P9-BIT-375

VALUES OF MATERIAL AND PHYSICAL CONSTANTS

Name	Symbol	Value	Units
Room temperature	T	300 (= 27°C)	K
Boltzman constant	k	1.38×10^{-23}	J/K
Electron charge	q	1.6×10^{-19}	C
Thermal voltage	$\phi_T = kT/q$	26	mV (at 300 K)
Intrinsic Carrier Concentration (Silicon)	n_i	1.5×10^{10}	cm^{-3} (at 300 K)
Electrical permittivity of vacuum	ε_0	8.85×10^{-14}	F/cm
Permittivity of Si	ε_{si}	1.05×10^{-12}	F/cm
Permittivity of SiO_2	ε_{ox}	3.5×10^{-13}	F/cm
Magnetic permeability of vacuum (similar for SiO_2)	μ_0	12.6×10^{-7}	Wb/Am
Speed of light (in vacuum)	c_0	30	cm/nsec
Speed of light (in SiO_2)	c_{ox}	15	cm/nsec

FORMULAS AND EQUATIONS

Diode

$$I_D = I_S(e^{V_D/\phi_T} - 1) = Q_D/\tau_T$$

$$C_j = \frac{C_{j0}}{(1 - V_D/\phi_0)^m}$$

$$K_{eq} = \frac{-\phi_0^m}{(V_{high} - V_{low})(1-m)} \times [(\phi_0 - V_{high})^{1-m} - (\phi_0 - V_{low})^{1-m}]$$

MOS Transistor

$$V_T = V_{T0} + \gamma(\sqrt{|-2\phi_F + V_{SB}|} - \sqrt{|-2\phi_F|})$$

$$I_D = \frac{k'_n}{2}\frac{W}{L}(V_{GS} - V_T)^2(1 + \lambda V_{DS}) \text{(sat)}$$

$$I_D = k'_n \frac{W}{L}\left((V_{GS} - V_T)V_{DS} - \frac{V_{DS}^2}{2}\right) \text{ (triode)}$$

Bipolar Transistor

$$I_C = I_F - \frac{I_R}{\alpha_R} \qquad I_E = \frac{I_F}{\alpha_F} - I_R \qquad I_B = I_E - I_C$$

$$I_F = I_S\left(e^{V_{BE}/\phi_T} - 1\right) \qquad I_R = I_S\left(e^{V_{BC}/\phi_T} - 1\right)$$

$$B_F = \frac{I_C}{I_B} \qquad \alpha_F = \frac{\beta_F}{\beta_F + 1} \qquad I_C = \frac{Q_F}{\tau_F}$$

Inverter

$$V_{OH} = f(V_{OL})$$

$$V_{OL} = f(V_{OH})$$

$$V_M = f(V_M)$$

$$t_p = \frac{C_L(V_{swing}/2)}{I_{avg}}$$

$$P_{dyn} = C_L V_{DD} V_{swing} f$$

$$P_{stat} = V_{DD} I_{DD}$$

Static CMOS Inverter

$$V_{OH} = V_{DD}$$

$$V_{OL} = GND$$

$$V_M = \frac{r(V_{DD} - |V_{Tp}|) + V_{Tn}}{1 + r} \quad \text{with} \quad r = \sqrt{\frac{k_p}{k_n}}$$

$$t_p = \frac{C_L V_{DD}}{2(V_{DD} - V_T)^2}\left(\frac{1}{k_n} + \frac{1}{k_p}\right)$$

$$P_{av} = C_L V_{DD}^2 f$$

ECL Inverter

$$V_{OH} = V_{CC} - V_{be(on)}$$

$$V_{OL} = V_{OH} - I_{EE} R_C$$

$$V_M = V_{ref} = (V_{OH} + V_{OL})/2$$

$$V_{IH,IL} = V_{ref} \pm \ln\left(\frac{V_{swing}}{2\phi_T} - 1\right)$$

Propagation delay: Table 3.8

$$P_{stat} = (V_{CC} - V_{EE})\left(I_{EE} + \frac{I_{bias}}{N} + 2\frac{\frac{V_{OH} + V_{OL}}{2} - V_{EE}}{R_B}\right)$$

$$P_{dyn} = C_T(V_{CC} - V_{EE})V_{swing} f$$

Interconnect

Lumped RC: $t_p = 0.69\ RC$

Distributed RC: $t_p = 0.38\ RC$

RC-chain:

$$\tau_N = \sum_{i=1}^{N} R_i \sum_{j=i}^{N} C_j = \sum_{i=1}^{N} C_i \sum_{j=1}^{i} R_j$$

Transmission line reflection:

$$\rho = \frac{V_{refl}}{V_{inc}} = \frac{I_{refl}}{I_{inc}} = \frac{R - Z_0}{R + Z_o}$$

DIGITAL INTEGRATED CIRCUITS

A DESIGN PERSPECTIVE

Prentice Hall Electronics and VSLI Series

Charles S. Sodini, Series Editor

LEE, SHUR, FIELDLY, YTTERDAL *Semiconductor Devices Modeling for VLSI*

RABAEY *Digital Integrated Circuits: A Design Perspective*

DIGITAL INTEGRATED CIRCUITS

A DESIGN PERSPECTIVE

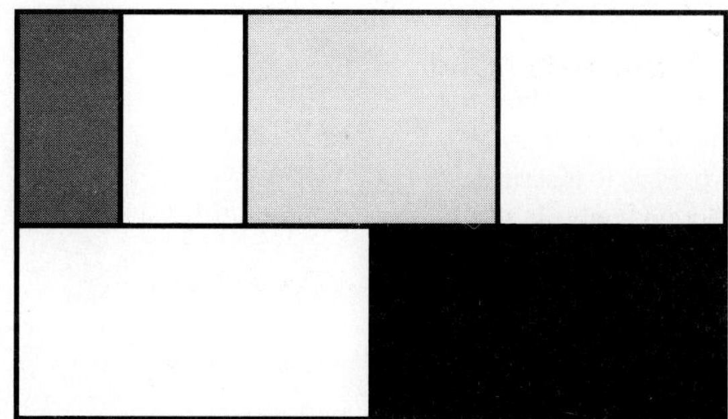

JAN M. RABAEY

PRENTICE HALL ELECTRONICS AND VLSI SERIES
CHARLES G. SODINI, SERIES EDITOR

Prentice-Hall International, Inc.

This edition may be sold only in those countries to which it is consigned by Prentice-Hall International. It is not to be re-exported and it is not for sale in the U.S.A., Mexico, or Canada.

Simon & Schuster/A Viacom Company
Upper Saddle River, New Jersey 07458

Printed in the United States of America

10 9 8 7 6 5 4 3 2 1

ISBN 0-13-394271-6

Prentice-Hall International (UK) Limited, *London*
Prentice-Hall of Australia Pty. Limited, *Sydney*
Prentice-Hall Canada Inc., *Toronto*
Prentice-Hall Hispanoamericana, S.A., *Mexico*
Prentice-Hall of India Private Limited, *New Delhi*
Prentice-Hall of Japan, Inc., *Tokyo*
Simon & Schuster Asia Pte. Ltd., *Singapore*
Editora Prentice-Hall do Brasil, Ltda., *Rio de Janeiro*
Prentice-Hall, Inc., *Upper Saddle River, New Jersey*

To Kathelijn

"Qu'est-ce que l'homme dans la nature?
Un néant a l'égard de l'infini,
un tout al l'égard du néant,
un milieu entre rien et tout."

"What is man in nature?
Nothing in relation to the infinite,
everything in relation to nothing,
a mean between nothing and everything."

— Blaise Pascal Pensées, no 4, 1670.

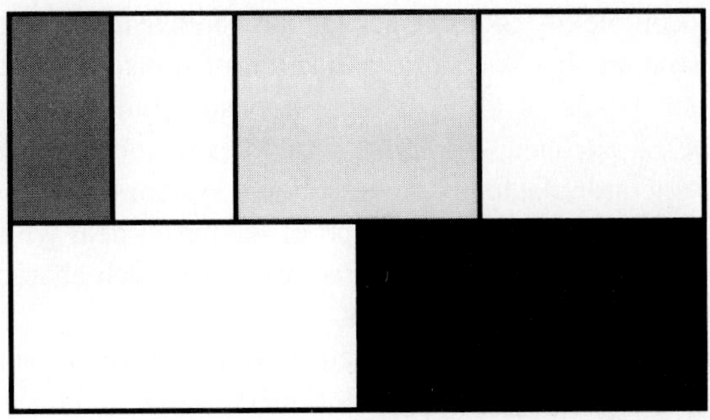

PREFACE

Why This Book?

Welcome to "*Digital Integrated Circuits: A Design Perspective.*" When you get this book in your hands, your first question very well might be, "With the already abundant number of text and reference books in the digital circuit and VLSI arenas, what is the value of this entry?" The answer is that a review of the existing material in the realm of digital circuit design reveals a major instructional gap between the **circuit and system visions** on digital design. This book merges both and provides a bridge between the top-down and bottom-up design approaches.

While starting from a solid understanding of the operation of electronic devices and an in-depth analysis of the nucleus of digital design—the inverter—we will gradually channel this knowledge into the design of more complex modules such as gates, adders, multipliers, registers, controllers, and memories. While doing so, we will identify the compelling questions that face the designer of today's complex circuits: What are the dominant design parameters, what section of the design should he focus on and what details could she ignore. Simplification is clearly the only approach to address the increasing complexity of the digital systems. However, oversimplification can lead to circuit failure since global circuit effects such as timing, interconnect, and power consumption are ignored. To avoid this pitfall it is important to design digital circuits with both a circuits and a systems perspective in mind. This is exactly the approach taken in this book, which brings the reader the knowledge and expertise needed to deal with complexity, using both analytical and experimental techniques.

Design from a System and a Circuit Perspective

The complexity of digital integrated circuits has grown dramatically over the past decades. Circuits with more than 1 million gates clocked at multiple 100 MHz combined with gigabit memories will emerge in the very near future. The sheer complexity of the design process has prompted the design community to respond with a dual solution: *design automation* and *design abstraction*. While the former helps the digital circuit designer in circuit generation, synthesis, analysis, and verification, the latter is the most important way to deal with the complexity issue. A design can be envisioned as a hierarchical composition of modules rather than a chaotic collection of transistors. At each design level, the complex, internal details of each of the composing modules are abstracted away and replaced by a *black box view,* or *model.* Examples of modeling levels are the transistor, gate, arithmetic operator, datapath, processor, and system levels. The impact of this *divide and conquer* approach is dramatic. Instead of having to deal with a myriad of elements, the designer considers only a handful of components, each characterized in performance and cost by a small number of parameters.

This design methodology, combined with an ever-increasing level of automation, suggests that the breed of digital circuit designers is soon to become extinct. This impression is reflected in the *VLSI design* instructional approach that addresses design in a **top-down** fashion, hiding the complex behavior of semiconductor devices.

Nothing is less true. While the top-down approach might work for a large number of circuits, the abstraction model suffers from major pitfalls and works only to a certain degree. In a high-performance circuit, for instance, the connection of one module to another influences the performance of both. The *interconnect wire* and associated parasitics become dominant factors in the circuit performance. The increasing *power dissipation* of high performance design translates into reduced reliability and increased packaging cost. Ensuring that the circuit operates correctly under these *high clock frequencies* is another challenge that faces the designer of advanced digital circuits. To address any of these issues requires an in-depth understanding of the underlying electrical concepts and constructs. This textbook covers these crucial concepts in detail and provides insights into factors that have a profound impact on reliability and performance.

In the ***bottom-up*** design philosophy, advocated in traditional digital circuit textbooks, the behavioral and performance model of a digital component is built starting from the transistor with all its peculiarities. While this approach results in an in-depth understanding of the component operation, it fails to translate this knowledge into a compact and simple model that can percolate upwards to help construct more complex modules. The prime talent of a good digital designer is to know when simplification is appropriate and when it is not. Acquiring this skill requires design experimentation and expertise. By taking a design-experimental approach, this book provides the student and professional the kind of hands-on experience that helps build that expertise.

It is my belief that bringing both circuit and systems views on design together results in a profound understanding of the design of complex digital circuits, while preparing the designer for new challenges that might be waiting around the corner. Only time will tell how successful this undertaking was.

Other Features This Book Offers

It is worth summarizing some other unique features we deem essential to accomplish the aforementioned goal and that form the underpinning of this textbook.

- Design-oriented perspectives are advocated throughout. Design challenges and guidelines are highlighted. Techniques introduced in the text are illustrated with real designs and complete SPICE analysis.

- It is the only current textbook that shows how to use the latest techniques to design complex high-performance, or low-power circuits.
- It covers crucial real-world system design issues such as signal integrity, power dissipation, interconnect, packaging, timing, and synchronization.
- It not only covers MOS but also addresses other high-performance technologies such as bipolar, BiCMOS, GaAs, and superconducting.
- It provides unique coverage of the latest design methodologies and tools, with a discussion of how to use them from a designers' perspective.
- It offers perspectives on how digital circuit technology might evolve in the future.
- The book features outstanding illustrations and a usable design-oriented four-color insert.
- An extensive instructional package is available over the internet from the author's web site at U.C. Berkeley. It includes design software, transparency masters, design problems, actual layouts, and hardware and software laboratories.

How to Use This Book

The core of the text is intended for use in a **senior-level digital circuit design class**. Around this kernel, we have included chapters and sections covering the more advanced topics. In the course of developing this book, it became obvious quickly that it is hard to define a subset of the digital circuit design domain that covers everyone's needs. On the one hand, a newcomer to the field needs detailed coverage of the basic concepts. On the other hand, feedback from early readers and reviewers indicated that an in-depth and extensive coverage of advanced topics and current issues is desirable and necessary. Providing this complete vision resulted in a text that exceeds the scope of a single-semester class. The more advanced material can be used as the basis for a **graduate class**. The wide coverage and the inclusion of state-of-the-art topics also makes the text useful as a reference work for professional engineers.

The organization of the material is such that the chapters can be taught or read in a great many ways, as long as a number of precedence relations are adhered to. An overview of these interdependencies is pictured in the chart below. The core of the text consists of Chapters 3, 4, 6, and 11. Chapters 1 and 2 can be considered as introductory. Students with a prior introduction to semiconductors can traverse quickly through Chapter 2. We urge

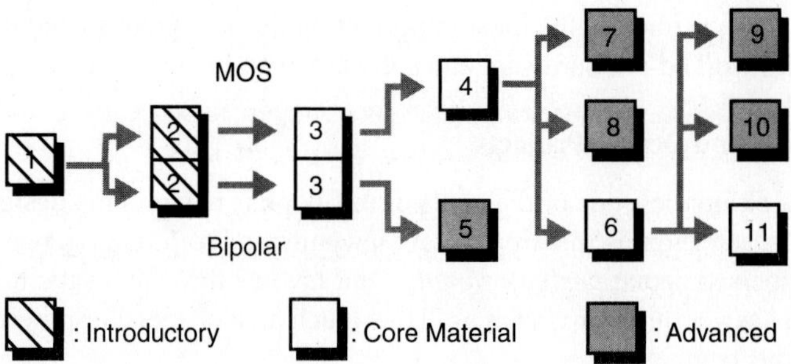

everyone to do at least that, as a number of important notations and foundations are introduced in that chapter. It is assumed that students taking this course are familiar with the basics of logic design.

Chapters 5, 7, 8, 9, and 10 are of a more advanced nature and can be used to provide a certain focus to the course. A course with a focus on the circuit aspects, for example, can supplement the core material with Chapters 5 and 10. A course focused on the digital system design should rather consider adding (parts of) Chapters 7, 8, and 9. All of these advanced chapters can be used to form the core of a graduate or a follow-on course. Other ordering options are possible. For instance, one might prefer to cover first the MOS parts of Chapters 2 and 3 before tackling the bipolar devices and gates. Sections considered advanced are marked with an asterisk in the text.

A number of possible paths through the material for a senior-level class are enumerated below. Many other variations are feasible. In the *instructor documentation,* provided with this book, we have included a number of complete syllabi similar to those used in school tests.

Basic circuit class (with minor prior device knowledge):
1, 2.1–3, 3.1–3, 4, 2.4–6, 3.4–5, 5.1–5.3, 6, 11.
Somewhat more advanced circuit coverage:
1, (2), 3.1–3, 4, 2.4–6, 3.4–5, 5.1–5.4, 6, 10, 11.
Course with systems focus:
1, (2.1–2.3), 3.1–3, 4, 6, 7, 8, 9, (10), 11.

We have chosen to separate layout-oriented discussions from the main text flow and to intersperse them through the text as *appendices*. Appendices A, C, and E cover design rules, cell layout, and datapath layout, respectively and are by preference covered in concurrence with the chapter they are attached to.

It was (and is) our intention to maintain a consistent flow through each of the chapters. The topics addressed in the chapter are *introduced* first, followed by a detailed and in-depth discussion of the ideas. A *Perspective* section discusses how the introduced concepts relate to real world designs and how they might be impacted by future evolutions. Each chapter finishes with a *Summary,* which briefly enumerates the topics covered in the text, followed by *To Probe Further* and *Reference* sections. These provide ample references and pointers for a reader interested in further details on some of the material. Each chapter ends with *exercises* and *design projects*, which range from the simple to the challenging. Solutions to the problem sets are provided in the *instructor's manual.* A *world-wide web companion* provides fully worked-out design problems and a complete set of overhead transparencies, extracting the most important figures and graphs from the text. A *laboratory book* with worked-out hardware and software experiments is in the making.

Problems, Exercises, and Design Projects

As the title of the book implies, one of the goals of this book is to stress the **design aspect** of digital circuits. To achieve this more practical viewpoint and to provide a real perspective, we have interspersed actual *design examples* and layouts throughout the text. These case studies help to answer questions, such as "How much area or speed or power is really saved by applying this technique?"

To mimic the real design process, we are making extensive use of design tools such as circuit- and switch-level simulation as well as layout editing and extraction. Computer analysis is used throughout to verify manual results, to illustrate new concepts, or to examine complex behavior beyond the reach of manual analysis.

Open-ended *design problems*, included at the end of most chapters, help to gain the all-important insight into design optimization and trade-off. The use of design editing, verification and analysis tools is recommended when attempting these design problems. Fully worked out versions of the problems can be found in the world-wide web page.

Finally, to facilitate the learning process, there are numerous examples included in the text. Each chapter contains a number of *problems or brain-teasers* (answers for which can be found in the back of the book), that provoke thinking and understanding while reading. Numerous challenging *exercises* (more than 200 of them) are included at the end of each chapter. Their goal is to provide the individual reader an independent gauge for his understanding of the material and to provide practice in the use of some of the design tools. Each problem is keyed to the text sections it refers to (e.g. <1.3>), the design tools that must be used when solving the problem (e.g., SPICE) and a rating, ranking the problems on difficulty: (E) easy; (M) moderate and (C) challenging. Problems marked with a (D) include a design or research elements.

The Worldwide Web Page — A Dynamic Companion

With the advent of modern networking technology, a textbook does not have to be a static entity anymore, but can become a dynamically evolving document. For this reason, we have established a worldwide web page for the book that contains the latest updates, new problem sets and design projects as well as extensive instructor material. Downloadable postscript files are available for complete transparency sets, covering all the material. Even more, we have provided a library of downloadable software tools useful in the context of this book (SPICE, MAGIC, IRSIM). It is our hope that this home page will become a forum for digital circuit design. Instructor contributions in terms of new problem sets or design projects are greatly welcomed and can help to keep your course more attractive and interesting for the students. We already received extra slide sets, addressing more advanced topics even before going to press. Please have a look at the following address: **http://infopad.eecs.berkeley.edu/~icdesign**. Comments and feedback are appreciated.

The Contents at a Glance

A quick scan of the table-of-contents shows how the ordering of chapters and the material covered are consistent with the advocated design methodology. Starting from a model of the semiconductor devices, we will gradually progress upwards, covering the inverter, the complex logic gate (NAND, EXOR, Flip-Flop), the functional (adder, multiplier, shifter, register) and the system module (datapath, controller, memory) levels of abstraction. For each of these layers, the dominant design parameters are identified and simplified models are constructed, abstracting away the nonessential details. While this layered modeling approach is the designer's best handle on complexity, it has some pitfalls. This is illustrated in Chapters 8 and 9, where topics with a global impact, such as interconnect parasitics and chip timing, are discussed. To further express the dichotomy between circuit and system design visions, we have divided the book contents into two parts: Part I (Chapters

2–6) addresses mostly the circuit perspective of digital circuit design, while Part II (Chapters 7–11) presents a more system oriented vision.

Chapter 1 serves as a global *introduction*. After a historical overview of digital circuit design, the concepts of hierarchical design and the different abstraction layers are introduced.

Chapter 2 contains a summary of the primary design building blocks, *the semiconductor devices*. The main goal of this chapter is to provide an intuitive understanding of the operation of the MOS and bipolar silicon transistors as well as to introduce the device models, which will be used extensively in the later chapters. Some of the artifacts of modern submicron devices are also discussed. Readers with prior device knowledge can traverse this material rather quickly.

Chapter 3 deals with the nucleus of digital design, the *inverter*. First, a number of fundamental properties of digital gates are introduced. These parameters, which help to quantify the performance and reliability of a gate, are derived in detail for two representative inverter structures: the static complementary CMOS and the bipolar ECL inverter. The techniques and approaches introduced in this chapter are of crucial importance, as they are repeated over and over again in the analysis of other gate structures and more complex gate structures.

This fundamental knowledge is extended in Chapter 4 to address the design of *simple digital CMOS gates*, such as NOR and NAND structures. It is demonstrated that, depending upon the dominant design constraint (reliability, area, performance, or power), other CMOS gate structures besides the complementary static gate can be attractive. The properties of a number of contemporary gate-logic families are analyzed and compared. Techniques to optimize the performance and power consumption of complex gates are introduced.

While CMOS gates are achieving ever higher speeds, other technologies are a necessity when even *higher performance* is required; for example, BiCMOS, GaAs, and superconducting technologies. While circuits, implemented in one of these processes, represent only a small portion of the total digital design market, an analysis of some representative gates for each technology is definitely worthwhile. Design automation is likely to turn the design of low- to medium-performance circuits into a chore rather than a challenge. The contributions of the digital designer will, rightly so, be focused on the implementation of the highest-performance components of the design. A large number of issues, currently raised in the design of high performance bipolar or GaAs circuits, might soon carry over to CMOS as well. Finally, the analysis of how the choice of a different device affects the nature and the performance of a gate structure is revealing and is instrumental in the building of a fundamental understanding of digital circuit design. High-speed gate design is the topic of Chapter 5 (which is optional for undergraduate courses).

All chapters prior to Chapter 6 deal exclusively with combinational circuits, that is circuits without a sense of the past history of the system. *Sequential logic circuits*, in contrast, can remember and store the past state. Chapter 6 discusses how this memory function can be accomplished using either positive feedback or charge storage. Besides analyzing the traditional bistable flip-flops, other sequential circuits such as the mono- and astable multivibrators are also introduced.

All previous chapters present a circuit-oriented approach towards digital design. The analysis and optimization process has been constrained to the individual gate. Starting from Chapter 7, we take our approach one step further and analyze how gates can be connected together to form the building blocks of a system. The design of a variety of complex *arithmetic building blocks* such as adders, multipliers, and shifters, is discussed first. This chapter is crucial because it demonstrates how the design techniques introduced in chapters 3 and 4 are extended to the next abstraction layer. The concept of the critical path is introduced and used extensively in the performance analysis and optimization. Higher-level performance models are derived. These help the designer to get a fundamental insight into the operation and quality of a design module, without having to resort to an in-depth and detailed analysis of the underlying circuitry.

Chapter 8 discusses the impact of *interconnect wiring* on the functionality and performance of a digital gate. A wire introduces parasitic capacitive, resistive, and inductive effects, which are becoming ever more important with the scaling of the technology. Techniques to efficiently model these parasitic effects are presented. Approaches to minimize the impact of the interconnect parasitics on both performance and circuit reliability are introduced. The impact of packaging technology on circuit operation and performance is discussed.

In order to operate sequential circuits correctly, a strict ordering of the switching events has to be imposed. Without these *timing* constraints, wrong data might be written into the memory cells. Most digital circuits use a synchronous, clocked approach to impose this ordering. In Chapter 9, the different approaches to digital circuit timing and clocking are discussed. The impact of important effects such as clock skew on the behavior of digital synchronous circuits is analyzed. The synchronous approach is contrasted with alternative techniques, such as self-timed circuits. The chapter concludes with a short introduction to synchronization and clock-generation circuits.

Whenever large amounts of data storage are needed, the digital designer resorts to special circuit modules, called *memories*. Semiconductor memories achieve very high storage density by compromising on some of the fundamental properties of digital gates. Chapter 10 discusses in depth the different memory classes and their implementation. Instrumental in the design of reliable and fast memories is the implementation of the peripheral circuitry, such as the decoders, sense amplifiers, drivers, and control circuitry, which are extensively covered. Finally, as the primary issue in memory design is to ensure that the device works consistently under all operating circumstances, the chapter concludes with a detailed discussion of memory reliability. This chapter as well as the previous one are optional for undergraduate courses.

The book concludes with a discussion of *design methodologies*. Design automation is the only way to cope with the ever-increasing complexity of digital designs. In Chapter 11, the prominent ways of producing large designs in a limited time are discussed. The chapter presents the common design representation and analysis approaches as well as the different implementation methodologies. The chapter ends with a short discussion of manufacturing tests, an often overlooked component of the digital design process.

Acknowledgments

The author would like to thank all those who contributed to the emergence, creation and correction of this manuscript. First of all, the graduate students of the University of California at Berkeley, who were an invaluable help in providing examples and improving the early drafts: Paul Landman, David Lidsky, Anthony Stratakos, Andy Burstein, Anantha Chandrakasan, Renu Mehra, Lisa Guerra, Ole Bentz, Tom Burd, Alfred Yeung, Scarlett Wu, Arthur Abnous, and Steve Lo. Thanks also to the students of the EE141 and EE241 courses at Berkeley, who suffered through many of the experimental class offerings based on this book. I am grateful to the students of the E506 class at Waseda University, Tokyo who represented the first foreign exposure of this material. I also would like to acknowledge the help and advice of a substantial number of people, who helped to review and improve this text: Ingrid Verbauwhede (UCB), Andrew Neureuther (UCB), Kevin Kornegay (IBM and Purdue), Charlie Sodini (MIT), Doug Hoy (Tektronix), Randy Allmon (DEC), Greg Uehara (Univ. of Hawaii), Massoud Pedram (USC), A. Srivastava (Louisiana State Univ.), C. Mastrangelo (Univ. Of Michigan), S. Embabi (Texas A&M), D. Ioannou (George Mason Univ.), T. DeMassa (Arizona State), B. Biderman (DEC), T. Knight (MIT), M. Elmasry (Waterloo), and many others.

I am extremely grateful to the staff at Prentice Hall, who have been instrumental in turning a rough manuscript into an enjoyable book. First of all, I would like to acknowledge the help and constructive feedback of the publishing editor, Alan Apt. The editorial development help of Sondra Chavez was greatly appreciated. Nick Murray was the copy editor and Mona Pompili made it all come together as production editor. Shirley McGuire helped us assemble the instructor's manual.

A special word of thanks for the two persons who taught me more than anyone else about the world of electronics (and much beyond), namely Hugo De Man and Robert Brodersen.

I would like to highlight to role of computer aids in developing this manuscript. All drafts were completely developed on the FrameMaker publishing system (Frame Technology Corporation). Graphs were generated using the *xvgr* tool (thanks, Paul Turner), Mathematica, and Matlab. For circuit simulations, I used both SPICE 3 (U.C. Berkeley) and HSPICE (Meta-Software). All layouts were generated using the MAGIC layout editor and the LAGER silicon compiler (both from U.C. Berkeley). The IRSIM tool (from Stanford University) was used extensively for the switch-level verification of more complex designs and is also suggested for use in a number of design projects.

Finally, I would like to express my extreme gratitude to my wife and companion Kathelijn for enduring these "lost years" of our life. She has been a constant support, help and encouragement during the writing of this manuscript. Fortunately, we will get some more time for skiing or hiking as of now.

Jan M. Rabaey
Berkeley, California

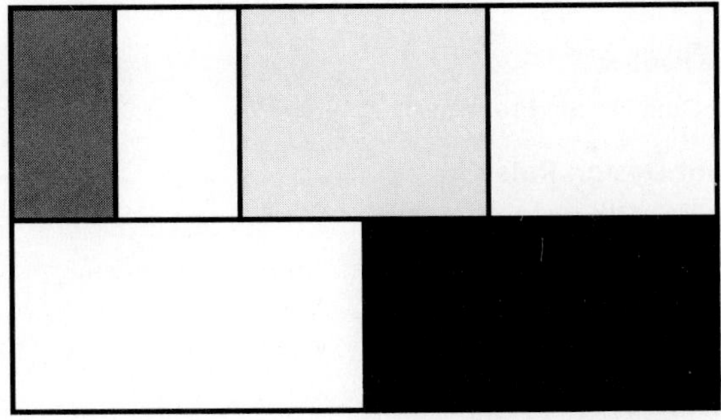

CONTENTS

CHAPTER

1

INTRODUCTION

The evolution of digital circuit design

Compelling issues in digital circuit design

Valuable references

1.1 A Historical Perspective

The concept of digital data manipulation has made a dramatic impact on our society. One has long grown accustomed to the idea of digital computers. Evolving steadily from mainframe and minicomputers, personal and laptop computers have proliferated into daily life. More significant, however, is a continuous trend towards digital solutions in all other areas of electronics. Instrumentation was one of the first noncomputing domains where the potential benefits of digital data manipulation over analog processing were recognized. Other areas such as control were soon to follow. Only recently have we witnessed the conversion of telecommunications and consumer electronics towards the digital format. Increasingly, telephone data is transmitted and processed digitally over both wired and wireless networks. The compact disk has revolutionized the audio world, and digital video is following in its footsteps.

The idea of implementing computational engines using an encoded data format is by no means an idea of our times. In the early nineteenth century, Babbage envisioned large-scale mechanical computing devices, called *Difference Engines* [Swade93]. Although these engines use the decimal number system rather than the binary representation now common in modern electronics, the underlying concepts are very similar. The Analytical Engine, developed in 1834, was perceived as a general-purpose computing machine, with features strikingly close to modern computers. Besides executing the basic repertoire of operations (addition, subtraction, multiplication, and division) in arbitrary sequences, the machine operated in a two-cycle sequence, called "store" and "mill" (execute), similar to current computers. It even used pipelining to speed up the execution of the addition operation! Unfortunately, the complexity and the cost of the designs made the concept impractical. For instance, the design of Difference Engine I (part of which is shown in Figure 1.1) required 25,000 mechanical parts at a total cost of £17,470 (in 1834!).

Figure 1.1 Working part of Babbage's Difference Engine I (1832), the first known automatic calculator (from [Swade93], courtesy of the Science Museum of London).

The electrical solution turned out to be more cost effective. Early digital electronics systems were based on magnetically controlled switches (or relays). They were mainly used in the implementation of very simple logic networks. Examples of such are train safety systems, where they are still being used at present. The age of digital electronic computing only started in full with the introduction of the vacuum tube. While originally used almost exclusively for analog processing, it was realized early on that the vacuum tube was useful for digital computations as well. Soon complete computers were realized. The era of the vacuum tube based computer culminated in the design of machines such as the ENIAC (intended for computing artillery firing tables) and the UNIVAC I (the first successful commercial computer). To get an idea about *integration density*, the ENIAC was 80 feet long, 8.5 feet high and several feet wide and incorporated 18,000 vacuum tubes. It became rapidly clear, however, that this design technology had reached its limits. Reliability problems and excessive power consumption made the implementation of larger engines economically and practically infeasible.

All changed with the invention of the *transistor* at Bell Telephone Laboratories in 1947 [Bardeen48], followed by the introduction of the bipolar transistor by Schockley in 1949 [Schockley49][1]. It took till 1956 before this led to the first bipolar digital logic gate, introduced by Harris [Harris56], and even more time before this translated into a set of integrated-circuit commercial logic gates, called the Fairchild Micrologic family [Norman60]. The first truly successful IC logic family, *TTL (Transistor-Transistor Logic)* was pioneered in 1962 [Beeson62]. Other logic families were devised with higher performance in mind. Examples of these are the current switching circuits that produced the first subnanosecond digital gates and culminated in the *ECL (Emitter-Coupled Logic)* family [Masaki74], which is discussed in more detail in this textbook. TTL had the advantage, however, of offering a higher integration density and was the basis of the first integrated circuit revolution. In fact, the manufacturing of TTL components is what spear-headed the first large semiconductor companies such as Fairchild, National, and Texas Instruments. The family was so successful that it composed the largest fraction of the digital semiconductor market until the 1980s.

Ultimately, bipolar digital logic lost the battle for hegemony in the digital design world for exactly the reasons that haunted the vacuum tube approach: the large power consumption per gate puts an upper limit on the number of gates that can be reliably integrated on a single die, package, housing, or box. Although attempts were made to develop high integration density, low-power bipolar families (such as *I^2L—Integrated Injection Logic* [Hart72]), the torch was gradually passed to the MOS digital integrated circuit approach.

The basic principle behind the MOSFET transistor (originally called IGFET) was proposed in a patent by J. Lilienfeld (Canada) as early as 1925, and, independently, by O. Heil in England in 1935. Insufficient knowledge of the materials and gate stability problems, however, delayed the practical usability of the device for a long time. Once these were solved, MOS digital integrated circuits started to take off in full in the early 1970s. Remarkably, the first MOS logic gates introduced were of the CMOS variety [Wanlass63], and this trend continued till the late 1960s. The complexity of the manufacturing process

---

[1] An intriguing overview of the evolution of digital integrated circuits can be found in [Murphy93]. (Most of the data in this overview has been extracted from this reference). It is accompanied by some of the historically ground-breaking publications in the domain of digital IC's.

delayed the full exploitation of these devices for two more decades. Instead, the first practical MOS integrated circuits were implemented in PMOS-only logic and were used in applications such as calculators. The second age of the digital integrated circuit revolution was inaugurated with the introduction of the first microprocessors by Intel in 1972 (the 4004) and 1974 (the 8080) [Shima74]. These processors were implemented in NMOS-only logic, that has the advantage of higher speed over the PMOS logic. Simultaneously, MOS technology enabled the realization of the first high-density semiconductor memories. For instance, the first 4Kbit MOS memory was introduced in 1970 [Hoff70].

These events were at the start of a truly astounding evolution towards ever higher integration densities and speed performances, a revolution that is still in full swing right now. The road to the current levels of integration has not been without hindrances, however. In the late 1970s, NMOS-only logic started to suffer from the same plague that made high-density bipolar logic unattractive or infeasible: power consumption. This realization, combined with progress in manufacturing technology, finally tilted the balance towards the CMOS technology, and this is where we still are today. Interestingly enough, power consumption concerns are rapidly becoming dominant in CMOS design as well, and this time there does not seem to be a new technology around the corner to alleviate the problem.

Although the large majority of the current integrated circuits are implemented in the MOS technology, other technologies come into play when very high performance is at stake. An example of this is the BiCMOS technology that combines bipolar and MOS devices on the same die. BiCMOS is effectively used in high-speed memories and gate arrays. When even higher performance is necessary, other technologies emerge besides the already mentioned bipolar silicon ECL family—Gallium-Arsenide, Silicon-Germanium and even superconducting technologies. While these circuits only fill in a small niche in the overall digital integrated circuit design scene, it is worth examining some of the issues emerging in the design of these circuits. With the continuing increase in performance of digital MOS circuits, design problems currently encountered in these high-speed technologies might come to haunt CMOS as well in the foreseeable future.

1.2 Issues in Digital Integrated Circuit Design

Integration density and performance of integrated circuits have gone through an astounding revolution in the last couple of decades. In the 1960s, Gordon Moore, then with Fairchild Corporation and later cofounder of Intel, predicted that the number of transistors that can be integrated on a single die would grow exponentially with time. This prediction, later called *Moore's law*, has proven to be amazingly visionary. Its validity is best illustrated with the aid of a set of graphs. Figure 1.2 plots the integration density of both logic ICs and memory as a function of time. As can be observed, integration complexity doubles approximately every 1 to 2 years. As a result, memory density has increased by more than a thousandfold since 1970.

An intriguing case study is offered by the microprocessor. From its inception in the early seventies, the microprocessor has grown in performance and complexity at a steady and predictable pace. The number of transistors and the clock frequency for a number of landmark designs are collected in Figure 1.3. The million-transistor/chip barrier was crossed in the late eighties. Clock frequencies double every three years and have reached

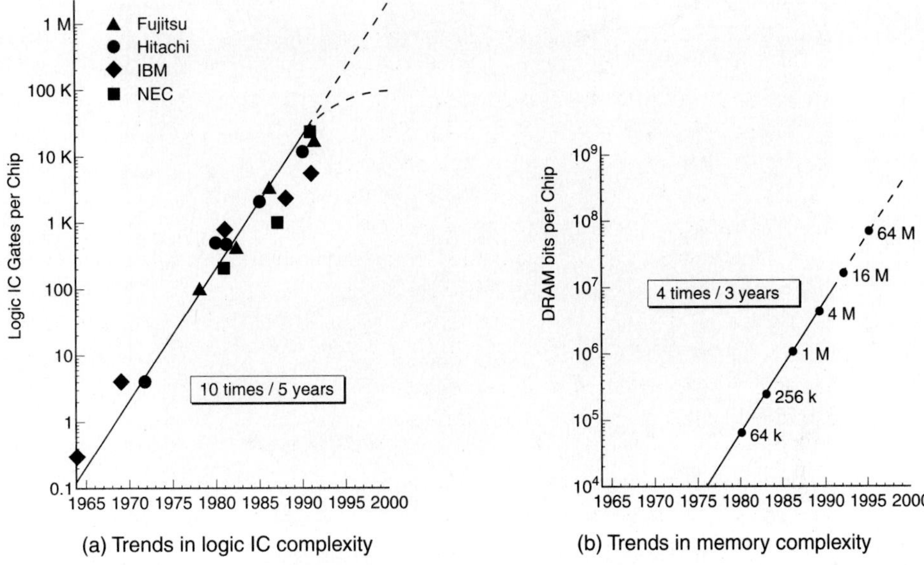

(a) Trends in logic IC complexity (b) Trends in memory complexity

Figure 1.2 Evolution of integration complexity of logic ICs and memories as a function of time (from [Masaki92]).

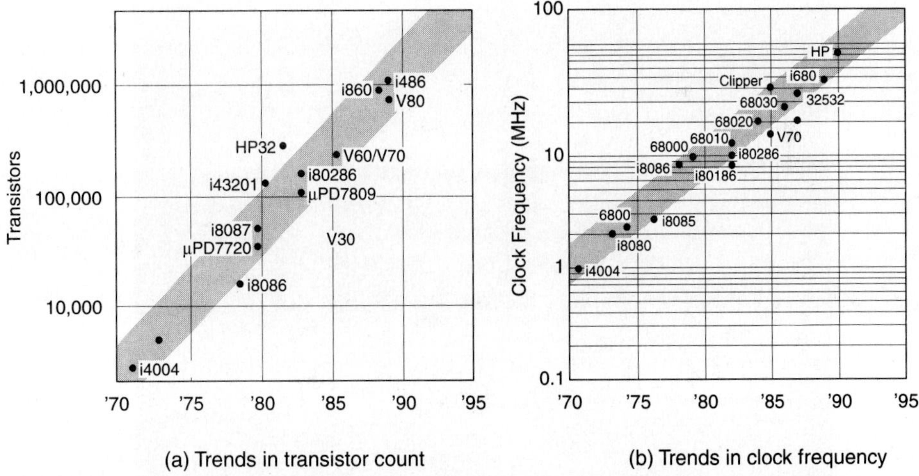

(a) Trends in transistor count (b) Trends in clock frequency

Figure 1.3 Evolution of microprocessor transistor count and clock frequency (from [Sasaki91]).

into the 100 MHz range. An even more important observation is that, as of now, these trends have not shown any signs of a slow-down.

It should be no surprise to the reader that this revolution has had a profound impact on how digital circuits are designed. Early designs were truly hand-crafted. Every transistor was laid out and optimized individually and carefully fitted into its environment. This is adequately illustrated in Figure 1.4a, which shows the design of the Intel 4004 microprocessor. This approach is, obviously, not appropriate when more than a million devices have to be created and assembled. With the rapid evolution of the design technology, time-

(a) The 4004 microprocessor (see also back cover)

(b) The Pentium™ microprocessor (see also back cover)

Figure 1.4 Comparing the design methodologies of the Intel 4004 (1971) and Pentium™ (1994) microprocessors (reprinted with permission from Intel).

to-market is one of the crucial factors in the ultimate success of a component. Designers have, therefore, increasingly adhered to rigid design methodologies and strategies that are more amenable to design automation. The impact of this approach is apparent from the layout of one of the later Intel microprocessors, the Pentium, shown in Figure 1.4b. Instead of the individualized approach of the earlier designs, a circuit is constructed in a hierarchical way: a processor is a collection of modules, each of which consists of a number of cells on its own. Cells are reused as much as possible to reduce the design effort and to enhance the chances for a first-time-right implementation. The fact that this hierarchical approach is at all possible is the key ingredient for the success of digital circuit design and also explains why, for instance, very large scale analog design has never caught on.

The obvious next question is why such an approach is feasible in the digital world and not (or to a lesser degree) in analog designs. The crucial concept here, and the most important one in dealing with the complexity issue, is *abstraction*. At each design level, the internal details of a complex module can be abstracted away and replaced by a *black box view* or *model*. This model contains virtually all the information needed to deal with the block at the next level of hierarchy. For instance, once a designer has implemented a multiplier module, its performance can be defined very accurately and can be captured in a model. The performance of this multiplier is in general only marginally influenced by the way it is utilized in a larger system. For all purposes, it can hence be considered a black box with known characteristics. As there exists no compelling need for the system designer to look inside this box, design complexity is substantially reduced. The impact of this *divide and conquer* approach is dramatic. Instead of having to deal with a myriad of elements, the designer has to consider only a handful of components, each of which are characterized in performance and cost by a small number of parameters.

This is analogous to a software designer using a library of software routines such as input/output drivers. Someone writing a large program does not bother to look inside those library routines. The only thing he cares about is the intended result of calling one of those modules. Imagine what writing software programs would be like if one had to fetch every bit individually from the disk and ensure its correctness instead of relying on handy "file open" and "get string" operators.

Typically used abstraction levels in digital circuit design are, in order of increasing abstraction, the device, circuit, gate, functional module (e.g., adder) and system levels (e.g., processor), as illustrated in Figure 1.5. A semiconductor device is an entity with a very complex behavior. No circuit designer will ever seriously consider the solid-state physics equations governing the behavior of the device when designing a digital gate. Instead he will use a simplified model that adequately describes the input-output behavior of the transistor. For instance, an AND gate is adequately described by its Boolean expression ($Z = A.B$), its bounding box, the position of the input and output terminals, and the delay between the inputs and the output.

This design philosophy has been the enabler for the emergence of elaborate *computer-aided design* (CAD) frameworks for digital integrated circuits; without it the current design complexity would not have been achievable. Design tools include simulation at the various complexity levels, design verification, layout generation, and design synthesis. An overview of these tools and design methodologies is given in Chapter 11 of this textbook.

Furthermore, to avoid the redesign and reverification of frequently used cells such as basic gates and arithmetic and memory modules, designers most often resort to *cell librar-*

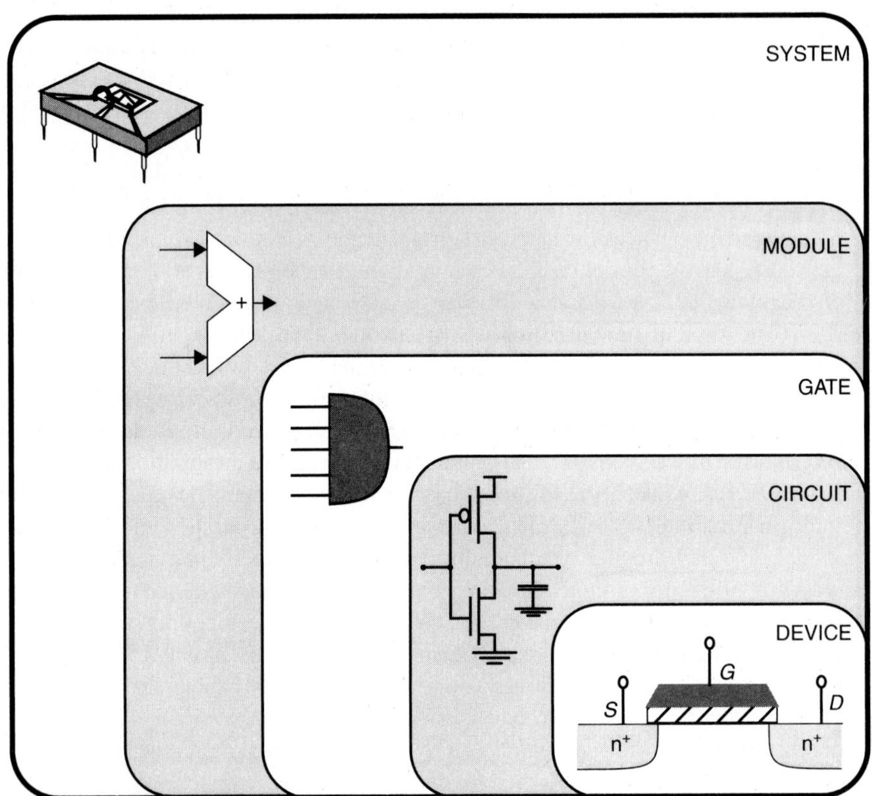

Figure 1.5 Design abstraction levels in digital circuits.

ies. These libraries contain not only the layouts, but also provide complete documentation and characterization of the behavior of the cells. The use of cell libraries is, for instance, apparent in the layout of the Pentium processor (Figure 1.4b). The integer and floating-point unit, just to name a few, contain large sections designed using the so-called *standard cell approach*. In this approach, logic gates are placed in rows of cells of equal height and interconnected using routing channels. The layout of such a block can be generated automatically given that a library of cells is available.

The preceding analysis demonstrates that design automation and modular design practices have effectively addressed some of the complexity issues incurred in contemporary digital design. This leads to the following pertinent question. If design automation solves all our design problems, why should we be concerned with digital circuit design at all? Will the next-generation digital designer ever have to worry about transistors or parasitics, or is the smallest design entity he will ever consider the gate and the module?

The truth is that the reality is more complex, and various reasons exist as to why an insight into digital circuits and their intricacies will still be an important asset for a long time to come.

- First of all, someone still has to *design and implement the module libraries*. Semiconductor technologies continue to advance from year to year, as demonstrated in

Figure 1.2, where the minimum MOS device dimensions are plotted as a function of time. Until one has developed a fool-proof approach towards "porting" a cell from one technology to another, each change in technology—which happens approximately every two years—requires a redesign of the library.

- Creating an adequate *model* of a cell or module requires an in-depth understanding of its internal operation. For instance, to identify the dominant performance parameters of a given design, one has to recognize the critical timing path first.
- The library-based approach works fine when the design constraints (speed, cost or power) are not stringent. This is the case for a large number of *application-specific designs*, where the main goal is to provide a more integrated system solution, and performance requirements are easily within the capabilities of the technology. Unfortunately for a large number of other products such as microprocessors, success hinges on high performance, and designers therefore tend to push technology to its limits. At that point, the hierarchical approach tends to become somewhat less attractive. To resort to our previous analogy to software methodologies, a programmer tends to "customize" software routines when execution speed is crucial; compilers—or design tools—are not yet to the level of what human sweat or ingenuity can deliver.
- Even more important is the observation that the abstraction-based approach is only correct to a certain degree. The performance of, for instance, an adder can be substantially influenced by the way it is connected to its environment. The interconnection wires themselves contribute to delay as they introduce parasitic capacitances, resistances and even inductances. The impact of the *interconnect parasitics* is bound to increase in the years to come with the scaling of the technology.
- Scaling tends to emphasize some other deficiencies of the abstraction-based model. Some design entities tend to be *global or external* (to resort anew to the software analogy). Examples of global factors are the clock signals, used for synchronization in a digital design, and the supply lines. Increasing the size of a digital design has a profound effect on these global signals. For instance, connecting more cells to a supply line can cause a voltage drop over the wire, which, in its turn, can slow down all the connected cells. Issues such as clock distribution, circuit synchronization, and supply-voltage distribution are becoming more and more critical. Coping with them requires a profound understanding of the intricacies of digital circuit design.
- Another impact of technology evolution is that *new design issues* and constraints tend to emerge over time. A typical example of this is the periodical reemergence of power dissipation as a constraining factor, as was already illustrated in the historical overview. Another example is the changing ratio between device and interconnect parasitics. To cope with these unforeseen factors, one must at least be able to model and analyze their impact, requiring once again a profound insight into circuit topology and behavior.
- Finally, when things can go wrong, they do. A fabricated circuit does not always exhibit the exact waveforms one might expect from advance simulations. Deviations can be caused by variations in the fabrication process parameters, or by the induc-

tance of the package, or by a badly modeled clock signal. *Troubleshooting* a design requires circuit expertise.

For all the above reasons, it is my belief that an in-depth knowledge of digital circuit design techniques and approaches is an essential asset for a digital-system designer. Even though she might not have to deal with the details of the circuit on a daily basis, the understanding will help her to cope with unexpected circumstances and to determine the dominant effects when analyzing a design.

Example 1.1 Clocks Defy Hierarchy

To illustrate some of the issues raised above, let us examine the impact of deficiencies in one of the most important global signals in a design, the *clock*. The function of the clock signal in a digital design is to order the multitude of events happening in the circuit. This task can be compared to the function of a traffic light that determines which cars are allowed to move. It also makes sure that all operations are completed before the next one starts—a traffic light should be green long enough to allow a car or a pedestrian to cross the road. Under ideal circumstances, the clock signal is a periodic step waveform with abrupt transitions between the low and the high values (Figure 1.6a).

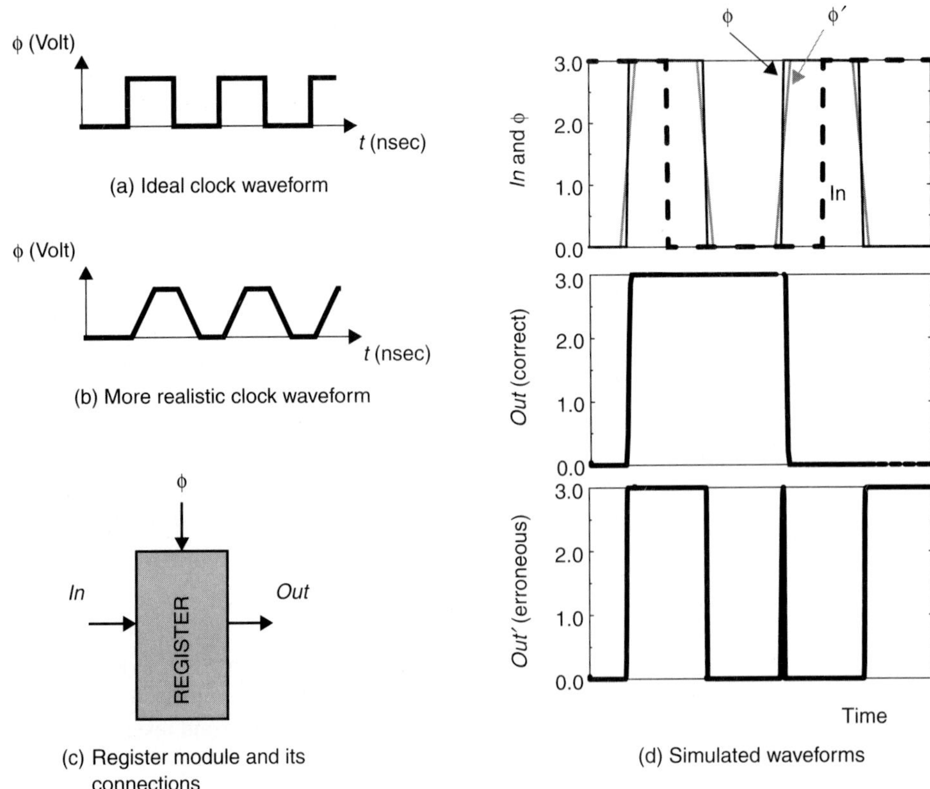

Figure 1.6 Reduced clock slopes can cause a register circuit to fail.

Consider, for instance, the circuit configuration of Figure 1.6c. The *register* module samples the value of the input signal at the rising edge of the clock signal ϕ. This sampled value is preserved and appears at the output until the clock rises anew and a new input is sampled. Under normal circuit operating conditions, this is exactly what happens, as demonstrated in the simulated response of Figure 1.6d. On the rising edge of clock ϕ, the input *In* is sampled and appears at the output *Out.*

Assume now that, due to added loading on the clock signal (for instance, connecting more latches), the clock signal is degenerated, and the clock slopes become less steep (clock ϕ' in Figure 1.6d). When the degeneration is within bounds, the functionality of the latch is not impacted. When these bounds are exceeded the latch suddenly starts to malfunction as shown in Figure 1.6d (signal *Out'*). The output signal makes unexpected transitions at the falling clock edge, and extra spikes can be observed as well. Propagation of these erroneous values can cause the digital system to go into a unforeseen mode and crash. This example clearly shows how global effects, such as adding extra load to a clock, can change the behavior of an individual module. Observe that the effects shown are not universal, but are a property of the register circuit used.

Besides the requirement of steep edges, other constraints must be imposed on clock signals to ensure correct operation. A second requirement related to *clock alignment*, is illustrated in Figure 1.7. The circuit under analysis consists of two cascaded registers, both operating on the rising edge of the clock ϕ. Under normal operating conditions, the input *In* gets sampled into the first register on the rising edge of ϕ and appears at the output exactly one clock period later. This is confirmed by the simulations shown in Figure 1.7b (signal *Out*).

Due to delays associated with routing the clock wires, it may happen that the clocks become misaligned with respect to each other. As a result, the registers are interpreting time

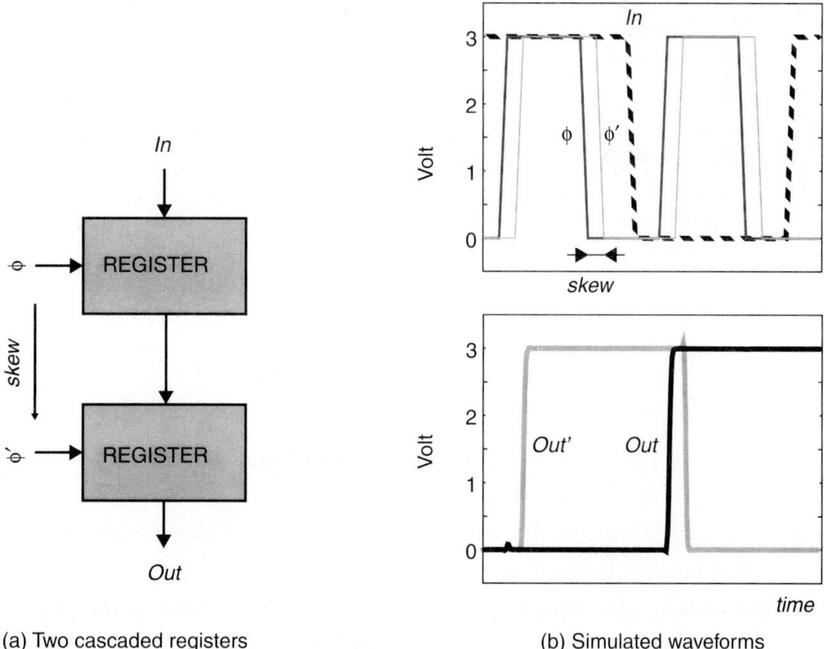

(a) Two cascaded registers

(b) Simulated waveforms

Figure 1.7 Impact of clock misalignment.

indicated by the clock signal differently. Consider the case that the clock signal for the second register is delayed—or skewed—by a value δ. The rising edge of the delayed clock ϕ' will postpone the sampling of the input of the second register. If the time it takes to propagate the output of the first register to the input of the second is smaller than the clock delay, the latter will sample the wrong value. This causes the output to change prematurely, as clearly illustrated in the simulation, where the signal *Out′* goes high at the first rising edge of ϕ' instead of the second one.

Clock misalignment, or *clock skew*, as it is normally called, is another example of how global signals may influence the functioning of a hierarchically designed system. Clock skew is actually one of the most critical design problems facing the designers of large, high-performance systems.

The purpose of this textbook is to provide *a bridge between the abstract vision of digital design and the underlying digital circuit and its peculiarities.* While starting from a solid understanding of the operation of electronic devices and an in-depth analysis of the nucleus of digital design—the inverter—we will gradually channel this knowledge into the design of more complex entities, such as complex gates, datapaths, registers, controllers, and memories. The persistent quest for a designer when designing each of the mentioned modules is to identify the dominant design parameters, to locate the section of the design he should focus his optimizations on, and to determine the specific properties that make the module under investigation (e.g., a memory) different from any others.

The text also addresses other compelling (global) issues in modern digital circuit design such as *power dissipation, interconnect, timing, and synchronization.* While most of the text concentrates on CMOS design, it is worthwhile exploring the boundaries of what can be achieved with semiconductor digital circuit technology. A short discussion on bipolar, BiCMOS, GaAs, and cryogenic circuit approaches is therefore included.

1.3 To Probe Further

The design of digital integrated circuits has been the topic of a multitude of textbooks and monographs. To help the reader find more information on some selected topics, an extensive list of reference works is listed below. The state-of-the-art developments in the area of digital design are generally reported in technical journals or conference proceedings, the most important of which are listed.

JOURNALS AND PROCEEDINGS

IEEE Journal of Solid-State Circuits

IEICE Transactions on Electronics (Japan)

Proceedings of The International Solid-State and Circuits Conference (ISSCC)

Proceedings of the Integrated Circuits Symposium

European Solid-State Circuits Conference (ESSCIRC)

REFERENCE BOOKS

General

H. Bakoglu, *Circuits, Interconnections, and Packaging for VLSI*, Addison-Wesley, 1990.

J. Buchanan, *CMOS/TTL Digital Systems Design*, McGraw-Hill, 1990.

J. Di Giacoma, *VLSI Handbook*, McGraw-Hill, 1989.

E. Friedman, ed., *Clock Distribution Networks in VLSI Circuits and Systems,* IEEE Press, 1995.

H. Haznedar, *Digital Micro-Electronics*, Benjamin/Cummings, 1991.

D. Hodges, *Semiconductor Memories*, IEEE Press, 1972.

D. Hodges and H. Jackson, *Analysis and Design of Digital Integrated Circuits*, 2nd ed., McGraw-Hill, 1988.

R. K. Watts, *Submicron Integrated Circuits*, Wiley, 1989.

Design Tools and Methodologies

W. Bhanzhaf, *Computer Aided Circuit Analysis Using SPICE*, Prentice Hall, 1992.

G. De Micheli, *Synthesis and Optimization of Digital Circuits*, McGraw-Hill, 1994.

E. Elmasry, ed., *Digital VLSI Systems*, IEEE Press, 1985.

S. Rubin, *Computer Aids for VLSI Design*, Addison-Wesley, 1987.

P. Tuinenga, *A Guide to Circuit Simulation & Analysis using PSpice*, Prentice Hall, 1988.

W. Wolf, *Modern VLSI Design: A Systems Approach*, Prentice Hall, 1994.

MOS

M. Annaratone, *Digital CMOS Circuit Design*, Kluwer, 1986.

T. Dillinger, *VLSI Engineering*, Prentice Hall, 1988.

E. Elmasry, ed., *Digital MOS Integrated Circuits*, IEEE Press, 1981.

E. Elmasry, ed., *Digital MOS Integrated Circuits II*, IEEE Press, 1992.

Glasser and Dopperpuhl, *The Design and Analysis of VLSI Circuits*, Addison-Wesley, 1985.

Mead and Conway, *Introduction to VLSI Systems*, Addison-Wesley, 1980.

D. Pucknell and K. Eshraghian, *Basic VLSI Design*, Prentice Hall, 1988.

M. Shoji, *CMOS Digital Circuit Technology*, Prentice Hall, 1988.

J. Uyemura, *Fundamentals of MOS Digital Integrated Circuits*, Addison-Wesley, 1988.

J. Uyemura, *Circuit Design for CMOS VLSI*, Kluwer, 1992.

H. Veendrick, *MOS IC's: From Basics to ASICS*, VCH, 1992.

Weste and Eshraghian, *Principles of CMOS VLSI Design*, Addison-Wesley, 1985, 1993.

Bipolar and BiCMOS

A. Alvarez, *BiCMOS Technology and Its Applications*, Kluwer, 1989.

M. Elmasry, *Digital Bipolar Integrated Circuits*, Wiley, 1983.

M. Elmasry, ed., *BiCMOS Integrated Circuit Design,* IEEE Press, 1994.

S. Embabi, A. Bellaouar, and M. Elmasry, *Digital BiCMOS Integrated Circuit Design*, Kluwer, 1993.

Lynn et al., eds., *Analysis and Design of Integrated Circuits*, McGraw-Hill, 1967.

Gallium Arsenide

N. Kanopoulos, *Gallium Arsenide Digital Integrated Circuits*, Prentice Hall, 1989.

V. Milutinovic, ed., *Microprocessor Design for GaAs Technology*, Prentice Hall, 1990.

S. Long and S. Butner, *Gallium Arsenide Digital Integrated Circuit Design*, McGraw-Hill, NY, 1990.

Low-Temperature Digital Electronics

R. Kisselman, *Low-Temperature Electronics*, IEEE Press, 1986.

REFERENCES

[Bardeen48] J. Bardeen and W. Brattain, "The Transistor, a Semiconductor Triode," *Phys. Rev.*, vol. 74, p. 230, July 15, 1948.

[Beeson62] R. Beeson and H. Ruegg, "New Forms of All Transistor Logic," *ISSCC Digest of Technical Papers*, pp. 10–11, Feb. 1962.

[Harris56] J. Harris, "Direct-Coupled Transitor Logic Circuitry in Digital Computers," *ISSCC Digest of Technical Papers*, p. 9, Feb. 1956.

[Hart72] C. Hart and M. Slob, "Integrated Injection Logic—A New Approach to LSI," *ISSCC Digest of Technical Papers*, pp. 92–93, Feb. 1972.

[Hoff70] E. Hoff, "Silicon-Gate Dynamic MOS Crams 1,024 Bits on a Chip," *Electronics*, pp. 68–73, August 3, 1970.

[Masaki74] A. Masaki, Y. Harada and T. Chiba, "200-Gate ECL Master-Slice LSI," *ISSCC Digest of Technical Papers*, pp. 62–63, Feb. 1974.

[Masaki92] A. Masaki, "Deep-Submicron CMOS Warms Up to High-Speed Logic," *Circuits and Devices Magazine*, Nov. 1992.

[Murphy93] B. Murphy, "Perspectives on Logic and Microprocessors," *Commemorative Supplement to the Digest of Technical Papers, ISSCC Conf.*, pp. 49–51, San Francisco, 1993.

[Norman60] R. Norman, J. Last and I. Haas, "Solid-State Micrologic Elements," *ISSCC Digest of Technical Papers*, pp. 82–83, Feb. 1960.

[Sasaki91] H. Sasaki, H. Abe, T. Enomoto, and Y. Yano, "Prospect for the Chip Architecture in Sub-Halh-Micron ULSI Era," *IEICE Transactions*, Vol. E 74, No. 1, pp. 119–129, January 1991.

[Schockley49] W. Schockley, "The Theory of pn Junctions in Semiconductors and pn-Junction Transistors," *BSTJ*, vol. 28, p. 435, 1949.

[Schutz94] J. Schutz, "A 3.3V, 0.6 mm BiCMOS Superscaler Microprocessor," *ISSCC Digest of Technical Papers*, pp. 202–203, Feb. 1994.

[Shima74] M. Shima, F. Faggin and S. Mazor, "An N-Channel, 8-bit Single-Chip Microprocessor," *ISSCC Digest of Technical Papers*, pp. 56–57, Feb. 1974.

[Swade93] D. Swade, "Redeeming Charles Babbage's Mechanical Computer," *Scientific American*, pp. 86–91, February 1993.

[Wanlass63] F. Wanlass, and C. Sah, "Nanowatt logic Using Field-Effect Metal-Oxide Semiconductor Triodes," *ISSCC Digest of Technical Papers*, pp. 32–32, Feb. 1963.

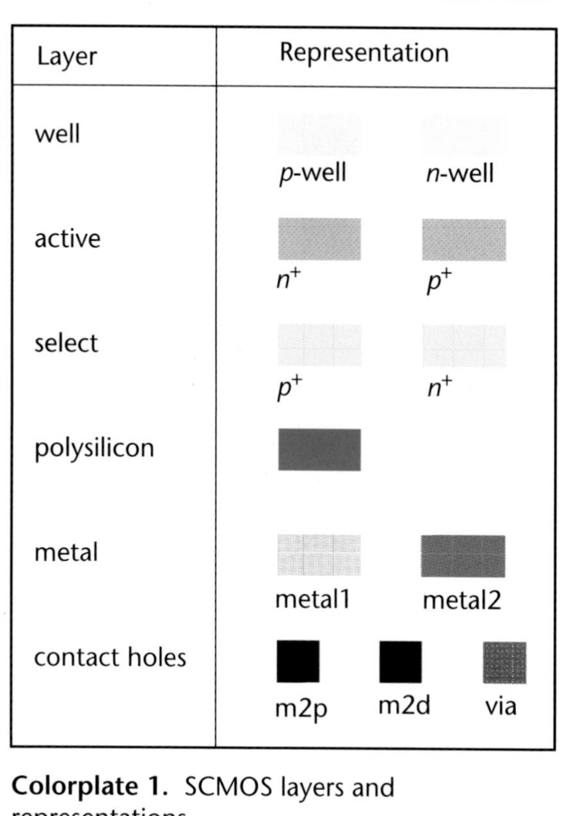

Layer
Representation
well
p-well
n-well
active
n+
p+
select
p+
n+
polysilicon
metal
metal1
metal2
contact holes
m2p
m2d
via

Colorplate 1. SCMOS layers and representations.

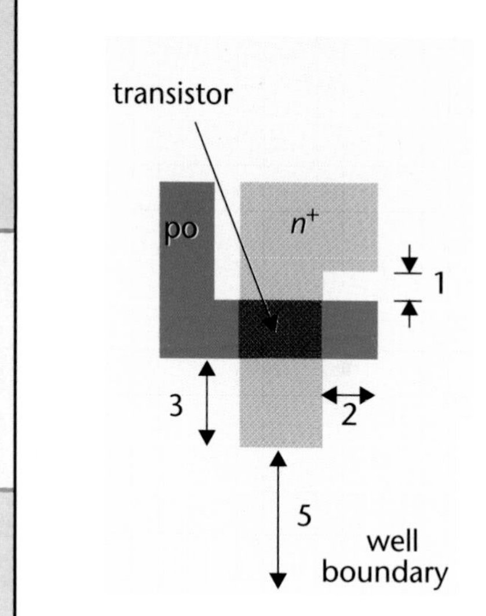

Colorplate 3. Design rules concerning transistor layout.

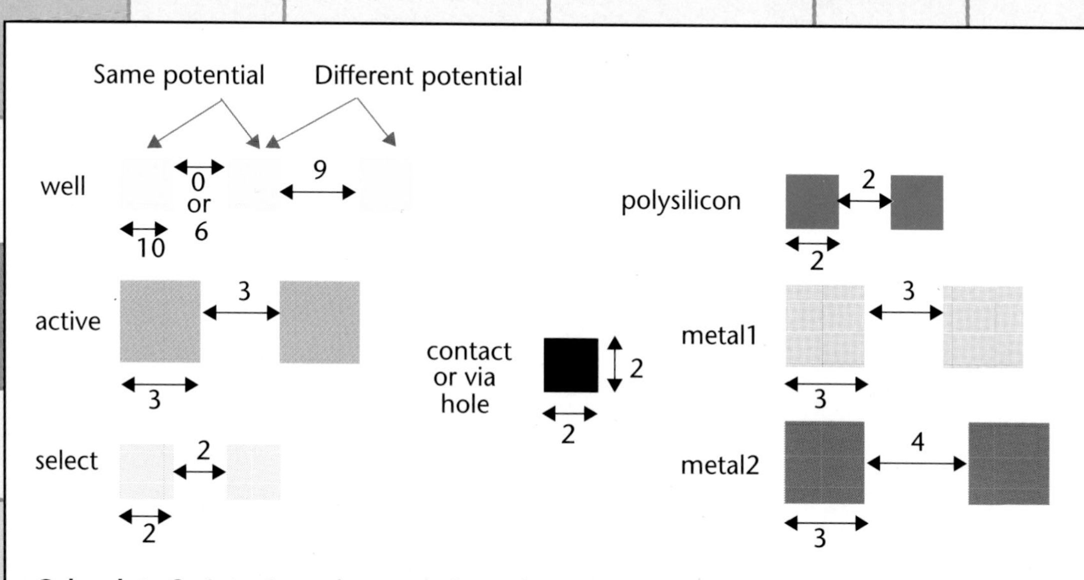

Colorplate 2. Intra-Layer layout design rules: minimum dimensions and spacings.

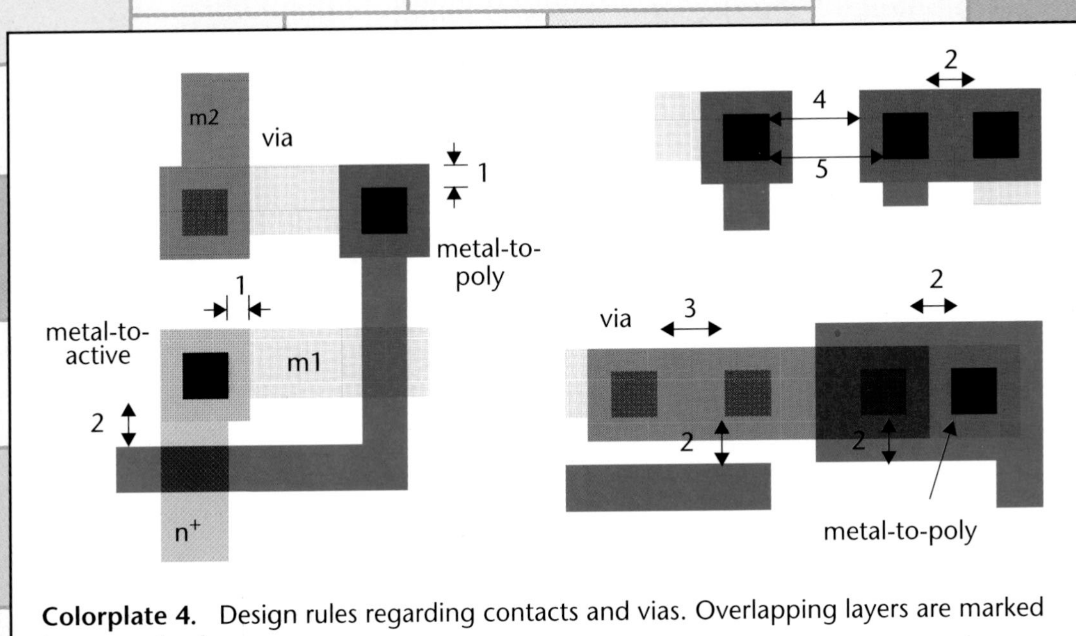

Colorplate 4. Design rules regarding contacts and vias. Overlapping layers are marked by merged colors.

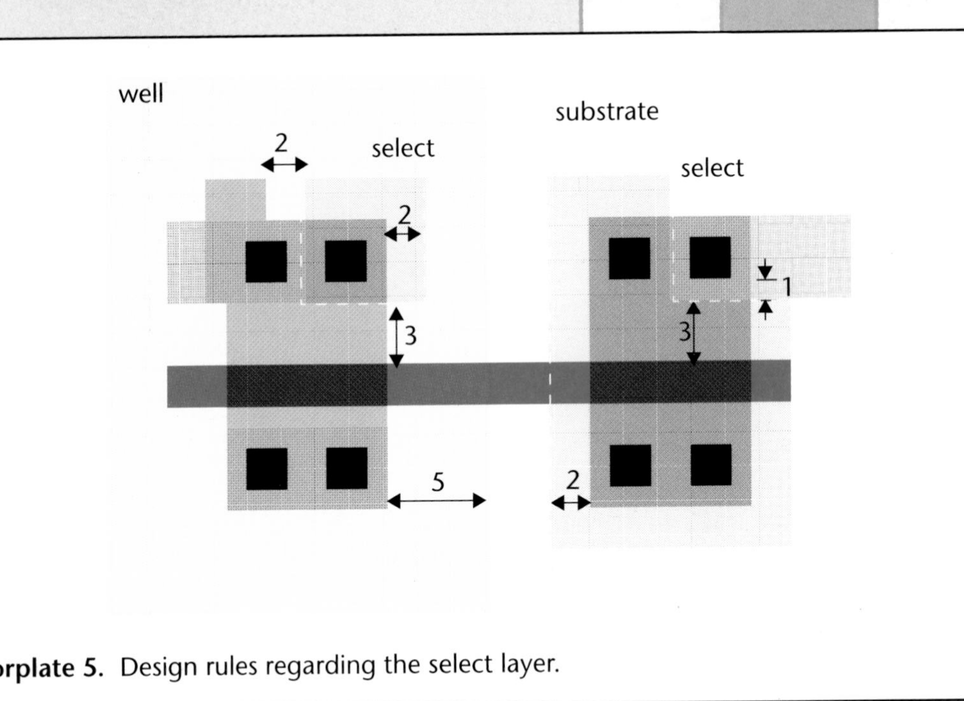

Colorplate 5. Design rules regarding the select layer.

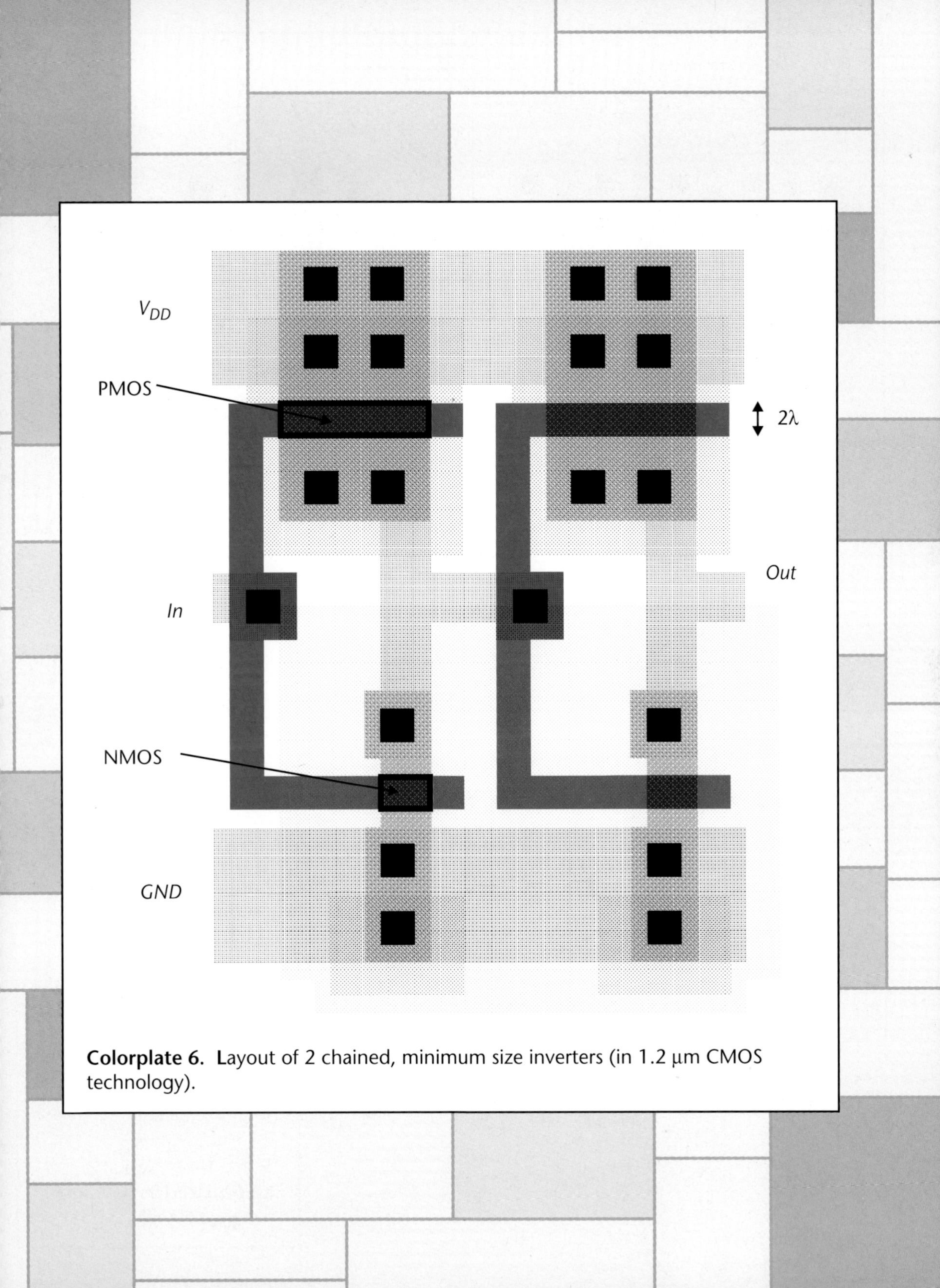

Colorplate 6. Layout of 2 chained, minimum size inverters (in 1.2 μm CMOS technology).

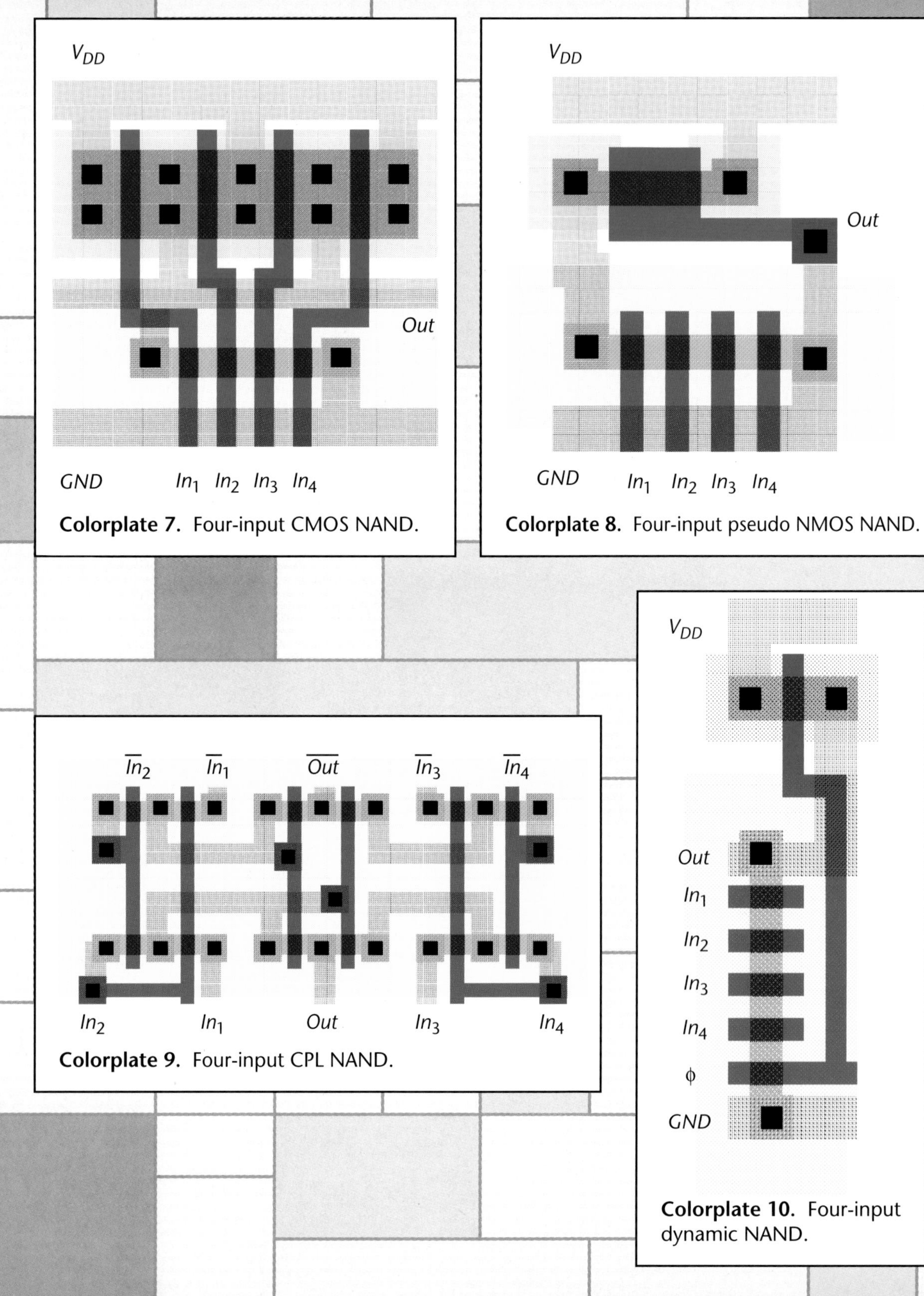

Colorplate 7. Four-input CMOS NAND.

Colorplate 8. Four-input pseudo NMOS NAND.

Colorplate 9. Four-input CPL NAND.

Colorplate 10. Four-input dynamic NAND.

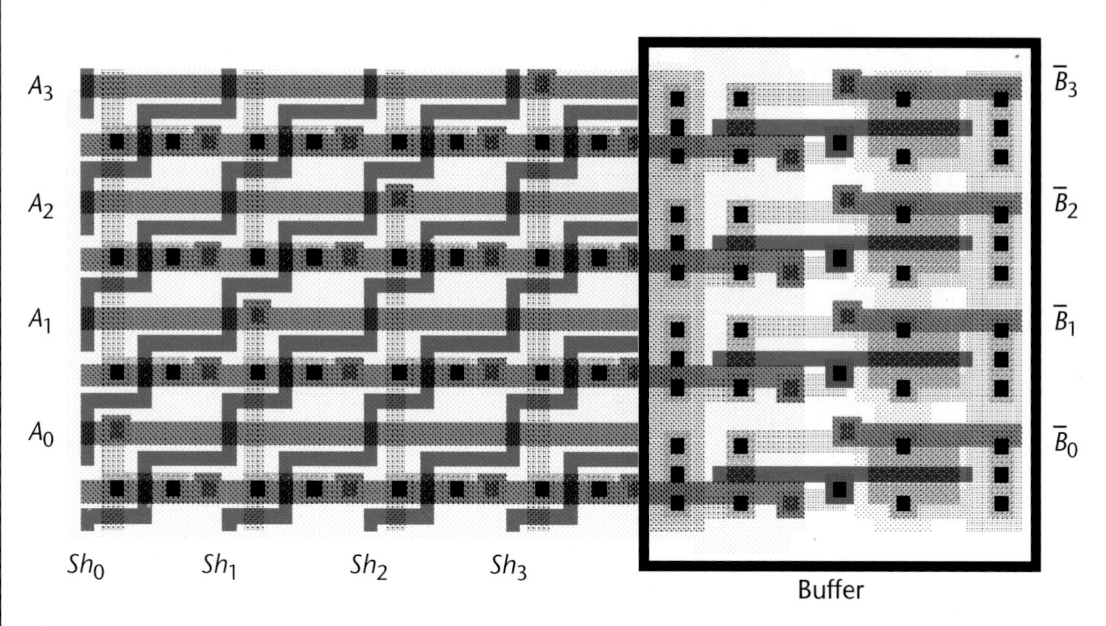

Colorplate 11. Four bit (by 4) barrel shifter—layout.

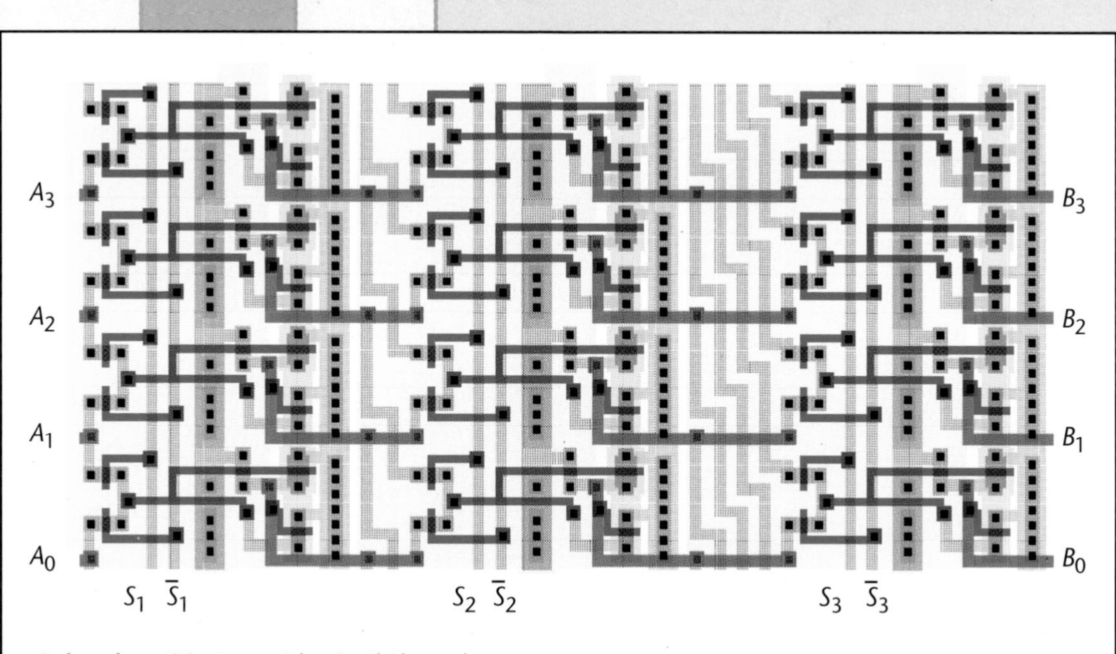

Colorplate 12. Logarithmic Shifter—layout.

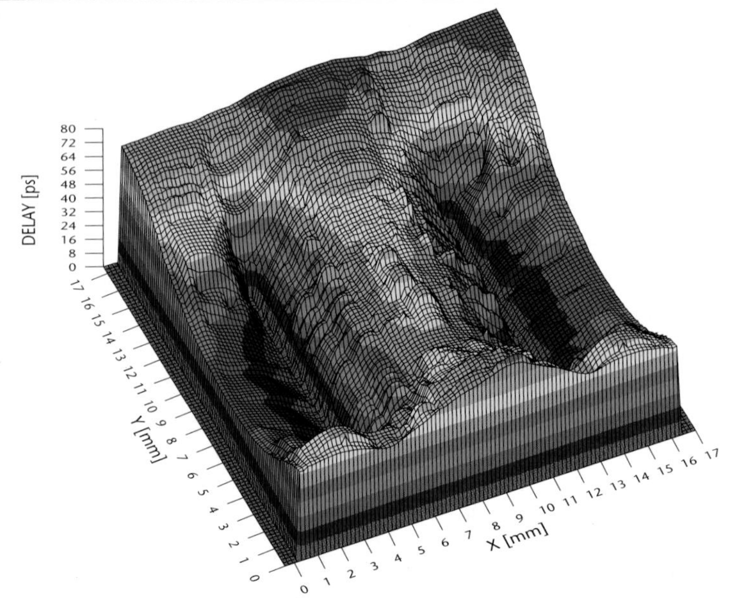

Colorplate 13. Clock skew in Digital Equipment 21164 Alpha microprocessor.

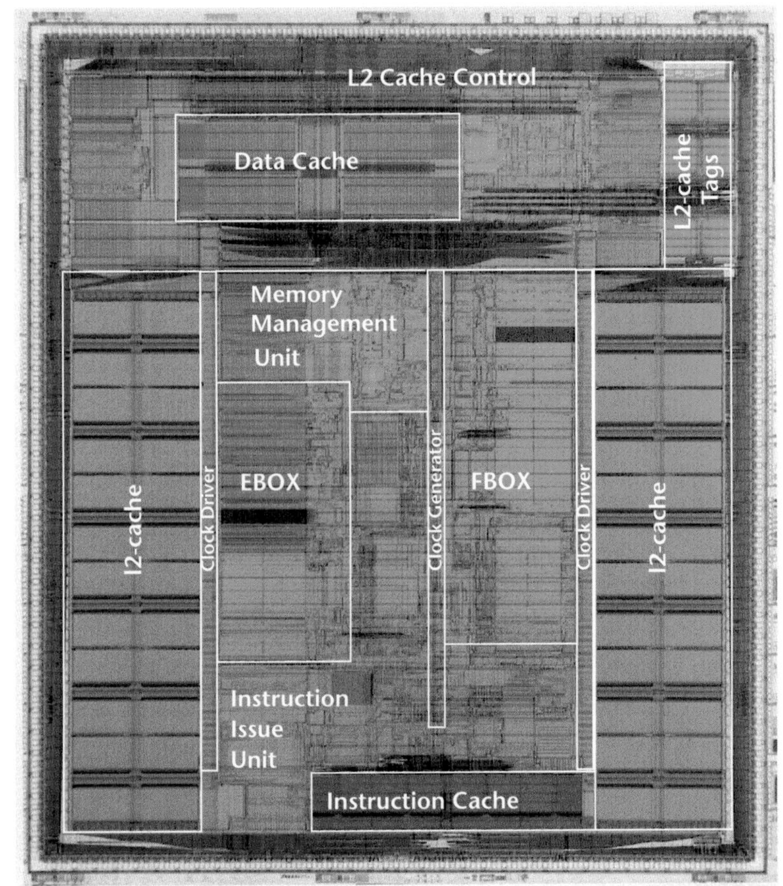

Colorplate 14. Digital Equipment Corporation 21164 Alpha microprocessor.

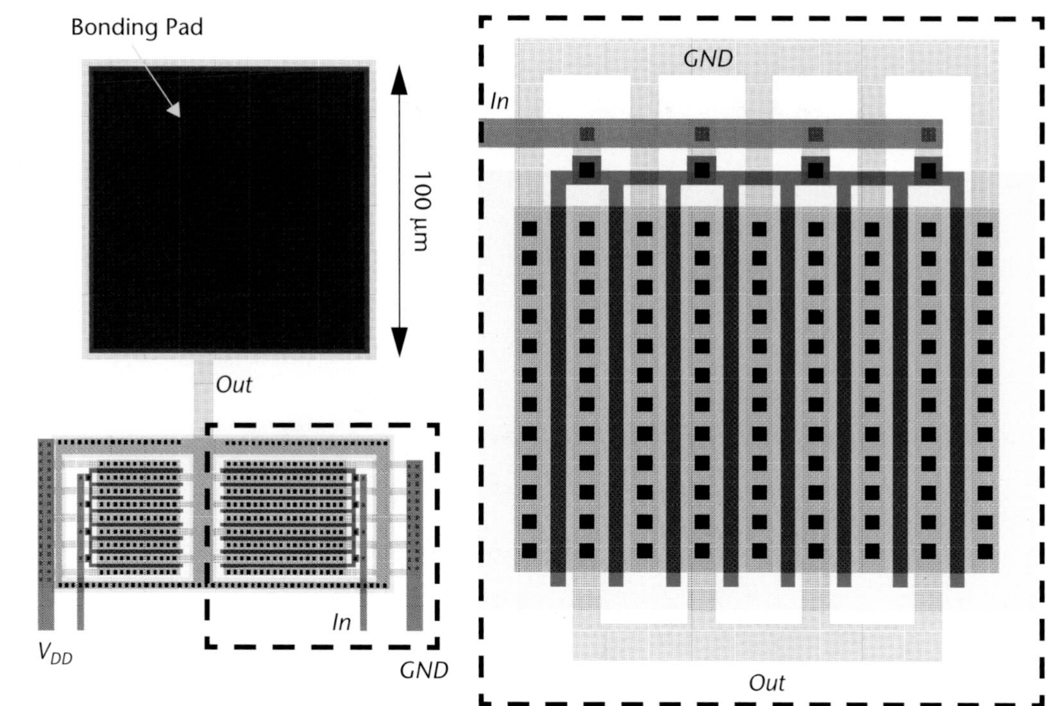

Colorplate 15. Layout of final stage of bonding-pad driver, including magnification of NMOS transistor.

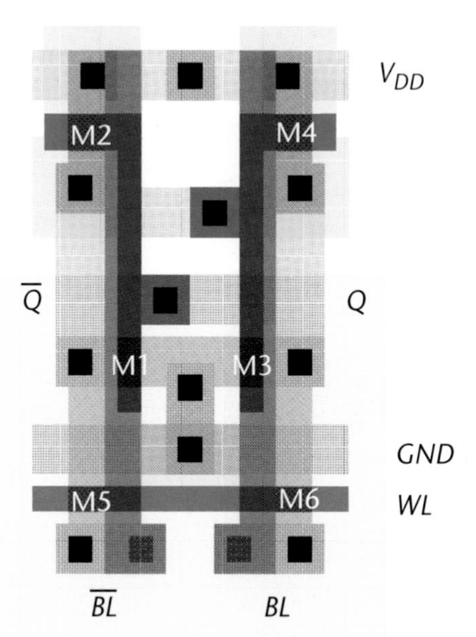

Colorplate 16. Layout of 6-transistor CMOS SRAM memory cell.

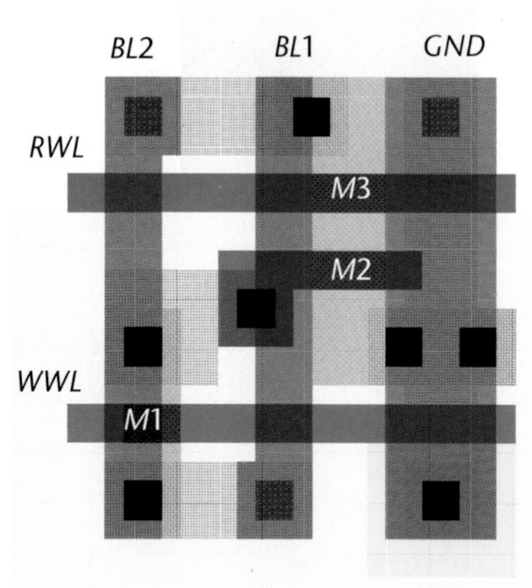

Colorplate 17. Layout of 3-transistor dynamic memory cell.

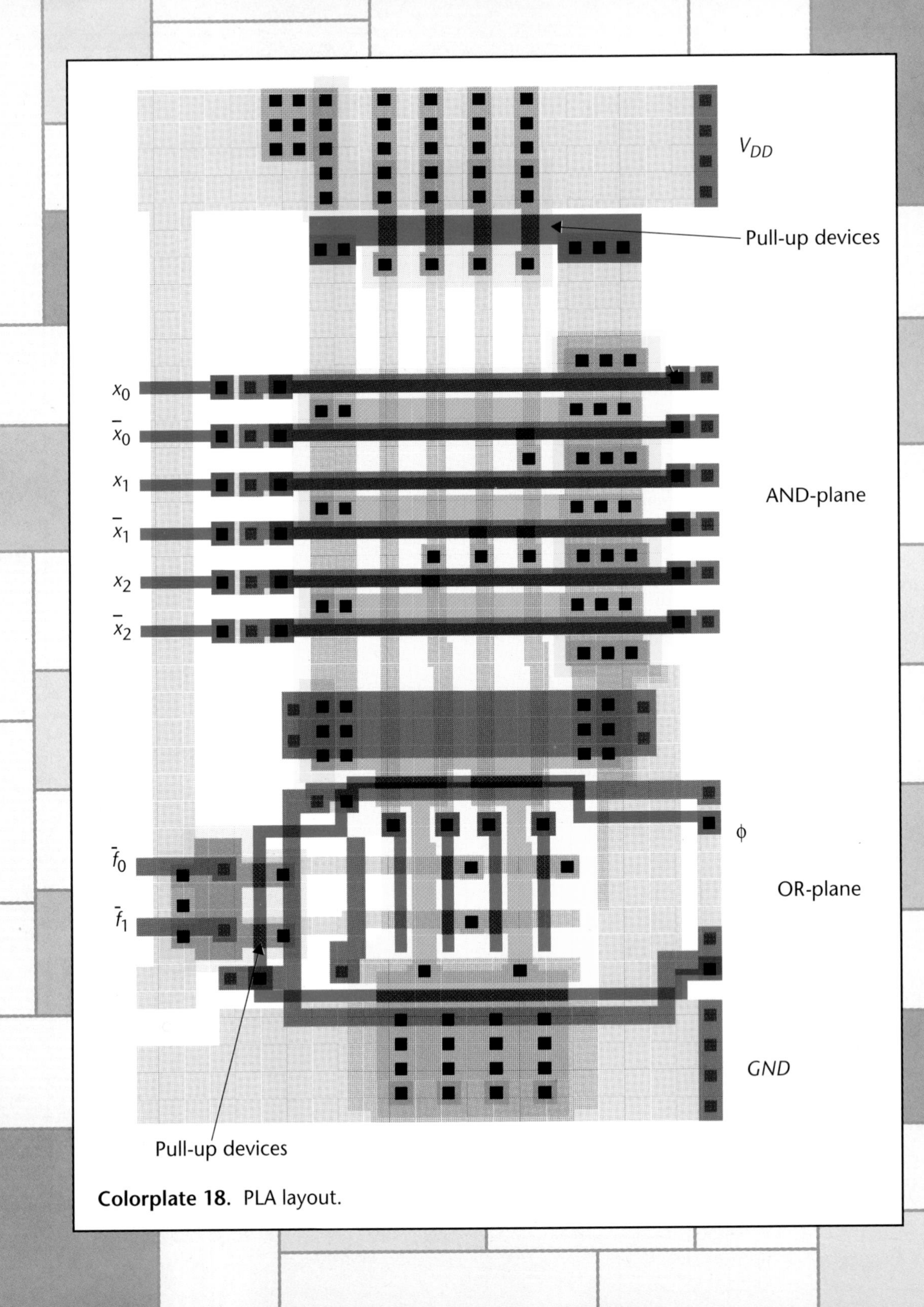

Colorplate 18. PLA layout.

1.4 Exercises

1. [E, None, 1.2] Based on the evolutionary trends described in the chapter, predict the integration complexity and the clock speed of a microprocessor in the year 2010. Determine also how much DRAM should be available on a single chip at that point in time, if Moore's law would still hold.
2. [D, None, 1.2] By scanning the literature, find the leading-edge devices at this point in time in the following domains: microprocessor, SRAM, and DRAM. Determine for each of those, the number of integrated devices, the overall area and the maximum clock speed. Evaluate the match with the trends predicted in section 1.2.
3. [D, None, 1.2] Find in the library the latest November issue of the *Journal of Solid State Circuits.* For each of the papers, determine its application class (such as microprocessor, signal processor, DRAM, SRAM, Gate Array), the type of manufacturing technology used (MOS, bipolar, etc.), the minimum feature size, the number of devices on a single die, and the maximum clock speed. Tabulate the results along the various application classes.
4. [E, None, 1.2] Provide at least three examples for each of the abstraction levels described in Figure 1.5.

PART

I

A CIRCUIT PERSPECTIVE

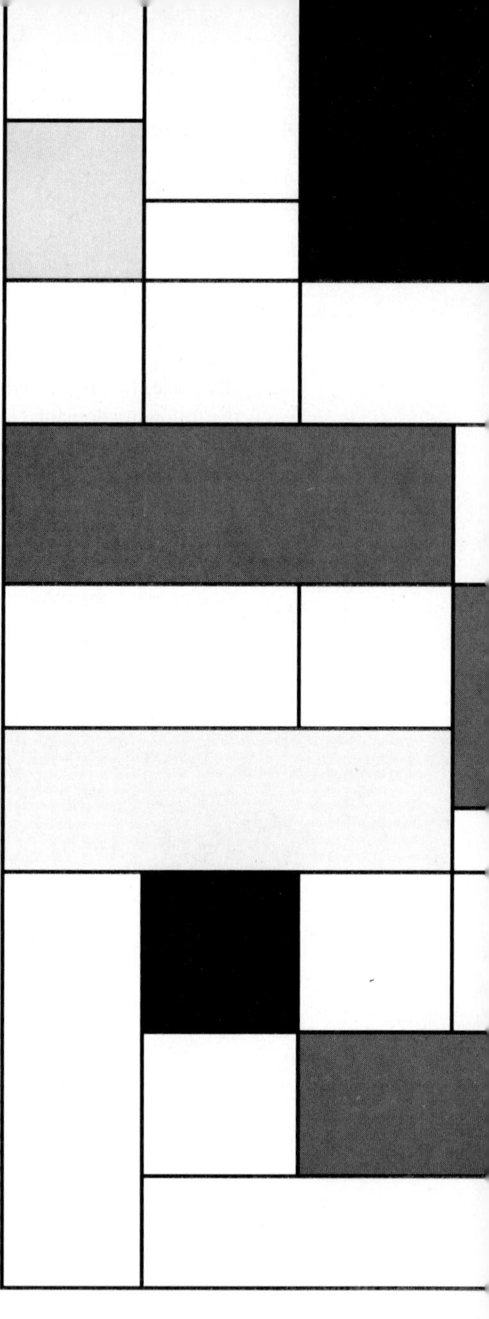

Philosophy is written
in this grand book
—I mean the universe—
which stands continually open
to our gaze,
but it cannot be understood
unless one first learns
to comprehend the language
and interpret the characters
in which it is written.
It is written in the language of
mathematics, and its characters are
triangles, circles and other geometrical
figures, without which
it is humanly impossible
to understand a single word of it.

Galileo Galilei
Il Saggiatore, 1623

CHAPTER 2

THE DEVICES

Qualitative understanding of MOS and bipolar device operation

Simple device models for manual analysis

Detailed device models for SPICE

Impact of process variations

2.1 Introduction

It is a well-known premise in engineering that the conception of a complex construction without a prior understanding of the underlying building blocks is a sure road to failure. This surely holds for digital circuit design as well. The basic building blocks in this engineering domain are the silicon semiconductor devices, more specifically the diodes, and the MOS and bipolar transistors.

Giving the reader the necessary *knowledge and understanding of these devices* is the prime motivation for this chapter. It is not our intention to present an in-depth discussion (we assume that the reader has some prior familiarity with electronic devices). The goal is rather to refresh the memory, to introduce some notational conventions, and to highlight a number of properties and parameters that are particularly important in the design of digital gates. We further identify the fundamental differences between bipolar and MOS transistors that helps to explain the differences in the topology of digital circuits manufactured in those technologies.

Another important function of this chapter is the introduction of the *device models*. Taking all the physical aspects of each device into account when designing complex digital circuits leads to an unnecessary complexity that quickly becomes intractable. Such an approach is similar to considering the molecular structure of concrete when constructing a bridge. To deal with this issue, an abstraction of the device behavior called a *model* is typically employed. A range of models can be conceived for each device presenting a trade-off between accuracy and complexity. A simple first-order model is useful for manual analysis. It has limited accuracy but helps us to understand the operation of the circuit and its dominant parameters. When more accurate results are needed, complex, second- or higher-order models are employed in conjunction with computer-aided simulation. In this chapter, we present both first-order models for manual analysis as well as higher-order models for simulation for each device of interest.

Designers tend to take the device parameters offered in the models for granted. They should be aware, however, that these are only nominal values, and that the actual parameter values vary with operating temperature, over manufacturing runs, or even over a single wafer. To highlight this issue, a short discussion on *process variations* and their impact is included in the chapter.

Since this text focuses on the *design aspect* of digital integrated circuits, a mere presentation of an analytical model of a device is not sufficient. Turning a conceived circuit into an actual implementation also requires a knowledge of the manufacturing process and its constraints. The interface between the design and processing world, is captured as a set of *design rules* that act as prescriptions for preparing the masks used in the fabrication process of integrated circuits. The design rules for a representative IC process are introduced in Appendix A to this chapter. A detailed description of IC fabrication processes is beyond the scope of this textbook.

2.2 The Diode

Although diodes rarely occur directly in the schematic diagrams of present-day digital gates, they are still omnipresent. For instance, each MOS transistor implicitly contains a number of reverse-biased diodes. Diodes are used to protect the input devices of an IC against static charges. Also, a number of bipolar gates use diodes as a means to adjust voltage levels. Therefore, a brief review of the basic properties and device equations of the diode is appropriate.

2.2.1 A First Glance at the Device

The *pn*-junction diode is the simplest of the semiconductor devices. Figure 2.1a shows a cross-section of a typical *pn*-junction. It consists of two homogeneous regions of *p*- and *n*-type material, separated by a region of transition from one type of doping to another, which is assumed thin. Such a device is called a *step* or *abrupt junction*. The *p*-type material is doped with *acceptor* impurities (such as boron), which results in the presence of holes as the dominant or majority carriers. Similarly, the doping of silicon with *donor* impurities (such as phosphorus or arsenic) creates an *n*-type material, where electrons are the majority carriers. Aluminum contacts provide access to the *p*- and *n*-terminals of the device. The circuit symbol of the diode, as used in schematic diagrams, is introduced in Figure 2.1c.

To understand the behavior of the *pn*-junction diode, we often resort to a one-dimensional simplification of the device (Figure 2.1b). Bringing the *p*- and *n*-type materials together causes a large concentration gradient at the boundary. The electron concentration changes from a high value in the *n*-type material to a very small value in the *p*-type

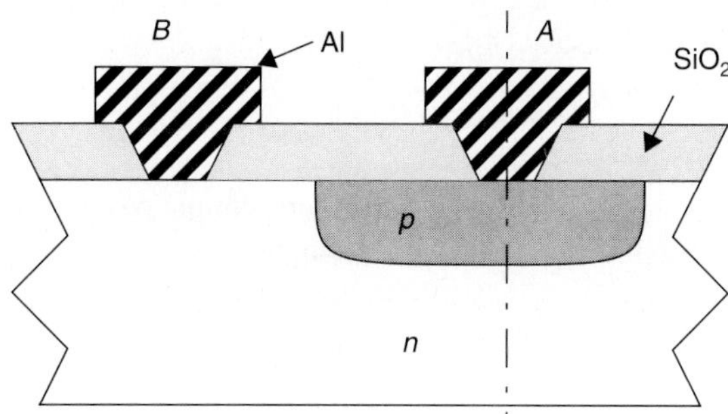

(a) Cross-section of *pn*-junction in an IC process

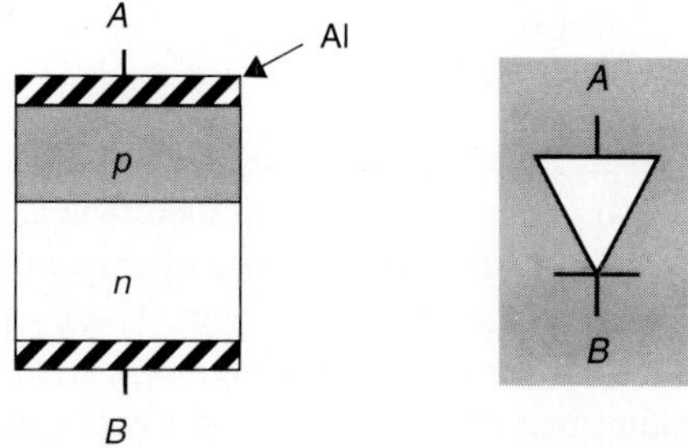

(b) One-dimensional representation

(c) Diode symbol

Figure 2.1 Abrupt *pn*-junction diode and its schematic symbol.

material The reverse is true for the hole concentration. This gradient causes electrons to *diffuse* from n to p and holes to diffuse from p to n. When the holes leave the p-type material, they leave behind immobile acceptor ions, which are negatively charged. Consequently, the p-type material is negatively charged in the vicinity of the pn-boundary. Similarly, a positive charge builds up on the n-side of the boundary as the diffusing electrons leave behind the positively charged donor ions. The region at the junction, where the majority carriers have been removed, leaving the fixed acceptor and donor ions, is called the *depletion* or *space-charge region.* The charges create an electric field across the boundary, directed from the n to the p-region. This field counteracts the diffusion of holes and electrons, as it causes electrons to *drift* from p to n and holes to drift from n to p. Under equilibrium, the depletion charge sets up an electric field such that the drift currents are equal and opposite to the diffusion currents, resulting in a zero net flow.

The above analysis is summarized in Figure 2.2 that plots the current directions, the charge density, the electrical field, and the electrostatic field of the abrupt pn-junction under zero-bias conditions. In the device shown, the p material is more heavily doped than the n, or $N_A > N_D$, with N_A and N_D the acceptor and donor concentrations, respectively. Hence, the charge concentration in the depletion region is higher on the p-side of the junction. Figure 2.2 also shows that under zero bias, there exists a voltage ϕ_0 across the junction, called the *built-in potential*. This potential has the value

$$\phi_0 = \phi_T \ln\left[\frac{N_A N_D}{n_i^2}\right] \tag{2.1}$$

where ϕ_T is the *thermal voltage*

$$\phi_T = \frac{kT}{q} = 26\,\text{mV at 300 K} \tag{2.2}$$

The quantity n_i is the intrinsic carrier concentration in a pure sample of the semiconductor and equals approximately $1.5 \times 10^{10}\ \text{cm}^{-3}$ at 300 K for silicon.

Example 2.1 Built-in Voltage of *pn*-junction

An abrupt junction has doping densities of $N_A = 10^{15}$ atoms/cm^3, and $N_D = 10^{16}$ atoms/cm^3. Calculate the built-in potential at 300 K.

From Eq. (2.1),

$$\phi_0 = 26\ln\left[\frac{10^{15}\times 10^{16}}{2.25\times 10^{20}}\right]\ \text{mV} = 638\ \text{mV}$$

Assume now that a forward voltage V_D is applied to the junction or, in other words, that the potential of the p-region is raised with respect to the n-zone. The applied potential lowers the potential barrier. Consequently, the flow of mobile carriers across the junction increases as the diffusion current dominates the drift component. These carriers traverse the depletion region and are injected into the neutral n- and p-regions, where they become minority carriers. Under the assumption that no voltage gradient exists over the neutral regions, which is approximately the case for most modern devices, these minority carriers will diffuse through the region as a result of the concentration gradient until they get

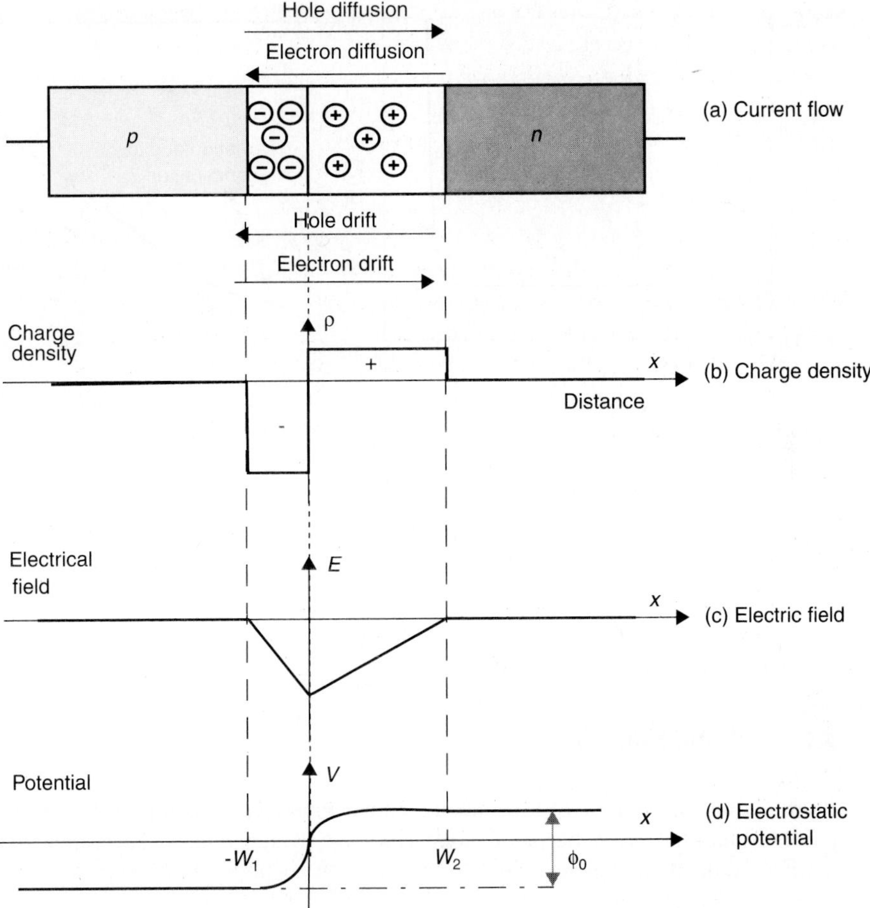

Figure 2.2 The abrupt *pn*-junction under equilibrium bias.

recombined with a majority carrier. The net result is a current flowing through the diode from the *p*-region to the *n*-region. The most important property of this current is its *exponential dependence* upon the applied bias voltage.

On the other hand, when a reverse voltage V_D is applied to the junction or when the potential of the *p*-region is lowered with respect to the *n*-region, the potential barrier is raised. This results in a reduction in the diffusion current, and the drift current becomes dominant. A current flows from the *n*-region to the *p*-region. Since the number of minority carriers in the neutral regions (electrons in the *p*-zone, holes in the *n*-region) is very small, this drift current component is virtually ignorable. It is fair to state that in the reverse-bias mode the diode operates as a nonconducting, or blocking, device. The diode thus acts as a one-way conductor. This is illustrated in Figure 2.3, which plots the diode current I_D as a function of the bias voltage V_D. The exponential behavior for positive-bias voltages is shown in Figure 2.3b, where the current is plotted on a logarithmic scale. The current increases by a factor of 10 for every extra 60 mV (= 2.3 ϕ_T) of forward bias. At small voltage levels ($V_D < 0.15$ V), a deviation from the exponential dependence can be observed,

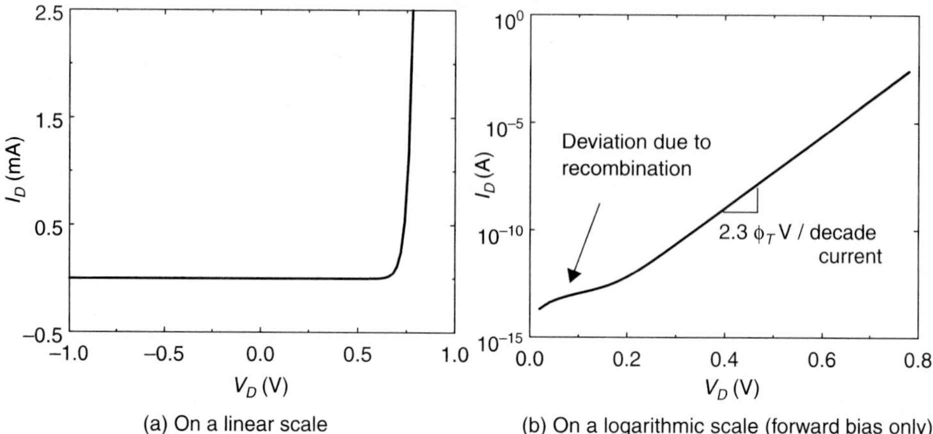

Figure 2.3 Diode current as a function of the bias voltage V_D.

which is due to the recombination of holes and electrons in the depletion region as discussed in more detail later in the chapter.

After this intuitive introduction, we present analytical expressions for the behavior of the *pn*-junction. A distinction is made between the *static* (or steady-state) and the *dynamic* (or transient) behavior of the device.

2.2.2 Static Behavior

From earlier encounters with semiconductor devices [e.g., Sedra87], the reader is most probably familiar with the *ideal diode equation*, which relates the current through the diode I_D to the diode bias voltage V_D

$$I_D = I_S(e^{V_D/\phi_T} - 1) \tag{2.3}$$

I_S represents a constant value, called the *saturation current* of the diode. Under reverse-bias conditions, where $V_D << 0$, $I_D \approx -I_S$ and equals the reverse-bias leakage current. ϕ_T is the thermal voltage of Eq. (2.2) and is equal to 26 mV at room temperature. The remainder of this section presents a physical background for this equation.

Forward Bias

When a positive voltage is applied to the junction, mobile carriers drift through the depletion region and are injected into the neutral regions, where they become *excess minority carriers* and diffuse in the direction of the terminal connections. It is the distribution of these excess minority carriers that dictates the static behavior of the *pn*-diode. Figure 2.4 shows the minority carrier concentrations in the neutral region near a *pn*-junction for the forward-bias condition. Observe that the majority carrier concentrations have to proceed along the same line, because charge neutrality dictates that any local increase in the electron (hole) concentration is matched by a similar increase in the hole (electron) density.

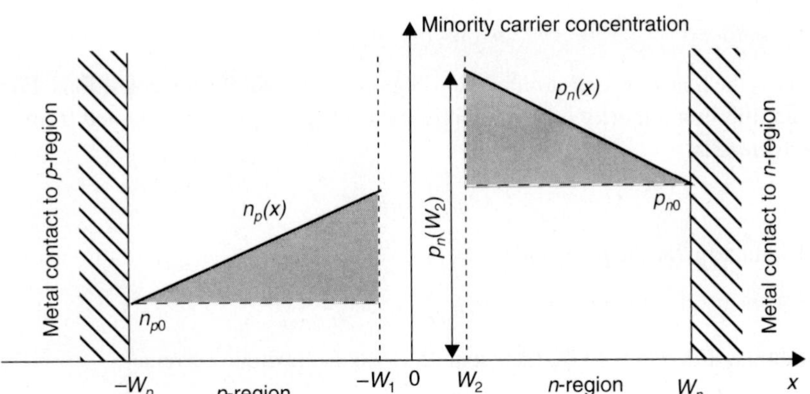

Figure 2.4 Minority carrier concentrations in the neutral region near an abrupt *pn*-junction under forward-bias conditions.

While the fractional increase of the minority carrier concentration is substantial, it is largely ignorable for the majority carriers.

Figure 2.4 shows a linear decay in the minority carrier concentrations when moving away from the junction. At the metal contacts (which can be assumed to be infinite sources or sinks of holes or electrons), the minority carrier concentrations are at their equilibrium values (n_{p0} and p_{n0}), independent of the applied bias. The linear decay model is valid under the assumption that the width of the *n*- or *p*-regions is sufficiently small so that injected minority carriers diffuse to the metal contact before recombining with majority carriers. This operation condition is called the *short-base diode model* and is valid for most contemporary semiconductor diodes.

The gradient in the minority concentrations causes a diffusion current in the neutral (also called *bulk*) regions that is proportional to that gradient. The constant of proportionality is called the *diffusion coefficient* (D_p and D_n for holes and electrons, respectively). Based on these observations, an expression for the diode current can be derived. In this derivation, we initially consider the *n*-region only. Similar expressions can be derived for the *p*-region.

$$I_{D,p} \sim \frac{dp_n}{dx} = -qA_D D_p \frac{dp_n}{dx} \tag{2.4}$$

with

$$p_n(x) = -\left[\frac{p_n(W_2) - p_{n0}}{W_n - W_2}\right]x + \left[\frac{p_n(W_2)W_n - p_{n0}W_2}{W_n - W_2}\right] \tag{2.5}$$

where $p_n(x)$ represents the hole concentration in the *n*-region as a function of the position x, A_D is the junction area, and q *is* the electron (hole) charge. Combining the two equations results in an expression of the diode current as a function of the minority carrier concentration at the boundary of the depletion region. The latter is determined by the *law of the junction,* which states that the concentration at the edge of the depletion region is an exponential function of the applied bias voltage

$$p_n(W_2) = p_{n0}e^{V_D/\phi_T} \tag{2.6}$$

with p_{n0} the hole concentration in the n-region under equilibrium conditions. For $N_A >> n_i$, the equilibrium minority hole concentration (in the n-region) is obtained from the following expression

$$p_{n0} \approx n_i^2/N_D \tag{2.7}$$

and, similarly, for the p-region

$$n_{p0} \approx n_i^2/N_A \tag{2.8}$$

Combining Equations (2.4), (2.5) and (2.6) yields the diode current,

$$I_{Dp} = qA_D D_p \frac{p_{n0}}{W_n - W_2}(e^{V_D/\phi_T} - 1) \tag{2.9}$$

Repeating the same analysis for the p-region and summing the p and n current-contributions produces the ideal diode current expression of Eq. (2.3). It also yields an expression for the saturation current I_S

$$I_S = qA_D\left(\frac{D_p p_{n0}}{W_n - W_2} + \frac{D_n n_{p0}}{W_p - W_1}\right) \tag{2.10}$$

Keep in mind that the above equation is based on a number of assumptions, which might not be valid for all actual devices. First of all, it is assumed that the length of the neutral regions is short enough that recombination does not occur (*short-base diode* model). For this to be valid, the widths of the p- and n-regions must be substantially smaller than a material constant called the *diffusion length* (denoted as L_p and L_n for holes and electrons, respectively). If this is not the case, the diode becomes of the *long-base* type. Minority carriers diffusing through the neutral region gradually recombine with majority carriers. This affects the minority-carrier concentration as illustrated in Figure 2.5. Instead of a linear decay, the concentration drops in an exponential fashion. In one diffusion length, the excess minority concentration drops to $1/e$ (= 0.37) of its original value. After a few diffusion lengths, virtually all injected carriers have recombined, and the minority carrier concentration reaches its thermal equilibrium value. The current equation remains essentially unchanged. The only modification is that the W_n and W_p parameters in the saturation current (Eq. (2.10)) are replaced by the diffusion lengths L_p and L_n.

Other assumptions are that the resistance of the neutral regions is negligible, and that the minority carrier injection levels are substantially lower than the majority concentration levels (*low-injection condition*). Later in the chapter we discuss how violating these conditions affects the device operation.

Eq. (2.10) clearly shows that the diode current is the composite result of a hole and an electron current. In most practical diodes, one of the sides has a substantially lighter doping level and hence produces a larger number of minority carriers. The corresponding current component dominates the overall value. For instance, in the example of Figure 2.2, the p-region has a heavier doping than the n-region. Consequently, $p_{n0} >> n_{p0}$, and the hole current dominates.

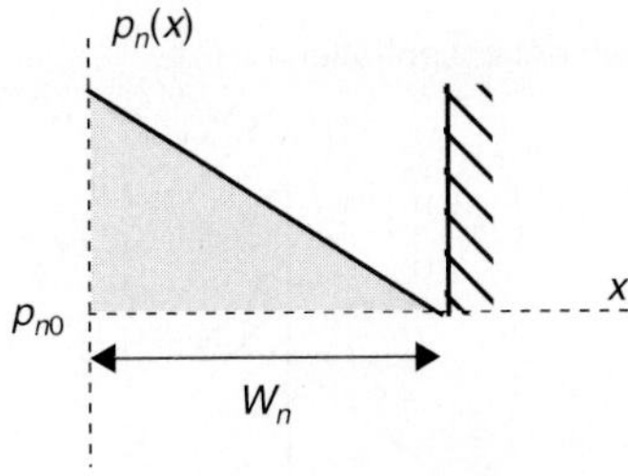

(a) Short-base diode: $W_n << L_p$

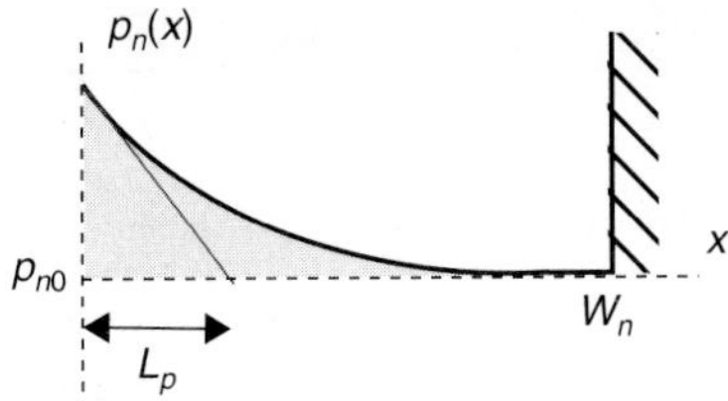

(b) Long-base diode: $W_n >> L_p$

Figure 2.5 Minority carrier concentrations in the *n*-region near a *pn*-junction.

Problem 2.1 Diode Current

For a diode with the following properties, compute the saturation current I_S. Also, solve V_D for $I_D = 0.1$ mA, assuming that $\phi_T = 26$ mV.

$A_D = 9\ \mu m^2$,
$N_D = 5 \times 10^{15}\ cm^{-3}$,
$N_A = 2.5 \times 10^{16}\ cm^{-3}$,
$\phi_0 = 0.795$ V,
$D_n = 25\ cm^2/sec$,
$D_p = 10\ cm^2/sec$,
$W_n = 5\ \mu m$ and $W_p = 0.7\ \mu m$,
$W_2 = 0.15\ \mu m$ and $W_1 = 0.03\ \mu m$.
$n_i = 1.5 \times 10^{10}\ cm^{-3}$ and $q = 1.6 \times 10^{-19}$ C.
Also, $L_n = 5\ \mu m$ and $L_p = 31\ \mu m$.

Reverse Bias

When applying a reverse-bias voltage to the junction, the ideal diode equation predicts that the diode current I_D approaches $-I_S$ for $|V_D| >> \phi_T$. This is readily understood when analyzing the minority carrier concentration distribution under reverse-bias conditions, as shown in Figure 2.6.[1] From the law of the junction (which is equally valid under reverse-bias conditions), it can be derived that the concentration of minority carriers at the depletion-region boundaries is small and actually approaches 0 when sufficient reverse bias is applied. At the metallic contacts, the concentration is restored to the thermal equilibrium value.

[1] It is worth observing that all equations derived for the forward-bias apply just as well under reverse-bias conditions.

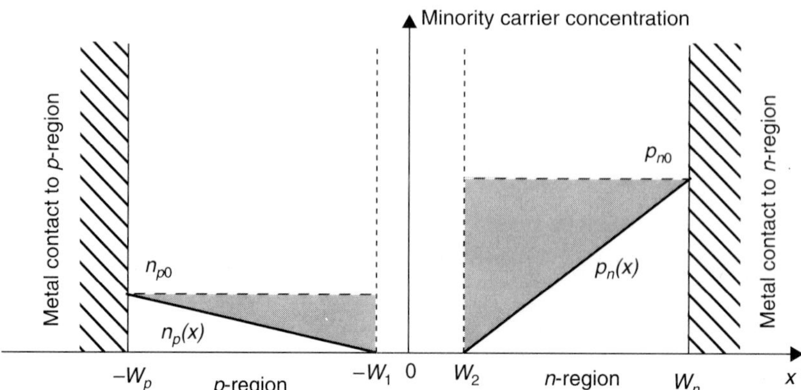

Figure 2.6 Minority carrier concentration in the neutral regions near the *pn*-junction under reverse-bias conditions.

The resulting gradient causes a diffusion of minority carriers towards the junction. Once they reach the depletion region they are swept across the junction by the electric field of the depletion region (which is actually increased by the reversed bias) and transported to their majority zone (holes to the p-region, electrons to the n-region). This reverse current is restricted by two factors: the limited availability of minority carriers (p_{n0}, n_{p0}) and the fact that the concentration gradient does not change much once the reverse-bias voltage is sufficiently large (which typically means $> 4\ \phi_T$), as is obvious when taking the derivative of Eq. (2.6) as well as from Figure 2.6.

It is worth mentioning that in actual devices, the reverse currents are substantially larger than the saturation current I_S. This is due to the thermal generation of hole and electron pairs in the depletion region. The electric field present sweeps these carriers out of the region, causing an additional current component. For typical silicon junctions, the saturation current is nominally in the range of 10^{-17} A/μm^2, while the actual reverse currents are approximately three orders of magnitude higher. Actual device measurements are, therefore, necessary to determine realistic values for the reverse diode leakage currents.

Models for Manual Analysis

The derived current-voltage equations can be summarized in a set of simple models that are useful in the manual analysis of diode circuits. A first model, shown in Figure 2.7a, is based on the ideal diode equation Eq. (2.3). While this model yields accurate results, it has the disadvantage of being strongly nonlinear. This prohibits a fast, first-order analysis of the dc-operation conditions of a network. An often-used, simplified model is derived by inspecting the diode current plot of Figure 2.3. For a "fully conducting" diode, the voltage drop over the diode V_D lies in a narrow range, approximately between 0.6 and 0.8 V. To a first degree, it is reasonable to assume that a conducting diode has a fixed voltage drop V_{Don} over it. Although the value of V_{Don} depends upon I_S, a value of 0.7 V is typically assumed. This gives rise to the model of Figure 2.7b, where a conducting diode is replaced by a fixed voltage source.

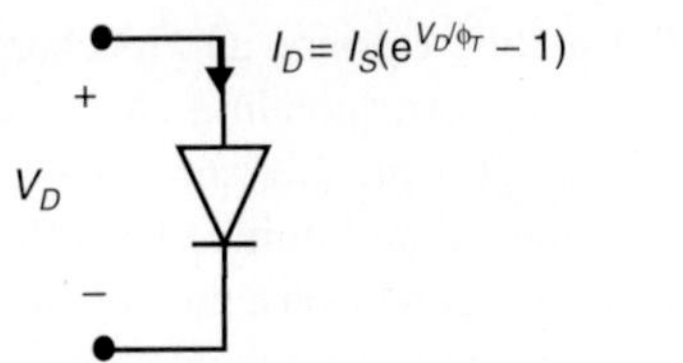

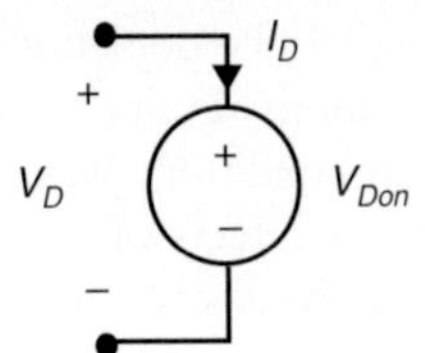

(a) Ideal diode model

(b) First-order diode model

Figure 2.7 Diode models.

Example 2.2 Analysis of Diode Network

Consider the simple network of Figure 2.8 and assume that $V_S = 3$ V, $R_S = 10$ kΩ and $I_S = 0.5 \times 10^{-16}$ A. The diode current and voltage are related by the following network equation

$$V_S - R_S I_D = V_D$$

Inserting the ideal diode equation and (painfully) solving the nonlinear equation using either numerical or iterative techniques yields the following solution: $I_D = 0.224$ mA, and $V_D = 0.757$ V. The simplified model with $V_{Don} = 0.7$ V produces similar results ($V_D = 0.7$ V, $I_D = 0.23$ A) with far less effort. It hence makes considerable sense to use this model when determining a first-order solution of a diode network.

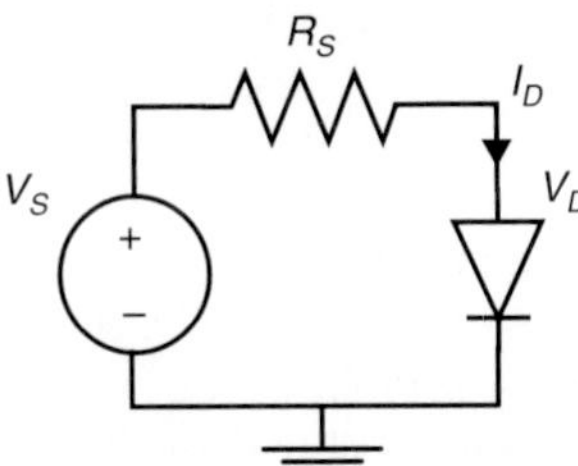

Figure 2.8 A simple diode circuit.

2.2.3 Dynamic, or Transient, Behavior

So far, we have mostly been concerned with the static, or steady-state, characteristics of the diode. Just as important in the design of digital circuits is the response of the device to changes in its bias conditions. The transient, or dynamic, response determines the maximum speed at which the device can be operated. Because the operation mode of the diode is a function of the amount of charge present in both the neutral and the space-charge regions, its dynamic behavior is strongly determined by how fast charge can be moved around. An accurate model of the charge distribution in a diode is, therefore, essential and will be presented first.

Depletion-Region Capacitance

In the ideal model, the depletion region is void of mobile carriers, and its charge is determined by the immobile donor and acceptor ions. The corresponding charge distribution under zero-bias conditions was plotted in Figure 2.2. This picture can be easily extended to incorporate the effects of forward or reverse biasing. At an intuitive level the following observations can be easily verified—under forward-bias conditions, the potential barrier is reduced, which means that less space charge is needed to produce the potential difference. This corresponds to a reduced depletion-region width. On the other hand, under reverse conditions, the potential barrier is increased corresponding to an increased space charge and a wider depletion region. These observations are confirmed by the well- known depletion-region expressions given below (a derivation of these expressions, which are valid for abrupt junctions, is either simple or can be found in any textbook on devices such as [Sedra87]). One observation is crucial—due to the global charge neutrality requirement of the diode, the total acceptor and donor charges must be numerically equal.

1. Depletion-region charge (V_D is positive for forward bias).

$$Q_j = A_D\sqrt{\left(2\varepsilon_{si}q\frac{N_A N_D}{N_A + N_D}\right)(\phi_0 - V_D)} \tag{2.11}$$

2. Depletion-region width.

$$W_j = W_2 - W_1 = \sqrt{\left(\frac{2\varepsilon_{si}}{q}\frac{N_A + N_D}{N_A N_D}\right)(\phi_0 - V_D)} \tag{2.12}$$

3. Maximum electric field.

$$E_j = \sqrt{\left(\frac{2q}{\varepsilon_{si}}\frac{N_A N_D}{N_A + N_D}\right)(\phi_0 - V_D)} \tag{2.13}$$

In the preceding equations ε_{si} stands for the electrical permittivity of silicon and equals 11.7 times the permittivity of a vacuum, or 1.053×10^{-12} F/cm. The ratio of the n- versus p-side of the depletion-region width is determined by the doping-level ratios: $\mathbf{W}_2/(-W_1) = N_A/N_D$.

From an abstract point of view, it is possible to visualize the depletion region as a capacitance, albeit one with very special characteristics. Because the space-charge region contains few mobile carriers, it can be conceived as an insulator with a dielectric constant ε_{si} of the semiconductor material. The n- and p-regions act as the capacitor plates. A small change in the voltage applied to the junction dV_D causes a change in the space charge dQ_j. Hence, a depletion-layer capacitance can be defined

$$\begin{aligned} C_j &= \frac{dQ_j}{dV_D} = A_D\sqrt{\left(\frac{\varepsilon_{si}q}{2}\frac{N_A N_D}{N_A + N_D}\right)(\phi_0 - V_D)^{-1}} \\ &= \frac{C_{j0}}{\sqrt{1 - V_D/\phi_0}} \end{aligned} \tag{2.14}$$

where C_{j0} is the capacitance under zero-bias conditions and is only a function of the physical parameters of the device.

$$C_{j0} = A_D \sqrt{\left(\frac{\varepsilon_{si} q}{2} \frac{N_A N_D}{N_A + N_D}\right) \phi_0^{-1}} \tag{2.15}$$

Notice that the same capacitance value is obtained when using the standard parallel-plate capacitor equation $C_j = \varepsilon_{si} A_D / W_j$ (with W_j given in Eq. (2.12)) Typically, the A_D factor is omitted, and C_j and C_{j0} are expressed as a capacitance/unit area.

The resulting junction capacitance is plotted in the function of the bias voltage in Figure 2.9 for a typical silicon diode found in MOS circuits. A strong *nonlinear dependence* can be observed. Note also that the capacitance decreases with an increasing reverse bias: a reverse bias of 5 V reduces the capacitance by more than a factor of two.

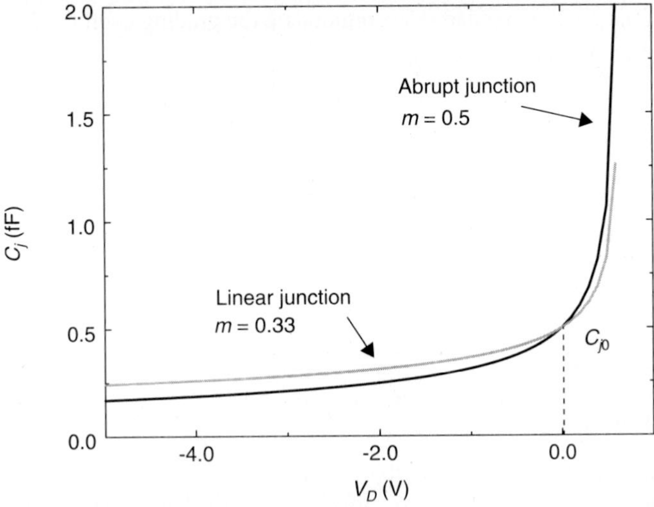

Figure 2.9 Junction capacitance (in fF/μm²) as a function of the applied bias voltage.

Example 2.3 Junction Capacitance

Consider the following silicon junction diode: $C_{j0} = 0.5$ fF/μm², $A_D = 12$ μm² and $\phi_0 = 0.64$ V. A reverse bias of −5 V results in a junction capacitance of 0.17 fF/μm², or, for the total diode, a capacitance of 2.02 fF.

Equation (2.14) is only valid under the condition that the *pn*-junction is an *abrupt junction*, where the transition from *n* to *p* material is instantaneous. This is often not the case in actual integrated-circuit *pn*-junctions, where the transition from *n* to *p* material can be gradual. In those cases, a linear distribution of the impurities across the junction is a better approximation than the step function of the abrupt junction. An analysis of the *linearly-graded junction* shows that the junction capacitance equation of Eq. (2.14) still holds, but with a variation in order of the denominator. A more generic expression for the junction capacitance can hence be provided,

$$C_j = \frac{C_{j0}}{(1 - V_D/\phi_0)^m} \tag{2.16}$$

where m is called the *grading coefficient* and equals 1/2 for the abrupt junction and 1/3 for the linear or graded junction. Both cases are illustrated in Figure 2.9.

The reader should be aware that the junction capacitance is actually a small-signal parameter whose value varies over bias points. In digital circuits, operating voltages tend to move rapidly over wide ranges. Under those circumstances, it is more attractive to replace the voltage-dependent, nonlinear capacitance C_j by an equivalent, linear capacitance C_{eq}. C_{eq} is defined such that, for a given voltage swing from voltages V_{high} to V_{low}, the same amount of charge is transferred as would be predicted by the nonlinear model

$$C_{eq} = \frac{\Delta Q_j}{\Delta V_D} = \frac{Q_j(V_{high}) - Q_j(V_{low})}{V_{high} - V_{low}} = K_{eq} C_{j0} \tag{2.17}$$

Combining Eq. (2.11) (extended to accommodate the grading coefficient m) and Eq. (2.17) yields the value of K_{eq}.

$$K_{eq} = \frac{-\phi_0^m}{(V_{high} - V_{low})(1-m)}[(\phi_0 - V_{high})^{1-m} - (\phi_0 - V_{low})^{1-m}] \tag{2.18}$$

Example 2.4 Average Junction Capacitance

The diode of Example 2.3 is switched between 0 and –5 V. Compute the average junction capacitance ($m = 0.5$).

For the defined voltage range and for $\phi_0 = 0.64$ V, K_{eq} evaluates to 0.502. The average capacitance hence equals 0.25 fF/μm^2.

Diffusion Capacitance

Under forward bias, the *pn*-junction exhibits a capacitive effect much larger than just the junction capacitance. This extra capacitive effect is due to the excess minority carrier charge stored at the boundaries of the depletion region.

It now turns out that this *excess charge is directly related to the current flowing through the diode*. The total excess minority charge stored in a region can be derived by integrating Eq. (2.5) over the complete region and the total diode area and taking into account that each carrier carries a charge q (= 1.6×10^{19} C). For the n-region, for instance, this results in the following relation:

$$\begin{aligned} Q_p &= qA_D \int_{W_2}^{W_n} (p_n(x) - p_{n0})dx \\ &= qA_D \frac{(W_n - W_2)p_{n0}(e^{V_D/\phi_T} - 1)}{2} \\ &= \frac{(W_n - W_2)^2}{2D_p} I_{Dp} \approx \frac{W_n^2}{2D_p} I_{Dp} \end{aligned} \tag{2.19}$$

with I_{Dp} the hole component of the diode current. The ratio of the squared width of the neutral region and the diffusion coefficient is another important device parameter called the *mean transit time*.

$$\tau_{Tp} = \frac{W_n^2}{2D_p} \text{ sec}$$

and (2.20)

$$\tau_{Tn} = \frac{W_p^2}{2D_n} \text{ sec}$$

The total diode current can now be expressed as a function of the excess minority carrier charge

$$I_D = \frac{Q_p}{\tau_{Tp}} + \frac{Q_n}{\tau_{Tn}} = \frac{Q_D}{\tau_T} \tag{2.21}$$

This equation simply states that, in the steady state, the current I_D is inversely proportional to the time it takes a carrier to transport from the junction to the metallic contact.

Example 2.5 Mean Transit Times

For the diode of Problem 2.1, the mean transit times evaluate to the following values:

$$\tau_{Tp} = (5\ \mu\text{m} - 0.15\ \mu\text{m})^2 / (2 \times 10\ \text{cm}^2/\text{sec}) = 11.7\ \text{nsec}$$

$$\tau_{Tn} = (0.7\ \mu\text{m} - 0.03\ \mu\text{m})^2 / (2 \times 25\ \text{cm}^2/\text{sec}) = 0.09\ \text{nsec}$$

Similar expressions can be derived for the long-base diode. In that case, the transit time is replaced by the *excess minority carrier lifetime* parameter, which indicates the mean time it takes for an injected minority carrier to recombine with a majority carrier.

$$\tau_p = L_p^2 / D_p \text{ sec}$$

and (2.22)

$$\tau_n = L_n^2 / D_n \text{ sec}$$

In silicon, typical values of L_p and L_n range from 1 to 100 μm, and the corresponding values of the lifetime are in the range of 1 to 10,000 nsec.

Under transient conditions, a change in current translates into a change in the excess minority carrier charge. In correspondence with the approach used for the depletion region, we model the effect of this charge by an equivalent capacitance called the *diffusion capacitance* C_d. The value of C_d is easily derived

$$C_d = \frac{dQ_D}{dV_D} = \tau_T \frac{dI_D}{dV_D} \approx \frac{\tau_T I_D}{\phi_T} \tag{2.23}$$

which shows a linear dependence upon I_D (as could be expected). For reverse bias, it is fair to assume that C_d is ignorable. Observe, once again, that **C_d is a small-signal capacitance** and is only valid around a given bias voltage.

Similar to the junction capacitance, an average diffusion capacitance can be defined for a voltage range of interest.

$$C_D = \frac{\Delta Q_D}{\Delta V_D} = \tau_T \frac{(I_D(V_{high}) - I_D(V_{low}))}{V_{high} - V_{low.}} = \phi_T \frac{(C_{d(high)} - C_{d(low)})}{V_{high} - V_{low}} \tag{2.24}$$

Example 2.6 Diffusion Capacitance

For $I_S = 0.5 \times 10^{-16}$A, $\tau_T = 1$ nsec, and $\phi_T = 26$ mV, C_d evaluates to a capacitance of 6.5 pF for a forward bias of 0.75V.

Diode Switching Time: A Case Study

The presented models can now be employed to determine the time it takes to switch a diode between two different states. Consider the circuit of Figure 2.10a. Before time 0, the voltage source V_{src} provides a strong forward bias for the diode ($V_1 >> V_{D,on}$). At time $t = 0$, the voltage source switches to a negative voltage, such that the diode goes into reverse bias. At time $t = T$, the voltage source is reversed again, turning on the diode. We simplify the analysis by replacing the voltage source and its resistance by the Norton equivalent circuit (Figure 2.10b). It is assumed that the source resistance R_{src} is large enough so that virtually all current flows through the diode in forward-bias conditions.

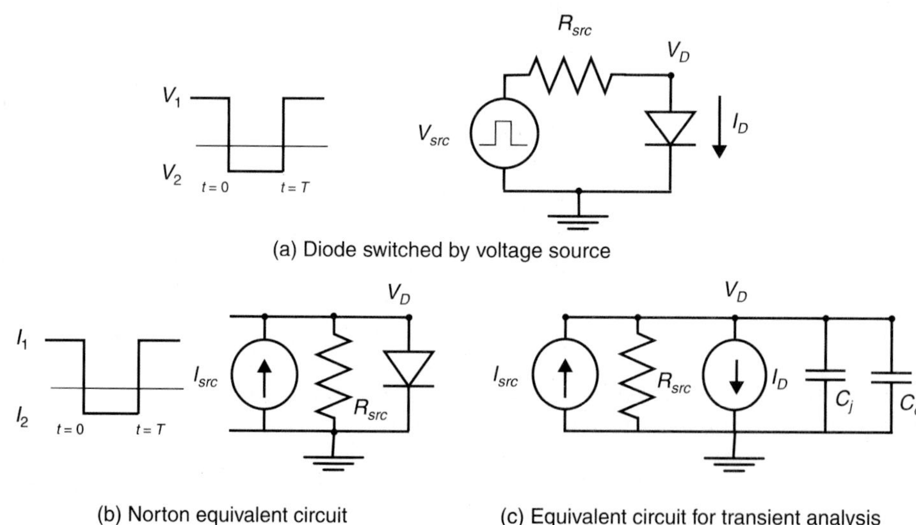

Figure 2.10 Diode switching time.

To determine how fast the circuit will move to a new steady state, we analyze the equivalent circuit of Figure 2.10c, where the diode is replaced by a nonlinear current source (representing the ideal diode equation) and two capacitances, representing the space charge (C_j) and the excess minority carrier charge (C_d), respectively. Once again, observe that C_j and C_d are small signal-capacitances. To be applicable to large-signal analysis, averaged capacitance values must be used. The model for the reverse-bias operation mode is obtained by simply eliminating the current source I_D and the diffusion capacitance C_D.

Deriving the transient response of this network seems simple, as it requires the "mere" solution of the following differential equation:

$$I_{src} = I_D(t) + (C_d + C_j)\frac{dV_D}{dt} = I_S(e^{V_D(t)/\phi_T} - 1) + (C_d + C_j)\frac{dV_D}{dt} \quad (2.25)$$

Unfortunately, this equation is heavily nonlinear again, and finding an analytical solution is a daunting task, which is easily solved by a computer but hard to perform manually. Observe that C_d and C_j are nonlinear functions of V_D as well. Some simplifications are hence at hand. A glimpse of how to tackle this is offered by an inspection of a simulated response, shown in Figure 2.11.

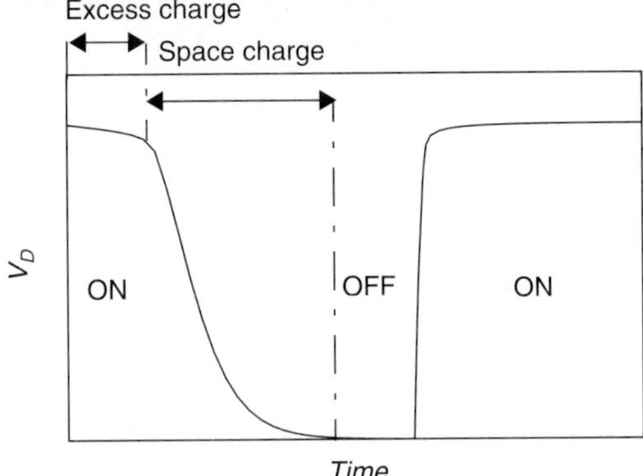

Figure 2.11 Simulated transient response of diode.

The turn-off transient, for instance, clearly displays two operation intervals:

1. Initially, the reverse current I_2 is used to remove the excess minority carrier charge from the neutral regions. During that time, the diode remains on, and the voltage over the diode is approximately constant. This is easily understood: a linear drop in voltage requires an exponential drop in current (and equivalently in excess charge). The constant voltage means that the space charge remains approximately constant as well. The effect of C_j can be ignored during this interval.

2. Once the diode has been turned off ($I_D \approx 0$), the circuit evolves towards steady state. While building a reverse bias over the diode, the space charge changes. In this region, the junction capacitance C_j dominates the performance.

The reader should be aware that the partitioning of the transient into two intervals is somewhat artificial and that both intervals overlap. For instance, in the later phases of the diode turn-off, not all excess charge is removed, yet the voltage over the diode starts to drop, changing the space charge. It is fair to assume that the impact of the space charge is dominant at that time, as the excess charge has been reduced to minuscule amounts (once again, this is a consequence of the exponential relationship between excess charge and diode voltage). The assumption is therefore very reasonable.

Based on these observations, we can derive the duration of both intervals. We first address the turn-off transient.

Removal of the excess charge. Instead of trying to solve Eq. (2.25), a more tangible way of monitoring the excess minority carrier charge is to operate in the charge domain. Using the charge-control expression of the diode current (Eq. (2.21)) we can derive the following expression

$$I_{src} = \frac{Q_D(t)}{\tau_T} + \frac{dQ_D}{dt} \tag{2.26}$$

This equation states that in transient operation, the current supplied to the diode splits in two fractions, as illustrated in Figure 2.12. A first component sustains the normal diffusion current, which is proportional to the excess minority carrier charge present. The second component adds (turn-on) or removes (turn-off) excess carrier charge. This component obviously drops to zero when the steady-state condition is reached.

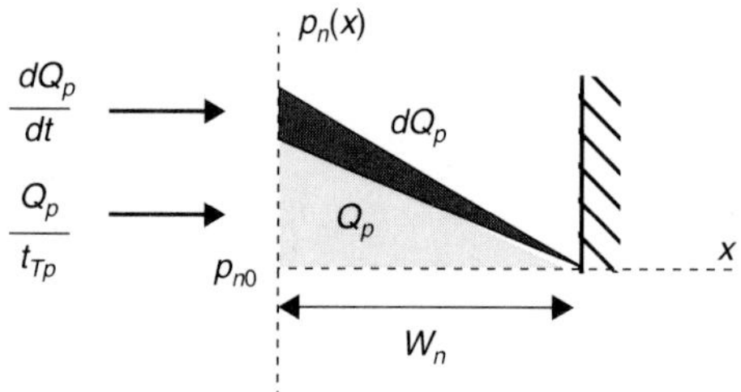

Figure 2.12 Incremental change in minority carrier charge during diode turn-on (showing *n*-region only) and the resulting current components.

Solving the differential equation, while taking into account that the initial value of Q_D equals $I_1 \times \tau_T$ and $I_{src} = I_2$, yields $Q_D(t)$

$$Q_D(t) = \tau_T[I_2 + (I_1 - I_2)e^{-t/\tau_T}] \tag{2.27}$$

The turn-off time is derived by solving for the time $t = t_1$, for which Q_D evaluates to 0.

$$t_1 = \tau_T \ln\left(\frac{I_1 - I_2}{-I_2}\right) \tag{2.28}$$

Changing the Space Charge. Once the diode is turned off, the circuit will evolve to a steady-state solution, where all the source current is flowing through the resistor R_{src}, or $V_D(t = \infty) = I_{src}R_{src} = I_2R_{src}$. A reverse voltage is built over the diode, which means that extra space charge has to be provided. The circuit operation during this period is modeled by the equivalent circuit of Figure 2.10b, where the diode current I_d is set to zero (when the diode is off, the reverse current can generally be ignored). Since the change in excess minority carrier charge in reverse mode is negligible, C_d is ignored as well. This remaining circuit, which is a simple *RC* circuit) is described by the following differential equation:

$$I_{src} = \frac{V_D(t)}{R_{src}} + C_j\frac{dV_D}{dt} \tag{2.29}$$

where C_j is the average junction capacitance over the voltage region of interest. Assuming that the value of V_D at time $t = t_1$ equals 0, the solution of this equation is the well-known exponential expression[2]

$$V_D(t) = I_2 R_{src}\left(1 - e^{-\frac{t-t_1}{R_{src}C_j}}\right) \quad (2.30)$$

The diode voltage reaches its final value in an asymptotic fashion. For such a waveform, the 90% point is often used to determine the end of the transition (as the 100% point would take infinitely long). It is easily determined that this 90% point is reached after 2.3 time-constants $R_{src}C_j$.

Turn-on Transient. Similar considerations hold for the turn-on transient, as is illustrated in Figure 2.11. Before the diode can be turned on, it is necessary to change the space charge first. The transient waveform for the diode voltage is easily derived (using a similar approach as above, and assuming that the transient starts at time $t = 0$):

$$V_D(t) = R_{src}\left(I_1 - (I_2 - I_1)e^{-\frac{t}{R_{src}C_j}}\right) \quad (2.31)$$

Solving this equation for $V_D = 0$ (the time the diode starts to turn on) yields

$$t_3 = R_{src}C_j \ln\left(\frac{I_1 - I_2}{I_1}\right) \quad (2.32)$$

The build-up of the space-charge is still described by the differential equation (2.26). With the proper initial condition ($Q_D(t = t_3) = 0$), solving this equation shows that the excess charge increases in an exponential fashion and asymptotically approaches its final value

$$Q_D(t) = I_1\tau_T\left[1 - e^{-\frac{t-t_3}{\tau_T}}\right] \quad (2.33)$$

It takes approximately 2.2 time constants τ_T for Q_D to reach 90% of its final value.

The procedures used above are representative of the analysis and derivation techniques to be used in the chapters to come—using a number of approximations and linearizations, a simple, tractable model is constructed of a complex, nonlinear circuit. Although inaccurate, this model fosters insight into the circuit operation and identifies the dominant parameters. The first-order solution obtained from this analysis can then be further optimized, or fine-tuned, using computer-aided tools.

Example 2.7 Diode Transient Response

Consider the diode circuit of Figure 2.10 for the following parameters: R_{src} = 50 kΩ, I_1 = 1 mA, I_2 = –0.1 mA. The following parameters are used for the diode: $I_S = 2 \times 10^{-16}$A, C_{j0} = 0.2 pF, τ_T = 5 nsec, and ϕ_0 = 0.65.

---

[2] One can argue that the junction capacitance starts to dominate at the point where $V_D = \phi_0$ and that this should be the starting point of the second interval. Adopting this assumption does not affect the final result in a major way.

The steady-state voltages over the diode are easily derived,

$$V_D \text{(diode on)} = \phi_T \ln (I_D/I_S) = 0.75 \text{ V}$$

$$V_D \text{(diode off)} = I_2 R_{src} = -0.1 \text{ mA} \times 50 \text{ k}\Omega = -5 \text{ V}.$$

Using the expressions derived above, we can further estimate the lengths of the various intervals in the turn-off and turn-on transients:

$$t_1 = 5 \text{ nsec} \times \ln\frac{1+0.1}{0.1} = 12 \text{ nsec}$$

Finding an approximation of $(t_2 - t_1)$ requires first of all a value for the average junction capacitance C_j, which can be obtained with the aid of Eq. (2.18),

$$C_j = K_{eq} C_{j0} = 0.51 \times 0.2 = 0.102 \text{ pF}$$

assuming a voltage swing from 0 to –5V. Using this information we can easily derive the 90% transition point

$$t_2 - t_1 = 2.2 \times 50 k\Omega \times 0.102 pF = 11.2 \text{ nsec}$$

This yields a total turn-off time of 23.2 nsec. For the turn-on, we obtain the following delays

$$t_3 = 50 \text{ k}\Omega \times 0.102 \text{ pF} \times \ln\frac{1+0.1}{1} = 0.49 \text{ nsec}$$

and

$$t_4 - t_3 = 2.2 \times 5 \text{ nsec} = 11 \text{ nsec}$$

which translates into a turn-on time of approximately 11.5 nsec. The faster response is due to larger current (1 mA) available for turning on the device (versus the 0.1 mA for turn-off).

A SPICE simulation of the same circuit produces a transient response similar to the one shown in Figure 2.11 and the following numerical results: t_1 (0 voltage crossing) = 13.2 nsec, t_2 (–4.5 V or 90% of the final value) = 23.6 nsec and t_3 (0 voltage crossing) = 0.57 nsec. Determining the correct value of t_4 is hard as there no direct way to derive the value of Q_D (and its 90% point) from the simulation results.

The obtained numbers are in perfect agreement with the simulated ones. The closeness of the match is, however, partially due to good luck. While building the estimation model, we made a number of approximations and simplifications, which obviously cause deviations from the actual result. The art of approximation is to keep the errors within reasonable bounds.

2.2.4 The Actual Diode—Secondary Effects

In practice, the diode current is less than what is predicted by the ideal diode equation. Not all applied bias voltage appears directly across the junction, as there is always some voltage drop over the neutral regions. Fortunately, the resistivity of the neutral zones is generally small (between 1 and 100 Ω, depending upon the doping levels) and the voltage drop only becomes significant for large currents (>1 mA). This effect can be modeled by adding a resistor in series with the n- and p-region diode contacts.

In the discussion above, it was assumed that under sufficient reverse bias, the reverse current reaches a constant value, which is essentially zero. When the reverse bias exceeds a certain level, called the *breakdown voltage*, the reverse current shows a dramatic increase as shown in Figure 2.13. In the diodes found in typical MOS and bipolar pro-

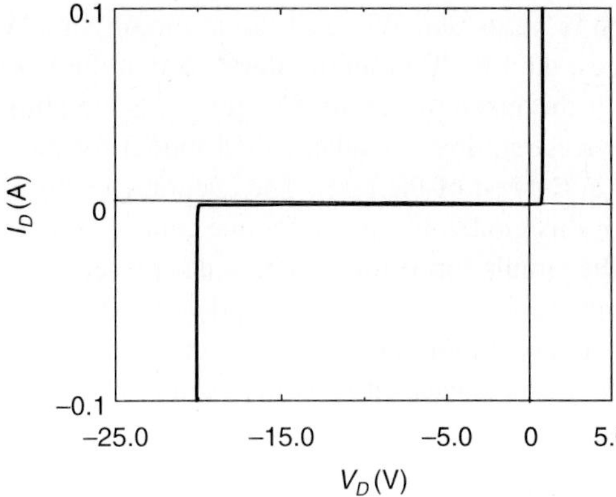

Figure 2.13 *I-V* characteristic of junction diode, showing breakdown under reverse-bias conditions (Breakdown voltage = 20 V).

cesses, this increase is caused by the *avalanche breakdown*. The increasing reverse bias causes the magnitude of the electrical field across the junction to increase. Consequently, carriers crossing the depletion region are accelerated to high velocity. At a critical field E_{crit}, the carriers reach a high enough energy level that electron-hole pairs are created on collision with immobile silicon atoms. These carriers create, in turn, more carriers before leaving the depletion region. The value of E_{crit} is approximately 2×10^5 V/cm for impurity concentrations of the order of 10^{16} cm^{-3}. While avalanche breakdown in itself is not destructive and its effects disappear after the reverse bias is removed, maintaining a diode for a long time in avalanche conditions is not recommended as the high current levels (and the associated power dissipation) might cause permanent damage to the structure. Observe that avalanche breakdown is not the only breakdown mechanism encountered in diodes. For highly doped diodes, another mechanism, called Zener breakdown, can occur. Discussion of this phenomenon is beyond the scope of this text.

Finally, it is worth mentioning that the diode current is affected by the operating *temperature* in a dual way:

1. The thermal voltage ϕ_T, which appears in the exponent of the current equation, is linearly dependent upon the temperature. An increase in ϕ_T causes the current to drop.
2. The saturation current I_S is also temperature-dependent, as the thermal equilibrium carrier concentrations increase with increasing temperature. Theoretically, the saturation current approximately doubles every 5 °C. Experimentally, the reverse current has been measured to double every 8 °C.

This dual dependence has a significant impact on the operation of a digital circuit. First of all, current levels (and hence power consumption) can increase substantially. For instance, for a forward bias of 0.7 V at 300 K, the current increases approximately 6%/°C, and doubles every 12 °C. Secondly, integrated circuits rely heavily on reverse-biased diodes as isolators. Increasing the temperature causes the leakage current to increase and decreases the isolation quality.

2.2.5 The SPICE Diode Model

In the preceding sections, we have presented a model for manual analysis of a diode circuit. For more complex circuits, or when a more accurate modeling of the diode that takes into account second-order effects is required, manual circuit evaluation becomes intractable, and computer-aided simulation is necessary. While different circuit simulators have been developed over the last decades, the SPICE program, developed at the University of California at Berkeley, is definitely the most successful [Nagel75]. Simulating an integrated circuit containing active devices requires a mathematical model for those devices (which is called the *SPICE model* in the rest of the text). The accuracy of the simulation depends directly upon the quality of this model. For instance, one cannot expect to see the result of a second-order effect in the simulation if this effect is not present in the device model. Creating accurate and computation-efficient SPICE models has been a long process and is by no means finished. Every major semiconductor company has developed their own proprietary models, which it claims have either better accuracy or computational efficiency and robustness.

The standard SPICE model for a diode is simple, as shown in Figure 2.14. The steady-state characteristic of the diode is modeled by the nonlinear current source I_D, which is a modified version of the ideal diode equation

$$I_D = I_S(e^{V_D/n\phi_T} - 1) \tag{2.34}$$

The extra parameter n is called the *emission coefficient*. It equals 1 for most common diodes but can be somewhat higher than 1 for others. The resistor R_s models the series resistance contributed by the neutral regions on both sides of the junction. For higher current levels, this resistance causes the internal diode V_D to differ from the externally applied voltage, hence causing the current to be lower than what would be expected from the ideal diode equation.

The dynamic behavior of the diode is modeled by the nonlinear capacitance C_D, which combines the two different charge-storage effects in the diode: the excess minority carrier charge and the space charge.

$$C_D = \frac{\tau_T I_S}{\phi_T} e^{V_D/n\phi_T} + \frac{C_{j0}}{(1 - V_D/\phi_0)^m} \tag{2.35}$$

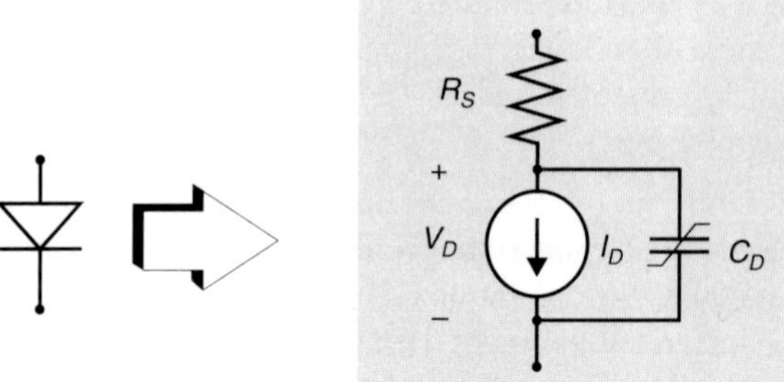

Figure 2.14 SPICE diode model.

We can verify that this equation is, aside from the introduction of the emission coefficient, nothing else than a combination of Eq. (2.23) and Eq. (2.16). The parameter τ_T is called the *transit time* and represents, depending upon the diode type, the excess minority carrier lifetime (τ_n, τ_p) for long-base diodes, or the mean transit time τ_T for short-base diodes.

A listing of the parameters used in the diode model is given in Table 2.1. Besides the parameter name, symbol, and SPICE name, the table contains also the default value used by SPICE in case the parameter is left undefined. Observe that this table is by no means complete. Other parameters are available to govern second-order effects such as breakdown, high-level injection, and noise. To be concise, we chose to limit the listing to the parameters of direct interest to this text. For a complete description of the device models (as well as the usage of SPICE), we refer to the numerous textbooks devoted to SPICE (e.g., [Banhzaf92], [Thorpe92]).

Table 2.1 First-order SPICE diode model parameters.

Parameter Name	Symbol	SPICE Name	Units	Default Value
Saturation current	I_S	IS	A	1.0 E–14
Emission coefficient	n	N	–	1
Series resistance	R_S	RS	Ω	0
Transit time	τ_T	TT	sec	0
Zero-bias junction capacitance	C_{j0}	CJ0	F	0
Grading coefficient	m	M	–	0.5
Junction potential	ϕ_0	VJ	V	1

2.3 The MOS(FET) Transistor

The metal-oxide-semiconductor field-effect transistor (MOSFET or MOS, for short) is certainly the workhorse of contemporary digital design. Its major assets are its integration density and a relatively simple manufacturing process, which make it possible to produce large and complex circuits in an economical way.

We restrict ourselves in this section to a general overview of the device and its parameters, as we did for the diode—after a generic overview of the device, we present an analytical description of the transistor from a static (steady-state) and dynamic (transient) viewpoint. The discussion concludes with an enumeration of some second-order effects and the introduction of the SPICE MOS transistor models.

2.3.1 A First Glance at the Device

A cross section of a typical *n*-channel MOS transistor (NMOS) is shown in Figure 2.15. Heavily doped *n*-type *source* and *drain* regions are implanted (or diffused) into a lightly doped *p*-type substrate (often called the *body*). A thin layer of silicon dioxide (SiO_2) is grown over the region between the source and drain and is covered by a conductive material, most often polycrystalline silicon (or polysilicon, for short). The conductive material forms the *gate* of the transistor. Neighboring devices are insulated from each other with the aid of a thick layer of SiO_2 (called the *field oxide*) and a reverse-biased *np*-diode, formed by adding an extra p^+ region, called the *channel-stop implant* (or *field implant*).

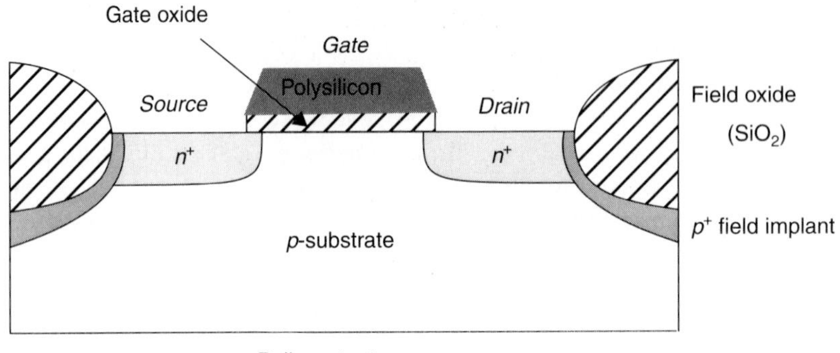

Figure 2.15 Cross section of NMOS transistor.

At the most superficial level, the NMOS transistor can be considered to act as a switch. When a voltage is applied to the gate that is larger than a given value called the *threshold voltage* V_T, a conducting channel is formed between drain and source. In the presence of a voltage difference between drain and source, current flows between the two. The conductivity of the channel is modulated by the gate voltage—the larger the voltage difference between gate and source, the smaller the channel resistance and the larger the current. When the gate voltage is lower than the threshold, no such channel exists, and the switch is considered open.

In an NMOS transistor, current is carried by electrons moving through an *n*-type channel between source and drain. This is in contrast with the *pn*-junction diode, where current is carried by both holes and electrons. MOS devices can also be made by using an *n*-type substrate and p^+ drain and source regions. In such a transistor, current is carried by holes moving through a *p*-type channel. Such a device is called a *p*-channel MOS, or PMOS transistor. In a complementary MOS technology (CMOS), both devices are present. In a pure NMOS or PMOS technology, the substrate is common to all devices and invariably connected to dc power supply voltage. In CMOS technology, PMOS and NMOS devices are fabricated in separate isolated regions called *wells* that are connected to different power supplies. Figure 2.16 shows a cross-section of a CMOS device, where PMOS transistors are implemented in a *n*-type area embedded in a *p*-type substrate. For obvious reasons, such a fabrication approach is called an *n-well* technology.

Circuit symbols for the various MOS transistors are shown in Figure 2.17. In general, the device is considered to be a three-terminal one with gate, drain, and source ports. In reality, the MOS transistor has a fourth terminal, the substrate. Since the substrate is generally connected to a dc supply that is identical for all devices of the same type (GND for NMOS, V_{dd} for PMOS), it is most often not shown on the schematics. In case a design deviates from that concept, a four-terminal symbol is also available as shown in Figure 2.17c. **If the fourth terminal is not shown, it is assumed that the substrate is connected to the appropriate supply.**

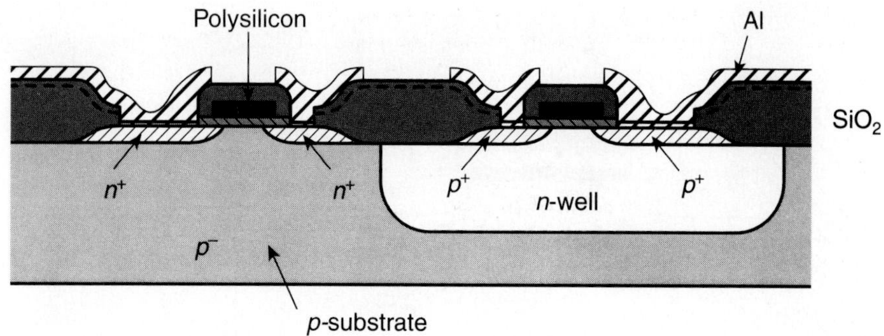

Figure 2.16 Cross section of CMOS *n*-well technology.

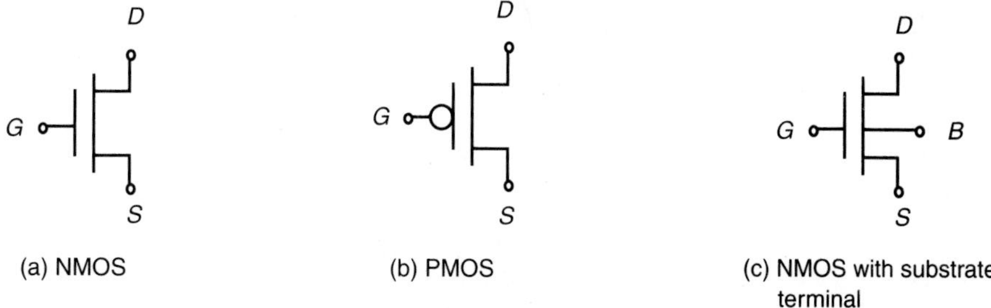

Figure 2.17 Circuit symbols for MOS transistors.

2.3.2 Static Behavior

In the derivation of the static model of the MOS transistor, we concentrate on the NMOS device. All the arguments made are valid for PMOS devices as well as will be discussed at the end of the section.

The Threshold Voltage

Consider first the case where $V_{GS} = 0$ and drain, source, and bulk are connected to ground. The drain and source are connected by back-to-back *pn*-junctions (substrate-source and substrate-drain). Under the mentioned conditions, both junctions have a 0 V bias and can be considered off, which results in an extremely high resistance between drain and source.

Assume now that a positive voltage is applied to the gate (with respect to the source), as shown in Figure 2.18. The gate and substrate form the plates of a capacitor with the gate oxide as the dielectric. The positive gate voltage causes positive charge to accumulate on the gate electrode and negative charge on the substrate side. The latter manifests itself initially by repelling mobile holes. Hence, a depletion region is formed below the gate. This depletion region is similar to the one occurring in a *pn*-junction diode. Consequently, similar expressions hold for the width and the space charge per unit area. Compare these expressions to Eq. (2.11) and Eq. (2.12).

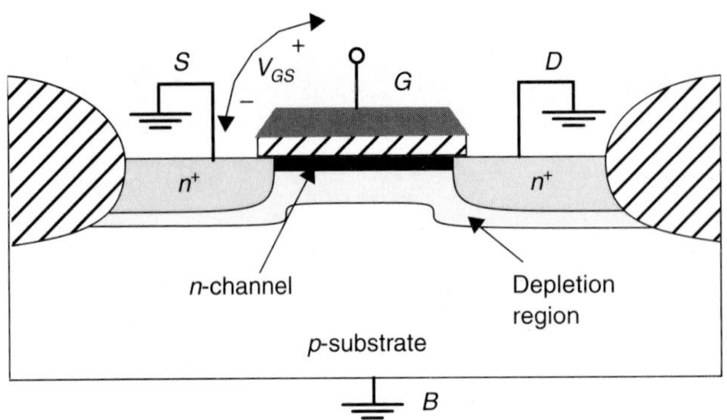

Figure 2.18 NMOS transistor for positive V_{GS}, showing depletion region and induced channel.

$$W_d = \sqrt{\frac{2\varepsilon_{si}\phi}{qN_A}} \tag{2.36}$$

and

$$Q_d = \sqrt{2qN_A\varepsilon_{si}\phi} \tag{2.37}$$

with N_A the substrate doping and ϕ the voltage across the depletion layer (i.e., the potential at the oxide-silicon boundary).

As the gate voltage increases, the potential at the silicon surface at some point reaches a critical value, where the semiconductor surface inverts to n-type material. This point marks the onset of a phenomenon known as *strong inversion* and occurs at a voltage equal to twice the *Fermi Potential* ($\phi_F \approx -0.3$ V for typical p-type silicon substrates).

Further increases in the gate voltage produce no further changes in the depletion-layer width, but result in additional electrons in the thin inversion layer directly under the oxide. These are drawn into the inversion layer from the heavily doped n+ source region. Hence, a continuous n-type channel is formed between the source and drain regions, whose conductivity is modulated by the gate-source voltage.

In the presence of an inversion layer, the charge stored in the depletion region is fixed and equals

$$Q_{B0} = \sqrt{2qN_A\varepsilon_{si}|-2\phi_F|} \tag{2.38}$$

In the presence of a substrate bias voltage V_{SB} (V_{SB} is normally positive for n-channel devices), the surface potential required for strong inversion increases and becomes $|-2\phi_F + V_{SB}|$. The charge stored in the depletion region then is expressed by Eq. (2.39)

$$Q_B = \sqrt{2qN_A\varepsilon_{si}(|-2\phi_F + V_{SB}|)} \tag{2.39}$$

The value of V_{GS} where strong inversion occurs is called the *threshold voltage* V_T. The expression of V_T consists of several components:

1. A flat-band voltage V_{FB} that represents the built-in voltage offset across the MOS structure. It consists of the work-function difference ϕ_{ms}, which exists between the gate polysilicon and the silicon, and some extra components to compensate for the (undesired) fixed charge Q_{ox}, sitting at the oxide-silicon interface, and the threshold-adjusting implanted impurities Q_I.

2. A second term V_B represents the voltage drop across the depletion region at inversion and equals $-2\phi_F$.

3. A final component V_{ox} stands for the potential drop across the gate oxide and is equal to Q_B / C_{ox}, with C_{ox} representing the gate-oxide capacitance per unit area.

$$V_T = V_{FB} + V_B + V_{ox} = \left(\phi_{ms} - \frac{Q_{ox}}{C_{ox}} - \frac{Q_I}{C_{ox}}\right) - 2\phi_F - \frac{Q_B}{C_{ox}} \tag{2.40}$$

While most of the terms in this expression are pure material or technology parameters, Q_B is a function of V_{SB}. It is therefore customary to reorganize the threshold equation in the following manner,

$$V_T = V_{T0} + \gamma\left(\sqrt{|-2\phi_F + V_{SB}|} - \sqrt{|-2\phi_F|}\right)$$

with

$$V_{T0} = \phi_{ms} - 2\phi_F - \frac{Q_{B0}}{C_{ox}} - \frac{Q_{ox}}{C_{ox}} - \frac{Q_I}{C_{ox}} \tag{2.41}$$

and

$$\gamma = \frac{\sqrt{2q\varepsilon_{si}N_A}}{C_{ox}}$$

where V_{T0} is the threshold voltage for $V_{SB} = 0$, and the parameter γ (gamma) is called the *body-effect coefficient*. The gate capacitance per unit area C_{ox} is expressed by Eq. (2.42).

$$C_{ox} = \frac{\varepsilon_{ox}}{t_{ox}} \tag{2.42}$$

with $\varepsilon_{ox} = 3.97 \times \varepsilon_o = 3.5 \times 10^{-13}$ F/cm the oxide permittivity. The expression t_{ox} stands for the oxide thickness, which is 20 nm (= 200 Å) or smaller for contemporary processes. This translates into an oxide capacitance of 1.75 fF/μm^2.

Observe that the threshold voltage has a **positive** value for a typical **NMOS** device, while it is **negative** for a normal **PMOS** transistor.

Example 2.8 Threshold Voltage of an NMOS Transistor

An MOS transistor has a threshold voltage of 0.75 V, while the body-effect coefficient equals 0.54. Compute the threshold voltage for V_{SB} = 5 V. $2\phi_F$ = – 0.6 V.

Using Eq. (2.41), we obtain V_T(5 V) = 0.75 V + 0.86 V = 1.6 V, which is more than twice the threshold under zero-bias conditions!

Current-Voltage Relations

Assume now that $V_{GS} > V_T$. A voltage difference V_{DS} causes a current I_D to flow from drain to source (Figure 2.19). Using a simple first-order analysis, an expression of the current as a function of V_{GS} and V_{DS} can be obtained.

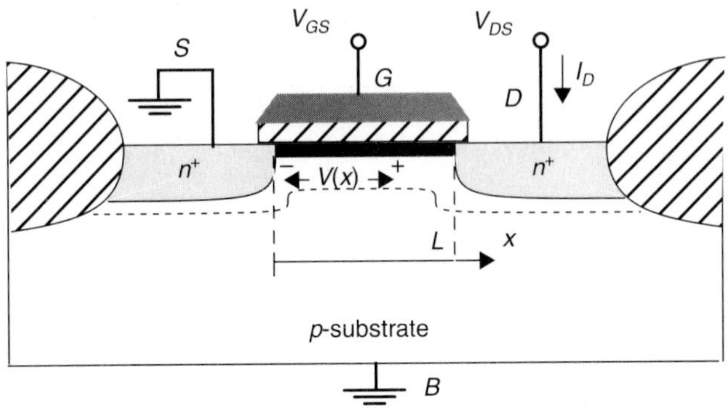

Figure 2.19 NMOS transistor with bias voltages.

At a point x along the channel, the voltage is $V(x)$, and the gate-to-channel voltage at that point equals $V_{GS} - V(x)$. Under the assumption that this voltage exceeds the threshold voltage all along the channel, the induced channel charge per unit area at point x can be computed.

$$Q_i(x) = -C_{ox}[V_{GS} - V(x) - V_T] \tag{2.43}$$

The current is given as the product of the drift velocity of the carriers υ_n and the available charge. Due to charge conservation, it is a constant over the length of the channel. W is the width of the channel in a direction perpendicular to the current flow.

$$I_D = -\upsilon_n(x)Q_i(x)W \tag{2.44}$$

The electron velocity is related to the electric field through a parameter called the *mobility* μ_n (expressed in $cm^2/V{\cdot}sec$). The mobility is a complex function of crystal structure, local fields, and so on. In general, an empirical value is used.

$$\upsilon_n = -\mu_n E(x) = \mu_n \frac{dV}{dx} \tag{2.45}$$

Combining Eq. (2.43) – Eq. (2.45) yields

$$I_D dx = \mu_n C_{ox} W(V_{GS} - V - V_T)dV \tag{2.46}$$

Integrating the equation over the length of the channel L yields the voltage-current relation of the NMOS transistor.

$$I_D = k'_n \frac{W}{L}\left[(V_{GS} - V_T)V_{DS} - \frac{V_{DS}^2}{2}\right] = k_n\left[(V_{GS} - V_T)V_{DS} - \frac{V_{DS}^2}{2}\right] \tag{2.47}$$

k'_n is called the *process transconductance parameter* and equals

$$k'_n = \mu_n C_{ox} = \frac{\mu_n \varepsilon_{ox}}{t_{ox}} \tag{2.48}$$

For a typical *n*-channel device with $t_{ox} = 20$ nm, k'_n equals 80 $\mu A/V^2$. The product of the process transconductance k'_n and the (W/L) ratio of an (NMOS) transistor is called the *gain factor* k_n of the device.

As the value of the drain-source voltage is further increased, the assumption that the channel voltage is larger than the threshold all along the channel ceases to hold. This happens when $V_{GS} - V(x) < V_T$. At that point, the induced charge is zero, and the conducting channel disappears or is *pinched off*. This is illustrated in Figure 2.20, which shows (in an

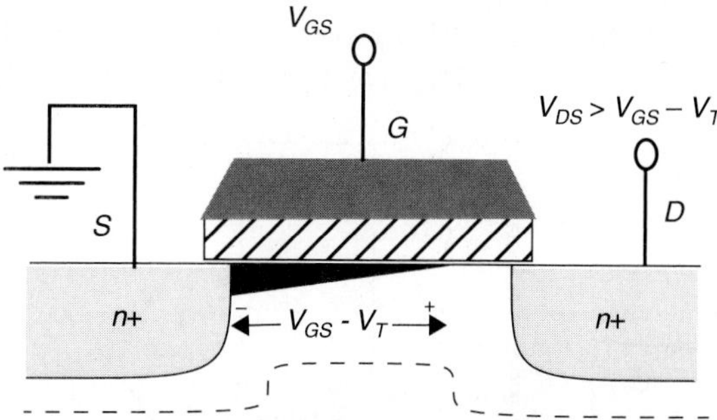

Figure 2.20 NMOS transistor under pinch-off conditions.

exaggerated fashion) how the channel thickness gradually is reduced from source to drain until pinch-off occurs. No channel exists in the vicinity of the drain region. Obviously, for this phenomenon to occur, it is essential that the pinch-off condition be met at the drain region, or

$$V_{GS} - V_{DS} \leq V_T. \tag{2.49}$$

Under those circumstances, the transistor is in the *saturation* region. When a continuous channel exists between source and drain, the device operates in the *triode* (or *linear*) mode.

In the saturation region, Eq. (2.47) no longer holds. The voltage difference over the induced channel (from the pinch-off point to the source) remains fixed at $V_{GS} - V_T$, and consequently, the current remains constant (or saturates). Replacing V_{DS} by $V_{GS} - V_T$ in Eq. (2.47) yields the drain current for the saturation mode.

$$I_D = \frac{k'_n}{2} \frac{W}{L} (V_{GS} - V_T)^2 \tag{2.50}$$

This equation is not entirely correct. The position of the pinch-off point, and hence the effective length of the conductive channel, is modulated by the applied V_{DS}. As can be observed from Eq. (2.50), the current increases when the length factor L is decreased. A more accurate description of the current of the saturated MOS transistor is given in Eq. (2.51).

$$I_D = \frac{k'_n}{2}\frac{W}{L}(V_{GS} - V_T)^2(1 + \lambda V_{DS}) \tag{2.51}$$

with λ an empirical constant parameter, called the *channel-length modulation.*[3]

Figure 2.21 plots I_D versus V_{DS} (with V_{GS} as a parameter) for an NMOS transistor. In the triode region, the transistor behaves like a voltage-controlled resistor, while in the saturation region, it acts as a voltage-controlled current source (when the channel-length modulation effect is ignored). Also shown is a plot of $\sqrt{I_D}$ as a function of V_{GS} (with V_{DS} a constant). As expected a linear relationship is observed for values of $V_{GS} >> V_T$. Notice also how the current does not drop abruptly to 0 at $V_{GS} = V_T$. At that point, the device goes into *subthreshold operation.* To turn the device completely off, the gate-source voltage has to be substantially lower than V_T. Subthreshold conduction is discussed in more detail later in the chapter, when we discuss some second-order effects in MOS transistors.

All the derived equations hold for the PMOS transistor as well. The only difference is that for PMOS devices, the polarities of all voltages and currents are reversed.

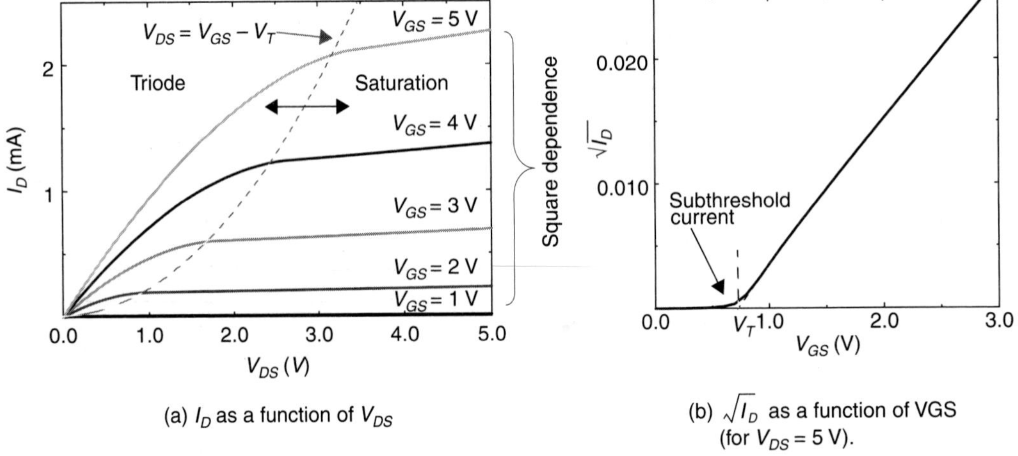

(a) I_D as a function of V_{DS}

(b) $\sqrt{I_D}$ as a function of VGS (for $V_{DS} = 5$ V).

Figure 2.21 *I-V* characteristics of NMOS transistor ($W = 100$ µm, $L = 20$ µm in a 1.2 µm CMOS technology).

Problem 2.2 PMOS *I-V* Characteristic

Draw the *I-V* curves of a PMOS transistor. Derive expressions for the PMOS drain current in the saturation and triode regions, which take the polarities of voltages and currents into account.

A Model for Manual Analysis

The derived equations can be combined into a simple device model, which we will employ for the manual analysis of MOS circuits in the rest of the book. It is summarized in Figure 2.22.

[3] Analytical expressions for λ have proven to be complex and inaccurate. Device experiments and simulations indicate that λ varies roughly with the inverse of the channel length.

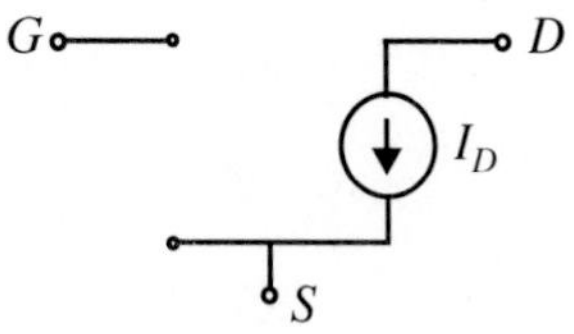

$V_{DS} > V_{GS} - V_T$

$$I_D = \frac{k'_n}{2}\frac{W}{L}(V_{GS} - V_T)^2(1 + \lambda V_{DS})$$

$V_{DS} < V_{GS} - V_T$

$$I_D = k'_n \frac{W}{L}\left((V_{GS} - V_T)V_{DS} - \frac{V_{DS}^2}{2}\right)$$

with

$$V_T = V_{T0} + \gamma(\sqrt{|-2\phi_F + V_{SB}|} - \sqrt{|-2\phi_F|})$$

Figure 2.22 An MOS model for manual analysis.

2.3.3 Dynamic Behavior

The transient behavior of the *pn*-junction diode is dominated by the moving of the excess minority carrier charge in the neutral zones and, to a second degree, by the space charge in the depletion region. Since the MOSFET is a majority carrier device, its dynamic response is solely determined by the time to (dis)charge the capacitances between the device ports and from the interconnecting lines. An accurate analysis of the nature and behavior of these capacitances is essential when designing high-performance digital circuits. They originate from three sources: the basic MOS structure, the channel charge, and the depletion regions of the reverse-biased *pn*-junctions of drain and source. Aside from the MOS structure capacitances, all capacitors are nonlinear and vary with the applied voltage. We discuss each of the components in turn.

MOS Structure Capacitances

The gate of the MOS transistor is isolated from the conducting channel by the gate oxide that has a capacitance per unit area equal to $C_{ox} = \varepsilon_{ox} / t_{ox}$. From the *I-V* equations, we learned that it is useful to have C_{ox} as large as possible, or to keep the oxide thickness very thin. The total value of this capacitance is called the *gate capacitance* C_g and equals $C_{ox}WL$. This gate capacitance can be decomposed into a number of elements, each with a different behavior. Obviously, one part of C_g contributes to the channel charge, and is discussed in a subsequent section. Another part is solely due to the topological structure of the transistor. This component is the subject of the remainder of this section.

Consider the transistor structure of Figure 2.23. Ideally, the source and drain diffusion should end right at the edge of the gate oxide. In reality, both source and drain tend to extend somewhat below the oxide by an amount x_d, called the *lateral diffusion*. Hence, the effective channel of the transistor L_{eff} becomes shorter than the drawn length (or the length the transistor was originally designed for) by a factor of $2x_d$. It also gives rise to a parasitic capacitance between gate and source (drain) that is called the *overlap capacitance*. This capacitance is strictly linear and has a fixed value

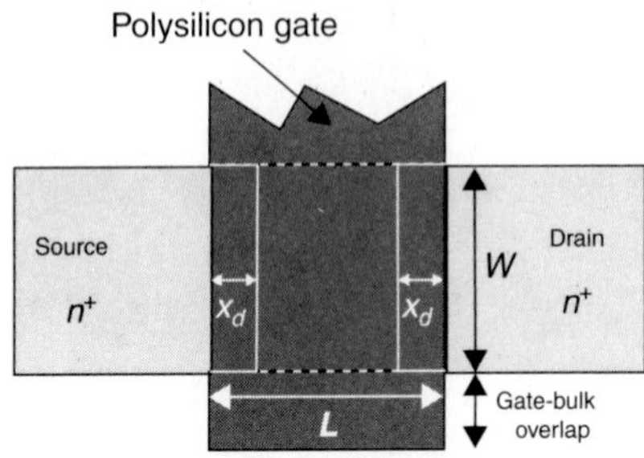

(a) Top view

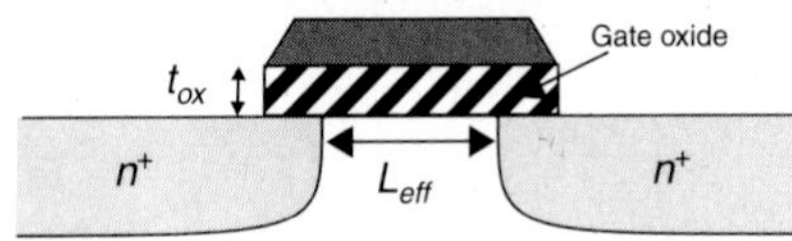

(b) Cross section

Figure 2.23 MOSFET overlap capacitance.

$$C_{gsO} = C_{gdO} = C_{ox}x_dW = C_OW \tag{2.52}$$

Since x_d is a technology-determined parameter, it is customary to combine it with the oxide capacitance to yield the overlap capacitance per unit transistor width C_O.

Channel Capacitance

The gate-to-channel capacitance can be decomposed into three components: C_{gs}, C_{gd}, and C_{gb}, being the capacitance between the gate and the source, drain, and bulk regions, respectively. All those components are nonlinear, and their value depends upon the operation region. To simplify the analysis, estimated and average values are used. For instance, in the cut-off mode, no channel exists, and the total capacitance $C_{ox}WL_{eff}$ appears between gate and bulk. In the triode region, an inversion layer is formed, which acts as a conductor between source and drain. Consequently, $C_{gb} = 0$ as the bulk electrode is shielded from the gate by the channel. Symmetry dictates that $C_{gs} \approx C_{gd} \approx C_{ox}W\,L_{eff}/2$. Finally, in the saturation mode, the channel is pinched off. The capacitance between gate and drain is thus approximately zero, and so is the gate-bulk capacitance. A careful analysis of the channel charge, taking into account the potential variations over the channel, indicates that C_{gs} averages 2/3 $C_{ox}WL_{eff}$. Although these expressions are approximations, they are adequate for the initial design estimates. The derived values are summarized in Table 2.2.

Table 2.2 Average channel capacitances of MOS transistor for different operation regions.

Operation Region	C_{gb}	C_{gs}	C_{gd}
Cutoff	$C_{ox}WL_{eff}$	0	0
Triode	0	$C_{ox}WL_{eff}/2$	$C_{ox}WL_{eff}/2$
Saturation	0	$(2/3)C_{ox}WL_{eff}$	0

Junction Capacitances

A final capacitive component is contributed by the reverse-biased source-bulk and drain-bulk *pn*-junctions. The depletion-region capacitance is nonlinear and decreases when the reverse bias is raised as discussed earlier. To understand the components of the junction capacitance (often called the *diffusion capacitance*), we must look at the source (drain) region and its surroundings. The detailed picture, shown in Figure 2.24, shows that the junction consists of two components:

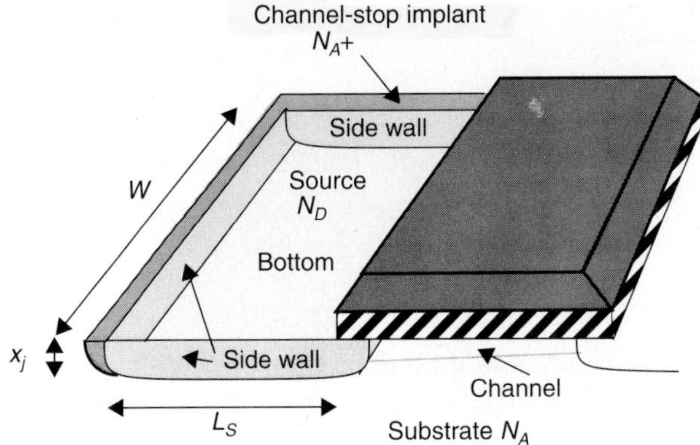

Figure 2.24 Detailed view of source junction.

- The *bottom-plate* junction, which is formed by the source region (with doping N_D) and the substrate with doping N_A. The total depletion region capacitance for this component equals $C_{bottom} = C_j W L_S$, with C_j the junction capacitance per unit area as given by Eq. (2.16). As the bottom-plate junction is typically of the abrupt type, the grading coefficient m is set to 0.5.
- The *side-wall* junction, formed by the source region with doping N_D and the p^+ channel-stop implant with doping level $N_A^{\ +}$. The doping level of the stopper is usually larger than that of the substrate, resulting in a larger capacitance per unit area. The side-wall junction resembles the graded type, which sets m to 1/3. Its total value equals $C_{sw} = C'_{jsw} x_j \,(W + 2 \times L_s)$. Notice that no side-wall capacitance is counted for the fourth side of the source region, as this represents the conductive channel. Since x_j, the junction depth, is a technology parameter, it is normally combined with C'_{jsw} into a capacitance per unit perimeter $C_{jsw} = C'_{jsw} x_j$.

An expression for the total junction capacitance can now be derived,

$$\begin{aligned} C_{diff} &= C_{bottom} + C_{sw} = C_j \times AREA + C_{jsw} \times PERIMETER \\ &= C_j L_S W + C_{jsw}(2L_S + W) \end{aligned} \qquad (2.53)$$

Since all these capacitances are small-signal capacitances, we normally linearize them and use average capacitances along the lines of Eq. (2.17).

Capacitive Device Model

All the above contributions can be combined in a single capacitive model for the MOS transistor, which is shown Figure 2.25. Its components are readily identified on the basis of the preceding discussions.

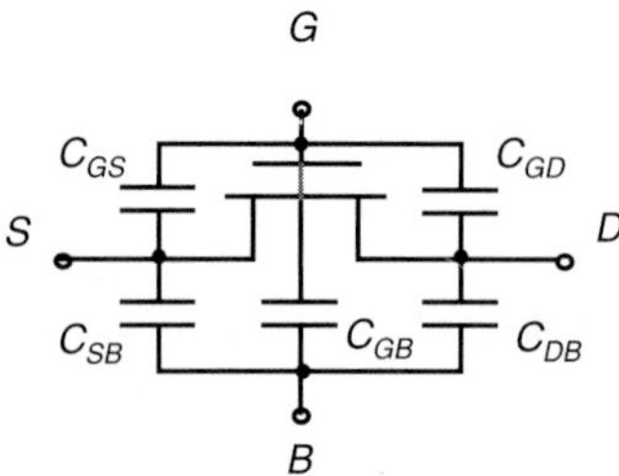

Figure 2.25 MOSFET capacitance model.

$$C_{GS} = C_{gs} + C_{gsO};\ C_{GD} = C_{gd} + C_{gdO};\ C_{GB} = C_{gb}$$

$$C_{SB} = C_{Sdiff};\ C_{DB} = C_{Ddiff} \tag{2.54}$$

A good understanding of this model as well as of the relative values of its components is essential in the design and optimization of high performance digital circuits. The dynamic performance of these circuits is directly proportional to the capacitance.

Example 2.9 MOS Transistor Capacitances

Consider an NMOS transistor with the following parameters: t_{ox} = 20 nm, L = 1.2 μm, W = 1.8 μm, $L_D = L_S$ = 3.6 μm, x_d = 0.15 μm, $C_{j0} = 3 \times 10^{-4}$ F/m^2, $C_{jsw0} = 8 \times 10^{-10}$ F/m. Determine the zero-bias value of all relevant capacitances.

The gate capacitance per unit area is easily derived as $(\varepsilon_{ox}/t_{ox})$ and equals 1.75 fF/μm^2. The total gate capacitance C_g is equal to WLC_{ox} = 3.78 fF. This is divided into the overlap capacitances ($C_{GSO} = C_{GDO} = Wx_dC_{ox}$ = 0.47 fF) and the channel capacitance, which splits between source, drain, and bulk terminals, dependent upon the operation region, and equals $3.78 - 2 \times 0.47 = 2.84$ fF.

The diffusion capacitance consists of the bottom and the side-wall capacitances. The former is equal to $C_{j0}\, L_DW$ = 1.95 fF, while the side-wall capacitance under zero-bias conditions evaluates to $C_{jsw0}\,(2L_D + W)$ = 7.2 fF.

The diffusion capacitance seems to dominate the gate capacitance. This is a worst-case condition, however. When increasing the value of the reverse bias, the diffusion capacitance is substantially reduced (to about 50% of its value). In general, it can be stated that both contributions are virtually similar in value. Observe also the large value of the side-wall versus bottom-plate capacitance for this particular process. Advanced processes reduce the diffusion capacitances by using materials such as SiO_2 to isolate the devices. This approach is called *trench isolation.*

2.3.4 The Actual MOS Transistor—Secondary Effects

Up to this point, we have discussed the behavior of an ideal MOS device. The operation of an actual transistor can deviate substantially from this model. This is especially true when the dimensions of the device reach the μm-range or below. At that point, the channel length becomes comparable to other device parameters such as the depth of drain and source junctions, and the width of their depletion regions. Such a device is called a *short-channel* transistor, in contrast to the *long-channel* devices discussed so far. The behavior of a long-channel device is adequately described by a one-dimensional model, where it is assumed that all current flows on the surface of the silicon and the electrical fields are oriented along that plane. In short-channel devices, those assumptions are no longer valid and a two-dimensional model is more appropriate. This results in important deviations from the ideal model.

The understanding of some of these second-order effects and their impact on the device behavior is essential in the design of contemporary digital circuits and therefore merits some discussion. One word of warning, though. Trying to take all those effects into account in a manual, first-order analysis results in intractable and opaque circuit models. It is therefore advisable to analyze and design MOS circuits first using the ideal model. The impact of the nonidealities can be studied in a second round using computer-aided simulation tools with more precise transistor models.

Threshold Variations

Eq. (2.41) states that the threshold voltage is only a function of the manufacturing technology and the applied body bias V_{SB}. The threshold can therefore be considered as a constant over all NMOS (PMOS) transistors in a design. As the device dimensions are reduced, this model becomes inaccurate, since the threshold potential becomes a function of L, W, and V_{DS}. For instance, in the derivation of V_{T0} it was assumed that all depletion charge beneath the gate originates from the MOS field effects. This ignores the depletion regions of the source and reverse-biased drain junction that become relatively more important with shrinking channel lengths. Since a part of the region below the gate is already depleted (by the source and drain fields), a smaller threshold voltage suffices to cause strong inversion. In other words, V_{T0} decreases with L for short-channel devices (Figure 2.26a). A similar

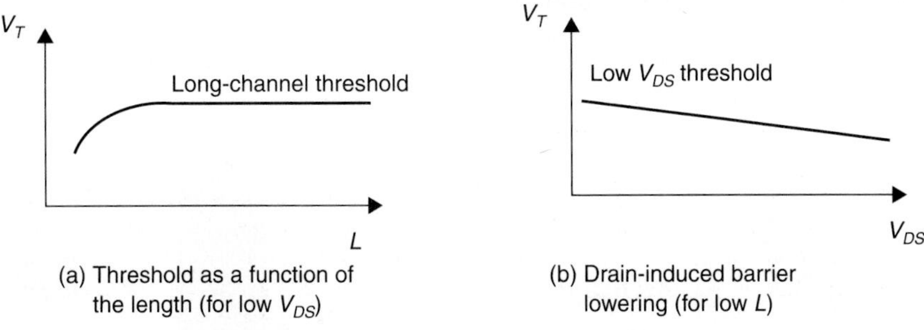

Figure 2.26 Threshold variations.

effect can be obtained by raising the drain-source (bulk) voltage, as this increases the width of the drain-junction depletion region. Consequently, the threshold decreases with increasing V_{DS}. This effect, called the *drain-induced barrier lowering,* or *DIBL*, causes the threshold potential to be a function of the operating voltages (Figure 2.26b). For high enough values of the drain voltage, the source and drain regions can even be shorted together, and normal transistor operation ceases to exist. This effect is called *punchthrough*.

Since the majority of the transistors in a digital circuit are designed at the minimum channel length, the variation of the threshold voltage as a function of the length is almost uniform over the complete design, and is therefore a minor issue. More troublesome is the DIBL, as this effect varies with the operating voltage. This is, for instance, a problem in dynamic memories, where the leakage current of a cell (being the subthreshold current of the access transistor) becomes a function of the voltage on the data-line, which is shared with many other cells.

Besides varying over a design, threshold voltages in short-channel devices also have the tendency to *drift over time*. This is the result of the *hot-carrier* effect [Hu92]. Over the last decades, device dimensions have been scaled down continuously, while the power supply and the operating voltages were kept constant. The resulting increase in the electrical field strength causes an increasing velocity of the electrons, which can leave the silicon and tunnel into the gate oxide upon reaching a high enough energy level. Electrons trapped in the oxide change the threshold voltage, typically increasing the thresholds of NMOS devices, while decreasing the V_T of PMOS transistors. For an electron to become hot, an electrical field of at least 10^4 V/cm is necessary. This condition is easily met in devices with channel lengths around or below 1 μm. The hot-electron phenomenon can lead to a long-term reliability problem, where a circuit might degrade or fail after being in use for a while.

Source-Drain Resistance

When transistors are scaled down, their junctions are shallower, and the contact openings become smaller. This results in an increase in the parasitic resistance in series with the drain and source regions, as shown in Figure 2.27a. The resistance of the drain (source) region can be expressed as

$$R_{S,D} = \frac{L_{S,D}}{W} R_{\square} + R_C \tag{2.55}$$

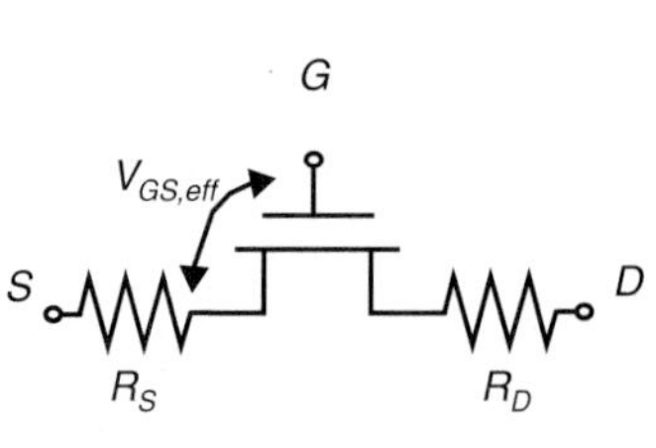

(a) Modeling the series resistance

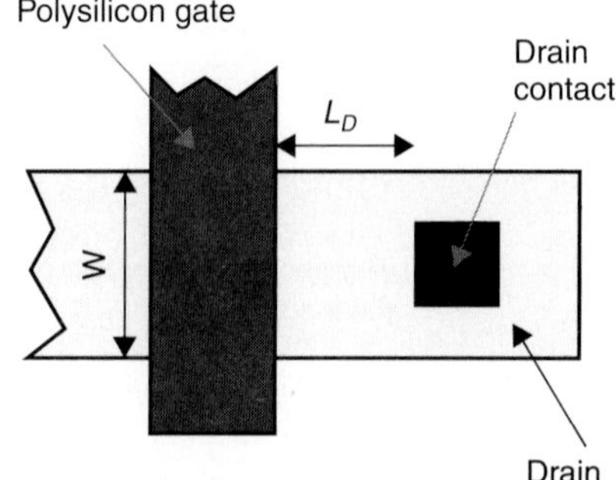

(b) Parameters of the series resistance

Figure 2.27 Series drain and source resistance.

with R_C the contact resistance, W the width of the transistor, and $L_{S,D}$ the length of the source or drain region (Figure 2.27b). $R_\square$ is the sheet resistance per square of the drain-source diffusion, which ranges from 50 Ω/□ to 1 kΩ/□. Observe that the resistance of a square of material is constant, independent of its size (see Chapter 8). The series resistance causes a deterioration in the device performance, as it reduces the drain current for a given control voltage. Keeping its value as small as possible is thus an important design goal (for both the device and the circuit engineer). One option is to cover the drain and source regions with a low-resistivity material, such as titanium or tungsten. This process is called *silicidation* and effectively reduces the parasitic resistance. Silicidation is furthermore used to reduce the resistance of the polysilicon gate. Making the transistor wider than needed is another possibility as should be obvious from Eq. (2.55).

Variations in *I-V* Characteristics

The voltage-current relations of a short-channel device deviate considerably from the ideal expressions of Eq. (2.47) and Eq. (2.51). The most important reasons for this difference are the *velocity saturation* and the *mobility degradation* effects. In Eq. (2.45), it was stated that the velocity of the carriers is proportional to the electrical field, independent of the value of that field. In other words, the carrier mobility is a constant. However, when the electrical field along the channel reaches a critical value E_{sat}, the velocity of the carriers tends to saturate as illustrated in Figure 2.28a.

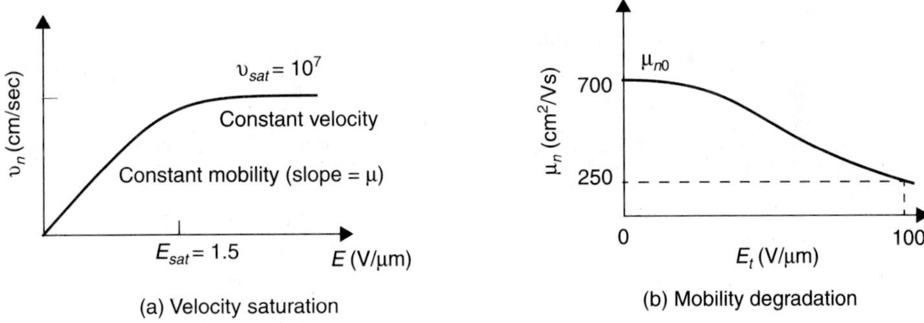

Figure 2.28 Effect of electrical field on electron velocity and mobility.

For *p*-type silicon, the critical field at which electron saturation occurs is 1.5×10^4 V/cm (or 1.5 V/μm), and the saturation velocity υ_{sat} equals 10^7 cm/sec. This means that in an NMOS device with a channel length of 1 μm, only a couple of volts between drain and source are needed to reach the saturation point. This condition is easily met in current short-channel devices. Holes in a *n*-type silicon saturate at the same velocity, although a higher electrical field is needed to achieve saturation ($\geq 10^5$ V/cm).

This effect has a profound impact on the operation of the transistor. This is easily realized when computing the transistor currents under the velocity-saturated condition. Combining Eq. (2.43) and Eq. (2.44) and setting υ_n to υ_{sat} yields a revised current expression,

$$I_{DSAT} = \upsilon_{sat} C_{ox} W(V_{GS} - V_{DSAT} - V_T) \tag{2.56}$$

with V_{DSAT} the drain-source voltage at which velocity saturation comes into play. Observe the *linear dependence* of the saturation current with respect to the gate-source voltage V_{GS}, which is in contrast with the squared dependence in the long- channel device. Consequently, reducing the operating voltage does not have such a significant effect in submicron devices as it would have in a long-channel transistor. Furthermore, I_D is independent of L in velocity-saturated devices, suggesting that current drive cannot be further improved by decreasing the channel length as was the case in long-channel transistors. Observe that this is only true to a first degree, as it ignores the influence of L on V_{DSAT}.

Reducing the channel length has another important impact on the transistor current: even at normal electric field levels, a reduction in the electron mobility can be observed. This effect, called *mobility degradation*, can be attributed to the vertical component of the electrical field, which is no longer ignorable in these small devices. This is illustrated in Figure 2.28b, where the electron mobility is plotted as a function of the transversal electrical field.

Both effects can be combined into an approximate but manageable model for the short-channel MOSFET transistor (proposed in [Toh88])

$$\begin{aligned} I_D &= \kappa \upsilon_{sat} C_{ox} W(V_{GS} - V_T) \text{ for } V_{DS} \geq V_{DSAT} \text{ (saturated region)} \\ &= \mu_n C_{ox} \frac{W}{L}\left((V_{GS} - V_T)V_{DS} - \frac{V_{DS}^2}{2}\right) \text{ for } V_{DS} \leq V_{DSAT} \text{ (triode region)} \end{aligned} \tag{2.57}$$

with V_{DSAT} the saturation voltage, given by

$$V_{DSAT} = (1 - \kappa)(V_{GS} - V_T) \tag{2.58}$$

where κ is a measure of the velocity-saturation degree (with E the longitudinal electrical field):

$$\kappa = \frac{1}{1 + E_{sat}/E} = \frac{1}{1 + E_{sat}L/(V_{GS} - V_T)} \tag{2.59}$$

Be aware of the fact that the mobility μ_n in Eq. (2.57) is not a constant, but a function of the applied electrical field as well due to the mobility degradation. Also, it is easily shown that for $E << E_{sat}$, Eq. (2.57) reverts to the long-channel model, taking into account that the electric field at saturation is related to the maximum velocity by the following expression: $E_{sat} = 2\,\upsilon_{sat}/\mu_n$ [Toh88].

From Eq. (2.58), it can be observed that a short-channel MOS has an extended saturation region as compared to a long-channel transistor (as $0 < \kappa < 1$). Both the extended saturation region and the linear dependence upon V_{GS} are apparent in the simulated *I-V* response of a short-channel device, shown in Figure 2.29a and b. The simulated device uses the same technology as the example of Figure 2.21, but the channel length is set to the minimum allowed value of 1.2 μm.

For detailed, excellent descriptions of short-channel effects in MOS transistors, please refer to [Muller86, Chen90, Sze81, Ko89].

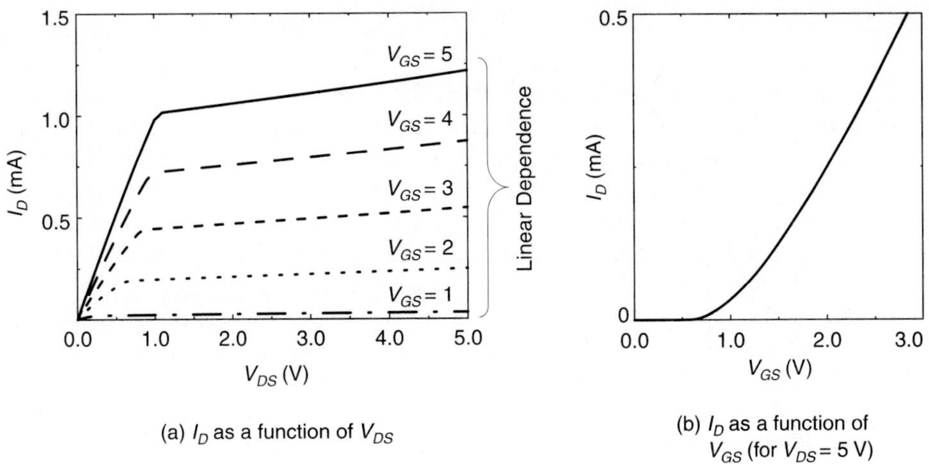

(a) I_D as a function of V_{DS}

(b) I_D as a function of V_{GS} (for V_{DS} = 5 V)

Figure 2.29 Short-channel NMOS transistor *I-V* characteristic (W = 4.6 μm, L = 1.2 μm in a 1.2 μm CMOS technology).

Example 2.10 Impact of Velocity Saturation

The devices of Figure 2.21 and Figure 2.29 have identical effective (*W/L*) ratios (when taking into account the lateral diffusion). If the long-channel model were valid, both devices would produce identical *I-V* characteristics. The latter transistor (W = 4.6 μm, L = 1.2 μm), however, suffers from velocity saturation, while this is not the case for the former device with its very long channel (W = 100 μm, L = 20 μm).

This results in a substantial drop in current drive for high voltage levels. For instance, at (V_{GS} = 5 V, V_{DS} = 5 V), the drain current of the small transistor is only 53% of the corresponding value of the longer transistor (1.2 mA versus 2.3 mA). At lower values of the drain-source voltage, the effect of velocity saturation is less significant. For instance, at (V_{GS} = 5 V, V_{DS} = 1 V) both devices yield approximately the same current of 0.95 mA.

Subthreshold Conduction

From inspection of Figure 2.21b and Figure 2.29b, the reader is probably aware that the MOS transistor is already partially conducting for voltages below the threshold voltage. This effect is called *subthreshold* or *weak-inversion* conduction. The onset of strong inversion means that ample carriers are available for conduction, but by no means implies that no current at all can flow for gate-source voltages below V_T. However, the current levels are small under those conditions. The transition from the on- to the off-condition is not abrupt, but gradual.

To study this effect in somewhat more detail, we redraw the I_D versus V_{GS} curve of Figure 2.29b on a logarithmic scale as shown in Figure 2.30. This clearly demonstrates that the current does not drop to zero immediately for $V_{GS} < V_T$, but actually decays in an exponential fashion, similar to the operation of a bipolar transistor (which will be discussed in Section 2.4). In the absence of a conducting channel, the n^+ (source) - p (bulk) -

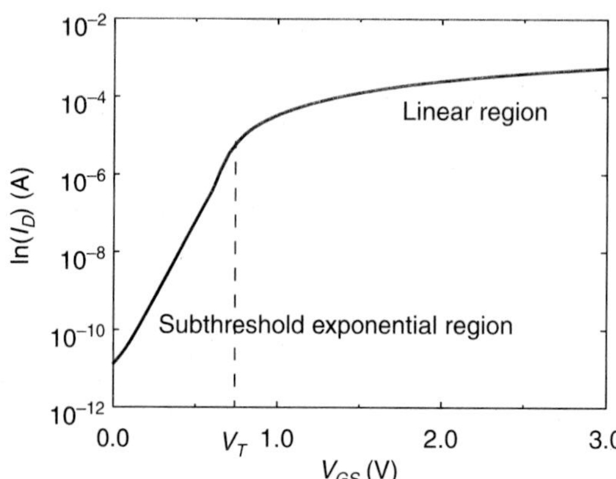

Figure 2.30 I_D current versus V_{GS} (on logarithmic scale), showing the exponential characteristic of the subthreshold region.

n^+ (drain) terminals actually form a parasitic bipolar transistor. In this operation region, the (inverse) rate of decrease of the current with respect to V_{GS} is approximated as stated in [Sze81]:

$$\left(\frac{d}{dV_{GS}}\ln(I_D)\right)^{-1} = \frac{kT}{q}\ln 10(1+\alpha) \tag{2.60}$$

The expression $(kT/q)\ln(10)$ evaluates to 60 mV/decade at room temperature. α equals 0 for an ideal device which means that at room temperature the subthreshold current drops by a factor of 10 for a reduction in V_{GS} of 60 mV. Unfortunately, α is larger than 1 for actual devices and the current drops at a reduced rate. The current drop is further affected in a negative sense by an increase in the operating temperature (most integrated circuits operate at temperatures considerably beyond room temperature). Note that α is a function of the transistor capacitances. Reducing it requires advanced and expensive technologies, such as silicon-on-insulator.

The presence of the subthreshold current detracts from the ideal switch model that we like to assume for the MOS transistor. In general, we want the current to be as close as possible to zero at $V_{GS} = 0$. This is especially important in the so-called *dynamic circuits*, which rely on the storage of charge on a capacitor and whose operation can be severely degraded by subthreshold leakage. This observation puts a firm lower bound on the value of the threshold voltage of the devices.

Example 2.11 Subthreshold Slope

For the example of Figure 2.30, a slope of 121 mV/decade is observed. This is equivalent to an α-factor of 1.

CMOS Latchup

The MOS technology contains a number of intrinsic bipolar transistors. These are especially troublesome in CMOS processes, where the combination of wells and substrates results in the formation of parasitic *n-p-n-p* structures. Triggering these thyristor-like devices leads to a shorting of the V_{DD} and V_{SS} lines, usually resulting in a destruction of the chip, or at best a system failure that can only be resolved by power-down.

Consider the *n*-well structure of Figure 2.31a. The *n-p-n-p* structure is formed by the source of the NMOS, the *p*-substrate, the *n*-well and the source of the PMOS. A circuit equivalent is shown in Figure 2.31b. When one of the two bipolar transistors gets forward biased (e.g., due to current flowing through the well, or substrate), it feeds the base of the other transistor. This positive feedback increases the current until the circuit fails or burns out.

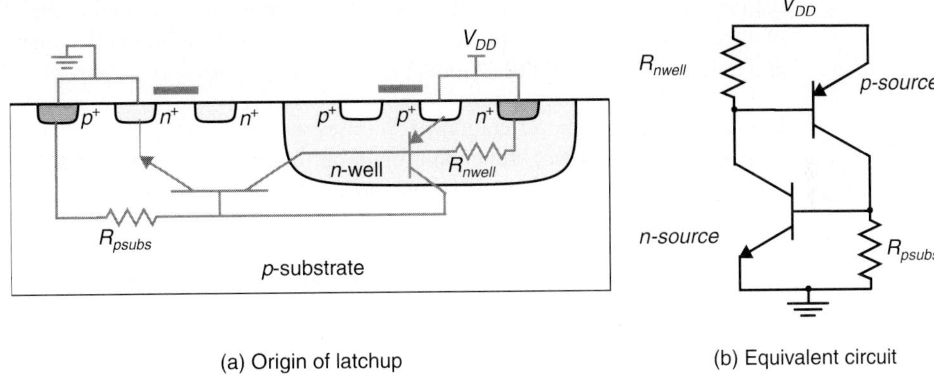

(a) Origin of latchup

(b) Equivalent circuit

Figure 2.31 CMOS latchup.

From the above analysis the message to the designer is clear—to avoid latchup, the resistances R_{nwell} and R_{psubs} should be minimized. This can be achieved by providing numerous well and substrate contacts, placed close to the source connections of the NMOS/PMOS devices. Devices carrying a lot of current (such as transistors in the I/O drivers) should be surrounded by *guard rings*. These circular well/substrate contacts, positioned around the transistor, reduce the resistance even further and reduce the gain of the parasitic bipolars. For an extensive discussion on how to avoid latchup, please refer to [Weste93]. The latchup effect was especially critical in early CMOS processes. In recent years, process innovations and improved design techniques have all but eliminated the risks for latchup.

2.3.5 SPICE Models for the MOS Transistor

The complexity of the behavior of the short-channel MOS transistor and its many parasitic effects has led to the development of a wealth of models for varying degrees of accuracy and computing efficiency. In general, more accuracy also means more complexity and, hence, an increased run time. In SPICE, the choice of the device model is set by the LEVEL parameter. In this section, we introduce some of the most commonly used SPICE MOSFET models as well as their parameters. We also revisit briefly our simple model for manual analysis to accommodate some of the second-order effects.

SPICE Models

- The LEVEL 1 SPICE model implements the *Shichman-Hodges model*, which is based on the square law long-channel expressions, derived earlier in this chapter. As it does not handle short-channel effects, it is not very appropriate for contemporary devices. Its main purpose is to verify a manual analysis.
- The LEVEL 2 model is a geometry-based model, which uses detailed device physics to define its equations. It handles effects such as velocity saturation, mobility degradation, and drain-induced barrier lowering. Unfortunately, including all 3D-effects of an advanced submicron process in a pure physics-based model becomes complex and inaccurate. The LEVEL 2 model is, therefore, virtually obsolete.
- LEVEL 3 is a semi-empirical model. It relies on measured device data to determine its parameters. However, depending upon on the specified parameters, the designer can use mixed models to calculate the threshold voltage and the drain current.
- LEVEL 4 (the Berkeley Short-Channel IGFET Model, or BSIM) provides a model that is analytically simple and is based on a small number of parameters, which are normally extracted from experimental data. Its accuracy and efficiency make it one of the most popular SPICE MOSFET models at present.
- A large number of other models are available, both from SPICE vendors and semiconductor manufacturers. A complete description of all those models would take the remainder of this book, which is, obviously, not the goal. For a good description of SPICE MOSFET models, please refer to [Antognetti88].

Table 2.3 lists the main SPICE model parameters (as used in LEVELS 1 to 3). The parameters covering the parasitic resistive and capacitive effects have been transferred to a separate table for the sake of clarity (Table 2.4). Whenever possible, we have correlated the SPICE parameter to the symbol used in this book. Observe that some of the defined parameters are redundant. For instance, PHI (ϕ_0) can be computed from process parameters such as the substrate doping. User-defined parameters always preside over the analytical value, however. The list is by no means complete, but is sufficient to cover the requirements of this and later chapters.

To conclude this lengthy enumeration, the parameters that can be defined an individual transistor have to be mentioned (Table 2.5). Not all these parameters have to be defined for each transistor. SPICE assumes default values (which are often zero!) for the missing factors. When accuracy is an issue, it is essential to painstakingly define the value of parameters such as the drain and source area or resistance. The NRS and NRD values multiply the sheet resistance RSH specified in the transistor model for an accurate representation of the parasitic series source and drain resistance of each transistor.

Example 2.12 SPICE MOSFET Model

The LEVEL 2 model for a 1.2 μm CMOS process is included. Models are provided for both the NMOS and PMOS devices. This process will serve as the generic CMOS technology in

Table 2.3 Main SPICE MOSFET model parameters.

Parameter Name	Symbol	SPICE Name	Units	Default Value
SPICE Model Index		LEVEL	–	1
Zero-Bias Threshold Voltage	V_{T0}	VT0	V	0
Process Transconductance	k'	KP	A/V^2	2.E-5
Body-Bias Parameter	γ	GAMMA	$V^{0.5}$	0
Channel Modulation	λ	LAMBDA	1/V	0
Oxide Thickness	t_{ox}	TOX	m	1.0E-7
Lateral Diffusion	x_d	LD	m	0
Metallurgical Junction Depth	x_j	XJ	m	0
Surface Inversion Potential	$2\|\phi_F\|$	PHI	V	0.6
Substrate Doping	N_A, N_D	NSUB	cm^{-3}	0
Surface-State Density	Q_{ss}/q	NSS	cm^{-3}	0
Fast Surface-State Density		NFS	cm^{-3}	0
Total Channel Charge Coefficient		NEFF	–	1
Type of Gate Material		TPG	–	1
Surface Mobility	μ_0	U0	cm^2/V-sec	600
Maximum Drift Velocity	υ_{max}	VMAX	m/s	0
Mobility Critical Field	E_{crit}	UCRIT	V/cm	1.0E4
Critical Field Exponent in Mobility Degradation		UEXP	–	0
Transverse Field Exponent (mobility)		UTRA	–	0

Table 2.4 SPICE Parameters for parasitics (resistances, capacitances).

Parameter Name	Symbol	SPICE Name	Units	Default Value
Source Resistance	R_S	RS	Ω	0
Drain Resistance	R_D	RD	Ω	0
Sheet Resistance (Source/Drain)	$R_\square$	RSH	Ω/□	0
Zero-Bias Bulk Junction Cap	C_{j0}	CJ	F/m^2	0
Bulk Junction Grading Coeff.	m	MJ	-	0.5
Zero-Bias Side-Wall Junction Cap	C_{jsw0}	CJSW	F/m	0
Side-Wall Grading Coeff.	m_{sw}	MJSW	-	0.3
Gate-Bulk Overlap Capacitance	C_{gbO}	CGBO	F/m	0
Gate-Source Overlap Capacitance	C_{gsO}	CGSO	F/m	0
Gate-Drain Overlap Capacitance	C_{gdO}	CGDO	F/m	0
Bulk Junction Leakage Current	I_S	IS	A	0
Bulk Junction Leakage Current Density	J_S	JS	A/m^2	1E-8
Bulk Junction Potential	ϕ_0	PB	V	0.8

Table 2.5 SPICE transistor parameters.

Parameter Name	Symbol	SPICE Name	Units	Default Value
Drawn Length	*L*	L	m	–
Effective Width	*W*	W	m	–
Source Area	*AREA*	AS	m^2	0
Drain Area	*AREA*	AD	m^2	0
Source Perimeter	*PERIM*	PS	m	0
Drain Perimeter	*PERIM*	PD	m	0
Squares of Source Diffusion		NRS	–	1
Squares of Drain Diffusion		NRD	–	1

the rest of the book (in both examples, problem sets, and design problems). The presented models are assumed to be the defaults, unless otherwise specified.[4]

```
* SPICE LEVEL 2 Model for 1.2 mm CMOS Process

.MODEL NMOS NMOS LEVEL=2 LD=0.15U TOX=200.0E-10
+ NSUB=5.37E+15 VTO=0.74 KP=8.0E-05 GAMMA=0.54
+ PHI=0.6 U0=656 UEXP=0.157 UCRIT=31444
+ DELTA=2.34 VMAX=55261 XJ=0.25U LAMBDA=0.037
+ NFS=1E+12 NEFF=1.001 NSS=1E+11 TPG=1.0 RSH=70.00
+ CGDO=4.3E-10 CGSO=4.3E-10 CJ=0.0003 MJ=0.66
+ CJSW=8.0E-10 MJSW=0.24 PB=0.58

.MODEL PMOS PMOS LEVEL=2 LD=0.15U TOX=200.0E-10
+ NSUB=4.33E+15 VTO=-0.74 KP=2.70E-05 GAMMA=0.58
+ PHI=0.6 U0=262 UEXP=0.324 UCRIT=65720
+ DELTA=1.79 VMAX=25694 XJ=0.25U LAMBDA=0.061
+ NFS=1E+12 NEFF=1.001 NSS=1E+11 TPG=-1.0 RSH=121
+ CGDO=4.3E-10 CGSO=4.3E-10 CJ=0.0005 MJ=0.51
+ CJSW=1.35E-10 MJSW=0.24 PB=0.64
```

Examples of a typical NMOS and PMOS transistor definition are given below. Transistor M1 is an NMOS device with its drain, gate, source, and bulk terminals connected to nodes 2, 1, 0, and 0, respectively. Its gate length is the minimum allowed in this technology (1.2 μm). This leads to an effective gate length L_{eff} of 0.9 μm, as the lateral diffusion equals 0.15 μm (LD in the transistor model). The PMOS device, connected between nodes 2,1, 5, and 5 (D, G, S, and B, respectively) is three times wider, which reduces the series resistance, but increases the parasitic diffusion capacitances as the area and perimeter of the drain and source regions go up.

```
M1 2 1 0 0 NMOS W=1.8U L=1.2U NRS=0.333 NRD=0.333
+ AD=6.5P PD=9.0U AS=6.5P PS=9.0U
M2 2 1 5 5 PMOS W=5.4U L=1.2U NRS=0.111 NRD=0.111
+ AD=16.2P PD=11.4U AS=16.2P PS=11.4U
```

[4] Be aware that this represents a hypothetical model. The LEVEL 2 parameters were derived from the LEVEL 4 model of an actual process, and some deviations were introduced in the translation.

Another Look at the Manual-Analysis Model

Even the simplified model for the velocity-saturated device (Eq. (2.57) to Eq. (2.59)) is too complex to be considered for manual analysis and would make the evaluation of any circuit with more than one transistor prohibitive. On the other hand, ignoring the short-channel effects completely leads to grossly inaccurate and overly optimistic results. Relying only on SPICE simulations leads to an *ad hoc* design strategy, where optimizations are made in a random order without a real understanding of the basic operation of the circuit and its prime parameters. As already established in the discussion on diodes, manual analysis is useful to derive a first-order solution and to build a conceptual understanding. A simple, yet reasonably accurate, model of the short-channel transistor is therefore essential.

A number of approaches to that goal are conceivable. A first one, adopted in this book, is to use the simple, long-channel equations for the transistor current (Eqs. (2.47) to (2.51)), but to adjust the dominant transistor parameters (k' and λ) so that a reasonable approximation of the current is obtained in the regions that count the most. The performance of an MOS digital circuit is primarily determined by the maximum available current (i.e., the current obtained for $V_{GS} = V_{DS}$ = supply voltage). A good matching in this region is therefore essential.

This idea is illustrated in Figure 2.32 for a supply voltage of 5 V. Drawn is the I-V curve of the short-channel device for V_{GS} = 5 V. An approximative long-channel device is conceived, which yields the same current at $V_{GS} = V_{DS}$ = 5 V and whose current slope approximates the actual one in that particular region. This results in heuristic values for k' and λ, which on the average will yield reasonable approximations. Obviously, this model results in substantial errors when used in other regions such as for small values of V_{GS} or when the supply voltage is changed. For the latter case, a new set of heuristic parameters can be derived. Observe, also, that this model tends to yield pessimistic results, which is in general desirable.

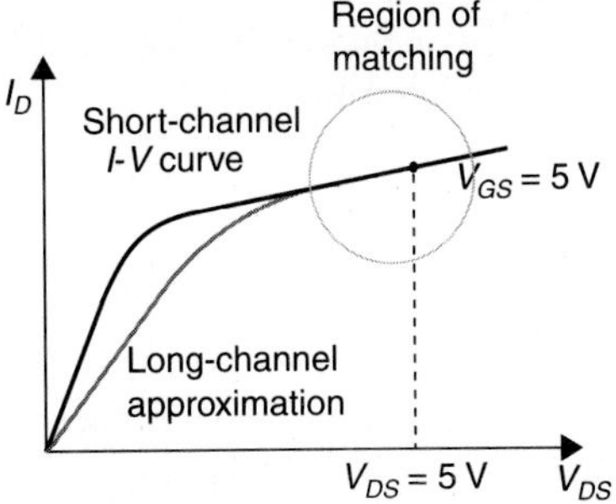

Figure 2.32 Retrofitting the long-channel model to short-channel devices (for a supply voltage of 5 V).

Example 2.13 Manual Analysis Model for 1.2 μm CMOS Process

Using the LEVEL 2 SPICE model of Example 2.12, the drain-source current values are obtained from SPICE for both an NMOS and a PMOS device. Both devices are identical in size (W = 2.0 μm, L = 1.2 μm).

$$I_{DN}\,(V_{GS} = 5\text{ V}, V_{DS} = 5\text{ V}) = 0.514\text{ mA}$$

$$I_{DN}\,(V_{GS} = 5\text{ V}, V_{DS} = 4.5\text{ V}) = 0.502\text{ mA}$$

$$I_{DP}\,(V_{GS} = -5\text{ V},\, V_{DS} = -5\text{ V}) = -0.212\text{ mA}$$

$$I_{DP}\,(V_{GS} = -5\text{ V},\, V_{DS} = -4.5\text{ V}) = -0.202\text{ mA}$$

Plugging those numbers into the long-channel transistor model for the saturated region yields a set of simultaneous equations (for both NMOS and PMOS transistors) from which k' and λ can be determined.

$$I_{DN} = \frac{k'_n}{2}\left(\frac{W_n}{L_{eff}}\right)(V_{GS} - V_{T0})^2(1 + \lambda V_{DS}) = \frac{k'_n}{2}\left(\frac{2}{0.9}\right)(5 - 0.743)^2(1 + \lambda V_{DS})$$

and

$$I_{DP} = -\frac{k'_p}{2}\left(\frac{W_p}{L_{eff}}\right)(|V_{GS}| - |V_{T0}|)^2(1 + \lambda|V_{DS}|) = -\frac{k'_p}{2}2.22(5 - 0.739)^2(1 + \lambda|V_{DS}|)$$

Observe that L_{eff} (the effective transistor length, equal to $L - 2\,x_d$) is used instead of the drawn length. The obtained parameters (which are equivalent to a SPICE LEVEL 1 model) are summarized in Table 2.6. Compare these values with their corresponding value in the LEVEL 2 model.

Table 2.6 Retrofitted LEVEL 1 parameters for 1.2 μm CMOS process.

	V_{T0} (V)	k' (A/V^2)	λ (V^{-1})
NMOS	0.743	19.6×10^{-6}	0.06
PMOS	–0.739	5.4×10^{-6}	0.19

Another reasonable approach is to use the model of Eqs. 2.57 to 2.59, but replace the value of κ, which is normally a function of V_{GS} (Eq. (2.59)) by a constant value. This is acceptable if the voltage-transition range is restricted [Toh88]. Others have modeled the effect of velocity saturation as an extra series resistance on the source side [Gray93]. However, the first approach is simple and produces acceptable results.

2.4 The Bipolar Transistor

MOS transistors took over the digital integrated circuit market in the 1970s, mainly as a consequence of their high integration density. Before that time, most digital gates were implemented in the bipolar technology. The dominance of the bipolar approach to digital design was exemplified in the wildly and widely successful TTL (Transistor-Transistor Logic) logic series, which persisted until the late 1980s. Although bipolar digital designs occupy only a small portion of the digital market at present, they still are the technology of choice when very high performance is required and are, therefore, worth studying.

In line with the previous sections on diodes and MOSFETs, we discuss the static and the dynamic behavior of the bipolar transistor after a brief introduction to the device. We conclude with the presentation of second-order effects and simulation models.

2.4.1 A First Glance at the Device

Figure 2.33a shows a cross section of a typical *npn* bipolar (junction) transistor structure. The heart of the transistor is the region between the dashed lines and consists of two *np* junctions, connected back to back. In the following analysis, we will consider the idealized transistor structure of Figure 2.33b. The transistor is a three-terminal device, where the two *n*-regions, called the *emitter* and the *collector*, sandwich the *p*-type *base* region. In contrast to the source and drain regions of the MOSFET, the emitter and collector regions are not interchangeable, as the emitter is much more heavily doped than the collector.

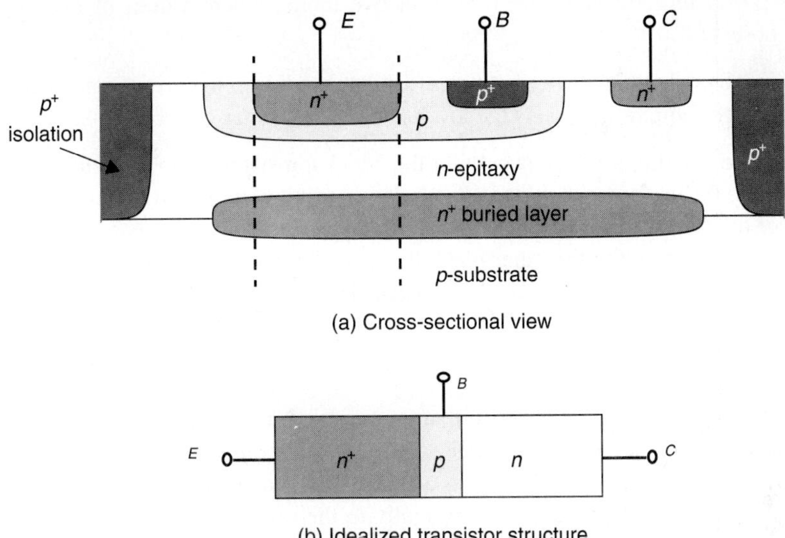

Figure 2.33 *npn* bipolar transistor.

Depending upon the voltages applied over the device terminals, the emitter-base and collector-base junctions are in the *forward-* or *reverse-biased* condition. Enumeration of all possible combinations results in Table 2.7, which summarizes the operation modes of the bipolar device. In digital circuits, the transistor is operated by preference in the cut-off or forward-active mode. Operation in the saturation or reverse regions is, in general, avoided as the circuit performance in those regions tends to deteriorate.

Table 2.7 Modes of operation of the bipolar transistor.

Mode	Emitter Junction	Collector Junction
Cut-off	Reverse	Reverse
Forward-active	Forward	Reverse
Reverse-active	Reverse	Forward
Saturation	Forward	Forward

In a superficial way, the operation of the device can be summarized as follows:

- As both junctions are reverse biased in the *cut-off* mode, no current flows into or out of the device. It can be considered *off*.
- In the *active* mode, the transistor acts as a current amplifier. The current flowing into the base results in a collector current that is β times larger. Furthermore, there exists an exponential relationship between the emitter-base voltage and the collector current. This relation is similar to the forward-bias condition of the junction diode. In the reverse active condition, this current gain is small and virtually nonexistent ($\beta \approx 1$) in contrast to the forward-active mode, where values of over 100 can be observed.
- Finally, when the device saturates, a substantial drop in current gain occurs. Typical for the saturation mode is the low value of the emitter-collector voltage.

Some important differences with the MOS transistor jump immediately into view. First of all, the *exponential relationship* between input voltage and output current makes it possible to drive large currents with small voltage excursions. This has a beneficial impact on the performance. On the other hand, the control terminal (i.e., the base) of the bipolar transistor carries an input current when the device is in active or forward mode. This means that the *input resistance* of the device is small compared to the MOS transistor. As will become apparent later, this feature makes the device not as amenable to high-density digital design.

It should come as no surprise that a complementary device, called the *pnp* bipolar junction transistor, can be conceived as well. They have only been used sparingly in digital design—adding a high-quality *pnp* to a bipolar process tends to raise the cost substantially—and are therefore treated superficially in the rest of this book. Schematic symbols for both *npn* and *pnp* devices, as well as sign conventions for voltage and currents are pictured in Figure 2.34.

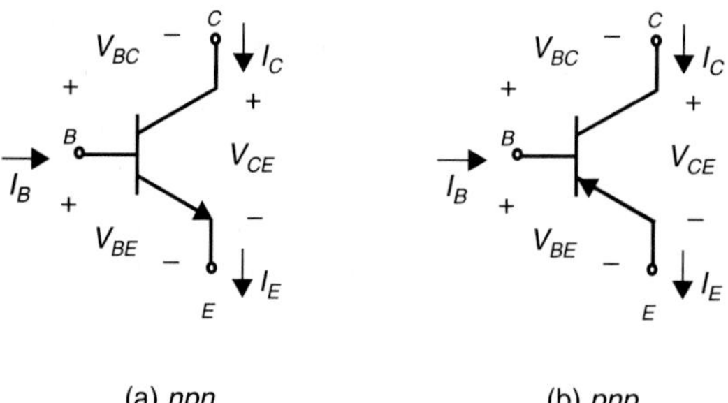

Figure 2.34 Bipolar transistor schematic symbols and sign conventions.

2.4.2 Static Behavior

Forward-Active Region

Figure 2.35 shows a cross section of the idealized transistor structure of Figure 2.33b as well as the minority carrier concentrations in the emitter, base, and collector regions. The concentrations are plotted for the forward-active operation mode. That is, the base-emitter (*be*) junction is forward-biased, while the base-collector (*bc*) junction is in reverse-bias condition. The subscripts *e, b,* and *c* are used to denote the various regions. As we know from our diode study, the forward bias causes excess minority carriers on the *be* side, while the reverse bias at the *bc* end causes the minority concentration to approach zero. We assume (without loss of generality) that the short-base diode model is valid for all junctions.

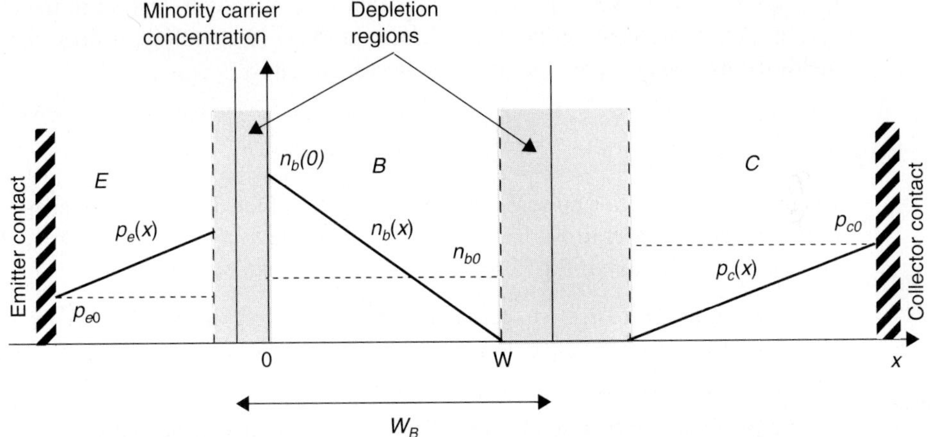

Figure 2.35 Minority carrier concentration profile in an *npn* transistor under forward-active conditions.

The *law of the junction,* Eq. (2.6), is still valid and can be used to evaluate the minority carrier concentrations at the boundaries of the base region

$$\begin{aligned} n_b(0) &= n_{b0}e^{V_{BE}/\phi_T} \\ n_b(W) &= n_{b0}e^{V_{BC}/\phi_T} \approx 0 \end{aligned} \tag{2.61}$$

Obviously, a concentration gradient exists within the base region causing minority electrons to diffuse from the emitter to the collector end. The width of the base region, which is substantially below 1 μm in contemporary technologies, is sufficiently smaller than diffusion length L_n. Consequently, the minority carriers in the base region display a linear gradient, similar to the case of the short-base diode shown in Figure 2.5. The value of this current is readily computed

$$\begin{aligned} I_{diff} &= qA_E D_b \frac{dn_b}{dx} = qA_E D_b\left(\frac{n_b(0) - n_b(W)}{W}\right) = \frac{qA_E D_b}{W} n_b(0) \\ &= \frac{qA_E D_b n_{b0}}{W} e^{V_{BE}/\phi_T} \end{aligned} \tag{2.62}$$

where A_E is the area of the of the transistor under the emitter, and D_b is the diffusion coefficient for minority carriers in the base region.

The following picture now emerges. The forward bias of the *be* junction reduces the potential barrier. Electrons, the majority carriers in the emitter, diffuse from the emitter to the base, where they become minority carriers. As the emitter is more heavily doped than the base, we assume for the moment that the *be* junction is a one-sided junction and that the hole current can be ignored. The concentration gradient in the base causes the injected (or emitted) electrons to diffuse towards the collector. From Eq. (2.62), it follows that this diffusion current is an exponential function of the applied base-emitter bias-voltage. Once arrived at the reverse-biased *bc* junction, these electrons are swept towards the collector by the local electrical field (the collector potential is positive with respect to the base). The current across the junction (or the collector current) is, therefore, a drift current. This yields the following expression for the collector current

$$I_C = \frac{qA_E D_b n_{b0}}{W} e^{V_{BE}/\phi_T} = I_S e^{V_{BE}/\phi_T} \tag{2.63}$$

where I_S is the saturation current. Since the thickness of the depletion regions is, generally, negligible with respect to the base region W_b, it is safe to assume that $W \approx W_b$. Be aware also that the derivation of Eq. (2.63) is based on some extra assumptions:

- The minority carrier concentration in the base region is substantially below the majority carrier concentration. This is called the *low-level injection* criterion.
- All voltage drops occur over the depletion regions. This assumes that the neutral regions are perfect conductors. This simplification will be accounted for later by adding series resistances.

The main assumption until now, however, was that all electrons make it safely from the emitter to the collector, or that $I_C = I_E$. In reality, this is clearly not the case. A more accurate picture of the currents flowing in the bipolar *npn* transistor is pictured in Figure 2.36.

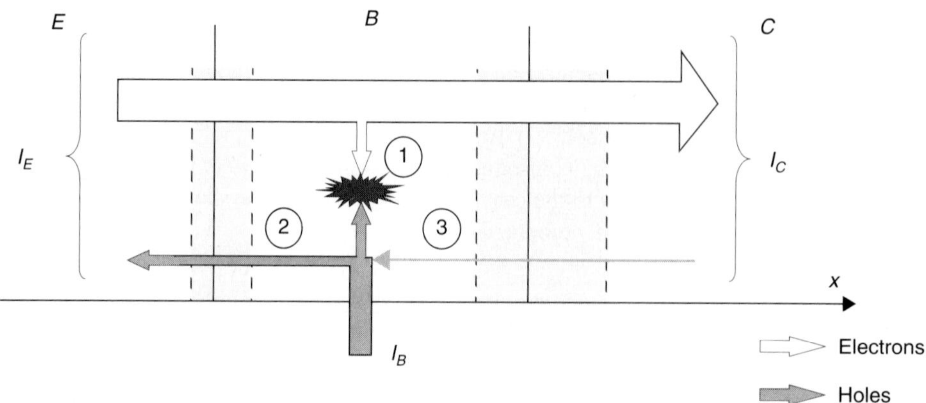

Figure 2.36 Components of the base current.

As shown, holes are flowing into the base to supply the following currents:

- *Recombination current* ①—Occasionally, minority electrons diffusing through the base recombine with majority hole carriers. To maintain charge neutrality in the base region, the vanished holes have to be replaced.
- *be-junction hole current* ②—The forward-biased base-emitter junction carries an electron as well as a hole current. Although the latter is substantially smaller due to the one-sided nature of the *be* junction, a small base current is still required to supply the carriers.
- *bc-junction hole current* ③—This current actually flows from the collector to the base and equals the saturation current of a reverse-biased junction. From the discussion on *pn*-junction diodes, we know that this component is negligible.

This analysis demonstrates that the base current I_B is relatively small; the smaller the better, actually. It relates to the collector current I_C by a constant ratio, called the *forward current gain* β_F:

$$\beta_F = \frac{I_C}{I_B} \tag{2.64}$$

For typical digital bipolar processes, β_F varies between 50 and 100. This means that a small hole current into the base sustains a large electron current at the collector; hence the current gain.

The relationship between I_E and I_B is obtained by enforcing the current conservation law

$$I_E = I_B + I_C = I_B(\beta_F + 1) \tag{2.65}$$

Finally, by combining Eq. (2.64) and Eq. (2.65), we can relate I_C and I_E.

$$\frac{I_C}{I_E} = \frac{\beta_F}{\beta_F + 1} = \alpha_F \leq 1 \tag{2.66}$$

where α_F is called the *forward common-base current gain.*

Reverse-Active Region

In this operation region, the situation is reversed—the base-collector junction is forward biased, while the base-emitter junction is in reverse mode. The picture is essentially the same as in the forward active mode, as shown in Figure 2.37. The major difference is that the gradient of the base minority carrier distribution is reversed, so that the diffusion current now is directed from collector to emitter. We use a similar approach to derive the current expression

$$I_E = -\frac{qA_C D_b n_{b0}}{W_B} e^{V_{BC}/\phi_T} \tag{2.67}$$

where A_C is the collector area. Notice that we observe the current conventions of Figure 2.34, which explains the negative sign of I_E. While this expression looks similar to Eq.

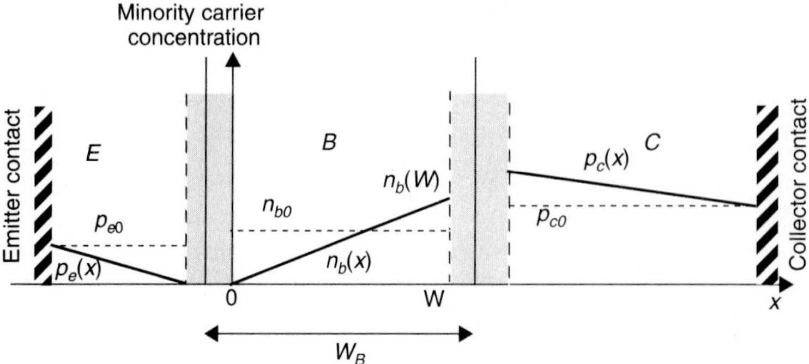

Figure 2.37 Minority carrier concentration profile in reverse-active operation mode.

(2.63), some major differences come to light when examining the components of the base current. While the recombination and base-emitter fractions stay at roughly the same level, the hole current from the base to the collector is substantial and actually exceeds the flow of electrons over the junction. The latter is explained by the fact that the base has a higher doping level than the collector, so that its current in forward bias is dominated by the flow of holes (Eq. (2.10)). These holes have to be provided by the base current, which becomes comparable to the emitter current. The *reverse current gain* β_R is, therefore, small

$$\beta_R = -\frac{I_E}{I_B} \approx 1 \tag{2.68}$$

and so is the *reverse common-base current gain*

$$\alpha_R = \frac{I_E}{I_C} = \frac{\beta_R}{\beta_R + 1} \ll 1 \tag{2.69}$$

Saturation Region

In this mode of operation, both junctions are forward-biased (V_{BE} and $V_{BC} > 4\,\phi_T$). We only consider the case where the emitter junction has a stronger bias (or $V_{BE} > V_{BC}$). The opposite case is called the *reverse-saturation condition* and occurs rarely in digital circuits.

Under this condition, excess minority carriers are present at both the emitter and collector boundaries of the base region, although a somewhat higher concentration is present at the emitter end. The resulting minority carrier concentration is plotted in Figure 2.38. Due to the short base width, a linear carrier gradient is still appropriate, and a diffusion current flows from emitter to collector, albeit of a smaller value than in the forward-active region. An important increase in the base charge is furthermore apparent. This gives rise to an increase in the recombination component of the base current. The combination of both factors indicates that the current gain is reduced in the saturation region, and is substantially smaller than β_F. A simple application of Kirchoff's voltage law also shows that saturation condition corresponds to a small value of V_{CE}.

$$V_{CE} = V_{BE} - V_{BC} \tag{2.70}$$

with $V_{BE} > V_{BC} > 4\phi_T$. When the device is in deep saturation, $V_{CE(sat)}$ generally ranges between 0.1 to 0.2V.

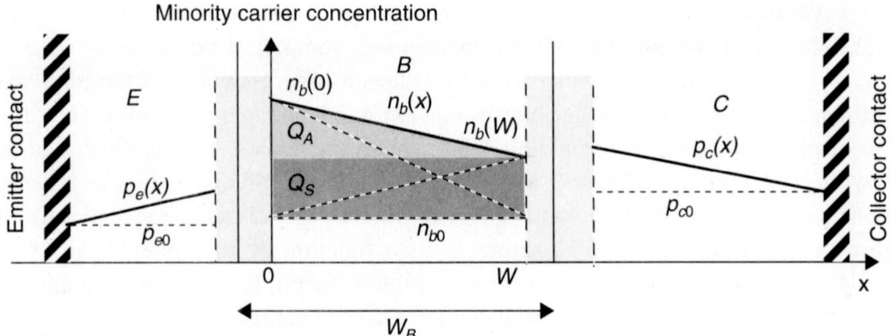

Figure 2.38 Minority carrier concentration in saturation region ($V_{BE} > V_{BC} > 4\phi_T$).

Cut-off Mode

Finally, in cut-off mode, both diodes are reverse-biased. The corresponding concentration profile is shown in Figure 2.39. No excess base charge is present. The currents into the terminals are limited to the saturation currents of the reverse-biased diodes and are extremely small. The transistor is considered to be in the off-state.

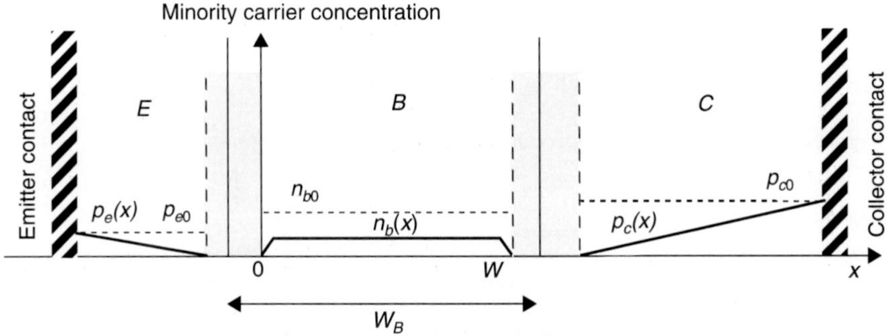

Figure 2.39 Minority carrier profile in cut-off mode.

A Global View

The different operation modes of the bipolar transistors can be unified into a single set of equations, called the *Ebers-Moll model*. The model combines both forward and reverse currents, and in its most general form is expressed by the following set of equations:

$$I_C = I_F - \frac{I_R}{\alpha_R} \qquad I_E = \frac{I_F}{\alpha_F} - I_R \qquad I_B = I_E - I_C \tag{2.71}$$

where

$$I_F = I_S(e^{V_{BE}/\phi_T} - 1) \qquad I_R = I_S(e^{V_{BC}/\phi_T} - 1) \tag{2.72}$$

and where I_S, α_F, and α_R are defined by Eqs. (2.63), (2.66), and (2.69) respectively. This form is often used for the computer representation of transistor large-signal behavior.

We now have all the information needed to plot the *I-V* characteristics of the bipolar transistor. The behavior of an MOS transistor was completely described by the variable set (I_D, V_{GS}, and V_{DS}), two of which could be chosen independently. In the bipolar case, the presence of the base current introduces one extra independent parameter. Therefore, a dual set of plots is needed to completely characterize the device. This is illustrated in the following pair of graphs. Figure 2.40 plots I_C as a function of V_{CE} with I_B as a parameter for both the reverse and the forward operation modes. Observe the reduced current gain in reverse operation. Figure 2.41 shows I_C as a function of V_{BE}, plotted on a logarithmic scale. The reduced slope of the currents at higher current levels is due to a number of secondary effects, which will be introduced later in the chapter.

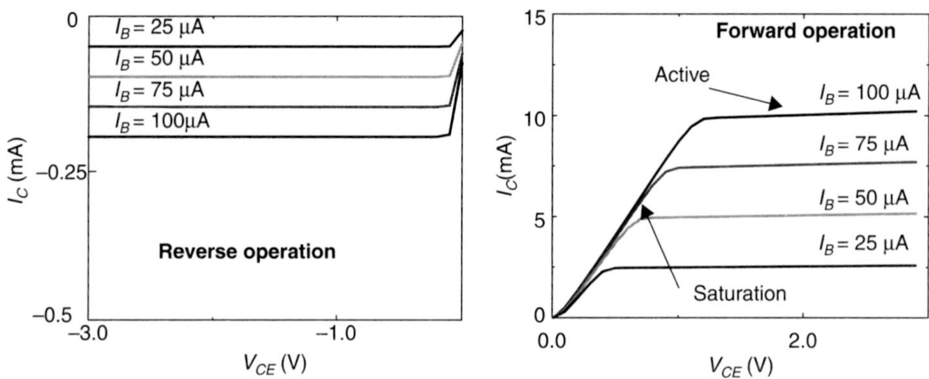

Figure 2.40 Collector current as a function of V_{CE} with I_B as a parameter.

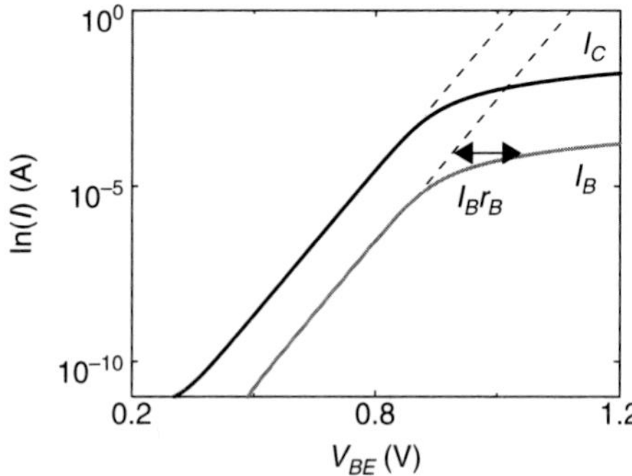

Figure 2.41 Collector and base current as a function of V_{BE} (for a constant V_{CE} of 3 V).

Problem 2.3 Current Gain of Bipolar Transistor.

Determine β_F and β_R of the device plotted in Figure 2.40.

Problem 2.4 *pnp* Transistor Characteristics.

Plot, approximately, the *I-V* characteristics of a *pnp* transistor.

Model for Manual Analysis

The Ebers-Moll model, while being accurate, tends to be too complex to be useful in the first-order analysis of a bipolar circuit. A simplified model of the bipolar transistor, intended for manual analysis, is therefore at hand. A summary of such a model, categorized per operation region, is given below (similar models can be established for the reverse operation modes).:

- **Cut-off**—$V_{BE} < V_{BE(on)}$, $V_{BC} < V_{BC(on)}$

  All currents are zero: $I_C = I_B = 0$

- **Forward-active**—$V_{BE} \geq V_{BE(on)}$, $V_{BC} < V_{BC(on)}$

  A first-order model for large signal computations consists of a base-emitter diode and a current source (Figure 2.42a). An even simpler version of this model, which is often useful, replaces the input diode by a voltage source with a fixed voltage $V_{BE(on)}$, which ranges between 0.6 and 0.75 V. This represents the fact that the base-emitter voltage varies only a little in the forward-active region due to the steep exponential characteristic (Figure 2.42b).

- **Forward-saturation**—$V_{BE} \geq V_{BE(sat)}$, $V_{BC} \geq V_{BC(on)}$, $V_{BE} > V_{BC}$

  During saturation, the collector-emitter voltage is assumed to be fixed at a voltage $V_{CE(sat)}$, which produces the model of Figure 2.42c.

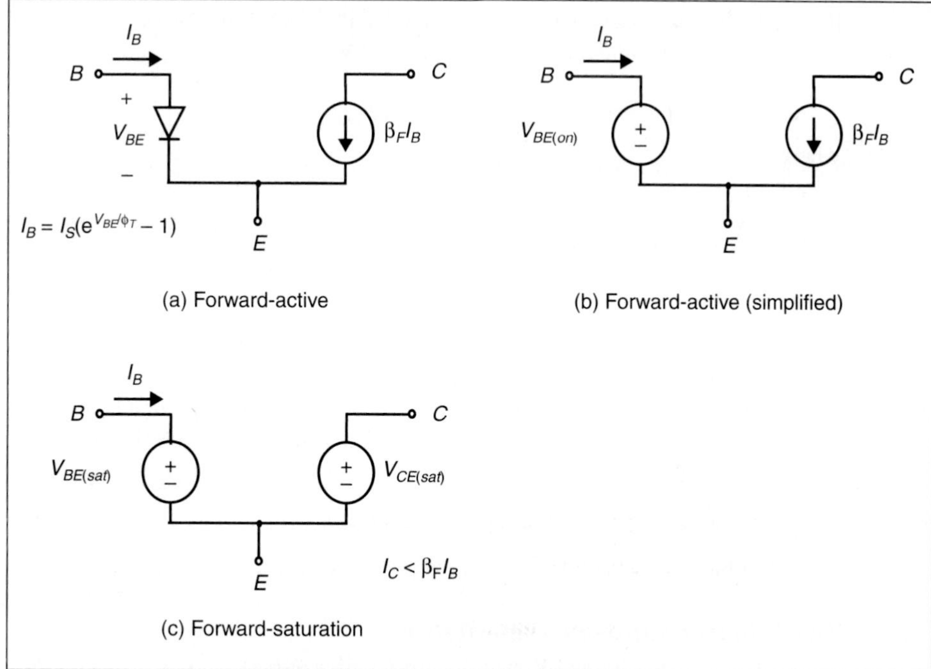

Figure 2.42 First-order large signal models for *npn* bipolar transistor

Example 2.14 Static *npn* Model for Manual Analysis

The *npn* transistor of Example 2.12 is to a first degree modeled by the following parameters:

$$\beta_F = 100, I_S = 1.E{-}17\text{ A}, V_{BE(on)} = 0.7\text{ V}, V_{BE(sat)} = 0.8\text{ V}, V_{CE(sat)} = 0.1\text{ V}.$$

2.4.3 Dynamic Behavior

As we have done for diodes and MOS devices, we model the dynamic behavior of the bipolar transistor by a set of capacitances. In the bipolar device, these capacitive effects originate from three sources: the base-emitter and base-collector depletion regions, the collector-substrate diode, and the excess minority carrier charge in the base. We discuss each of them individually.

The Base-Emitter and Base-Collector Depletion Capacitances

Depending upon the operation region, the *be* and *bc* junctions are either forward- or reverse-biased. This has an important impact on charge storage in the depletion region and, hence, on the equivalent capacitance. The nature of the junction capacitance and its model should be familiar by now, as it has been discussed extensively in previous sections. To combat the nonlinearity of the junction capacitance, the linearized large-scale model of Eqs. (2.17)–(2.18) (repeated below for clarity) is often used. In determining the value of the K_{eq} multiplier factor, one should be careful in delineating exactly the range of the voltage swing over the diode, since this affects the accuracy.

$$C_{eq} = K_{eq} C_{j0} \tag{2.73}$$

where

$$K_{eq} = \frac{-\phi_0^m}{(V_{high} - V_{low})(1-m)}[(\phi_0 - V_{high})^{1-m} - (\phi_0 - V_{low})^{1-m}] \tag{2.74}$$

Both junction capacitances are thus adequately modeled by the following parameters:

- Their zero-bias value (C_{be0} and C_{bc0}, respectively). Although the value of the collector capacitance is typically a trifle larger than its counterpart at the emitter side, both values are comparable. For instance, for an *npn* device in an advanced bipolar process with emitter size of 0.6 μm × 2.4 μm, C_{be0} = 6.7 fF, and C_{bc0} = 7.5 fF [Yamaguchi88].
- The junction grading coefficients m_{be} and m_{bc}.
- The built-in potentials.

The Collector-Substrate Capacitance

As can be observed from the cross-sectional view of the bipolar transistor (Figure 2.33), the *n*-type collector is isolated from the substrate by an inverse-biased diode. The model is complicated by the presence of the n^+ buried layer and the p^+ isolation regions. This results in a parasitic capacitance, which is, once again, a nonlinear junction capacitance. Since the area of the collector-substrate junction is considerably larger than the emitter area, the zero-bias value of this parasitic capacitance can be substantial and might dominate the performance of the device. The value of C_{cs0} typically ranges between 1 and 2.5 times the base-emitter capacitance. Advanced processes reduce its impact by providing trench isolation between devices (i.e., a vertical plug of SiO_2 inserted between transistors), or even by using isolating substrates.

The Base Charge—An Introduction

When deriving the diode current model, we were able to relate the diode current to the excess minority charge in the neutral regions. This proved to be beneficial, especially when studying the dynamic characteristics of the device. A similar approach works for the bipolar transistor as well. The important parameter here is the *excess minority carrier charge in the base region*, which is a strong function of the operation mode.

1. Forward-active

Assuming that the minority carriers in the base region display a linear gradient, as shown in Figure 2.35, the total excess minority charge in the base is readily computed.

$$Q_F = qA_E W_B\left(\frac{n_b(0) + n_b(W) - 2n_{b0}}{2}\right) \approx qA_E W_B\frac{n_b(0)}{2} \tag{2.75}$$

this yields the following relationship between the collector current of Eq. (2.63) and the excess base charge,

$$I_C = \frac{Q_F}{\tau_F} \tag{2.76}$$

with τ_F the *mean forward-transit time*, which can be interpreted as the mean time for the minority carriers to diffuse from the emitter to the collector.

$$\tau_F = \frac{W_B^2}{2D_B} \tag{2.77}$$

Observe the similarity between the forward transit time and the mean transit time of the short-base diode, Eq. (2.20). Keeping this transit time as short as possible is clearly a good idea, as it results in a larger current for a smaller base charge. This can be accomplished be reducing the base width to a maximal extent. This trend is apparent in contemporary bipolar processes, where W_B varies between 50 and 100 nm. Values of τ_F range between 5 and 30 psec.

In a similar fashion, we can relate the base current I_B to the base charge

$$I_B = \frac{Q_F}{\tau_{BF}} \tag{2.78}$$

with τ_{BF} the *minority carrier lifetime* in the base region during forward-active operation. It accounts for the three factors shown in Figure 2.36. Taking into account that the base and collector currents are related by the forward current gain, a relationship between the base transit and lifetimes is easily derived.

$$\beta_F = \frac{\tau_{BF}}{\tau_F} \tag{2.79}$$

2. Reverse-active

A similar analysis yields the following charge-driven model for the reverse currents.

$$I_E = -\frac{Q_R}{\tau_R} \qquad I_C = -\frac{Q_R}{\tau_R} - \frac{Q_R}{\tau_{BR}} \qquad \text{and} \qquad \beta_R = \frac{\tau_{BR}}{\tau_R} \tag{2.80}$$

3. Forward-saturation

This excess base charge in the saturation region can be divided into a forward and a reverse component, as shown by the dotted lines in Figure 2.38. This results in the following expression for I_C,

$$I_C = \frac{Q_F}{\tau_F} - \frac{Q_R}{\tau_R}\frac{1}{\alpha_R} = \frac{Q_F}{\tau_F} - Q_R\left(\frac{1}{\tau_R} + \frac{1}{\tau_{BR}}\right) \tag{2.81}$$

It is conceptually more interesting to consider a different subdivision as illustrated by the shadings of grey in Figure 2.38.

The first component, called the *critical base charge* Q_A, has a triangular shape. It represents the charge needed to bring the transistor to the *edge-of-saturation (eos)*, which is the boundary condition between forward-active and saturation. At the *eos*, the equations governing the forward-active mode are still valid.

$$I_{C(eos)} = \frac{Q_A}{\tau_F} = \beta_F I_{B(eos)} \tag{2.82}$$

where $I_{B(eos)}$ is the corresponding base current.

The second component, Q_S, is called the *overdrive charge* and is the result of both junctions being in forward bias. Being rectangular, it does not contribute to the gradient of the excess minority carriers, nor to the collector current. It arises from the pushing of more current into the base than is required to reach the saturation condition. The current in excess of the $I_{B(eos)}$ is called the *excess base current* I_{BS} and is related to the overdrive charge by a time constant τ_S, called the *saturation time constant*,

$$I_{BS} = \frac{Q_S}{\tau_S} \tag{2.83}$$

which is a weighted average of τ_F and τ_R [Hodges88].

$$\tau_S = \frac{\alpha_F(\tau_F + \alpha_R\tau_R)}{1 - \alpha_F\alpha_R} \approx \frac{\alpha_R}{1 - \alpha_R}\tau_R \tag{2.84}$$

(assuming that $\tau_F << \tau_R$ and $\alpha_F \approx 1$).

With the transistor in saturation, an expression for the (static) base current is now readily derived

$$I_B = I_{B(eos)} + I_{BS} = \frac{Q_A}{\tau_{BF}} + \frac{Q_S}{\tau_S} \tag{2.85}$$

It is, especially, the overdrive charge that makes the saturation region an operation mode to avoid in digital design. Recall from the discussion of diode dynamics that it takes time to remove or provide this minority excess charge. Bringing a transistor in and out of saturation is a slow operation that is inconsistent with the high performance requirements we impose on digital circuits.

Changing the Base Charge—The Diffusion Capacitance

From the above, it becomes apparent that a considerable amount of excess minority carrier charge is stored in the base region of an active or saturated bipolar transistor. Changing the operation mode of the device requires the addition or the removal of this charge.

As with the diode, this minority excess charge can be modeled by capacitances in parallel with the *be* junction (representing the forward charge Q_F) and the *bc* junction (representing the reverse charge Q_R). The value of these capacitors is a strong function of the transistor operation mode and the current levels. A large-signal approach, in the sense of Eqs. (2.23)–(2.24), is therefore appropriate. Observe that the base charge is an exponential function of the voltage over the junctions.

As an example, during forward-active operation Q_R is ignorable. Only a single capacitor between base and emitter, representing $Q_F = I_C\tau_F$, has to be considered. This capacitor is often called *the diffusion capacitance* C_d, as it represents the carriers diffusing through the base region. Similar to the approach used for the diode, the **small-signal value** of C_d is readily computed.

$$C_d = \tau_F\frac{I_C}{\phi_T} \tag{2.86}$$

The large-signal value (called C_D) is comparable to Eq. (2.24). During saturation, the base charge can be represented as the sum of the forward and reverse charges. For manual analysis, it is often more convenient to divide the base charge into the *overdrive* and the *edge-of-saturation* (eos) charge in correspondence with the model discussed in the previous section. Bringing a transistor out of saturation simply requires the removal of the overdrive charge.

Bipolar Transistor Capacitive Model

The effects of all the above contributions are combined into a single capacitive model for the bipolar transistor, as shown in Figure 2.43. Its components can be identified based on the preceding discussion. Having an intuitive feeling for what components are important for each operation condition is what makes a good bipolar designer.

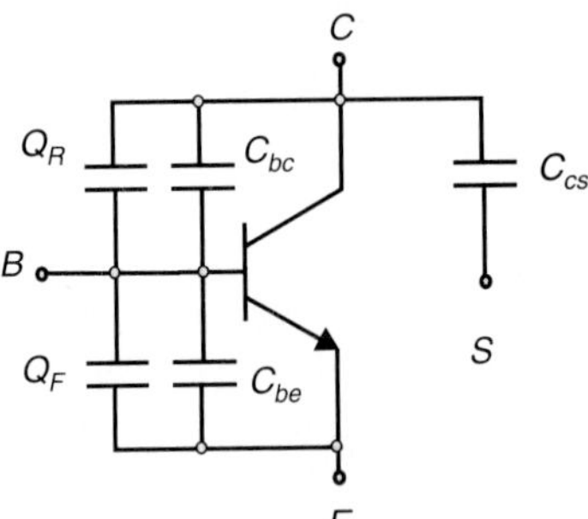

Figure 2.43 Capacitive model of an *npn* bipolar transistor.

Bipolar Transistor Switching Time—A Case Study

Consider the simple circuit of Figure 2.44. We analyze the case where the transistor is initially on, or $V_0 > V_{BE(on)}$, and in forward-active operation mode. At time $t = 0$, the input voltage drops to 0, and the device is turned off.

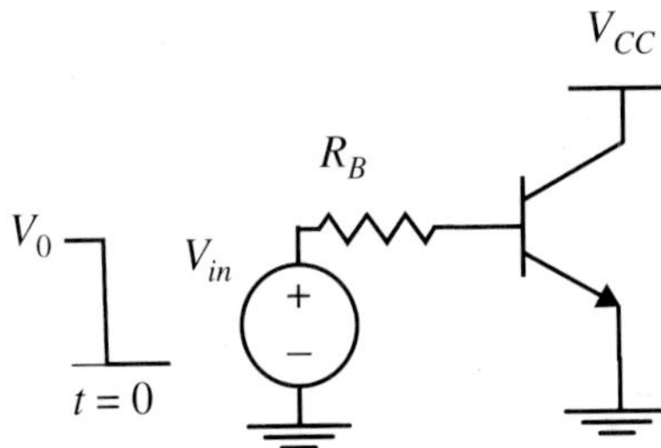

Figure 2.44 Example of *npn*-transistor-based circuit.

Similar to the diode case, the turn-off process proceeds in two phases. In a first step, the base charge is removed, turning the transistor off. During this phase, the base-emitter voltage remains relatively constant. In the second phase, the base-emitter voltage drops to zero. This requires the (dis)charging of the junction capacitances. We analyze each of these phases individually. Again, be aware that this break-up is a simplification.

1. Removal of the base charge

Before time $t = 0$, the excess minority-carrier charge in the base equals

$$Q_F(t = 0) = Q_{F0} = I_C(t = 0) \times \tau_F$$

where $I_C(t = 0)$ is expressed by the following equation:

$$I_C(t = 0) = \beta_F \times I_B(t = 0) = \beta_F(V_0 - V_{BE(on)})/R_B$$

We may assume that during the base-charge removal the base-emitter voltage remains relatively constant and is approximately equal to $V_{BE(on)}$. This means the space charge of the *be* and *bc* junctions is constant also, and not a factor at this point. The length of this interval is thus uniquely defined by the time it takes to remove the base charge. Under the assumption of a fixed base-emitter voltage, the base current that accomplishes this is constant and equals

$$I_B = (0 - V_{BE(on)})/R_B$$

This current actually serves two purposes: (1) to sustain the recombination in the base and (2) to remove the base charge. This dual purpose is expressed by the following equation,

$$I_B = \frac{Q_F}{\tau_{BF}} + \frac{dQ_F}{dt} \tag{2.87}$$

Solving the differential equation, taking the initial condition for Q_F into account, yields $Q_F(t)$,

$$Q_F(t) = \tau_{BF}[I_B - (I_B(t=0) - I_B)e^{-t/\tau_{BF}}] \tag{2.88}$$

and the time to remove the excess charge ($Q_F = 0$).

$$t_{base} = \tau_{BF}\ln\left[\frac{I_B(t=0) - I_B}{-I_B}\right] \tag{2.89}$$

The recombination factor in Eq. (2.87) is often small compared to I_B and actually drops off very rapidly. It can be ignored for a first-order analysis, which means that all base current is assumed to be used for base-charge removal. This yields a simplified expression for t_{base}.

$$t_{base} = \frac{\Delta Q_F}{I_B} = -\frac{Q_{F0}}{I_B} = -\tau_{BF}\frac{I_B(t=0)}{I_B} = \frac{C_D \Delta V}{I_B} \tag{2.90}$$

Observe especially the last component of Eq. (2.90). C_D stands for the large-signal equivalent capacitance (defined over the range ΔV). This equation states that, if the recombination can be ignored, the base charge removal can be modeled as the discharging of a capacitor.

2. Changing the space charge of the *be* and *bc* junctions

Once the transistor is off, the rest of the transient is devoted to discharging the junction capacitances to 0 volt. During this time span, the circuit is modeled as a first-order RC-circuit. The time to reach 90% of the final value is a simple function of the time constant of the network

$$t_{space} = 2.2R_B(C_{be} + C_{bc}) \tag{2.91}$$

Example 2.15 Dynamic Behavior of an *npn* Transistor

Assume that the bipolar transistor in Figure 2.44 is characterized by the following parameters: $\beta_F = 100$, $V_{BE(on)} = 0.75$ V, $\tau_F = 10$ psec, $C_{be0} = 20$ fF, $C_{bc0} = 22$ fF, $C_{cs0} = 47$ fF, $m = 0.33$, $\phi_0 = 0.75$ V. The following circuit parameters are further assumed: $V_{CC} = 2$ V, $V_0 = 1$ V, and $R_B = 5$ kΩ.

First we compute the initial values of the currents. Since the V_{CE} of the transistor is equal to 2 V, the transistor is, obviously, in the forward-active mode. The base current is expressed as

$$I_B\,(t = 0) = (1 - 0.75)\,/\,(5 \times 10^3) = 50\ \mu\text{A}$$

After the switching of the input, the base current changes direction, or

$$I_B = (-0.75)\,/\,(5 \times 10^3) = -150\ \mu\text{A}$$

The time to remove the base charge can now be computed with the aid of Eq. (2.89).

$$t_{base} = (100 \times 10\ \text{psec})\ \ln\,(200\ \mu\text{A}\,/\,150\ \mu\text{A}) = 288\ \text{psec}$$

The simplified expression of Eq. (2.90), on the other hand, yields 333 psec. This is sufficiently close to be useful as a first-order approximation.

As mentioned, the circuit behaves as an *RC*-equivalent network during the space-charge removal. To approximate the time constant of the network, it is necessary first to linearize the junction capacitances over the voltage range of interest. Observe that the collector voltage remains fixed during the transient, so that C_{bc} can be considered as a grounded capacitor. The voltage over the base-emitter junctions decreases from 0.75 V to 0 V, while the voltage over the *bc* junction increases from 1.25 V to 2 V. Injecting those values into Eq. (2.18) yields the following values of K_{eq}: $K_{eq(be)} = 1.5$ and $K_{eq(bc)} = 0.68$. The value of t_{depl} can now be computed

$$t_{depl} = 2.2 \times 5\ \text{k}\Omega \times (1.5 \times 20\ \text{fF} + 0.68 \times 22\ \text{fF}) = 495\ \text{psec}$$

which is considerably larger than the base-charge removal period.

These results are validated by the SPICE simulations shown in Figure 2.45. During the removal of the base charge, a fast decrease in the collector current can be observed. At the end of the base-charge removal phase, the collector current drops to 0, the transistor goes into the cut-off mode, and the base-emitter voltage decays to 0 V in an exponential manner. The simulated values of t_{base} and t_{depl} equal 302 psec and 429 psec (the transition between both regions is somewhat fuzzy and is determined here by the point where the extrapolated linear collector current crosses the zero-axis). The discrepancy between the predicted and simulated values of t_{depl} is due to the linearization, which is more accurate in the reverse- than the forward-biased regions of the diodes.

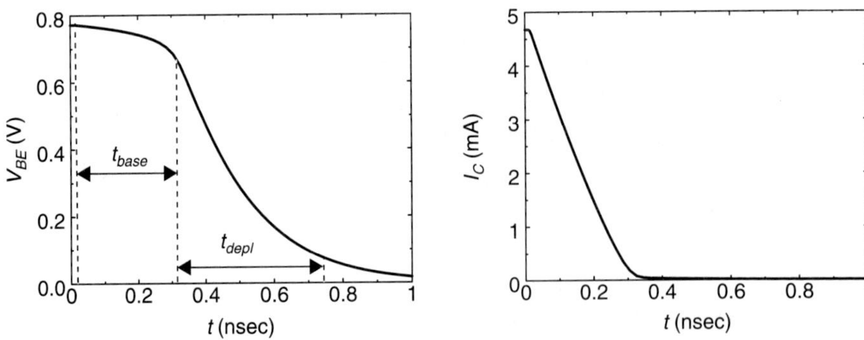

Figure 2.45 Simulated transient response of the circuit of Figure 2.44.

Problem 2.5 Turning on the Bipolar Transistor

Derive the time it takes to turn on the *npn* transistor in the example of Figure 2.44, assuming the same device and circuit parameters and assuming a step from 0 V to 1 V at the input.

2.4.4 The Actual Bipolar Transistor—Secondary Effects

In the previous sections, we have sketched a model of an ideal *npn* device. As always, the actual transistor is somewhat more complex, and some second-order parameters have to be considered.

The Early Voltage

In the analysis of the forward-active operation mode, we assumed the base-collector junction to be reverse-biased and to have no effect on the value of the collector current (Eq. (2.63)). This is a useful approximation for first-order analysis, but is not strictly true in practice. The reason for this is easily understood. Until now, we ignored the width of the depletion regions with respect to the width of the base region W_B. In other words, we assumed that the width of the neutral base region W is equal to the base width W_B. For thin base regions, this is not really the case. Furthermore, an increase in the collector-base voltage increases the reverse bias and, consequently, the width of the depletion region. Since this decreases the effective base width, it reduces the forward transit time τ_F and raises the collector current I_C.

The collector current therefore increases as a function of the collector voltage, as is shown in Figure 2.46, which plots the collector voltage as a function of V_{CE} with V_{BE} as a parameter. It is interesting to observe that all characteristics extrapolate to a single point on the V_{CE}-axis, called the *early voltage* V_A. For integrated circuit transistors, V_A varies from 15 V to 100 V. The inverse of the early voltage is analogous to the channel-length modulation parameter λ in the MOS transistor.

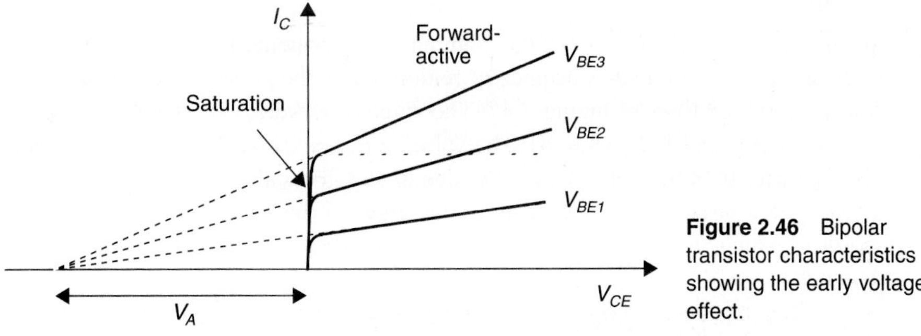

Figure 2.46 Bipolar transistor characteristics showing the early voltage effect.

The collector-current Eq. (2.63) can be adapted to compensate for the early effect. Due to additional computational complexity this model is rarely used in manual calculations and is reserved for computer models.

$$I_C = I_S\left(1 + \frac{V_{CE}}{V_A}\right)e^{V_{BE}/\phi_T} \tag{2.92}$$

Transistor Breakdown Voltages

The mechanism of avalanche breakdown in a *pn*-junction was described in section 2.3.4 Similar effects occur at the *be* and *bc* junctions of the bipolar transistor. This limits the voltages that can be applied to the device. The breakdown voltages depend upon the transistor configuration and are commonly expressed as BV_{CBO} (common base configuration breakdown voltage), V_{CEO} (common emitter configuration) and BV_{EBO} (base-emitter breakdown). The values of the breakdown voltages vary from 6–8 V (BV_{EBO}) to several tens of volts for the others. Since breakdown creates excessive power dissipation and might lead to device destruction when sustained, it is avoided in digital bipolar design.

Parasitic Resistances

Parasitic resistances are produced by the finite resistance of the neutral regions of the transistor, as shown in Figure 2.47. While r_E is normally very small (1–5 Ω), r_B and r_C can be substantial and have a significant impact on the device performance due to the high doping level of the emitter region.

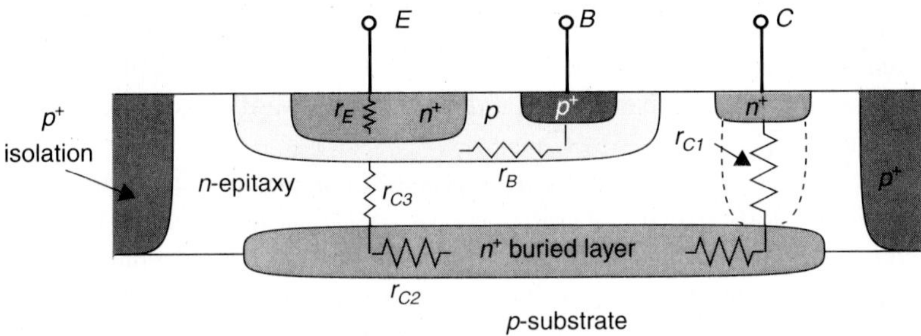

Figure 2.47 *npn* transistor structure, showing parasitic resistances.

The collector resistance consists of three components, labeled r_{C1}, r_{C2}, and r_{C3}. Of these, the latter is in general the dominant component. In advanced processes, r_{C3} is reduced by adding a low-resistance n^+ region below the collector (called a *deep collector*), shown in dotted lines in Figure 2.47. The value of r_C varies between 20 Ω (with deep collector) to almost 1 kΩ (without). The value of r_B ranges from 50 Ω to 500 Ω and unfortunately varies with the collector current due to an effect called *current crowding*. This effect causes the transistor action to occur at the edges of the emitter area, instead of the central portion, which results in a varying distance between the active base region and the base contact.

The presence of these parasitic resistances is particularly noticed under high collector-current conditions. The voltage drop over the resistances reduces the voltage differences at the internal device terminals and, as a result, the transistor currents. Simulation accuracy is strongly contingent upon a careful modeling of the resistances. The designer can, if necessary, reduce the parasitic effects by modifying the transistor structure. Available options include increasing the emitter area, and providing multiple base and emitter terminals.

β_F Variations

The ideal model states that in the forward-active region, the current gain β_F is a constant. This parameter, in fact, does vary with the operating conditions of the device as shown in Figure 2.48, which plots the values of $\ln(I_C, I_B)$ as a function of V_{BE}. It can be observed that for intermediate values, a constant value of β_F can indeed be observed. At low current values, a degradation of the current gain occurs. This is attributed to an increase in the base current, caused by the recombination of carriers in the *be* depletion region. This effect is present at all current levels, but only has an impact under very low current conditions. At the other end of the spectrum, the collector current drops below the ideal current as a result of *high-level injection* effects—as the injected minority carrier density in the base approaches the majority carrier density, the hole current from base to emitter becomes substantial leading to a decline in the collector current.

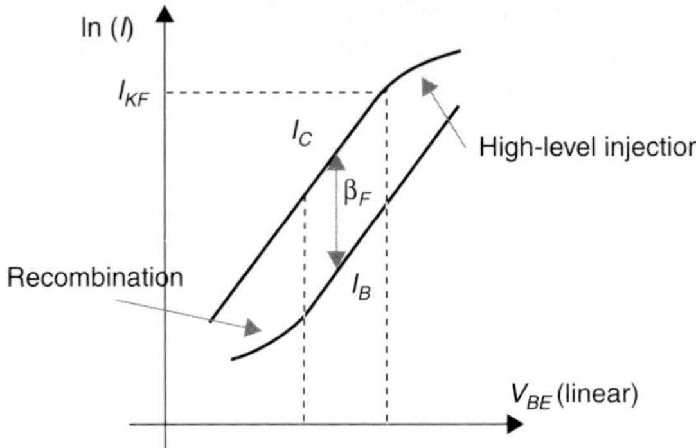

Figure 2.48 Deterioration of the current gain at low and high current levels.

In addition to high-level injection, the β_F at high currents is also affected by the onset of the *Kirk effect* or *base stretching*. This occurs when the level of injected electrons into the collector becomes comparable to the donor-atom doping density of the collector region. This causes a collapse of the space-charge region, and the base stretches out into the collector regions, resulting once again in a current-gain degradation.

The effects of the low- and high-level injection can modeled by modulating the emission coefficient n in the base and collector currents Eq. (2.93). For instance, the high-level injection effect is modeled by modifying n from the standard value of 1 to 2. The point of onset of the high-level injection effects is called the *knee current* (I_{KF}).

$$I_C = I_S e^{V_{BE}/n\phi_T} \tag{2.93}$$

The impact of some of the high-current and parasitic resistance effects is observed in the *I-V* plots of Figure 2.41. Notice especially how the voltage drop over the base resistance causes both the collector and base currents to degrade for high values of V_{BE}.

2.4.5 SPICE Models for the Bipolar Transistor

Earlier versions of SPICE supported two separate models for the bipolar transistor: the *Ebers-Moll (E-M)* and the charge-based *Gummel-Poon (G-P)* model. Both were merged in later versions into the *modified Gummel-Poon* model which incorporates various extensions to model high-bias conditions [Getreu76]. Under default conditions for certain parameters, this model automatically reduces to the simple E-M model.

The complete model for an *npn* transistor, as implemented in SPICE, is shown in Figure 2.49 (The model of a *pnp* device is obtained by reversing the polarities of V_{BE}, V_{BC},

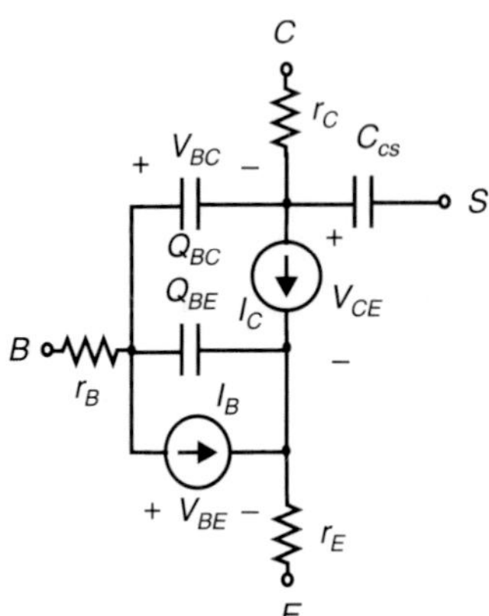

Figure 2.49 SPICE model for an *npn* transistor.

V_{CE}, I_C and I_B). Under normal operating conditions, the values of the current sources are given by the modified Ebers-Moll equations

$$I_C = I_S(e^{V_{BE}/(n_F\phi_T)} - e^{V_{BC}/(n_R\phi_T)})\left(1 - \frac{V_{BC}}{V_A}\right) - \frac{I_S}{\beta_R}(e^{V_{BC}/(n_R\phi_T)} - 1) \tag{2.94}$$

and

$$I_B = \frac{I_S}{\beta_F}(e^{V_{BE}/(n_F\phi_T)} - 1) + \frac{I_S}{\beta_R}(e^{V_{BC}/(n_R\phi_T)} - 1) \tag{2.95}$$

where I_S is the saturation current, V_A the early voltage, and n_F and n_R the forward and reverse current-emission coefficients. As can be observed, the model superimposes the forward- and reverse-operation conditions. The *base-width modulation* effect is incorporated in the collector current through the *early voltage* V_A.

This model does not take into account the second-order effects occurring at low current levels (depletion-layer recombination) or high injection levels. These are properly accounted for in the Gummel-Poon model, which relates all terminal voltages and currents to the base charge. A complete description of the model is beyond the scope of this text.

The intrinsic capacitances and resistances are modeled by the expressions presented in Sections 2.2.4 and 2.4.3. Q_{BE} and Q_{BC} represent the combined effects of the depletion and base charge for the *be* and the *bc* junctions, respectively.

Table 2.8 lists the main SPICE model parameters. The parameters covering the parasitic resistive and capacitive effects have been transferred to a separate table for the sake of clarity (Table 2.9).

In contrast to the MOS transistor case, not many parameters can be defined for the individual transistor. The only parameter that can be defined at the instantiation time of a device is the AREA factor, which determines how many of the bipolar junction transistors of type MODNAME are put in parallel to create that particular device. The default value is 1. This parameter is not used very often. The common practice is to provide a different model for every possible device dimension, which may be used. This is acceptable, because a typical bipolar digital design only uses a small range of device dimensions, contrary to the MOS practice, where transistor dimensions can vary widely, even in a single circuit.

Table 2.8 Main SPICE bipolar junction transistor model parameters.

Parameter Name	Symbol	SPICE Name	Units	Default Value
Transport Saturation Current	I_S	IS	A	1.0E–16
Maximum Forward Current Gain	β_F	BF	–	100
Forward Current-Emission Coefficient	n_F	NF	–	1
Forward Early Voltage	V_{AF}	VAF	V	∞
Maximum Reverse Current Gain	β_R	BR	–	1
Reverse Current-Emission Coefficient	n_R	NR	–	1
Reverse Early Voltage	V_{AR}	VAR	V	∞
Corner for Forward Beta High-Current Roll-off[a]	I_{KF}	IKF	A	∞
be Junction Leakage Saturation Current[a]		ISE	A	1.0E–13
be Junction Leakage Emission Coeff. (low-current condition)[a]		NE	–	1.5
Corner for Reverse Beta High Current Roll-off[a]	I_{KR}	IKR	A	∞
bc junction leakage saturation current[a]		ISC	A	1.0E–13
bc junction leakage emission coeff. (low-current condition)[a]		NC	–	2
Ideal Forward Transit Time	τ_F	TF	sec	0
Ideal Reverse Transit Time	τ_R	TR	sec	0

a. Gummel-Poon Model Parameter

Table 2.9 SPICE Parameters for parasitics (resistances, capacitances).

Parameter Name	Symbol	SPICE Name	Units	Default Value
Emitter Resistance	r_E	RE	Ω	0
Collector Resistance	r_C	RC	Ω	0
Zero-Bias Base Resistance	r_B	RB	Ω	0
Minimum Base Resistance		RBM	Ω	RB
Current where RB falls halfway to RBM		IRB	A	∞
Zero-Bias be-Junction Capacitance	C_{be0}	CJE	F	0
be-Junction Grading Coeff.	m_{be}	MJE	–	0.33
be-Junction Built-in Voltage	ϕ_{be}	VJE	V	0.75
Zero-Bias bc-Junction Capacitance	C_{bc0}	CJC	F	0
bc-Junction Grading Coeff.	m_{bc}	MJC	–	0.33
bc-Junction Built-in Voltage	ϕ_{bc}	VJC	V	0.75
Zero-Bias Collector-Substrate Cap.	C_{cs0}	CJS	F	0
cs-Junction Grading Coeff.	m_{cs}	MJS	–	0
cs-Junction Built-in Voltage	ϕ_{cs}	VJS	V	0.75

Example 2.16 SPICE Bipolar Junction Transistor Model

The model for a bipolar *npn* transistor with an emitter area of 2 μm × 3.75 μm, implemented in a process technology with a minimum line-width of 1 μm, is included. We use this as the generic bipolar transistor in the rest of the text, unless specified otherwise.

```
* SPICE Model for 2 x 3.75 npn transistor

.MODEL BF=100 BR=1 IS=1.E-17 VAF=50
+ TF=10E-12 TR=5E-9 IKF=2E-2 IKR=0.5
+ RE=20 RC=75 RB=120
+ CJE=20E-15 VJE=0.8 MJE=0.5 CJC=22E-15 VJC=0.7 MJC=0.33
+ CJS=47E-15 VJS=0.7 MJS=0.33
```

An example of an actual transistor instantiation is given below. Transistor *Q*1 is an NPN device with its collector, base and emitter terminals connected to nodes 2, 1, 0, respectively. Transistor *Q*2 consists of four transistors of type NPN, connected in parallel, and has its substrate connected to node 5.

```
Q1 2 1 0 NPN
Q2 2 1 0 5 NPN 4
```

2.5 A Word on Process Variations

The preceding discussion has assumed that a device is adequately modeled by a single set of parameters. In reality, the parameters of an identical transistor vary from wafer to wafer or even between transistors on the same die, depending upon the position. The observed random distribution between supposedly identical devices is primarily the result of two factors:

1. Variations in the process parameters, such as impurity concentration densities, oxide thicknesses, and diffusion depths. These are caused by nonuniform conditions during the deposition and/or the diffusion of the impurities. This introduces variations in the sheet resistances and transistor parameters such as the threshold voltage.

2. Variations in the dimensions of the devices, mainly resulting from the limited resolution of the photolithographic process. This causes (W/L) variations in MOS transistors and mismatches in the emitter areas of bipolar devices.

These variations can cause considerable deviations from the nominal or expected behavior of a circuit (in both positive and negative directions), which can have a serious impact on the design process. Assume, for instance, that you are supposed to design a microprocessor running at a clock frequency of 200 MHz. It is economically important that the majority of the manufactured dies meet that performance requirement. One way to achieve this goal is to design the circuit assuming worst-case values for all possible device parameters. While safe, this approach is prohibitively conservative and would result in severely overdesigned and hence uneconomical circuits. Quite a number of the design parameters are totally uncorrelated. For instance, variations in the length of an MOS transistor are unrelated to variations of the oxide thickness as both are set by different process steps.

Example 2.17 MOS Transistor Process Variations

To illustrate the possible impact of process variations on the performance of an MOS device, consider for instance a minimum-size NMOS device in our generic 1.2 μm CMOS process. A later chapter will establish that the speed of the device is proportional to the drain current that can be delivered.

Assume that $V_{GS} = V_{DS} = 5$ V. The transistor operates, obviously, in the saturation mode. The nominal current of the minimal-size device ($W = 1.8$ μm, $L_{eff} = L - 2x_d = 0.9$ μm) then equals

$$I_D = \frac{k'_n}{2}\left(\frac{W}{L_{eff}}\right)(V_{GS} - V_T)^2$$
$$= (1/2)19.6 \times 10^{-6} \times (2)(5 - 0.75)^2 = 354\ \mu A$$

The following parameters in this expression are subject to process variations:

1. The *threshold voltage* V_T can vary for numerous reasons: changes in oxide thickness, substrate, poly and implant impurity levels, and the surface charge. Accurate control of the threshold voltage is an important goal for many reasons, Where in the past thresholds could vary by as much as 50%, state-of-the-art digital processes manage to control the thresholds to within 25 mV. Let us assume for this example a 25% variation on V_T.
2. k'_n: The main cause for variations in the process transconductance is changes in oxide thickness. Variations can also occur in the mobility but to a lesser degree. Assume a 10% variation in k'_n.
3. Variations in W and L. These are mainly caused by the lithographic process. Observe that variations in W and L are totally uncorrelated since the first is determined in the field-oxide step, while the second is defined by the polysilicon definition and the source and drain diffusion processes. Assume that the distribution of both W and L has a 3σ value of 0.15 μm. For a Gaussian distribution of the transistor dimensions, this means that 96% fall within this bound.

Finally, the supply voltage is not a constant either, as a result of resistive voltage drops over the supply lines on the printed circuit board as well as noise. Assume that, for this example, a variation of 0.5 V is possible in the supply voltage and, consequently, in V_{GS}.

Merging all those factors results in the following worst- and best-case values for I_D:

$$I_{D,max} = \frac{(19.6 + 1.96)}{2}\left(\frac{1.8 + 0.15}{0.9 - 0.15}\right)((5 + 0.5) - (0.75 - 0.1875))^2 = 683\ \mu A$$

$$I_{D,min} = \frac{(19.6 - 1.96)}{2}\left(\frac{1.8 - 0.15}{0.9 + 0.15}\right)((5 - 0.5) - (0.75 + 0.1875))^2 = 176\ \mu A$$

The current levels and the circuit performance can thus vary by 100% in the extreme cases. To guarantee that all fabricated circuits meet the performance requirements, it is necessary to make the transistor width twice as wide as would be required in the nominal case. This translates into a severe area penalty.

Fortunately, these worst- (or best-) case conditions occur only very rarely in reality. The probability that all parameters assume their worst-case values simultaneously is very low, and most designs will display a performance centered around the nominal design. The art of the *design for manufacturability* is to center the nominal design so that the majority of the fabricated circuits (e.g., 99%) will meet the performance specifications, while keeping the area overhead minimal.

Specialized design tools to help meet this goal are available. For instance, the Monte Carlo analysis approach [Jensen91] simulates a circuit over a wide range of randomly chosen values for the device parameters. The result is a distribution plot of design parameters (such as the speed or the sensitivity to noise) that can help to determine if the nominal design is economically viable. Examples of such distribution plots, showing the impact of variations in the effective transistor channel length and the PMOS transistor thresholds on the speed of an adder cell, are shown in Figure 2.50. As can be observed, technology variations can have a substantial impact on the performance parameters of a design.

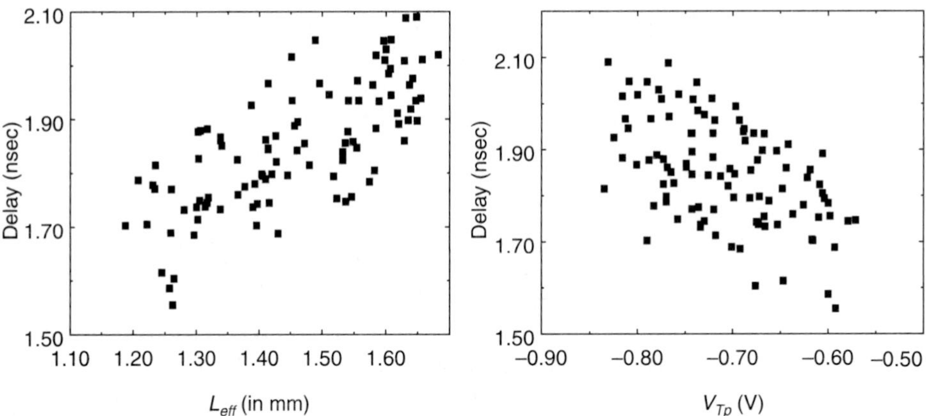

Figure 2.50 Distribution plots of speed of adder circuit as a function of varying device parameters, as obtained by a Monte Carlo analysis. The circuit is implemented in a 2 μm (nominal) CMOS technology (*courtesy of Eric Boskin, UCB, and ATMEL corp.*).

One important conclusion from the above discussion is that SPICE simulations should be treated with care. The device parameters presented in a model represent average values, measured over a batch of manufactured wafers. Actual implementations are bound to differ from the simulation results, and for reasons other than imperfections in the modeling approach. Be furthermore aware that temperature variations on the die can present another source for parameter deviations. Optimizing an MOS circuit with SPICE to a resolution level of picoseconds is clearly a waste of effort.

2.6 Perspective: Future Device Developments

Over the last decades, we have observed a dramatic reduction in the dimensions of both MOS and bipolar transistors. This has resulted in a spectacular increase in integration density, combined with an improved speed performance. Most interesting though, is that no signs of a slow-down can be observed at present. To illustrate this point, Table 2.10 shows a potential scenario of how MOSFET technology might evolve in the foreseeable future. Observe how the operating voltages have to be reduced with diminishing device dimensions. This is due to both reliability and power-consumption considerations. Notice also how the current density remains approximately constant.

Table 2.10 MOSFET technology projection (from [Hu93]).

Year of Introduction	1994	1997	2000	2003	2006	2009
Channel length (μm)	0.4	0.3	0.25	0.18	0.13	0.1
Gate oxide (nm)	12	7	6	4.5	4	4
V_{DD} (V)	3.3	2.2	2.2	1.5	1.5	1.5
V_T(V)	0.7	0.7	0.7	0.6	0.6	0.6
NMOS I_{Dsat} (mA/μm) (@ $V_{GS} = V_{DD}$)	0.35	0.27	0.31	0.21	0.29	0.33
PMOS I_{Dsat} (mA/μm) (@ $V_{GS} = V_{DD}$)	0.16	0.11	0.14	0.09	0.13	0.16

From the above, it is reasonable to conclude that both integration density and performance will continue to increase. The obvious question is for how long? Experimental sub-0.1 micron MOS devices have proven to be operational in the laboratories and to display current characteristics that are surprisingly close to present-day transistors. Current thinking projects that the MOS structure can probably survive to approximately 0.03 μm, although getting there will require substantial *device engineering*. From this perspective, integrated circuits integrating over a 100 million transistors clocked at speeds of more than 1 GHz seem to be well under way.

Whether this will actually happen is an open question. Even though it might be technologically feasible, other parameters have an equal impact on the feasibility of such an undertaking. A first doubt is if such a part can be manufactured in an economical way. Current semiconductor plants cost over $1 billion, and this price is expected to rise substantially with smaller feature sizes. Design considerations also play a role. Power consumption of such a component might be prohibitive. The growing role of interconnect parasitics might put an upper bound on performance. Finally, system considerations might determine what level of integration is really desirable. All in all, it is obvious that the design of semiconductor circuits still faces an exciting future.

2.7 Summary

In this chapter, we have presented a number of devices, including junction diodes, MOSFET, and bipolar junctions transistors. Besides an intuitive understanding of their behavior, we have presented a variety of modeling approaches, ranging from simple models, useful for a first-order manual analysis of the circuit operation, to complex SPICE models. These models will be used extensively starting in the next chapter, where we look at the fundamental building blocks of digital circuits.

- The static behavior of the junction diode is well described by the ideal diode equation that states that the current is an *exponential function of the applied voltage bias*. The dynamic behavior is governed by the storage of charge in both the space-charge and the neutral regions of the diode. The switching time of the diode is dominated by the time it takes to provide or remove these charges.

- The MOS(FET) transistor is a voltage-controlled device, where the controlling gate terminal is insulated from the conducting channel by a SiO_2 capacitor. Based on the value of the gate-source voltage with respect to a threshold voltage V_T, three operation regions have been identified: *cut-off, linear,* and *saturation*. At best, *the output current is a quadratic function of the control voltage*. One of the most enticing properties of the MOS transistor, which makes it particularly amenable to digital design, is that it approximates a voltage-controlled switch: when the control voltage is low, the switch is nonconducting (open); for a high control voltage, a conducting channel is formed, and the switch can be considered closed. This two-state operation matches the concepts of binary digital logic.

- The dynamic operation of the MOS transistor is dominated by the *device capacitors*. The main contributors are the gate capacitance and the capacitance formed by the depletion regions of the source and drain junctions. The minimization of these capacitances is the prime requirement in high-performance MOS design.

- The continuing reduction of the device dimensions to the submicron range has introduced some substantial deviations from the traditional long-channel MOS transistor model. The most important one is the *velocity saturation* effect, which changes the dependence of the transistor current with respect to the controlling voltage from quadratic to linear. Models for this effect as well as other second-order parasitics have been introduced.

- The bipolar junction transistor consists of two back-to-back *pn*-junctions, sharing a common region called the base. Depending upon the biasing of the junctions, four regions of operation can be defined: *forward- and reverse-active, saturation,* and *cut-off.* The most important property to remember is *the exponential relationship between the base-emitter voltage and the collector current.* As a result, a small voltage change causes a large current variation. This results in fast digital circuits with large current-driving capabilities.

- The dynamic behavior of the bipolar transistor is dominated by the *excess minority carrier charge*, stored in the base. Also important is the space charge of the base-emitter and base-collector depletion regions. Due to the large amount of base charge, operation in the saturation region is slow and should be avoided in digital circuits.

- The actual bipolar transistor operation diverges from the ideal behavior due to a number of parasitic effects. Especially important are the *early effect, the parasitic resistances, and the high-current effects.*

- SPICE models and their parameters have been introduced for all devices. It was observed that these models represent an average behavior and can vary over a single wafer or die.

2.8 To Probe Further

Semiconductor devices have been discussed in numerous books, reprint volumes, tutorials, and journal articles. The *IEEE Journal on Electron Devices* is one of the prime journals, where most of the state-of-the-art devices and their modeling are discussed. The books and journals referenced below contain excellent discussions of the semiconductor devices of interest or refer to specific topics brought up in the course of this chapter.

REFERENCES

[Antognetti88] P. Antognetti and G. Masobrio (eds.), *Semiconductor Device Modeling with SPICE*, McGraw-Hill, 1988.

[Banzhaf92] W. Bhanzhaf, *Computer Aided Analysis Using PSPICE*, 2nd ed., Prentice Hall, 1992.

[Chen90] J. Chen, *CMOS Devices and Technology for VLSI*, Prentice Hall, 1990.

[Getreu76] I. Getreu, "Modeling the Bipolar Transistor," Tektronix Inc., 1976.

[Gray69] P. Gray and C. Searle, *Electronic Principles*, John Wiley and Sons, 1969.

[Gray93] P. Gray and R. Meyer, *Analysis and Design of Analog Integrated Circuits*, 3rd ed., John Wiley and Sons, 1993.

[Haznedar91] H. Haznedar, *Digital Microelectronics*, Benjamin/Cummings, 1991.

[Hodges88] D. Hodges and H. Jackson, *Analysis and Design of Digital Integrated Circuits*, 2nd ed., McGraw-Hill, 1988.

[Howe95] R. Howe and S. Sodini, *Microelectronics: An Integrated Approach*, forthcoming, Prentice Hall, 1995.

[Hu92] C. Hu, "IC Reliability Simulation," *IEEE Journal of Solid State Circuits*, vol. 27, no. 3, pp. 241–246, March 1992.

[Hu93] C. Hu, "Future CMOS Scaling and Reliability," *IEEE Proceedings*, vol. 81, no. 5, May 1993.

[Jensen91] G. Jensen et al., "Monte Carlo Simulation of Semiconductor Devices," *Computer Physics Communications*, 67, pp. 1–61, August 1991.

[Ko89] P. Ko, "Approaches to Scaling," in *VLSI Electronics: Microstructure Science*, vol. 18, chapter 1, pp. 1–37, Academic Press, 1989.

[Muller86] R. Muller and T. Kamins, *Device Electronics for Integrated Circuits*, 2nd ed., John Wiley and Sons, 1986.

[Nagel75] L. Nagel, "SPICE2: a Computer Program to Simulate Semiconductor Circuits," Memo ERL-M520, Dept. Elect. and Computer Science, University of California at Berkeley, 1975.

[Sedra87] A. Sedra and K. Smith, *Microelectronic Circuits*, 2nd ed., Holt, Rinehart and Winston, 1987.

[Sheu87] B. Sheu, D. Scharfetter, P. Ko, and M. Jeng, "BSIM: Berkeley Short-Channel IGFET Model for MOS Transistors," *IEEE Journal of Solid-State Circuits*, vol. SC-22, no. 4, pp. 558–565, August 1987.

[Sze81] S. Sze, *Physics of Semiconductor Devices*, 2nd ed., John Wiley and Sons, 1981.

[Thorpe92] T. Thorpe, *Computerized Circuit Analysis with SPICE*, John Wiley and Sons, 1992.

[Toh88] K. Toh, P. Koh, and R. Meyer, "An Engineering Model for Short-Channel MOS Devices," *IEEE Journal of Solid-Sate Circuits*, vol. 23. no. 4, pp 950–957, August 1988.

[Tsividis87] Y. Tsividis, *Operation and Modeling of the MOS Transistor*, McGraw-Hill, 1987.

[Yamaguchi88] T. Yamaguchi et al., "Process and Device Performance of a High-Speed Double Poly-Si Bipolar Technology Using Borsenic-Poly Process with Coupling-Base Implant," *IEEE Trans. Electron. Devices*, vol. 35, no 8, pp. 1247–1255, August 1988.

[Weste93] N. Weste and K. Eshragian, *Principles of CMOS VLSI Design: A Systems Perspective*, Addison-Wesley, 1993.

2.9 Exercises and Design Problems

For all problems, use the device parameters provided in Chapter 2 (Sections 2.2.5, 2.3.5, and 2.4.5) and the inside back book cover, unless otherwise mentioned. Also assume T = 300 K by default.

1. [E,SPICE,2.23]

a. Consider the circuit of Figure 2.51. Using the simple model, with $V_{Don} = 0.7$ V, solve for I_D.

b. Find I_D and V_D using the ideal diode equation. Use $I_s = 10^{-14}$ A and $T = 300$ K.

c. Solve for V_{D1}, V_{D2}, and I_D using SPICE.

d. Repeat parts *b* and *c* using $I_S = 10^{-16}$ A, $T = 300$K, and $I_S = 10^{-14}$A, $T = 350$ K.

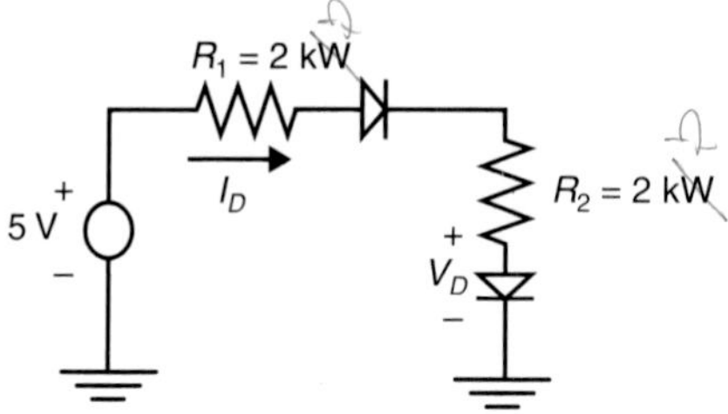

Figure 2.51 Resistor diode circuit.

2. [M, None, 2.2.3] For the circuit in Figure 2.52, $V_s = 3.3$ V. Assume $A_D = 12\ \mu m^2$, $\phi_0 = 0.65$ V, and $m = 0.5$. $N_A = 2.5$ E16 and $N_D = 5$ E15.

a. Find I_D and V_D.

b. Is the diode forward- or reverse-biased?

c. Find the depletion region width, W_j, of the diode.

d. Use the parallel-plate model to find the junction capacitance, C_j.

e. Set $V_s = 1.5$ V. Again using the parallel-plate model, explain qualitatively why C_j increases.

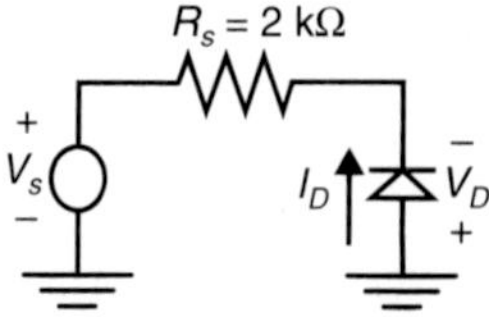

Figure 2.52 Series diode circuit

3. [C, None, 2.2.3] For the circuit in Figure 2.53, sketch $i_R(t)$, $v_d(t)$, and $v_o(t)$ showing quantitative values for the asymptotes and time constants. Assume $V_{D(on)} = 0.7$ V and the minority carrier transit time $\tau_T = 50$ nsec. Further assume a short-base model with a high hole-injection efficiency. Neglect the effects of junction capacitance.

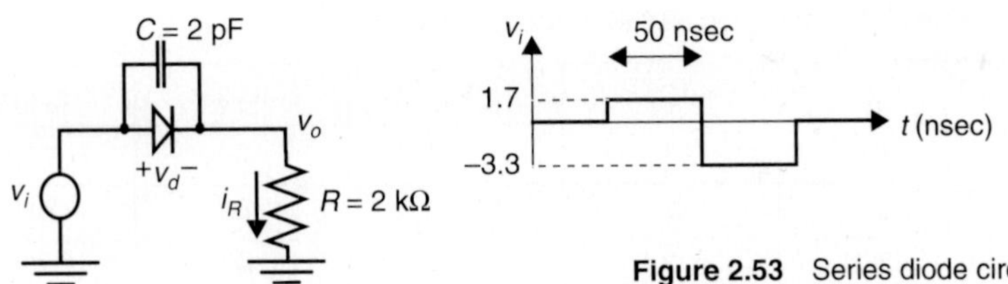

Figure 2.53 Series diode circuit

4. [E, SPICE, 2.3.2] Figure 2.54 shows NMOS and PMOS devices with drains, source, and gate ports annotated. Determine the mode of operation (saturation, triode, or cutoff) and drain current I_D for each of the biasing configurations given below. Verify with SPICE. Use the following transistor data: NMOS: k'_n = 60 μA/V^2, V_{T0} = 0.7 V, λ = 0.1 V^{-1}, PMOS: k'_p = 20 μA/V^2, V_{T0} = −0.8 V, λ = 0.1 V^{-1}. Assume (W/L) = 1.

a. NMOS: V_{GS} = 3.3 V, V_{DS} = 3.3 V. PMOS: V_{GS} = −0.5 V, V_{DS} = −1.5 V.

b. NMOS: V_{GS} = 3.3 V, V_{DS} = 2.2 V. PMOS: V_{GS} = −3.3 V, V_{DS} = −2.6 V.

c. NMOS: V_{GS} = 0.6 V, V_{DS} = 0.1 V. PMOS: V_{GS} = −3.3 V, V_{DS} = −0.5 V.

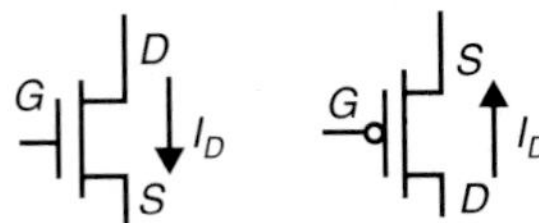

Figure 2.54 NMOS and PMOS devices.

5. [E, SPICE, 2.3] Using SPICE plot the I-V characteristics for the following devices.

a. NMOS W = 2.4 μm, L = 0.6 μm

b. NMOS W = 12.8 μm, L = 3.6 μm

c. PMOS W = 2.4 μm, L = 0.6 μm

d. PMOS W = 12.8 μm, L = 3.6 μm

6. [E, SPICE, 2.3] Indicate on the plots from problem 5.

a. the regions of operation.

b. the effects of channel length modulation.

c. Which of the devices are in velocity saturation? Explain how this can be observed on the I-V plots.

7. [M, None, 2.3] Repeat problem 5 using hand analysis using the LEVEL 1 model. Explain the differences with the results of problem 5 and suggest approaches to reduce the discrepancies.

8. [E, None, 2.3.2] You are given an NMOS device with the following process parameters: N_A = 10^{16} cm^{-3} (substrate), N_D = 10^{20} cm^{-3} (gate), N_{SS} = 10^{11} cm^{-3} (surface states), and x_j = 0.5 μm. Assume there has been no threshold channel implant.

a. Due to process variations in t_{ox} you observe values of V_{T0} between −0.1 V and 0 V. What is the range of oxide thicknesses?

b. You desire to make a low-V_T process with $V_{T0,min}$ = 0.3 V. What channel implant dose is required?

9. [M, None, 2.3.2] Given the data in Table 2.11 for an NMOS transistor with k' =20 μA/V^2, calculate V_{t0}, γ, λ, 2|ϕ_f|, and W/L:

10. [M, None, 2.3] The data points in Table 2.12 were measured for an MOS transistor:

a. What type of device is this: PMOS or NMOS, V_T > 0 or < 0?

b. Is this device velocity saturated? Why or why not.

Table 2.11 Measured NMOS transistor data

	V_{GS}	V_{DS}	V_{BS}	I_D (μA)
1	3	5	0	1210
2	5	5	0	4410
3	5	10	0	5292
4	5	5	–2	3265
5	5	5	–5	2381

c. Derive the values of $k = k'(W/L)$, V_{T0}, and λ. Determine the operation region for each of the rows in the table.

d. Suppose that t_{ox} is reduced. How do k, V_{to} and λ change?

Table 2.12 Measured MOS transistor data.

	V_{GS}(V)	V_{DS}(V)	V_{SB}(V)	I_D (μA)
1	–2	3	0	0
2	0	1	0	47.86
3	0	5	0	62.11
4	1	3.5	0	192.52
5	1	5	0	208.76
6	5	5	0	1182

11. [M, None, 2.3.2] Consider the circuit configuration of Figure 2.55.

a. Write down the equations (and only those) which you need to determine the voltage at node X. Do NOT plug in any values yet. Assume that $\lambda_p = 0$.

b. Draw the (approximative) load lines for both MOS transistor and resistor. Mark some of the significant points.

c. Determine the required width of the transistor (for $L_{eff} = 1.2$ μm) such that X equals 1.5 V.

d. We have, so far, assumed that M_1 is a long-channel device. Redraw the load lines assuming that M_1 is velocity-saturated. Will the voltage at X rise or fall?

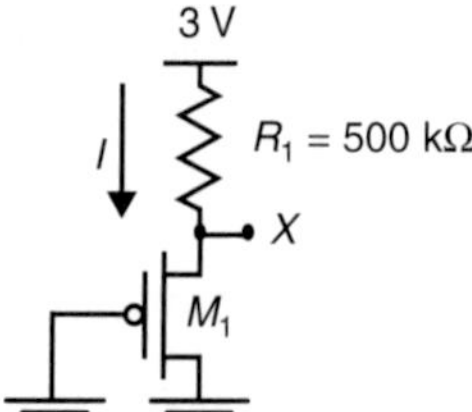

Figure 2.55 MOS circuit.

12. [M, None, 2.3.2] The circuit of Figure 2.56 is known as a *source-follower* configuration. It achieves a DC level shift between the input and output. The value of this shift is determined by the current I_0. Assume $L_D = 0.15$ μm, $\gamma = 0.543$, $2|\phi_f| = 0.6$ V, $V_{T0} = 0.74$ V, $k' = 19.6$ μA/V^2, and $\lambda = 0$.

a. Derive an expression giving V_i as a function of V_o and $V_T(V_o)$. If we neglect body effect, what is the nominal value of the level shift performed by this circuit.

b. The NMOS transistor experiences a shift in V_T due to the body effect. Find V_T as a function of V_o for V_o ranging from 0 to 3 V with 0.5 V intervals. Plot V_T vs. V_o.

c. Plot V_o vs. V_i as V_o varies from 0 to 3 V with 0.5 V intervals. Plot two curves: one neglecting the body effect and one accounting for it. How does the body effect influence the operation of the level converter? At V_o (body effect) = 3 V, find V_o (ideal) and, thus, determine the maximum error introduced by body effect.

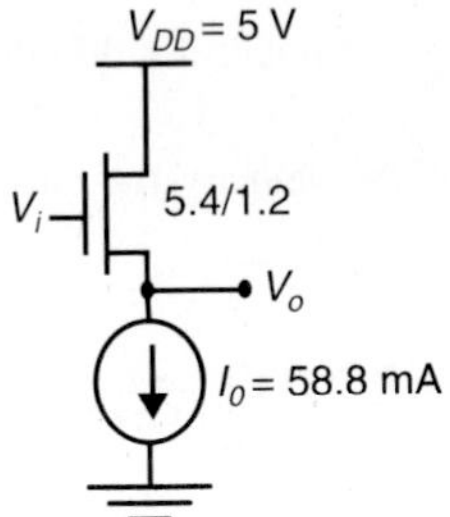

Figure 2.56 Source-follower level converter.

13. [M, SPICE, 2.3.2] Problem 13 uses the MOS circuit of Figure 2.57.

a. Plot V_{out} vs. V_{in} with V_{in} varying from 0 to 5 volts (use steps of 1 V). V_{DD} = 5 V.

b. Repeat *a* using SPICE.

c. Repeat *a* and *b* using a MOS transistor with $(W / L) = 4.8 / 1.2$. Is the discrepancy between manual and computer analysis larger or smaller. Explain why.

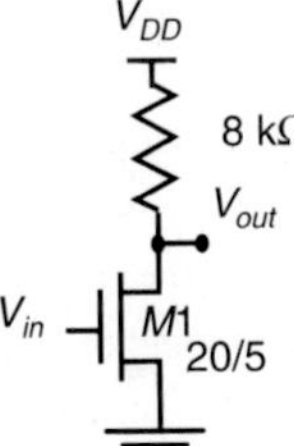

Figure 2.57 MOS circuit.

14. [M, None, 2.3.3] Compute the gate and diffusion capacitances for transistor $M1$ of Figure 2.57. Assume that drain and source areas are rectangular, and are 20 µm wide and 5 µm long. Use the parameters of Example 3.5 to determine the capacitance values. Assume $m_j = 0.5$ and $m_{jsw} = 0.33$. Also compute the total charge stored at node *In*, for the following conditions:

a. V_{in} = 5 V, V_{out} = 5 V, 2.5 V, and 0 V.

b. V_{in} = 0 V, V_{out} = 5 V, 2.5 V, and 0 V.

15. [C, SPICE, 2.5] Though impossible to quantify exactly by hand, it is a good idea to understand process variations and be able to at least get a rough estimate for the limits of their effects.

a. For the circuit of Figure 2.57, calculate nominal, minimum, and maximum currents in the NMOS device with V_{in} = 0 V, 2.5 V and 5 V. Assume 3σ variations in V_{T0} of 25 mV, in k' of 15%, and in lithographic etching of 0.15 µm.

b. Analyze the impact of these current variations on the output voltage. Assume that the load resistor also can vary by 10%. Verify the result with SPICE.

16. [M, None, 2.3.4] Short-channel effects:

a. Use the fact that current can be expressed as the product of the carrier charge per unit length and the velocity of carriers ($I_{DS} = Qv$) to derive I_{DS} as a function of W, C_{ox}, $V_{GS} - V_T$, and carrier velocity v.

b. For a long-channel device, the carrier velocity is the mobility times the applied electric field. The electrical field, which has dimensions of V/m, is simply $(V_{GS} - V_T) / 2L$. Derive I_{DS} for a long-channel device.

c. From the equation derived in a, find I_{DS} for a short-channel device in terms of the maximum carrier velocity, v_{max}.

d. Based on the results of b and c describe the most important differences between short-channel and long-channel devices.

17. [C, None, 2.3.4] Another equation, which models the velocity-saturated drain current of an MOS transistor is given by

$$I_{dsat} = \frac{1}{1 + (V_{GS} - V_t)/(E_{sat}L)}\left(\frac{\mu_0 C_{ox}}{2}\right)\frac{W}{L}(V_{GS} - V_T)^2$$

Using this equation it is possible to see that velocity saturation can be modeled by a MOS device with a source-degeneration resistor (see Figure 2.58).

a. Find the value of R_S such that $I_{DSAT}(V_{GS}, V_{DS})$ for the composite transistor in the figure matches the above velocity-saturated drain current equation. *Hint: the voltage drop across R_S is typically small.*

b. Given $E_{sat} = 1.5$ V/μm and $k' = \mu_0 C_{ox} = 20$ μA/V^2, what value of R_S is required to model velocity saturation. How does this value depend on W and L?

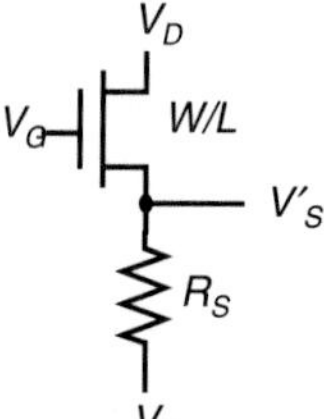

Figure 2.58 Source-degeneration model of velocity saturation.

18. [C, SPICE, 2.2.3] The circuit in Figure 2.59 shows an NMOS device with its parasitic diode. The NMOS transistor has $k_n = \mu_n C_{ox} W / L = 6000$ μA / V^2, $V_{T0} = 0.7$ V, and $\gamma = 0$, and is off ($V_g = 0$ V) until time t_1 when the gate voltage steps to 5 V. The parasitic diode has $A_D = 100$ μm^2, $\phi_0 = 0.65$ V, and $m = 0.33$. $I_{BIAS} = 500$ μA.

a. Solve for the initial conditions on I_D, V_D, and I_{M1} (at time t_{1-}).

b. Find the minority carrier charge in the diode, Q_D.

c. Find the time for the NMOS device to remove this excess minority carrier charge from the diode. During this interval, assume that V_D is constant, and model the NMOS device as a constant current source. Verify your results with SPICE.

d. Find the final values of I_D, V_D, and I_{M1} (after the NMOS device has been on for a long time).

e. Find the diode junction capacitance, C_j, linearized over the values of V_D found in parts a and d.

f. Find the time to reach 90% of the final value of V_D. During the diode space-charge removal, model the NMOS device as a current source of value equal to the average current at the interval endpoints. Compare your results with those of SPICE.

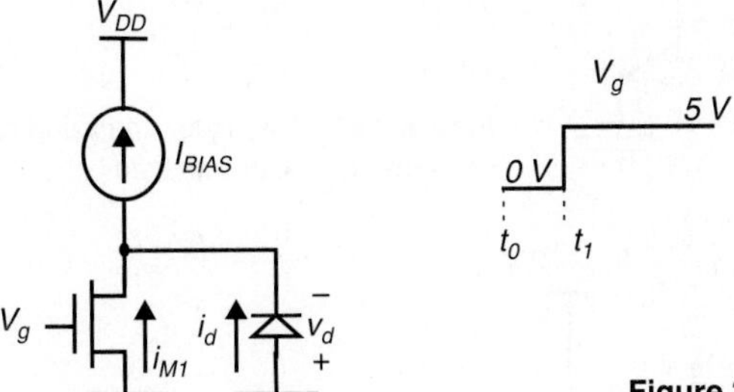

Figure 2.59 NMOS with parasitic diode.

19. [E, SPICE, 2.3.2] Figure 2.60 shows *npn* and *pnp* devices biased with bias resistors R_B and R_C. Determine the mode of operation (forward-active, saturation, or cut-off) and collector current I_C for the resistor values given below. Verify with SPICE. Use the following transistor data. *npn:* $\beta_F = 100$, $V_{BE(on)} = 0.7$ V, $V_{BE(sat)} = 0.8$ V, $V_{CE(sat)} = 0.2$ V. *pnp:* $\beta_F = 30$, $V_{BE(on)} = -0.7$ V, $V_{BE(sat)} = -0.8$ V, $V_{CE(sat)} = -0.2$ V.

a. $R_B = 5\ \text{k}\Omega$, $R_C = 1\ \text{k}\Omega$

b. $R_B = 50\ \text{k}\Omega$, $R_C = 0.5\ \text{k}\Omega$

c. $R_B = 10\ \text{k}\Omega$, $R_C = 250\ \Omega$

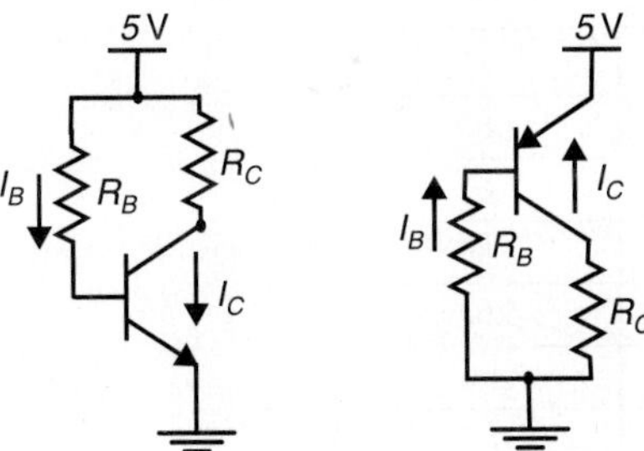

Figure 2.60 *npn* and *pnp* devices with their biasing schemes.

20. [M, None, 2.5] Assuming 3σ process variations of 5% in R_C, 10% in R_B, and 20% in β_F, repeat Problem 19. For each case, find the region of operation for both devices, and the minimum and maximum values for I_B, I_C, and V_{CE}.

21. [M, None, 2.4.3] In this problem, we consider the delays involved in turning on the bipolar transistor. Sketch $v_{BE}(t)$, $Q_F(t)$, and $i_C(t)$ for the circuit of Figure 2.61. Annotate the critical delay components on your graphs. Use the transistor data of Example 2.15.

22. [C, None, 2.4.3] Consider adding a capacitor in parallel with the base resistor in the circuit of Problem 21 (Figure 2.62). The capacitor is referred to as a speed-up capacitor, and it allows the output current to achieve its steady-state value almost instantaneously. How? What value of C is required to achieve this effect. Repeat problem 21 using this value for C.

23. [M, None, 2.4.4] The deviation in the actual collector current of a bipolar transistor, I_{CA}, from the ideal value, $I_{CI} = I_S e^{+V_{BE}/\phi_T}$, due to parasitic and high-level injection effects can be modeled as a base resistance, R_B.

a. Derive an expression for R_B as a function of I_{CI} and I_{CA}.

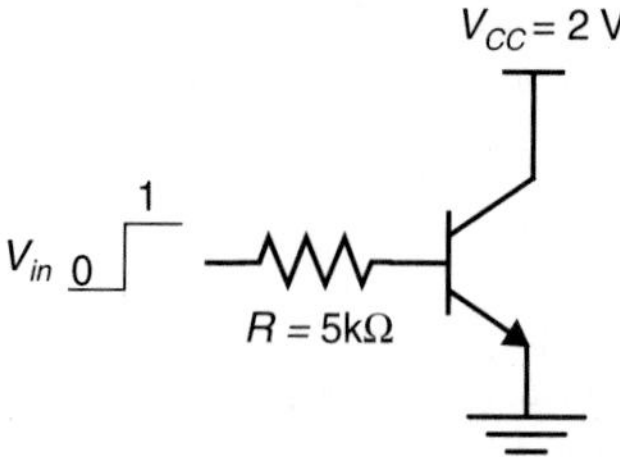

Figure 2.61 Circuit for analyzing turn-on transient of bipolar transistor.

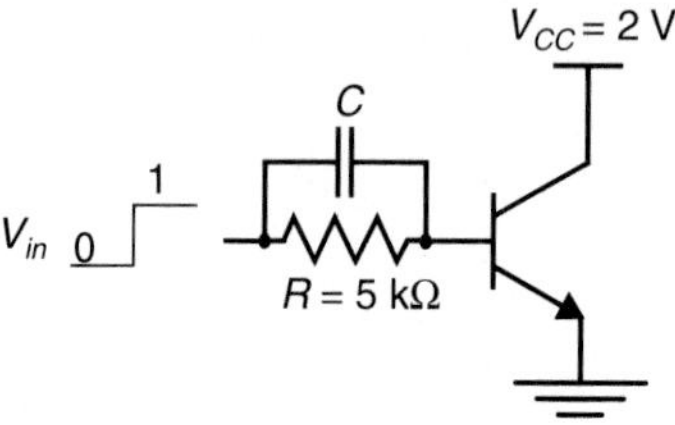

Figure 2.62 Bipolar transistor circuit with speed-up capacitor.

b. Plot R_B versus I_{CA} for the measured current values of Table 2.13. Use a log scale for I_{CA}. Assume $I_S = 10^{-15}$A, and $\beta_F = 100$.

Table 2.13 Table of measured collector currents

	V_{BE} (V)	I_{CA} (mA)
1	0.72	1.25
2	0.73	1.83
3	0.74	2.67
4	0.75	3.89
5	0.76	5.65
6	0.78	11.6
7	0.81	31.5
8	0.85	90.6

DESIGN PROBLEM

Measure the *I-V* characteristics of discrete MOS and bipolar transistors, and derive the first-order SPICE parameters. Compare the simulated characteristics with the measured ones.

APPENDIX

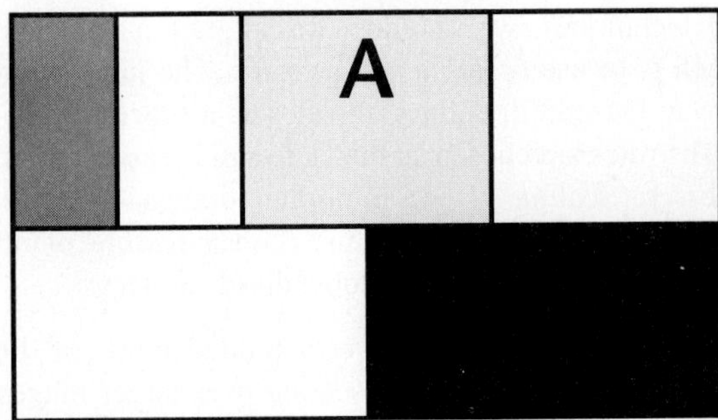

LAYOUT DESIGN RULES

Creating a manufacturable layout

The goal of defining a set of design rules is to allow for a ready translation of a circuit concept into an actual geometry in silicon. The design rules act as the interface between the circuit designer and the process engineer. As processes become more complex, requiring the designer to understand the intricacies of the fabrication process and interpret the relations between the different masks is a sure road to trouble.

Circuit designers in general want tighter, smaller designs, which lead to higher performance and higher circuit density. The process engineer, on the other hand, wants a reproducible and high-yield process. Design rules are, consequently, a compromise that attempts to satisfy both sides.

The design rules provide a set of guidelines for constructing the various masks needed in the patterning process. They consist of minimum-width and minimum-spacing constraints and requirements between objects on the same or on different layers.

The fundamental unity in the definition of a set of design rules is the *minimum line width*. It stands for the minimum mask dimension that can be safely transferred to the semiconductor material. In general, the minimum line width is set by the resolution of the patterning process, which is most commonly based on optical lithography. More advanced approaches use electron-beam or X-ray sources that offer a finer resolution, but are less attractive from an economical viewpoint.

Even for the same minimum dimension, design rules tend to differ from company to company, and from process to process. This makes porting an existing design between different processes a time-consuming task. One approach to address this issue is to use advanced CAD techniques, which allow for migration between compatible processes. Another approach is to use *scalable design rules*. The latter approach, made popular by Mead and Conway [Mead80], defines all rules as a function of a single parameter, most often called λ. The rules are chosen so that a design is easily ported over a cross section of industrial processes. Scaling of the minimum dimension is accomplished by simply changing the value of λ. This results in a *linear scaling* of all dimensions. Such an approach, while attractive, suffers from some disadvantages:

1. Linear scaling is only possible over a limited range of dimensions (for instance, between 3 μm and 1 μm). When scaling over larger ranges (for instance, into the submicron range), the relations between the different layers tend to vary in a nonlinear way that cannot be adequately covered by the linear scaling rules.

2. Scalable design rules are conservative. As they represent a cross section over different technologies, they have to represent the worst-case rules for the whole set. This results in overdimensioned and less-dense designs.

For these reasons, scalable design rules are normally avoided by industry. As circuit density is a prime goal in industrial designs, most semiconductor companies tend to use *micron rules*, which express the design rules in absolute dimensions and can therefore exploit the features of a given process to a maximum degree. Scaling and porting designs between technologies under these rules is more demanding and has to be performed either manually or using advanced CAD tools.

For small projects, fast prototyping, or educational use, on the other hand, the scalable design rules present a flexible and versatile design methodology. For this textbook, we have selected the MOSIS SCMOS (Scalable CMOS) scalable design rules as our preferred design medium for CMOS [Mosis]. The rest of this appendix is devoted to a short introduction and overview of the SCMOS rules (which we have slightly simplified for the sake of clarity).

As mentioned, in a λ-based design-rule set, all design rules are expressed as a function of λ. For a given process, λ is set to a specific absolute value, and all design dimensions are consequently translated into absolute numbers. Typically, the minimum line width of a process is set to 2λ. For instance, for a 1.2 μm process (i.e., a process with a minimum line width of 1.2 μm), λ equals 0.6 μm.

A design-rule set now consists of the following entities: a set of interconnect layers, relations between objects on the same layer, and relations between objects on different layers. We discuss each of them in sequence.

Layer Representation

The layer concept translates the intractable set of masks currently used in CMOS into a simple set of conceptual layout levels that are easier to visualize by the circuit designer. From a designer's viewpoint, all CMOS designs are based on the following entities:

- *Substrates* and/or *wells*, being p-type (for NMOS devices) and n-type (for PMOS)
- *Diffusion regions* (n^+ and p^+) defining the areas where transistors can be formed. These regions are often called the *active areas*. Diffusions of an inverse type are

needed to implement contacts to the wells or to the substrate. These are called *select regions*.

- One or more *polysilicon* layers, which are used to form the gate electrodes of the transistors (but serve as interconnect layers as well).
- One or more *metal interconnect* layers (typically Al).
- *Contact* layers to provide interlayer connections.

A layout consists of a combination of polygons, each of which is attached to a certain layer. The functionality of the circuit is determined by the choice of the layers, as well as the interplay between objects on different layers. For instance, an MOS transistor is formed by the cross section of the diffusion layer and the polysilicon layer. An interconnection between two metal layers is formed by a cross section between the two metal layers and an additional contact layer. To visualize these relations, each layer is assigned a standard color (or stipple pattern for a black-and-white representation). The different layers used in the SCMOS process are represented in Figure A.1 (gray scale) or Colorplate 1 (color insert).

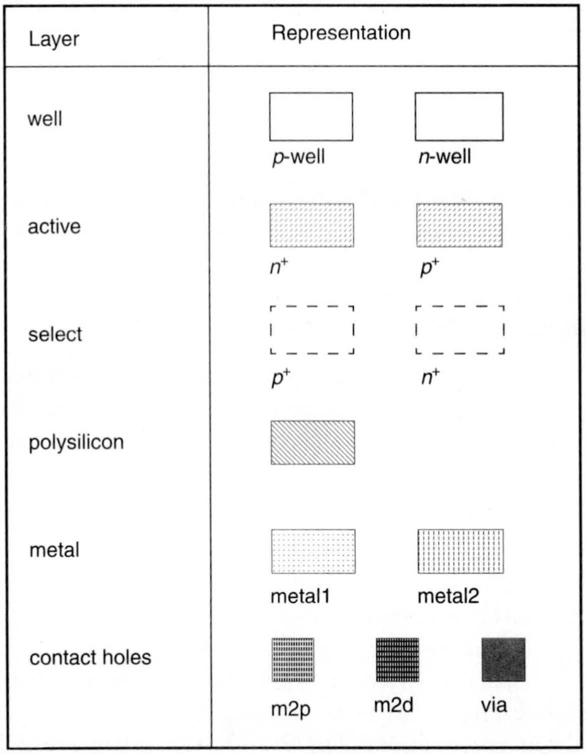

Layer	Representation		
well	*p*-well	*n*-well	
active	n^+	p^+	
select	p^+	n^+	
polysilicon			
metal	metal1	metal2	
contact holes	m2p	m2d	via

Figure A.1 SCMOS layers and representations. m2p and m2d stand for metal1-to-poly and metal1-to-diffusion, respectively. See also Colorplate 1.

Intralayer Constraints

A first set of rules defines the minimum dimensions of objects on each layer, as well as the minimum spacings between objects on the same layer. All distances are expressed in λ. These constraints are presented in a pictorial fashion in Figure A.2.

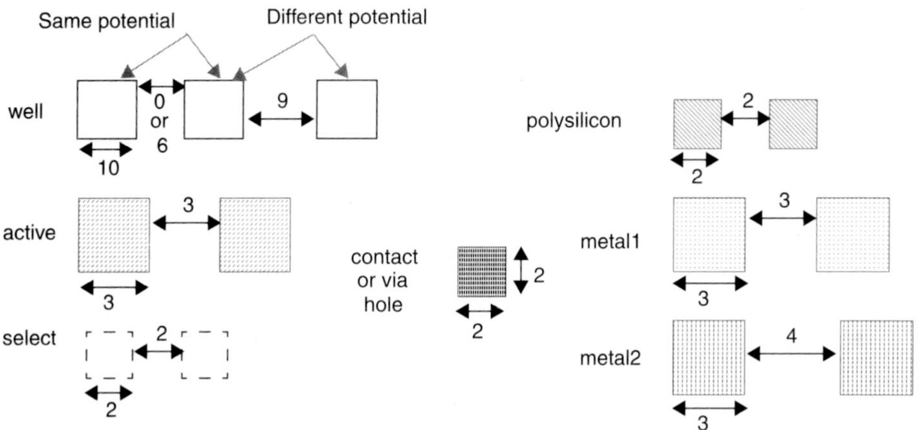

Figure A.2 Intralayer layout design rules: minimum dimensions and spacings. See also Colorplate 2.

Interlayer Constraints

These rules tend to be more complex. The fact that multiple layers are involved makes it harder to visualize their meaning or functionality. Understanding layout requires the capability of translating the two-dimensional picture of the layout drawing into the three-dimensional reality of the actual device. This takes some practice.

We present these rules in a set of separate groupings.

1. *Transistor Rules* (Figure A.3). A transistor is formed by the overlap of the active and the polysilicon layers. From the intralayer design rules, it is already clear that the minimum length of a transistor equals 2λ (the minimum width of polysilicon), while its minimum width is equal to 3λ (the minimum width of diffusion). Extra rules include the spacing between the active area and the well boundary, the gate overlap of the active area, and the active overlap of the gate.

2. *Contact and Via Rules* (Figure A.4). A contact (which forms an interconnection between metal1 and active or polysilicon) or a via (which connects metal1 and metal2) is formed by overlapping the two interconnecting layers and providing a contact hole, filled with metal, between the two. In the SCMOS rules, both interconnecting layers have to extend at least one λ beyond the area of the contact hole, which sets the minimum size of a contact to 4λ × 4λ. This is larger than the dimensions of a minimum-size transistor! Multiple changes between interconnect layers are thus to be avoided. The figure, furthermore, points out the minimum spacings between contact and via holes, as well as their relationship with layers.

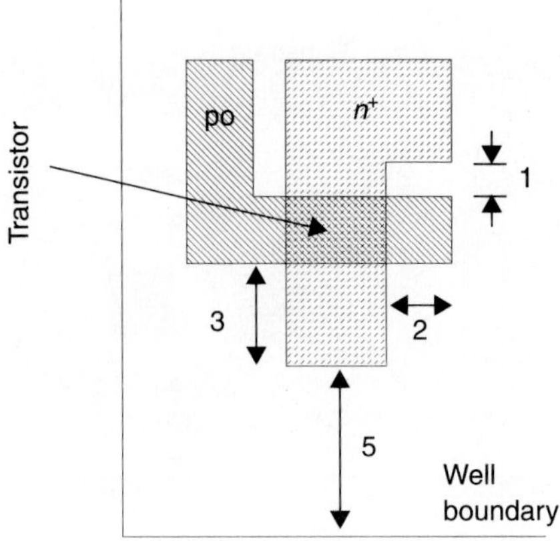

Figure A.3 Design rules concerning transistor layout. See also Colorplate 3.

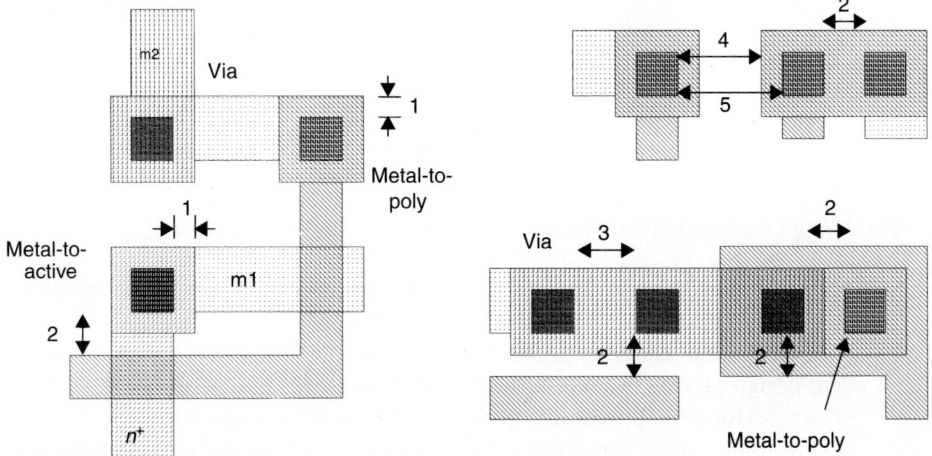

Figure A.4 Design rules regarding contacts and vias. Overlapping layers are marked by shaded regions. See also Colorplate 4.

Well and Substrate Contacts (Figure A.5). For robust digital circuit design, it is important for the well and substrate regions to be adequately connected to the supply voltages. Failing to do so results in a resistive path between the substrate contact of the transistors and the supply rails, and can lead to possibly devastating parasitic effects, such as latchup. It is therefore advisable to provide numerous substrate (well) contacts spread over the complete region. To establish an ohmic contact between a supply rail, implemented in metal1, and a p-type material, a p^+ diffusion region must be provided. This is enabled by the *select* layer, which reverses the type of diffusion. A number of rules regarding the use of the *select layer* are illustrated in Figure A.5.

Consider an n-well process, which implements the PMOS transistors into an n-type well diffused in a p-type material. The nominal diffusion is p^+. To invert the polarity of the diffusion, an n-select layer is provided that helps to establish the n^+ diffusions for the well-

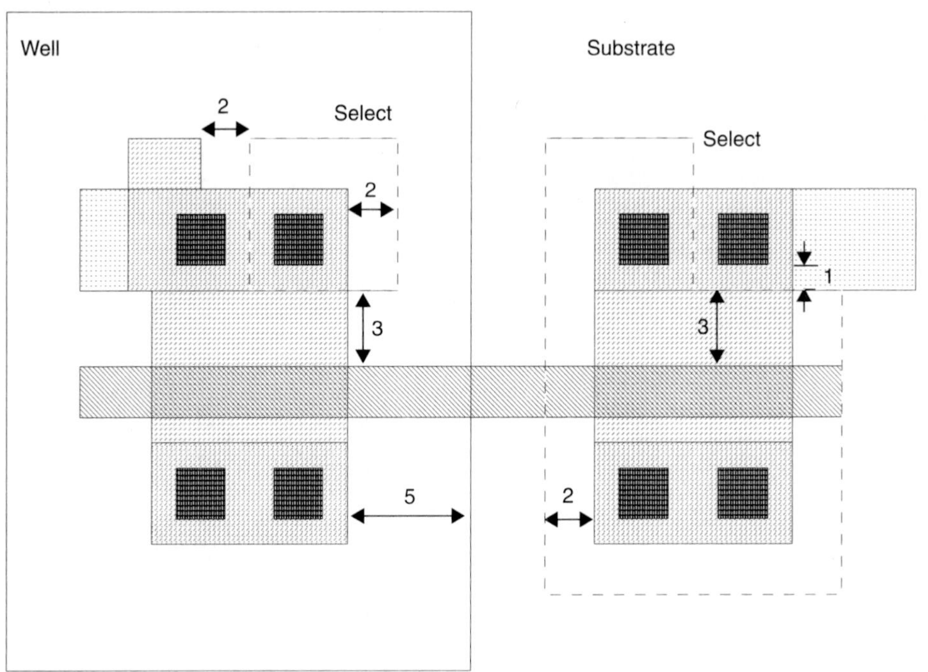

Figure A.5 Design rules regarding the select layer. See also Colorplate 5.

contacts in the n-region as well as the n^+ source and drain regions for the NMOS transistors in the substrate.

Ensuring that none of the design rules is violated is a fundamental requirement of the design process. Failing to do so will almost surely lead to a nonfunctional design. Doing so for a complex design that can contain millions of transistors is no sinecure, especially when taking into account the complexity of some design-rule sets. While design teams used to spend numerous hours staring at room-size layout plots, most of this task is now done by computers. Computer-aided *Design-Rule Checking* (called *DRC*) is an integral part of the design cycle for virtually every chip produced today. A number of layout tools even perform *on-line DRC* and check the design in the background during the time of conception.

Example A.1 Layout Example

An example of a complete layout is shown in Figure A.6. To help the visualization process, a vertical cross section of the process along the design center is included as well as a circuit schematic.

It is left as an exercise for the reader to determine the sizes of both the NMOS and the PMOS transistor, as well as the maximum current they can carry. Assume a supply voltage of 5 V. Also assume the design is implemented in the generic 1.2 μm CMOS process described in Chapter 2.

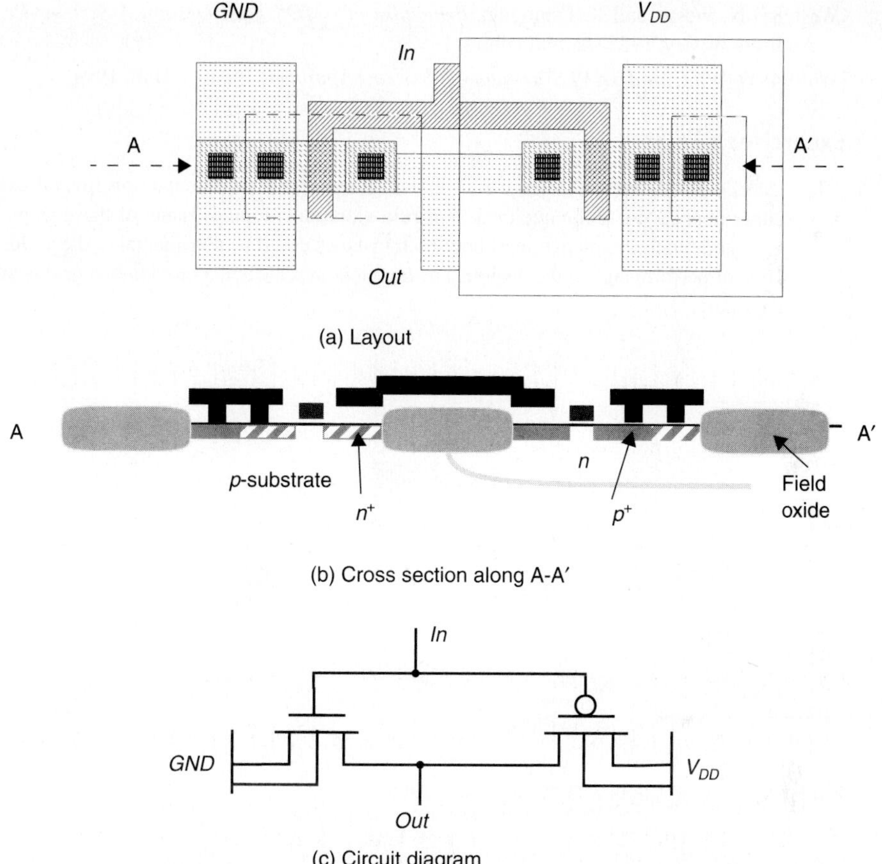

Figure A.6 A detailed layout example, including vertical process cross section and circuit diagram.

To Probe Further

Complete information on the MOSIS SCMOS design rules is available via the Internet by sending electronic mail to mosis@mosis.edu. On the worldwide web, the information is available at the following address: http://info.broker.isi.edu/1/mosis. A more detailed version is also included in the instructor's manual and the www-page of the book.

For novice designers, extensive introductions on design rules can be found in the following references; [Wolf94] offers an especially comprehensive and well-illustrated discussion.

REFERENCES

[Mead80] C. Mead and L. Conway, *Introduction to VLSI Systems*, Addison-Wesley, 1980.

[Mosis] *MOSIS Scalable and Generic CMOS Design Rules*, Rev.6, ISI, February 1988.

[Weste85] N. Weste, and K. Eshragian, *Principles of CMOS VLSI Design: A Systems Perspective*, Addison-Wesley, 1985 (second edition 1993).

[Wolf94] W. Wolf, *Modern VLSI Design—A Systems Approach*, Prentice Hall, 1994.

Excercises

1. [M&D, None, App. A] Figure A.7 shows an inverter layout containing several examples of either poor layout technique or design-rule violations. Find as many of these as possible and give a qualitative explanation of how each instance could be detrimental to the yield, functionality, or performance of the design. The layout is to scale and a one-lambda grid is provided in the figure.

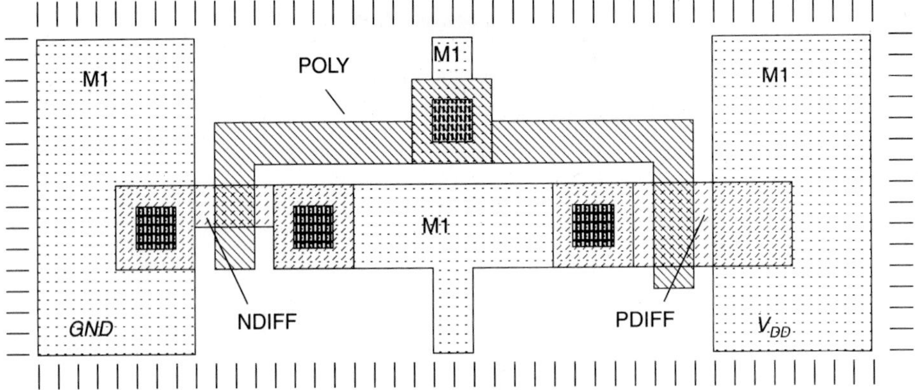

Figure A.7 Inverter layout with design-rule violations

APPENDIX

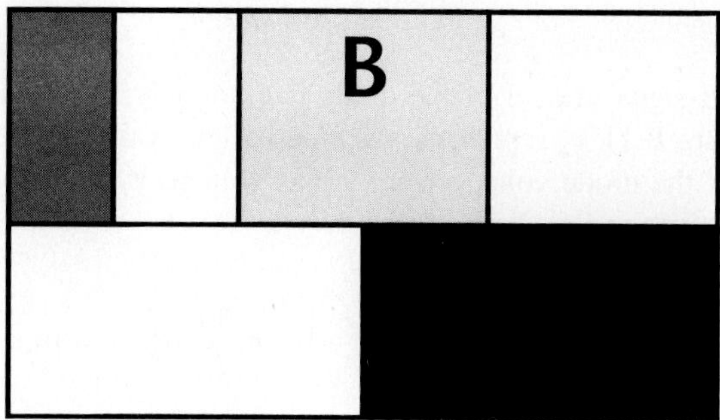

SMALL-SIGNAL MODELS

Small-signal models of diodes, MOS, and bipolar transistors

Circuits often operate with signal levels that are small compared to the bias currents and voltages. Under those conditions, a small-signal model can be employed that linearizes the device behavior and allows for the calculation of the circuit gain, terminal impedances, and frequency response without including the bias quantities.

Such models are especially useful in analog circuits, where signal excursions tend to be small compared to the supply voltage. In digital circuits, signal swings span a large fraction of the supply voltage, and the linearized, small-signal model is of a limited value. In spite of this observation, the linearized model comes in handy when analyzing some specific properties of a digital gate, for example, the calculation of the gain of a gate in the transient region.

A short overview of the small-signal models of the devices, which were introduced in Chapter 2, is therefore appropriate. The models are presented without derivation. We refer the interested reader to other textbooks such as [Sedra87] and [Gray93]. It suffices to state that the small-signal parameters represent the derivative of the device-current equations with respect to the controlling voltages (V_D for a diode, V_{GS} and V_{DS} for an MOS transistor, and V_{BE} and V_{CE} for a bipolar device). Observe also that the presented models are intended for the static analysis and do not include the capacitive effects. A dynamic model is obtained by adding the linearized small-signal capacitances that were derived for each device in Chapter 2.

Diode

The static small-signal model of the diode is extremely simple and consists of a single resistance (Figure B.1). v_d represents the small-signal value of the diode voltage, that is, the deviation of the diode voltage from a bias voltage V_D. The value of the small-signal resistance r_o is given in Eq. (B.1).

$$r_o = \left(\frac{dI_D}{dV_D}\right)^{-1} = \frac{\phi_T}{I_D} \tag{B.1}$$

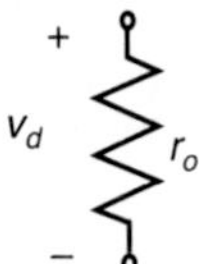

Figure B.1 Static small-signal model of diode.

MOS Transistor

The (static) small-signal model of the MOS transistor is given in Figure B.2. v_{gs} (lowercase characters) stands for the small-signal value of the gate-source voltage, that is, the deviation of the gate-source voltage from a bias voltage V_{GS}. The values of the transconductance g_m and the output resistance r_o depend upon the operation region and the bias conditions. They are summarized in Table B.1. Observe that the presented model ignores the small-signal dependence on the substrate voltage (node *B*), as this is not an issue in the context of this textbook.

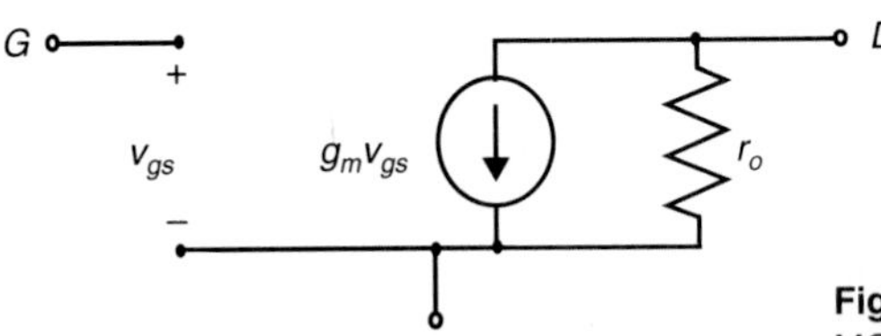

Figure B.2 Static small-signal model of MOS transistor.

Table B.1 Small-signal parameters of an MOS transistor.

	g_m	r_o
Linear	kV_{DS}	$[k(V_{GS} - V_T - V_{DS})]^{-1}$
Saturation	$k(V_{GS} - V_T)$	$1/\lambda I_D$

Bipolar Transistor

A simple static model, ignoring the parasitic resistances, is given in Figure B.3, where v_{be} (lowercase characters) stands for the small-signal value of the base-emitter voltage. The

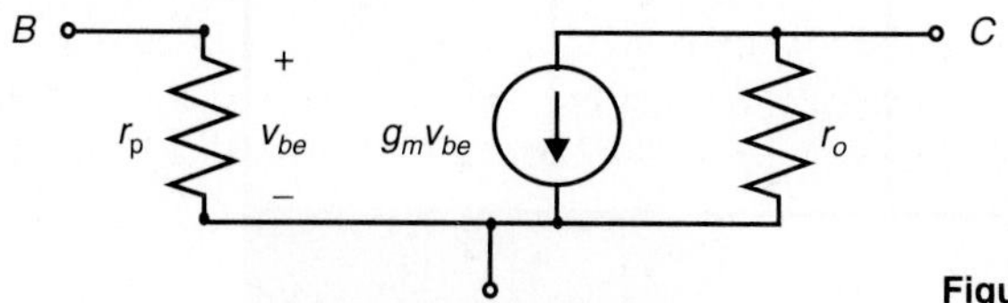

Figure B.3 Static small-signal model of bipolar transistor.

values of the transconductance g_m, the input resistance r_π, and the output resistance r_o depend upon the operation region and the bias conditions, and are summarized in Table B.2 for the forward-active region (the only region we will really use). Once again, the dynamic small-signal model can be obtained by adding the small signal-capacitances derived in Chapter 2.

Table B.2 Basic small-signal parameters of a bipolar transistor.

	g_m	r_π	r_o
Forward-active	I_C/ϕ_T	β_F/g_m	$(V_A/\phi_T)/g_m$

REFERENCES

[Gray93] P. Gray and R. Meyer, *Analysis and Design of Analog Integrated Circuits*, 3rd ed., John Wiley and Sons, 1993.

[Sedra87] A. Sedra and K. Smith, *Microelectronic Circuits*, 2nd ed., Holt, Rinehart and Winston, 1987.

Exercises

1. [E, None, App. B] Table B.1 presents equations for the small-signal values $g_m = \partial I_D/\partial V_{GS}$ and $r_o = (\partial I_D/\partial V_{DS})^{-1}$ for an MOS device without derivation. Derive these expressions both for $\lambda = 0$ and for nonzero λ's.

2. [E, None, App. B] Do the same as in Exercise 1 for the small-signal parameters of the bipolar device given in Table B.2.

CHAPTER

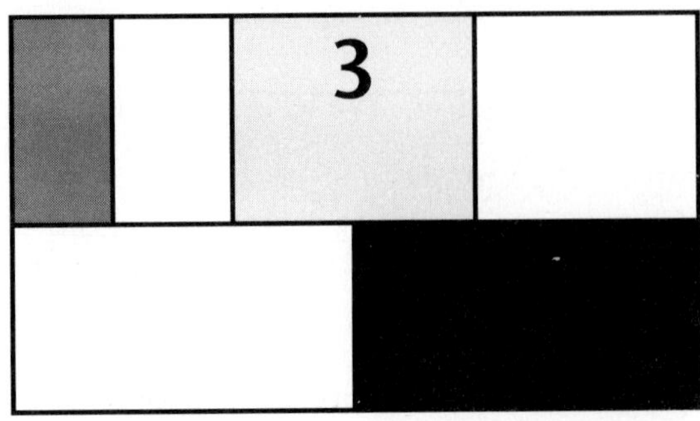

THE INVERTER

Quality measures of a digital gate: area, robustness, speed, and energy consumption

Analyzing and optimizing an inverter design

Two contrasting approaches: static CMOS and bipolar ECL

3.1 Introduction

The inverter is truly the nucleus of all digital designs. Once its operation and properties are clearly understood, designing more intricate structures such as NAND gates, adders, multipliers, and microprocessors is greatly simplified. The electrical behavior of these complex circuits can be almost completely derived by extrapolating the results obtained for inverters. The analysis of inverters can be extended to explain the behavior of more complex gates such as NAND, NOR, or XOR, which in turn form the building blocks for modules such as multipliers and processors.

The choice of a technology or a design style dramatically affects the density, performance, and the power consumption of a design. To illustrate this, we discuss in detail the behavior of static complementary CMOS and bipolar ECL inverters, which are representative gates for both MOS and bipolar technologies. Although these are not the only gate topologies in use (see Chapters 4 and 5), they are certainly the most popular at present. For each gate, we analyze the following fundamental properties:

- *robustness*, expressed by the static (or steady-state) behavior
- *performance*, determined by the dynamic (or transient) response
- *heat dissipation and supply capacity requirements,* set by the power consumption

The first section provides precise definitions for each of the above properties. While each of these parameters can be easily quantified for a given technology, we also discuss how they are affected by *scaling of the technology*. Finally, the properties of the presented gates are summarized, and some suggestions are provided on selecting a technology.

3.2 Definitions and Properties

This section defines a set of basic properties of a digital gate. These properties help to quantify the behavior of a gate from different perspectives: complexity, functionality, robustness, performance, and energy consumption. Although we concentrate here on the behavior of the simplest of all gates, the inverter, similar properties can also be defined for more complex components, such as NAND, NOR, and XOR gates.

3.2.1 Area and Complexity

Having a small area is, obviously, a desirable property for a digital gate. The smaller the gate, the higher the integration density and the smaller the die size. Die size directly relates to the *fabrication cost* of a design. Smaller gates tend also to be faster, as the total gate capacitance—which is one of the dominant performance parameters—often scales with the area.

The *number of transistors* in a gate is indicative for the expected implementation area. Other parameters may have an impact, though. For instance, a complex interconnect pattern between the transistors can cause the wiring area to dominate.

3.2.2 Functionality and Robustness: The Static Behavior

A prime requirement for a digital gate is, obviously, that it perform the digital function it is designed for. The measured behavior of a manufactured gate normally deviates from the expected response. One reason for this aberration are the variations in the manufacturing process. As was explained in Chapter 2, the dimensions, threshold voltages, and currents of an MOS transistor vary between runs or even on a single wafer or die. The electrical behavior of a circuit can be profoundly affected by those variations. The presence of disturbing noise sources on or off the chip is another source of deviations in circuit response. The word *noise* in the context of digital circuits means *"unwanted variations of voltages and currents at the logic nodes."* Noise signals can enter a circuit in many ways. Some examples of digital noise sources are depicted in Figure 3.1. For instance, two wires placed side by side in an integrated circuit form a coupling capacitor and a mutual inductance. Hence, a voltage or a current change on one of the wires can influence the signals on the neighboring wire. Noise on the power and ground rails of a gate also influences the signal levels in the gate. How to cope with all these disturbances is one of the main challenges in the design of high-performance digital circuits and is a recurring topic in this book.

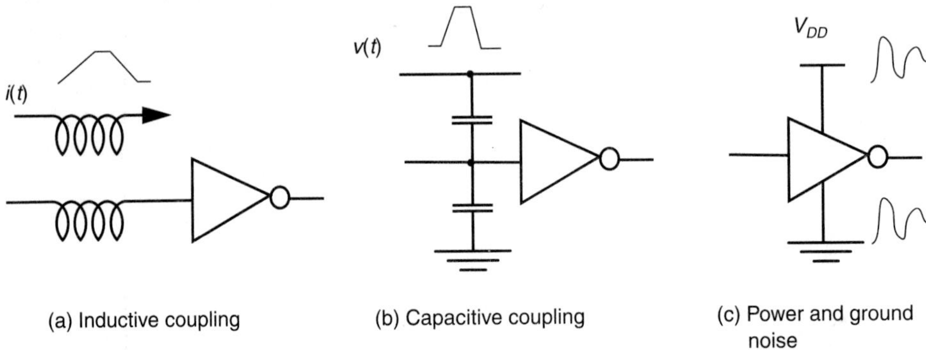

Figure 3.1 Noise sources in digital circuits.

The steady-state parameters of a gate measure how robust the structure is with respect to both variations in the manufacturing process and noise disturbances. The definition and derivation of these parameters requires a prior understanding of how digital signals are represented in the world of electronic circuits.

Digital circuits (DC) perform operations on *logical* (or *Boolean*) variables. A logical variable x can only assume two discrete values:

$$x \in \{0,1\}$$

As an example, the inversion (i.e., the function that an inverter performs) implements the following compositional relationship between two Boolean variables x and y:

$$y = \bar{x}: \{x = 0 \Rightarrow y = 1; x = 1 \Rightarrow y = 0\} \tag{3.1}$$

A logical variable is, however, a mathematical abstraction. In a physical implementation, such a variable is represented by an electrical quantity. This is most often a node voltage that is not discrete but can adopt a continuous range of values. It is necessary to turn the electrical voltage into a discrete variable by associating *a nominal voltage level* with each logic state: 1 ⇔ V_{OH}, 0 ⇔ V_{OL}, where V_{OH} and V_{OL} represent the *high* and the *low* logic levels, respectively. Applying V_{OH} to the input of the gate yields V_{OL} at the output and vice versa. The difference between the two is called the *logic swing*.

$$
\begin{aligned}
V_{OH} &= \overline{(V_{OL})} \\
V_{OL} &= \overline{(V_{OH})}
\end{aligned}
\tag{3.2}
$$

The Voltage-Transfer Characteristic

Assume now that a logical variable *in* serves as the input to an inverting gate that produces the variable *out*. The electrical function of a gate is best expressed by its *voltage-transfer characteristic* (VTC) (sometimes called the *DC transfer characteristic*), which plots the output voltage as a function of the input voltage $V_{out} = f(V_{in})$. An example of an inverter VTC is shown in Figure 3.2. The high and low nominal voltages, V_{OH} and V_{OL}, can readily be identified—$V_{OH} = f(V_{OL})$ and $V_{OL} = f(V_{OH})$. Another point of interest of the VTC is the *gate or switching threshold voltage* V_M (not to be confused with the threshold voltage of a transistor), that is defined as $V_M = f(V_M)$. V_M can also be found graphically at the intersection of the VTC curve and the line given by $V_{out} = V_{in}$. The gate threshold voltage presents the midpoint of the switching characteristics, which is obtained when the output of a gate is short-circuited to the input. This point will prove to be of particular interest when studying circuits with feedback (also called *sequential circuits*).

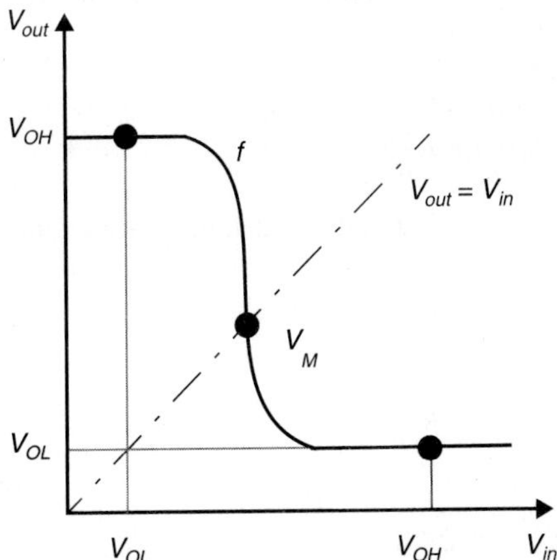

Figure 3.2 Inverter voltage-transfer characteristic.

Even if an ideal nominal value is applied at the input of a gate, the output signal often deviates from the expected nominal value. These deviations can be caused by noise or by the loading on the output of the gate (i.e., by the number of gates connected to the output signal). Figure 3.3a illustrates how a logic level is represented in reality by a range of acceptable voltages, separated by a region of uncertainty, rather than by nominal levels alone. The regions of acceptable high and low voltages are delimited by the V_{IH} and V_{IL} voltage levels, respectively. These represent by definition the points where the gain

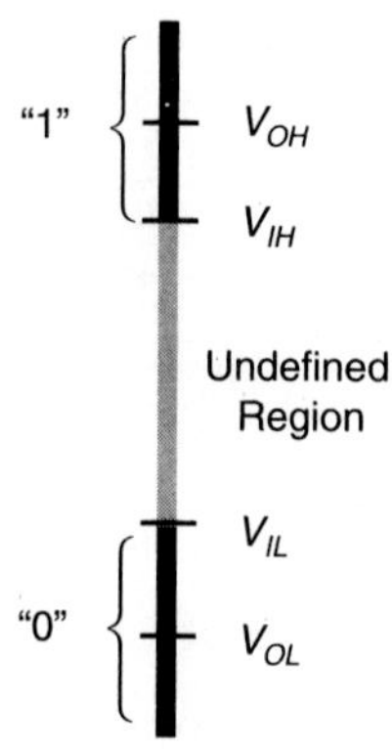

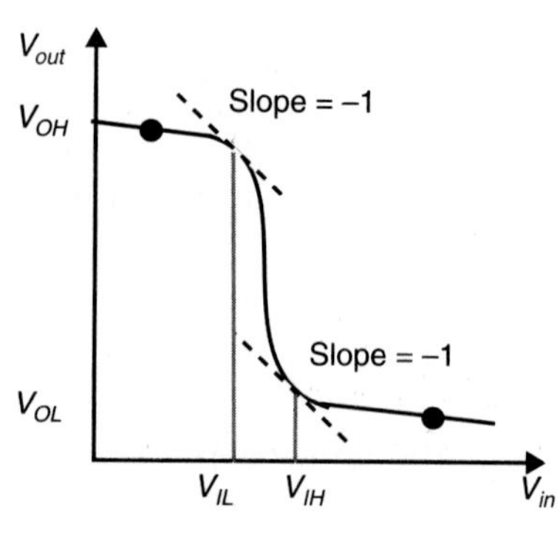

(a) Relationship between voltage and logic levels (b) Definition of V_{IH} and V_{IL}

Figure 3.3 Mapping logic levels to the voltage domain.

(= dV_{out} / dV_{in}) of the VTC equals –1 as shown in Figure 3.3b. The region between V_{IH} and V_{IL} is called the *undefined region* (sometimes also referred to as *transition width,* or *TW*). Steady-state signals should avoid this region if proper circuit operation is to be ensured.

Noise Margins

For a gate to be robust and insensitive to noise disturbances, it is essential that the "0" and "1" intervals be as large as possible. A measure of the sensitivity of a gate to noise is given by the noise margins NM_L (*noise margin low*) and NM_H (*noise margin high*), which quantize the size of the legal "0" and "1", respectively:

$$\begin{aligned} NM_L &= V_{IL} - V_{OL} \\ NM_H &= V_{OH} - V_{IH} \end{aligned} \tag{3.3}$$

The noise margins represent the levels of noise that can be sustained when gates are cascaded as illustrated in Figure 3.4. It is obvious that the margins should be larger than 0 for a digital circuit to be functional and by preference should be as large as possible.

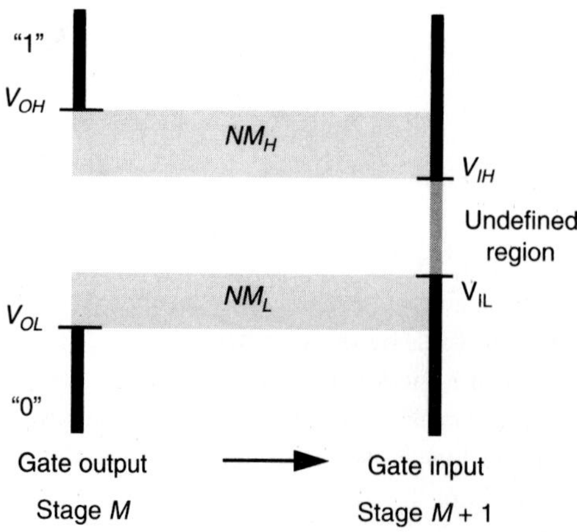

Figure 3.4 Cascaded inverter gates: definition of noise margins.

Regenerative Property

A large noise margin is a desirable, but not sufficient requirement. Assume that a signal is disturbed by noise and differs from the nominal voltage levels. As long as the signal is within the noise margins, the following gate continues to function correctly, although its output voltage varies from the nominal one. This deviation is added to the noise injected at the output node and passed to the next gate. The effect of different noise sources may accumulate and eventually force a signal level into the undefined region. This, fortunately, does not happen if the gate possesses the *regenerative property*, which ensures that a disturbed signal gradually converges back to one of the nominal voltage levels after passing through a number of logical stages. This property can be understood as follows:

An input voltage v_{in} ($v_{in} \in$ "0") is applied to a chain of N inverters (Figure 3.5a). Assuming that the number of inverters in the chain is even, the output voltage v_{out} ($N \rightarrow \infty$) will equal V_{OL} if and only if the inverter possesses the regenerative property. Similarly, when an input voltage v_{in} ($v_{in} \in$ "1") is applied to the inverter chain, the output voltage will approach the nominal value V_{OH}.

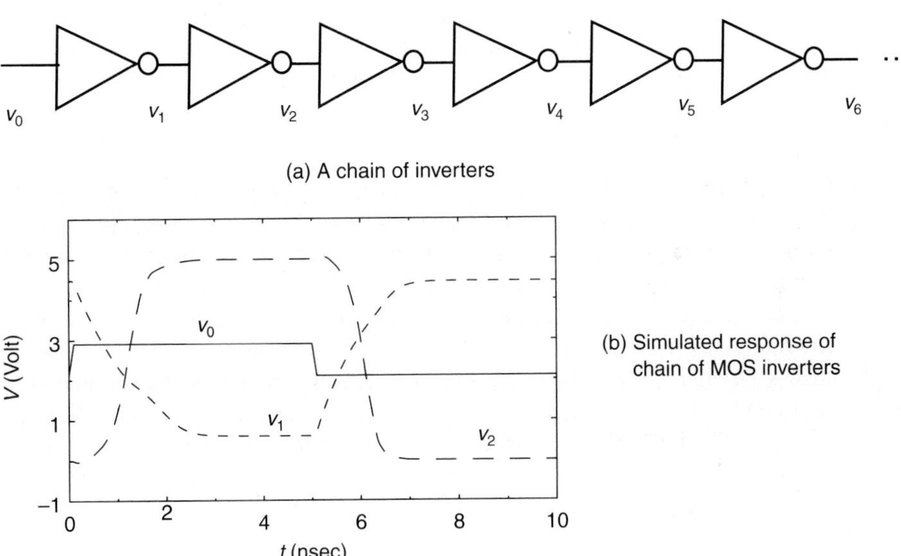

Figure 3.5 The regenerative property.

Example 3.1 Regenerative property

The concept of regeneration is illustrated in Figure 3.5b, which plots the simulated transient response of a chain of CMOS inverters. The input signal to the chain is a step-waveform with a degraded amplitude, which could be caused by noise. Instead of swinging from rail to rail, v_0 only extends between 2.1 and 2.9 V. From the simulation, it can be observed that this deviation rapidly disappears, while progressing through the chain; v_1, for instance, extends from 0.6 V to 4.45 V. Even further, v_2 already swings between the nominal V_{OL} and V_{OH}. The inverter used in this example clearly possesses the regenerative property.

The conditions under which a gate is regenerative can be intuitively derived by analyzing a simple case study. Figure 3.6(a) plots the VTC of an inverter $V_{out} = f(V_{in})$ as well as its inverse function *finv*(), which reverts the function of the *x*- and *y*-axis and is defined as follows:

$$in = f(out) \Rightarrow in = finv(out) \tag{3.4}$$

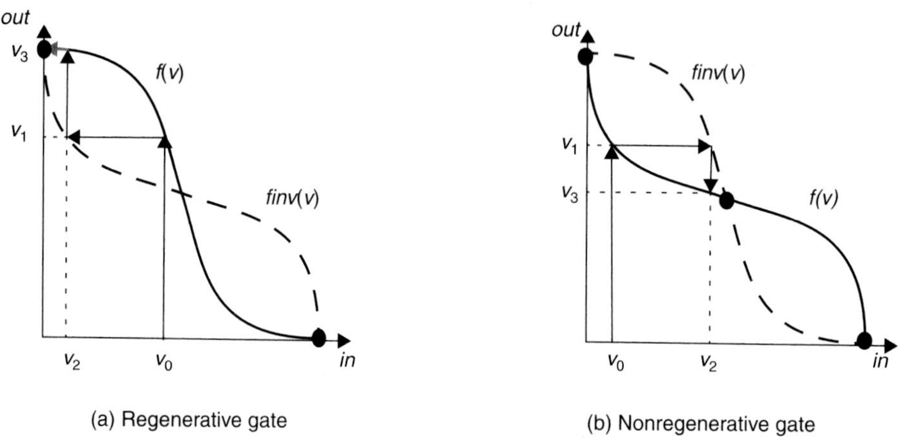

Figure 3.6 Conditions for regeneration.

Assume that a voltage v_0, deviating from the nominal voltages, is applied to the first inverter in the chain. The output voltage of this inverter equals $v_1 = f(v_0)$ and is applied to the next inverter. Graphically this corresponds to $v_1 = finv(v_2)$. The signal voltage gradually converges to the nominal signal after a number of inverter stages, as indicated by the arrows. In Figure 3.6(b) the signal does not converge to any of the nominal voltage levels but to an intermediate voltage level. Hence, the characteristic is nonregenerative. The difference between the two cases is due to the gain characteristics of the gates. To be regenerative, the VTC should have a transient region (or undefined region) with a gain *greater than* 1 in absolute value, bordered by the two legal zones, where the gain should be *smaller than* 1. Such a gate has two stable operating points. This clarifies the definition of the V_{IH} and the V_{IL} levels that form the boundaries between the legal and the transient zones.

Directivity

The directivity property requires a gate to be *unidirectional*, that is, changes in an output level should not appear at any unchanging input of the same circuit. If not, an output-signal transition reflects to the gate inputs as a noise signal, affecting the signal integrity.

In real gate implementations, full directivity can never be achieved. Some feedback of changes in output levels to the inputs cannot be avoided. Capacitive coupling between inputs and outputs is a typical example of such a feedback. It is important to minimize these changes so that they do not affect the logic levels of the input signals.

Fan-In and Fan-Out

The *fan-out* denotes *the number of load gates N that are connected to the output of the driving gate* (Figure 3.7). Increasing the fan-out of a gate can affect its logic output levels. From the world of analog amplifiers, we know that this effect is minimized by making the input resistance of the load gates as large as possible (minimizing the input currents) and by keeping the output resistance of the driving gate small (reducing the effects of load currents on the output voltage). When the fan-out is large, the added load can deteriorate the dynamic performance of the driving gate. For these reasons, many generic and library components define a *maximum fan-out* to guarantee that the static and dynamic performance of the element meet specification.

The *fan-in* of a gate is defined as the *number of inputs* to the gate (Figure 3.7b). Gates with large fan-in tend to be more complex, which often results in inferior static and dynamic properties.

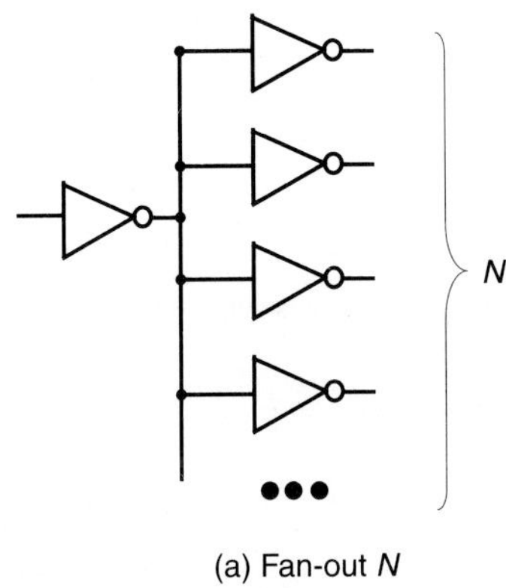

(a) Fan-out N

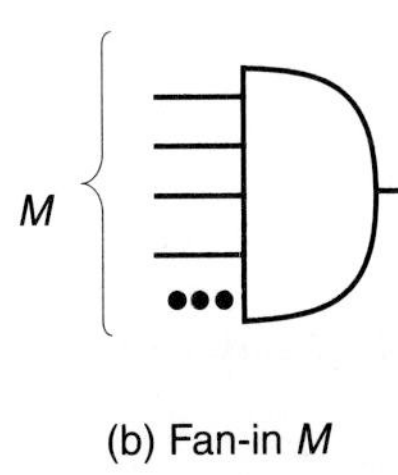

(b) Fan-in M

Figure 3.7 Definition of fan-out and fan-in of a digital gate.

The Ideal Digital Gate

Based on the above observations, we can define the *ideal* digital gate from a static perspective. The ideal inverter model is important because it gives us a metric by which we can judge the quality of actual implementations.

Its VTC is shown in Figure 3.8 and has the following properties: infinite gain in the transition region, and gate threshold located in the middle of the logic swing, with high and low noise margins equal to half the swing. The input and output impedances of the ideal gate are infinity and zero, respectively (i.e., the gate has unlimited fan-out). While this ideal VTC is unfortunately impossible in real designs, some implementations, such as the static CMOS inverter, come close.

Example 3.2 Voltage-Transfer Characteristic

Figure 3.9 shows an example of a voltage-transfer characteristic of an actual, but outdated gate structure (as produced by SPICE in the DC analysis mode). The values of the dc-parameters are derived from inspection of the graph.

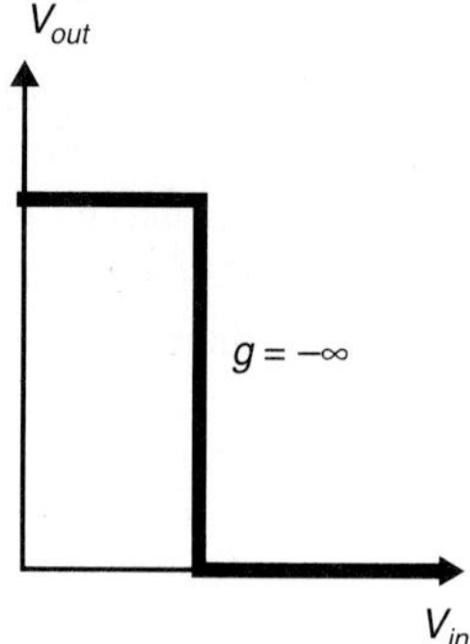

Figure 3.8 Ideal voltage-transfer characteristic.

$$V_{OH} = 3.5\text{ V}; \quad V_{OL} = 0.45\text{ V}$$

$$V_{IH} = 2.35\text{ V}; \quad V_{IL} = 0.66\text{ V}$$

$$V_M = 1.64\text{ V}$$

$$NM_H = 1.15\text{ V}; \quad NM_L = 0.21\text{ V}$$

The observed transfer characteristic, obviously, is far from ideal: it is asymmetrical, has a very low value for NM_L, and the voltage swing of 3.05 V is substantially below the maximum obtainable value of 5 V (which is the value of the supply voltage for this design).

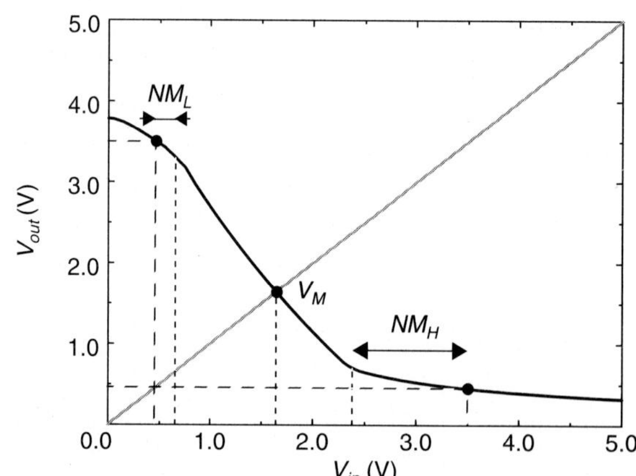

Figure 3.9 Voltage-transfer characteristic of an NMOS inverter of the 1970s.

3.2.3 Performance: The Dynamic Behavior

The *propagation delay* t_p of a gate defines how quickly it responds to a change at its input and relates directly to the speed and performance metrics. The propagation delay expresses *the delay experienced by a signal when passing through a gate*. It is measured between the 50% transition points of the input and output waveforms, as shown in Figure 3.10 for an inverting gate.[1] Because a gate displays different response times for rising or falling input waveforms, two definitions of the propagation delay are necessary. The t_{pLH} defines the response time of the gate for a *low to high* (or positive) output transition, while t_{pHL} refers to a *high to low* (or negative) transition. The overall propagation delay t_p is defined as the average of the two,

$$t_p = \frac{t_{pLH} + t_{pHL}}{2} \tag{3.5}$$

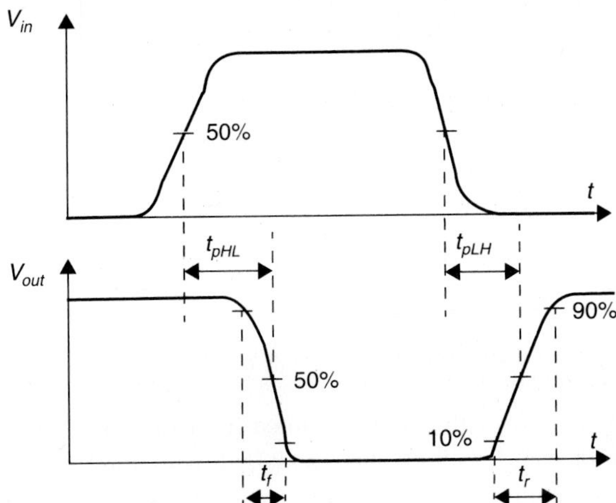

Figure 3.10 Definition of propagation delays and rise and fall times.

Knowledge of t_p is, however, not sufficient to completely characterize circuit performance. The power consumption, noise behavior, and, indirectly, the speed of a gate are also strong functions of the *signal slopes* (as will become clear later in this chapter). This can be quantified with the *rise and fall time* measures t_r and t_f, which are defined between the 10% and 90% points of the waveforms (Figure 3.10).

The propagation delay of a gate is a function of its fan-in and fan-out. Fan-out gates present an increased load (mostly capacitive) to the driving gate and slow its performance. The increased complexity of a gate due to a large fan-in also has a negative influence on the performance. When comparing the performance of gates in different technologies, it is important not to confuse the picture by including second-order parameters such as fan-in and fan-out. It is therefore useful to find a uniform way of measuring the t_p of a gate, so that technologies can be judged on an equal footing. The de-facto standard circuit for delay measurement is the *ring oscillator*, which consists of an odd number of inverters connected in a circular chain (Figure 3.11). Due to the odd number of inversions, this circuit does not have a stable operating point and oscillates. The period T of the oscillation is determined by the propagation time of a signal transition through the complete chain, or

[1] The 50% definition is inspired the assumption that the switching threshold V_M is typically located in the middle of the logic swing.

$T = 2 \times t_p \times N$ with N the number of inverters in the chain. The factor 2 results from the observation that a full cycle requires both a low-to-high and a high-to-low transition. Note that this equation is only valid for $2Nt_p >> t_f + t_r$. If this condition is not met, the circuit might not oscillate—one "wave" of signals propagating through the ring will overlap with a successor and eventually dampen the oscillation. Typically, a ring oscillator needs a least five stages to be operational.

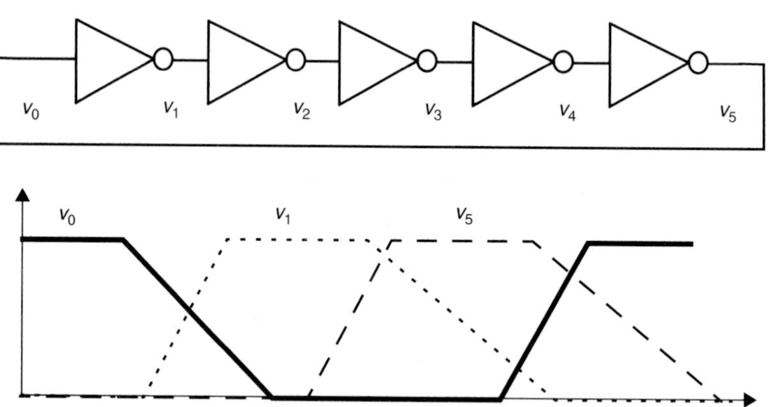

Figure 3.11 Ring oscillator circuit for propagation-delay measurement.

We must be extremely careful with results obtained from ring oscillator measurements. A t_p of 100 psec by no means implies that a circuit built with those gates will operate at 10 GHz. The oscillator results are primarily useful for quantifying the differences between various manufacturing technologies and gate topologies. The oscillator is an idealized circuit where each gate has a fan-in and fan-out of exactly one and parasitic loads are minimal. In more realistic digital circuits, fan-ins and fan-outs are higher, and interconnect delays are non-negligible. The gate functionality is also substantialy more complex than a simple invert operation. As a result, the achievable clock frequency on average is 50 to a 100 times slower than the frequency predicted from ring oscillator measurements. This is an average observation; carefully optimized designs might approach the ideal frequency more closely.

Example 3.3 Propagation Delay of First-Order *RC* Network

Digital circuits are often modeled as first-order *RC* networks of the type shown in Figure 3.12. The propagation delay of such a network is thus of considerable interest.

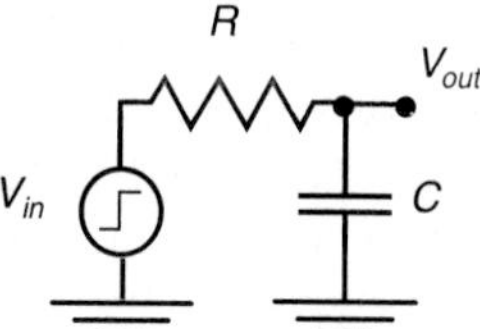

Figure 3.12 First-order *RC* network.

When applying a step input (with V_{in} going from 0 to V), the transient response of this circuit is known to be an exponential function, and is given by the following expression (where $\tau = RC$, the time constant of the network):

$$V_{out}(t) = (1 - e^{-t/\tau})\, V \tag{3.6}$$

The time to reach the 50% point is easily computed as $t = \ln(2)\tau = 0.69\tau$. Similarly, it takes $t = \ln(9)\tau = 2.2\tau$ to get to the 90% point. It is worth memorizing these numbers, as they are extensively used in the rest of the text.

3.2.4 Power and Energy Consumption

The power consumption of a gate determines how much heat the circuit dissipates and how much energy is consumed per operation. These factors influence a great number of critical design decisions, such as the packaging and cooling requirements, supply-line sizing, power-supply capacity, and, most important, the number of circuits that can be integrated onto a single chip. For instance, it is mainly power considerations that prevent the development of very large bipolar digital integrated circuits. Therefore, power dissipation is an important property of a gate that affects feasibility, cost, and reliability. Depending upon the design problem at hand, different dissipation measures have to be considered. For instance, the peak power P_{peak} is important when studying supply-line sizing. When addressing cooling or battery requirements, one is predominantly interested in the average power dissipation P_{av}. Both measures are defined in equation Eq. (3.7):

$$\begin{aligned} P_{peak} &= i_{peak}V_{supply} = max[p(t)] \\ P_{av} &= \frac{1}{T}\int_0^T p(t)dt = \frac{V_{supply}}{T}\int_0^T i_{supply}(t)dt \end{aligned} \tag{3.7}$$

where i_{supply} is the current being drawn from the supply voltage V_{supply} over the interval $t \in [0,T]$, and i_{peak} is the maximum value of i_{supply} over that interval. The dissipation can further be decomposed into static and dynamic components. The latter occurs only during transients, when the gate is switching. It is due to the charging of capacitors and temporary current paths between the supply rails, and is, therefore, proportional to the switching frequency: *the higher the number of switching events, the higher the power consumption.* The static component on the other hand is present even when no switching occurs and is caused by static conductive paths between the supply rails or by leakage currents. It is always present, even when the circuit is in stand-by. Minimization of this consumption source is a worthwhile goal.

The propagation delay and the power consumption of a gate are related—the propagation delay is mostly determined by the speed at which a given amount of energy can be stored on the gate capacitors. The faster the energy transfer (or the higher the power consumption), the faster the gate. For a given technology and gate topology, the product of power consumption and propagation delay is generally a constant. This product is called the *power-delay product* (or PDP) and can be considered as a quality measure for a switching device. The PDP is simply the *energy* consumed by the gate *per switching event.* The ring oscillator is again the circuit of choice for measuring the PDP of a logic family.

3.3 The Static CMOS Inverter

In the following sections, we will derive the static and dynamic parameters for two popular gates, i.e., the CMOS and the ECL inverter. For each structure, we will initiate the discussion with an intuitive analysis of the gate operation.

3.3.1 A First Glance

Figure 3.13 shows the circuit diagram of a static CMOS inverter. Its operation is readily understood from a simplified circuit model. Qualitatively, an MOS transistor can be modeled as *a switch with a finite on-resistance* R_{on}. When $|V_{GS}| < |V_T|$, the switch is open; when $V_{GS} > V_T$, the transistor behaves as a finite resistance (Figure 3.14a). This leads to the following interpretation of the inverter. When V_{in} is high (or equal to V_{DD}), the NMOS transistor is on, while the PMOS is off. This yields the equivalent circuit of Figure 3.14b. A direct path exists between V_{out} and the ground node, resulting in a steady-state value of 0 V. On the other hand, when the input voltage is low (0 V), NMOS and PMOS transistors are off and on, respectively. The equivalent circuit of Figure 3.14c shows that a path exists between V_{DD} and V_{out}, yielding a high output voltage. Notice that no path exists between the supply and ground in steady-state operation. Consequently, the inverter does not consume any static power (ignoring leakage).

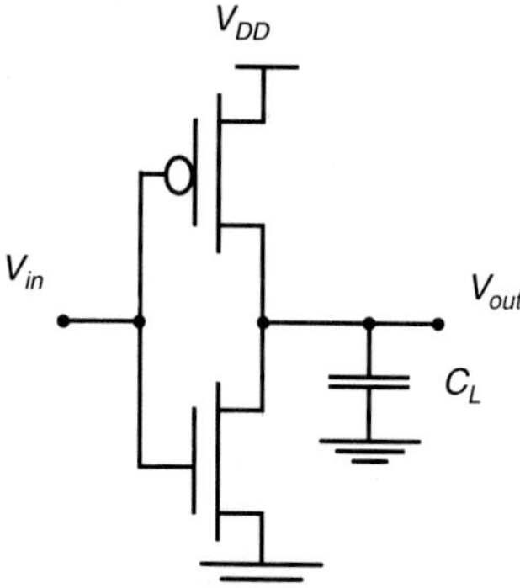

Figure 3.13 Static CMOS inverter. V_{DD} stands for the supply voltage.

SIDELINE: The above observation is one of the prime reasons CMOS superceded NMOS as the digital technology of choice. NMOS was very popular in the 1970s and early 1980s. All early microprocessors, such as the Intel 4004, were pure NMOS designs. The lack of complementary devices (such as the NMOS and PMOS transistor) in a pure NMOS technology makes the realization of inverters with zero static power nontrivial. This static power consumption of the basic NMOS gate puts an upper bound on the number of gates that can be integrated on a single chip; hence the move to CMOS in the 1980s.

A number of other important properties of static CMOS can be derived from this switch-level view:

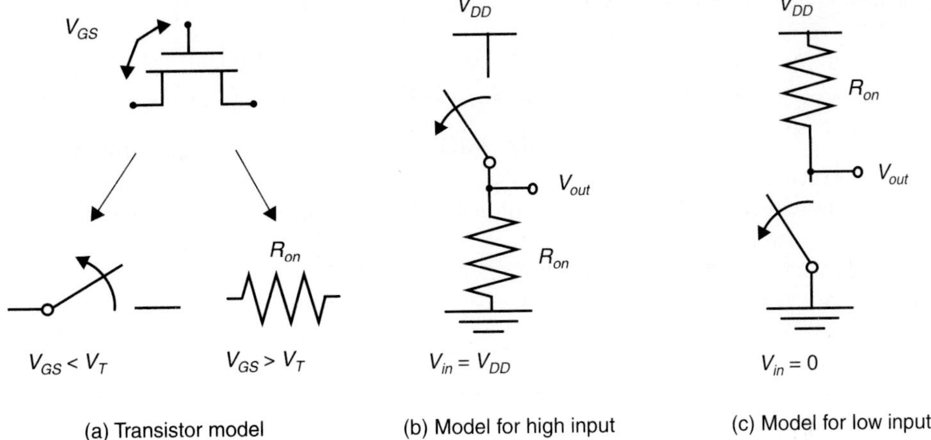

Figure 3.14 Switch models of CMOS inverter.

- The high and low output levels equal V_{DD} and *GND*, respectively; in other words, the voltage swing is equal to the supply voltage. This results in high noise margins.
- The logic levels are not dependent upon the relative device sizes, so that the transistors can be minimum size. Gates with this property are called *ratioless*. This is in contrast with *ratioed logic*, where logic levels are determined by the relative dimensions of the composing transistors.
- In steady state, there always exists a path with finite resistance between the output and either V_{DD} or *GND*. A well-designed CMOS inverter, therefore, has a *low output impedance*, which makes it less sensitive to noise and disturbances. Typical values of the output resistance are in the range of 10 kΩ (for the technology under consideration).
- The *input resistance* of the CMOS inverter is extremely high, as the gate of an MOS transistor is a virtually perfect insulator and draws no dc input current. Since the input node of the inverter only connects to transistor gates, the steady-state input current is nearly zero. A single inverter can theoretically drive an infinite number of gates (or have an infinite fan-out) and still be functionally operational; however, increasing the fan-out also increases the propagation delay, as will become clear below. So, although fan-out does not have any effect on the steady-state behavior, it degrades the transient response.

The nature and the form of the voltage-transfer characteristic (VTC) can be graphically deduced from the load-line plots, which superimpose the current characteristics of the NMOS and the PMOS devices. Creating such a graph requires that the *I-V* curves of the NMOS and PMOS devices are transformed onto a common coordinate set. We have selected the output voltage and the NMOS drain current I_{DN} as the independent variables. The PMOS *I-V* relations can be translated into this variable space by the following relations (the subscripts *n* and *p* denote the NMOS and PMOS devices, respectively).

$$I_{DSp} = -I_{DSn}$$
$$V_{GSn} = V_{in} \; ; \; V_{GSp} = V_{in} - V_{DD} \tag{3.8}$$
$$V_{DSn} = V_{out} \; ; \; V_{DSp} = V_{out} - V_{DD}$$

The load-line curves of the PMOS device are obtained by a mirroring around the x-axis and a horizontal shift over V_{DD}. This procedure is illustrated in Figure 3.15, where the subsequent steps to adjust the original PMOS I-V curves to the common coordinate set V_{in}, V_{out} and I_{Dn} are enumerated.

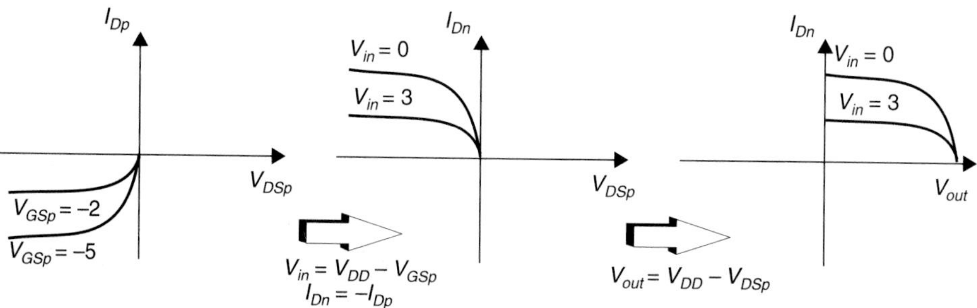

Figure 3.15 Transforming PMOS I-V characteristic to a common coordinate set.

The resulting load lines are plotted in Figure 3.16.[2] For valid dc operating points, the currents through the NMOS and PMOS devices must be equal. Graphically, this means that the dc points must be located at the intersection of corresponding load lines. A number of those points (for V_{in} = 0, 1, 2, 3, 4, and 5 V) are marked on the graph. As can be observed, all operating points are located either at the high or low output levels. The VTC

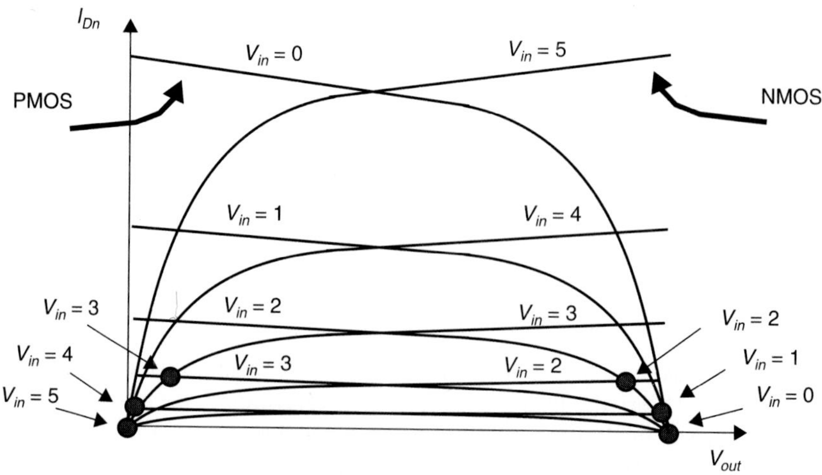

Figure 3.16 Load curves for NMOS and PMOS transistors of the static CMOS inverter (V_{DD} = 5 V). The dots represent the dc operation points for various input voltages.

[2] If not familiar with the concept of load lines, please refer to [Sedra87].

of the inverter hence exhibits a very narrow transition zone. This results from the high gain during the switching transient, when both NMOS and PMOS are simultaneously on. The gain of the circuit is determined by the transconductances of the transistors and their output resistances. The latter are large in the transition region as both devices are in saturation. This translates into the VTC of Figure 3.17.

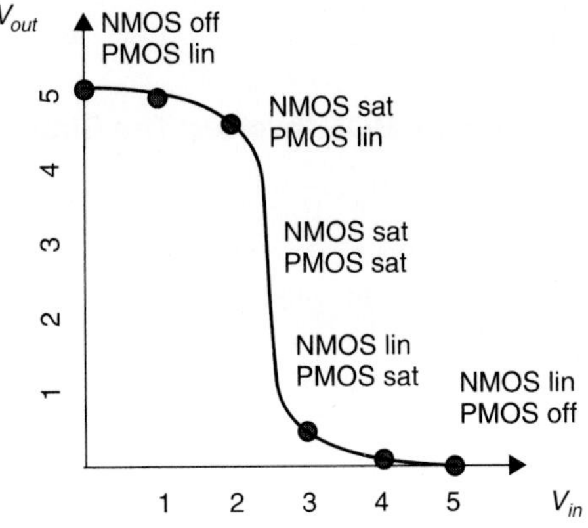

Figure 3.17 VTC of static CMOS inverter, derived from Figure 3.16 (V_{DD} = 5 V). For each operation region, the modes of the transistors are annotated.

Before going into the analytical details of the operation of the CMOS inverter, a qualitative analysis of the transient behavior of the gate is appropriate. This response is dominated mainly by the output capacitance of the gate, C_L, which is composed of the diffusion capacitances of the NMOS and PMOS transistors, the capacitance of the interconnect wires, and the input capacitance of the fan-out. Assuming temporarily that the transistors switch instantaneously, the transient response can be approximated again using the simplified switch model (Figure 3.18). Let us consider the low-to-high transition (Figure 3.18a). The gate response time is determined by the time it takes to charge the capacitor C_L through the resistor R_{on} (i.e., by the time constant $C_L R_{on}$). A fast gate is built either

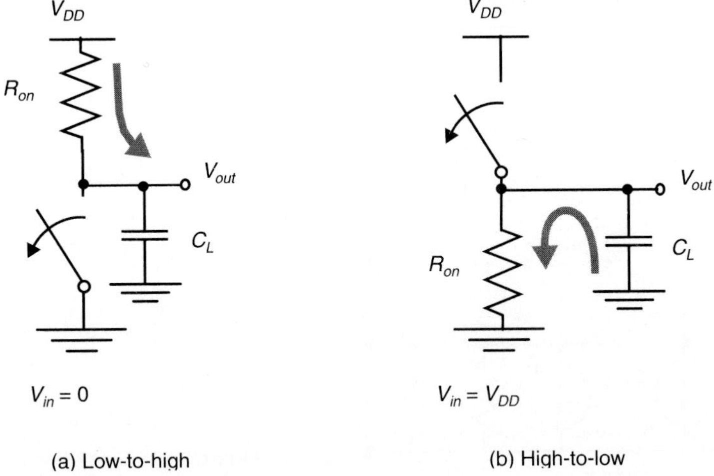

Figure 3.18 Switch model of dynamic behavior of static CMOS inverter.

by keeping the output capacitance small or by decreasing the on-resistance of the PMOS transistor. The latter is achieved by increasing the *W/L* ratio of the device, as is apparent from an inspection of the current equations of an MOS device. A similar consideration is true for the high-to-low transition Figure 3.18(b). One should be aware that the on-resistance of the NMOS and PMOS transistor is not constant, but is a nonlinear function of the voltage across the transistor. This complicates the exact determination of the propagation delay.

3.3.2 Evaluating the Robustness of the CMOS Inverter: The Static Behavior

In the qualitative discussion above, the overall shape of the voltage-transfer characteristic of the static CMOS inverter was derived, as were the values of V_{OH} and V_{OL} (V_{DD} and *GND*, respectively). It remains to determine the precise values of V_M, V_{IH}, and V_{IL} as well as the noise margins. By definition, the values of V_{IH} and V_{IL} are the points at which

$$\frac{\partial V_{out}}{\partial V_{in}} = -1 \, .$$

In the terminology of the analog circuit designer, these are the points where the small-signal gain *g* of the amplifier, formed by the inverter, is equal to –1.

SIDELINE: Surprisingly (or not so surprisingly), an inverter can also be used as an analog amplifier. The static CMOS inverter/amplifier actually has a very high gain in its transition region (although the rest of its amplifier properties such as supply noise rejection are rather poor). This observation can be used to demonstrate one of the major differences between analog and digital design: where the analog designer would bias the amplifier in the middle of the transient region, so that a maximum linearity is obtained, the digital designer will operate the device in the regions of extreme nonlinearity, resulting in well-defined and well-separated high and low signals.

The small-signal model of the MOS transistor and expressions for its parameters were introduced in Appendix B. By inserting the model for the NMOS and PMOS transistor and by shorting all dc voltage sources, the small-signal model of the inverter is obtained as shown in Figure 3.19. The gain of this circuit is determined by inspection of the circuit and is set to –1 for $V_{in} = V_{IH}$ and V_{IL}

$$g = \frac{v_{out}}{v_{in}} = -(g_{mn} + g_{mp}) \times (r_{on} \parallel r_{op}) = -1 \tag{3.9}$$

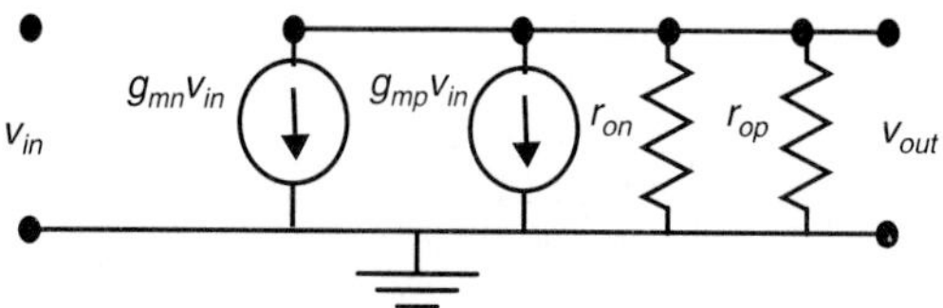

Figure 3.19 Small-signal model of a CMOS inverter.

The parameters of the model depend upon the operating modes of the transistors. Consider for instance V_{IH}. For $V_{in} = V_{IH}$, the PMOS and NMOS transistors can be assumed to be in the *saturation* and *linear* regions, respectively (Figure 3.17). Using the expressions presented in Table B.1, this results in the following values for the small-signal parameters. To simplify the analysis, the channel-length modulation is ignored, or $\lambda_p = 0$.

$$\begin{aligned} g_{mn} &= k_n V_{out} \\ g_{mp} &= k_p(V_{DD} - V_{IH} - |V_{Tp}|) \\ r_{on} &= \frac{1}{k_n(V_{IH} - V_{out} - V_{Tn})} \\ r_{op} &= \infty \end{aligned} \tag{3.10}$$

Inserting those expressions into the gain formula Eq. (3.9) results in a first relation between V_{IH} and V_{out}. A second relation is obtained by noting that the static currents through the NMOS and PMOS transistors must be identical.

$$g = -(g_{mn} + g_{mp}) \times (r_{on} \| r_{op}) = -\frac{k_n V_{out} + k_p(V_{DD} - V_{IH} - |V_{Tp}|)}{k_n(V_{IH} - V_{out} - V_{Tn})} = -1 \tag{3.11}$$

and

$$k_n\left[(V_{IH} - V_{Tn})V_{out} - \frac{V_{out}^2}{2}\right] = \frac{k_p}{2}(V_{DD} - V_{IH} - |V_{Tp}|)^2 \tag{3.12}$$

Solving V_{out} from Eq. (3.11) and substituting the result into Eq. (3.12) produces a second-order equation with one root between 0 and V_{DD}. An analytic expression for V_{IH} is complex and, therefore, not included.

Similar equations can be derived for the $V_{in} = \mathbf{V_{IL}}$. In this case however, the NMOS operates in *saturation*, while the PMOS device is in the *linear* mode.

$$\begin{aligned} g_{mn} &= k_n(V_{IL} - V_{Tn}) \\ g_{mp} &= k_p(V_{DD} - V_{out}) \\ r_{on} &= \infty \\ r_{op} &= \frac{1}{k_p(v_{out} - V_{IL} - |V_{Tp}|)} \end{aligned} \tag{3.13}$$

so that

$$g = -(g_{mn} + g_{mp}) \times (r_{on} \| r_{op}) = -\frac{k_n(V_{IL} - V_{Tn}) + k_p(V_{DD} - V_{out})}{k_p(V_{out} - V_{IL} - |V_{Tp}|)} = -1 \tag{3.14}$$

and

$$k_p\left[(V_{DD} - V_{IL} - |V_{Tp}|)(V_{DD} - V_{out}) - \frac{(V_{DD} - V_{out})^2}{2}\right] = \frac{k_n}{2}(V_{IL} - V_{Tn})^2 \tag{3.15}$$

The inverter threshold, V_M, is defined as the point where $V_{in} = V_{out}$. Its value can be obtained graphically from the intersection of the VTC with the line given by $V_{in} = V_{out}$ as shown in Figure 3.2. In this region, both PMOS and NMOS are saturated, and the expression for V_M is obtained by equating the currents through the transistors.

$$\frac{k_n}{2}(V_M - V_{Tn})^2 = \frac{k_p}{2}(V_{DD} - V_M - |V_{Tp}|)^2 \tag{3.16}$$

which yields

$$V_M = \frac{r(V_{DD} - |V_{Tp}|) + V_{Tn}}{1 + r} \quad \text{with} \quad r = \sqrt{\frac{k_p}{k_n}} \tag{3.17}$$

This leads to the conclusion that V_M is situated in the middle of the available voltage swing (or at $V_{DD}/2$) when $k_n = k_p$ (assuming that the threshold voltages of the PMOS and NMOS transistors are comparable, which is generally the case). This is equivalent to sizing the PMOS device so that $(W/L)_p = (W/L)_n \times (k'_n/k'_p)$. For the 1.2 μm CMOS process presented in Chapter 2 (Example 2.12), this means making the PMOS approximately three times wider than the NMOS device.

This observation can be generalized:

When designing static CMOS circuits, it is necessary to make the PMOS section (k'_n/k'_p) times wider than the NMOS section, if one wants to maximize the noise margins and obtain symmetrical characteristics.

Having the threshold of the inverter situated right in the middle of the voltage swing results in comparable values for the low and high noise margins, which is in general considered beneficial. Figure 3.20 plots the simulated value of V_M for various ratios of k_p/k_n. An analysis of this curve generates some interesting observations:

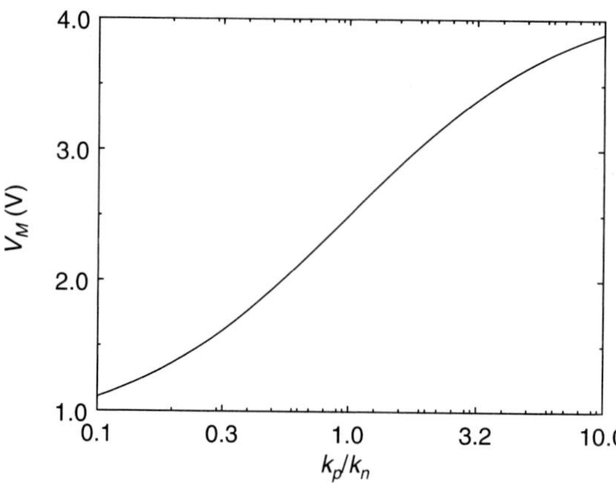

Figure 3.20 Inverter threshold V_M in function of the k_p/k_n ratio (plotted on a logarithmic scale). (V_{DD} = 5V, $V_{Tn} = |V_{Tp}|$ = 0.8 V).

1. V_M is relatively insensitive to variations of the k_p/k_n ratio around the center point. This means that small variations of the ratio (e.g., making it 0.7 or 1.5) do not disturb the transfer characteristic too much. This observation is one of the reasons why it is an accepted practice in industrial designs to set the width of the PMOS transis-

tor to only two times the width of the NMOS, saving some valuable area. Secondary effects such as channel-length modulation and velocity saturation further help to make this a reasonable decision.

2. The effect of changing the k_p/k_n ratio is to shift the transient region of the VTC. Increasing the width of the PMOS or the NMOS moves V_M towards V_{DD} or GND respectively. This property can be very useful, as asymmetrical transfer characteristics are actually desirable in some designs. This is demonstrated by the example of Figure 3.21. The incoming signal V_{in} has a very noisy zero value. Passing this signal through a symmetrical inverter would lead to erroneous values (Figure 3.21a). This can be addressed by raising the threshold of the inverter, which results in a correct response (Figure 3.21b).

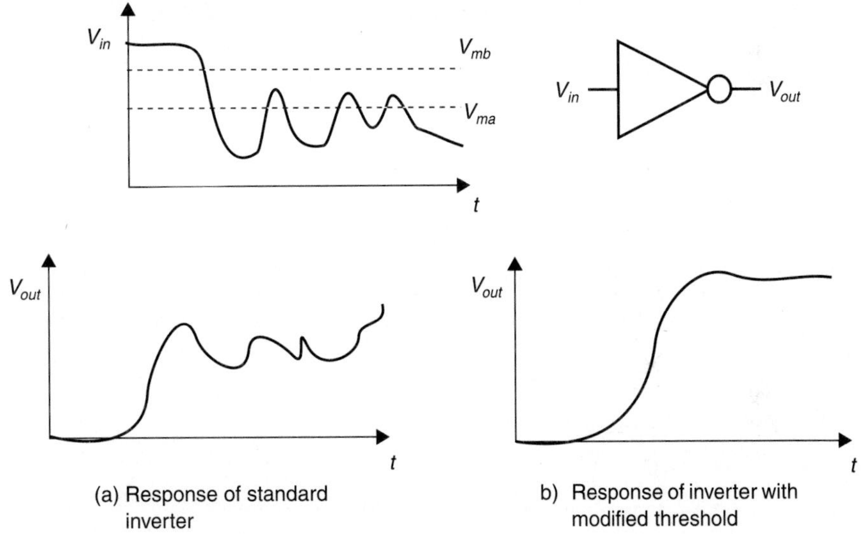

Figure 3.21 Changing the inverter threshold can improve the circuit reliability.

Example 3.4 VTC of a CMOS Inverter

As an example, we analyze and plot the VTC of an inverter designed in the 1.2 μm CMOS technology (Example 2.12). The transistors are chosen to be of minimal length. In order to make the response approximately symmetrical, we make the PMOS transistor three times wider than the NMOS device, as $k'_p = 27\ \mu A/V^2$ and $k'_n = 80\ \mu A/V^2$ [3]

$$V_{DD} = 5\ V$$

$$(W/L)_n = 1.8 / 1.2$$

$$(W/L)_p = 5.4 / 1.2$$

[3] Observe that the model-2 parameters are used. This proves to be more accurate for this dc analysis. Remember that the level-1 parameters were derived for the (V_{GS} = 5 V, V_{DS} = 5 V) operation point, which is meaningful for transient analysis, but tends to produce large errors in the transition region (see section 2.3.5).

The dc parameters of the inverter can be found using the expressions derived above:

$$V_{OL} = 0\text{ V},\ V_{OH} = 5\text{ V}.$$

To obtain V_{IH}, we solve the following equations (derived from Eq. (3.11) and Eq. (3.12)).

$$80 \times [2 \times (V_{IH} - 0.74) \times V_{out} - V^2_{out}] = 3 \times 27 \times (5 - 0.74 - V_{IH})^2$$

$$80 \times V_{out} + 3 \times 27 \times (5 - 0.74 - V_{IH}) = 80 \times (V_{IH} - V_{out} - 0.74)$$

This yields $V_{IH} = 2.92$ V, $V_{out} = 0.42$ V, and $NM_H = 5\text{ V} - V_{IH} = 2.08$ V.

Similarly the values of V_{IL} and NM_L can be computed.

$$3 \times 27 \times [2 \times (5 - 0.74 - V_{IL}) \times (5 - V_{out}) - (5 - V_{out})^2] = 80 \times (V_{IL} - 0.74)^2$$

$$80 \times (V_{IL} - 0.74) + 3 \times 27 \times (5 - V_{out}) = 3 \times 27 \times (V_{out} - V_{IL} - 0.74)$$

which yields

$$V_{IL} = NM_L = 2.06\text{ V}$$

Finally, V_M can be computed with the aid of Eq. (3.17), given that $r = 1.01$

$$V_M = 2.51\text{ V}$$

A plot of the VTC of the inverter, obtained using SPICE, is shown in Figure 3.22. The values of the dc parameters extracted from this plot ($V_{IL} = 2.17$ V, $V_{IH} = 2.80$ V, and $V_M = 2.57$ V) are in close agreement with the manually obtained values.

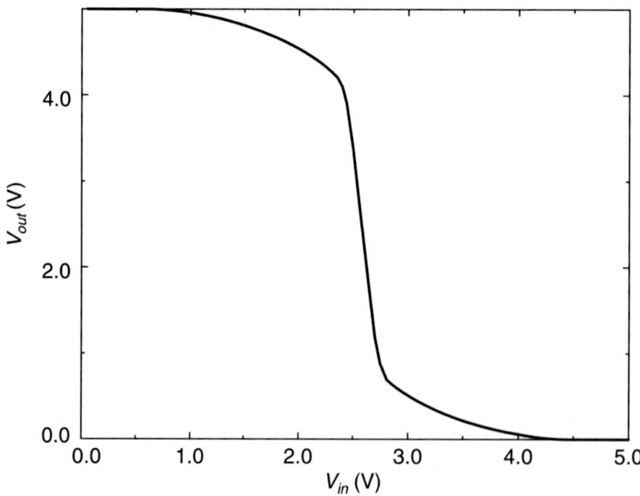

Figure 3.22 Simulated voltage-transfer characteristic of inverter.

3.3.3 Performance of CMOS Inverter: The Dynamic Behavior

As explained in the qualitative discussion, the propagation delay of the CMOS inverter is determined by the time it takes to charge and discharge the load capacitor C_L through the PMOS and NMOS transistors, respectively. This observation suggests that **getting C_L as small as possible is crucial to the realization of high-performance CMOS circuits.** Before deriving detailed expressions for the propagation delay, it is hence worthwhile to study the load capacitance in detail. The next section analyzes the components of C_L and proposes techniques for manual estimation.

Computing the Capacitances

The propagation-delay model assumes that all capacitances are lumped together in one single capacitor C_L, located between V_{out} and *GND*. Manual analysis of MOS circuits where each capacitor is considered individually is virtually impossible and is exacerbated by the many nonlinear capacitances in the MOS transistor model. We now describe an approach to derive a lumped model. Be aware, however, that this is a considerable simplification of the actual situation, even in the case of a simple inverter.

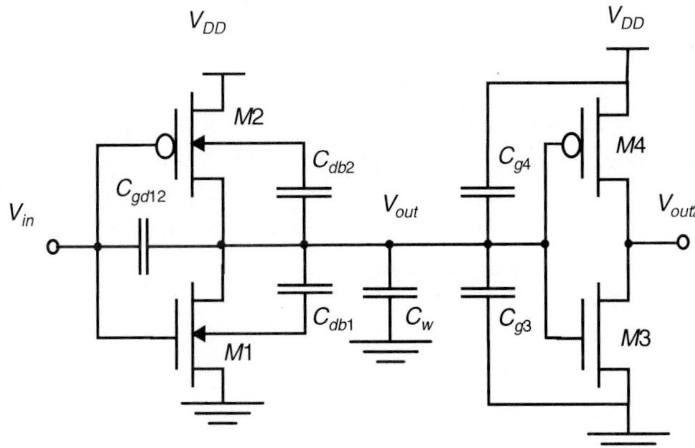

Figure 3.23 Parasitic capacitances, influencing the transient behavior of the cascaded inverter pair.

Figure 3.23 shows the schematic of a cascaded inverter pair. It includes all the capacitances influencing the transient response of node V_{out}. It is assumed that the input V_{in} is driven by an *ideal voltage source* with fixed rise and fall times. The following components can now be discerned in C_L (observe that only capacitances connected to the output node are considered):

- $\boldsymbol{C_{gd12}}$—As $M1$ and $M2$ are either in cut-off or in the saturation mode in the steady-state operation condition, it is reasonable to assume that the only contributions to this capacitance are the *overlap capacitances* of both M1 and M2. This is a consequence of the observation that in those operation modes the gate capacitance is either completely between gate and bulk (cut-off) or gate and source (saturation) as was discussed in Chapter 2. In the lumped capacitor model, the gate-drain capacitor is replaced by a capacitor to ground. The contribution of this capacitance should, therefore, be counted twice, due to the so-called *Miller effect:* the effective voltage change over the gate-drain capacitor during a low-high or high-low transition is actually twice the output voltage swing, as the terminals of the capacitor are moving in opposite directions (Figure 3.24). We use the following equation for the gate-drain capacitors: $C_{gd} = 2\ C_{GD0}W$ (with C_{GD0} the overlap capacitance per unit width as used in the SPICE model). For an in-depth discussion of the Miller effect, please refer to textbooks such as Sedra and Smith ([Sedra87], p. 57).

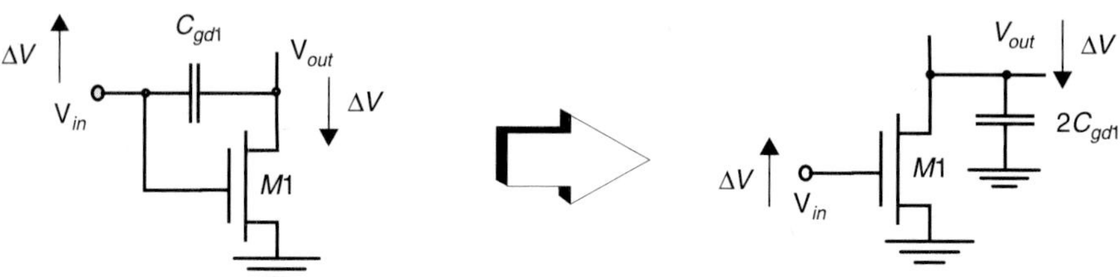

Figure 3.24 The Miller effect—A capacitor experiencing identical but opposite voltage swings at both its terminals can be replaced by a capacitor to ground, whose value is two times the original value.

- **C_{db1} and C_{db2}**—The capacitance between drain and bulk is due to the reverse-biased *pn*-junction. Such a capacitor is, unfortunately, quite nonlinear and depends heavily on the applied voltage. As shown in Chapter 2, the best approach to simplify the hand analysis is to replace the nonlinear capacitor by a linear one with the same change in charge for the voltage range of interest. A multiplication factor K_{eq} is introduced to relate the linearized capacitor to the value of the junction capacitance under zero-bias conditions.

$$C_{eq} = K_{eq}C_{j0} \tag{3.18}$$

with C_{j0} the junction capacitance per unit area under zero-bias conditions. The values of the bottom-plate and side-wall junction capacitances under zero-bias conditions can be found in the SPICE MOS model card as the CJ and CJSW parameters (for the bottom junction and the side-wall junctions of the drain or source diffusions, respectively). An expression for K_{eq} was derived in Eq. (2.18) and is repeated here for convenience

$$K_{eq} = \frac{-\phi_0^m}{(V_{high} - V_{low})(1-m)}[(\phi_0 - V_{high})^{1-m} - (\phi_0 - V_{low})^{1-m}] \tag{3.19}$$

with ϕ_0 the built-in junction potential (PB in the SPICE model) and m the grading coefficient of the junction. Observe that the junction voltage is defined to be negative for reverse-biased junctions.

Example 3.5 K_{eq} for a 5 V CMOS Inverter

Consider the inverter of Figure 3.23 designed in the 1.2 μm CMOS technology. ϕ_0 equals 0.6 V for both NMOS and PMOS transistors. Assume m = 0.5 (abrupt junction).

Consider first the NMOS transistor (C_{db1} in Figure 3.23). The propagation delay is defined by the time between the 50% transitions of the input and the output. For the CMOS inverter, this is the time-instance where V_{out} reaches 2.5 V, as the output voltage swing goes from rail to rail or equals 5 V. We, therefore, linearize the junction capacitance over the interval {5 V, 2.5 V} for the high-to-low transition and {0, 2.5 V} for the low-to-high transition.

During the high-to-low transition at the output, V_{out} initially equals 5 V. Because the bulk of the NMOS device is connected to *GND*, this translates into a reverse voltage of 5 V over the drain junction or $V_{high} = -5$ V. At the 50% point, $V_{out} = 2.5$ V or $V_{low} = -2.5$ V. Plugging those values into Eq. (3.19) yields $K_{eq} = 0.375$. During the low-to-high transition, V_{low} and V_{high} equal 0 V and –2.5 V, respectively, resulting in a higher value for K_{eq}, equaling 0.611.

The PMOS transistor has reverse values, as its substrate is connected to 5 V. Hence, for the high-to-low transition $K_{eq} = 0.611$ ($V_{low} = 0$, $V_{high} = -2.5$ V), and $K_{eq} = 0.375$ for the low-to-high transition ($V_{low} = -2.5$ V, $V_{high} = -5$ V).

Using this approach, the junction capacitance can be replaced by a linear component and treated as any other device capacitance.The result of the linearization is a minor distortion of the voltage waveforms. The logic delays are not significantly influenced by this simplification.

- C_w—The capacitance due to the wiring is negligible in this particular case (< 1 fF). This is not necessarily the case for more complex designs or for future technologies, as we will discuss extensively in later chapters.

- C_{g3} **and** C_{g4}—The analysis is simplified here by assuming that the total gate capacitance of the loading gates (M3 and M4) is connected between V_{out} and *GND* (or V_{DD}). Overlap and gate capacitance are clustered into a single component $C_g = C_{ox}WL$. This is only partially true as it ignores the Miller effect on the gate-drain capacitances. This has a relatively minor effect on the accuracy, since we can safely assume that the connecting gate does not switch before the 50% point is reached, and V_{out2}, therefore, remains constant in the interval of interest. A second approximation is that the channel capacitance of the connecting gate is constant over the interval of interest and situated between gate and *GND*. Once again, this assumption introduces only a minor error.

Example 3.6 Capacitances of a 1.2 μm CMOS Inverter

A minimum-size, symmetrical CMOS inverter has been designed in the 1.2 μm CMOS technology. The layout is shown in Figure 3.25. The supply voltage V_{DD} is set to 5 V. From the layout, we derive the transistor sizes, diffusion areas, and perimeters. This data is summarized in Table 3.1. As an example, we will derive the drain area and perimeter for the NMOS transistor. The drain area is formed by the metal-diffusion contact, which has an area of $4 \times 4\ \lambda^2$, and the rectangle between contact and gate, which has an area of $3 \times 1\ \lambda^2$. This results in a total area of 19 λ^2, or 6.84 μm^2 (as $\lambda = 0.6$ μm). The perimeter of the drain area is rather involved and consists of the following components (going counterclockwise): $5 + 4 + 4 + 1 + 1 = 15\lambda$ or PD = $15 \times 0.6 = 9.0$ μm. Notice that the gate side of the drain perimeter is not included, as this is not a part of the side-wall. The drain area and perimeter of the PMOS transistor are derived similarly (the rectangular shape makes the exercise considerably simpler).

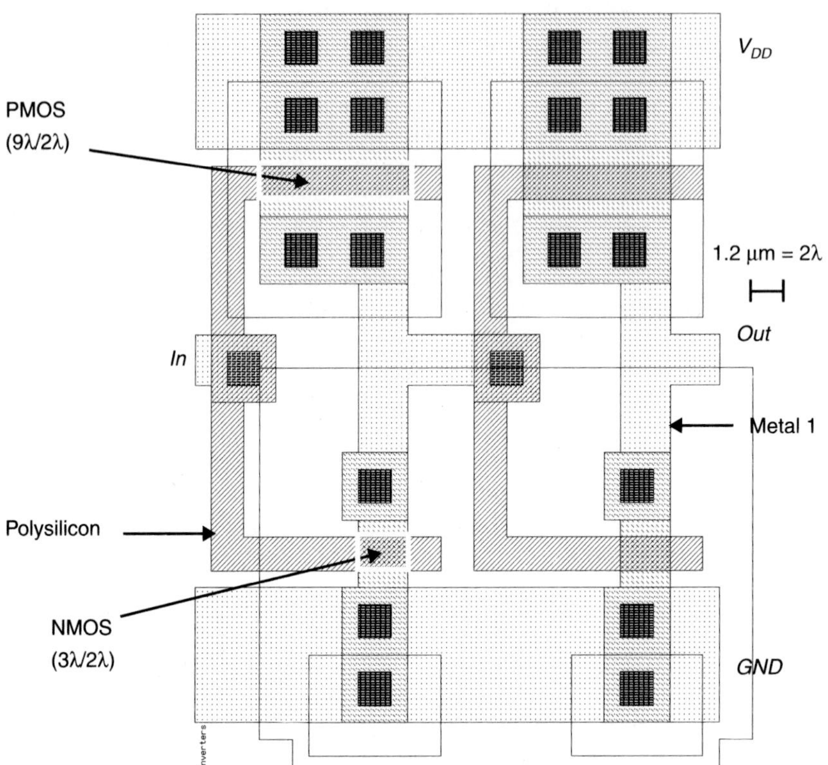

Figure 3.25 Layout of two chained, minimum-size inverters (1.2 μm CMOS technology) (see also Color-plate 6).

Table 3.1 Inverter transistor data.

	W/L	AD (μm²)	PD (μm)	AS (μm²)	PS (μm)
NMOS	1.8/1.2	6.84 (19 λ^2)	9.0 (15λ)	6.8 (19 λ^2)	9.0 (15λ)
PMOS	5.4/1.2	3*5.4 = 16.2(45 λ^2)	3*2 + 5.4 = 11.4 (19λ)	16.2 (45 λ^2)	11.4 (19λ)

This physical information can be combined with the approximations derived above to come up with an estimation of C_L. From the SPICE model, the following capacitor parameters are obtained:

Overlap capacitance: CGD0(NMOS) = 0.43 fF/μm; CGDO(PMOS) = 0.43 fF/μm

Bottom junction capacitance: CJ(NMOS) = 0.3 fF/μm^2; CJ(PMOS) = 0.5 fF/μm^2

Side-wall junction capacitance: CJSW(NMOS) = 0.8 fF/μm; CJSW(PMOS) = 0.135 fF/μm

Gate capacitance: C_{ox}(NMOS) = C_{ox}(PMOS) = ε_{ox}/t_{ox} = 1.76 fF/μm^2

Bringing all the components together results in Table 3.2. Notice that the load capacitance is almost evenly split between its two major components: the internal capacitance (diffusion and overlap capacitances) and the gate capacitance of the connecting gate. Technically speaking, we should use different K_{eq} values for the bottom-plate and side-wall capacitances (as the latter is of the linear-gradient type). To simplify the derivation, we consider both of them to be abrupt and use the values of Example 3.5.

Table 3.2 Components of C_L (for high-to-low and low-to-high transitions).

Capacitor	Expression	Value (fF) (H→L)	Value (fF) (L→H)
C_{gd1}	$2\ \text{CGD0}\ W_n$	1.55	1.55
C_{gd2}	$2\ \text{CGD0}\ W_p$	4.65	4.65
C_{db1}	$K_{eqn}\ (AD_n\ CJ + PD_n\ CJSW)$	3.45	5.6
C_{db2}	$K_{eqp}\ (AD_p\ CJ + PD_p\ CJSW)$	5.9	3.6
C_{g3}	$C_{ox}\ W_n\ L_n$	3.8	3.8
C_{g4}	$C_{ox}\ W_p\ L_p$	11.4	11.4
C_w	From Extraction	2	2
C_L	Σ	32.75	32.6

Propagation Delay: First-Order Analysis

The propagation delay can be computed by integrating the capacitor (dis)charge current, which results in Eq. (3.20).

$$t_p = C_L \int_{v_1}^{v_2} \frac{dv}{i(v)} \tag{3.20}$$

with i the (dis)charging current, v the voltage over the capacitor, and v_1 and v_2 the initial and final voltage. An exact computation of this equation is complex, as $i(v)$ is a nonlinear function of v. A reasonable approximation of the propagation delay, adequate for manual analysis, can be obtained by replacing the time-varying charging current by a fixed current I_{av}, which is the average of the currents at the end points of the voltage transition. This simplification transforms Eq. (3.20) into a more tractable expression.

$$t_p = \frac{C_L(v_2 - v_1)}{I_{av}} \tag{3.21}$$

Remember that the propagation delay is defined as the time it takes for the output to reach the 50% point. For the low-to-high transition, $v_1 = V_{OL}$ and $v_2 = (V_{OH}+V_{OL})/2$. For the high-to-low transition, $v_1 = V_{OH}$, and $v_2 = (V_{OH} + V_{OL})/2$. As a result, the following holds for both t_{pLH} and t_{pHL}

$$t_p = C_L \frac{(V_{OH} - V_{OL})/2}{|I_{av}|} \quad (3.22)$$

Consider first the t_{pLH} of the CMOS inverter. To simplify the analysis, we assume that the input signal changes abruptly from V_{DD} to 0. In this case, the NMOS transistor turns off immediately, and only the PMOS device contributes to the current. It is easily verified that this transistor will be in saturation as long as $V_{out} < |V_{Tp}|$ and in linear mode for the rest of the output range.

$$V_{OH} - V_{OL} = V_{DD}$$

$$I(V_{out} = 0) = \frac{k_p}{2}(V_{DD} - |V_{Tp}|)^2 \quad (3.23)$$

$$I\left(V_{out} = \frac{V_{DD}}{2}\right) = k_p\left[(V_{DD} - |V_{Tp}|)\frac{V_{DD}}{2} - \frac{V_{DD}^2}{8}\right]$$

This leads to the following expression for I_{av}:

$$I_{av} = \frac{I(V_{out} = 0) + I\left(V_{out} = \frac{V_{DD}}{2}\right)}{2} = \frac{k_p}{2}\left(\frac{7V_{DD}^2}{8} + \frac{|V_{Tp}|^2}{2} - \frac{3V_{DD}|V_{Tp}|}{2}\right) \quad (3.24)$$

Using the above expression in a manual analysis is somewhat cumbersome. A simpler expression is obtained when it is assumed that the PMOS stays in saturation between $V_{out} = 0$ and $V_{out} = V_{DD}/2$ and hence acts as a current source. Although not entirely exact, this approximation introduces only a minor error (8% and 5% for a V_{DD} of 5 and 3 V, respectively). The average charging current then equals the saturation current and is given by

$$I_{av} = \frac{k_p}{2}(V_{DD} - |V_{Tp}|)^2 \quad (3.25)$$

and using Eq. (3.22)

$$t_{pLH} = \frac{C_L V_{DD}}{k_p(V_{DD} - |V_{Tp}|)^2} \quad (3.26)$$

When $V_{DD} >> |V_{Tp}|$ (which is a reasonable approximation for $V_{DD} = 5$ or 3.3 V), Eq. (3.26) simplifies to a straightforward relation, which will prove to be useful for first-order calculations.

$$t_{pLH} \approx \frac{C_L}{k_p V_{DD}} \quad (3.27)$$

A similar relation can be derived for t_{pHL}

$$t_{pHL} = \frac{C_L V_{DD}}{k_n(V_{DD} - V_{Tn})^2} \approx \frac{C_L}{k_n V_{DD}} \quad (3.28)$$

and finally,

$$t_p \approx \frac{1}{2}(t_{pLH} + t_{pHL}) = \frac{C_L}{2V_{DD}}\left(\frac{1}{k_p} + \frac{1}{k_n}\right) \tag{3.29}$$

Very often, it is desirable for a gate to have identical propagation delays for both rising and falling inputs. This condition can be achieved by making k_p and k_n approximately equal. Remember that this condition is identical to the requirement for a symmetrical VTC.

Design Challenges

From the above, we deduce that the propagation delay of a gate can be minimized in the following ways:

- *Reduce C_L.* Remember that three factors contribute to the load capacitance: the internal diffusion capacitance of the gate itself, the interconnect capacitance, and the fan-out. Careful layout can help to reduce the diffusion and interconnect capacitances.
- *Increase k_p and k_n*, or, equivalently, increase the W/L ratio of the transistors. This might seem to be a straightforward and desirable solution. A word of caution is necessary. Increasing the transistor sizes also increases the diffusion capacitance (and C_L) as well as the gate capacitance. The latter will increase the fan-out factor of the driving gate and adversely affect its speed. This issue will be treated in more detail when we discuss the driving of large capacitors in Chapter 8.
- *Increase V_{DD}.* Normally the designer does not have too much control over this factor, as the supply voltage is determined by system and technology considerations. In fact, with the scaling of the technology to feature sizes of 0.5 μm and below, an opposite trend will be observed. More and more, the nominal supply voltage for commodity devices is reduced to either 3.3 V or 3 V. Further decreases can even be expected, mainly to cope with the increasing electrical fields. Reducing the supply voltage also helps to alleviate power-consumption problems arising from the ever-increasing number of gates on an IC (as will be discussed in a coming section).

Example 3.7 Propagation Delay of a 1.2 μm CMOS Inverter

To derive the propagation delays of the CMOS inverter of Figure 3.25, we make use of Eq. (3.22). The load capacitance C_L has been computed in Example 3.6. Next, we need to estimate the average (dis)charging currents.

When computing these currents, it is important to realize that the effective length of the transistors equals the drawn length (1.2 μm), reduced by twice the lateral diffusion LD (2 × 0.15 μm) or L_{eff} = 0.9 μm. As an example, we derive the current through the PMOS at the beginning of the charging period. The PMOS is in saturation at that point. Hence,

$$I_{PMOS}(V_{out} = 0) = (1/2)\,(5.4 \times 10^{-6})\,(5.4/0.9)\,(5 - 0.74)^2\,(1 + 0.19 \times 5) = 0.57 \text{ mA}$$

$$I_{PMOS}(V_{out} = 2.5 \text{ V}) = (5.4 \times 10^{-6})\,(5.4/0.9)\,((5 - 0.74)2.5 - 2.5^2/2) = 0.24 \text{ mA}$$

Notice that the LEVEL-1 model parameters are used in the above equations (as derived in Example 2.13). To achieve adequate matching between estimated and simulated results, the channel-length modulation factor λ is included in the computations. The NMOS currents are computed similarly. The results are summarized in Table 3.3.

$$I_{NMOS}(V_{out} = 5 \text{ V}) = (1/2)\,(19.6 \times 10^{-6})\,(1.8/0.9)\,(5 - 0.74)^2\,(1{+}0.06 \times 5) = 0.46 \text{ mA}.$$

$$I_{NMOS}(V_{out} = 2.5 \text{ V}) = (19.6 \times 10^{-6})\,(1.8/0.9)\,((5 - 0.74)2.5 - 2.5^2/2) = 0.29 \text{ mA}$$

Table 3.3 Values of (dis)charge currents for operation points of interest.

	V_{out} (Volt)	Operation Mode	I (mA)	I_{av} (mA)
For t_{pLH}	0	PMOS sat.	0.57	
	2.5	PMOS lin.	0.24	0.41
For t_{pHL}	5	NMOS sat.	0.46	
	2.5	NMOS lin.	0.29	0.38

The propagation delays of the gate can now be computed by plugging the results of the previous sections into Eq. (3.22), which produces the results of Table 3.4.

Table 3.4 Propagation delays for the cascaded inverter pair (manual & SPICE).

	t_{pHL} (psec)	t_{pLH} (psec)	t_p (psec)
Manual	215.5	198.8	207.1
SPICE	250.0	199.5	224.8

The accuracy of this analysis is checked using SPICE (using the LEVEL-2 model). The computed transient response of the circuit is plotted in Figure 3.26, while the derived propagation delays are compared to the manual ones in Table 3.4. These results are good considering the simplifications used both in the computations of the average (dis)charge current and the lumped linearized load capacitor. On the simulated transient response, notice the overshoots on the output signals. These are caused by the capacitive coupling between the gate and the drain of the inverter transistors.

WARNING: This example might give the impression that manual analysis always leads to close approximations of the actual response. This is not necessarily the case. Large deviations can often be observed between first- and higher-order models. The purpose of the manual analysis is to get a basic insight in the behavior of the circuit and to determine the dominant parameters. A detailed simulation is indispensable when quantitative data is required. Consider the example above a stroke of good luck.

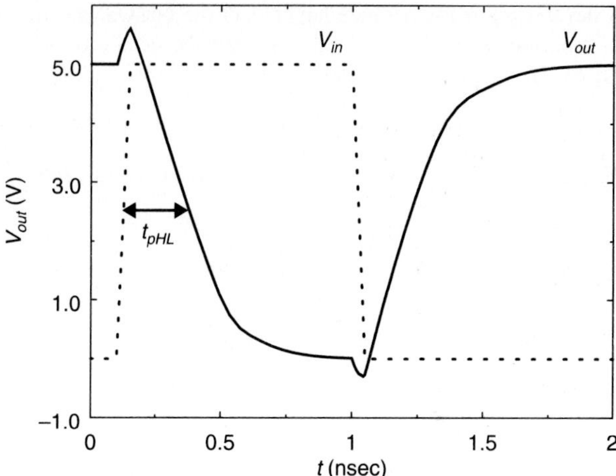

Figure 3.26 Simulated transient response of the inverter of Figure 3.25.

A couple of interesting observations can be drawn from this detailed analysis of the transient response of the complementary MOS inverter:

- In the layout of Figure 3.25, the PMOS transistor was made approximately three times larger than the NMOS device to equate the high-to-low and low-to-high propagation delays and create a symmetrical inverter. This up-sizing has a negative effect on the load capacitance of the gate, as becomes clear from Table 3.2: the diffusion, overlap and gate capacitances of the PMOS device are approximately three times larger than the corresponding values of the NMOS device. If symmetry and reduced noise margins are not of prime concern, it is possible to speed up the inverter by actually reducing the width of the PMOS device! While this has a negative effect on t_{pLH} (the charging current through the PMOS device is reduced), it substantially improves t_{pHL}. When two contradictory effects are present, there must exist a transistor ratio that optimizes the propagation delay of the inverter. It is easily proven that this optimum point is reached when the ratio between PMOS and NMOS devices is set to $\sqrt{\mu_n/\mu_p}$ (see Example 3.8). For our example, this reduces the propagation delay to 217 nsec, while making the design area smaller! On the negative side, this also reduces the low noise margin NM_L to 1.81 V (down from 2.06 V). This value is still acceptable for most designs.

Example 3.8 Sizing of CMOS Inverter Loaded by an Identical Gate

Consider two (identical) cascaded CMOS inverters. The load capacitance of the first gate equals approximately

$$C_L = (C_{dp1} + C_{dn1}) + (C_{gp2} + C_{gn2}) + C_W \tag{3.30}$$

where C_{dp1} and C_{dn1} are the equivalent drain diffusion capacitances of PMOS and NMOS transistors of the first inverter, while C_{gp2} and C_{gn2} are the gate capacitances of the second gate. C_W represents the wiring capacitance.

If the PMOS devices are made α times larger than the NMOS ones ($\alpha = (W/L)_p$ / $(W/L)_n$), all transistor capacitances will scale in approximately the same way, or $C_{dp1} \approx \alpha\, C_{dn1}$, and $C_{gp2} \approx \alpha\, C_{gn2}$. Eq. (3.30) can then be rewritten:

$$C_L = (1+\alpha)(C_{dn1} + C_{gn2}) + C_W \tag{3.31}$$

An expression for the propagation delay can then be derived, based on Eq. (3.29).

$$t_p = \left(\frac{(1+\alpha)(C_{dn1} + C_{gn2}) + C_W}{2V_{DD}k_n}\right)\left(1 + \frac{\mu_n}{\mu_p\alpha}\right) \tag{3.32}$$

The optimal value of α can be found by setting $\frac{\partial t_p}{\partial \alpha}$ to 0, which yields

$$\alpha_{opt} = \sqrt{\frac{\mu_n}{\mu_p}\left(1 + \frac{C_w}{C_{dn1} + C_{gn2}}\right)} \tag{3.33}$$

This means that when the wiring capacitance is negligible (C_{dn1}+C_{gn2} >> C_W), α_{opt} equals $\sqrt{\mu_n/\mu_p} \approx 1.73$, in contrast to the factor three normally used in the noncascaded case. If the wiring capacitance dominates, larger values of α should be used.

- We can also observe in Table 3.2 that approximately half of the load capacitance and the propagation delay is due to the inverter itself (diffusion and overlap capacitances), while the other half is attributable to the fan-out gate (gate capacitance). The latter factor, called the *extrinsic capacitance*, dominates the performance of the gate for larger fan-outs. In fact, the propagation delay increases linearly with the fan-out N

$$t_p(N) = t_p(0) + N(t_p(1) - t_p(0)) \tag{3.34}$$

with $t_p(0)$ and $t_p(1)$ respectively the propagation delays of the inverter with no fan-out and with a fan-out of a single gate. The derivation of this equation is rather simple and is based on the linearity between propagation delay and load capacitance (This is left as an exercise for the reader).

Second-Order Performance Issues

Although the above derivations give a rather accurate picture of the dynamic behavior of the CMOS inverter, we should mention that the effects of a number of second-order parameters and parasitics were not included. These extra factors can have an adverse effect on the performance of the gate.

- **The rise/fall time of the input signal**—All the above expressions were derived under the assumption that the input signal to the inverter abruptly changed from 0 to V_{DD} or vice-versa. Only one of the devices is assumed to be on during the (dis)charging process. In reality, the input signal changes gradually and, temporarily, PMOS and NMOS transistors conduct simultaneously. This affects the total current available for (dis)charging and impacts the propagation delay. Figure 3.27 plots the propagation delay (t_{pHL}) of a minimum-size inverter as a function of the input signal rise time (as obtained from SPICE). It can be observed that t_{pHL} increases (approximately) linearly with increasing rise-time values $t_r > t_{pHL}$. It has been shown in analytical studies [Hodges88] that the effect of the nonzero rise time is adequately modeled by a mean-square dependence of the following form

$$t_{pHL(actual)} = \sqrt{t^2_{pHL(step)} + (t_r/2)^2} \tag{3.35}$$

This is in accordance with the simulation results of Figure 3.27, where, for $t_r > t_{pHL}$, the following dependence was observed: $t_{pHL} = 0.84\ t_r$. Similar results hold for the low-to-high transition as well.

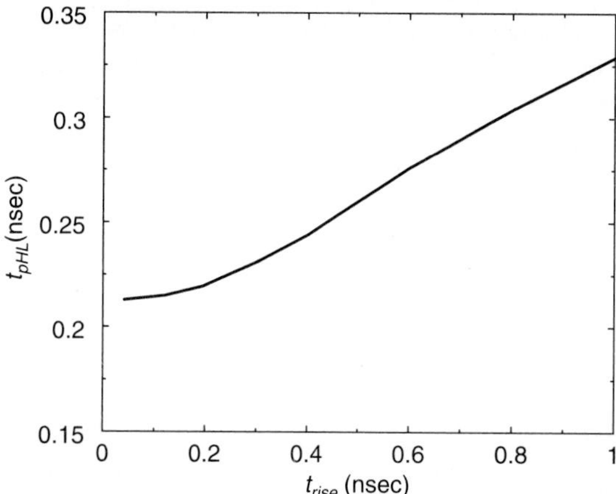

Figure 3.27 t_{pHL} as a function of the input signal rise time (minimum-size inverter with fan-out of a single gate).

Design Challenge

It is advantageous to keep the signal rise times smaller than or equal to the gate propagation delays. This proves not only to be advantageous for the speed, but also for power consumption, as will be discussed later. Keeping the rise and fall times of the signals small (and of approximately equal values) is one of the major challenges in high-performance design.

- **Velocity Saturation**—Throughout this discussion, we have assumed that the maximum (dis)charge current is equal to the saturation current of the transistors and hence is proportional to V_{DD}^2. With the onset of velocity saturation in small-geometry devices, this no longer holds. I_{av} tends to be proportional to V_{DD} instead (Eq. (2.57)). As a result, for $V_{DD} >> V_T$, Eq. (3.36) better describes the propagation delay of the gate (as obtained by modifying Eq. (3.28) and Eq. (3.29)):

$$t_p \approx \frac{C_L}{2}\left(\frac{1}{k_p} + \frac{1}{k_n}\right) \tag{3.36}$$

No first-order dependence of t_p with respect to V_{DD} can be noticed. This is demonstrated in Figure 3.28, where the propagation delay is plotted as a function of V_{DD} for a minimal-size inverter for both long- and short-channel devices. For larger values of V_{DD}, not much influence is apparent, and t_p remains approximately constant (for $V_{DD} > 4\ V_T$). The effect of lowering V_{DD} becomes more prominent once the supply voltage approaches the threshold of the transistors. A sharp increase in delay can be observed below $2\ V_T$. At that point, the simplifications used to derive Eq. (3.29) are no longer valid.

In general, it can be stated that running velocity-saturated devices at high voltages is not beneficial. For submicron devices, it turns out that lowering the supply voltage from 5 V to 3.3 V has only a marginal effect on the performance (between 10% and 30%, depending upon the values of the thresholds).

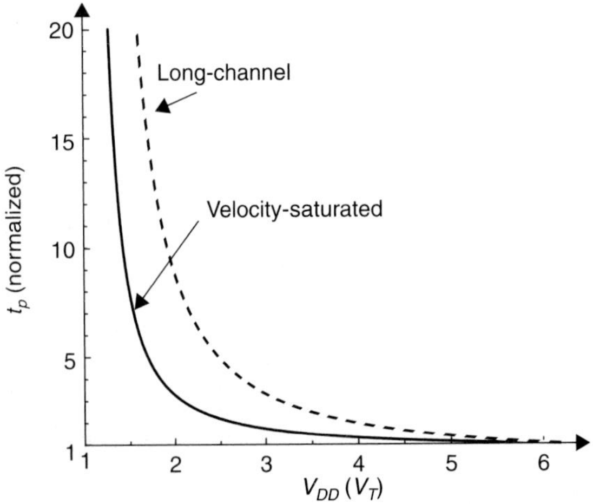

Figure 3.28 Inverter delay as a function of the supply voltage (expressed as a multiple of threshold voltages). The delay is normalized to 1 for $V_{DD} = 6.25\ V_T$, which equals 5 V for V_T = 0.8 V.

- **Source Resistance**—In the discussion of the MOS transistor, it was mentioned that a more accurate model of the device includes source and drain resistances R_S and R_D in series with the transistor. Until now, this resistance was ignored. As shown in Figure 3.29, this resistance can have a detrimental effect on the performance of the CMOS inverter, especially for wide devices with large switching currents. In the ideal case, the maximum discharge current equals

$$I_{sat,ideal} = \frac{k_n}{2}(V_{DD} - V_{Tn})^2 \tag{3.37}$$

The source resistance affects this maximum current in two ways. First of all, the V_{GS} of the discharge transistor is reduced, hence lowering the current. Secondly, as the source of the transistor is no longer grounded, the threshold of the transistor increases due to body effect. This further decreases the discharge current.

$$I_{sat,R} = \frac{k_n}{2}(V_{DD} - V_S - V_{Tn}(V_S))^2 \tag{3.38}$$

with $V_S = R_S I_{sat,R}$. As the above equation is implicit, it must be solved iteratively.

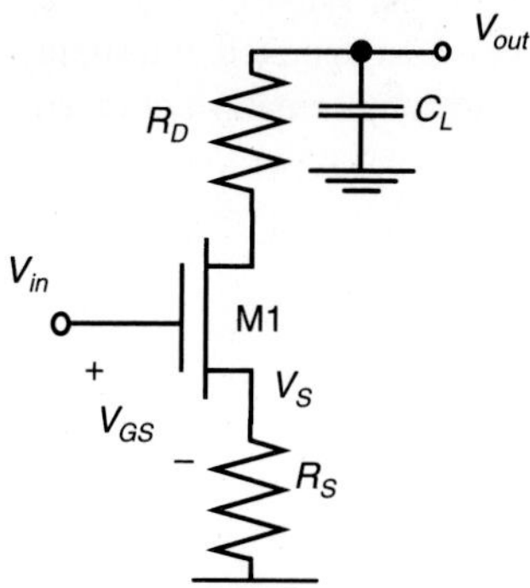

Figure 3.29 Effect of series source (and drain) resistance on performance.

The value of R_S tends to range from tens of ohms to several kΩ, depending upon the manufacturing process and the device width. For example, assume that the transistor M1 of Figure 3.29 is a minimal-size device. For our standard 1.2 μm CMOS process, the values of R_S and R_D equal 70 Ω, approximately. Only a minor impact of these parasitic resistors can be observed: the saturation current is reduced by 1.3% for V_{GS} and V_{DS} equal 5 V. Consider, on the other hand, a process that uses the *lightly doped drain* (LDD) approach, where the values of R_S and R_D can reach approximately 1 to 1.5 kΩ. This number is comparable to the equivalent channel resistance of the minimum-size NMOS transistor, which is typically around 10 kΩ. For the same voltage levels, the series resistances reduce the saturation current by 20%!

Problem 3.1 Propagation Delay as a Function of On-Resistance

In this section, we have expressed the propagation delay as a function of the average (dis)charge current Eq. (3.21). Another approach would be to replace the (dis)charging transistor by its average on-resistance R_{on} and model the circuit as an *RC*-network (as proposed in the introduction to the CMOS inverter). Derive an expression of the propagation delay using this alternative approach.

3.3.4 Power Consumption and Power-Delay Product

By now, it has become fairly clear that the static CMOS inverter has an almost ideal VTC—symmetrical shape, full logic swing, and high noise margins. The main reason, however, that the majority of contemporary high-complexity designs are implemented in static CMOS is the almost complete absence of power consumption in steady-state mode.

Static Consumption

Ideally, the static power consumption of the CMOS inverter is equal to zero, as the PMOS and NMOS devices are never on simultaneously in steady-state operation. There is, unfortunately, always a leakage current present, flowing through the reverse-biased diode junctions of the transistors located between the source or drain and the substrate, as shown in Figure 3.30. This contribution is, in general, very small and can be ignored. For the device sizes under consideration, the leakage current typically ranges between 0.1 nA and 0.5 nA at room temperature. For a die with 1 million devices, operated at a supply voltage of 5 V, this results in a power consumption of 0.5 mW, which is clearly not much of an issue.

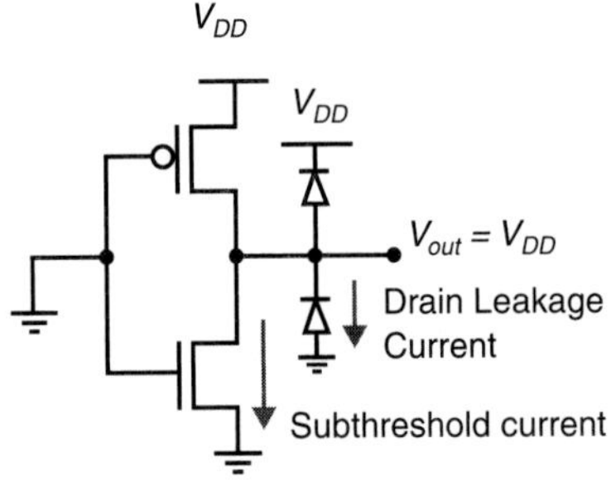

Figure 3.30 Sources of leakage currents in CMOS inverter (for $V_{in} = 0$ V).

A more important source of leakage current is potentially the subthreshold current of the transistors. As discussed in Chapter 2, an MOS transistor can experience a drain-source current, even when V_{GS} is smaller than the threshold voltage (Figure 2.30). The closer the threshold voltage is to zero volts, the larger the leakage current at $V_{GS} = 0$ V and the larger the static power consumption. To offset this effect, the threshold voltage of the device is generally kept above 0.5 V in current technologies.

For both sources of leakage, the resulting static power dissipation is expressed by the following equation

$$P_{stat} = I_{leakage}V_{DD} \tag{3.39}$$

Finally, it is worth mentioning that the junction leakage currents are caused by thermally generated carriers. Therefore, their value increases with increasing junction temperature, and this occurs in an exponential fashion. At 85°C (a common junction temperature limit for commercial hardware), the leakage currents increase by a factor of 60 over their room-temperature values.

The majority of the power is consumed during switching. This dynamic power can be divided into two components:

Dynamic Consumption Due to the Load Capacitance C_L

Each time the capacitor C_L gets charged through the PMOS transistor, its voltage rises from 0 to V_{DD}, and a certain amount of energy is drawn from the power supply. Part of this energy is dissipated in the PMOS device, while the remainder is stored on the load capacitor. During the high-to-low transition, this capacitor is discharged, and the stored energy is dissipated in the NMOS transistor.

A precise measure for this energy consumption can be derived. Let us first consider the low-to-high transition. We assume, initially, that the input waveform has zero rise and fall times, or, in other words, that the NMOS and PMOS devices are never on simultaneously. Therefore, the equivalent circuit of Figure 3.31 is valid. The values of the energy

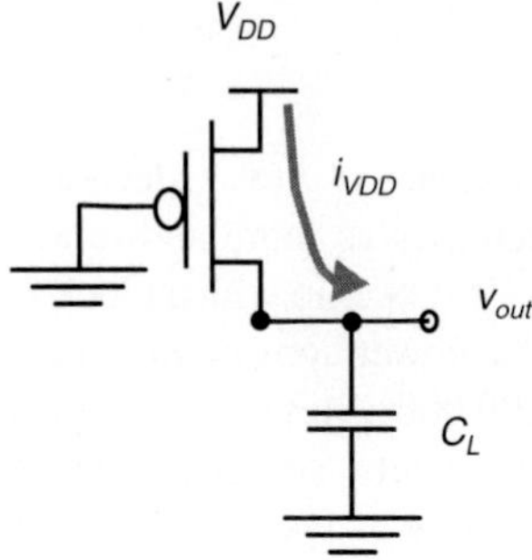

Figure 3.31 Equivalent circuit during the low-to-high transition.

E_{VDD}, taken from the supply during the transition, as well as the energy E_C, stored on the capacitor at the end of the transition, can be derived by integrating the instantaneous power of the period of interest. The corresponding waveforms of $v_{out}(t)$ and $i_{VDD}(t)$ are pictured in Figure 3.32.

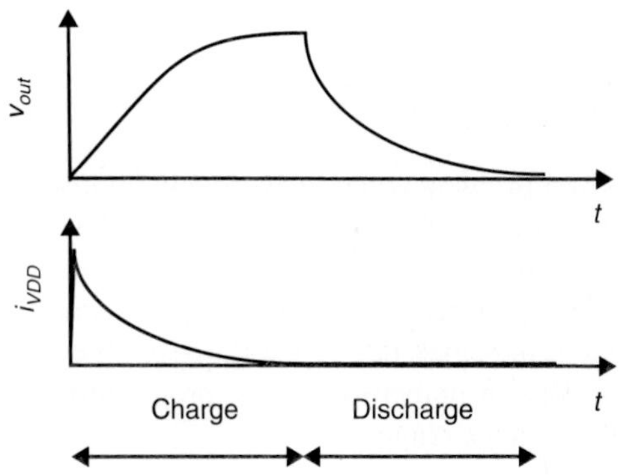

Figure 3.32 Output voltages and supply current during (dis)charge of C_L.

$$E_{VDD} = \int_0^\infty i_{VDD}(t)V_{DD}dt = V_{DD}\int_0^\infty C_L\frac{dv_{out}}{dt}dt = C_LV_{DD}\int_0^{V_{DD}} dv_{out} = C_LV_{DD}^2 \quad (3.40)$$

$$E_C = \int_0^\infty i_{VDD}(t)v_{out}dt = \int_0^\infty C_L\frac{dv_{out}}{dt}v_{out}dt = C_L\int_0^{V_{DD}} v_{out}dv_{out} = \frac{C_LV_{DD}^2}{2} \quad (3.41)$$

These results can also be derived by observing that during the low-to-high transition, C_L is loaded with a charge C_LV_{DD}. Providing this charge requires an energy from the supply equal to $C_LV_{DD}^2$ $(= Q \times V_{DD})$. The energy stored on the capacitor equals $C_LV_{DD}^2/2$. This means that only half of the energy supplied by the power source is stored on C_L. The other half has been dissipated by the PMOS transistor. Notice that this energy dissipation is independent of the size (and hence the resistance) of the PMOS device! During the discharge phase, the charge is removed from the capacitor, and its energy is dissipated in the

NMOS device. Once again, there is no dependence on the size of the device. In summary, each switching cycle (consisting of an L→H and an H→L transition) takes a fixed amount of energy, equal to $C_L V_{DD}^2$. In order to compute the power consumption, we have to take into account how often the device is switched. If the gate is switched **on and off** f times per second, the power consumption equals

$$P_{dyn} = C_L V_{DD}^2 f \tag{3.42}$$

Advances in technology result in ever-higher of values of f (as t_p decreases). At the same time, the total capacitance on the chip (C_L) increases as more and more gates are placed on a single die. Consider for instance a 1.2 μm CMOS chip with a clock rate of 100 Mhz and an average load capacitance of 30 fF/gate. The power consumption per gate for a 5 V supply then equals 75 μW. For a design with 200,000 gates, this would result in a power consumption of 15 W! This evaluation is, fortunately, somewhat pessimistic. In reality, not all gates in the complete IC switch at the full rate of 100 Mhz. The actual activity in the circuit is substantially lower. This can be accommodated in Eq. (3.42) by replacing C_L by $C_{EFF} = \alpha C_L$, where C_{EFF} (called the *effective capacitance*) is the average capacitance switched every clock cycle, and α is the activity factor of the circuit. For our example, an activity factor of 20% reduces the average consumption to 3 W.

On the other hand, the above number does not take into account the driving of the output pins of the package, which normally consumes a substantial amount of the power budget. Suppose that the above chip contains 100 output pins, each loaded with 20 pF (typical) and switched at the rate of 20 Mhz between 0 and 5 V. This results in an additional consumption of 1 W.

It should be obvious that the power problem will only worsen in future technologies. This is one of the reasons that lower supply voltages (e.g., 3 V or 3.3 V) are becoming more and more attractive. Remember that reducing V_{DD} has a quadratic effect on P_{dyn}. For instance, reducing V_{DD} from 5 V to 3 V for our example drops the power dissipation from 4 W to 1.44 W. This assumes that the same clock rate can still be sustained. Figure 3.28 demonstrates that this assumption is not that unrealistic, as the propagation delay of the CMOS gate is rather constant over this voltage range.

Consumption Due to Direct-Path Currents

In reality, the assumption of the zero rise and fall times of the input wave forms is not correct. As a result, a direct current path between V_{DD} and GND exists for a short period of time during switching, while the NMOS and the PMOS transistors are conducting simultaneously. This is illustrated in Figure 3.33. Under the reasonable assumption that the resulting current spikes can be approximated as triangles and that $V_{DD} >> |V_T|$, we can compute the energy consumed per switching period,

$$E_{dp} = V_{DD}\frac{I_{peak}t_r}{2} + V_{DD}\frac{I_{peak}t_f}{2} = \frac{t_r + t_f}{2} V_{DD} I_{peak} \tag{3.43}$$

as well as the average power consumption.

$$P_{dp} = \frac{t_r + t_f}{2} V_{DD} I_{peak} f \tag{3.44}$$

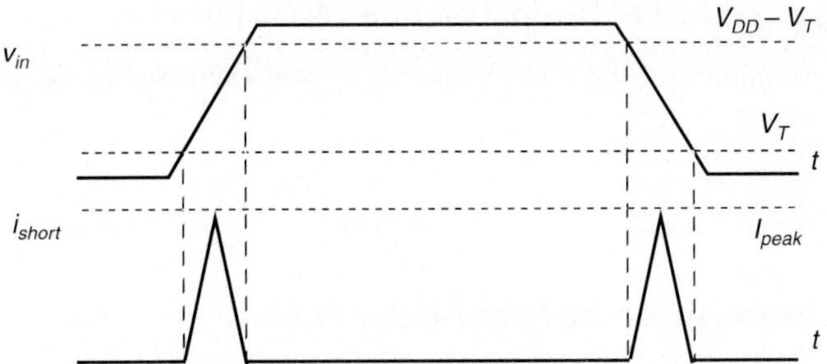

Figure 3.33 Current spiking during transients.

I_{peak} is determined by the saturation current of the devices and is hence directly proportional to the sizes of the transistors. To minimize $P_{dp,}$ slowly changing input wave forms with large values of t_f and t_r should be avoided. Unfortunately, this is not always possible, such as when driving large capacitive loads (as will be discussed in Chapter 8). In general, P_{dp} is substantially smaller than P_{dyn} and can be ignored in first-order calculations. For instance, it has been derived [Veendrick84] that if each inverter in a string of inverters is designed so that the input and output rise and fall times are equal, the short-circuit dissipation will be less than 20% of the dynamic dissipation. This topic is treated more extensively in Chapter 4.

Example 3.9 Direct Path Current in a CMOS Inverter

For example, in the inverter of Figure 3.25, the current flowing at the midpoint of the input transition (V_{in} = 2.5 V) equals 0.14 mA. This value can be extracted from Figure 3.34, where the dc value of the static current is plotted as a function of the input voltage (using SPICE). For a rise and fall time of 300 psec (equivalent to taking the input from a similar inverter), E_{spike} equals 0.21 pJ. This is about 3.5 times smaller than the dynamic energy consumed per cycle (0.75 pJ = 30 fF × 5^2).

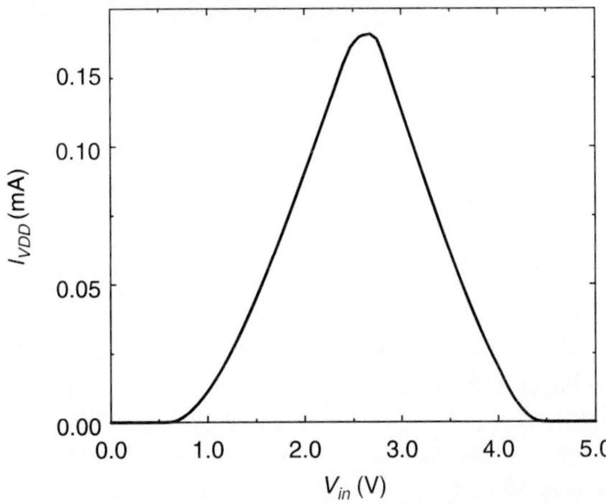

Figure 3.34 The dc supply current as a function of input voltage V_{in}.

Putting It All Together: Power Dissipation of a CMOS Inverter

The total power consumption of the CMOS inverter is now expressed as the sum of its three components:

$$P_{tot} = P_{dyn} + P_{dp} + P_{stat} = C_L V_{DD}^2 f + V_{DD} I_{peak}\left(\frac{t_r + t_f}{2}\right) f + V_{DD} I_{leak} \tag{3.45}$$

The Energy per Operation, or the Power-Delay Product

As discussed in Section 3.2.4, the power-delay product (PDP) is a quality measure for a logic gate and measures the *energy* consumed by the gate *per switching event*. A switching event is defined here as the sequence of a $0 \rightarrow 1$ and a $1 \rightarrow 0$ transition. Other textbooks consider a *single transition*, which results in a PDP of $C_L \times VDD^2/2$. The frequency f', in that case, is equal to $2f$, which results in the same average dynamic power consumption. The advantage of the definitions used here is that the high-to-low and low-to-high transitions are lumped together, and only one type of transition (typically the $0 \rightarrow 1$) has to be considered. When ignoring the contributions of the static and direct-path currents to the power consumption, this energy consumption was derived above, in Eq. (3.40), and equals

$$PDP = C_L V_{DD}^2 \tag{3.46}$$

The PDP of our example gate for a fan-out of 1 equals 0.75 pJ.

Design Challenge

It is interesting to note that the PDP of a CMOS gate is only a function of the load capacitance and the supply voltage. As it is a desirable feature to minimize the PDP, one should keep both C_L and V_{DD} as small as allowable or possible.

3.3.5 A Look into the Future: Effects of Technology Scaling

Integrated-circuit fabrication technology is continually improving. As a result, the internal dimensions of semiconductor devices are steadily decreasing. Early integrated circuits in the mid-1960s used internal dimensions in the range of 10 to 20 μm. Current state-of-the-art designs employ minimal feature sizes between 0.5 and 1.2 μm, while advanced memory structures already use 0.3 μm devices. This trend towards ever-smaller dimensions, illustrated in Figure 3.35a, is likely to continue and has not yet shown any signs of slowing down, although this is likely to happen in the future as fundamental physical (or economical!) limits are reached.

At the same time, the size of circuits that can be fabricated economically continues to increase. While the maximum size of early integrated circuits was limited to a couple of mm^2, circuits of over 400 mm^2 are currently on the market! As a result, the number of gates that can be integrated on a single chip has grown dramatically as illustrated in Figure 3.35b.

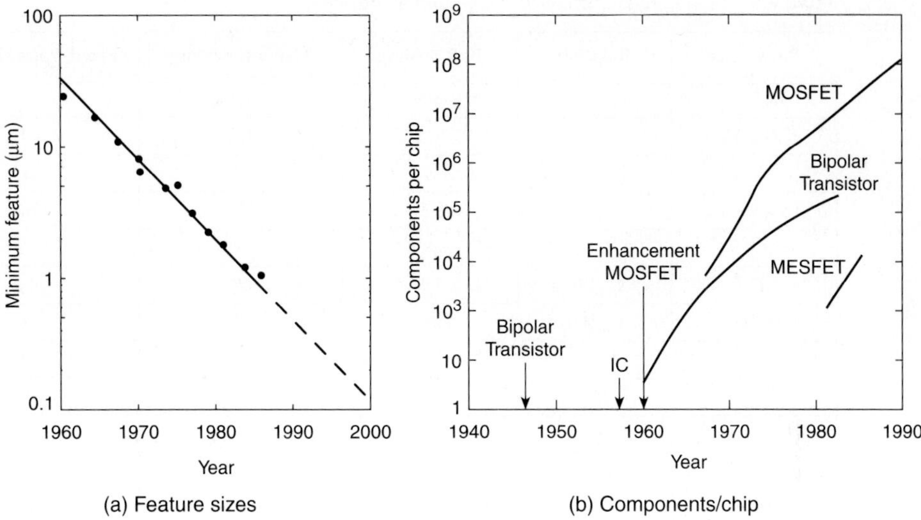

Figure 3.35 Trends in feature sizes (a) and components per chip (b) on the most advanced commercial integrated circuits (from S. M. Sze in [Watts89]).

Besides increasing the number of devices per IC, scaling has a profound effect on the performance of a circuit in terms of speed and power. In this section, we will examine first-order changes in the CMOS gate characteristics as a function of the internal device dimensions. Two different models can be used to study the effects of scaling [Dennard74, Baccarani84].

Full Scaling (Constant Electrical Field Scaling)

In this ideal model, all the dimensions of the MOS devices, the voltage supply level and the depletion widths are scaled by the same factor S. The goal is to keep the electrical field patterns in the scaled device identical to those in the original device. Keeping the electrical fields constant ensures the physical integrity of the device and avoids breakdown or other secondary effects. This scaling leads to greater device density, higher speed, and reduced power consumption. The performance is improved due to the reduced capacitances as well as the reduced voltage swing. The effects of full scaling on the device and circuit parameters are summarized in the third column of Table 3.5. The results clearly demonstrate the beneficial effects of scaling—the speed of the circuit increases in a linear fashion, while the power-delay product is reduced in a cubic fashion![4]

[4] Some assumptions were made when deriving this table:

1. It is assumed that the carrier mobilities are not affected by the scaling.
2. The substrate doping N_{sub} is scaled so that the maximum depletion-layer width is reduced by a factor S.
3. It is furthermore assumed that the propagation delay of the device is mainly determined by the intrinsic capacitance (the gate capacitance) and that other device capacitances, such as the diffusion capacitances, scale appropriately. This assumption is approximately true for the full-scaling case, but not for fixed-voltage scaling, where C_{diff} scales as $1/\sqrt{S}$.

Table 3.5 Scaling relationships for long-channel devices.

Parameter	Relation	Full Scaling	General Scaling	Fixed-Voltage Scaling
W, L, t_{ox}		$1/S$	$1/S$	$1/S$
V_{DD}, V_T		$1/S$	$1/U$	1
N_{SUB}	V/W_{depl}^2	S	S^2/U	S^2
Area/Device	WL	$1/S^2$	$1/S^2$	$1/S^2$
C_{ox}	$1/t_{ox}$	S	S	S
C_L	$C_{ox}WL$	$1/S$	$1/S$	$1/S$
k_n, k_p	$C_{ox}W/L$	S	S	S
I_{av}	$k_{n,p}V^2$	$1/S$	S/U^2	S
J_{av}	I_{av}/Area	S	S^3/U^2	S^3
t_p (intrinsic)	$C_L V / I_{av}$	$1/S$	U/S^2	$1/S^2$
P_{av}	$C_L V^2 / t_p$	$1/S^2$	S/U^3	S
PDP	$C_L V^2$	$1/S^3$	$1/SU^2$	$1/S$

In reality, full scaling is not a feasible option. First of all, to keep new devices compatible with existing components, voltages cannot be scaled arbitrarily. Providing multiple voltage supplies is expensive. As a result, voltages have not been scaled down along with feature sizes, and designers adhere to well-defined standards for supply voltages and signal levels. The majority of integrated circuit are still being designed with a 5 V supply, although new standards such as 3 V and 3.3 V are making rapid inroads. Some of the most advanced devices already use voltages around 1.5 V (for a 0.4 μm feature size; these voltages are only used internally on the chip, not for input-output signals). Secondly, some of the device voltages such as the silicon bandgap and the built-in junction potential, are material parameters and cannot be scaled. Finally, the scaling potential of the transistor threshold is limited. Making the threshold too low makes it difficult to turn off the devices completely. This is aggravated by the large process variation of the value of the threshold, even on the same wafer. Therefore, a more general scaling model is needed, where dimensions and voltages are scaled independently.

General Scaling

The general scaling model is shown in the fourth column of Table 3.5. Here, device dimensions are scaled by a factor S, while voltages are scaled by a factor U. When the voltage is held constant (which has been the standard scenario in the last decade), $U = 1$, and the scaling model is reduced to the model shown in the last column of Table 3.5. Note that in this case the currents are increased by a factor of S. This results in a rather impressive speed-up ($1/S^2$) at the expense of increased power consumption (S). The latter observation is one of the key reasons why lower supply voltages are unavoidable.

Scaling for Submicron Devices

It should be emphasized that the above model is an oversimplification and that a number of important second-order effects are ignored. For example, mobility degradation, velocity saturation, drain-induced barrier lowering, and series resistances are ignored. Velocity saturation, in particular, can have a profound effect on the scaling behavior, as it reduces the voltage dependency of I_{av} from quadratic to linear. As a result, the decrease in propagation delay with S will not be as large as predicted in Table 3.5. The scaling picture for short-channel devices is summarized in Table 3.6. Only the entries that differ from the long-channel case are provided. Notice that the propagation delay (to a first order) is independent of the voltage.

Table 3.6 Scaling relationships for short-channel devices.

Parameter	Relation	Full Scaling	General Scaling	Fixed-Voltage Scaling
I_{av}	$C_{ox}WV$	$1/S$	$1/U$	1
J_{av}	I_{av}/Area	S	S^2/U	S^2
t_p (intrinsic)	C_LV / I_{av}	$1/S$	$1/S$	$1/S$
P_{av}	C_LV^2 / t_p	$1/S^2$	$1/U^2$	1

Example 3.10 Scaling of a CMOS Inverter

Consider the inverter introduced in Example 3.7. The propagation delay of the gate was determined to be 225 psec, while the PDP was evaluated at 0.75 pJ. When scaling the technology to 0.6 μm (factor 2), the propagation delay will scale to 56 nsec, and the PDP will reduce to 0.375 pJ, if the supply voltage is kept at 5 V. When the voltage is reduced to 3 V, on the other hand, the propagation delay only reduces to 94 nsec, but a reduction of the PDP by a factor of 5.5 is obtained! Velocity-saturation effects might make this picture less optimistic, though.

Some of the conclusions of this section are summarized in Figure 3.36, where the propagation delays and PDPs for a number of industrial processes are plotted as functions of minimum feature lengths.

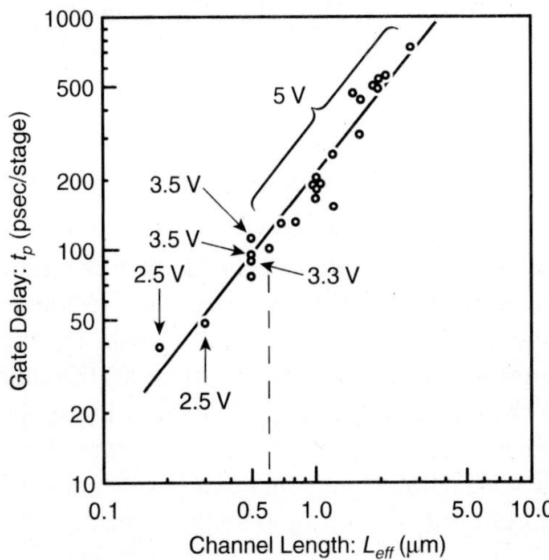

Figure 3.36 Propagation delay as a function of minimum feature length (from [Kakumu90]). It should be noted that the supply voltage is reduced for feature lengths below 0.6 μm.

3.4 The Bipolar ECL Inverter

While we learned in the previous section that the CMOS inverter has an almost ideal dc characteristic and low power consumption, it also became clear that the speed of the inverter is restricted by the fact that the maximum current is proportional to V_{DD}^2 (or V_{DD} under velocity-saturation conditions). This puts a firm upper bound on the achievable clock speeds. The bipolar device, on the other hand, is known to have an exponential relationship between collector current and base-emitter voltage, which means that a small change in voltage can provide a large current. It is therefore reasonable to assume that bipolar gates should be capable of achieving superior propagation delays.

Although the TTL gate (transistor-transistor-logic) has long been the flagship of the bipolar digital logic gates, its importance has been reduced dramatically in the last decade as CMOS has become more and more competitive. Most modern bipolar digital designs are implemented in a circuit style called emitter-coupled logic (or ECL). We will, therefore, concentrate on this logic family. Following the scheme of the CMOS discussion, we will first describe the ECL gate in a qualitative fashion, followed by a quantitative analysis of its static and dynamic behavior, as well as the power dissipation. A discussion of the effects of technology scaling concludes the section.

3.4.1 Issues in Bipolar Digital Design: A Case Study

Before engaging in a discussion of the ECL gate, some issues in the design of bipolar gates should be highlighted. The particular nature of the bipolar junction transistor and the lack of a good complementary device in most standard bipolar manufacturing processes translate into gate topologies that are dramatically different from the CMOS structures. The most important features can be summarized as follows:

- The high transconductance of the bipolar devices translates to a large variation in collector current for a small change in the input voltage.
- The input resistance of the transistor is finite. This means that a fan-out gate presents a current-sink for the driving gate. This is in contrast to CMOS designs, where fan-out presents only a capacitive load.
- The presence of an excessive amount of base-charge makes the transistor very slow when operated in the saturation mode. Saturating the device should, therefore, be avoided as far as possible.

To illustrate these observations, let us first examine a simple bipolar gate, called the *RTL* (*resistor-transistor logic*) gate. A schematic of such a structure is shown in Figure 3.37. It can be readily seen from the circuit that with the input voltage V_{in} less than the turn-on voltage $V_{BE(on)}$ of the transistor, the transistor is in cut-off mode, the collector current I_C essentially equals zero, and the output voltage V_{out} is equal to the supply voltage V_{CC}. When the input voltage is increased above $V_{BE(on)}$, the transistor turns on and enters the forward-active mode. Further increasing the input voltage causes the output voltage to drop due to the increasing collector current I_C. This drop is swift, as the bipolar transistor is characterized by a *large gain in the forward-active mode*. With sufficient input voltage, when the output voltage has fallen sufficiently, the transistor will enter the saturation region. The output voltage V_{out} remains fixed at $V_{CE(sat)} \approx 0.1$ V, the saturation collector-emitter voltage. To enter this mode of operation, we require (as a first approximation) that $V_{in} \geq V_{BE(sat)}$ (≈ 0.8 V). Observe that the above analysis uses the simple model for manual analysis of the bipolar transistor, presented in Figure 2.42.

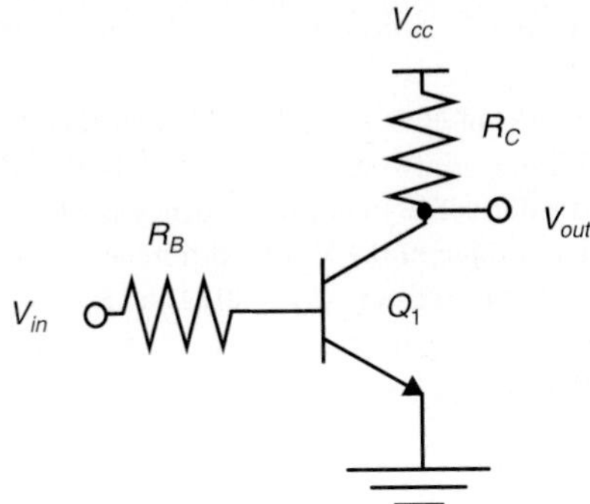

Figure 3.37 Resistor-transistor logic (RTL) inverter.

From the above, it becomes obvious that the gate of Figure 3.37 acts as an inverter. A rough sketch of the voltage-transfer characteristic is shown in Figure 3.38, where linear interpolations are used to join the two major breakpoints of the graph.

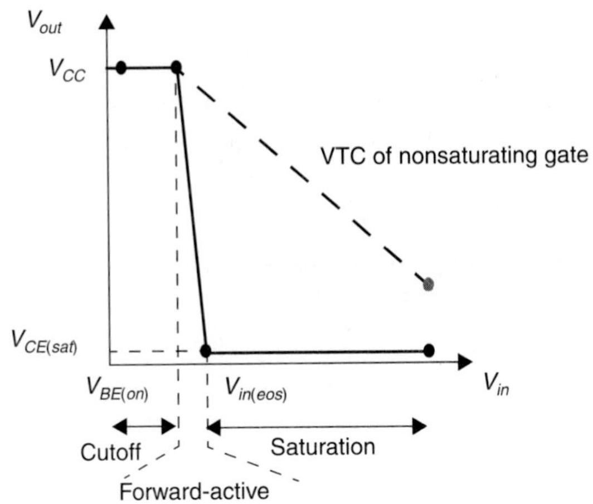

Figure 3.38 Approximated voltage transfer characteristic of inverter.

Example 3.11 VTC of an RTL Inverter

The VTC of an RTL inverter is derived for the following component and parameter values: $V_{CC} = 5$ V, $R_C = 1$ kΩ, and $R_B = 10$ kΩ. The bipolar transistor model presented in Chapter 2 is used for Q_1. Initially, it is assumed that the fan-out of the inverter is set to 0.

- For $V_{in} < V_{BE(on)}$, Q_1 is in cut-off, $I_C = 0$, and $V_{out} = V_{OH} = V_{CC}$.
- In the low output mode, Q_1 is in saturation (or $V_{out} = V_{CE(sat)} \approx 0.1$ V), if the condition $V_{BE} \geq V_{BE(sat)}$ is fulfilled. The boundary condition on V_{in} for this to be valid can be derived by analyzing the circuit with Q_1 at the edge of saturation. This translates to the following condition on V_{in},

$$I_B = \frac{I_C}{\beta_F} = \frac{V_{CC} - V_{CE(sat)}}{R_C \beta_F}$$

$$V_{in(eos)} = V_{BE(eos)} + I_B R_B = V_{BE(eos)} + \frac{R_B}{R_C}\left(\frac{V_{CC} - V_{CE(sat)}}{\beta_F}\right) \tag{3.47}$$

Assuming $V_{CE(sat)} = 0.1$ V, $V_{BE(eos)} = 0.8$V, and $\beta_F = 100$, Eq. (3.47) evaluates to $V_{in(eos)} =$ 1.29 V.

The simulated VTC of the RTL inverter is plotted in Figure 3.39 (consider only the case of zero fan-out at present). The observed breakpoints equal approximately 0.7 V ($\sim V_{BE(on)}$) and 1.4 V, which are close to the predicted values. The major discrepancy is the value of V_{OL}, which equals 0.5 V in contrast to the predicted value of 0.1 V. This difference is caused by the nonzero value of the series collector and emitter resistances ($R_E = 20\ \Omega$ and $R_C = 75\ \Omega$). In the saturation mode, the collector current I_C equals approximately $V_{CC}/R_C = 5$ mA. This causes an extra voltage drop of $5 \times 10^{-3}(75 + 20) = 0.48$ V over the collector-emitter terminals of Q_1.

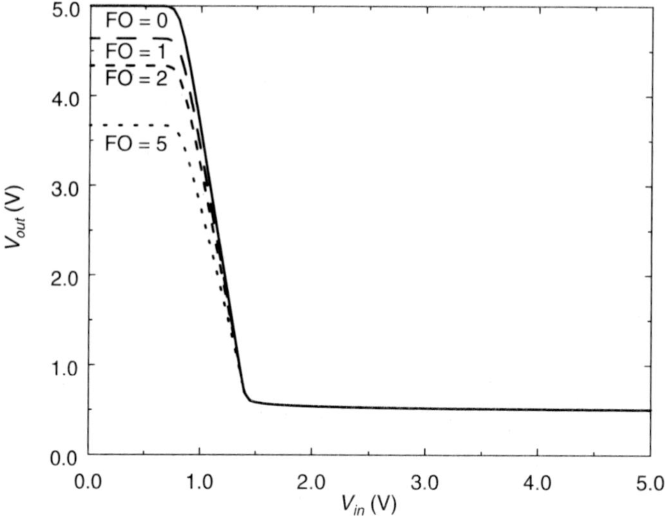

Figure 3.39 Simulated voltage-transfer characteristic of RTL inverter for different fan-out configurations.

It can be observed that the RTL inverter displays some major **dc problems**:

1. The VTC is *asymmetrical*. The NM_L is less than or equal to 0.2 V. Noise and disturbances on the *GND* line or input signal could easily result in faulty operation.

2. To obtain a reasonable low value of V_{OL}, it is necessary to drive the *transistor into saturation*. From the earlier remarks, this can be labeled as unfortunate as the excessive base charge has a significant impact on switching performance. One option is to choose the resistor values so that saturation never occurs for the voltage range of interest, which is equivalent to requiring that $V_{in(eos)} > V_{CC}$ in Eq. (3.47). This

approach causes a deterioration of V_{OL} and reduces the gain in the transient region. The corresponding VTC is plotted in shaded lines in Figure 3.38.

3. The *output impedance* of the gate is equal to R_C when the output is *high*. Reducing R_C is not a reasonable option, as this raises the value of the collector current I_C and increases power consumption and switching noise.[5] Furthermore, in contrast to the CMOS inverter, the gate has a finite input impedance. When Q_1 is on, current flows into its base through R_B. The V_{OH} of the inverter (and the NM_H) is, hence, sensitive to loading as is illustrated in Figure 3.40. This effectively restricts the fan-out of the inverter to approximately three gates.

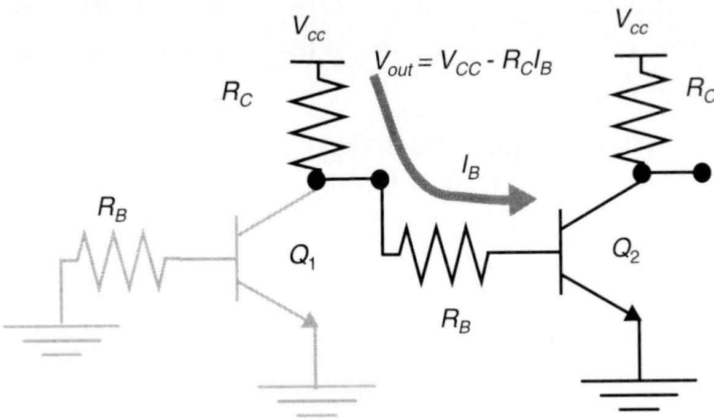

Figure 3.40 Effect of fan-out on V_{OH}.

Example 3.12 Effect of Fan-out on the VTC of the RTL Inverter

Suppose that the output of the RTL gate of the previous example is connected to N identical gates and that its output is high. In this mode, each of the connecting gates will sink an amount of current I_{in} into its base as illustrated in Figure 3.40.

$$I_{in} = \frac{V_{out} - V_{BE(sat)}}{R_B} \tag{3.48}$$

This results in the following value of V_{out} as a function of the fan-out N:

$$\begin{aligned} V_{out} &= V_{CC} - N \cdot R_C \cdot I_{in} = V_{CC} - N \cdot R_C \cdot \frac{V_{out} - V_{BE(sat)}}{R_B} \\ &= \frac{V_{CC} + N(R_C/R_B)V_{BE(sat)}}{1 + N(R_C/R_B)} \end{aligned} \tag{3.49}$$

For large values of N, V_{out} ($= V_{OH}$) will eventually approach $V_{BE(sat)} \approx 0.8$ V, which means that the NM_H is reduced to zero (or is even negative)! For smaller values of N, the situation is not as bad, however. For instance, $N = 5$ results in an output voltage of $V_{OH} = 3.6$ V, as is confirmed by the simulations of Figure 3.39.

---

[5] *Switching noise* is the noise present on the supply rails resulting from the switching of large currents in the presence of parasitic resistances and inductors.

The main problem of the RTL gate is, however, evident in its **transient behavior**. Consider, for instance, the t_{pLH}. The propagation delay can be decomposed into two elements:

1. Initially, Q_1 is in deep saturation. Turning off the device requires the removal of the base charge (through R_B). This is a slow operation.
2. As in the CMOS case, a major part of the switching time consists of charging up the load capacitance C_L from V_{OL} to V_{OH}. In the RTL case, this has to happen through the load resistance R_C, and the delay is determined by the time-constant $R_C C_L$. Making R_C small increases $I_{C(sat)}$ [$= (V_{CC} - V_{CE(sat)})/R_C$]. This raises both the power consumption as well as the first component of the propagation delay. Indeed, the charge stored in the base region is proportional to the collector current ($I_C = Q_F/\tau_F$, as discussed in Chapter 2).

Similar arguments can be made regarding t_{pHL}. The situation is not as bad in this case, as C_L is discharged by Q_1 instead of through a resistor. This is considerably faster. The dominant part of the propagation delay is the buildup of the base charge to bring Q_1 into saturation. Observe that avoiding saturation is not an option, since it results in a substantial degradation of the dc behavior.

Finally, note that the gate consumes static power when the output is low, because a direct path between V_{CC} and GND exists through the resistor and transistor.

Example 3.13 Transient Response of the RTL Inverter

The transient response of the RTL inverter of the previous example (for a fan-out of 1) is simulated using SPICE and is shown in Figure 3.41 (a manual analysis of the transient behavior of a bipolar circuit is complex and is, therefore, delayed until later in the chapter). The

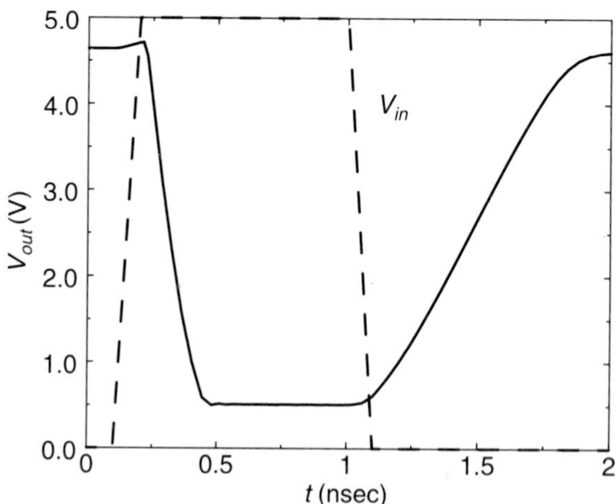

Figure 3.41 Transient response of an RTL inverter (for a fan-out of 1).

response is clearly asymmetrical: $t_{pLH} = 420$ psec, and $t_{pHL} = 165$ psec. The resulting propagation delay is equal to 292.5 psec, which is inferior to the performance of the CMOS inverter! An important part of this delay is due to the base-charge buildup and removal. A fast bipolar gate should avoid having its transistors going into saturation, since this is where the major base-charge buildup happens.

In summary, a more effective bipolar gate should address the dc and ac issues raised above. In the next section, we discuss how this is accomplished in the ECL gate.

3.4.2 The Emitter-Coupled Logic (ECL) Gate at a Glance

The ECL gate differs fundamentally from both CMOS and RTL inverters. It has been conceived with *utmost performance* in mind. One of the means to accomplish this is to keep the logic swing low. The ECL gate, hence, operates typically with a swing of only 0.5 V (compare this to the 3 to 5 V CMOS structures). Ensuring a reasonable noise margin under those conditions is nontrivial and requires careful circuit design.

The resulting gate is rather complex and is a composition of three components: *the differential pair, the output driver, and the bias network*. Each of these affects one particular aspect of the gate functionality and performance.

The Differential Pair (or Current Switch)

The function of the input stage of the ECL gate is to provide maximal noise margins, low noise sensitivity, and fast switching in the presence of a *small logic swing*. A schematic of this portion of an ECL gate is plotted in Figure 3.42. This structure is well known in the

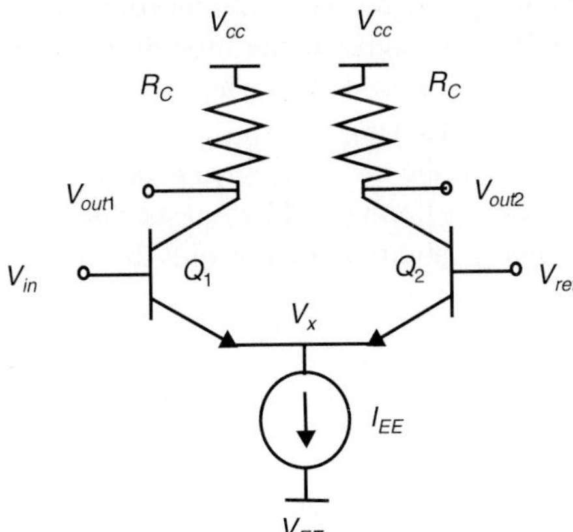

Figure 3.42 The differential pair.

analog world, where it is called an *emitter-coupled pair*. One of the two inputs is connected to a reference voltage V_{ref} that is generated by the bias network (discussed later). Assuming that both transistors Q_1 and Q_2 are identical, the circuit is biased in a completely symmetrical fashion if V_{in} is set equal to V_{ref} as typically would be the case in an analog design: $V_{be1} = V_{be2}$, $I_{C1} = I_{C2} = I_{EE}/2$ and $V_{out1} = V_{out2} = V_{CC} - R_C I_{EE}/2$. This bias condition turns out to be rather useless for digital purposes, where a key requirement is that two easily distinguishable operation modes are produced.

Let us now increase (decrease) V_{in} by a small amount with respect to V_{ref}. This changes the current balance in the differential pair and increases (decreases) the fraction of

I_{EE} that is routed through Q_1. A critical observation is that this current fraction is an *exponential function* of the voltage difference between V_{in} and V_{ref}. This is expressed in Eq. (3.50), which is based on the familiar collector-current equation of the bipolar transistor (2.63) (with ϕ_T equal to kT/q, the thermal voltage).

$$\frac{I_{C1}}{I_{C2}} = \frac{e^{\frac{V_{in} - V_x}{\phi_T}} - 1}{e^{\frac{V_{ref} - V_x}{\phi_T}} - 1} \approx e^{\frac{V_{in} - V_{ref}}{\phi_T}} \tag{3.50}$$

At room temperature, **a 60 mV increase in V_{in} causes a tenfold increase of I_{C1}/I_{C2}**, while a 120 mV increase is sufficient to set I_{C1} to 100 I_{C2}. With I_{C2} only 1% of I_{C1}, it can be stated that, for all practical purposes, Q_2 is essentially off, and that Q_1 is carrying all of I_{EE}. As a result, approximate values for V_{out1} and V_{out2} can be derived.

$$\begin{aligned} V_{out1} &= V_{CC} - I_{EE}R_C \quad \text{(logic low)} \\ V_{out2} &= V_{CC} \quad \text{(logic high)} \end{aligned} \tag{3.51}$$

A similar argument can be used to analyze the case where V_{in} drops below V_{ref}. Here, all the current is diverted to Q_2, resulting in a high V_{out1} and a low V_{out2}. A number of conclusions can be drawn from this simple analysis:

1. The ECL gate provides *differential (or complementary) outputs*. Both the output signal (V_{out1}) and its inverted value (V_{out2}) are simultaneously available. This is a distinct advantage, as it eliminates the need for an extra inverter to produce the complementary signal. This also prevents some of the time-differential problems introduced by additional inverters. For example, in logic design it often happens that both a signal and its complement are needed simultaneously. When the complementary signal is generated using an inverter, the inverted signal is delayed with respect to the original (Figure 3.43a). This causes timing problems, especially in very high-speed designs. The differential output capability of ECL avoids this problem (Figure 3.43b).

2. Another advantage of the differential operation is an *increased noise immunity*, as common mode noise signals—i.e., signal disturbances common to both V_{in} and V_{ref}—are rejected to a large degree.

3. The *transition region* from one logic state to the other is *centered around* V_{ref} and is, to the first order, independent of the transistor parameters. To determine the boundaries of the transition region, an often-used simplification is to place them at the points where the current ratio drops to 1%. This is only a qualitative approximation, which does not strictly follow the definitions of V_{IH} and V_{IL} (gain = −1). A more detailed analysis is needed to produce precise values and will follow later. Using the simplified definition, the transition width evaluates to approximately 240 mV.

$$\begin{aligned} V_{IL} &= V_{ref} - 120mV \\ V_{IH} &= V_{ref} + 120mV \end{aligned} \tag{3.52}$$

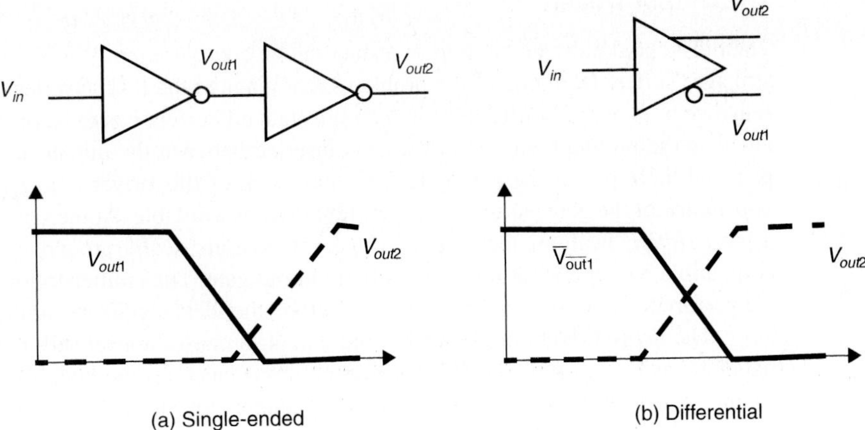

Figure 3.43 Advantage of differential (b) over single-ended (a) gate.

In the case of the static CMOS and the RTL gates, the logic swing approximates the voltage difference between the supply rails. This is not so for the ECL gate, where the difference between V_{OH} and V_{OL} is determined by the bias current I_{EE} and the load resistance R_C, as was illustrated in (3.51). The reduced voltage swing results in a faster operation, as the time to (dis)charge the load capacitor is reduced.

4. The *current drawn from the supply is always constant* and equal to I_{EE}. Therefore, the differential pair consumes a constant amount of static power equal to $I_{EE}(V_{CC} - V_{EE})$. This static power consumption precludes the use of ECL for the design of very complex VLSI circuits (> 100,000 gates), as the removal of the excessive heat becomes either impossible or excessively expensive. Therefore, the use of ECL-style circuits is mostly confined to the design of very high-speed building blocks such as CPUs for mainframe computers.

   On the other hand, the fact that the current drawn from the supply is constant and independent of the switching operation reduces the amount of switching noise on the supply lines. This is further aided by the low logic swing, which reduces the current spikes caused by charging and discharging the load capacitance. As a result, ECL circuits can operate reliably, even with small noise margins.

   Furthermore, it can be observed that the output voltages (V_{out1} and V_{out2}) are defined with respect to V_{CC} and are, to a first degree, not influenced by the value of V_{EE}. Noise on the latter supply therefore has virtually no effect on the signal levels. On the other hand, V_{CC} has to be as clean as possible.

5. With a proper choice of the bias current, the input voltage levels, and the resistor values, it can be assured that *neither Q_1 nor Q_2 ever goes into saturation*. Actually, in normal operation, both transistors are always forward-active. This speeds up the device operation, as the time to alter the base charge is drastically reduced.

The Output Driver

The differential pair can be used in a stand-alone configuration. While it is very fast, this structure suffers from one of the problems that plagued the RTL inverter. Its VTC is very sensitive to *output loading*, since in the high state all current has to be provided by R_C. To alleviate the problem, an output driver is inserted between the outputs of the differential pair and the input of the next gate. The main task of this driver is to *reduce the output impedance* of the gate, so that ample current drive is available. At the same time, it acts as a *level shifter*. It aligns the output logic levels symmetrically around V_{ref}, making them compatible with the input logic levels of a fan-out gate. The emitter-follower circuit module performs all these functions and is, therefore, the ideal candidate for the output driver.

The revised circuit is shown in Figure 3.44. It causes a level shift on the output signal equal to $V_{BE(on)}$. It also reduces the output impedance for the high logic level from R_C to approximately $R_C/(\beta_F+1)$, thus providing ample current-drive capability (with β_F the forward-current gain of the bipolar device). To complete the circuit, a resistor R_B has been added at the input terminal. R_B serves as a current path to V_{EE} from the base of Q_1 or Q_2. This helps to speed up the removal of the base charge Q_F during the switching of the differential pair. For a very high speed design, R_B also acts as termination for the transmission line formed by the interconnect. We will discuss this in depth in Chapter 8.

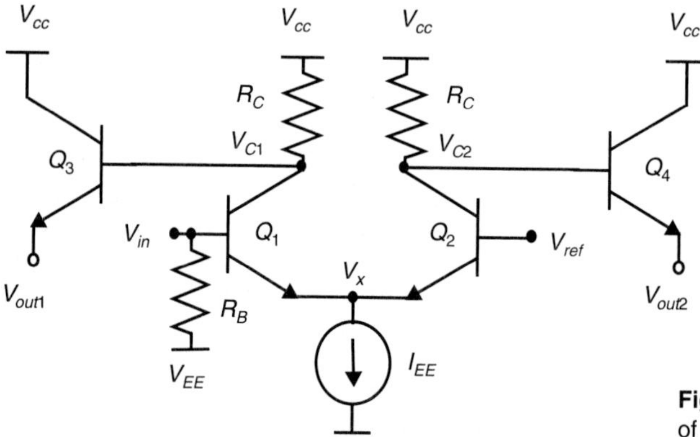

Figure 3.44 ECL gate consisting of differential pair and emitter-followers.

Design Consideration

The differential pair draws a constant current from the supply. This is obviously not true for the emitter-followers, whose current levels are substantially different between the high and the low output states. Large current peaks occur during the switching events, while (dis)charging the load capacitances. This induces large fluctuations in the supply voltages and can cause malfunctioning of the circuit, given the small noise margins. To avoid this cross-coupling, a common practice is to provide a *separate V_{CC} supply for the output drivers.*

The Bias Network

The purpose of the bias is to generate the reference voltage V_{ref}. The prime requirement for this reference is that it is to be centered as much as possible in the middle of the logic swing, independent of the operating temperature. As was mentioned in Chapter 2, the voltage drop over a forward-biased base-emitter junction is a strong function of temperature. Typically, V_{BE} changes linearly with the temperature, or $\Delta V_{BE} = -k\Delta T$, with k approximately equal to 1.5 mV/°C. This drift causes the voltage-transfer characteristic of the ECL gate and its noise margins to vary as a function of temperature as well. The ambient temperature in ECL devices can be quite high, due to the large power dissipation of the gate. Therefore, the bias network has to be designed so that V_{ref} tracks those changes and the noise margins are kept symmetrical over a wide range of temperatures. An example of such a network is shown in Figure 3.45. The value of the reference voltage is easily derived.

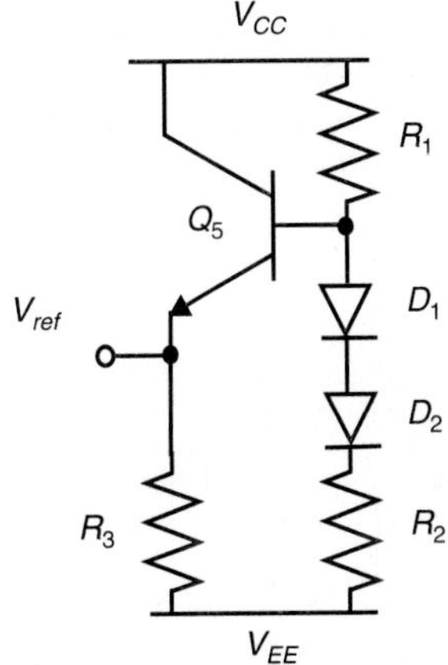

Figure 3.45 Voltage reference circuit (as used in the ECL 10K series from Motorola)

$$
\begin{aligned}
V_{ref} &= V_{B5} - V_{BE(on)} \\
V_{B5} &= V_{CC} - \frac{R_1}{R_1 + R_2}(V_{CC} - 2V_D - V_{EE})
\end{aligned}
\tag{3.53}
$$

Transistor Q_5 is connected in the emitter-follower configuration, reducing the output impedance of the reference network. Changes in temperature affect the $V_{BE(on)}$ of Q_5 and Q_2. These changes are tracked to a first order by the change in the voltage drops across diodes D_1 and D_2. Actual implementations of this circuit have shown that it ensures the centering of V_{ref} between V_{OH} and V_{OL} for a temperature range from –30°C to +85°C.

In advanced ECL configurations, the bias network is also used to generate other reference voltages, such as the bias voltage for the current source. Fortunately, a single bias network can serve multiple gates, reducing the overall power consumption and the area overhead.

After this qualitative analysis of the structure and properties of the basic ECL gate, we will now derive the quantitative dc and ac characteristics of the gate. Before proceeding, we would like to mention that there exist many different ECL families. The most famous are the ECL 10K and 100K series and the MECL I, II, and III circuits [Hodges88]. The major differences between these are the values of the resistors, the type of current source used and the construction of the bias network. Other variants of the gate are used in modern, integrated ECL circuits used in instrumentation circuits, CPUs, and even microprocessors (see, for example, [Jouppi93]). Instead of studying all those versions separately, we discuss only one single gate structure in detail. Similar approaches can then be employed to analyze the characteristics of related structures or even to design custom ECL gates, whose current levels are optimized for the required performance or power consumption.

3.4.3 Robustness and Noise Immunity: The Steady-State Characteristics

This section will analyze, in detail, how a typical ECL gate structure, as shown in Figure 3.46, can operate with a low voltage swing, and yet maintain sufficient noise immunity. The gate topology is similar to the one presented before, with the exception of the current source. The latter is implemented as a resistance R_S in series with a transistor Q_6, whose base is dc-biased at V_{CS}. To a first order, this produces the following bias current,

$$I_{EE} = \frac{(V_{CS} - V_{BE(on)} - V_{EE})}{R_S} \tag{3.54}$$

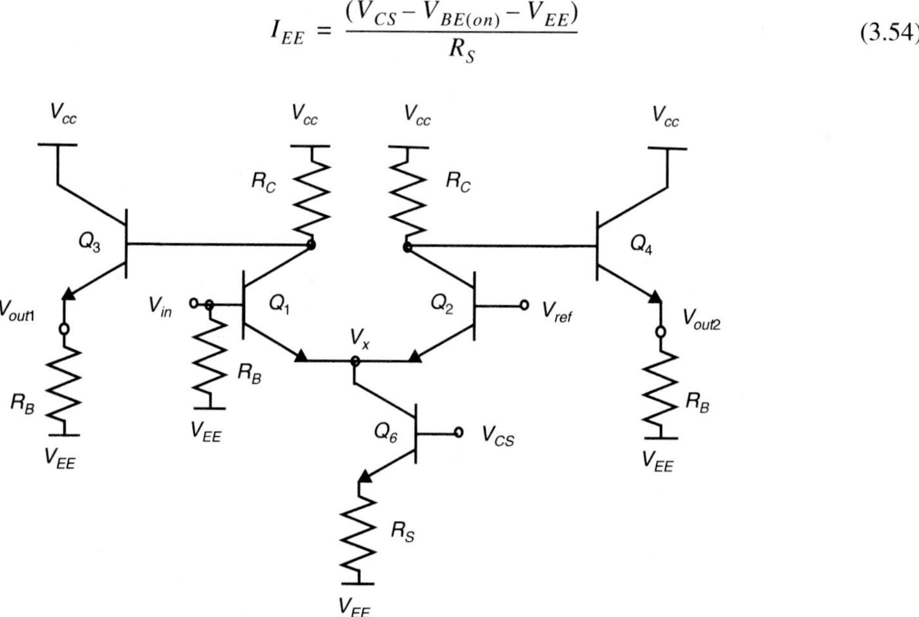

Figure 3.46 ECL gate with transistor/resistor current source.

For a manual analysis, we may assume that I_{EE} is constant over the range of input voltages. We will derive the VTC assuming a fan-out of 1, or, in other words, with the output stage loaded with the resistance R_B.

Suppose first of all that all transistors are operating in the forward-active region. Under those conditions, the following circuit equations are valid:

$$\frac{I_{C1}}{I_{C2}} = e^{\frac{V_{in} - V_{ref}}{\phi_T}}$$

$$\begin{aligned} I_{C1} + I_{C2} &= I_{EE}\frac{\beta_F}{\beta_F + 1} \approx I_{EE} \\ V_{C1} &= V_{CC} - R_C(I_{C1} + I_{B3}) \\ V_{out1} &= V_{C1} - V_{BE(on)} \\ I_{B3} &= \frac{V_{o1} - V_{EE}}{(\beta_F + 1)R_B} \end{aligned} \tag{3.55}$$

These equations can be solved to yield the following equations for I_{C1} and V_{out1} (similar equations can be derived for I_{C2} and V_{out2}).

$$I_{C1} = \frac{I_{EE}e^x}{1 + e^x} \tag{3.56}$$

with $x = (V_{in} - V_{ref})/\phi_T$, and

$$\begin{aligned} V_{out1} &= \frac{(\beta_F + 1)R_B}{(\beta_F + 1)R_B + R_C}\left[V_{CC} - V_{BE(on)} - R_C I_{C1} + \frac{V_{EE}R_C}{(\beta_F + 1)R_B}\right] \\ &\approx V_{CC} - V_{BE(on)} - R_C I_{C1} \qquad (\text{for } R_B(\beta_F + 1) \gg R_C) \end{aligned} \tag{3.57}$$

The important parameters of the voltage-transfer characteristic can now be derived.

Nominal Voltage Levels

For $V_{in} >> V_{ref}$, $I_{C1} = I_{EE}$, and $V_{OL} \approx V_{CC} - V_{BE(on)} - R_C I_{EE}$. For $V_{in} << V_{ref}$, $I_{C1} \approx 0$, and $V_{OH} \approx V_{CC} - V_{BE(on)} - V_b$. V_b is the voltage drop over R_C, caused by the base current of Q_3 (= $R_C I_{B3}$). This factor can normally be ignored for manual analysis. The logic swing of the gate hence equals

$$V_{swing} = R_C I_{EE} \tag{3.58}$$

Noise Margins

Deriving exact expressions for V_{IH} and V_{IL} is somewhat more involved, due to the heavily nonlinear current relations. As mentioned earlier, one common approach to circumvent this analysis is to use revised definitions of V_{IH} and V_{IL}. Specifically, V_{IL} is defined as the point where transistor Q_1 carries 1% of I_{EE}, while at V_{IH} Q_1 carries 99% of the total current. The corresponding values of V_{in} can be derived from Eq. (3.56):

$$\begin{aligned} \frac{I_{C1}}{I_{EE}} &= \frac{e^x}{1 + e^x} = \alpha = 0.01 \\ V_{IL, IH} &= V_{ref} \pm \phi_T ln\left(\frac{\alpha}{1 - \alpha}\right) \end{aligned} \tag{3.59}$$

At room temperature ($\phi_T = 26$ mV), we find that $V_{IL} = V_{ref} - 120$ mV and $V_{IH} = V_{ref} +$ 120 mV. This results in a very narrow transition region of 240 mV.

The results ensuing from this approach tend to differ considerably from those obtained using the traditional unit gain definition. Consequently, it is worthwhile to derive more accurate expressions. In correspondence to earlier derivations, we use the small-signal approach. A number of simplifications have to be considered to make the analysis tractable.

- Combining the small-signal models of the differential pair and the emitter-follower yields rather intractable equations. The model can be simplified in an important way by observing that the input impedance of the emitter-follower is approximately equal to its load impedance, multiplied by ($\beta_F + 1$). This technique, illustrated in Figure 3.47, is called the *resistance reflection rule*[6] and will be quite useful for the computation of the propagation delay of the ECL gate later in the chapter. Furthermore, the gain of the emitter-follower is approximately equal to 1 and will have virtually no effect on the overall gain of the gate. Using this approach, we can eliminate the emitter-follower from the small-signal model and replace it by a resistor $R'_B = R_B(\beta_F + 1)$ in parallel with R_C.

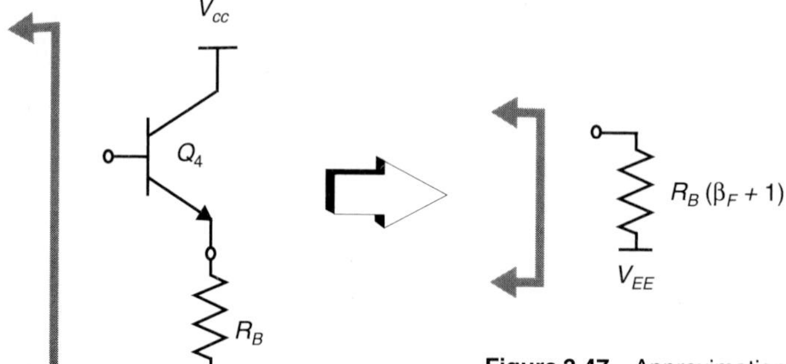

Figure 3.47 Approximation of the loading effect of the emitter-follower using the reflection rule.

- We assume that, at V_{IH} and V_{IL}, the differential amplifier is sufficiently balanced so that the small-signal gain of the amplifier approximates the gain of the amplifier under symmetrical biasing conditions. Although the actual situation deviates substantially from this assumption (at V_{IH} and V_{IL}, an important imbalance exists between I_{C1} and I_{C2}), it produces a reasonable approximation useful for manual analysis.

Under those assumptions the gain of the amplifier equals $g = -g_m R_L/2$, with R_L the load resistance of the amplifier ($= R_C \parallel R_B(\beta_F + 1)$) and g_m the transconductance of Q_1

[6] For an in-depth derivation of this rule, please refer to textbooks on analog circuit design, such as *Microelectronic Circuits* by Sedra and Smith [Sedra87, p. 457].

$(= I_{C1}/V_T)$.[7] The values of V_{IH} and V_{IL} can now be found by setting g to –1 and plugging in the appropriate value of I_{C1} (Eq. (3.56)).

$$V_{IH,IL} = V_{ref} \pm V_T \ln\left(\frac{I_{EE}(R_C||R_B(\beta_F + 1))}{2\phi_T} - 1\right) \approx V_{ref} \pm \phi_T \ln\left(\frac{V_{swing}}{2\phi_T} - 1\right) \quad (3.60)$$

As expected, V_{IH} and V_{IL} differ only a couple of ϕ_T (= 26 mV at room temperature) from V_{ref} (for V_{swing} varying between 250 mV and 1V). Their values depend on the value of the logic swing $I_{EE}R_C$ in a logarithmic fashion, or, in other words, they are rather insensitive to voltage variations. It is worth observing that a more precise analysis, which does not employ the second assumption, yields only a slightly different equation:

$$V_{IH,IL} = V_{ref} \pm \phi_T \operatorname{acosh}\left(\frac{V_{swing}}{2\phi_T} - 1\right) \quad (3.61)$$

Switching Threshold

Finally, V_M can be determined by setting V_{in} equal to V_{out} in Eq. (3.57). This leads, once again, to a hopelessly complex expression that is difficult to solve analytically. Some interesting information can be obtained by realizing that the preferred value of V_M equals V_{ref}. Under this circumstance, I_{EE} is split equally over both branches of the current switch, and a dc bias condition for the gate can be derived:

$$V_{ref} = V_M \approx V_{CC} - V_{BE(on)} - \frac{R_C I_{EE}}{2} \quad (3.62)$$

Other dc bias conditions can be derived that help to define the values of the resistors and the current source. For example, R_C and I_{EE} have to be chosen so that transistors Q_1 and Q_2 are biased in the forward-active region over the complete range of input voltages.

The results of the above analysis are summarized in Figure 3.48, where an asymptotic picture of the ECL voltage transfer characteristic is drawn. Notice that for large values of the input signal V_{in} transistor Q_1 eventually saturates, causing output V_{out1} to track V_{in} in a linear fashion (with $V_{out1} = V_{in} - V_{BE(sat)} + V_{CE(sat)}$). This operation mode is of no concern as it is out of the normal input voltage range.

Example 3.14 VTC of ECL Gate

An ECL gate was designed with the following parameters: V_{CC} = 0 V, V_{EE} = –5 V, V_{ref} = –0.95 V, I_{EE} = 0.5 mA, R_C = 1 kΩ, and R_B = 50 kΩ. The β_F of the bipolar devices equals 100, while I_S is set to 10^{-17}A. The bipolar transistors are fabricated using the bipolar process described in Chapter 2. All transistors are minimum-size devices with $A_{emitter}$ = 2.0 μm × 3.75 μm. The VTC of the device is computed using both manual analysis and SPICE.

First, the exact value of the $V_{BE(on)}$ of the emitter-follower transistors has to be derived. This voltage varies with the collector current. A reasonable approximation is obtained by setting I_C to (5 V – 0.7 V) / 50 kΩ ≈ 0.1 mA, which equals the collector current of the emitter-follower in the high output state. This results in the following value of V_{BE} (for ϕ_T = 26 mV):

---

7 For more information on the small-signal model of the differential pair, please refer to standard analog design handbooks, such as [Sedra87, pp. 494–496].

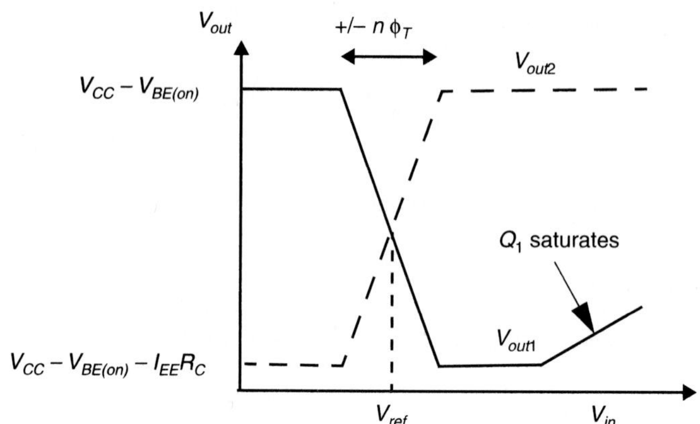

Figure 3.48 Approximate VTC of ECL gate.

$$V_{BE(on)} = \phi_T \ln\left(\frac{I_C}{I_S}\right) = 0.778\,\text{V}\,.$$

The important parameters of the voltage-transfer characteristic can now be computed.

$$\begin{aligned}
V_{OH} &= V_{CC} - V_{BE(on)} = -0.78\,\text{V} \\
V_{OL} &= V_{CC} - V_{BE(on)} - I_{EE}R_C = -1.28\,\text{V} \\
V_{IH} &= -0.95 + 0.026\ln\left(\frac{0.5}{2 \times 0.026} - 1\right) = -0.89\,\text{V} \\
V_{IL} &= -0.95 - 0.026\ln\left(\frac{0.5}{2 \times 0.026} - 1\right) = -1.01\,\text{V} \\
V_M &= V_{ref} = -0.95\,\text{V}
\end{aligned} \tag{3.63}$$

The VTC has also been computed using SPICE and is plotted in Figure 3.49. The important parameters have been extracted and are compared with the manual results in Table 3.7. An excellent correspondence is observed[.

Table 3.7 DC-parameters of ECL inverter (manual versus computed).

	Manual	SPICE
V_{OH}	–0.78	–0.77
V_{OL}	–1.28	–1.26
V_{IH}	–0.89	–0.87
V_{IL}	–1.01	–1.03
V_M	–0.95	–0.96
NM_L	0.27	0.23
NM_H	0.11	0.1

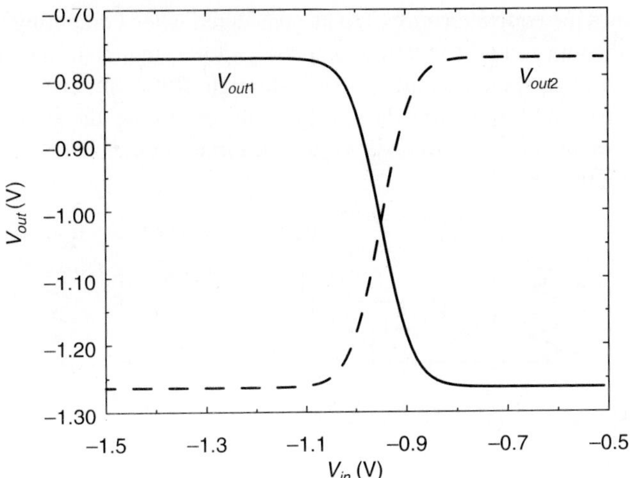

Figure 3.49 Simulated voltage-transfer characteristic of ECL gate.

Design Considerations

Compared to the CMOS inverter, the small logic swing (0.49 V) and noise margins are apparent. The smaller swing has the advantage of increased switching speed. The smaller noise margins can be tolerated because the logic levels are robust and not particularly dependent on *process parameters*. For instance, the V_{BE} of a bipolar process is far more predictable than the threshold of an MOS transistor. The reduced voltage swing also helps to keep the noise levels down. In fast ECL circuits, however, with many simultaneous switching actions, it is essential to keep the noise within bounds.

At this point, it is worth observing that all voltage levels (such as V_{OH} and V_{OL}) in an ECL gate are defined with respect to V_{CC} and are only affected in a minor way by V_{EE}. It makes sense to assign V_{CC} the best available fixed potential, which is generally the ground potential of the power supply, which is, in general, the "most clean" supply rail. Therefore, ECL circuits will normally set V_{CC} to 0 V. This requires that V_{EE} be at a negative potential, which explains the choice of the supply voltages in the example (V_{CC} = 0 V, V_{EE} = –5 V). The standard value used for the 10k and 100k series was –5.2 V. Current ECL circuits operate at lower supply voltages ranging from –3 V to –4.5 V.

Finally, it is worth mentioning that the dc characteristics of the above gate are virtually unaffected by fan-out. This is due to the very low output impedance of the emitter-follower.

3.4.4 ECL Switching Speed: The Transient Behavior

Studying the propagation delay of the ECL gate is a considerably more involved task than the analysis of the CMOS inverter. The structure is inherently more complex, as it contains internal nodes, each of which could dominate the performance. This is illustrated in Figure

3.50, which shows the equivalent circuit to be considered when computing the propagation delay between the input and the inverting output V_{out1} for a single fan-out. Getting an accurate prediction of the response is only possible through SPICE simulation. Unfortunately, this does not give an insight into the mechanisms governing the transitions. Such an insight can only be obtained by studying a tractable circuit model, obtained through significant simplifications:

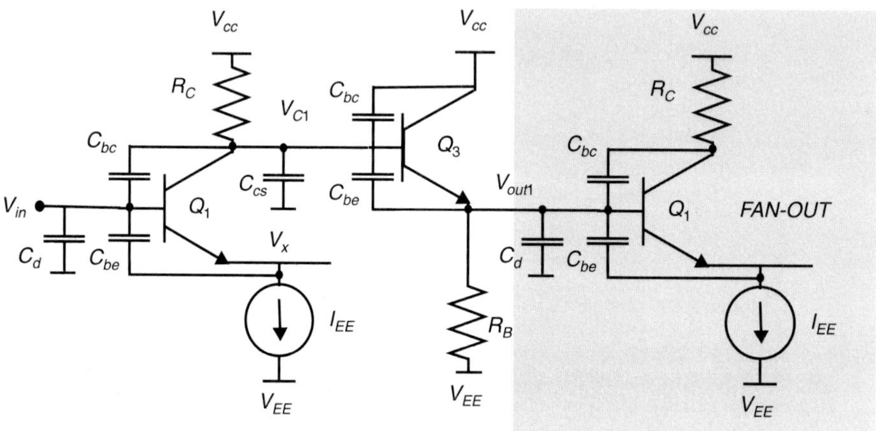

Figure 3.50 Circuit model of ECL gate with single fan-out, used in propagation delay analysis.

- All capacitances are *linearized* over the appropriate voltage range and *lumped* together into a limited number of capacitances.
- Internal nodes are *eliminated* (as much as possible) using the *reflection rule* (introduced in the dc analysis).
- The transient response is *decomposed into several steps,* based on the dominant effects governing the response at each step. Although in reality **these steps do partially overlap**, this approach yields reasonable accuracy, while simplifying the analysis substantially. In particular, we isolate the switching of the differential pair from the (dis)charging of the load capacitors. This important assumption dramatically simplifies the analysis, while enabling the identification of the dominant performance parameters.

Switching the Differential Pair

The analysis of the large-signal transient response of the current switch is intricate and yields complex expressions.[8] Complicating the derivation is the impact of the base and emitter resistances, the feedback effect offered by the coupled emitters, and the coupling between the transistor terminals. To avoid this multiple-page numerical art-work, we use a simple heuristic model. In Chapter 2, it was derived that the transient response of a bipolar transistor is dominated by the base and depletion-region charge. The same is true for the differential pair. To illustrate this statement, let us study the simplified network of Figure

[8] For a detailed analysis, please refer to [Embabi93, pp.215–224]

3.51. Some of the most important transient parameters are annotated on the diagram. Notice that in this model, the collectors are kept at V_{CC}, which effectively eliminates the impact of the discharging of the load capacitances.

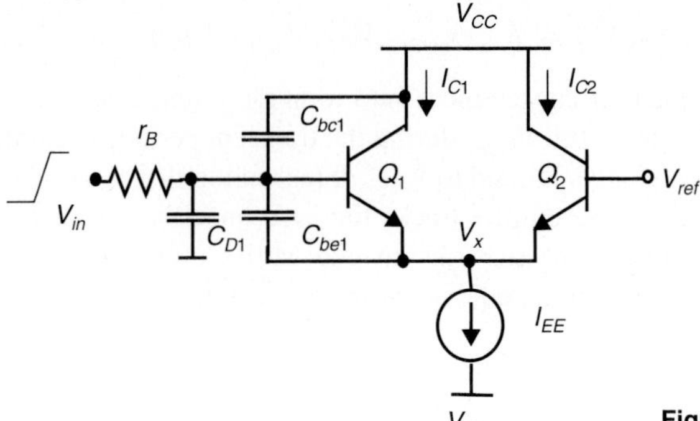

Figure 3.51 Switching the differential pair.

The transient response of the current switch is simulated for varying values of the current source I_{EE}. The resulting collector currents for both branches are plotted in Figure 3.52. The switching time is a strong function of I_{EE} for large values of the current source, which suggests that the base charge is the dominant factor under those circumstances. On the other hand, the delay is independent of I_{EE} for smaller current values. This means that the depletion charge of the *be* and *bc* junctions has become the most important factor. Varying the value of the junction capacitances in the simulation confirms this assumption. Observe the large spikes in the collector currents during the switching of the input signal. This is due to the capacitive coupling between base and emitter of Q_1. A simple model can now be constructed for both mechanisms.

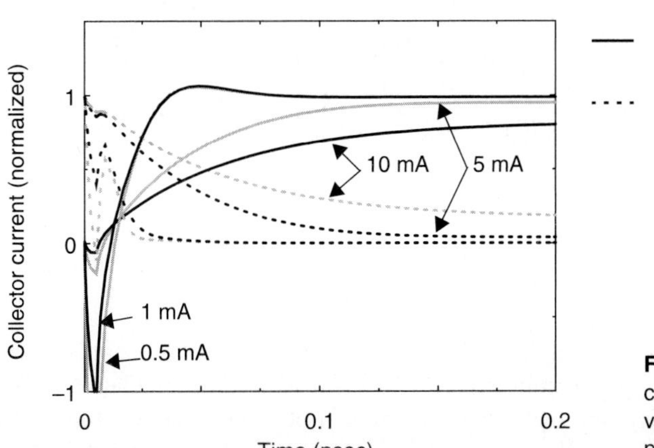

Figure 3.52 Simulated collector currents of differential pair (final value normalized to 1), with I_{EE} as a parameter.

(Dis)charging the Depletion Capacitances (space charge). Based on our common practice, we first replace all nonlinear junctions capacitors by linearized equivalents. This is accomplished by multiplying the zero-bias capacitance with the K_{eq} factor (Eq. (2.18)), which is a function of the applied voltage range. The voltage at the base of Q_1 swings between V_{OL} to V_{OH}, while the emitter voltage moves between $V_{ref} - V_{BE(on)}$ to

$V_{OH} - V_{BE(on)}$. The collector voltage stays constant at V_{CC}. Knowledge of these voltage ranges allows for the computation of the K_{eq} values:

$$\begin{aligned} K_{eq,be} &= K_{eq}(V_1, V_2) = K_{eq}(V_{OL} - V_{ref} + V_{BE(on)}, V_{BE(on)}) \\ K_{eq,bc} &= K_{eq}(V_1, V_2) = K_{eq}(V_{OL} - V_{CC}, V_{OH} - V_{CC}) \end{aligned} \tag{3.64}$$

The lumping of the junction capacitances into a single component requires some extra caution. The C_{bcI} undergoes a full V_{swing} during the transient period. The total voltage swing over C_{beI}, on the other hand, is limited to $V_{swing}/2$ (assuming that V_{ref} is placed in the middle of the voltage swing), as the emitter tracks the base once the transistor is on. Its impact is thus half of the other capacitor (an inverse Miller effect). Taking this into account, the following expression for the equivalent input capacitance can be derived over the V_{swing} voltage range.

$$C_{in,j} = K_{eq,bc}C_{bc1} + K_{eq,be}C_{be1}/2 \tag{3.65}$$

The circuit diagram of Figure 3.51 suggests that the (dis)charging of the depletion capacitances is dominated by the $r_B C_{in,j}$ time-constant. Achieving 90% of the final value requires approximately 2.2 time-constants, if the voltage waveform at the base is exponential (which is approximately the case). Under these conditions, the time to change the space charge is adequately modeled by the following expression:

$$t_{space} = 2.2 r_B C_{in,j} = 2.2 r_B (K_{eq,bc}C_{bc1} + K_{eq,be}C_{be1}/2) \tag{3.66}$$

(Dis)charging the Diffusion Capacitance (base charge). The diffusion capacitance C_{DI} represents the excess-charge storage in the transistor base (introduced in Chapter 2). C_{D1} is placed at the base of the transistor and stores an amount of charge equal to ΔQ_F over the voltage range of interest. From Chapter 2, recall the following expression,

$$C_D = \frac{\Delta Q_F}{\Delta V} \tag{3.67}$$

During a single transition, the current through Q_1 evolves from virtually nonexisting to I_{EE}, or $\Delta Q_F = Q_F = \tau_F I_{EE}$. The voltage swing at the base during that period is, obviously, V_{swing}. Hence,

$$C_{D1} = \frac{\Delta Q_F}{\Delta V} = \frac{\tau_F I_{EE}}{V_{swing}} \tag{3.68}$$

It is worth observing that the diffusion capacitance is proportional to the value of the current source. This proportionality is clearly manifested in the simulation results of Figure 3.52. Remembering that the voltage swing can be expressed as a function of I_{EE} as well (Eq. (3.58)) yields another interesting relation. Eq. (3.69) states that the diffusion capacitance of Q_1 is only a function of the forward transit time τ_F and the collector resistance R_C (a design parameter!).

$$C_{D1} = \frac{\tau_F I_{EE}}{R_C I_{EE}} = \frac{\tau_F}{R_C} \tag{3.69}$$

Deriving an exact expression of the time to change the base charge is difficult, especially due to the impact of the emitter-coupling (and other factors, such as emitter resistance). The time is proportional, though, to the time-constant of the input circuit, formed by r_B and C_{D1}. This suggests the following model for t_{base}:

$$t_{base} = \alpha r_B C_{D1} \tag{3.70}$$

where α is an empirical factor, and depends upon circuit and device parameters. From simulations, we derive that α approximately equals 2 and 5 for the 50% and 90% points, respectively.

Propagation Delay of the Differential Pair. The overall switching time of the differential pair can now be expressed as the sum of the space- and base-charge components (assuming that both mechanisms happen consecutively). Remember from the simulations that the response is typically dominated by a single mechanism.

$$t_{dp} = t_{space} + t_{base} = r_B(2.2C_{in,j} + \alpha C_{D1}) \tag{3.71}$$

Example 3.15 Switching the Differential Pair

Consider the ECL gate of Example 3.14. The bipolar transistor model defined in Chapter 2 provides the extra parameter values needed for the transient analysis:

$$r_B = 120\ \Omega,\ r_E = 20\ \Omega, \text{ and } r_C = 75\ \Omega;\ C_{be} = 20\text{ fF},\ C_{bc} = 22\text{ fF, and } C_{cs} = 47\text{ fF}.$$
$$m_{be} = 0.5,\ m_{bc} = m_{cs} = 0.33,\ \tau_F = 10\text{ psec.}$$

With $I_{EE} = 0.5$ mA (or $R_c = 1$ kΩ) and $\tau_F = 10$ psec, C_{D1} equals 10 fF. C_{be} and C_{bc} are defined to be 20 and 22 fF, while their K_{eq}-factors evaluate to 3.35 and 0.75, respectively (with $\phi_0 = 0.7$ V).[9] This yields the following capacitance values:

$$C_{D1} = 10\text{ psec/ }1\text{k}\Omega = 10\text{ fF};\ C_{in,j} = 3.35 \times 20 / 2 + 0.75 \times 22 = 50\text{ fF}$$

The time it takes for the collector currents to reach 90% of their value is now approximated using Eq. (3.71) (and assuming that $\alpha = 5$).

$$t_{dp} = 120\ \Omega \times (2.2 \times 50\text{ fF} + 5 \times 10\text{ fF}) = 20\text{ psec}$$

(Dis)Charging the Load Capacitances

To derive the second component of the transient response, we assume that the current switch has already reverted state. This means that the current-switch transistors are either off, or carry the complete current I_{EE}. Before addressing the analysis of the propagation delay, it is good practice to derive the capacitance values first, as we did for the MOS inverter.

Deriving the Load Capacitances. As illustrated in Figure 3.50, the capacitance model of the ECL inverter is complex and consists of many contributions. To simplify the analysis, we gather all the individual contributions into two lumped capacitors: the collector capacitance C_C and the load capacitance C_L.

[9] In the computation of K_{eq} for C_{be}, it is assumed that the maximum forward bias is equal to ϕ_0.

1. The **collector capacitance** $C_C = C_{cs1} + C_{bc1} + C_{bc3} + C_{D3}$.

The first three factors in C_C represent the depletion capacitances of the collector-substrate junction of Q_1 and the base-collector junctions of Q_1 and Q_3. Since the emitter-follower transistor Q_3 is assumed to be always on and in forward-active mode, the voltage over the base-emitter junction is constant, and its depletion charge remains unchanged. This explains the absence of C_{be3} in the list. The equivalent junction capacitances are easily computed once the voltage excursions are known. The base of Q_1 is assumed to stay constant, while its collector voltage V_C swings between V_{CC} to $V_{CC} - V_{swing}$. The collector of Q_3 stays at V_{CC}. No Miller effect occurs, and the capacitances can simply be added, weighted with the appropriate K_{eq} factors.

C_{D3} models the change in base charge of Q_3 during a transition:

$$\Delta Q_F = \tau_F (V_{OH} - V_{OL})/R_B = \tau_F V_{swing}/R_B,$$

where V_{OH}/R_B and V_{OL}/R_B approximate the collector currents of Q_3 in the high and low output state, respectively. The equivalent diffusion capacitance is derived by dividing ΔQ_F by the voltage swing of interest:

$$C_{D3} = \frac{\Delta Q_F}{\Delta V} = \frac{\tau_F V_{swing}}{R_B V_{swing}} = \frac{\tau_F}{R_B} \tag{3.72}$$

Observe the similarity to Eq. (3.69).

2. The **load capacitance** $C_L = C_W + C_{in}$

The load of the gate equals the sum of the wiring capacitance C_W and the input capacitance of the fan-out gates. The latter is a complex combination of space- and base-charge capacitance, and varies with the direction of the transition. To simplify the analysis, we just assume a fixed, average value in this study.

Example 3.16 Load Capacitances of an ECL Gate

Consider, again, the ECL gate of Example 3.14. The values of the load capacitances are readily computed.

1. C_C—Due to the high value of R_B, C_{d3} evaluates to 0.2 fF and can be ignored. The K_{eq} factors for C_{bc1}, C_{bc3}, and C_{cs1} equal 0.84, 0.91, and 0.51, respectively (the low value for C_{CS} is due to the strong reverse bias on that junction, with the substrate at –5 V). This yields the following value for the collector capacitor.

$$C_C = 0.51 \times 47 + 0.84 \times 22 + 0.91 \times 22 + 0.2 \text{ fF} = 62.7 \text{ fF}$$

2. C_L—We assume a value of 60 fF here, which approximates the average input capacitance of a similar ECL gate.

Discharging the Load Capacitances. The time it takes to discharge the collector and output nodes is determined by two competing mechanisms: the discharging of the load capacitance C_L through the load resistor R_B, and the discharging of the collector capacitance by the current source I_{EE}. Both processes are illustrated in the (simplified) transistor network of Figure 3.53. Depending upon the values of the resistances, capacitances, and current levels, either one of them can be dominant.

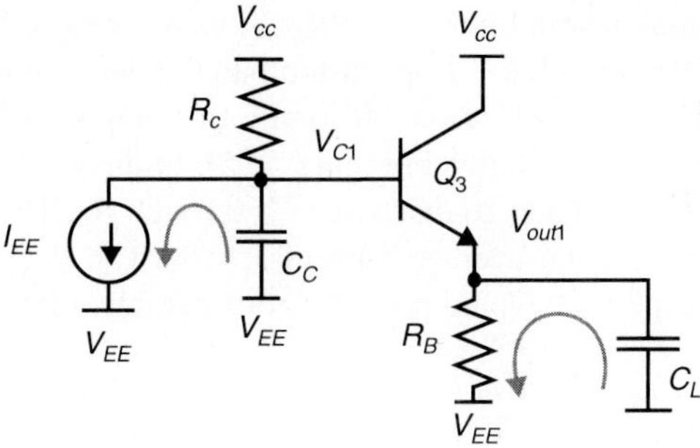

Figure 3.53 Discharging the collector and load capacitances.

When $R_BC_L << R_CC_C$, the discharging of the collector node through I_{EE} dominates the performance, and the discharge time is proportional to the time-constant R_CC_C. Observe that value of I_{EE} sets the dc voltage levels, but does not influence the discharge time. If this argument does not seem obvious, a useful exercise is to construct the Thevenin equivalent circuit of the network, consisting of I_{EE}, V_{CC}, and R_C.

A more accurate derivation takes the base current of Q_3 into account. Even for very small values of $R_B\,C_L$, V_{out1} can never fall faster than V_{c1}, since this causes the emitter-follower to turn on. The nodes, therefore, discharge in unison. The collector and output nodes are tightly coupled as the output node follows the collector node with a (fixed) voltage difference of $V_{BE(on)}$. This operation is adequately modeled by the equivalent circuit of Figure

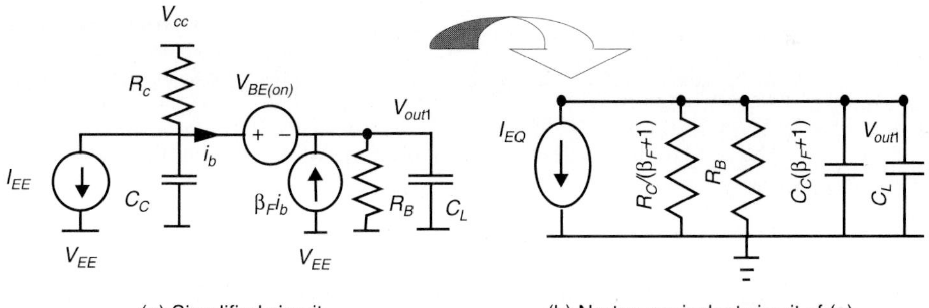

Figure 3.54 Equivalent circuits for the computation of t_{pHL}, for $R_BCL << R_CC_C$, with $I_{EQ} = I_{EE}(\beta_F + 1) - V_{EE}/R_B - (V_{CC} - V_{BE(on)})(\beta_F + 1)/R_C$.

3.54a (where the base-emitter diode is modeled as a voltage source), or by the even simpler Norton equivalent circuit model of Figure 3.54b. The same circuit could have been derived in a single step by using the reflection rule to eliminate the internal node. Divide all impedances at the base of Q_3 by ($\beta_F + 1$) and move them to the emitter node. Notice that this means that C_c is multiplied with (β_F + 1). The discharge time is approximated by the time-constant of the resulting circuit:

$$t_{discharge1} = 0.69\left(R_B \parallel \frac{R_C}{\beta_F + 1}\right)[C_L + C_C(\beta_F + 1)] \tag{3.73}$$

For large values of $R_C C_C$, this expression simplifies to $0.69R_C C_C$, as was projected.

When $\boldsymbol{R_B C_L >> R_C C_C}$, the collector voltage drops faster than the output node, which means that the emitter-follower transistor Q_3 shuts off (until the output voltage gets clamped to its final value by the follower, as shown in Figure 3.55). In this case, the propagation delay is dominated by the time it takes to discharge C_L through R_B. The effect of the clamping makes it necessary to use the average current approach to determine the delay, as the *RC*-technique used in Eq. (3.73) would predict a considerably different result.

$$I_{av} = \frac{1}{2}\left(\frac{V_{OH} - V_{EE}}{R_B} + \frac{V_{OH} - V_{swing}/2 - V_{EE}}{R_B}\right) = \frac{1}{R_B}\left(V_{OH} - V_{EE} - \frac{V_{swing}}{4}\right)$$

$$t_{discharge2} = \frac{C_L(V_{swing}/2)}{I_{av}} = \frac{0.5C_L R_B V_{swing}}{V_{OH} - V_{EE} - \frac{V_{swing}}{4}} \approx 0.5C_L R_B\left(\frac{V_{swing}}{V_{CC} - V_{EE}}\right) \tag{3.74}$$

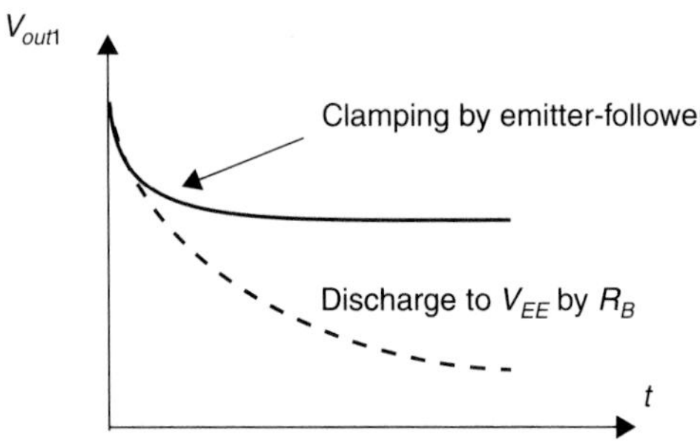

Figure 3.55 Effect of voltage clamping on discharge waveform. When only the *RC*-network would be considered, the output would discharge to V_{EE}. This is prevented by the emitter-follower.

The actual discharge time is approximated by taking the worst case of both scenarios:

$$t_{discharge} \approx \max(t_{discharge1}, t_{discharge2}) \tag{3.75}$$

with $t_{discharge2}$ the time it takes to discharge the collector node and $t_{discharge1}$ the time to discharge the load capacitance through R_B.

Charging the Load Capacitances. A similar approach can be taken to estimate the time it takes to charge the load capacitances C_C (through R_C) and C_L (through Q_3). The process is illustrated in the circuit model of Figure 3.56. To obtain accurate expressions for the delay, we resort again to the reflection rule and the use of equivalent circuits, as shown in Figure 3.57a and b. The propagation delay of the resulting circuit is easily derived:

$$t_{charge} = 0.69\left(\frac{R_C}{\beta_F + 1} \parallel R_B\right)[C_C(\beta_F + 1) + C_L] \tag{3.76}$$

which is similar to the expression obtained for $t_{discharge2}$ above.

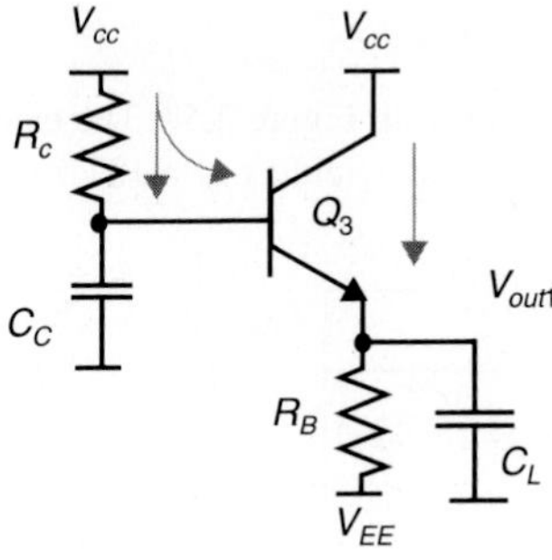

Figure 3.56 Equivalent circuit for the computation of t_{pLH}.

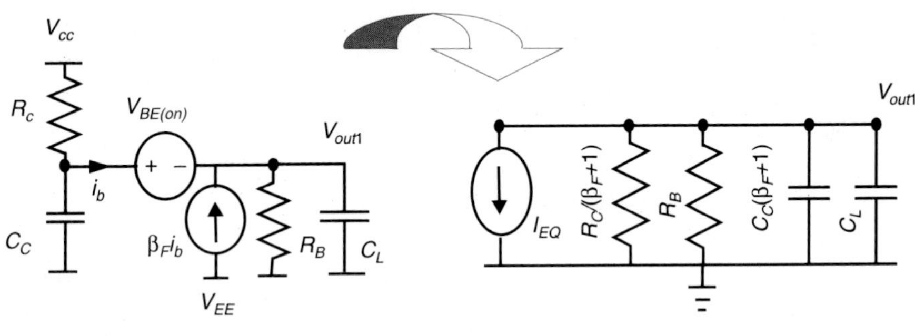

Figure 3.57 Equivalent circuits for the computation of t_{pLH}, with $I_{EQ} = -V_{EE}/R_B - (V_{CC} - V_{BE(on)}) \times (\beta_F + 1)/R_C$.

Example 3.17 ECL Transient Response

The total propagation delay of the ECL gate of Example 3.14 can now be computed (recall that $R_C = 1\text{k}\Omega$, and $R_B = 50$ kW).

Consider first the t_{pHL}. The time to switch the differential pair was computed in Example 3.15 and equals 20 psec. The second factor of the delay is the time to discharge the node capacitance. Since $R_B C_L >> R_C C_C$, the output node time-constant dominates, and the discharge time is given by Eq. (3.74).

$$t_{discharge} = (0.5 \times 60 \text{ fF} \times 50 \text{ k}\Omega \times 0.5 \text{ V}) / (-0.7 + 5 - 0.125 \text{ V}) = 180 \text{ psec}.$$

Adding the two components leads to the following approximation:

$$t_{pHL} = 20 \text{ psec} + 180 \text{ psec} = 200 \text{ psec}.$$

The low-to-high transition consists of turning off the current switch and charging the load capacitances. The charge time is approximated by Eq. (3.76), with the collector node presenting the largest delay.

$$t_{charge} = 0.69\, R_C C_C = 0.69 \times 1 \text{ k}\Omega \times 62.7 \text{ fF} = 43.3 \text{ psec}.$$

Combining this with the delay of the differential pair yields

$$t_{pLH} = 20 + 43.4 = 63.3 \text{ psec}$$

and

$$t_p = (200 + 63.3)/2 = 132 \text{ psec}$$

The simulated transient response of the circuit is shown in Figure 3.58. The extracted values of t_{pHL} and t_{pLH} respectively equal 163 psec and 92 psec, yielding t_p = 127.5 psec, which is consistent with the estimated results.

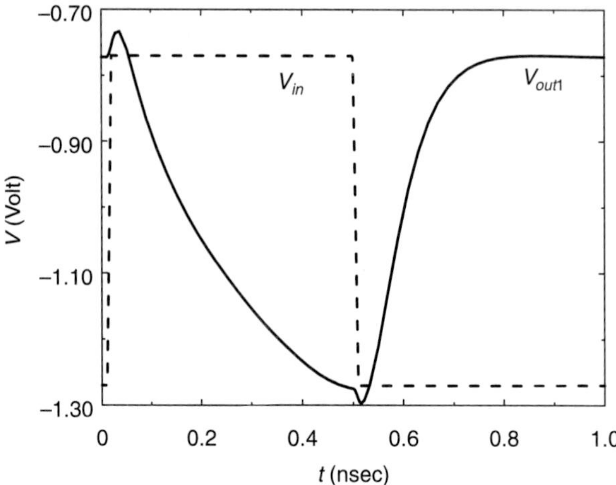

Figure 3.58 Simulated transient response of ECL inverter.

A Global View of the Transient Response

The results of this extensive analysis are summarized in Table 3.8. Observe again that these expressions are first-order models and are by no means intended to be accurate. SPICE simulations are indispensable if precise numbers are required. The most important function of the presented model is to provide a clear insight into the parameters dominating the performance of the ECL structure.

The analysis shows that the ECL gate exhibits asymmetrical high-to-low and low-to-high transitions. The difference between charge and discharge times is most obvious for large load values. In that case, the rise time is determined by the $C_L R_C/\beta_F$ time constant,

Table 3.8 Propagation delay of ECL gate: summary.

		Switching the differential pair	**Capacitive (dis)charge**
t_{pHL}	$R_B C_L >> R_C C_c$	$r_B(2.2C_{in,j} + \alpha C_{D1})$	$0.5 C_L R_B \left(\frac{V_{swing}}{V_{CC} - V_{EE}}\right)$
	$R_B C_L << R_C C_c$		$0.69\left(\frac{R_C}{\beta_F + 1} \parallel R_B\right)(C_C(\beta_F + 1) + C_L)$
t_{pLH}		$r_B(2.2C_{in,j} + \alpha C_{D1})$	$0.69\left(\frac{R_C}{\beta_F + 1} \parallel R_B\right)(C_C(\beta_F + 1) + C_L)$

while the fall time at the output is set by $C_L R_B$. For C_L smaller or comparable to C_C, t_{pHL} and t_{pLH} are approximately equal, and the delays are dominated by the $C_C R_C$ time constant. The former case occurs most often in discrete ECL gates or in gates driving large capacitive loads.The latter situation occurs in gates internal to modules such as adders or CPUs. In these circumstances, the propagation delay is mainly dominated by the intrinsic capacitances, modeled by C_C.

Design Considerations

In the presented example, the propagation delay is dominated by the capacitive (dis)charge time. This causes the propagation delay to *increase linearly with the capacitive load*, as in the CMOS case. It is important to notice, however, that the ECL gate is less sensitive to capacitive loading than the CMOS counterpart. During a low-to-high transition, C_L is charged through the emitter-follower, which means that a substantial charging current is available (or the effective capacitive load is divided by the current gain of the emitter-follower). The discharge time for large load capacitances is dominated by the $R_B C_L$ time-constant, which can be made very small by reducing the value of R_B.

Compared to the propagation delay of the CMOS inverter (225 psec), a speedup of 1.8 is obtained for a single fan-out. Considerable improvements over this result can be attained. The gate design presented in the example is far from optimal. Its delay can be lowered by increasing the current levels in both the differential pair and the emitter-follower. This effectively *trades off an improvement in performance against an increased power consumption.* For a given logic swing, increased current levels mean reduced values of resistors R_C and R_B. From Table 3.8, it is apparent that this translates to a reduction in capacitive (dis)charge times.

At this point, you probably wonder how low a delay can be obtained and if an optimum current level can be defined. In effect, increasing the collector current also increases the base charge in the bipolar transistors, and, consequently, the turn-on and turn-off times of the devices. This is clearly demonstrated in Eqs. (3.69) and (3.72), which state that the diffusion capacitances are inversely proportional to resistor values. At some current level, the diffusion capacitance becomes dominant, and the delay levels off and even starts to grow. Other second-order effects, such as the Kirk effect and the collector resistance come into play at higher current levels and cause a further degradation. To illustrate this behavior, we have plotted the simulated t_{pLH} of our ECL gate as a function of I_{EE} (we have kept R_B constant for this analysis). A sharp decrease in the delay can initially be observed due to a reduction in the dominant time-constant $R_C C_C$. For higher current levels, the time to turn-off Q_1 starts to dominate, and the delay rises again (in accordance with $r_B C_{D1}$).

In summary, careful optimization of the gate structure can substantially reduce the delay. Combining this with a state-of-the-art bipolar technology (the parasitic capacitances can be as low as 10 fF in a 0.8 μm bipolar process) leads to ECL propagation delays ranging from 40 to 70 psec. This is between 3 to 4 times faster than its CMOS equivalent.

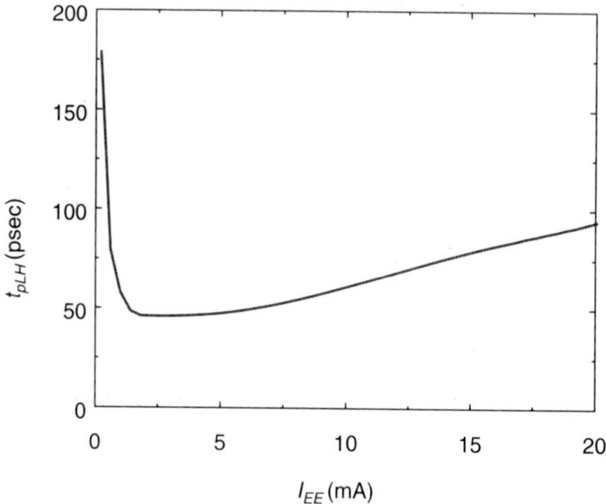

Figure 3.59 t_{pLH} of ECL inverter as a function of the I_{EE} bias current.

3.4.5 Power Consumption

The ECL gate clearly consumes static power. Sources of this static consumption are the emitter-coupled pair, the bias network, and the output stage. The dissipation of the latter depends upon the value of the pull-down resistor and the termination mechanism used. The power consumption of the bias network, on the other hand, can be distributed over multiple gates. These observations are summarized in the following equation,

$$P_{stat} = (V_{CC} - V_{EE})\left(I_{EE} + \frac{I_{bias}}{N} + 2\frac{\frac{V_{OH}+V_{OL}}{2} - V_{EE}}{R_B}\right) \quad (3.77)$$

where N equals the number of gates serviced by a single bias network. The factor 2 in the power-consumption expression for the emitter-follower is due to the complementary outputs.

Furthermore, dynamic power is consumed during switching. During each switching event, both internal and external capacitances are (dis)charged, requiring a charge transfer equal to $V_{swing}(C_C + C_L)$. In contrast to the CMOS inverter, the voltage swing is considerably smaller than the supply voltage. This results in a slight modification in the expression for the dynamic power consumption (compare this to the expression in Eq. (3.42), derived for the CMOS inverter).

$$P_{dyn} = C_T(V_{CC} - V_{EE})V_{swing}f \quad (3.78)$$

with C_T the sum of all capacitances switched.

The static power factor generally dominates the overall dissipation. In fact, it can be shown that for the dynamic power to become the dominating factor, it is necessary to switch the gate faster than the propagation delay would allow.

Example 3.18 Power Dissipation of the ECL Inverter

The power dissipation of the inverter of Example 3.14 is analyzed. We ignore the power consumed in the bias network.The static power consumption is computed with the aid of Eq. (3.77).

$$P_{stat} = 5\ (0.5\ \text{mA} + 0.162\ \text{mA}) = 3.3\ \text{mW}$$

Even when switching the gate at the maximum allowable speed ($f = 1/t_p$), the dynamic consumption is still smaller,

$$P_{dyn} = (60\ \text{fF} + 72.8\ \text{fF}) \times 5\text{V} \times 0.5\text{V}\ /\ 127.5\ \text{psec} = 2.6\ \text{mW}$$

Combining the two factors yields a considerable consumption of approximately 6 mW. The power-delay product of the gate equals 760 fJ, which is almost identical to the number obtained for the CMOS inverter. The reduced switching delay is offset by the increased power consumption. Notice, however, that PDPs as low as 120 fJ have been reported in state-of-the-art 0.6 μm bipolar processes. The high power dissipation obviously constrains the number of gates that can be integrated on a single die.

Design Consideration

Finally, one more observation is worth mentioning. During the qualitative analysis of the gate, we have already noted that the differential pair draws a constant current from the supply, hence introducing virtually no switching noise on the supply lines. This is also true for the bias network, but is not the case for the emitter-follower. Depending upon the values of the terminating resistor and the load capacitance, the switching causes large stepwise current variations in the collectors of the emitter-follower transistors. As will be discussed in later chapters, parasitic inductances can translate those current fluctuations into supply noise (ringing, voltage spikes), which might cause the circuit to fail. Therefore, ECL circuits often employ two V_{CC}'s: a "clean" V_{CC}, which connects to the differential pairs and the bias networks, and a "dirty" V_{CC}, which feeds the emitter-followers. This separation avoids the feeding back of switching transients into differential pairs that could prove disastrous given the low noise margins.

3.4.6 Looking Ahead: Scaling the Technology

When discussing the influence of technology scaling on the performance of a CMOS process, we introduced the ideal scaling model. In this model, both dimensions and voltages are scaled in a similar way, keeping the electrical fields in the devices approximately constant.

Bipolar scaling is considerably more complex than MOS scaling, as more device parameters are involved in the design. Also, power-supply voltage and logic swing cannot be reduced much, since they already approach their lower limit at room temperature. The on-voltage of the base-emitter junction is a built-in parameter, which is virtually unaffected by scaling. The full-scaling model is therefore not appropriate for the ECL inverter, and the fixed-voltage scaling model must be used. Consequently, both electrical field strengths and current densities increase when scaling.

A typical bipolar scaling model is presented in Table 3.9. Voltages and currents are kept to a constant level. This means that the current density increases with a squared factor. Capacitances, consisting of junction and diffusion capacitances, are scaled linearly, which requires a scaling of the base width as well as the doping levels in base and collector regions. Thus, bipolar scaling means not only shrinking the lateral dimensions, but also the vertical profile of the transistor.

Table 3.9 Scaling model for bipolar inverter (from [Tang89]).

Parameter	Scaling Factor
A_E	$1/S^2$
w_B	$1/S^{0.8}$
V_{supply}, V_{swing}	1
J	S^2
I	1
C_d, C_j	$1/S$
t_p	$1/S$
P	1

This model is illustrated in Figure 3.60, where the projected gate delay of an ECL circuit is plotted as a function of the current level and feature size. The plot also provides insight in the number of gates that can be integrated on a single die (for a maximum consumption of 2 Watts).

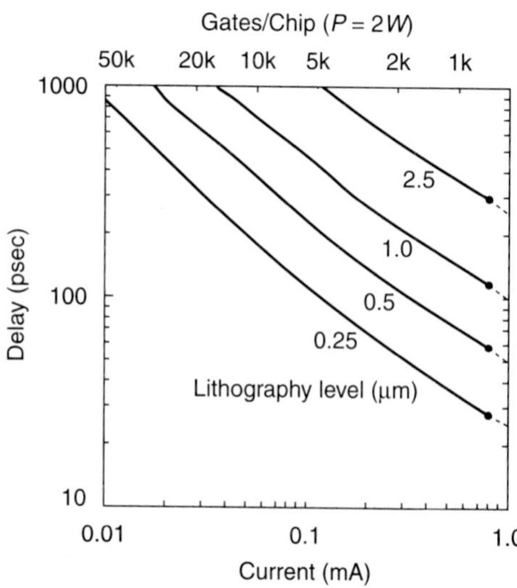

Figure 3.60 Projected gate delay of scaled ECL circuit in function of current level and feature size (from [Tang89]).

3.5 Perspective: Area, Performance, and Dissipation

This chapter has presented two radically different gate topologies. The nature and the structure of those circuits is a direct consequence of the employed technology. It is, for instance, entirely conceivable to implement a CMOS gate using the differential pair structure of the ECL gate. Due to the lower transconductance of the MOS device, this approach would require substantially larger voltage swings at the inputs or, alternatively, much larger input transistors. Similarly, the structure of the CMOS inverter with its complementary devices can be applied to the bipolar technology. Attempts in that direction result in complex gate topologies with reduced performance.

Overall, we conclude that the ECL technology offers an important *speed advantage* over CMOS by at least a factor of 2 and normally closer to 4 or 5. The ECL gate has, furthermore, the advantage of being far less sensitive to capacitive loading. These features come at the expense of a *large area/gate and an increased power consumption*. The layout density of an ECL inverter is approximately 3 to 10 times smaller than that of a CMOS inverter, realized in a comparable technology. Both these factors limit the number of gates that can be economically integrated on a single die.

To put this statement in perspective, let us compare the state of the art in CMOS and ECL designs: the CMOS Intel Pentium™ processor [Schutz94] clocks at 100 MHz (and has to use a limited number of bipolar gates to achieve that performance!), and integrates 3.3 million transistors on a 163 mm^2 die (0.6 μm technology) for a power consumption of 8 W. A DEC ECL microprocessor [Jouppi93], on the other hand, combines 486,000 transistors on a 194 mm^2 die (1 μm technology), clocks at 300 MHz and consumes 115 W.

ECL is therefore used extensively in the design of high-performance, medium-complexity components, where the speed performance attainable in CMOS is inadequate. Examples of such are advanced instrumentation applications, compute cores for mainframe computers and supercomputers, and, recently, also some microprocessors.

The ultimate question is how this relation will evolve in the future. A number of indicators suggest that CMOS will catch up with ECL in performance in the not-too-distant future [Masaki92]. This observation is predominantly based on the inferior scaling properties of bipolar devices and the ECL gate structure, especially with respect to the supply voltage. Only the future can confirm the truth of this prediction.

3.6 Summary

The following concepts were introduced in this chapter:

- A digital gate is characterized by a number of steady-state, dynamic, and energy parameters. These parameters categorize the gate in terms of robustness, performance and heat dissipation. Especially important are the voltage-transfer characteristic, the noise margins, the propagation delay, and the power-delay product.
- The static CMOS inverter has an almost ideal voltage-transfer characteristic. The logic swing is equal to the supply voltage and is not a function of the transistor sizes. The noise margins of a symmetrical inverter (where PMOS and NMOS transistor

have equal current-driving strength) approach $V_{DD}/2$. The steady-state response is not affected by fan-out.

- Its propagation delay is dominated by the time it takes to charge or discharge the load capacitor C_L. To a first order, it can be approximated as

$$\frac{C_L}{2V_{DD}}\left(\frac{1}{k_p} + \frac{1}{k_n}\right).$$

Keeping the load capacitance small is the most effective means of implementing high-performance circuits.

- The power dissipation is dominated by the dynamic power consumed in charging and discharging the load capacitor. It is given by $C_L V_{DD}^2 f$.
- Scaling the technology is an effective means of reducing the area, propagation delay and, power consumption of a gate. The impact is even more striking if the supply voltage is scaled simultaneously.
- The bipolar ECL gates consists of three components: the *differential pair*, which implements the logic function and achieves high switching speed by keeping the transistors from saturation; the *output stage*, which provides ample driving current and performs level shifting, and the *bias network*, which has to be designed to provide maximum insensitivity to temperature variations.
- It achieves high performance by reducing the logic swing (as low as 500 mV) and by operating at high current levels. This comes at the expense of the noise margins. To ensure correct operation at high speed, it is necessary to take special precautions with respect to supply-induced noise.
- The use of an emitter-follower as output stage ensures that the propagation delay is rather insensitive to variations in the output load. Loads of multiple pF's do not cause any major degradation in performance.
- A major disadvantage of the ECL gate is its static power dissipation. This prevents the implementation of very large scale integrated circuits in this technology.
- The bipolar technology does not scale as effectively as MOS. The presence of built-in, invariable voltages prevents a sustained scaling of the supply voltage.

3.7 To Probe Further

The operation of the inverter has been the topic of numerous publications and textbooks. Virtually every textbook on digital design devotes a substantial number of pages to the analysis of the basic inverter gate. An extensive list of references was presented in Chapter 1. Some references of particular interest that were explicitly quoted in this chapter are given below.

REFERENCES

[Baccarani84] G. Baccarani, M. Wordeman, and R. Dennard, "Generalized Scaling Theory and Its Application to 1/4 Micrometer MOSFET Design," *IEEE Trans. Electron Devices*, ED-31(4): p. 452, 1984.

[Brews89] J. Brews et al., "The Submicrometer Silicon MOSFET," in [Watts89].

[De Man87] De Man H., *Computer Aided Design of Digital Integrated Circuits*, Lecture Notes, Katholieke Universiteit Leuven, Belgium.

[Dennard74] R. Dennard et al., "Design of Ion-Implanted MOSFETS with Very Small Physical Dimensions," *IEEE Journal of Solid-State Circuits*, SC-9, pp. 256–258, 1974.

[Embabi93] S. Embabi, A. Bellaouar, M. Elmasry, *Digital BiCMOS Integrated Circuit Design*, Kluwer Academic Publishers, 1993.

[Hodges88] D. Hodges and H. Jackson, *Analysis and Design of Digital Integrated Circuits*, McGraw-Hill, 1988.

[Jouppi93] N. Jouppi et al., "A 300 MHz 115W 32b Bipolar ECL Microprocessor with On-Chip Caches," *Proc. IEEE ISSCC Conf.*, pp. 84–85, February 1993.

[Kakumu90] M. Kakumu and M. Kinugawa, "Power-Supply Voltage Impact on Circuit Performance for Half and Lower Submicrometer CMOS LSI," *IEEE Journal of Solid-State Circuits*, vol. 37, no. 8, pp. 1900–1908, August 1990.

[Lohstroh81] J. Lohstroh, "Devices and Circuits for Bipolar (V)LSI," *Proceedings of the IEEE*, vol. 69, pp. 812–826, July 1981.

[Masaki92] A. Masaki, "Deep-Submicron CMOS Warms Up to High-Speed Logic," *IEEE Circuits and Devices Magazine*, pp. 18–24, November 1992.

[Schutz94] J. Schutz, "A 3.3V 0.6 mm BiCMOS Superscaler Microprocessor," *ISSCC Digest of Technical Papers*, pp. 202–203, February 1994.

[Sedra87] Sedra and Smith, *MicroElectronic Circuits*, Holt, Rinehart and Winston, 1987.

[Shoji88] Shoji, M., *CMOS Digital Circuit Technology*, Prentice Hall, 1988.

[Tang89] D. Tang, "Scaling the Silicon Bipolar Transistor," in [Watts89].

[Veendrick84] H. Veendrick, "Short-Circuit Dissipation of Static CMOS Circuitry and its Impact on the Design of Buffer Circuits," *IEEE Journal of Solid-State Circuits*, vol. SC-19, no. 4, pp. 468–473, 1984.

[Watts89] Watts R., ed., *SubMicron Integrated Circuits*, Wiley, 1989.

3.8 Exercises and Design Problems

For all problems, use the device parameters provided in Chapter 2 (as well as the inside back cover), unless otherwise mentioned.

1. [E, None, 3.2] The VTC of a fictitious gate is shown in Figure 3.61. Determine (approximately) V_{OL}, V_{OH}, V_{IL}, V_{IH}, NM_L, and NM_H. Is this gate regenerative?
2. [M, None, 3.2] A diode-based digital gate is shown is shown in Figure 3.62.
   - **a.** What logic function does this gate implement? Assume that $V_{in,low} = 0$ V and $V_{in,high} = 3$ V.
   - **b.** Plot the VTC for V_{in} ranging from -5 V to 5 V. You may assume that the diode represents an open circuit when off and imposes a fixed voltage drop of 0.7 V when on (or is represented by an ideal diode).

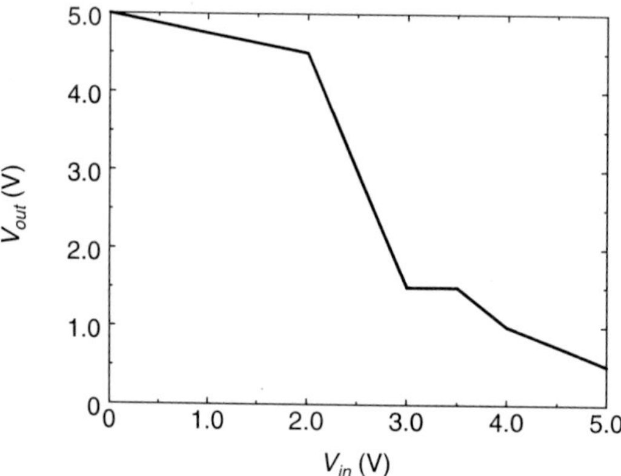

Figure 3.61 Voltage transfer characteristic of fictitious inverter.

c. Is the gate regenerative? Determine the input and output resistance for high and low inputs (assume $V_{in,low} = -3$ V and $V_{in,high} = 3$ V).

d. Determine the power consumption of the gate for $V_{in} = 0$ V and $V_{in} = 3$ V. Assume that all internal and external capacitances can be ignored.

e. Derive the VTC for a fan-out of 1 identical gate.

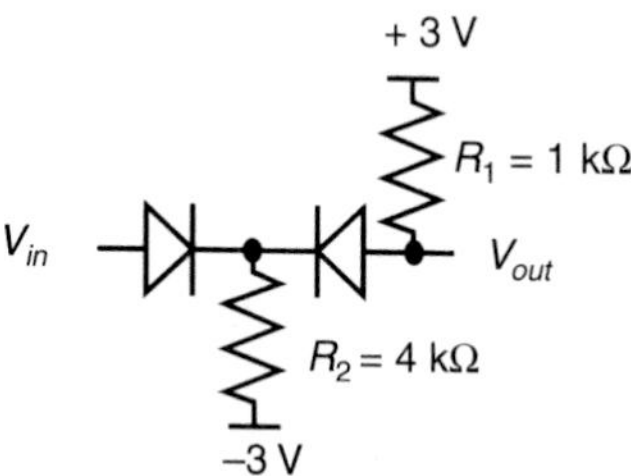

Figure 3.62 A diode-based digital gate.

3. [E, None, 3.3.2] The gate of Figure 3.63a uses a fictitious device as load. The device is characterized by the *I-V* curve of Figure 3.58b. Determine V_{OH} and V_{OL} of the gate. (Hint: use a graphical solution approach). Assume $V_T = 1$ V, $V_{DD} = 5$ V, $\lambda_n = 0$ V^{-1}, $k_n = 0.2286$ mA/V^2 for the NMOS transistor.

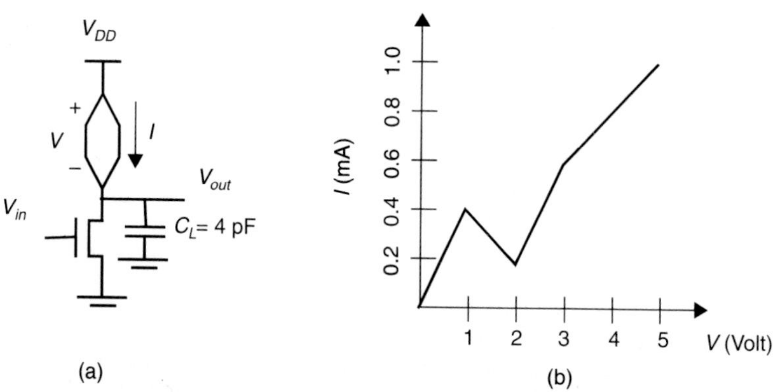

Figure 3.63 An inverter with a fictitious load, whose *I-V* characteristic is shown in (b).

4. [M, SPICE, 3.3.2] The layout of a static CMOS inverter is given in Figure 3.64. ($1\lambda = 0.6\ \mu m$).

a. Determine the sizes of the NMOS and PMOS transistor.

b. Derive the VTC and its parameters (V_{OH}, V_{OL}, V_M, V_{IH}, and V_{IL}).

c. Is the VTC affected when the output of the gates is connected to the inputs of 4 similar gates?

d. Compare your results with the VTC obtained by SPICE (using the LEVEL-2 model provided in Chapter 2).

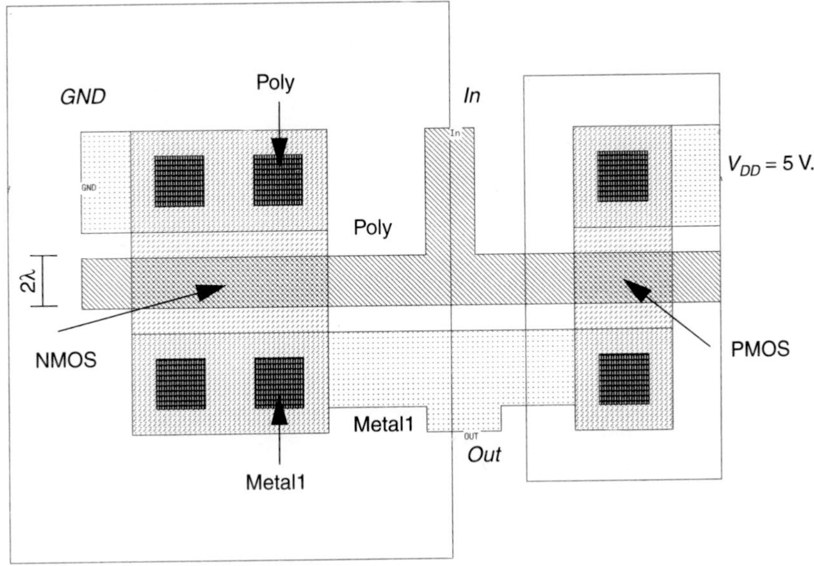

Figure 3.64 CMOS inverter layout.

5. [E, None, 3.3.2] Redesign the inverter of Figure 3.64 to achieve a switching threshold of approximately 1.5 V (only transistor sizes are needed). How are the noise margins affected by this modification?

6. The text defines the V_{IH} and V_{IL} points of the VTC as the points where the gain equals -1. This leads to complex expressions. A simpler approach is to use a piecewise linear approximation for the VTC, as shown in Figure 3.65. The transient region is approximated by a straight line, whose gain equals the gain in the mid-point V_M. The crossover with the V_{OH} and the V_{OL} lines is used to define V_{IH} and V_{IL} points.

a. Using this approach, derive expressions for the noise margins of the CMOS inverter. Assume that $k_n = k_p$, $V_M = V_{DD}/2$, and $V_{Tn} = |V_{Tp}|$. To solve this problem, you have to take into account the effect of channel-length modulation. This can be accounted for by using the following expression for the small-signal resistance in the saturation region: $r_{o,sat} = 1/(\lambda I_D)$ (instead of the value of ∞, used in the text).

b. Compare the obtained results with the values derived using the gain $=-1$ approach. For the comparison, assume $V_{DD} = 5V$, $V_{Tn} = |V_{Tp}| = 0.75$ V, $k_n = k_p = 40\ \mu A/V^2$, $\lambda_n = 0.06$, $\lambda_p = 0.2$.

7. [M, SPICE, 3.3.2] The noise margins of a CMOS inverter are highly dependent on the sizing ratio, $r = k_p/k_n$, of the NMOS and PMOS transistors. Use SPICE with $V_{Tn} = |V_{Tp}|$ to answer the following:

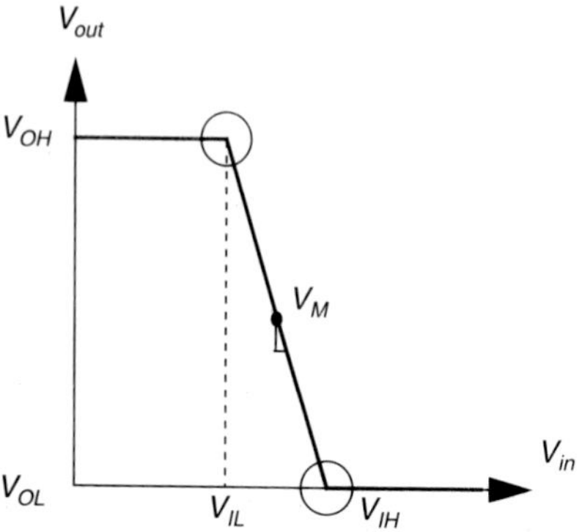

Figure 3.65 A different approach to derive V_{IL} and V_{IH}.

a. What value of *r* results in equal noise margins? Discuss qualitatively why.

b. As the ratio *r* is decreased to values below 1, discuss the effect on NM_H and NM_L. Give a qualitative explanation.

8. [M, SPICE, 3.3.2] Figure 3.66 shows an NMOS inverter with resistive load.

a. Discuss qualitatively why this circuit behaves as an inverter.

b. Find V_{OH} and V_{OL}.

c. Calculate V_{IH} and V_{IL}.

d. Find NM_L and NM_H, and plot the VTC using SPICE.

e. Compute the average power dissipation for: (i) $V_{in} = 0$ V and (ii) $V_{in} = 5$ V

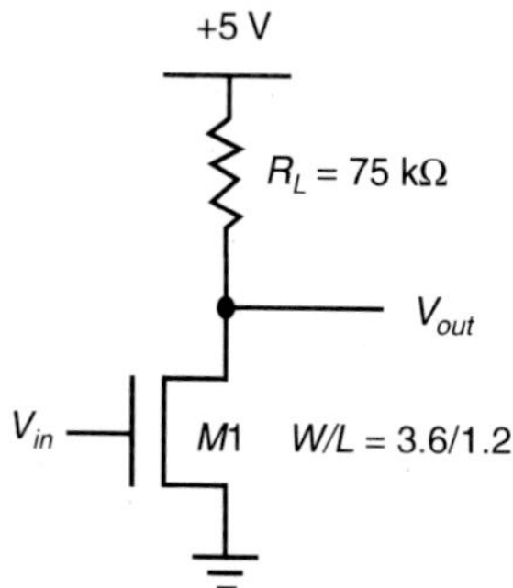

Figure 3.66 Resistive-load inverter

9. [E, SPICE, 3.3.2] For this problem refer to the inverter of Problem 8.

a. Sketch the VTCs for R_L = 37k, 75k, and 150k on a single graph (use SPICE).

b. Comment on the relationship between the critical VTC voltages (i.e., V_{OL}, V_{OH}, V_{IL}, V_{IH}) and the load resistance, R_L.

c. Do high or low impedance loads seem to produce more ideal inverter characteristics? Why?

10. [M, None, 3.3.2] Another MOS inverter, known as the *saturated enhancement load* inverter, is shown in Figure 3.67.

a. Find V_{OH} (assume V_{in} = 0.25 V) and V_{OL}.

b. Calculate the *W/L* required for M2 to achieve a V_{OL} of 0.25 V.

c. Calculate V_{IH} and V_{IL}.

d. Find NM_L and NM_H, and sketch the VTC.

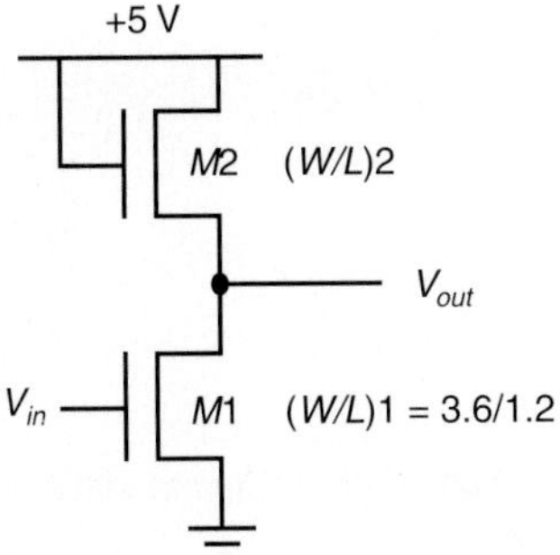

Figure 3.67 Saturated enhancement load inverter.

11. [M, SPICE, 3.3.3] For the inverter design of Example 3.6, do the following (assuming abrupt junctions and $\phi_0 = 0.6$ V).

a. Derive the values of all parasitic capacitances, assuming that the supply voltage is set at 3.3 V.

b. Calculate an effective output-loading capacitance based on these parasitics.

c. Find t_{plh} and t_{phl} for this basic inverter.

d. What is the maximum fan-out of identical inverters this gate can drive before its delay becomes larger than 2 ns.

e. Derive the propagation delay using SPICE.

12. [E, None, 3.3.3] For the inverter of Figure 3.66 and an output load of 3 pF,

a. Calculate t_{plh}, t_{phl}, and t_p.

b. Are the rising and falling delays equal? Why or why not?

c. Compute the static and dynamic power dissipation (assume that gate is clocked as fast as possible).

13. [M, SPICE, 3.3] Consider a CMOS inverter with a load capacitance of 60 fF. The NMOS device has a *W/L* ratio equal to 3.6/1.2, while the PMOS equals 10/1.2. The supply voltage V_{DD} is set to 5 V. Use SPICE to compute the energy consumed during a low-to-high transition followed by a high-to-low transition at the output.

a. Assume that the input waveform has a rise and fall time of 5 nsec. Determine how much of the energy consumption is due to direct path current.

b. Repeat part (a) for rise/fall times of 0.5 nsec and 20 nsec at the input. Determine which input waveform minimizes the dissipation due to direct path currents.

14. [M, None, 3.3.3] An NMOS transistor is used to charge up a large capacitor, as shown in Figure 3.68.

a. Determine the t_{pLH} of this circuit, assuming an ideal step from 0 to 10 V at the input node.

b. Assume that a resistor R_S of 5 kΩ is used to discharge the capacitance to ground. Determine t_{pHL}.

c. Determine how much energy is taken from the supply during the charging of the capacitor. Determine how much of this is dissipated in M1. How much is dissipated in the pull-down resistance during discharge? How does this change when R_S is reduced to 1 kΩ.

d. The NMOS transistor is replaced by a PMOS device, sized so that k_p is equal to the k_n of the original NMOS. Will the resulting structure be faster? Explain why or why not.

15. [M, None, 3.3.5] Consider scaling a CMOS technology by S > 1. In order to maintain compatibility with existing system components, you decide to use constant voltage scaling.

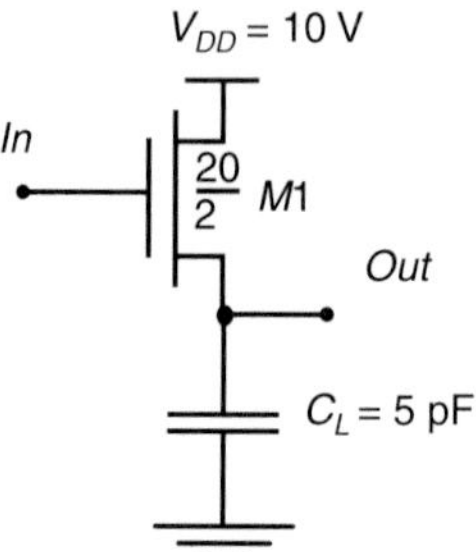

Figure 3.68 Circuit diagram with annotated *W/L* ratios

a. In traditional constant voltage scaling, transistor widths scale inversely with S, $W \propto 1/S$. To avoid the power increases associated with constant voltage scaling, however, you decide to change the scaling factor for *W*. What should this new scaling factor be to maintain approximately constant power. Assume long-channel devices (i.e., neglect velocity saturation).

b. How does delay scale under this new methodology?

c. Assuming short-channel devices (i.e., velocity saturation), how would transistor widths have to scale to maintain the constant power requirement?

16. [M, None, 3.4.3] Consider the ECL circuit shown in Figure 3.69. Let $V_{CC1} = V_{CC2} = 0$ V, $V_{EE} = -3.3$ V, $R_E = 400\ \Omega$, $V_{BE(on)} = 0.7$ V, $V_{CE(sat)} = 0.1$ V, $\beta_F = 70$, and $V_R = -1.5$ V.

a. Calculate the values of resistors R_B and R_C such that $V_{OH} = -1$ V and $V_{OL} = -2$ V at V_{out}.

b. Find the static power consumption for: (i) $V_{in} = V_{OL}$ and (ii) $V_{in} = V_{OH}$.

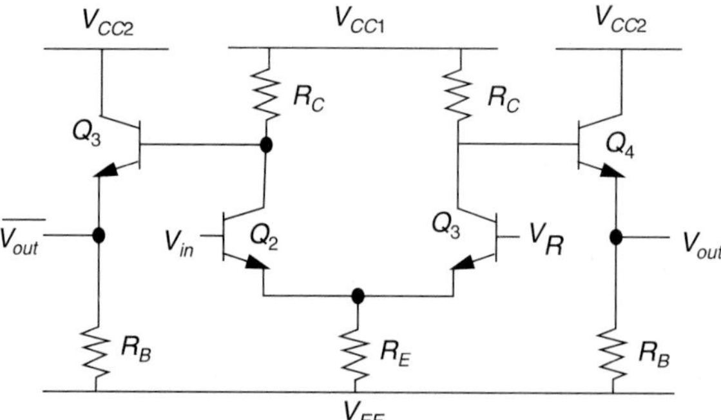

Figure 3.69 Simple ECL gate.

17. [E, SPICE, 3.4.3] Derive the voltage transfer characteristic of the ECL gate of Figure 3.69 using SPICE. Assume the parameter values of Problem 16 and use the following values for the resistances: $R_B = 5\ k\Omega$, $R_C = 250\ \Omega$. Determine also the value of the static power consumption for the various input conditions..

18. [C, None, 3.4.3] Shown in Figure 3.70 is an alternative implementation of an ECL device where $V_{CC} = 0$ V, $V_{EE} = -5$ V, $V_R = -2.4$ V, and $V_{BE(on)} = 0.7$ V. Neglect base currents.

a. Sketch the VTC, giving numerical values for V_{OH}, V_{OL}, V_{IH}, V_{IL}.

b. What is the maximum steady-state current drawn from the supply (with zero fan-out and with $V_{in} = 0$ V)?

c. This circuit is particularly sensitive to digital noise (i.e., V_{CC} bounce). Why? How would you change the circuit to reduce this effect?

d. What advantage does the two-transistor output stage formed by Q_3 and Q_4 have over the standard single-transistor ECL output stage?

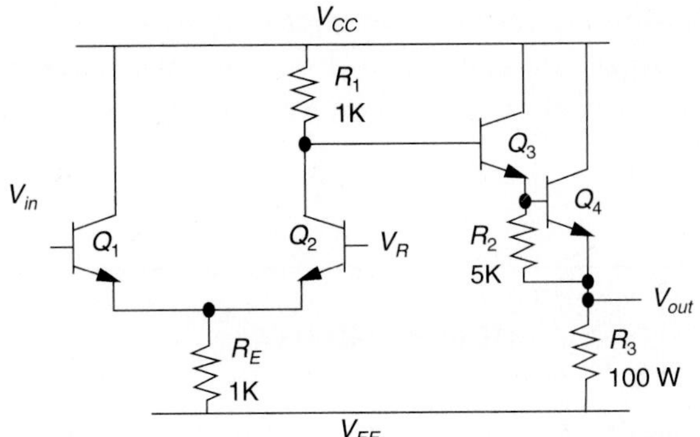

Figure 3.70 Alternative ECL gate

19. [M, None, 3.4.3] Consider the ECL gate of Figure 3.69. Let $V_{CC1} = V_{CC2} = 0$ V, $V_{EE} = -3.3$ V, $V_{BE(on)} = 0.7$ V, $V_{BE(sat)} = 0.8$ V, $V_{CE(sat)} = 0.1$ V, and $V_R = -1.2$ V. Finally, assume a logic swing of 0.8 V symmetrical about the reference voltage, and use compatible logic levels at inputs and outputs. Neglect base currents.

a. Calculate the values of resistors R_B, R_C, and R_E required to achieve a maximum emitter current of 3 mA for each transistor.

b. Sketch the VTC, showing the values of all the breakpoints.

c. Compute the noise margins.

20. [M, SPICE, 3.4.5] For this problem, use the circuit of Figure 3.69 with the parameters of Problem 19. Use $R_E = 250\ \Omega$, $R_C = 120\ \Omega$, and $R_B = 10\ \text{k}\Omega$. Assume a fan-out of one identical inverter.

a. Calculate the parasitic loading capacitances of the non-inverting output.

b. Compute t_{plh} and t_{phl} at V_{out}.

c. Verify your solution using SPICE.

21. [C, None, 3.4] A variant of an ECL gate is shown in Figure 3.71.

a. Determine R_E and R_C such that the logic swing at the output equals 0.8 V, while the maximum static power consumption of the gate equals 50 mW. You may assume that the input swings between 1.1 V and 1.9 V.

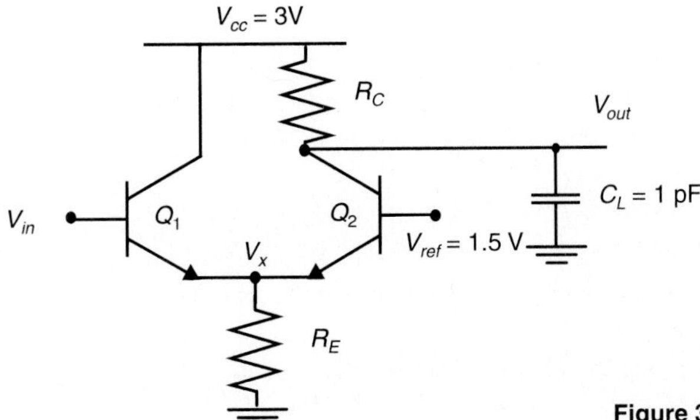

Figure 3.71 Single ended ECL gate.

b. Determine t_{pLH}. You may assume that all internal capacitances can be ignored and that C_L is the only capacitance of interest. Assume the following values: $V_{swing} = 1$ V, $V_{in(low)} = 1$ V, $V_{in(high)} = 2$ V, $R_E = 800\ \Omega$, $R_C = 1$ kΩ.

c. Determine t_{pHL}. Use the parameter values of part *b*.

d. Determine the dynamic energy dissipation during a low-to-high transition at the output.

e. Will any of the two bipolar transistors ever saturate (for V_{in} between 0 and 3 V)? Explain your answer.

DESIGN PROBLEM

Using the 1.2 μm CMOS introduced in Chapter 2, design a static CMOS inverter that meets the following requirements:

1. Matched pull-up and pull-down times (i.e., $t_{pHL} = t_{pLH}$).
2. $t_p = 5$ nsec (± 0.1 nsec).

The load capacitance connected to the output is equal to 4 pF. Notice that this capacitance is substantially larger than the internal capacitances of the gate.

Determine the *W* and *L* of the transistors. To reduce the parasitics, use minimal lengths ($L = 1.2$ μm) for all transistors. Verify and optimize the design using SPICE after proposing a first design using manual computations. Compute also the energy consumed per transition. If you have a layout editor (such as MAGIC) available, perform the physical design, extract the real circuit parameters, and compare the simulated results with the ones obtained earlier.

CHAPTER

DESIGNING COMBINATIONAL LOGIC GATES IN CMOS

In-depth discussion of logic families in CMOS—static and dynamic, pass-transistor, nonratioed and ratioed logic

Optimizing a logic gate for area, speed, or robustness

Low-power circuit-design techniques

4.1 Introduction

The preceding chapter deals extensively with the design of the inverter in both MOS and bipolar technologies. This knowledge is now extended to address the design of simple digital gates such as NOR, NAND, and XOR structures. Before discussing all possible digital gates, we first restrict our study to the class of the *combinational logic* or *non-regenerative* circuits. These gates have the property that at any point in time, the output of the circuit is related directly to its input signals by some Boolean expression (ignoring the short propagation delay of the composing gates). No intentional connection between outputs and inputs is present. This class of circuits is so important that it is discussed in both this chapter and the next.

In another class of circuits, known as *sequential* or *regenerative* circuits, the output is not only a function of the current input data, but also of previous values of the input signals. Circuits such as registers, counters, oscillators, and memory "remember" past events and hence have a sense of *history*. A common characteristic of sequential circuits is that one or more outputs are intentionally connected back to inputs. The difference between combinational and sequential circuits is illustrated in Figure 4.1.

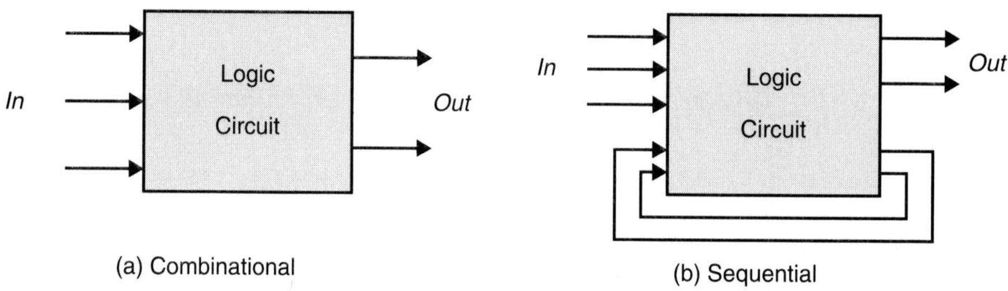

Figure 4.1 Classification of logic circuits.

Combinational logic forms the core of most digital integrated circuits such as fast arithmetic units and controllers. The design requirements imposed on the logic circuitry can vary widely. *Area* is often the prime concern, as it has a direct impact on cost. In many state-of-the-art designs, *speed* tends to be the dominating requirement. Contemporary microprocessors are excellent examples of designs in this class. For other applications, minimizing the *power consumption* is crucial, as in the design of portable applications such as mobile telephones.

These different design requirements generally translate into the use of different circuit styles, or even different manufacturing technologies. This chapter gives an overview of the most popular design techniques commonly used in CMOS technology. Chapter 5 extends this analysis to other technologies such as bipolar or GaAs. The different approaches are evaluated and compared using actual design examples. The initial discussions concentrate mainly on the minimization of either the area or the delay of a design. While power consumption used to be considered only as an afterthought, it is rapidly becoming an important performance criterion. Hence, a discussion of design techniques for low power appears at the end of the chapter.

4.2 Static CMOS Design

The static CMOS inverter discussed in Chapter 3 has excellent properties in many areas: low sensitivity to noise and process variations, excellent speed, and low power consumption. Most of those properties are carried over to more complex logic gates implemented using the same circuit topology. Unfortunately, complex static CMOS gates such as NAND gates with three or more inputs become large and slow. Other design styles have been devised to address this issue. In this section, we sequentially address the complementary, the ratioed, and the pass-transistor logic styles, all of which belong to the class of the *static* circuits. This means that at every point in time (except during the switching transients), each gate output is connected to either V_{DD} or V_{ss} via a low-resistance path. Also, the outputs of the gates assume at all times the value of the Boolean function implemented by the circuit (ignoring, once again, the transient effects during switching periods). This is in contrast to the *dynamic* circuit class, that relies on temporary storage of signal values on the capacitance of high-impedance circuit nodes. This approach has the advantage that the resulting gate is simpler and faster. On the other hand, its design and operation are more involved than those of its static counterpart, due to an increased sensitivity to noise. The design and analysis of dynamic gates is discussed in the Section 4.3.

4.2.1 Complementary CMOS

A static CMOS gate, as represented by the CMOS inverter of Chapter 3, is a combination of two networks, called the *pull-up network* (PUN) and the *pull-down* network (PDN) (Figure 4.2). The PUN consists solely of PMOS transistors and provides a conditional connection to V_{DD}. The PDN potentially connects the output to V_{SS} and contains only NMOS devices. The PUN and PDN networks should be designed so that, whatever the value of the inputs, *one and only one* of the networks is conducting in steady state. In this way, a path always exists between V_{DD} and the output F, realizing a high output ("one"), or, alternatively, between V_{SS} and F for a low output ("zero"). This is equivalent to stating that the output node is always a *low-impedance* node in steady state.

In constructing the PDN and PUN networks, the following observations should be kept in mind:

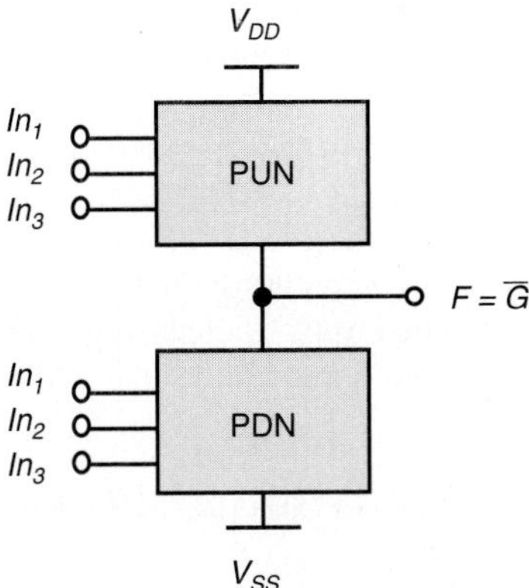

Figure 4.2 Complementary logic gate as a combination of a PUN (pull-up network) and a PDN (pull-down network).

- A transistor (both NMOS and PMOS) can be thought of as a switch controlled by its gate signal.
- An NMOS switch closes when the controlling signal is high. A PMOS transistor, on the other hand, acts as an inverse switch; that is, the switch closes when the controlling signal is low.
- The PDN is constructed of NMOS devices, while PMOS transistors are used in the PUN. The main reason for this choice is that NMOS transistors produce "strong zeros" and PMOS devices generate "strong ones". We can clarify this statement with the following simple example. Assume that we try to discharge capacitance C_L to *GND* through either an NMOS transistor (with the gate connected to V_{DD}) or a PMOS device (with the gate connected to *GND*). The NMOS transistor discharges the capacitor all the way to *GND* (hence producing a strong zero), while the PMOS device shuts off when $V_{out} = |V_{Tp}|$ is reached (producing a weak zero). The former case is clearly preferable. Similar considerations lead to the choice of PMOS transistors in the PUN.
- A series connection of switches corresponds to an *AND*-operation, and a parallel connection of switches is equivalent to an *OR*-ing of the inputs.
- The pull-up and pull-down networks are *dual* networks, which means that a parallel connection of transistors in the pull-up network corresponds to a series connection of the corresponding devices in the pull-down network and vice versa.[1]

  This property is understood from the following argument. Suppose that the pull-down network of a CMOS gate is known and implements the logic function *G*. Since the PDN connects to *GND*, the CMOS gate implements the inverse function $F = \overline{G}$. We wish to derive the structure of the corresponding PUN. Since the PUN connects to V_{DD}, it has to be conducting when F = TRUE (or in other words, it must implement *F*). Taking into account the above, as well as the fact that the PMOS transistors of the PUN are inverse switches, the following relation has to be valid:

$$\overline{G(In_1, In_2, In_3, \ldots)} \equiv F(\overline{In_1}, \overline{In_2}, \overline{In_3}, \ldots) \tag{4.1}$$

  This condition is met if (but not only if) *F* and *G* are dual equations, where each *AND* operation in *F* is replaced by an *OR* in *G* and vice-versa. This is a direct consequence of De Morgan's theorems, which state the following identities:

$$\begin{aligned} \overline{A+B} &= \overline{A}\,\overline{B} \\ \overline{AB} &= \overline{A}+\overline{B} \end{aligned} \tag{4.2}$$

- The complementary gate is *inverting* (implementing functions such as NAND, NOR, and XNOR). Implementing a noninverting Boolean function (such as AND OR, or XOR) in one stage is not possible and requires the addition of an extra inverter stage.

[1] The duality is a satisfying but not necessary requirement. Other valid PUN/PDN combinations can be envisioned, some of which will be illustrated in later chapters.

Example 4.1 Two-Input NAND Gate

Figure 4.3 shows a simple two-input NAND gate ($F = \overline{A \cdot B}$). The PUN consists of two parallel PMOS transistors. This means that F is 1 if $A = 0$ or $B = 0$, which is equivalent to $F = \overline{A} + \overline{B} = \overline{A \cdot B}$. The PDN, which consists of two series NMOS transistors, provides a connection to GND when both $A = 1$ and $B = 1$. Consequently, it implements $G = A \cdot B = \overline{F}$, which is consistent with the PUN network. It can be easily verified that the output F is always connected to either V_{DD} or GND, but never to both.

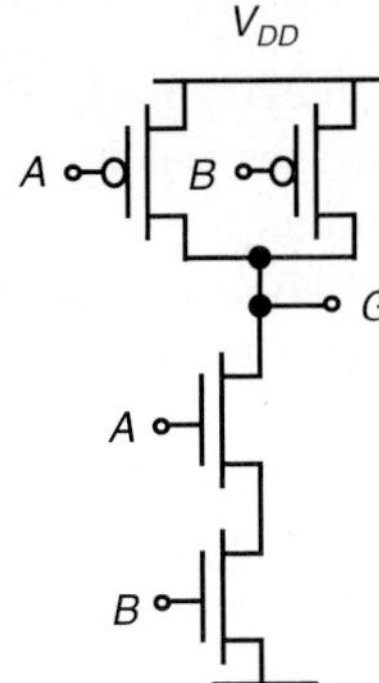

Figure 4.3 Two-input NAND gate in complementary static CMOS style.

Problem 4.1 Complex CMOS Gate

A more complex static CMOS gate is shown in Figure 4.4. The pull-up and pull-down circuits once again form dual networks. Derive the logic function of this gate and verify that for every possible input combination there always exists a path to either V_{DD} or GND.

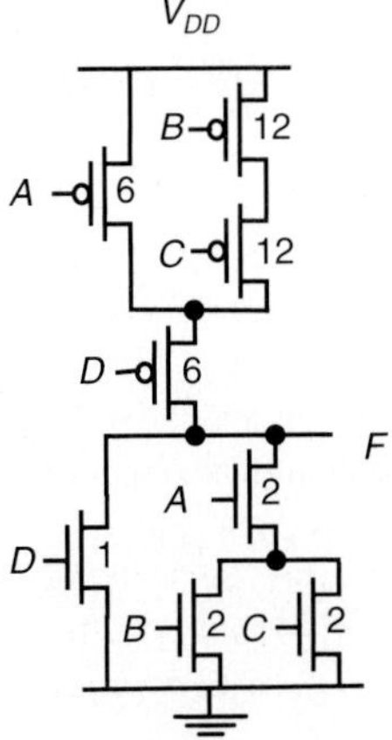

Figure 4.4 Complex complementary CMOS gate. The numbers indicate transistor sizes. Minimum-size transistors are denoted with a unit value. PMOS transistors are tripled in size with respect to NMOS devices.

Properties of Complementary CMOS Gates

Static CMOS gates inherit all the nice properties of the basic CMOS inverter as introduced in Chapter 3:

- *High noise margins.* V_{OH} and V_{OL} are at V_{DD} and *GND*, respectively.
- *No static power consumption*, as there is never a direct path between V_{DD} and V_{SS} (*GND*) in steady-state mode.
- *Comparable rise and fall times* (under the appropriate scaling conditions).

The last point requires some further analysis. When studying the CMOS inverter, we observed that identical rise and fall times are obtained under the condition that the PMOS (PUN) and NMOS (PDN) networks have identical current-driving capabilities. The smaller mobility of the PMOS transistor requires that those devices be widened by a factor μ_n/μ_p. Similar considerations are valid for the more complex gates as well. The analysis is complicated by the fact that the resistance of the PUN and PDN networks is a function of the value of the input signals. In general, we should consider the *worst-case condition*.

Studying the dynamic behavior of those complex gates using the full transistor model quickly becomes intractable. The switch model, already introduced in Figure 3.14, is often used to approximate the transient behavior of a complex gate. In this approach, the transistor is modeled as a switch with an infinite off-resistance and a fixed resistance R_{on} in the on-state. R_{on} is chosen so that the equivalent *RC*-circuit has a propagation delay identical to the original transistor-capacitor combination. Notice that R_{on} is inversely proportional to the *W/L* ratio of the transistor, which can therefore be considered as a conductivity factor.

The on-resistance of an MOS transistor depends upon the operation point and varies during the switching transient. Similar to the approach taken when computing the (dis)charge current (Section 3.3.3), a reasonable approximation is to use a fixed R_{on}, which is the average value of the resistances at the end point of the transitions. For instance, when computing t_{pHL} of a CMOS inverter, R_{on} can be approximated in the following way:

$$\begin{aligned} R_{on} &= \frac{1}{2}(R_{NMOS}(V_{out} = V_{DD}) + R_{NMOS}(V_{out} = V_{DD}/2)) \\ &= \frac{1}{2}\left[\left(\frac{V_{DS}}{I_D}\right)_{V_{out} = V_{DD}} + \left(\frac{V_{DS}}{I_D}\right)_{V_{out} = V_{DD}/2}\right] \end{aligned} \tag{4.3}$$

Example 4.2 Computing R_{on}

For example, for the 1.2 μm CMOS process and with V_{DD} = 5 V, the following resistance values can be computed (taking the data from Table 3.3 and normalizing it to W/L_{eff} = 1):

$R_n(W/L_{eff} = 2) = (5\text{ V} / 0.46\text{ mA} + 2.5\text{ V} / 0.29\text{ mA}) / 2 = 9.7\text{ k}\Omega$ (for t_{pHL})

$R_n(W/L_{eff} = 1) = 9.7 \times 2 = 19.4\text{ k}\Omega$ (for t_{pHL})

$R_p(W/L_{eff} = 6) = (5\text{ V} / 0.57\text{ mA} + 2.5\text{ V} / 0.24\text{ mA}) / 2 = 9.6\text{ k}\Omega$ (for t_{pLH})

$R_p(W/L_{eff} = 1) = 9.6 \times 6 = 57.6\text{ k}\Omega$ (for t_{pLH})

Divide these values by (W/L_{eff}) for larger transistor sizes.

Deriving the propagation delay now becomes identical to the analysis of the resulting *RC* network. The equivalent circuit for a CMOS inverter is shown in Figure 4.5a. R_n and R_p should clearly be made identical to achieve similar values for t_{pHL} and t_{pLH}.

Consider now the two-input NAND gate of Figure 4.5b. Assume first that $R_n = R_p =$ resistance of a minimum-size NMOS transistor. When analyzing the performance of a complex circuit, it is important to realize that the operation speed of the circuit is determined by the *worst-case delay* over the complete set of all possible input combinations. When sizing the transistors in a gate with multiple fan-ins, we should therefore pick the combination of inputs that triggers the worst-case conditions. Consider the low-to-high transition for the two-input NAND gate. The worst-case scenario is activated when only a single PMOS transistor is turned on. Activating the second PMOS device only reduces the propagation delay, as the resistances are connected in parallel. The worst-case value of t_{pLH} is therefore estimated as $0.69R_pC_L$ (which expresses the propagation delay of an *RC* network).

On the other hand, t_{pHL} equals $2 \times 0.69R_nC_L$, because in the worst (and only possible) scenario the two pull-down devices are connected in series. In order to make the pull-down network as fast as the pull-up, it is necessary to double the width of the NMOS devices. A similar analysis shows that the PMOS devices must be doubled in width to design a two-input NOR gate with similar worst-case rise and fall characteristics (Figure 4.5c).

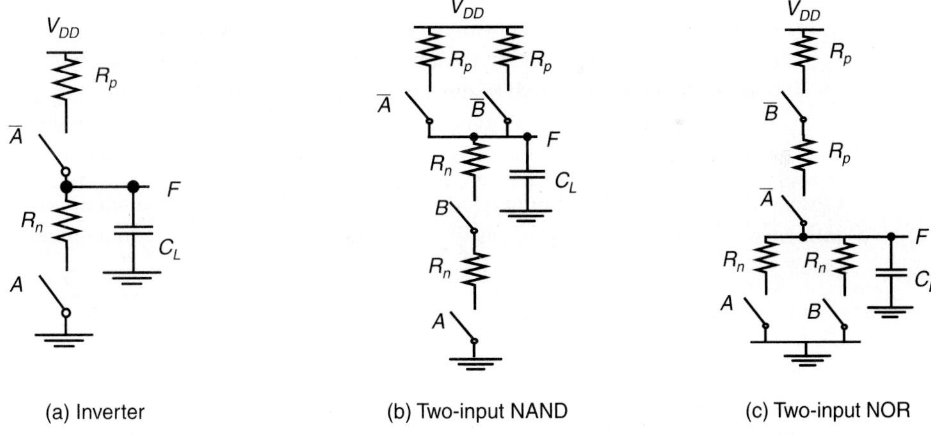

Figure 4.5 Switch-level models of complementary CMOS gates. It is assumed that the load capacitance C_L dominates.

Using those scaling rules, the proper device sizes for the complex gate of Figure 4.4 are derived. The resulting transistor sizes are annotated on the figure. It is assumed that the minimum PMOS device is three times wider than the minimum NMOS transistor to compensate for the reduced mobility.

Problem 4.2 Sizing of Transistors in Complementary CMOS Gates

Describe how the transistor sizes of Figure 4.4 were derived.

Notwithstanding the excellent properties of the complementary CMOS gates, the above analysis demonstrates that some important problems arise when implementing gates with large fan-in values. This makes this circuit style less interesting for the implementation of high-speed, complex gates.

1. A gate with N inputs requires $2N$ transistors (N in the PUN, N in the PDN). Other circuit styles (which will be introduced shortly) require at most $N + 1$ transistors. While this difference is not that significant for smaller gates (an inverter needs two transistors in both approaches), it becomes dominant for gates with a large fan-in—a complementary gate with a fan-in of 4 needs eight transistors compared to the potential minimum of five! This obviously has a substantial impact on the area.

2. The propagation delay of a complementary CMOS gate deteriorates rapidly as a function of the *fan-in*. First of all, the larger number of transistors ($2N$) increases the overall capacitance of the gate (both at the output node and at the internal nodes). Secondly, a series connection of transistors in either the PUN or PDN slows the gate as well, because the effective (dis)charging resistance is increased. At first it appears as if we can resolve this by widening the devices in a transistor chain, as was discussed above. This does, however, not improve the performance as much as expected from the first order model: widening a device also increases its gate and diffusion capacitances, which has an adverse affect on the gate performance.

3. Finally, *fan-out* has a larger impact on the gate delay in complementary CMOS than in some other logic styles. In the complementary circuit style, each input connects to both an NMOS and a PMOS device and presents a load to the driving gate equal to the sum of their gate capacitances. This is not the case for other circuit techniques, where an input only connects to a single NMOS (or PMOS) transistor.

Observations 2 and 3 are summarized by the following formula, which approximates the influence of fan-in and fan-out on the propagation delay of the complementary CMOS gate:

$$t_p = a_1 FI + a_2 FI^2 + a_3 FO \tag{4.4}$$

where FI and FO are the fan-in and fan-out of the gate, respectively, and a_1, a_2, and a_3 are weighting factors, which are a function of the technology. The linear dependence on the fan-out can be understood from the fact that the load capacitance increases linearly with the fan-out. The quadratic dependence on the fan-in is explained with the aid of Figure 4.5b. Increasing the fan-in raises both the capacitance C_L (as the capacitance is proportional to the number of transistors) and the (dis)charging resistance in a linear way, assuming that the transistors are not scaled. This statement is confirmed by simulation results, as shown in Figure 4.6, which plots the dependence of the propagation delay (for a NAND gate) as a function of the fan-in. The t_{pLH} increases linearly, as a result of the linearly increasing value of the output capacitance. The simultaneous increase in the values of the pull-down resistance and the load capacitance results in a quadratic curve for t_{pHL}. Gates with a fan-in greater than or equal to 4 become excessively slow and must be avoided.

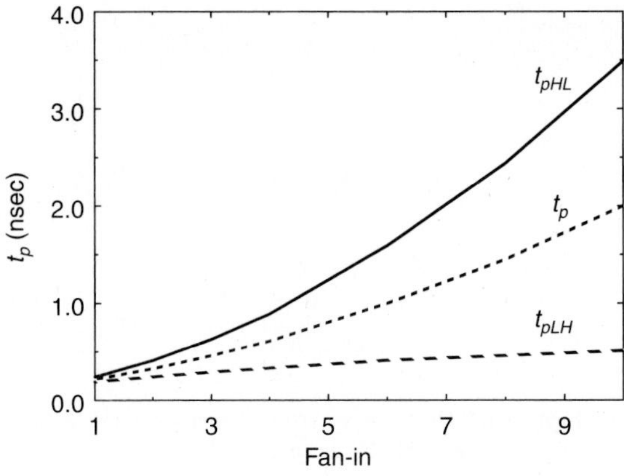

Figure 4.6 Propagation delay of CMOS NAND gate as a function of fan-in. (A fan-out of one inverter is assumed, and all pull-down transistors are minimal size)

Design Techniques for Large Fan-in

Several approaches may be used to alleviate this problem.

1. Transistor Sizing

Increasing the transistor sizes, as illustrated in Figure 4.4, increases the available (dis)charging current and decreases the second-order factor in Eq. (4.4). However, as mentioned earlier, widening the transistors results in larger parasitic capacitors, which do not only affect the propagation delay of the gate in question, but also present a larger load to the preceding gate. This technique should, therefore, be used with caution. If the load capacitance is dominated by the intrinsic capacitance of the gate, widening the device only creates a "self-loading" effect, and the propagation delay is unaffected.

2. Progressive Transistor Sizing

The previous analysis assumes that all intrinsic capacitances can be lumped into a single load capacitance C_L and that no capacitance is present at the internal nodes of the pull-up and pull-down networks. Under those assumptions, making all transistors in a series chain equal in size makes sense. This model is an over-simplification, however, that becomes more and more incorrect for increasing fan-in. In the latter case, it is more appropriate to consider the network of Figure 4.7 (only the pull-down network is shown here). Deriving the delay of this circuit requires us to solve a network of capacitors and resistive switches. We defer techniques for doing so until later. Here, we only present an intuitive approach for optimizing the delay by sizing the devices progressively.

While transistor M_N has to conduct the discharge current of the load capacitance C_L, M_1 has to carry the discharge current from the total capacitance $C_{tot} = C_L + \ldots + C_3 + C_2 + C_1$, which is substantially larger. Consequently, a progressive scaling of the transistors is beneficial: $M_1 > M_2 > M_3 > M_N$. This technique has, for instance, proven to be advantageous in the decoders of memories, where gates with a large fan-in are common.

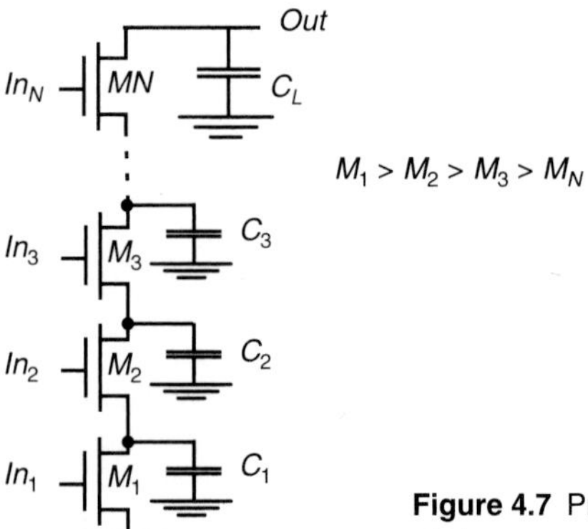

Figure 4.7 Progressive sizing of transistors in large transistor chains copes with the extra load of internal capacitances.

Example 4.3 Progressive Transistor Sizing

The effect of the progressive sizing approach is illustrated using the circuit of Figure 4.7 (with $N = 5$). C_L is set to 15 fF, while all internal capacitances are set to 10 fF. For the case where all NMOS transistors are minimum size, SPICE predicts a propagation delay (t_{pHL}) of 1.1 nsec. The transistors M_5 to M_1 are then made progressively wider in such a way that the width of the transistor is proportional to the total capacitor it has to discharge. M_5 is minimum size, $W_{M4} = W_{M5}(C_L + C_4)/C_L$, $W_{M3} = W_{M5}(C_L + C_4 + C_3)/C_L$, and so on. The resulting circuit has a t_{pHL} of 0.81 nsec, or a reduction with 26.5%! (This simulation takes into account that the intermediate capacitances increase with the widening of the transistors.)

For an excellent treatment on the optimal sizing of transistors in a complex network, we refer the interested reader to [Shoji88, pp. 131–143].

3. Transistor Ordering

Some signals in complex combinational logic blocks might be more critical than others. Not all inputs of a gate arrive at the same time (due, for instance, to the propagation delays of the preceding logical gates). An input signal to a gate is called *critical* if it is the last signal of all inputs to assume a stable value. The path through the logic which determines the ultimate speed of the structure is called the *critical path.*

Putting the critical-path transistors closer to the output of the gate can result in a speed-up. This is demonstrated in Figure 4.8. Signal In_1 is assumed to be a critical signal (or is part of the critical path of the combinational logic block). Suppose further that In_2 and In_3 are high and that In_1 undergoes a 0→1 transition. Assume also that C_L is initially charged high. In case (a), no path to *GND* exists until M_1 is turned on, which is unfortunately the last event to happen. The delay between the arrival of In_1 and the output is therefore determined by the time it takes to discharge $C_L + C_1 + C_2$. In the second case, C_1 and C_2 are already discharged when In_1 changes. Only C_L still has to be discharged, resulting in a faster response time.

Example 4.4 Transistor Ordering in a Four-input NAND gate

The t_{PHL} of an actual four-input NAND gate is simulated using SPICE for two extreme cases. The critical input is applied to the bottommost (case a) and the uppermost (case b) transistors

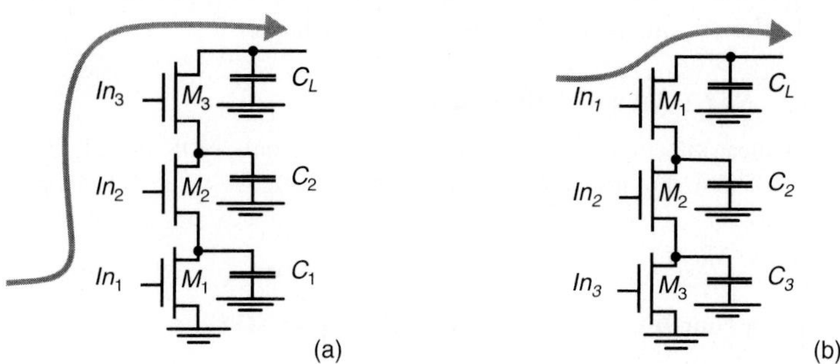

Figure 4.8 Influence of transistor ordering on propagation delay. Signal In_1 is the critical signal.

of the pull-down network. All other transistors are already enabled in advance. The resulting values of the propagation delay equal 717 nsec and 607 nsec, respectively. The ordering of the transistors yields a performance improvement of 15%.

---

4. Improved Logic Design

Manipulating the logic equations can reduce the fan-in requirements and hence reduce the gate delay, as illustrated in Figure 4.9. The quadratic dependency of the gate delay on fan-in makes the six-input NOR gate extremely slow. Partitioning the NOR-gate into two three-input gates results in a significant speed-up (as predicted by Eq. (4.4)), which offsets by far the extra delay incurred by turning the inverter into a two-input NAND gate.

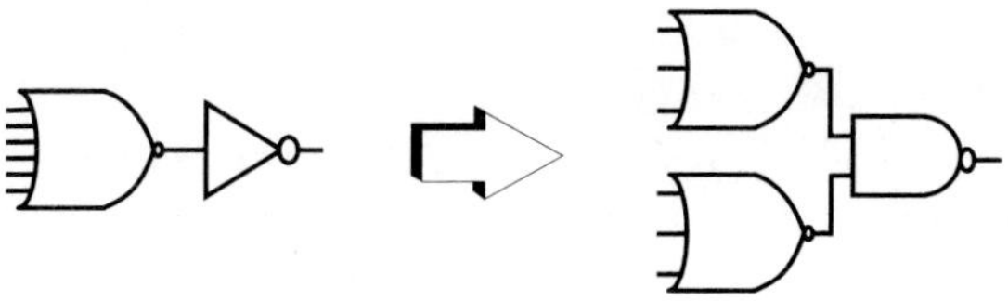

Figure 4.9 Logic restructuring can reduce the gate fan-in.

5. Use Another Circuit Style

To offset the negative impact of a large fan-in, designers have come up with a myriad of alternative CMOS design approaches. A number of representative styles are discussed in the subsequent sections.

The design techniques presented above deal with improving the performance of gates with a large fan-in. Often the speed of the gate is dominated by the fan-out factor. Driving large load capacitances with complex logic gates is expensive, because all transistors must be scaled upwards to provide sufficient current drive. For gates with a large fan-out, it is often beneficial to insert a buffer (i.e., an inverter or a cascade of inverters) between the complex gate and the fanout. The buffer can be scaled appropriately to drive the load capacitance, while the transistors of the complex gate can be of minimal size. The design of buffer circuits is treated extensively in Chapter 8.

Problem 4.3 NAND Versus NOR in Static CMOS

All Boolean functions can be implemented using only NOR or only NAND gates. Which approach (NAND-only or NOR-only) is more attractive when designing in complementary CMOS?

Example 4.5 A Four-Input Complementary CMOS NAND Gate

The layout of a four-input NAND gate is shown in Figure 4.10. No transistor sizing (besides the appropriate scaling of the PMOS devices for mobility) is applied. Hence all NMOS transistors have a (W/L) of (1.8/1.2), while the PMOS devices are set to (5.4/1.2).

To simplify the manual analysis, it is customary to replace the serial chain of NMOS transistors by a single device whose channel length is the sum of the lengths of the transistors in the chain. This is equivalent to stating that the resistance of the discharge network is similar to the series connection of the resistances of the individual transistors. The parasitic capacitances of this hypothetical transistor can be estimated as the sum of the capacitances of the individual devices (this is a worst-case scenario). Remember, however, that the merging of the series transistors into a single device is a simplification that ignores a number of second-order influences, such as the body effect and the distributed nature of the parasitic capacitors.

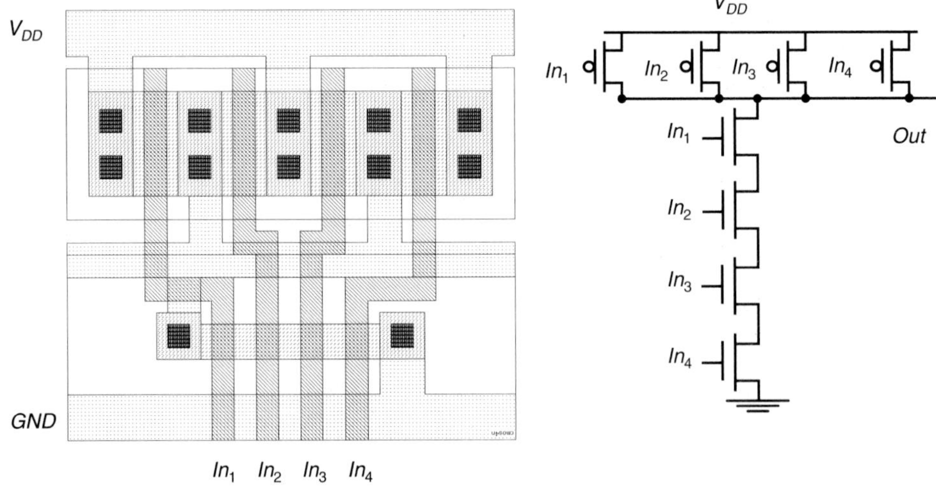

Figure 4.10 Layout and schematics of a four-input NAND gate in complementary CMOS. See also Colorplate 7.

Using techniques similar to those employed for the CMOS inverter in Chapter 3 (Table 3.2), we can compute the capacitance value at the output node from inspection of the layout. It is assumed that the output connects to a single, minimum-size inverter. The different contributions are summarized in Table 4.1. To determine the values of the diffusion capacitances, the areas and perimeters of the individual transistors are combined into a single contribution. From an inspection of the layout, we obtain the following numbers: Area(PMOS) =

$6 \times 9 \times 2 = 108\ \lambda^2$, Perimeter(PMOS) = $6 \times 2 \times 2 = 24\ \lambda$, Area(NMOS) = $4 \times 4 + 3 \times 1 + 2 \times 3 \times 3 = 37\ \lambda^2$, Perimeter(NMOS) = 15 (top transistor) + $2 \times 2 \times 3 = 27\ \lambda$. Determining the exact value of the gate-drain (and source!) capacitances is a lot more complex. In the PMOS section, only the switching transistor (one out of four in the worst case) is prone to the Miller effect and is multiplied by the factor 2. The other devices contribute a single overlap capacitance. The same is true for the PDN devices. Observe that the gate-source capacitances of the three uppermost transistors have to be accounted for as well ($C_{gd+sn} = 2C_{gd}$(Miller) + $3C_{gd} + 3C_{gs}$).

Table 4.1 Components of C_L (for high-to-low and low-to-high transitions).

Capacitor	Value (fF) (H→L)	Value (fF) (L→H)
C_{gd+sn}	$8 \times (1.55/2) = 6.2$	$8 \times (1.55/2) = 6.2$
C_{gdp}	$5 \times (4.65/2) = 11.63$	$5 \times (4.65/2) = 11.63$
C_{dbn}	$0.375 \times (0.3 \times 37 \times 0.36 + 0.8 \times 27 \times 0.6) = 6.36$	$0.611 \times (0.3 \times 37 \times 0.36 + 0.8 \times 27 \times 0.6) = 10.36$
C_{dbp}	$0.611 \times (0.5 \times 108 \times 0.36 + 0.135 \times 24 \times 0.6) = 13.07$	$0.375 \times (0.5 \times 108 \times 0.36 + 0.135 \times 24 \times 0.6) = 8.02$
C_{fanout}	15.2	15.2
C_L	52.46	51.41

The total load capacitance is about 150% higher than for the CMOS inverter discussed in the previous chapter. This increase is not as high as would be expected from a first-order inspection. When designing complex gates, it is often possible to reduce the diffusion area (and hence the capacitance) of a device by sharing it between two transistors. This approach is employed extensively in both the PDN and PUN networks of the design of Figure 4.10.

The values of the (dis)charge currents are determined using previously discussed techniques (cf. Section 3.3.4). For the pull-up network, we assume that only one transistor is active (worst case), while for the analysis of the pull-down network, we use the extra-long transistor equivalent. This results in an average pull-up current of 0.4 mA (identical to the value of Table 3.3), while the pull-down current equals 0.095 mA, a reduction by a factor of 4. This translates to estimated values of $t_{pLH} = 0.3$ nsec and $t_{pHL} = 1.4$ nsec. As expected, a major mismatch exists between the high-to-low and low-to-high transitions.

The simulated transient response of the network is plotted in Figure 4.11. This response represents the worst case, as input In_4 toggles the bottom transistor of the NMOS transistor chain. The gate parameters are summarized in Table 4.2. While the t_{pLH} is estimated accurately, a large deficiency can be observed for the t_{pHL}. This can be attributed to two factors. First of all, the modeling of the pull-down network by a single device is grossly inadequate. It

Table 4.2 The dc and ac parameters of a four-input NAND gate in complementary CMOS.

Area	Static Current	V_{OH}	V_{OL}	V_M	NM_H	NM_L	t_{pHL}	t_{pLH}	t_p
533 μm^2	0 A	5 V	0 V	2.63 V	1.28 V	2.33 V	0.89 nsec	0.33 nsec	0.61 nsec

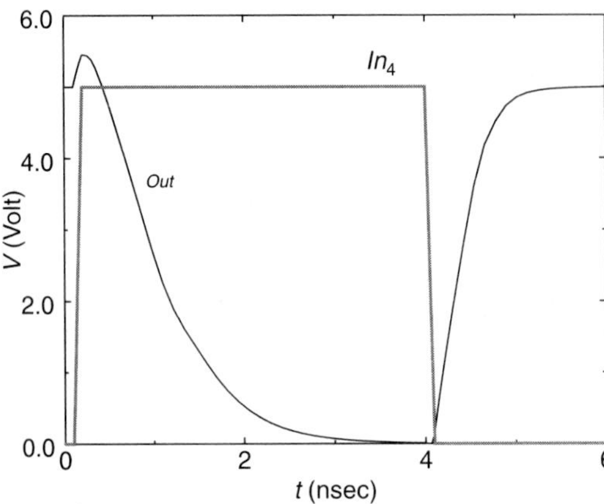

Figure 4.11 Simulated propagation delay of four-input NAND gate (for a fanout of one inverter).

ignores or misrepresents a number of parameters such as the body effect of the intermediate transistors as well as the operation modes of the individual devices. Lumping all capacitors into a single load capacitance is a worst-case scenario. Most importantly, the manual analysis used a SPICE level-1 model value for k_n, which was empirically derived for a channel length of 1.2 μm and for V_{GS} and V_{DS} = 5 V (as was discussed in Chapter 2). Connecting a number of transistors in series effectively increases the channel length, causing the transistors to operate in the long-channel mode (versus velocity-saturation mode). The discharge current is therefore higher than predicted by our simple transistor model. Replacing k_n with a more appropriate value would give a closer estimation. This demonstrates that when we are working with very small-geometry devices, manual analysis is mainly to be used for guidance. SPICE simulations are essential to obtain the actual results.

CAUTION: Although the circuit styles discussed in the following sections might sound very challenging and exciting, and might be superior to static CMOS in many respects, none of them has the *robustness and ease of design* of complementary CMOS. Therefore, use them sparingly and with caution. For designs that have no extreme area, complexity, or speed constraints, complementary CMOS is the recommended design style.

4.2.2 Ratioed Logic

Concept

One technique to reduce the circuit complexity in static CMOS is to revert to an approach similar to the one used in the bipolar RTL inverter, described in Chapter 3. Instead of a combination of active pull-down and pull-up networks, such a gate consists of an NMOS pull-down network that realizes the *logic function*, and a simple *load device* (Figure 4.12a). For an inverter, the PDN is no more than a single NMOS transistor. The load can be a passive device, such as a resistor, or an active element, such as a transistor. Let us first assume that both PDN and load can be represented as linearized resistors. The operation is as follows: For a low-input signal (or combination of inputs), the pull-down network is off,

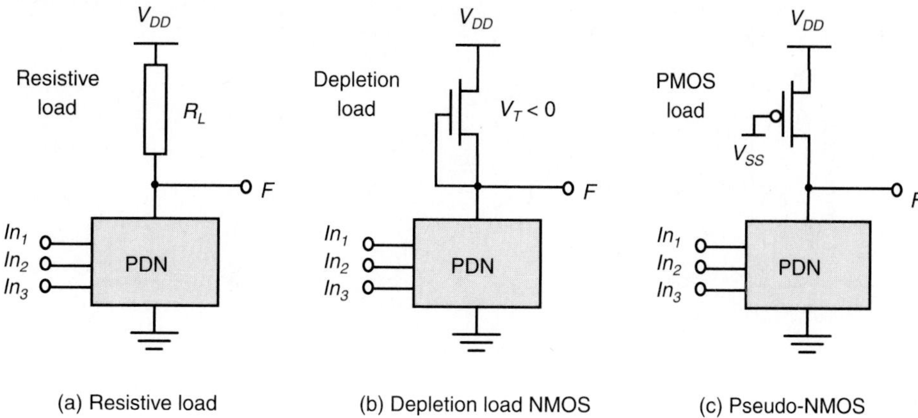

Figure 4.12 Ratioed logic gates.

and the output is pulled high by the load. When the input goes high, the driver transistor turns on, and the resulting output voltage is determined by the resistive division between the impedances of the pull-down and load networks:

$$V_{OL} = \frac{R_{PDN}}{R_L + R_{PDN}} V_{DD} \tag{4.5}$$

To keep the low noise margin high, it is important to choose $R_L >> R_{PDN}$. This style of logic is therefore called *ratioed*, because a careful scaling of the impedances (or transistor sizes) is required to obtain a workable gate. This is in contrast to the *ratioless* logic styles, such as complementary CMOS, where the low and high levels do not depend upon transistor sizes.

On the other hand, R_L should be capable of delivering as much current as possible during the transient operation to ensure fast switching action. Simplified expressions of the propagation delays of the ratioed gate are given in Eq. (4.6). These expressions are easily derived from the equivalent switch network.

$$\begin{aligned} t_{pLH} &= 0.69 R_L C_L \\ t_{pHL} &= 0.69 (R_L \parallel R_{PDN}) C_L \end{aligned} \tag{4.6}$$

The t_{pLH} is of particular concern. The requirement that R_L be simultaneously large (>> R_{PDN}, to maintain noise margins) and small (for performance as well as power dissipation considerations) has resulted in the creation of a wide variety of possible load configurations, some of which are presented below.

The first choice for the load device is a simple resistor (Figure 4.12a). Its load line, which describes the available charge current as a function of the output voltage of the gate, is plotted in Figure 4.13. As expected, a linear dependence is observed, as given by Eq. (4.7):

$$I_L = \frac{V_{DD} - V_{out}}{R_L} \tag{4.7}$$

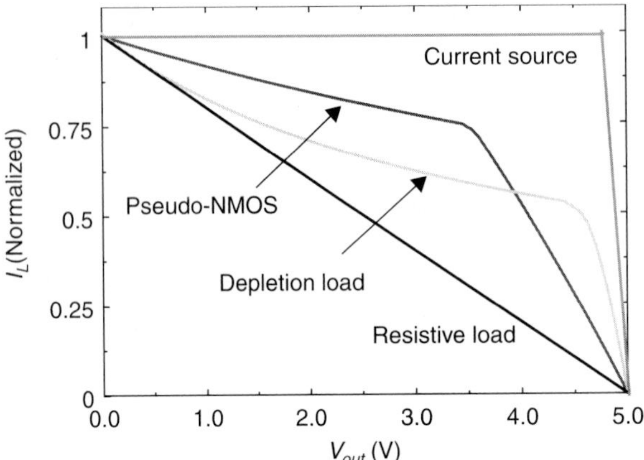

Figure 4.13 Load lines (I_L versus V_{out}) for various types of load devices. (The currents are normalized at V_{out} = 0 V.)

This linear relationship has the disadvantage that the charging current drops rapidly once V_{out} starts to rise. A more ideal load would be a current source, which has the property that available current is independent of the output voltage. One can easily determine that, for the same initial current level, the current source reduces t_{pLH} by 25%.

Problem 4.4 Resistive Versus Current-Source Load

Compare the delay of charging a capacitor from 0 to V_{DD} with a resistor R_L versus the delay using a current source. Assume that the value of the current source equals V_{DD}/R_L, which is the initial charging current available from the resistive load.

The *depletion load* inverter, shown in Figure 4.12b, was a first attempt to realize such a load and emerged as the most popular gate in the NMOS era.[2] The load consists of an NMOS depletion transistor with its gate connected to its source. A *depletion transistor* is a device with a negative threshold. In other words, the transistor is on for $V_{GS} = 0$. This is exactly the case in the circuit of Figure 4.12b, where the V_{GS} of the load device equals 0, and the transistor, being a depletion device, is always on. To a first approximation, the load acts as a current source, as its current in the saturation mode approximates

$$I_L = \frac{k_{n,\,load}}{2}(|V_{Tn}|)^2 \tag{4.8}$$

The actual load line is plotted in Figure 4.13. It deviates from the ideal current source characteristic for two reasons:

- The *channel length modulation* factor λ, which modulates the available current in the saturation mode.
- The *body effect*—The source of the load transistor is connected to the output of the inverter. The threshold of the load transistor therefore varies as a function of V_{out}.

[2] This period lasted until the early 1980s. Complementary devices were not readily or economically available before that time.

The body effect reduces the value of $|V_{Tn}|$, and hence also the available current I_L, for increasing values of V_{out}.

The latter effect in particular results in an important deterioration of the load characteristic compared to the first-order model. Even so, the depletion load outperforms the resistive load dramatically, as illustrated in Figure 4.13. The depletion load gate also requires far less area. Implementing the large resistor needed to obtain reasonable noise margins requires a substantial amount of silicon area. For instance, a resistor of 40 kΩ, as is typically needed, takes 3,200 μm^2 when implemented in n^+-diffusion. This is equivalent to more than one thousand minimum-size transistors!

Pseudo-NMOS

A grounded PMOS device presents an even better load (Figure 4.12c). This configuration, which is called *pseudo-NMOS* because it resembles the depletion NMOS load, is superior to the preceding approach. First of all, the PMOS transistor does not experience any body effect as its V_{SB} is constant and equal to 0. Secondly, the PMOS device is driven by a V_{GS} equal to $-V_{DD}$, resulting in a higher load-current level for similarly sized devices. The load current in saturation (ignoring the channel-length modulation) is given by Eq. (4.9):

$$I_L = \frac{k_p}{2}(V_{DD} - |V_{Tp}|)^2 \tag{4.9}$$

The simulated load line for the PMOS load is shown in Figure 4.13 as well. Be aware that the curves in this plot are normalized. The overall characteristic is substantially closer to the ideal current source, although the pseudo-NMOS load leaves the saturation mode earlier than the depletion load. This can be attributed to the larger V_{GS} value. When comparing similar-size depletion-NMOS and PMOS loads, we find that the latter supplies more current, even though its mobility is smaller. The superior load-line characteristic results in improved dc and ac characteristics.

Computing the dc transfer characteristic of the pseudo-NMOS proceeds along paths similar to those used for its complementary CMOS counterpart. In both cases, V_{OH} equals V_{DD}. The major difference here is that V_{OL} differs from *GND*, due to the ratioed nature of the pseudo-NMOS structure. The value of V_{OL} is obtained by equating the currents through the driver and load devices for $V_{in} = V_{DD}$. At this operation point, the NMOS driver resides in linear mode, while the PMOS load is saturated

$$k_n\left((V_{DD} - V_{Tn})V_{OL} - \frac{V_{OL}^2}{2}\right) = \frac{k_p}{2}(V_{DD} - |V_{Tp}|)^2 \tag{4.10}$$

Eq. (4.10) can now be solved for V_{OL}. For reasons of simplicity, we will assume that $V_{Tn} = |V_{Tp}|$.

$$V_{OL} = (V_{DD} - V_T)\left(1 - \sqrt{1 - \frac{k_p}{k_n}}\right) \text{ (assuming that } V_T = V_{Tn} = |V_{Tp}|) \tag{4.11}$$

Similarly, we can derive a value for V_M by solving the current equations for $V_{in} = V_{out}$, which yields Eq. (4.12). In this case, the NMOS and PMOS devices reside in saturation and linear mode, respectively.

$$V_M = V_T + (V_{DD} - V_T)\sqrt{\frac{k_p}{k_n + k_p}} \tag{4.12}$$

A designer using pseudo-NMOS has to cope with some important problems. First of all, from Eq. (4.12) it is clear that the VTC is asymmetrical, because V_M is not located in the middle of the voltage swing. Furthermore, the gate has asymmetrical rise and fall characteristics. During the low-to-high transition, C_L is charged by the load device, while during the high-to-low period, the propagation delay is determined by the discharging of the capacitor through the pull-down network. A final, but important disadvantage (which is actually common to all ratioed gates of Figure 4.12) is that they consume static power when the output is low, because a direct current path exists between V_{DD} and GND through the load and driver devices. The average power consumption in the low-output mode is easily derived

$$P_{av(low)} = V_{DD}I_{low} = \frac{k_p}{2}V_{DD}(V_{DD} - V_T)^2 \tag{4.13}$$

Problem 4.5 Propagation Delay of the Pseudo-NMOS Inverter

Derive expressions for the propagation delay of the pseudo-NMOS inverter. Use techniques like those employed for the complementary CMOS gate.

The trade-offs to be made during the design of a pseudo-NMOS inverter can best be summarized with the aid of the *RC* equivalent circuit of Figure 4.14. For the sake of simplicity, we assume here that the load can be approximated as a current source for the entire operation region. The following design constraints are valid:

1. To reduce static power consumption, I_L should be low.
2. To obtain a reasonable NM_L, $V_{OL} = I_L R_{PDN}$ should be low.
3. To reduce $t_{PLH} \approx (C_L V_{DD}) / (2I_L)$, I_L should be high.
4. To reduce $t_{pHL} \approx 0.69\ R_{PDN}\ C_L$, R_{PDN} should be kept small.

Condition 2 imposes a well-defined ratio $r = (W/L)_n/(W/L)_p$ of driver and load transistor sizes, as was made clear in Eq. (4.11). This ratio is a function of the technology and the

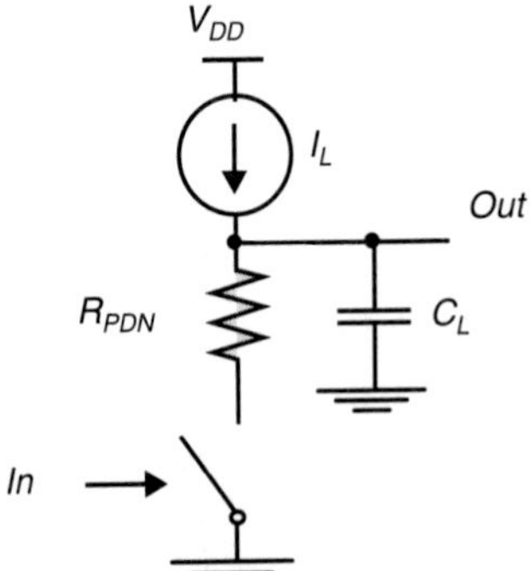

Figure 4.14 Switch model of pseudo-NMOS inverter.

required noise margins. For instance, to obtain a V_{OL} of 0.2 V for the 1.2 μm CMOS process (with V_{DD} = 5 V) requires a ratio r = 3. The wide NMOS driver automatically guarantees condition 4. On the other hand, conditions 1 and 3 are contradictory: realizing a faster gate means more static power consumption as well as reduced noise margins!

Design Considerations

The pseudo-NMOS logic style is particularly attractive when designing complex gates with a large fan-in. The pull-up network of the complementary CMOS gate is replaced by a single-load transistor. For example, a four-input NOR gate is shown in Figure 4.15. The structure has the following attractive features:

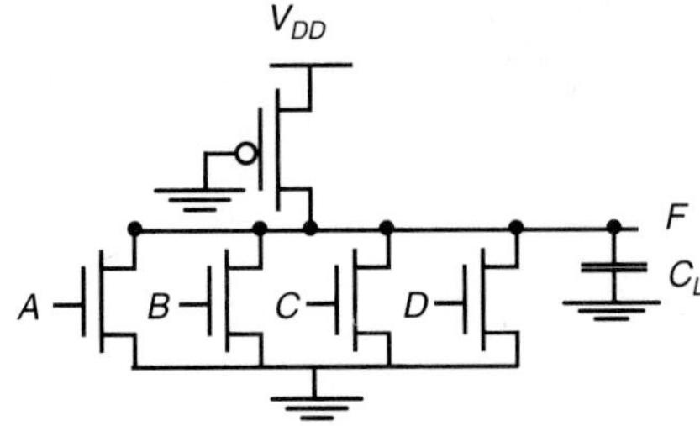

Figure 4.15 Four-input pseudo-NMOS NOR gate.

- A gate with a fan-in of N requires only N +1 transistors, resulting in a smaller area as well as smaller parasitic capacitances.
- Each input connects to only a single transistor, presenting a smaller load to the preceding gate.

This simplification comes at the expense of static power consumption, which eliminates the applicability of pseudo-NMOS for very large circuits. A minimal-size gate consumes approximately 1 mW of static power (ignoring dynamic power consumption)! A 100,000 gate circuit therefore consumes 50 W (assuming that half of the gates are in the low-output mode). Pseudo-NMOS gates, on the other hand, can be used quite effectively in small subcircuits, where speed is of a major importance, or in circuits where we know that the majority of the outputs are high and hence do not consume power. The latter is the case in address decoders, used in memories.

Example 4.6 A Pseudo-NMOS Four-input NAND Gate

To put the properties of the pseudo-NMOS gate in a perspective, we will design a four-input NAND gate. The gate schematics and layout are shown in Figure 4.16. The gate parameters (summarized in Table 4.3) can be compared with the complementary CMOS gate of Figure 4.10. To ensure a fair comparison, the NMOS transistors of the pull-down network were initially chosen to be of minimal size. The computed load capacitance varies between 23.8 fF and 24.8 fF (depending upon the direction of the transition). This is substantially smaller than the corresponding values for the complementary CMOS gate.

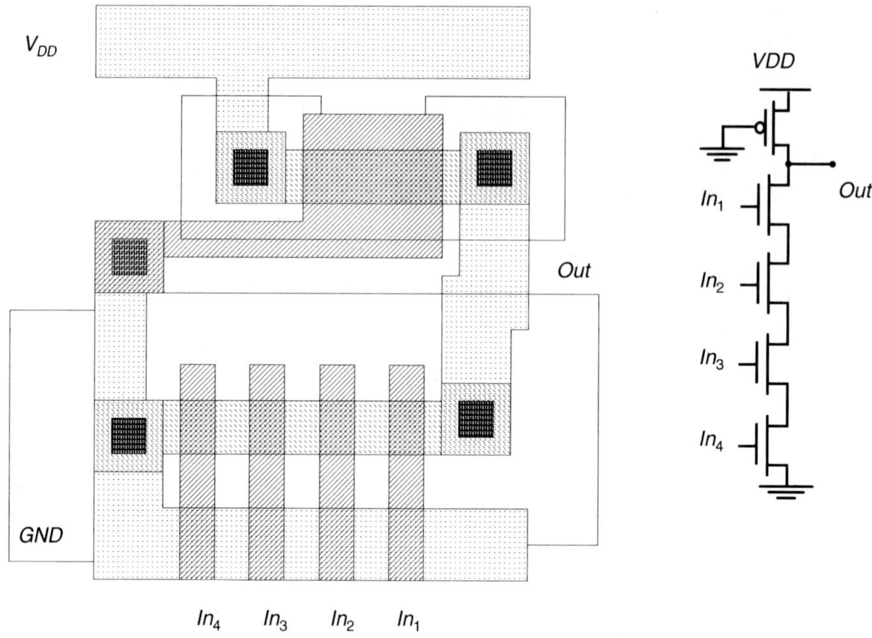

Figure 4.16 Layout and schematics of four-input pseudo-NMOS NAND. The following transistor sizes are used: $(W/L)_{NMOS}$ = (1.8/1.2); $(W/L)_{PMOS}$ = (1.8/4.8). See also Colorplate 8.

Table 4.3 The dc and ac parameters of 4-input NAND gate in pseudo-NMOS.

Area	Static Current	Transistors	V_{OH}	V_{OL}	V_M	NM_H	NM_L	t_{pHL}	t_{pLH}	t_p
288 μm^2	48.5 μA	5	5 V	0.48 V	1.62 V	3.26 V	0.27 V	0.6 nsec	2.37 nsec	1.49 nsec

Some important conclusions can be drawn from these results. While substantially smaller than the complementary gate, the pseudo-NMOS device suffers from a serious degradation in the lower noise margin. In the presented design, the ratio between each of the NMOS transistors and the PMOS load equals 4 (drawn lengths). Trying to improve the noise margin by making the PMOS-load longer has a detrimental effect on the t_{pLH}, which is already large. The asymmetrical nature of the propagation delay (a ratio of 4 between t_{pLH} and t_{pHL}) makes the pseudo-NMOS gate particularly well-suited for applications where transitions in one direction should be as fast as possible, while the speed in the other direction is not as crucial. When interpreting the results of Table 4.3, we should also take into account that the input capacitance of the gate is substantially reduced (to a single NMOS transistor gate capacitance). It is therefore reasonable to increase the size of the NMOS devices in an attempt to speed up the gate, while keeping the transistor ratio the same. Increasing the (W/L) of the NMOS devices to (7.2/1.2) and the of PMOS transistor to (1.8/1.2) increases the size of the gate somewhat (360 μm^2) as well as the static power consumption (I_{stat} = 0.18 mA), but reduces t_p to 0.82 nsec (t_{pHL} = 0.24 nsec; t_{pLH} = 1.4 nsec).

Problem 4.6 NAND Versus NOR in Pseudo-NMOS

Given the choice between NOR or NAND logic, which one would you prefer for implementation in pseudo-NMOS?

How to Build Even Better Loads

The grounded PMOS load is a good imitation of an ideal current-source load. For certain circuit configurations, some simple modifications can further improve either the speed or the power consumption.

Consider, for instance, the case where a gate is known to switch only during certain time periods. An example of such a circuit is the address decoder of a memory, which should only switch after a change in the address has been detected. Such a circuit should have a low power consumption and large noise margins in *standby mode*, but should switch fast after an address change has been detected (even at the expense of more power). This is achieved by the circuit of Figure 4.17. In standby mode, the large PMOS pull-up M_1 is turned off. The small PMOS device represents a large resistance, and hence a small static current and a low value for V_{OL}. After an address transition is detected, the large PMOS is turned on, resulting in a larger current drive and a fast low-to-high transition at the output.

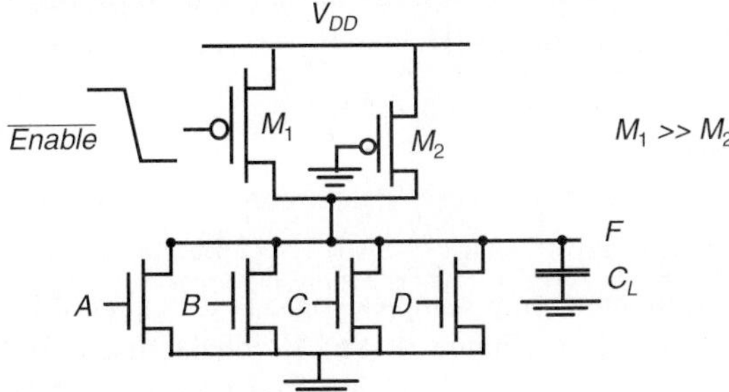

Figure 4.17 Pseudo-NMOS gate with adaptive load.

Another approach allows us to completely eliminate the static current. Let us assume that the complement of each signal is always available. This requires each gate to generate both polarities of the output signal. This is similar in concept to the ECL gate, discussed in Section 3.4. Such a gate, called *differential cascade voltage switch logic* (or DCVSL) is presented conceptually in Figure 4.18a. The pull-down networks PDN1 and PDN2 are complementary, and implement the required logic function and its inverse. Assume now that, for a given set of inputs, PDN1 conducts while PDN2 does not. Node *Out* is pulled down. This turns on load transistor M_2, pulling up $\overline{Out}$. This in turn cuts off load transistor M_1. The gate is clearly free of static current paths as only PDN1 and M_2 are conducting.

The DCVSL gate has the speed advantage of pseudo-NMOS—the reduction of the parasitic capacitances at the output nodes produces a faster response. At the same time, static power consumption is eliminated (although the current during switching increases). This comes at the expense of extra area, as each gate requires two pull-down networks.

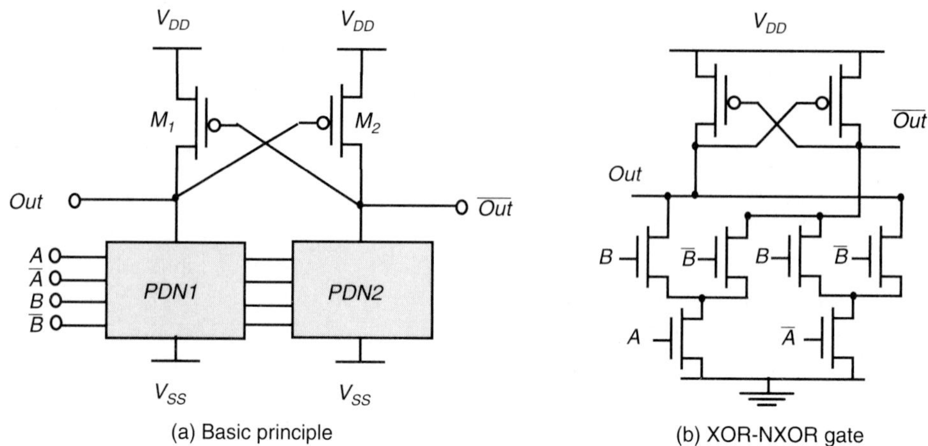

Figure 4.18 DCVSL logic gate.

This is not as bad as it seems at first sight. The availability of complementary (or differential) signals eliminates extra inverter stages, as was already noted for the ECL gate. Furthermore, the PDNs might have some logic in common that can be shared between the two networks. An example of the latter is shown in Figure 4.18b. This circuit simultaneously implements a two-input XOR and XNOR gate. The transistors connected to the *A*-inputs are shared between the two PDNs. DCVSL has, for instance, been used for the implementation of fast error-correcting logic in memories [Heller84].

4.2.3 Pass-Transistor Logic

Concept

Another promising approach to implementing complex logic is to realize it as a logical network of switches, or pass-transistors (Figure 4.19a) [Radhakrishnan85]. We have already observed that a series connection of two switches constitutes an AND function, while a parallel connection represents an OR. More complex gates are easily realized by a combination of the above. An example of an AND gate, implemented using this approach, is shown Figure 4.19b. The switch driven by $\overline{B}$ seems to be redundant at first glance. Its presence is essential to ensure that a low-impedance path exists to the supply rails under all circumstances, or, in this particular case, when *B* is low. Since the switch network itself is not driven by any of the supply rails, the insertion of a static buffer is occasionally needed to *regenerate* the signals.

The pass-transistor approach has the advantage of being simple and fast. Complex CMOS combinational logic is implemented with a minimal number of transistors. This reduces the parasitic capacitances and results in fast circuits. The static and transient performance of such a structure strongly depend upon the availability of a *high-quality switch with low parasitic resistance and capacitance*. Although the MOS transistor in itself is a switch of reasonable performance, some deficiencies will become apparent in the subsequent sections. Pass-transistor logic networks are, therefore, often constructed from *bidirectional transmission gates* (or *pass gates*). These gates are composed of an NMOS

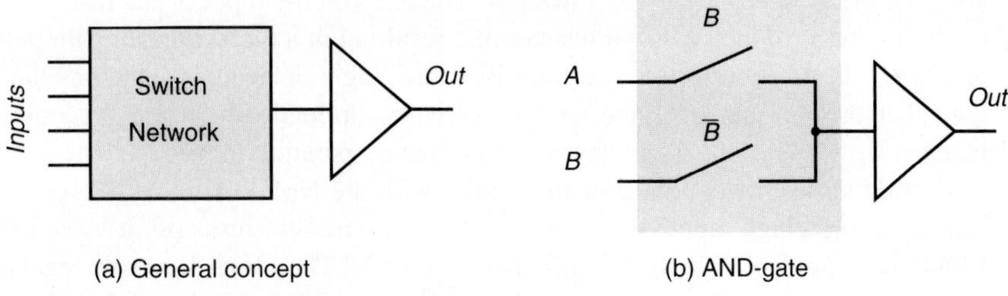

Figure 4.19 Pass-transistor logic.

transistor and a PMOS device in a parallel arrangement as shown in Figure 4.20a. The transmission gate acts as a bidirectional switch controlled by the gate signal C. When $C = 1$, both MOSFETs are on, allowing the signal to pass through the gate. In short,

$$A = B \quad \text{if} \quad C = 1 \tag{4.14}$$

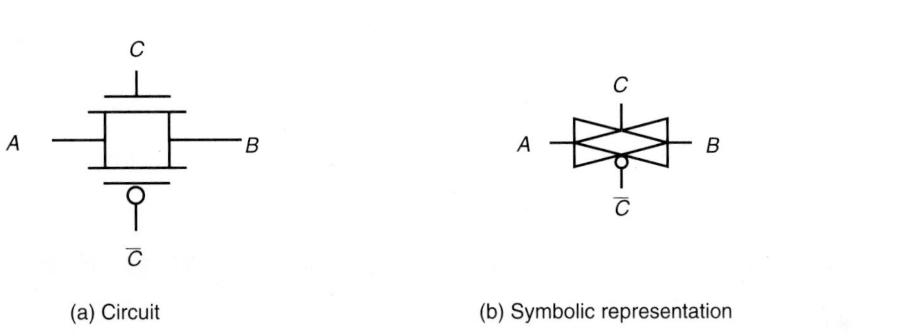

Figure 4.20 CMOS transmission gate.

On the other hand, $C = 0$ places both transistors in cutoff, creating an open circuit between nodes A and B. Figure 4.20b shows a commonly used transmission-gate symbol.

The reason that we select this particular configuration for the CMOS transmission gate becomes apparent with the aid of Figure 4.21a. Suppose that we want to charge capacitor C_L through a single NMOS pass-transistor M_n, controlled by a 5 V gate signal C. M_n turns off when V_B reaches $5 - V_{Tn}(V_B)$, or approximately 3.5 V. A full threshold voltage (increased to 1.5 V by the body effect) is lost. This reduces the noise margin and causes static power consumption, as illustrated in Figure 4.21b. Here, the output of the pass-tran-

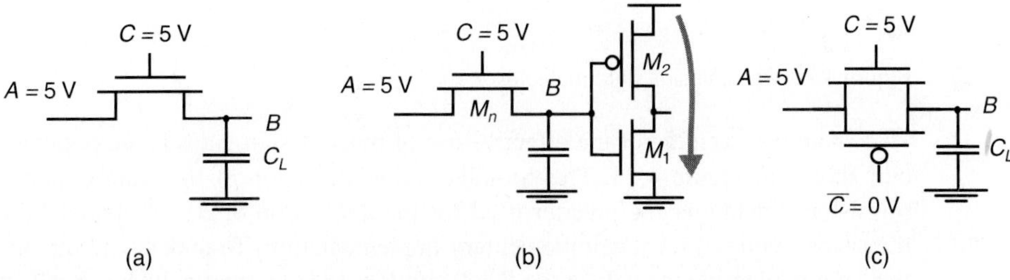

Figure 4.21 Threshold-loss problems in pass-transistor logic.

sistor connects to a static CMOS inverter. The 3.5 V at the input of the inverter is not high enough to turn off the PMOS transistor M_2, resulting in a direct current path between V_{DD} and *GND*. In addition to the threshold loss, the single-device pass gate has the disadvantage that the resistance of the switch increases dramatically when the output voltage reaches $V_{in} - V_{Tn}$, as the transistor goes into linear operation mode.

Adding a PMOS transistor in parallel with the NMOS (Figure 4.21c) solves these problems: for a high input ($V_A = 5$ V), the NMOS transistor turns off at $V_B \approx 3.5$ V, but the PMOS does not, because its V_{GSp} is constant at –5 V! Therefore, V_B easily reaches 5 V, and no voltage drop occurs across the transmission gate. The same is valid at the low end of the voltage scale ($V_A = 0$ V), where the NMOS stays on after the PMOS has turned off. In other words, an NMOS transistor provides a poor "1" and a PMOS a poor "0" level. The combination of the two results in an excellent switch. The disadvantage of this circuit is that the controlling signal as well as its complement must be available.

Transmission gates can be used to build some complex gates very efficiently. The simplest example of this type of circuit is the (inverting) two-input multiplexer shown in Figure 4.22. This gate either selects input *A* or *B* based on the value of the control signal *S*, which is equivalent to implementing the following Boolean function:

$$\overline{F} = (A \cdot S + B \cdot \overline{S}) \tag{4.15}$$

A complementary implementation of the gate requires eight transistors instead of six.

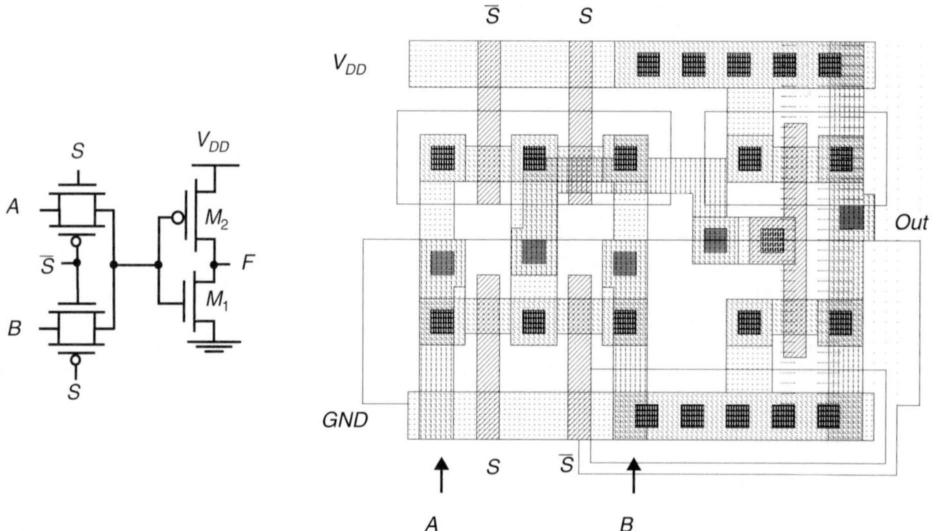

Figure 4.22 Transmission gate multiplexer and its layout.

Another example of the effective use of transmission gates is the popular XOR circuit shown in Figure 4.23. The complete implementation of this gate requires only six transistors (including the inverter used for the generation of $\overline{B}$), compared to the twelve transistors required for a complementary implementation. To understand the operation of this circuit, we have to analyze the $B = 0$ and $B = 1$ cases separately. For $B = 1$, transistors

M_1 and M_2 act as an inverter while the transmission gate M_3/M_4 is off; hence $F = \overline{A}B$. In the opposite case, M_1 and M_2 are disabled, and the transmission gate is operational, or $F = A\overline{B}$. The combination of both results in the XOR function. Notice that, regardless of the values of A and B, node F always has a connection to either V_{DD} or GND and is hence a low-impedance node. If this were not true, the circuit would be *dynamic*, and an occasional refresh would be required to counter the effects of charge leakage. When designing static-pass transistor networks, it is essential to adhere to the low-impedance rule under all circumstances. Other examples where transmission-gate logic is effectively used are fast adder circuits and registers. Both circuits will be discussed in later chapters.

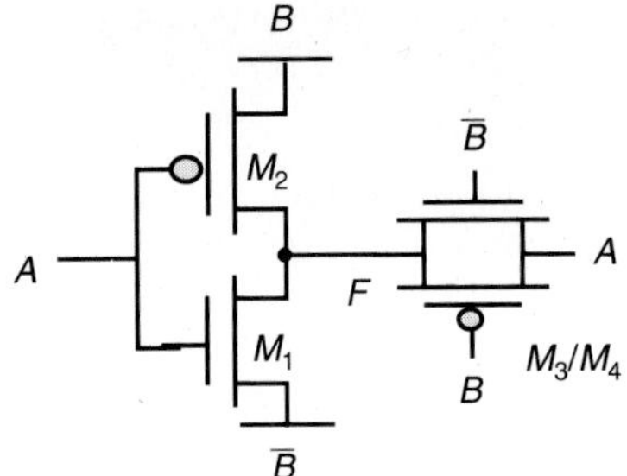

Figure 4.23 Transmission gate XOR.

Design Issues in Static Pass-Transistor Logic Design

When designing transmission-gate-based devices, one has to be aware of a number of design problems that are specific to that circuit class.

1. Resistance.

A transmission gate is, unfortunately, not an ideal switch, because it has a series resistance associated with it. To get an idea of the nature and value of this resistance, let us analyze the design instance of Figure 4.21c, when charging a capacitance C_L from 0 V to V_{DD}, that is, when passing a 1 from input to output. The resistance of the switch is modeled as a parallel connection of the resistances R_n and R_p of the NMOS and PMOS devices, defined as $(V_{DD} - V_{out})/I_n$ and $(V_{DD} - V_{out})/I_p$, respectively. The currents through the devices are obviously dependent on the value of V_{out} and the operating mode of the transistors. During the low-to-high transition, the pass-transistors traverse through a number of operation modes. As its V_{GS} is always equal to V_{DS}, the NMOS transistor is either in saturation or off. The V_{GS} of the PMOS is equal to V_{DD}, and the device changes from saturation to linear during the transient. When computing I_p and I_n, it is important to incorporate the body effect. The operating modes of the transistors for different ranges of V_{out} are summarized below.

- $V_{out} < |V_{Tp}|$: NMOS and PMOS saturated.
- $|V_{Tp}| < V_{out} < V_{DD} - V_{Tn}$: NMOS saturated, PMOS linear.
- $V_{DD} - V_{Tn} < V_{out}$: NMOS cutoff, PMOS linear.

The simulated value of $R_{eq} = R_p \parallel R_n$ as a function of V_{out} is plotted in Figure 4.24. It can be observed that R_{eq} is relatively constant ($\approx$ 10 kΩ in this particular case). The same is true in other design instances (for instance, when discharging C_L). When analyzing transmission-gate networks, the simplifying assumption that the switch has a constant resistive value is therefore acceptable. Deriving an analytical expression for R_{eq} is nontrivial, because the resistance depends on the operating region and is influenced by factors such as the body effect. A first-order approximation, which is often used, assumes that both devices are in the linear operating mode and ignores the body effect. This produces the following simplified model (with V_A and V_B the input and output nodes of the switch, respectively):

$$G_{eq} = \frac{1}{R_{eq}} = \frac{k_n(V_{DD} - V_B - V_{Tn})(V_A - V_B)}{(V_A - V_B)} + \frac{k_p(V_{DD} - |V_{Tp}|)(V_A - V_B)}{V_A - V_B} \approx k_n(V_{DD} - V_{Tn}) + k_p(V_{DD} - |V_{Tp}|) \tag{4.16}$$

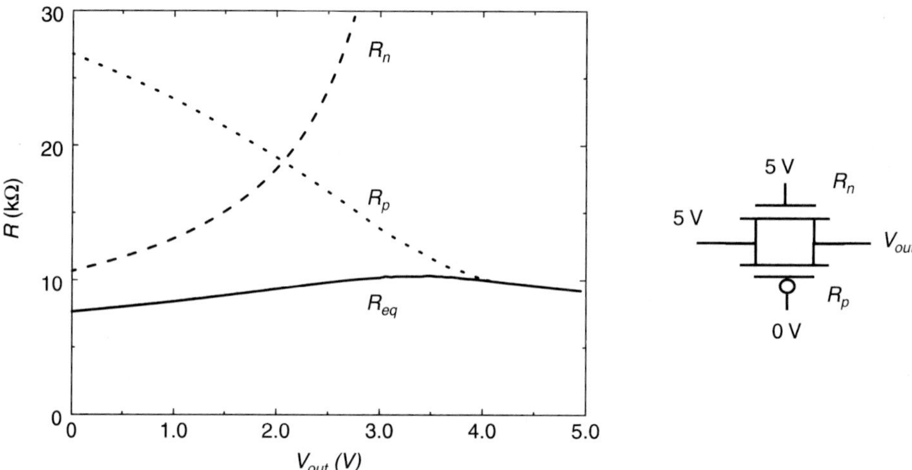

Figure 4.24 Simulated equivalent resistance of transmission gate for low-to-high transition (for $(W/L)_n = (W/L)_p = 1.8/1.2$).

More accurate expressions for both high and low transitions can be found in [Shoji88], pp. 168–170.

Example 4.7 Resistance of a Transmission Gate

The resistance of the transmission gate of Figure 4.24 is estimated first for $V_A = 5$ V and $V_B = 0$ V. Both transistors are in saturation mode at that time. The following expression is valid (ignoring channel-length modulation):

$$G_{eq} = \frac{I_n + I_p}{V_{DD}} = \frac{k_n(V_{DD} - V_{Tn})^2 + k_p(V_{DD} - |V_{Tp}|)^2}{2V_{DD}} = 90 \times 10^{-6} \text{ mho}$$

or $R_{eq} = 11$ kΩ. The simplified model given in Eq. (4.16) predicts a value of 9.4 kΩ.

Problem 4.7 Equivalent Resistance During Discharge

Determine the operation modes of the NMOS and PMOS transistors in the transmission gate as a function of V_{out} when discharging C_L.

2. Delay

Figure 4.25a shows a chain of n transmission gates. Such a configuration often occurs in circuits such as adders. Assume that all control signals are set to 1 and a step is applied at the input. To analyze the propagation delay of this network, the transmission gates are replaced by their equivalent resistances R_{eq}. This produces the network of Figure 4.25b.

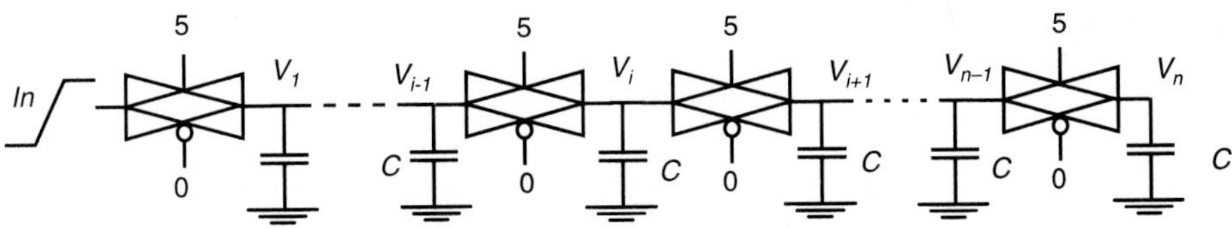

(a) A chain of transmission gates

(b) Equivalent RC network

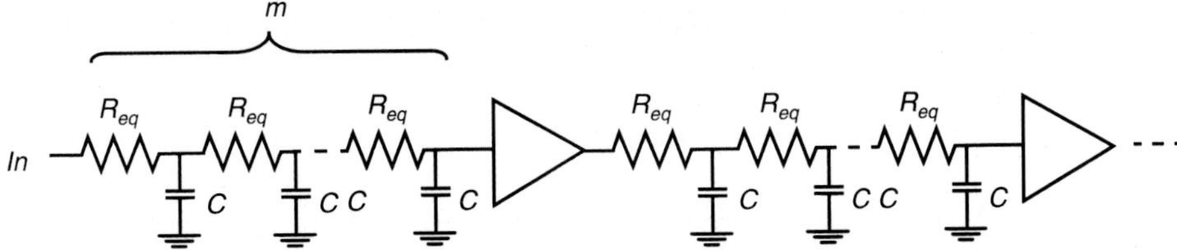

(c) Optimizing the speed by inserting buffers

Figure 4.25 Speed optimization in transmission-gate networks.

Finding the delay of this network requires us to solve a set of differential equations of the form

$$\frac{\partial V_i}{\partial t} = \frac{1}{R_{eq}C}(V_{i+1} + V_{i-1} - 2V_i) \tag{4.17}$$

Finding an analytical solution to this set is extremely hard, but we can estimate the dominant *time constant* at the output of a chain of n transmission gates as follows:[3]

$$\tau(V_n) = \sum_{k=0}^{n} CR_{eq}k = CR_{eq}\frac{n(n+1)}{2} \tag{4.18}$$

This means that the propagation delay is proportional to n^2 and increases rapidly with the number of switches in the chain. For large values of n, it is therefore recommended to break the chain every m switches and to insert buffers (Figure 4.25c). Assuming a propagation delay t_{buf} for each buffer, the overall propagation delay of the transmission-gate/buffer network can be computed as follows,

$$\begin{aligned} t_p &= 0.69\left[\frac{n}{m}CR_{eq}\frac{m(m+1)}{2}\right] + \left(\frac{n}{m} - 1\right)t_{buf} \\ &= 0.69\left[CR_{eq}\frac{n(m+1)}{2}\right] + \left(\frac{n}{m} - 1\right)t_{buf} \end{aligned} \tag{4.19}$$

The resulting delay exhibits only a linear dependence on the number of switches n, in contrast to the unbuffered circuit, which is quadratic in n. The optimal number of switches m_{opt} between buffers can be found by setting the derivative

$$\frac{\partial t_p}{\partial m}$$

to 0, which yields

$$m_{opt} = 1.7\sqrt{\frac{t_{pbuf}}{CR_{eq}}} \tag{4.20}$$

Obviously, the number of switches per segment grows with increasing values of t_{buf}. In current technologies, m_{opt} typically equals 3 or 4.

Example 4.8 Pass-Transistor Chain

Consider a design in our generic CMOS technology. From Figure 4.24, a value of $R_{eq} \approx 10$ kΩ can be derived for the on-resistance of the switch. Assuming $C = 10$ fF and $t_{pbuf} = 0.5$ nsec, an optimal value of m equal to 3.8 is obtained. This means that a buffer should be introduced for every four transmission gates.

3. Transistor Sizing in Transmission-Gate Logic

The transmission gate logic family is a member of the class of the ratioless logic gates. The dc characteristics are not affected by the sizes of the pass-transistors at all. On the other hand, one might wonder about the effect of the transistor dimensions on the performance. To a first order, increasing the (W/L) of the devices has no net impact on the

[3] An in-depth discussion on how to derive and interpret this expression (using the Elmore Delay technique) is presented in Chapter 8.

propagation delay: it reduces the resistance, while simultaneously increasing the diffusion capacitance. **Devices close to minimum-size should hence be used, unless an external load capacitance is the dominating factor.** When optimizing long chains of pass-transistors, such as the one shown in Figure 4.25, some benefit can be reaped by using a progressive sizing approach in the style of Figure 4.7. As in pull-down or pull-up networks, the first device in the chain has to (dis)charge the parasitic capacitances of all other devices in the chain. This device should therefore be made larger than the second one, and so forth. A good way to determine the exact values of the transistor sizes is to study the *RC*-equivalent network of the circuit (Figure 4.25b) and to pick the resistances so that the overall *RC* time constant is minimized. Techniques to analyze such a network are introduced in Chapter 8.

NMOS-Only Transmission-Gate Logic

Although the transmission gate of Figure 4.20 possesses some excellent properties, such as an almost constant resistance and no threshold loss, it has the disadvantage that it requires both an NMOS and a PMOS transistor, which have to be located in different wells. This reduces the layout efficiency of the design. Also, the control signal has to be presented in both polarities (true and complemented), which once again has a negative influence on the layout density. Furthermore, the parallel connection of PMOS and NMOS results in increased node capacitances and reduced performance. It would therefore be advantageous if we could implement transmission-gate logic using NMOS transistors only. Unfortunately, as demonstrated in Figure 4.21, NMOS-only pass-transistors are subject to voltage loss. This is not a real problem if the voltage levels are subsequently restored by a complementary CMOS inverter. Such a circuit suffers from two major drawbacks:

- Reduced noise margins, due to the threshold voltage drop. Notice that the NM_L on the controlling signal C (Figure 4.21b) is only equal to V_{Tn}.
- Static power consumption, as discussed above.

Several techniques have been proposed to get around this problem. One of them is presented in Figure 4.26. A single PMOS device, called a *level restorer*, is added between V_{DD} and the input of the inverter. The gate of the PMOS device is connected to the output of the inverter. This is, in effect, a *feedback* circuit. The restorer turns on when the output of the inverter goes low ($V_{out} < V_{DD} - |V_{tp}|$) and pulls the input of the inverter to V_{DD}. No static power is consumed by the inverter. Furthermore, no static current path can exist

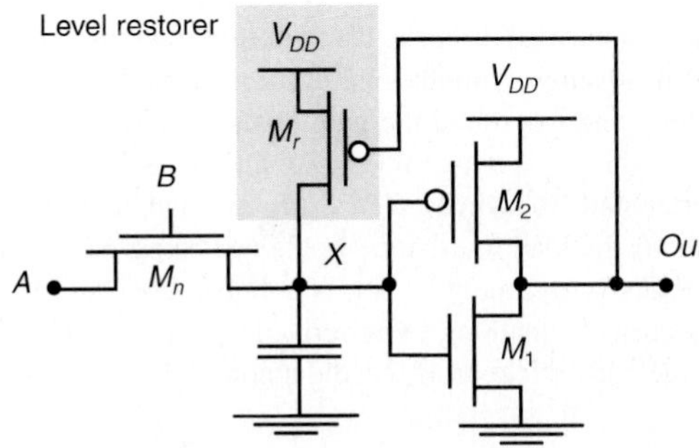

Figure 4.26 Level-restoring circuit.

through the level restorer and the pass-transistor, since the restorer is only active when *A* is high (and hence not connected to *GND*). In summary, this circuit has the advantage that all voltage levels are either at *GND* or V_{DD}, and no static power is consumed.

We must still determine how to size the level-restoring transistor. Strictly speaking, pass-transistor circuits are ratioless, since there never exists a conducting path between V_{DD} and *GND*. In the level-restoring circuit, no such path exists in steady-state operation mode either. However, in transient mode such a path may exist. Consider Figure 4.27. We use the notation R_1 to denote the equivalent on-resistance of transistor M_1, R_2 for M_2, and so on. During the switching of signal *B*, with node *A* in a low state and the storage node *X* initially at V_{DD}, a conducting path from V_{DD} to *GND* temporarily exists through M_r, M_n, and M_3. Some careful transistor sizing is necessary to make the circuit function correctly: when R_r is made too small, it is impossible to bring the voltage at node *X* below V_{Tn}. Hence, the inverter output never switches to V_{DD}, and the level-restoring transistor stays on. This sizing problem can be reformulated in the following way:

> There must exist a voltage $V_{sw} < V_{DD}$ at the *B* input to the pass-transistor M_n, such that for $V_B > V_{sw}$, the voltage at node *X* drops below the threshold of the inverter, $V_M = f(R_1, R_2)$. This condition is sufficient to guarantee a switching of the output voltage V_{out} to V_{DD} and a turning off of the level-restoring transistor.

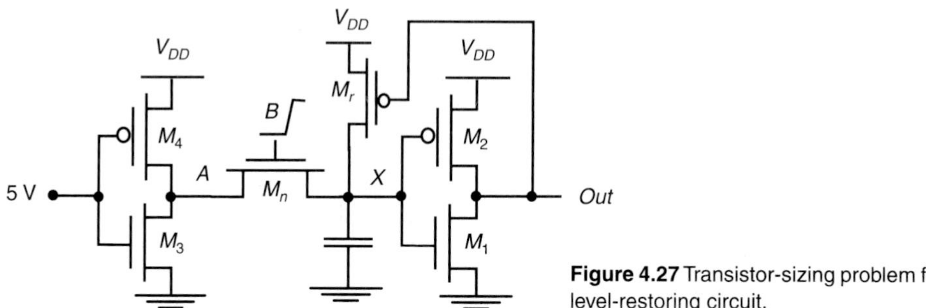

Figure 4.27 Transistor-sizing problem for level-restoring circuit.

The above problem occurs often in circuits with feedback, and is called *the writability problem*. It is discussed in more detail in Chapter 6, where we analyze sequential circuits based on feedback.

Example 4.9 Sizing of a Level Restorer

Analyzing the circuit as a whole is nontrivial, because the restoring transistor acts as a feedback device. We therefore recommend simplifying the circuit for manual analysis. One way to do so is to open the feedback loop and to ground the gate of the restoring transistor when determining the switching point (this is a reasonable assumption, as the feedback only becomes effective once the inverter starts to switch). Hence, M_r, M_n, and M_3 form a "pseudo-NMOS-like" configuration, with M_r the load transistor, M_n the switching device, and M_3 an additional resistor between *A* and *GND*. Assuming that M_r is in linear mode and M_n in saturation (for $V_X = V_M$), the following current equation can be derived. It is assumed here that V_A is sufficiently close to ground that the body effect on M_n can be ignored:

$$\begin{aligned} I &= k_3(V_{DD} - V_{Tn})V_A \\ &= \frac{k_n}{2}(V_B - V_A - V_{Tn})^2 \\ &= k_r\left[(V_{DD} - |V_{Tp}|)(V_{DD} - V_M) - \frac{(V_{DD} - V_M)^2}{2}\right] \end{aligned} \tag{4.21}$$

Solving this set of equations proceeds as follows. The last equation sets the value for I. From this we can derive V_A (using the first equation), after which an expression for V_B can be found as a function of the k-parameters. Imposing the condition $V_B < V_{DD}$ results in a bound on the transistor ratios. Assuming identical sizes for M_3 and M_n results in the following relation (for $V_{DD} = 5$ V, $V_{Tn} = |V_{Tp}| = 0.75$ V, and $V_M = 2.5$ V):

$$V_B = 3.87\sqrt{\frac{k_r}{k_n}} + 1.76\frac{k_r}{k_n} + 0.75 \le 5V \tag{4.22}$$

The boundary condition for this constraint to be valid is $m = k_n/k_r > 1.55$. This is confirmed in Figure 4.28, which shows the VTC of this circuit for different values of m (as obtained by SPICE). We observe that the gate does not switch for small values of m. Setting m to approximately 3 seems to be a reasonable design criterion. This corresponds to identical sizes for the NMOS pass-transistor and the PMOS restoring device.

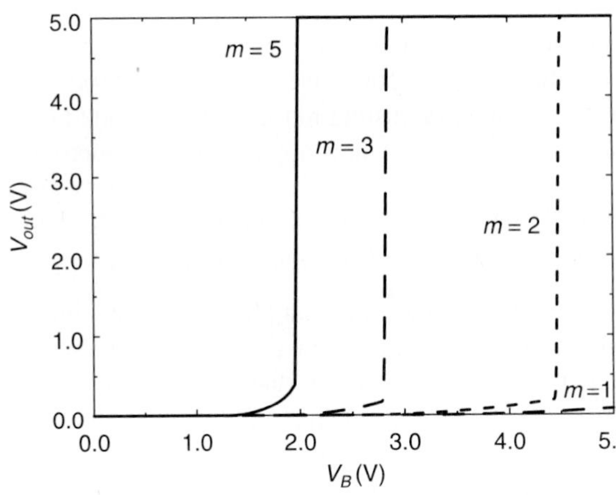

Figure 4.28 VTC simulation of NMOS-only pass-transistor gate for different values of $m = k_r/k_n$ (transistor M_3 is chosen to be minimum size).

Another concern is the influence of the level restorer on the switching speed of the device. Adding the restoring device increases the capacitance at the internal node X, slowing down the gate. The rise time of the gate is further negatively affected, since, the level-restoring transistor M_r fights the decrease in voltage at node X before being switched off. The resulting time-offset is obvious in the simulated transient response of the circuit, where the response of the circuit with a restoring device is compared to the case where M_r is eliminated (Figure 4.29a). On the other hand, the level restorer reduces the fall time, since the PMOS transistor, once turned on, speeds the pull-up action. This effect is especially significant when V_{out} approaches GND, and it offsets the slow-down caused by the increased capacitance. To clearly understand the behavior of the circuit, it is worthwhile to study the voltage response at the intermediate node X, shown in Figure 4.29b. Notice that when no restoring device is present, the intermediate node never reaches V_{DD}, resulting in a reduced current-driving capability.

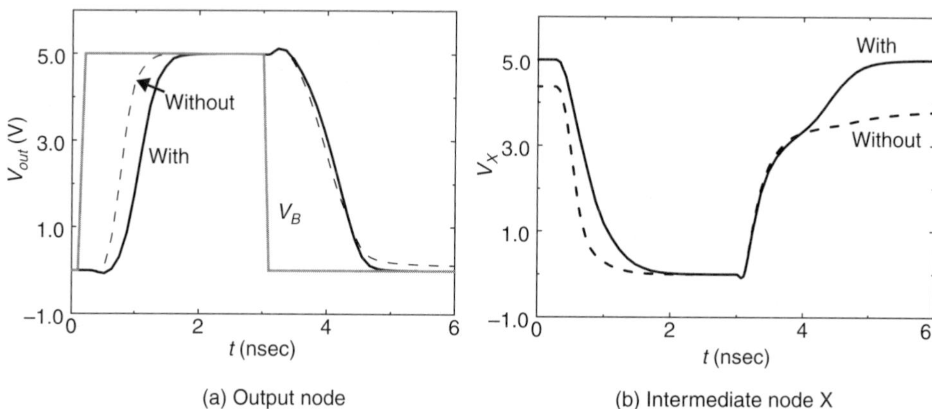

Figure 4.29 Transient response of NMOS-only pass-transistor network with and without level restorer: V_{out} (a) and V_X (b) versus t.

Another approach to implementing NMOS-only pass-transistor networks is to change the thresholds of the transistors using ion-implantations (if your manufacturer allows you to do so). We can either opt to increase the threshold of M_2 or to decrease the V_T of the pass-transistor M_n. For the latter, one could use *zero-threshold transistors*. These are NMOS transistors, where the threshold-adjusting ion-implant is omitted (called a *natural device*) or where a threshold-adjusting implant is applied. A V_T of approximately 0 V is obtained. This is the approach taken in a circuit style called *CPL (complementary pass-transistor logic)*. A number of CPL gates (AND/NAND, OR/NOR, and XOR/NXOR) are shown in Figure 4.30. These gates possess a number of interesting properties:

- The circuits are *differential*, which means that complementary data inputs and outputs are always available. Although generating the differential signals requires extra circuitry, the differential style has the advantage that some complex gates such as XORs and adders can be realized efficiently with a small number of transistors. Furthermore, the availability of both polarities of every signal eliminates the need for extra inverters, as is often the case in static CMOS or pseudo-NMOS. This results in an additional speed-up (similar to the ECL and DCVSL circuit styles).

- CPL belongs to the class of *static* gates, because the output-defining nodes are always tied to either V_{DD} or GND. This is advantageous for the noise resilience.

- The *threshold* of the pass-transistors (including the body effect) is often *reduced* to below $|V_{Tp}|$, eliminating static power consumption in the subsequent buffer unit. The reduced threshold value also increases the switching speed, because the current through the pass-transistor is proportional to $(V_{GS} - V_T)^2$ when the device is saturated.

- The design is very modular. In effect, all gates use exactly the same topology. Only the inputs are permutated. This makes the design of a library of gates very simple. More complex gates can be built by cascading the standard pass-transistor modules.

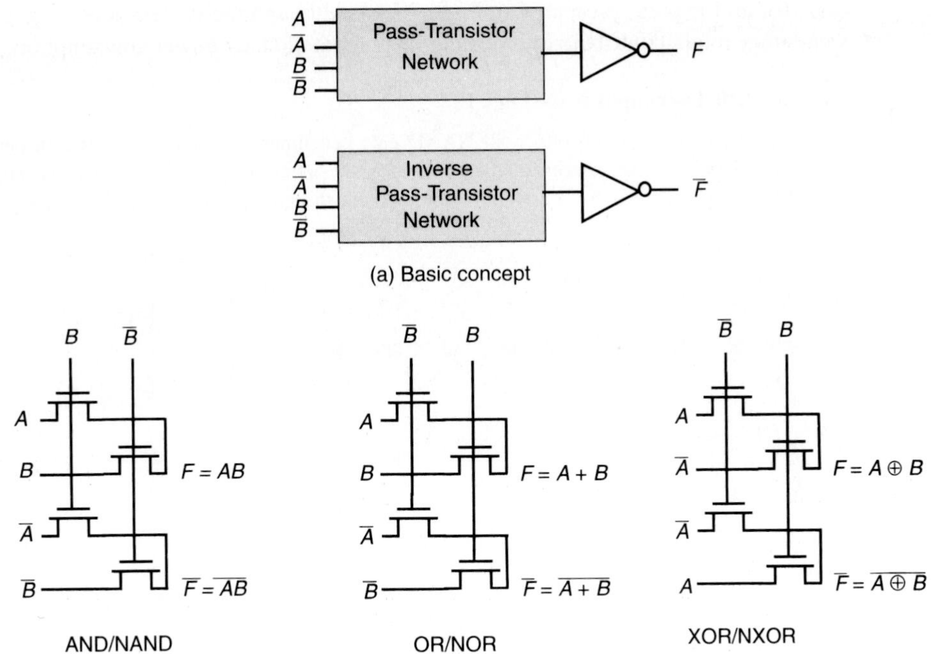

Figure 4.30 Complementary pass-transistor logic (CPL).

On the other hand, some disadvantages should be mentioned. The use of zero-threshold transistors can be dangerous due to the reduced noise margin. In addition, turning off a zero-V_T device is hard. Subthreshold currents can flow through the pass-transistors, even if V_{GS} is slightly below V_T. This is demonstrated in Figure 4.31, which points out a potential sneak dc-current path. A careful examination of the input patterns and the switching configurations reveals that such a dc path is present for 50% of the input condi-

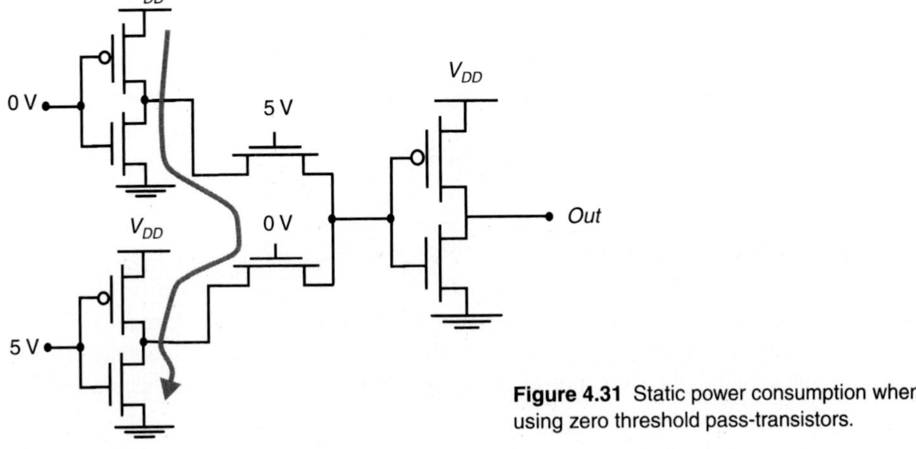

Figure 4.31 Static power consumption when using zero threshold pass-transistors.

tions for all the gates presented in Figure 4.30. Although the current levels are generally small, they might contribute to a substantial amount of static power consumption.

Example 4.10 Four-input NAND in CPL

The by now common four-input NAND gate benchmark is used to put CPL in perspective. Based on the associativity of the boolean AND operation [$A \cdot B \cdot C \cdot D$ = (A·B)·(C·D)], a two-stage approach has been adopted to implement the gate (Figure 4.32). The total number of

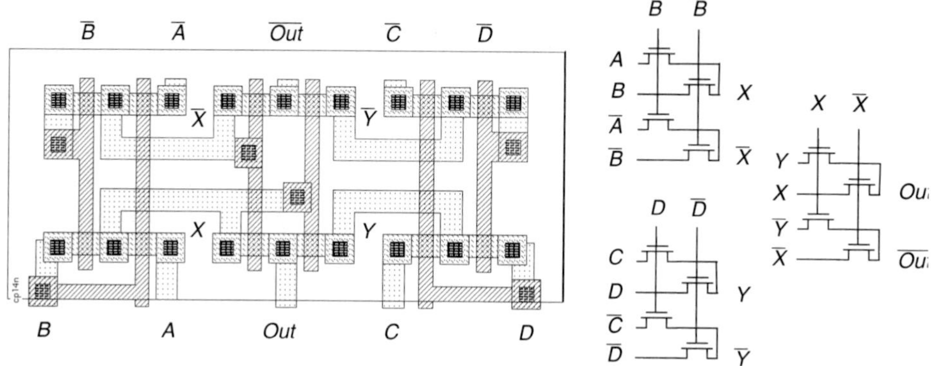

Figure 4.32 Layout and schematics of four-input NAND-gate using CPL (the final inverter stage is omitted). See also Colorplate 9.

transistors in the gate (including the final buffer) is 14. This is substantially higher than previously discussed gates. This factor, combined with the complicated routing requirements, makes this circuit style not particularly efficient for this gate. One should, however, be aware of the fact that the structure simultaneously implements the AND and the NAND functions, which might reduce the transistor count of the overall circuit. Circuit simulations project a propagation delay of 0.7 nsec (t_{pHL} = 1.05 nsec, t_{pLH} = 0.45 nsec) (assuming the availability of a zero-threshold device for the pass-transistor network).

In summary, CPL is a conceptually simple and modular logic style. Its applicability depends strongly upon the logic function to be implemented. The availability of a simple XOR as well of the ease of implementing some specific gate structures makes it attractive for structures such as adders and multipliers. Some extremely fast and efficient implementations have been reported in that application domain [Yano90]. When considering CPL, the designer should not ignore the implicit routing overhead of the complementary signals, which is apparent in the layout of Figure 4.32.

4.3 Dynamic CMOS Design

It was mentioned earlier that the static CMOS logic with fan-in N requires $2N$ devices versus only $N + 1$ for ratioed pseudo-NMOS logic. *Dynamic logic* obtains a similar result, while avoiding the static power consumption of pseudo-NMOS. It uses a sequence of *pre-*

charge and conditional *evaluation* phases to realize complex logic functions with less transistors and no static power.

4.3.1 Dynamic Logic: Basic Principles

Consider the circuit of Figure 4.33a, which realizes a so-called dynamic ϕn block. The PDN (pull-down network) is similar in composition to the ones encountered in complementary CMOS or pseudo-NMOS. The operation of this circuit can be divided into two major phases: *precharge* and *evaluation*. What mode the circuit is in is determined by a signal ϕ, called the *clock signal*.

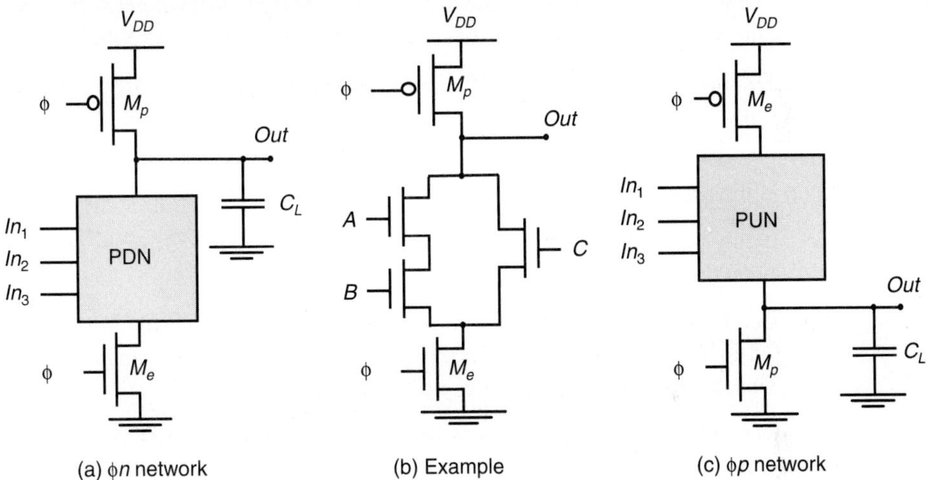

Figure 4.33 Basic concepts of dynamic circuits.

Precharge

When $\phi = 0$, the output node *Out* is precharged to V_{DD} by the PMOS transistor M_p. During that time, the NMOS transistor M_e is off, so that no dc current flows regardless of the values of the input signals.

Evaluation

When $\phi = 1$, the precharge transistor M_p is off, and the evaluation transistor M_e is turned on. Depending upon the values of the inputs and the composition of the PDN, a conditional path between *Out* and *GND* is created. If such a path exists, *Out* is discharged, and a low output signal is obtained. If not, the precharged value remains stored on the output capacitance C_L, which is a combination of the diffusion capacitors, the wiring capacitance, and the input capacitance of the fan-out gates. A high output value is obtained. During the evaluation phase, the only possible path between the output node and a supply rail is to *GND*. Consequently, once *Out* is discharged, it cannot be charged again! This is in contrast with the static gates, where the output node is low-impedance under all possible circumstances. The inputs to the gate can therefore make *at most one transition during evaluation.*

As an example, consider the circuit of Figure 4.33b. During $\phi = 0$, *Out* is precharged to V_{DD}. In the evaluation phase, a conducting path is created between *Out* and *GND* if (and only if) $A{\cdot}B{+}C$ is TRUE. If not, the output remains at V_{DD}. The following function is thus realized:

$$Out = \overline{A \cdot B + C} \quad (\text{when } \phi = 1) \tag{4.23}$$

Notice that during the precharge phase, the output equals 1, independent of the inputs.

A number of important properties can be derived for the dynamic gate circuit:

- The logic function is implemented by the NMOS pull-down network. The construction of the PDN proceeds just as it does for static CMOS and pseudo-NMOS.
- The *number of transistors* (for complex gates) is substantially lower than in the static case: $N + 2$ versus $2N$.
- It is *nonratioed*. The noise margin does not depend on transistor ratios, as is the case in the pseudo-NMOS family.
- It only consumes *dynamic power*. No static current path ever exists between V_{DD} and *GND* (besides the leakage currents).
- Due to the reduced number of transistors per gate and the single-transistor load per fan-in, the load capacitance for this gate is substantially lower than for static CMOS. This results in *faster switching speeds*.

Once we understand the concept of the ϕn block, we can also conceive a ϕp block, which consists of an NMOS precharge transistor and a PMOS pull-up network (PUN) (Figure 4.33c). This block is predischarged (to GND!) during $\phi = 1$ and evaluates during the $\phi = 0$ phase. Due to the lower mobility of the PMOS devices, a ϕp block is slower than a ϕn block. An appropriate scaling of the PMOS transistors is recommended.

Steady-State Behavior of Dynamic Logic

In the following sections, we concentrate on the behavior of the ϕn block. Similar considerations hold for the ϕp sections.

The low and high output levels V_{OL} and V_{OH} are easily identified as *GND* and V_{DD} and are not dependent upon the transistor sizes. The other VTC parameters are dramatically different from the static gates discussed above. Noise margins and switching thresholds have been defined as static quantities, which are not influenced by time. To be functional, a dynamic gate requires a periodic sequence of precharges and refreshes. Pure static analysis, therefore, does not apply. For instance, the value of the noise margins is a function of the length of the evaluation period. If the clock period is too long, the high output level is severely affected by charge leakage. On the other hand, extending the evaluation period results in a lower value of V_{OL}. Even so, some qualitative statements can be made.

The pull-down network of a dynamic inverter starts to conduct when the input signal exceeds the threshold voltage (V_{Tn}) of the NMOS pull-down transistor. If one waits long enough, the output eventually reaches *GND*. Therefore, it is reasonable to set the switching threshold (V_M) as well as V_{IH} and V_{IL} of the gate equal to V_{Tn}.[4] This translates to a low value for the NM_L.

In reality, the situation is even worse. Due to *subthreshold currents*, the PDN even starts to conduct for values of the inputs close to but smaller than V_{Tn}. Subthreshold conduction is one of the reasons for keeping the transistor threshold larger than 0.5 V.

In the high output state, the output impedance of the gate is very high, since the output node is floating. Hence, the output level is sensitive to noise and disturbances. For instance, capacitive coupling of other signals might cause a loss of charge, which cannot be recovered. Charge loss is discussed in more detail later in the section. Fortunately, the NM_H of the dynamic gate is high, so that a reasonable amount of noise can be tolerated.

4.3.2 Performance of Dynamic Logic

Besides the small area due to the small number of transistors, the most attractive property of the dynamic gate is its high switching speed. The simple construction of the gate and the small number of transistors results in a small value for the load capacitance C_L. The analysis of the switching behavior of the gate has some interesting peculiarities to it. After the precharge phase, the output is high. For a low input signal, no additional switching occurs. As a result, $t_{pLH} = 0$! Remember that the low-to-high transition is the weak point of the pseudo-NMOS gate. The high-to-low transition, on the other hand, requires the discharging of the output capacitance through the pull-down network. Therefore t_{pHL} is proportional to C_L and the current-sinking capabilities of the PDN. The presence of the evaluation transistor slows the gate somewhat, as it presents an extra series resistance to the pull-down network. Omitting this transistor, while functionally not forbidden, is dangerous, since its task is to prevent the presence of current paths between V_{DD} and *GND* during precharge.

The above analysis is somewhat unfair, because it ignores the influence of the precharge time on the switching speed of the gate. The precharge time is determined by the time it takes to charge C_L through the PMOS precharge transistor. During this time, the logic in the gate cannot be utilized. This is not necessarily a drawback. Very often, the overall digital system can be designed in such a way that the precharge time coincides with other system functions. For instance, the precharge of the arithmetic unit in a microprocessor can coincide with the instruction decode. The designer has to be aware of this "dead zone" in the use of dynamic logic, and should carefully consider the pros and cons of its usage, taking the overall system requirements into account.

The designer is free to choose the size of the PMOS precharge transistor at will, in contrast to the situation with pseudo-NMOS. Its size does not effect the dc operating levels. Making the transistor larger decreases the required precharge time. On the other hand, making the precharge transistor too large results in an increased t_{pHL} due to the increased value of C_L.

---

[4] We are ignoring charge leakage for now. In the presence of leakage, the outputs will always be eventually discharged to 0 if the evaluation period is extended long enough.

Example 4.11 A Four-Input Dynamic NAND Gate

The four-input NAND example was also designed using the dynamic-circuit style. The resulting layout is shown in Figure 4.34. Notice the size of the gate, which is approximately 2.5 times smaller than the equivalent complementary CMOS gate.

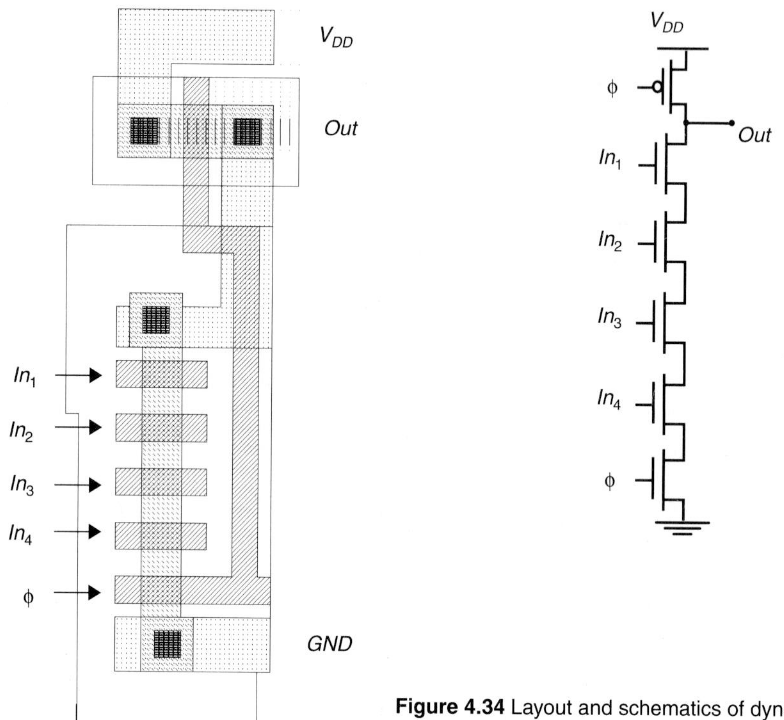

Figure 4.34 Layout and schematics of dynamic four-input NAND gate. See also Colorplate 10.

Due to the dynamic nature of the gate, the derivation of the voltage-transfer characteristic diverges from the traditional approach. We just assume that the switching threshold of the gate equals the threshold of the NMOS pull-down transistor. This results in asymmetrical noise margins, as shown in Table 4.4.

To analyze the dynamic behavior, we follow the common procedure. First we compute the node capacitances. Using an approach identical to the one described in previous examples, C_L is determined to equal 24.5 fF (during evaluation) and 25.8 fF (during precharge) for a fan-out of one static inverter. This is substantially lower than the C_L of the complementary CMOS inverter (~ 30fF), so that we can expect a fast response. The dynamic behavior of the gate is simulated with SPICE. It is assumed that all inputs are set high. At the advent of the clock, the output node is discharged. The resulting transient response is plotted in Figure 4.35. Notice the spikes on the output signals due to capacitive coupling between the clock and the output signal. The resulting propagation delays are summarized in Table 4.4. Once again, be aware that the t_p given in the table does not account for the time spent during precharge. The length of the precharge time can be adjusted by changing the size of the PMOS precharge transistor. Making the PMOS too large should be avoided, however, as it both slows down the gate and increases the capacitive load on the clock line. For large designs, the latter factor might become a major design concern because the clock load can become excessive and hard to drive.

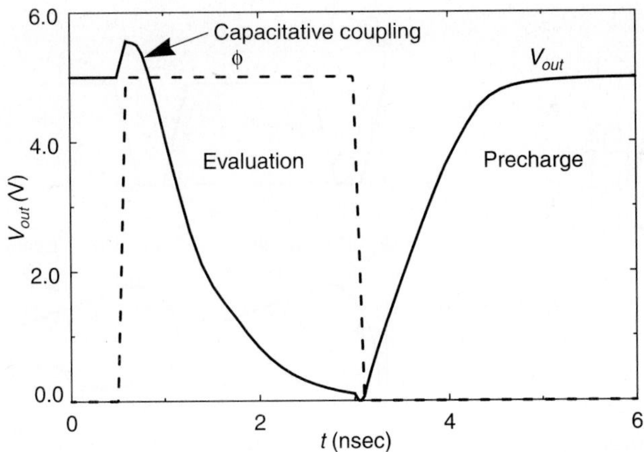

Figure 4.35 Transient response of four-input dynamic NAND gate.

Table 4.4 The dc and ac parameters of a four-input dynamic NAND.

Area	Static Current	Transistors	V_{OH}	V_{OL}	V_M	NM_H	NM_L	t_{pHL}	t_{pLH}	t_p
212 μm^2	0 μA	6	5 V	0 V	0.75V	4.25 V	0.75 V	0.74 nsec	0 nsec	0.37 nsec

4.3.3 Noise Considerations in Dynamic Design

The dynamic circuit concept results in simple and fast structures at the expense of a reduced robustness with regards to noise. This is aggravated by the fact that the gate has a number of inherent deficiencies that must be dealt with to guarantee functional operation.

Charge Leakage

The operation of a dynamic gate relies on the dynamic storage of the output value on a capacitor. Due to leakage currents, this charge gradually leaks away, resulting eventually in malfunctioning of the gate. Two sources of leakage can be identified, as illustrated in Figure 4.36a.

First of all, the storage capacitance C_L partly consists of the drain diffusion capacitance of the pull-down NMOS transistor. The charge stored on C_L will slowly leak away through the *reverse-biased diode* of the diffusion area. At room temperature (T = 300 K), the leakage-current density J_l is typically between 0.1 and 1 $\mu A/cm^2$. The density is temperature-dependent and doubles with every 10 K. For a diffusion with an area of 1 μm^2, the leakage current at room temperature equals approximately 10^{-15} to 10^{-14} A.

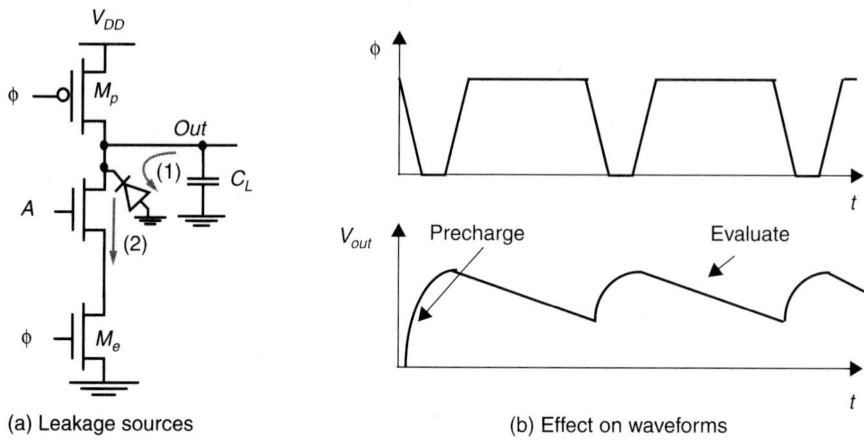

Figure 4.36 Leakage in dynamic circuits.

Secondly, although the input transistor is reportedly off when $A = 0$, some *sub-threshold current* can still flow from the drain to the source. This effect becomes more pronounced when V_A is not completely 0, but approaches V_{Tn} in the presence of noise. It is therefore advantageous to have V_{Tn} sufficiently high (≥ 0.5 V).

Charge leakage causes a degradation in the high level (Figure 4.36b). Dynamic circuits therefore require a minimal clock rate, which is typically between 250 Hz and 1 kHz. This makes the usage of dynamic techniques unattractive for certain products such as watches or low-end consumer products (e.g., toys), which are battery operated and normally run at very low speeds to conserve power. This also makes these circuits hard to test, because stalling of the clocks could result in a loss of the logic levels.

Charge Sharing

Consider the circuit of Figure 4.37. During the precharge phase, the output node is precharged to V_{DD}. Assume now that during precharge all inputs are set to 0 and that the capacitance C_a is discharged. Assume further that input B remains at 0 during evaluation, while input A makes a $0 \rightarrow 1$ transition, turning transistor M_a on. The charge stored originally on capacitor C_L is redistributed over C_L and C_a. This causes a drop in the output voltage, which cannot be recovered due to the dynamic nature of the circuit.

The influence on the output voltage is readily calculated. Under the above assumptions, the following initial conditions are valid: $V_{out}(t = 0) = V_{DD}$ and $V_X(t = 0) = 0$. Two cases must be considered:

1. $\Delta V_{out} < V_{Tn}$—In this case, the final value of V_X equals $V_{DD} - V_{Tn}(V_X)$. Charge conservation yields

$$C_L V_{DD} = C_L V_{out}(t) + C_a[V_{DD} - V_{Tn}(V_X)]$$

or

$$\Delta V_{out} = V_{out}(t) - V_{DD} = -\frac{C_a}{C_L}[V_{DD} - V_{Tn}(V_X)] \tag{4.24}$$

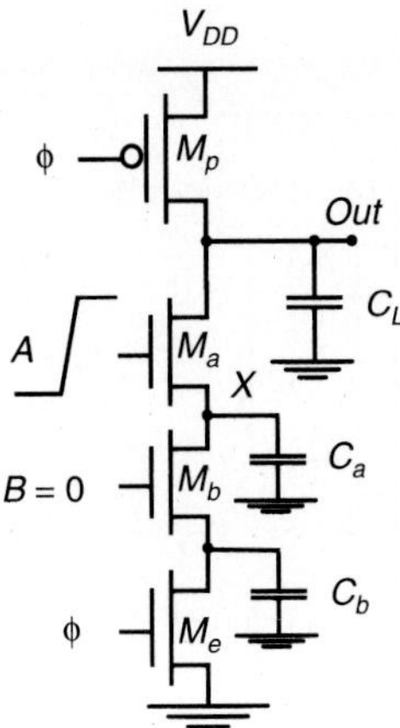

Figure 4.37 Charge sharing in dynamic networks.

2. $\Delta V_{out} > V_{Tn}$—V_{out} and V_X reach the same value:

$$\Delta V_{out} = -V_{DD}\left(\frac{C_a}{C_a + C_L}\right) \tag{4.25}$$

Overall, it is desirable to keep the value of ΔV_{out} below $|V_{Tp}|$. The output of the dynamic gate might be connected to a static inverter, in which case the low level of V_{out} would cause static power consumption. This translates into the following design constraint:

$$\frac{C_a}{C_L} < \frac{|V_{Tp}|}{V_{DD} - V_{Tn}} \approx 0.2 \tag{4.26}$$

C_a is normally smaller than C_L, since the latter includes wiring and fan-out capacitance, while C_a includes only the diffusion and the gate capacitance of the neighboring transistors.

Example 4.12 Charge Redistribution

For the four-input NAND gate of Figure 4.34, the internal capacitance equals 3.1 fF, compared to the 25 fF of the load capacitance. Multiple internal capacitances can be strung together, however; for instance, only the top three transistors of the PDN may be enabled. Under those circumstances, the total voltage drop, predicted by Eq. (4.25), equals 1.3 V. This is sufficient to disrupt the operation of a subsequent fan-out gate. The effect becomes even more pronounced in complex PDN, which contain multiple transistors in parallel.

One way to attack both the charge redistribution and the leakage problems is to make the logic block *pseudo-static*. This is achieved by adding a small, highly resistive, PMOS transistor (with its gate connected to *GND*) in parallel with the precharge transistor, creating in effect a pseudo-NMOS gate (Figure 4.38a). This transistor, also called a *bleeder*, reduces the impedance of the output node, but introduces static power consumption. Keeping the bleeder long and narrow minimizes this effect. Another solution is to precharge the internal nodes using a clock-driven transistor, as shown in Figure 4.38b. Both solutions unfortunately have an adverse effect on the area.

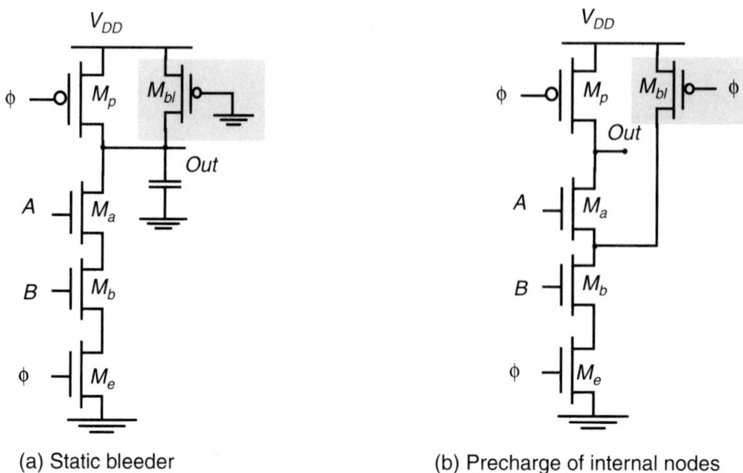

Figure 4.38 Charge redistribution and leakage remedies.

All the above considerations demonstrate that the design of dynamic circuits is rather tricky and requires extreme care. It should therefore only be attempted when high performance is required.

Clock Feedthrough

The clock signal is coupled to the storage node by the gate-source capacitance and the gate-overlap capacitance of the precharge device. The fast rising and falling edges of the clock couple into the signal node, as is adequately demonstrated in the simulation of Figure 4.39. The precise value of the coupling effect is hard to derive manually, although some rough estimation is possible using a capacitive divider model. Simulation is more appropriate, though.

The danger of clock feedthrough is that it causes the signal level to rise sufficiently above the supply voltage that the (normally reverse-biased) junction diodes become forward-biased. This causes electron injection into the substrate, which can be collected by a

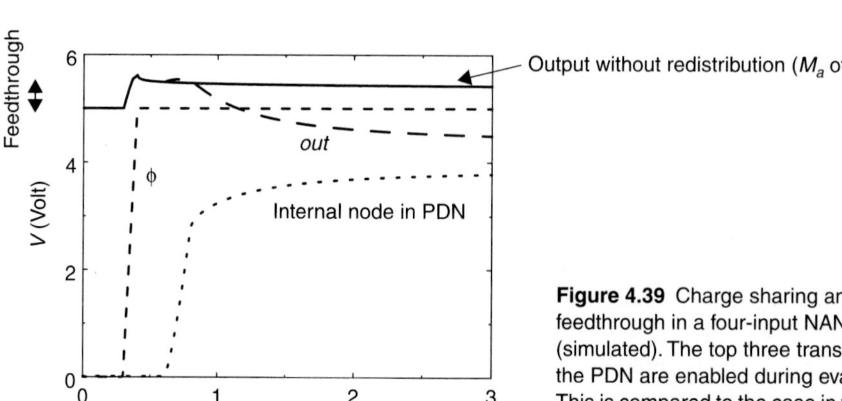

Figure 4.39 Charge sharing and clock feedthrough in a four-input NAND gate (simulated). The top three transistors of the PDN are enabled during evaluation. This is compared to the case in which M_a is left in cut-off mode.

nearby high impedance node in the 1 state, eventually resulting in faulty operation. CMOS latchup might be another result of this injection. This can be avoided by providing a sufficient number of well contacts close to the precharge device to collect the injected currents. For all purposes, high-speed dynamic circuits should be carefully simulated to ensure that clock feedthrough effects stay within bounds.

Example 4.13 Charge Redistribution and Clock Feedthrough

The combined impact of charge redistribution and clock feedthrough is illustrated for the example of the four-input NAND gate. The simulation results clearly demonstrate the effects of clock feedthrough. The magnitude of the feedthrough is limited by the on-voltage of the drain junction of M_p. The overshoot limits the damage of the charge redistribution in the evaluation phase, when the top three transistors in the PDN are turned on. This is contrasted to the case in which the top transistor is left off and no redistribution occurs.

4.3.4 Cascading Dynamic Gates

Trouble arises when we try to cascade a number of dynamic gates. Consider two simple ϕn inverters connected in series, as shown in Figure 4.40a. The potential problem is that all outputs (and hence the inputs to the next gate) are being precharged to 1. The PDN of the second gate is thus in a conducting state at the onset of the evaluation phase. Suppose now that *In* makes a $0 \rightarrow 1$ transition (Figure 4.40b). At the onset of the evaluation period ($\phi = 1$), output Out_1 starts to discharge. As long as Out_1 exceeds the switching threshold of the second gate, which approximately equals V_{Tn}, a conducting path exists between Out_2 and *GND*. Out_2 therefore discharges as well, and wrongly so, as the correct output of the gate equals 1. This conducting path is only turned off when Out_1 reaches V_{Tn} and shuts off the NMOS pull-down transistor. This leaves Out_2 at an intermediate voltage level. The correct level will not be recovered, since dynamic gates rely on capacitive storage, in contrast to static gates, which have dc restoration. The charge loss leads to reduced noise margins and eventual malfunctioning.

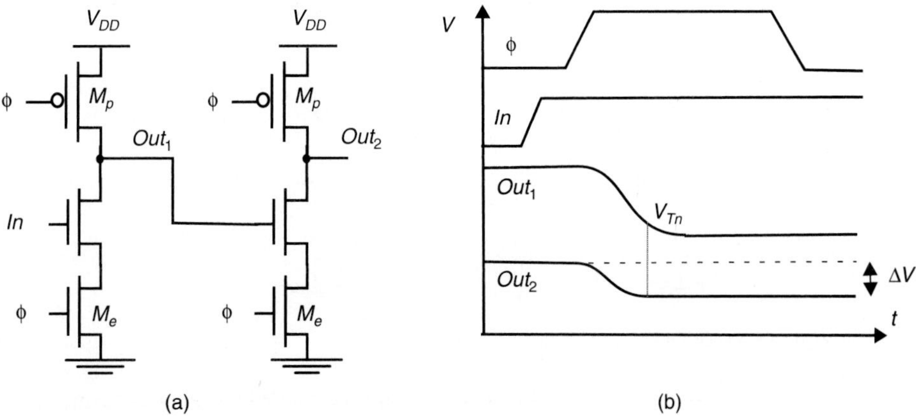

Figure 4.40 Cascade of dynamic ϕn blocks.

It is obvious that the cascading problems arise because the output (and hence the input to the next stage) is precharged to 1. Setting the inputs to 0 during precharge could solve this problem. In doing so, all logic transistors of the next function block are turned off after precharge, and no inadvertent discharging of the storage capacitors can occur during evaluation. In other words, correct operation is guaranteed (ignoring charge redistribution and leakage) as long as **the inputs can only make a single 0 → 1 transition during the evaluation period**. This eliminates the inadvertent discharging since transistors will only be turned on when needed and at most one time per cycle. There is only one exception to this rule. External inputs to the logic blocks should be stable during the evaluation phase and can only change during the precharge period. Note that a similar rule can be derived for a ϕp block. Here, the inputs are only allowed to make 1 → 0 transitions during the evaluation phase.

A number of design styles complying with the above rule have been developed. The most important ones are discussed below.

DOMINO Logic

A DOMINO logic module [Krambeck82] consists of a ϕn block followed by a static inverter (Figure 4.41). This ensures that all inputs to the next logic block are set to 0 after the precharge period. Hence, the only possible transition during the evaluation period is the 0 → 1 transition, so that the formulated rule is obeyed. The introduction of the static inverter has the additional advantage that the fan-out of the gate is driven by a static inverter with a low-impedance output, which increases noise immunity. The buffer furthermore reduces the capacitance of the dynamic output node by separating internal and load capacitances. Finally, the buffer itself can be optimized to drive the fan-out in an optimal way for high speed.

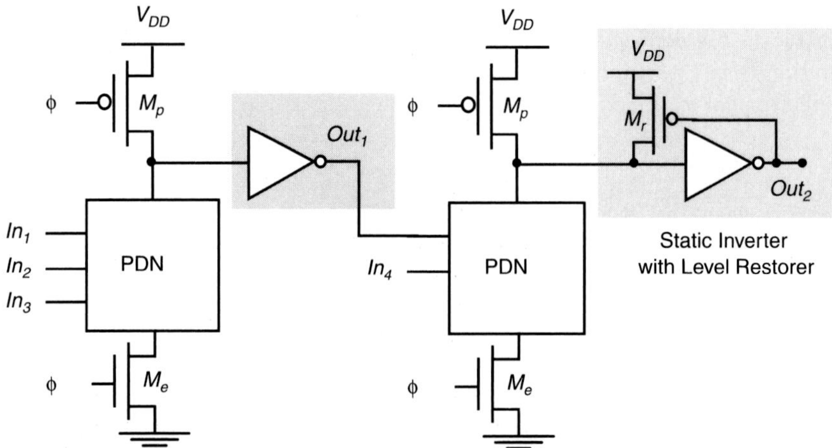

Figure 4.41 DOMINO CMOS logic.

Consider now the operation of a chain of DOMINO gates. During precharge, all inputs (except the external ones) are set to 0. During evaluation, the output of the first DOMINO block either stays at 0 (no delay!) or makes a 0 → 1 transition, affecting the second DOMINO. This effect might ripple through the whole chain, one after the other, as with a line of falling dominoes—hence the name.

DOMINO CMOS has the following properties:

- Only noninverting logic can be implemented. This might be a problem for the implementation of certain Boolean functions.
- It is appropriate for usage in complex, large fan-out circuits such as ALUs and complex control circuits.
- Very high speeds can be achieved: only a rising edge delay exists, while t_{pHL} equals zero (as the output node is precharged low). The static inverter can be optimized to match the fan-out, which is already much smaller than in the complimentary static CMOS case (only a single gate capacitance per input).

DOMINO can be made more immune to parasitic effects such as charge sharing and charge loss, by adding a level-restoring transistor to the static CMOS inverter (as shown in Figure 4.41). This PMOS operates in a similar way as the level-restoring device, discussed in the section on pass-transistor logic.

DOMINO CMOS has been used in the design of a number of very high speed integrated circuits. For instance, the first 32-bit microprocessor (the BellMAC 32, developed at AT&T [Murphy81]) was designed using this logic style. In recent years however, pure DOMINO designs have become rather rare. The noninverting property of the logic style makes its usage cumbersome.

np-CMOS

Instead of using a static inverter to ensure that only $0 \rightarrow 1$ transitions occur during precharge, one can exploit the duality between ϕn blocks and ϕp blocks. The precharge output value of a ϕn block equals 1, which is the correct value for the inputs of a ϕp block during precharge: all PMOS transistors of the PUN are turned off, and erroneous discharge at the onset of the evaluation phase is prevented. In a similar way, a ϕn block can follow a ϕp block without any problems, as the precharge value of inputs equals 0. To make the evaluation and precharge times of the ϕn and ϕp blocks coincide, one has to clock the ϕp blocks with an inverse clock $\overline{\phi}$. An example of such a circuit, called *np*-CMOS logic,[5] is shown in Figure 4.42 ([Goncalvez83, Friedman84, Lee86]). This logic style is the basis of a popular design style, called NORA, which is specifically oriented towards sequential circuits, and is discussed in Chapter 6. A disadvantage of the *np*-CMOS logic style is that the ϕp blocks are slower than the ϕn modules, due to the lower mobility of the PMOS transistors in the logic network. Equalizing the propagation delays requires extra area. The resulting layouts are very dense and can achieve extremely high speeds. Compared to DOMINO, *np*-CMOS is more than 20% faster due to the elimination of the static inverter and the smaller load capacitance. For example, the DEC alpha-processor, the first CMOS 250 Mhz microprocessor, made extensive use of *np*-CMOS logic [Dopperpuhl92].

[5] Also called *zipper CMOS*.

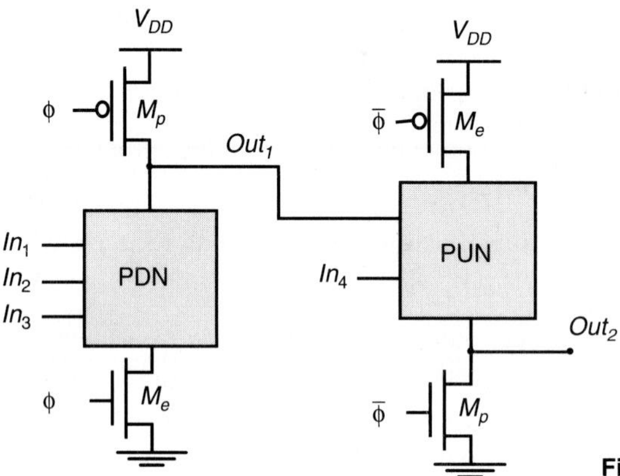

Figure 4.42 The *np*-CMOS logic circuit style.

Problem 4.8 Noise Margins of *np*-CMOS

A disadvantage of *np*-CMOS is a reduced noise tolerance. The noise margins of the dynamic blocks are respectively: $NM_H = |V_{Tp}|$ and $NM_L = V_{Tn}$. Try to determine why this is so.

4.4 Power Consumption in CMOS Gates

Until recently, power consumption was only an afterthought in the design process of CMOS circuits. As the density and size of chips and systems continues to increase, the difficulty of providing adequate cooling either adds significant cost to the system or limits the amount of functionality that can be provided on a single die. Techniques to reduce the power consumption of a design are therefore receiving more attention. The popularity of portable applications that prefer low power consumption to prolong the battery lifetime, has added intensity to this quest. Examples of the latter can be found in the worlds of audio, video, and laptop computing.

In Chapter 3, we discussed the sources of power consumption in the complementary CMOS inverter. These considerations generally remain valid for more complex gates, although some extra considerations have to be taken into account. This is the topic of this section. Most important, we will introduce the concept of *switching activity.* This concept, which is essential to determine the dynamic power consumption of a CMOS design, will be applied to both static and dynamic gates. Other sources of power consumption, such as glitching and direct-path current, are discussed as well. Finally, a number of techniques to reduce power consumption are introduced.

4.4.1 Switching Activity of a Logic Gate

Power in CMOS circuits is mainly consumed during the switching of the gates. The static power dissipation of most gate topologies (besides pseudo-NMOS) is limited to leakage. In Chapter 3, we derived an expression for the dynamic power consumption of an inverter:

$$P_{dyn} = C_L V_{DD}^2 f_{0 \to 1} \tag{4.27}$$

with $f_{0\to1}$ the frequency of energy-consuming transitions (or $0 \to 1$ transitions for static CMOS). It is easily realized that this expression also holds for more complex gates as the nature of the energy consumption remains identical: charging and discharging capacitors. Minimizing power consumption then boils down to reducing one or a number of factors of Eq. (4.27). The V_{DD} factor is, obviously, the most influential due to quadratic dependence.

Computing the dissipation of a complex gate is complicated by the $f_{0\to1}$ factor, also called the *switching activity*. While this factor is easily computed for an inverter, it turns out to be far more complex in the case of higher-order gates and circuits. One concern is that the switching activity of a network is a function of the nature and the statistics of the input signals: If the input signals remain unchanged, no switching happens, and the dynamic power consumption is zero! On the other hand, rapidly changing signals provoke plenty of switching and hence dissipation. Other factors influencing the activity are the circuit style (e.g., dynamic versus static), the function to be implemented, and the overall network topology. These factors can be incorporated by introducing a slight modification in Eq. (4.27):

$$P_{dyn} = C_L V_{DD}^2 f_{0 \to 1} = C_L V_{DD}^2 P_{0 \to 1} f \tag{4.28}$$

with f the average event rate of the inputs and $P_{0\to1}$ the probability that an input transition results in a $0 \to 1$ (or power-consuming) event.

Let us consider the case of a two-input NOR-gate, implemented in static, complementary CMOS. Assume that the inputs to the gate have a uniform distribution of high and low levels (once again, statistical information regarding the input signals is important when analyzing power dissipation). This means that the four possible input combinations for inputs A and B (00, 01, 10, and 11) are equally likely. From the truth table of Table 4.5, we can derive that the probability for the output to be low equals 3/4, while the output is high in 1/4 of the cases.

Table 4.5 Truth table for two- input NOR gate.

A	*B*	*Out*
0	0	1
0	1	0
1	0	0
1	1	0

The probability that a transition of one of the input signals results in energy consumption (or a $0 \to 1$ transition at the output) is equal to the probability that the gate is initially in the 0-output state (= 3/4) times the probability that the next output will be a 1 (= 1/4). The chances of a power-consuming transition are thus given by the following equation.

$$P_{0 \to 1} = P_0 P_1 = (1 - P_1)P_1 = \frac{3}{4} \times \frac{1}{4} = \frac{3}{16} \tag{4.29}$$

The state transition diagram annotated with the transition probabilities is shown in Figure 4.43. Note that the output probabilities are no longer uniform.

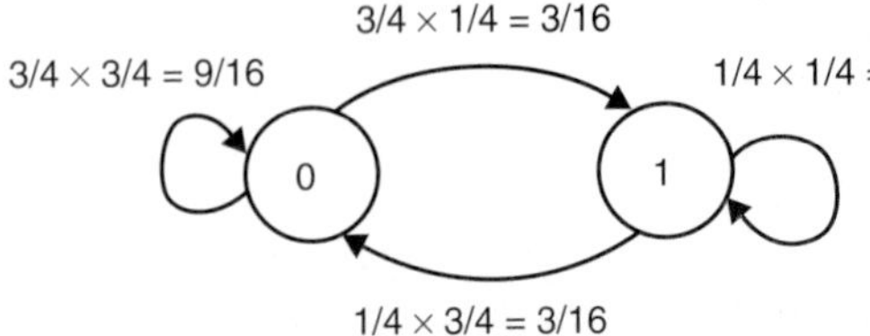

Figure 4.43 State transition diagram and state transition probabilities for two-input NOR gate.

The situation becomes more complicated when the input signals are not evenly distributed, which often occurs in logic circuits with multiple cascaded layers of combinational logic as is demonstrated later in this section. The probability that the output equals 1 (P_1) is then a function of the input distributions P_A and P_B (being the probabilities that the inputs A and B equal 1, respectively). For the two-input NOR gate, this relation is readily expressed as

$$P_1 = (1 - P_A)(1 - P_B) \tag{4.30}$$

which translates into a generalized expression for the transition probability:

$$P_{0 \to 1} = (1 - P_1)P_1 = [1 - (1 - P_A)(1 - P_B)][(1 - P_A)(1 - P_B)] \tag{4.31}$$

A plot of the output transition probability (which is proportional to the average power consumption) as a function of P_A and P_B is shown in Figure 4.44. Observe how this graph degrades into the simple inverter case when one of the input probabilities is set to 0. This picture demonstrates the impact of input probabilities on the dissipation. Ignoring signal statistics in power analysis obviously results in substantial errors.

Problem 4.9 Power Dissipation of Basic Logic Gates

Derive the $0 \to 1$ output transition probabilities for the basic logic gates (AND, OR, EXOR). The results to be obtained are given in Table 4.6. Derive also the transition probabilities of the inverse gates (NAND, NOR, NXOR).

Table 4.6 Output transition probabilities for static logic gates.

	$P_{0 \to 1}$
AND	$(1 - P_A P_B)P_A P_B$
OR	$(1 - P_A)(1 - P_B)[1 - (1 - P_A)(1 - P_B)]$
EXOR	$[1 - (P_A + P_B - 2P_A P_B)](P_A + P_B - 2P_A P_B)$

The evaluation of the switching activity becomes more involved in the case of complex logic networks. When a signal propagates through a number of logic layers, its signal and transition probabilities become "colored." Consider the circuit of Figure 4.45a, where both gates are implemented in static logic, and all inputs A, B, and C have equal probabilities of equaling 0 or 1. The probability that the node X undergoes a power-consuming tran-

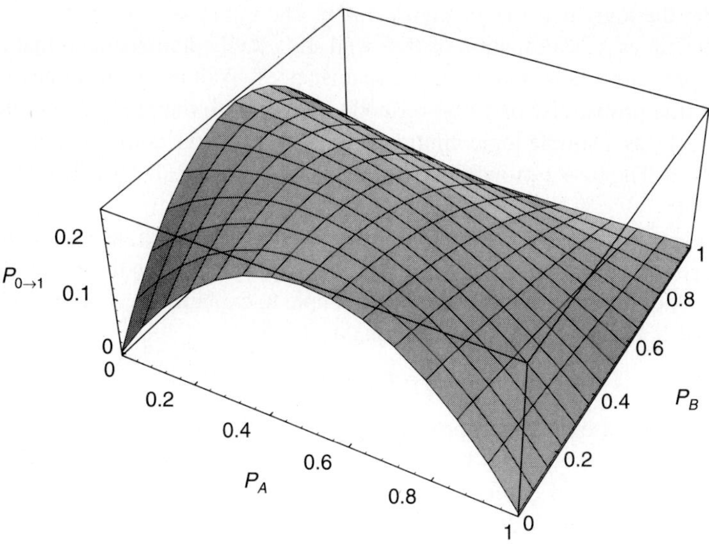

Figure 4.44 Transition probability (or equivalent energy dissipation) of a two-input NOR gate as a function of the input probabilities P_A and P_B.

sition has been derived above to equal 3/16, while its value equals 1 in 3 out of 4 cases. As a result, the input signals to the AND gate have an uneven distribution, and the extended expression of Table 4.7 must be used. Evaluation of the AND gate expression yields a transition probability for node Z of $(1 - 3/4 \times 1/2)\ (3/4 \times 1/2) = 15/64$.

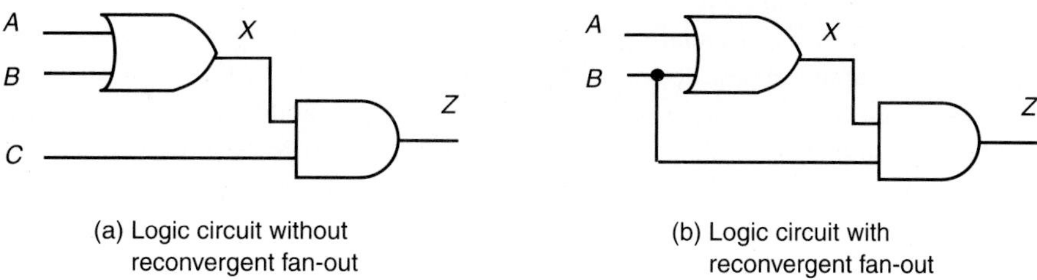

(a) Logic circuit without reconvergent fan-out

(b) Logic circuit with reconvergent fan-out

Figure 4.45 Examples of logical networks and transition probabilities.

The analysis technique used in the above example is straightforward. Signal and transition probabilities are evaluated in an ordered fashion, progressing from the input to the output node. This approach suffers, however, from some severe limitations. First, of all, it does not deal with networks with feedback (such as sequential circuits). Due to the cyclic nature of these circuits, finding an input-to-output traversal order is not possible. Secondly, it assumes that the signal probabilities at the input of a gate are independent. This is rarely the case in actual circuits, where *reconvergent fan-out* often causes intersignal dependencies: due to converging paths in the logic, a number of inputs to a logic gate can be functions of the same primary input.

Consider the logic network of Figure 4.45b. The inputs to the AND gate (X and B) are interdependent as X is a function of B as well. It is easily demonstrated that the analysis approach presented fails under these circumstances. Walking from inputs to outputs yields a transition probability of 15/64 for node Z (similar to the previous example). This proves to be false, as a simple logic minimization shows that the complete network can be reduced to $Z = B$. The $0 \rightarrow 1$ probability for this network equals $(1/2 \times 1/2) = 1/4$, which is the correct value.

To get the precise results in the progressive analysis approach, its is essential to take signal inter-dependencies into account. We can accomplish this with the aid of conditional probabilities. Once again, we use our simple example to explain this concept. For an AND gate, Z equals 1 if and only if B and X are equal to 1.

$$P_Z = P(Z = 1) = P(B = 1, X = 1) \tag{4.32}$$

$P(B = 1, X = 1)$ expresses the probability that B and X are equal to 1 simultaneously. If B and X are independent, $P(B = 1, X = 1)$ can be decomposed and yields the expression for the AND gate, derived earlier: $P_Z = P(B = 1).P(X = 1) = P_B P_X$. If a dependency between the two exists (as it does in Figure 4.45b), a conditional probability has to be employed.

$$P_Z = P(X{=}1|B{=}1).P(B{=}1|X{=}1) = P(X{=}1|B{=}1).P(B{=}1) \tag{4.33}$$

The first factor in Eq. (4.33) represents the probability that $X = 1$ given that $B = 1$. The extra condition is necessary because X is dependent upon B. Inspection of the network shows that this probability is equal to 1, resulting in the following signal probability for Z: $P_Z = P(B = 1) = P_B$. This conclusion comes as no surprise, because we have already determined that Z and B are identical signals.

Deriving those expressions in a structured way for large networks with reconvergent fan-out is difficult, especially when the networks are also sequential and contain feedback loops. Computer-aided design (CAD) tools are essential in this regard. The purpose of those tools is to estimate or analyze the transition probabilities at the nodes of a network. To be meaningful, they have to take in a typical sequence of input signals, as the power dissipation is a strong function of the statistics of those signals.

Problem 4.10 Switching Activity of Combinational Networks

Derive the average switching energy of the logic network, implementing the following logical functions: $S = A \cdot B$ and $C = A \oplus B$. Assume that static CMOS is the implementation style and that both gates are loaded with an equal capacitance C_L. Analyze the energy over the complete range of signal statistics for both inputs A and B. It is worth mentioning that the gate structure in question is nothing else than the *half-adder* circuit, which is discussed at length in Chapter 7.

The resulting transition probability as a function of the input statistics is plotted in Figure 4.46. A strong dependency on the input-signal statistics can once again be observed.

Finally, let us consider the trade-off between a static and a dynamic implementation of a given function. In a dynamic implementation, the signal is being precharged every clock-cycle (under the assumption that an NMOS pull-down network is employed). Power is therefore consumed during the precharge operation every time the output capacitor is

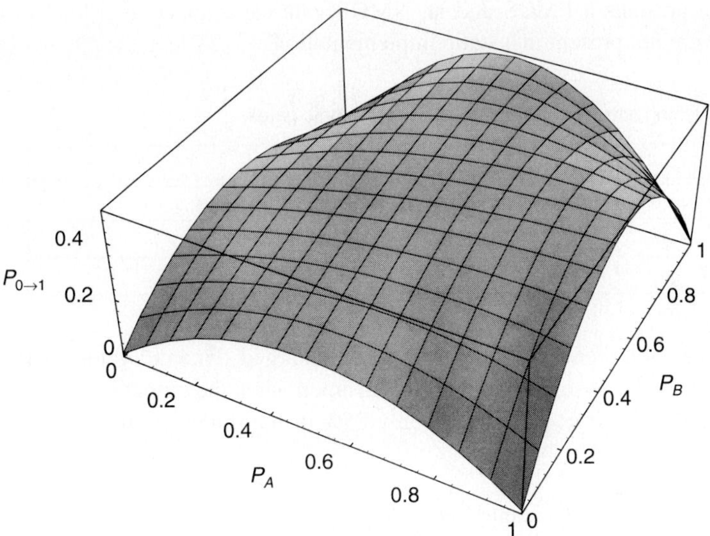

Figure 4.46 Switching activity of a static half-adder cell.

discharged in the preceding cycle. This is equivalent to stating that power is consumed every time the output equals 0, independent of the preceding or following values! The dynamic power consumption in a dynamic network is solely determined by the *signal-value probabilities*, not by the transition probabilities. It is easy to see that **this signal probability is always larger than the transition probability,** which is the product of two signal probabilities, each of which is smaller than 1.

If the inputs to a static CMOS gate do not change from the previous sample period, no power is consumed. This is not true in the case of dynamic logic, where power can be consumed even if the inputs remain constant. Unlike the dynamic case, in which the activity depends only on the signal probability, the transition probability in the static implementation depends on previous state.

Example 4.14 Activity in Dynamic Logic

For the two-input NOR gate, the 0-probability equals $(P_A + P_B - P_AP_B)$. Assuming that all inputs are equally probable, there is a 75% probability that the output node discharges after the precharge phase, implying that the activity for such a gate is 0.75 (i.e., $P_{NOR} = 0.75\ C_LV_{DD}^2 f_{clk}$). This is three times higher than the corresponding static implementation.

The switching activities for the basic dynamic gates (assuming precharge to V_{DD}) are summarized in Table 4.7. The activity is, again, influenced by the logic function. For an OR gate, for instance, the activity is 1/4 for a dynamic implementation (assuming equally probable inputs), which is virtually identical to the 3/16 activity of the static implementation. Be aware, however, that the capacitance being switched is substantially smaller in a dynamic implementation, as was demonstrated earlier in this chapter. On the negative side, the power analysis of a dynamic gate should include the power dissipated in driving the capacitance of the *clock lines*, which are being switched at the full rate. Each dynamic gate presents a PMOS and an NMOS gate capacitance as a load to the clock driver. These are not present in a static implementation.

Table 4.7 Switching activity for precharged dynamic logic gates.

	$P_{0\rightarrow 1}$
AND	$(1 - P_A P_B)$
OR	$(1 - P_A)(1 - P_B)$
EXOR	$(1 - [P_A + P_B - 2P_A P_B)]$

A similar table can be obtained for predischarged dynamic gates (with a PMOS PUN network). In this case, power is only consumed when the output equals 1, or $P_{0\rightarrow 1} = P_1$. For the two-input NOR-gate, this is only 25% of the time (for an even distribution of 0's and 1's at the inputs).

Problem 4.11 Activity in Dynamic Networks

Derive the activity factor for the output of the networks of Figure 4.45a, assuming that the gates are implemented using dynamic CMOS.

4.4.2 Glitching in Static CMOS Circuits

When analyzing the transition probabilities of complex, multistage logic networks in the preceding section, we ignored the fact that the gates have a nonzero propagation delay. In reality, the finite propagation delay from one logic block to the next can cause spurious transitions (called *glitches, critical races,* or *dynamic hazards*) to occur—a node can exhibit multiple transitions in a single clock cycle before settling to the correct logic level. For example, consider a cascade of two NOR gates, as shown in Figure 4.47. Assume that

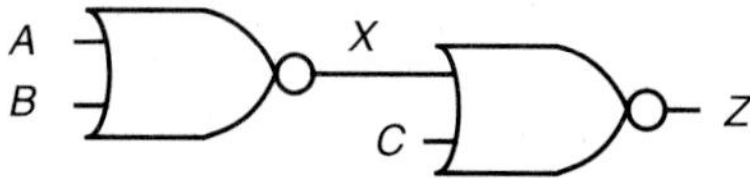

Figure 4.47 Glitching in static CMOS circuits.

each gate has an identical delay (equal to one "unit") and that inputs *A, B*, and *C* arrive at the same time. Since there is a finite propagation delay time through the first NOR gate, the second NOR gate initially evaluates with the previous value of *X*. When the correct value from the first gate finally ripples through, the same gate evaluates again. The first evaluation was redundant, consuming power without performing any "useful" function. For example, consider the transition from 101 to 000 for inputs *ABC* (Figure 4.48). The output should remain 0, but it makes an extra transition to 1 before settling to the correct value, dissipating additional energy.

A typical example of the effect of glitching is shown in Figure 4.49, which displays the simulated response of a chain of NAND gates for all inputs going simultaneously from 0 to 1. The output bits are shown as a function of time. A logic analysis of the circuit

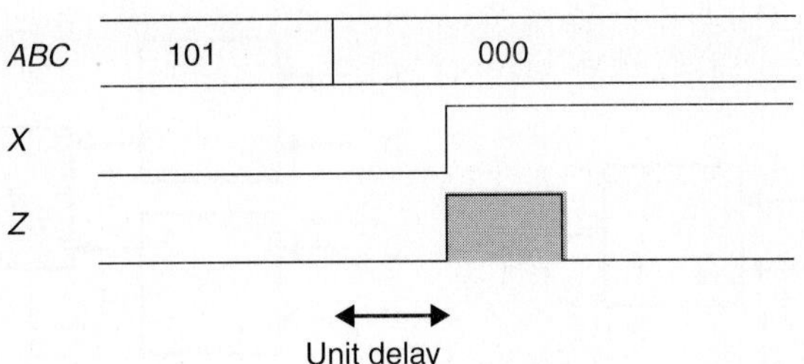

Figure 4.48 Glitching in multilevel logic networks.

shows the correct outputs to be a sequence of alternating 0's and 1's. As can be observed, all signals initially evolve in the direction of the *GND* rail. This trend is later reversed for half of the signals, as it takes some time for the correct outputs to propagate down the chain. Actually, the farther down the chain, the worse it gets. Although the glitches are only partial (i.e., not from rail to rail), they contribute significantly to the power dissipation. In later chapters, we will see that long chains of gates often occur in important logic structures such as adders and multipliers. The dissipation caused by the spurious transitions can reach up to 25% of the total dissipation for some of those circuits.

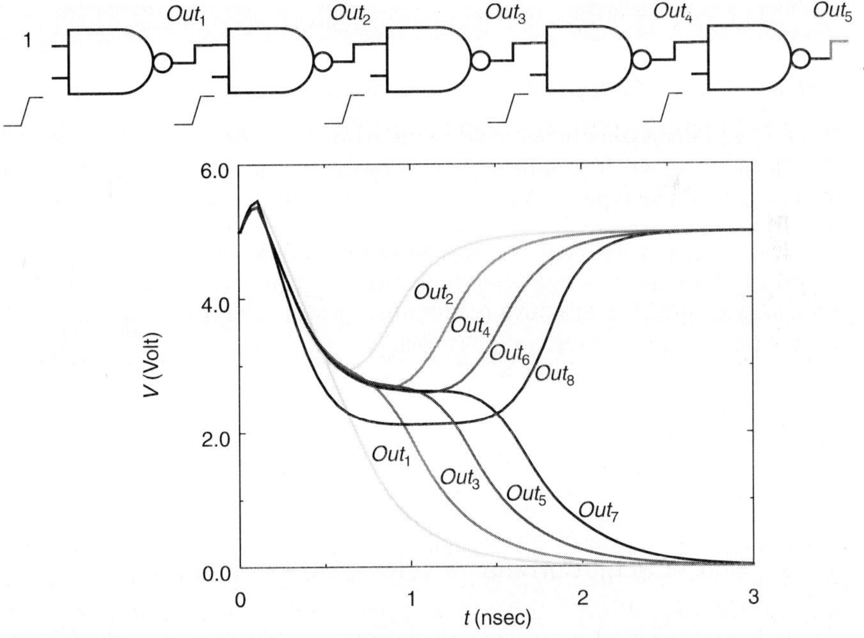

Figure 4.49 Glitching in a chain of NAND gates.

The occurrence of glitching in a circuit is mainly due to a mismatch in the path lengths in the network. If all input signals of a gate change simultaneously, no glitching occurs. On the other hand, if input signals change at different times, a dynamic hazard might develop. Such a mismatch in signal timing is typically the result of different path lengths with respect to the primary inputs of the network. This is illustrated in Figure 4.50. Assume that all operators *F* have the same unit delay. The first network (a) suffers from

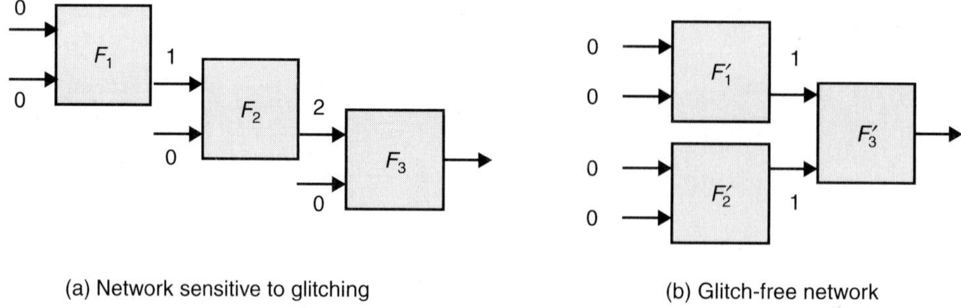

Figure 4.50 Glitching is influenced by matching of signal path lengths. The annotated numbers indicate the signal arrival times.

glitching as a result of the wide disparity between the arrival times of the input signals for a gate. For example, for gate F_3, one input settles at time 0, while the second one only arrives at time 2. Redesigning the network so that all arrival times are identical can dramatically reduce the number of transitions (network b).

Design Technique

The above observation can be translated into the following design rule:

> To eliminate the impact of dynamic hazards on the power dissipation of a static CMOS circuit, it is sufficient to equalize the arrival times of all signals at the inputs of each gate.

Strict adherence to this rule is hard in practice. Fortunately, making the path lengths to the inputs of a gate approximately the same is sufficient to virtually eliminate all glitching. The attentive reader observes that the stated condition is beneficial for the performance of the circuit as well.

Glitching is not an issue in dynamic logic, as the correct operation of such a circuit requires that each node undergo at most one transition per clock cycle.

4.4.3 Short-Circuit Currents in Static CMOS Circuits

As discussed in Chapter 3, another source of power dissipation in a static CMOS circuit is the flow of current from V_{DD} to GND during switching, when both NMOS and PMOS are conducting simultaneously. Such a path never exists in a dynamic circuit, as precharge and evaluate transistors should never be on simultaneously; if they were, it would lead to malfunction. Short-circuit currents are therefore encountered only in static designs. The total amount of energy dissipated in the short-circuit current is a function of the on-time of the transistors and the operation modes of the devices.

Consider a static CMOS inverter with a $0 \rightarrow 1$ transition at the input. Assume first that the load capacitance is very large, so that the output fall time is significantly larger than the input rise time (Figure 4.51a). Under those circumstances, the input moves through the transient region before the output starts to change. As the source-drain voltage of the PMOS device is approximately 0 during that period, the device shuts off without ever delivering any current. The short-circuit current is close to zero in this case. Consider now the reverse case, where the output capacitance is very small, and the output fall time is substantially smaller than the input rise time (Figure 4.51b). The drain-source voltage of the PMOS device equals V_{DD} for most of the transition period, guaranteeing the maximal short-circuit current (equal to the saturation current of the PMOS). This is clearly the worst case. This analysis leads to the conclusion that the short-circuit dissipation is minimized by making the output rise/fall time larger than the input rise/fall time.

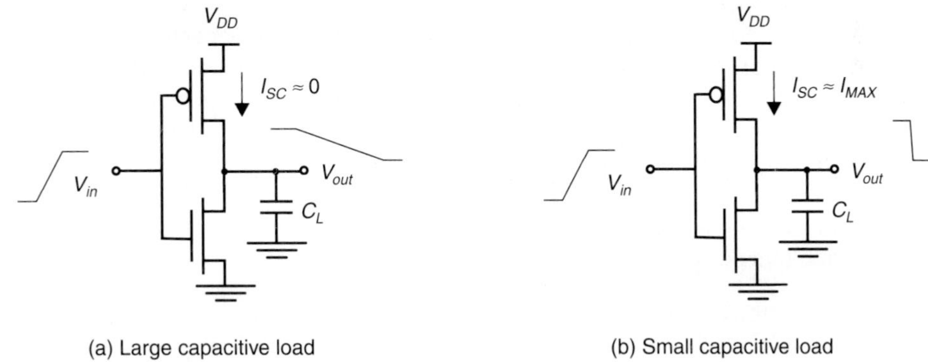

(a) Large capacitive load (b) Small capacitive load

Figure 4.51 Impact of load capacitance on short-circuit current.

On the other hand, making the output rise/fall time too large slows down the circuit and can cause short-circuit currents in the fan-out gates. This would be a perfect example of local optimization, forgetting the global picture.

Design Technique

A more practical rule, which optimizes the power consumption in a global way, can be formulated (Veendrick84]):

> The power dissipation due to short-circuit currents is minimized by matching the rise/fall times of the input and output signals. At the overall circuit level, this means that rise/fall times of all signals should be kept constant within a range.

Making the input and output rise times of a gate identical is not the optimum solution for that particular gate on its own, but keeps the overall short-circuit current within bounds. This is shown in Figure 4.52, which plots the short-circuit energy dissipation of an inverter (normalized with respect to the zero-input rise time dissipation) as a function of the ratio r between input and output rise/fall times. When the load capacitance is too small for a given inverter size ($r > 2...3$ for $V_{DD} = 5$ V), the power is

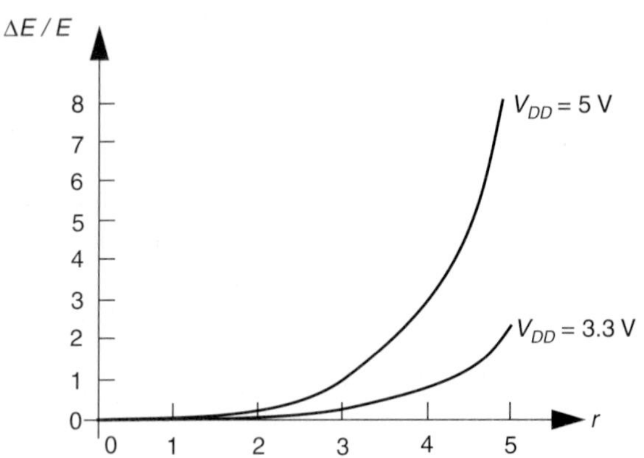

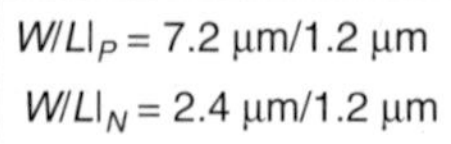

Figure 4.52 Short-circuit energy dissipation (normalized with respect to zero input rise time energy) for a static CMOS inverter as a function of the ratio between input and output rise/fall times (from [Burd94]).

dominated by the short-circuit current. For very large capacitance values, all power dissipation is devoted to charging and discharging the load capacitance. When the rise/fall times of inputs and outputs are equalized, most power dissipation is associated with the dynamic power and only a minor fraction (< 10%) is devoted to short-circuit currents. Notice also that the impact of short-circuit current is reduced when we lower the supply voltage. In the extreme case, when $V_{DD} < V_{Tn} + |V_{Tp}|$, short-circuit dissipation is completely eliminated, because both devices are never on simultaneously.

Adhering to the above concept seems simple enough, but it is often violated seriously in current circuit design practice. Consider a standard cell library. The standard-cell methodology is a semicustom design approach (discussed in Chapter 11) that uses a library of fixed cells to implement random logic functions. To ensure that the library cells meet the timing constraints for a wide variety of load capacitances, all cell transistors are overdimensioned. As a result, the output rise/fall times might easily be smaller than the input slopes, resulting in substantial short-circuit dissipation. The ratio of short-circuit dissipation versus overall consumption can exceed 50% for this approach. Careful transistor sizing (based on the fan-out of a gate) is necessary if power consumption is a main concern.

4.4.4 Analyzing Power Consumption Using SPICE

The average power consumption of a circuit is defined by Eq. (3.2), which is repeated here for the sake of convenience.

$$P_{av} = \frac{1}{T}\int_0^T p(t)dt = \frac{V_{DD}}{T}\int_0^T i_{DD}(t)dt \tag{4.34}$$

with T the period of interest, and V_{DD} and i_{DD} the supply voltage and current, respectively. Some implementations of SPICE provide built-in functions to measure the average value of a circuit signal. For instance, the HSPICE *.MEASURE TRAN AVG* command computes the area under a given transient response and divides it by the period of interest. This is identical to the definition given in Eq. (4.34). Other implementations of SPICE are, unfortunately, not as extensive. This is not as bad as it seems, as long as one realizes that SPICE is actually a differential equation solver. A small circuit can easily be conceived that acts as an integrator and whose output signal is nothing but the average power.

Consider, for instance, the circuit of Figure 4.53. The current delivered by the power supply is measured by the current-controlled current source and integrated on the capacitor C. The resistance R is only provided for DC-convergence reasons and should be chosen as high as possible to minimize leakage. A clever choice of the element parameter ensures that the output voltage P_{av} equals the average power consumption. The operation of the circuit is summarized in Eq. (4.35) under the assumption that the initial voltage on the capacitor C is zero.

$$C\frac{dP_{av}}{dt} = ki_{DD}$$

$$or \tag{4.35}$$

$$P_{av} = \frac{k}{C}\int_0^T i_{DD}dt$$

Equating Eq. (4.34) and Eq. (4.35) yields the necessary conditions for the equivalent circuit parameters: $k/C = V_{DD}/T$. Under these circumstances, the equivalent circuit shown presents a convenient means of tracking the average power in a digital circuit.

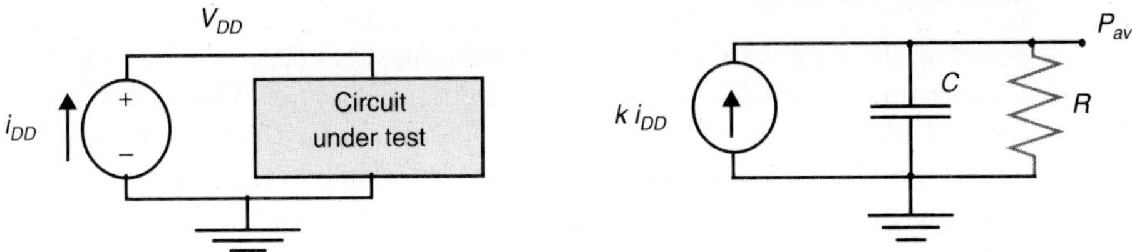

Figure 4.53 Equivalent circuit to measure average power in SPICE.

Example 4.15 Average Power of Inverter

The average power consumption of the inverter of Example 3.6 is analyzed using the above technique for a toggle period of 2 nsec. The resulting power consumption is plotted in Figure 4.54, showing an average power consumption of approximately 0.45 mW. This is equivalent to a PDP of 0.9 pJ (which is close to the 0.75 pJ derived in Chapter 3). Observe the slightly negative dip during the high-to-low transition. This is due to the injection of current into the supply, when the output briefly overshoots V_{DD} as a result of the capacitive coupling between input and output (as is apparent from the simulation of Figure 3.26).

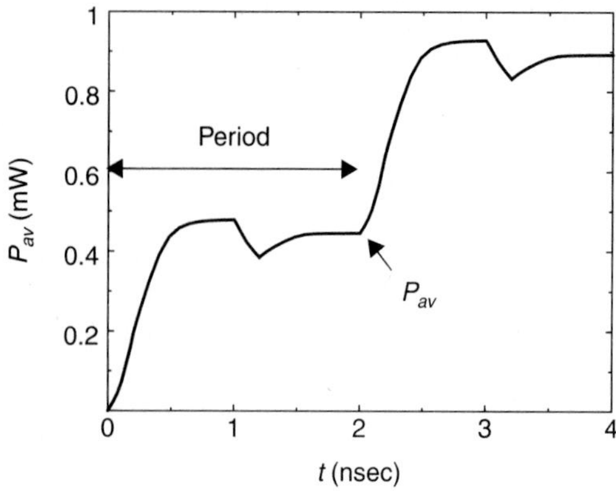

Figure 4.54 Deriving the power consumption using SPICE.

4.4.5 Low-Power CMOS Design

Assuming that short-circuit current, glitching, and leakage can be kept in bounds by employing the appropriate circuit design approaches, the dominant power source in CMOS is the dynamic power consumption, expressed by Eq. (4.29). Suppose further that the throughput or performance of the circuit is to be kept constant, which means that f is a fixed parameter. Under those circumstances, the power of a design can be reduced by manipulating two parameters: the supply voltage V_{DD} and the *effective capacitance* C_{eff}, which is the product of the physical capacitance C and the switching activity $P_{0 \to 1}$. Each of these factors is discussed below.

Reducing the Supply Voltage

Reducing the supply voltage is obviously the most effective way of reducing the power consumption, since the dissipation is proportional to V_{DD}^2. This is confirmed in Figure 4.55, which plots the measured power-delay product (which equals $C_{eff}V_{DD}^2$) of two experimental circuits. Therefore, reducing the supply voltage results in a *quadratic* improvement in the power-delay product of a logic family ([Chandrakasan92]). It is interesting to observe that the PDP (which is considered as one of the quality measures of a logic style) continues to improve when dropping the supply voltage.

Unfortunately, this simple solution to low-power design comes at a cost. The impact of reducing V_{DD} on the delay is shown in Figure 4.56 for a variety of different logic circuits, ranging in size from 56 to 44,000 transistors and spanning a variety of functions. All circuits exhibit in essence the same dependence. Clearly, we pay a speed penalty for a V_{DD} reduction. The performance loss becomes particularly significant when V_{DD} approaches the sum of the threshold voltages of the devices. This behavior is accurately predicted by Eq. (3.21) and Eq. (3.22) presented in Chapter 3 and repeated below.

$$t_p \sim \left(\frac{C_L V_{DD}}{2}\right)\left[\frac{1}{k_n(V_{DD} - V_{Tn})^2} + \frac{1}{k_p(V_{DD} - |V_{Tp}|)^2}\right] \tag{4.36}$$

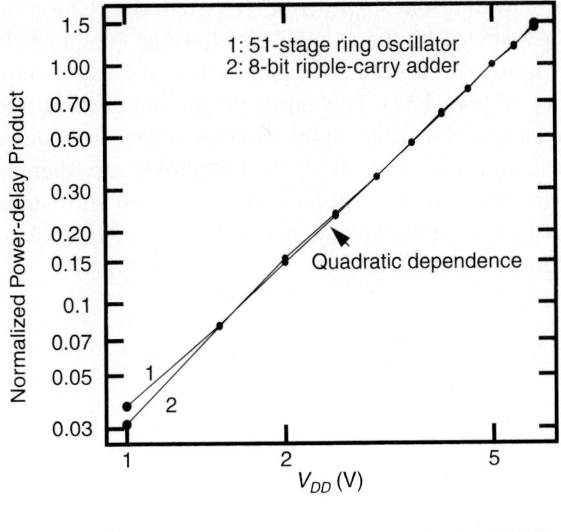

Figure 4.55 Normalized power-delay product of two experimental circuits as a function of V_{DD}.

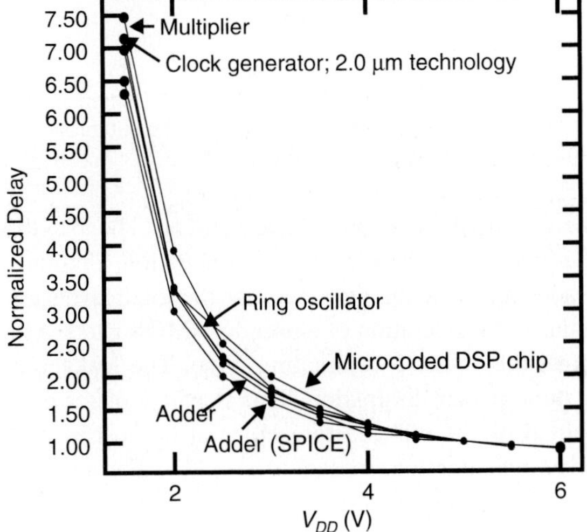

Figure 4.56 Normalized propagation delay as a function of supply voltage for a number of experimental circuits.

In summary, we see that the power-delay product improves as delays increase (through reduction of the supply voltage). To conserve energy, it is desirable to operate at the *slowest* possible speed. Maintaining the overall system throughput at the same time requires a compensation for the increased delays at low voltages.

Example 4.16 Effect of Supply-Voltage Reduction on Power Consumption and Speed

From Figure 4.56, we can see that the performance of a circuit degrades by a factor of 2 when dropping the supply voltage with a factor of 1.7 (from 5 V to 2.94 V). This reduction in supply voltage reduces the power-delay product by a factor of 2.89.

A first approach is to adopt a technology with *lower threshold voltages*. Reducing the thresholds of the devices moves the curves in Figure 4.56 to the left, which means that the performance penalty for lowering the supply voltage is reduced. Unfortunately, the threshold voltages are lower-bounded by the amount of allowable subthreshold leakage current. If the V_T becomes too low, even setting the gate-source voltage to zero does not completely turn off the device (Figure 4.57). This raises the minimum clock frequency in dynamic circuits (due to the increased leakage on the dynamic nodes) and causes standby currents and reduced noise margins in static designs. Thresholds are, therefore, never smaller than 0.5–0.6 V in standard processes (usually they are even substantially larger and go up to 1 V) and lowering the supply voltage below 1.2–2 V (2 times V_T) is excessively expensive in terms of lost performance.

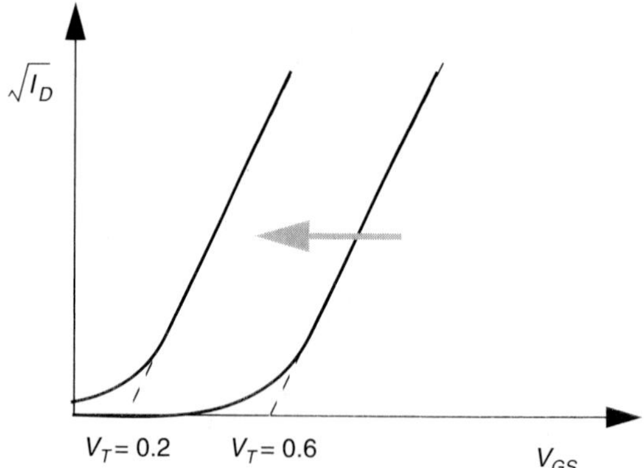

Figure 4.57 Decreasing the threshold increases the subthreshold current at $V_{GS} = 0$.

This lower bound on the thresholds is in some sense artificial. The idea that the leakage current in a static CMOS circuit has to be zero is a preconception. Certainly, the presence of leakage currents degrades the noise margins, because the logic levels are no longer equal to the supply rails (similar to the operation of a pseudo-NMOS gate). As long as the noise margins are within range, this is not a compelling issue. The leakage currents, of course, cause an increase in static power dissipation. This is partly offset by the drop in supply voltage, which is enabled by the reduced thresholds at no cost in performance.[6] Remember that the static power consumption is equal to $I_{static}V_{DD}$. For a 0.25 μm CMOS process, the following circuit configurations obtain the same performance: 3 V supply, 0.7 V V_T and 0.45 V supply, 0.1 V V_T. The power consumption of the latter is, however, 45 times smaller [Liu93]!

Similar considerations hold for dynamic circuits. The minimum allowed threshold voltage in a dynamic circuit is set by the minimum operating frequency. The potential power savings are not as impressive and are limited to approximately a factor of 8 when maintaining the performance level (for a supply voltage of 1 V and a minimum threshold of 0.3 V). This demonstrates that **static circuits are more amenable to supply scaling than dynamic ones.**

Changing the process parameters is not an option for most digital circuit designers. In that case, the loss in performance due to the supply reduction must be compensated by other means. There are ample opportunities to do so at the circuit or architectural design

[6] To avoid power dissipation in standby mode, techniques to shut off a circuit completely are a necessity in this approach.

levels. For instance, it is well known that high performance does not necessarily rely on the highest possible clock frequency. Parallel architectures that combine multiple slow computational elements can achieve exactly the same goal. This might result in an increase in area, but the reduction in power might well be worth it (as the cost of silicon is gradually being reduced with the scaling of the technology). Some of these ideas that amount to trading off *area for power* are explored in more detail in Chapter 7.

Reducing the Effective Capacitance

When a lower bound on the supply voltage is set by external constraints (as often happens in real-world designs), or when the performance degradation due to lowering the supply voltage is intolerable, the only means of reducing the dissipation is by lowering the effective capacitance. This can be achieved by addressing both of its components: the physical capacitance and the switching activity.

Lowering the physical capacitance is an overall worthwhile goal, which also helps to improve the performance of the circuit. For instance, some of the circuit styles discussed in this chapter come with a substantially reduced capacitance and can result in low-power operation. A CPL adder, for example, reportedly uses 30% less power compared to a conventional static CMOS adder (for a 4 V supply). This reduction, which is even more significant for lower supply voltages, is mainly due to the decrease in capacitance [Yano90].

As most of the capacitance in a combinational logic circuit is due to transistor capacitances (gate and diffusion), it makes sense to keep those contributions to a minimum when designing for low power. This means that transistors should be kept to *minimal size* whenever possible or reasonable. This definitely affects the performance of the circuit, but the effect can be offset by using logic or architectural speed-up techniques, as briefly discussed in the previous section. The only instances where transistors should be sized up is when the load capacitance is dominated by extrinsic capacitances (such as fan-out or wiring capacitance). Again, this is contrary to common design practices used in cell libraries, where transistors are generally made large to accommodate a range of loading and performance requirements.

To illustrate this point, let us analyze the simple case of a static inverter driving a load capacitance consisting of an intrinsic (C_{int}) and an extrinsic component (C_{ext}) (Figure 4.58a). While the former represents the diffusion capacitances, the latter stands for wiring capacitance and fan-out. It is assumed that the ratio between PMOS and NMOS transistors is constant. The factor N stands for the inverter sizing factor, where N is equal to 1 for an inverter constructed of minimum-size devices. We can see that the intrinsic capacitance of the scaled device is proportional to N (or $C_{int}(scaled) = NC_{int}$). Figure 4.58b plots the normalized energy (per transition) as a function of the scaling factor N with the ratio between the extrinsic and intrinsic capacitance as a parameter: $\alpha = C_{ext} / C_{int}$. The speed of all implementations is kept constant by appropriately adjusting the supply voltage: larger values of N normally mean lower values of the supply voltage.

When $\alpha = 0$ (or the load capacitance is zero), the lowest energy consumption is obtained when using minimum-size devices. Only when the extrinsic capacitances dominate ($\alpha > 1$) does it make sense to widen the devices. Notice, once again, that this analysis assumes a sustained performance.

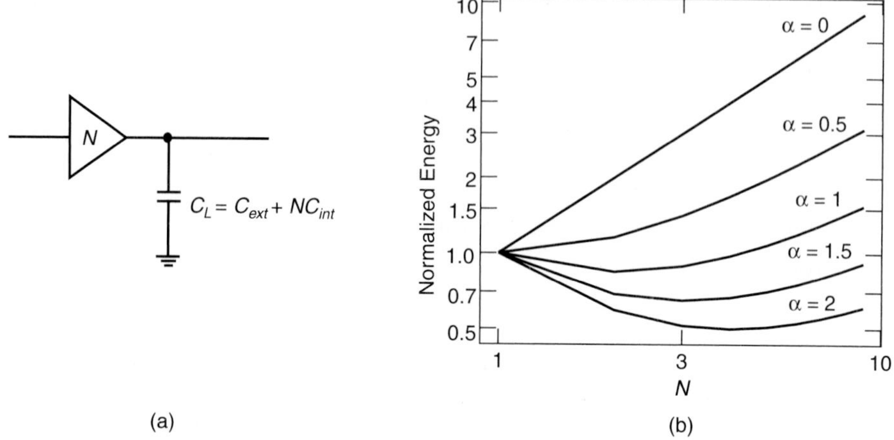

Figure 4.58 Normalized energy of a MOS inverter with load capacitance C_L, as a function of the inverter size N and the ratio between the extrinsic and intrinsic capacitance α (= C_{ext}/C_{int}).

Example 4.17 Transistor Sizing for Inverter

We derive a simplified expression for the normalized energy of the inverter of Figure 4.58 as a function of N and α. The energy is normalized with respect to the case for $N = 1$, which is called the *reference*.

We have already established that the propagation delay of an inverter can be approximated by the following expression:

$$t_p \approx \frac{C_L}{2V_{DD}}\left(\frac{1}{k_n} + \frac{1}{k_p}\right)$$

Observing that $C_L = (N + \alpha)\, C_{int}$ and that $k_n = Nk_{n,ref}$ and $k_p = Nk_{p,ref}$, we can express the propagation delay of the scaled inverter:

$$t_p = \frac{(N+\alpha)}{N} t_{p,\,ref}$$

Keeping the propagation delay of the scaled inverter constant with respect to the reference case means lowering the supply voltage:

$$V'_{DD} = \frac{(N+\alpha)}{N(1+\alpha)} V_{DD}$$

where V'_{DD} and V_{DD} are the supply voltages of the scaled and reference inverters, respectively.

The dissipated energy of the scaled inverter is now derived:

$$E' = C_L(V'_{DD})^2 = \frac{(N+\alpha)^3}{N^2(1+\alpha)^2} C_{int}(V_{DD})^2 = \frac{(N+\alpha)^3}{N^2(1+\alpha)} E_{ref} \tag{4.37}$$

The energy is minimized for $N = 2\alpha$ (for $N > 1$).

Another approach to reducing the physical capacitance is to avoid the extensive sharing of resources. Figure 4.59 shows a common bus architecture, where a single shared bus is connected to a number of drivers and receivers. This configuration results in a large bus capacitance, due to the large number of devices connected to the bus and also to the long wire-length of the bus (which probably stretches over a large part of the chip). From a power perspective, it is beneficial to avoid the extensive sharing and to use dedicated point-to-point buses instead. This probably has a negative effect on the chip area, but reduces the physical capacitance being switched per transition. Once again, this amounts to *trading off area versus power.*

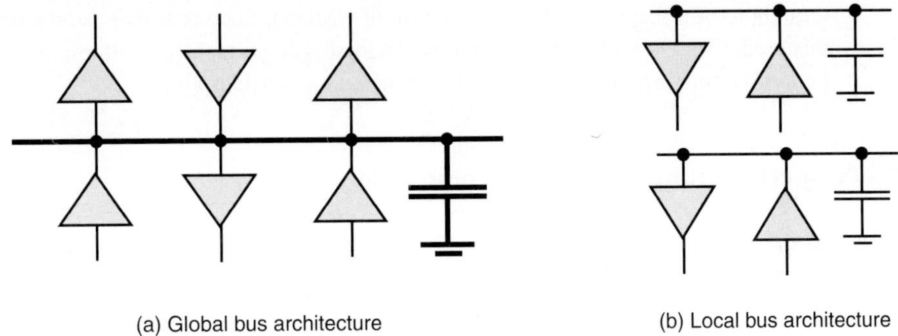

(a) Global bus architecture (b) Local bus architecture

Figure 4.59 Local bus architectures reduce the effective capacitance.

While reducing the physical capacitance helps, another way to reduce the dissipation is to switch that capacitor less often, or, in other words, *reduce the switching activity.* A good example in this class is the choice between static and dynamic circuits. Due to the periodic precharging, dynamic circuits exhibit a larger switching activity than their static counterparts. This is further aggravated by the necessary connection to the clock, which performs a transition every cycle.

We can reduce switching activity by many other means, ranging from the circuit to the logical and functional design levels (some of those are explored further in later chapters).

Example 4.18 Optimizing Switching Activity at the Logic Level

Consider the two static logic circuits of Figure 4.60. Which one consumes the least amount of power and why?

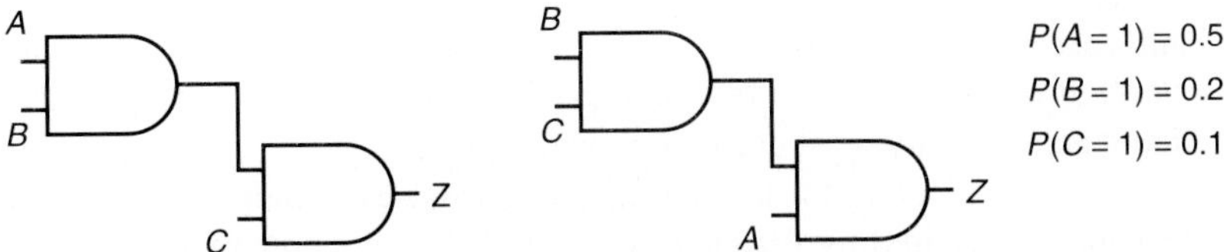

Figure 4.60 Reordering of inputs affects the circuit activity.

The answer is obtained by analyzing the transition probabilities at the circuit nodes. Since both circuits implement identical logic functionality, it is obvious that the activity at the output node Z is equal in both cases. The difference is the activity at the intermediate node. In the first circuit, this activity equals $(1 - 0.5 \times 0.2)\ (0.5 \times 0.2) = 0.09$. In the second case, the probability that a $0 \rightarrow 1$ transition occurs equals $(1 - 0.2 \times 0.1)(0.2 \times 0.1) = 0.0196$. This is substantially lower. From this we learn that it is beneficial to postpone the introduction of signals with a high transition rate (i.e., signals with a signal probability close to 0.5). A simple reordering of the input signals is often sufficient to accomplish that goal.

Although not a new problem, reducing the power consumption of an integrated circuit, or keeping it within bounds, is definitely becoming one of the most important issues in digital circuit designs. As this problem has only resurfaced in recent years (it was thought to be solved with the introduction of CMOS), there is ample use for innovation in this area. Circuits with reduced swings, smart supply voltages and alternative clock distribution schemes are only a couple of options under consideration.

4.5 Perspective: How to Choose a Logic Style

In the preceding sections, we have discussed several gate-implementation approaches using the CMOS technology. Each of the circuit styles has its advantages and disadvantages. Which one to select depends upon the primary requirement: ease of design, robustness, area, speed, or power dissipation. No single style optimizes all these measures at the same time. Even more, the approach of choice can vary from logic function to logic function.

The static approach has the advantage of being robust in the presence of noise. This makes the design process rather trouble-free and amenable to a high degree of automation. This ease-of-design does not come for free: for complex gates with a large fan-in, complementary CMOS becomes expensive in terms of area and performance. Alternative static logic styles have therefore been devised. Pseudo-NMOS is simple and fast at the expense of a reduced noise margin and static power dissipation. Pass-transistor logic is attractive for the implementation of a number of specific circuits, such as multiplexers and XOR-dominated logic such as adders.

Dynamic logic, on the other hand, makes it possible to implement fast and small complex gates. This comes at a price. Parasitic effects such as charge sharing make the design process a precarious job. Charge leakage forces a periodic refresh, which puts a lower bound on the operating frequency of the circuit.

The properties of the presented logic styles are summarized in Table 4.8.[7] Although the table can serve as a base for selecting a design style, bear in mind that this selection is influenced by a large number of elements, such as *ease of design, robustness, system clocking requirements, fan-out, functionality, and testing*. Deciding which parameters are dominant is exactly what makes a good designer.

The current trend is towards an increased use of complementary static CMOS. This tendency is inspired by the increased use of design-automation tools at the logic design level. These tools emphasize optimization at the logic rather than the circuit level and put a premium on robustness. Another argument is that static CMOS is more amenable to voltage scaling than some of the other approaches discussed in this chapter.

[7] This comparison should be treated cautiously. The four-input NAND gate might not be the right representative for a particular circuit style. For instance, CPL is treated rather unfairly by the comparison.

Table 4.8 Four-input NAND-gates: Summary.

Style	Ratioed	Static Power	Number of Transistors	Area (μm^2)	Propagation Delay (nsec)
Complementary	No	No	8	533	0.61
Pseudo-NMOS	Yes	Yes	5	288	1.49
CPL	No	No	14	800	0.75
Dynamic (NP)	No	No	6	212	0.37

4.6 Summary

In this chapter, we have extensively analyzed the behavior and performance of combinational CMOS digital circuits with regard to area, speed, and power.

- Static complementary CMOS combines dual pull-down and pull-up networks, only one of which is enabled at any time.
- The performance of a CMOS gate is a strong function of fan-in. Techniques to deal with fan-in include transistor sizing, input reordering, and partitioning. The speed is also a linear function of the fan-out. Extra buffering is needed for large fan-outs.
- The ratioed logic style consists of an active pull-down (up) network connected to a load device. This results in a substantial reduction in gate complexity at the expense of static power consumption and an asymmetrical response. Careful transistor sizing is necessary to maintain sufficient noise margins. The most popular approaches in this class are the pseudo-NMOS techniques and the differential DCVSL, which requires complementary signals.
- Pass-transistor logic implements a logic gate as a simple switch network. This results in very simple implementations for some logic functions. Long cascades of switches are to be avoided due to a quadratic increase in delay with respect to the number of elements in the chain. NMOS-only pass-transistor logic produces even simpler structures, but might suffer from static power consumption and reduced noise margins. This problem can be addressed by adding a level-restoring transistor.
- The operation of dynamic logic is based on the storage of charge on a capacitive node and the conditional discharging of that node as a function of the inputs. This calls for a two-phase scheme, consisting of a precharge followed by an evaluation step. Dynamic logic trades off noise margin for performance. It is sensitive to parasitic effects such as leakage, charge redistribution, and clock feedthrough. Cascading dynamic gates can cause hazards and should be addressed carefully.
- The power consumption of a logic network is strongly related to the switching activity of the network. This activity is a function of the input statistics, the network topology, and the logic style.
- Sources of power consumption such as glitches and short-circuit currents can be minimized by careful circuit design and transistor sizing.

- Power consumption is minimized by reducing the supply voltage, which increases the delay. Trading off area for power is a way to compensate for that performance loss.

4.7 To Probe Further

The topic of (C)MOS logic styles is treated extensively in the literature. Numerous texts have been devoted to the issue. Some of the most comprehensive treatments can be found in [Glasser85], [Annaratone86], [Elmasry91], [Uyemura92], and [Weste93]. Regarding the intricacies of high-performance design, [Shoji88] offers the most in-depth discussion of the optimization and analysis of digital MOS circuits. The topic of power minimization is relatively new. Excellent reference works are [Chandrakasan95] and [Rabaey95].

Innovations in the MOS logic area are typically published in the proceedings of the ISSCC Conference and the VLSI circuits symposium, as well as the *IEEE Journal of Solid State Circuits* (especially the November issue).

REFERENCES

[Annaratone86] M. Annaratone, *Digital CMOS Circuit Design*, Kluwer, 1986.

[Burd94] T. Burd, *Low Power CMOS Library Design Methodology*, M.S. thesis, University of California—Berkeley, December 1994.

[Chandrakasan92] A. Chandrakasan, S. Sheng, and R. Brodersen, "Low Power CMOS Digital Design," *IEEE Journal of Solid State Circuits*, vol. SC-27, no. 4, pp. 1082–1087, April 1992.

[Chandrakasan94] A. Chandrakasan, *Low Power Digital CMOS Design*, Ph.D. thesis, University of California—Berkeley, Memorandum No. UCB/ERL M94/65, August 1994.

[Chandrakasan95] A. Chandrakasan and R. Brodersen, *Low Power Digital CMOS Design*, Kluwer, 1995.

[Chu86] K. Chu and D. Pulfrey, "Design Procedures for Differential Cascade Logic," *IEEE Journal of Solid State Circuits*, vol. SC-21, no. 6 (Dec. 1986), pp. 1082–1087.

[Dopperpuhl92] D. Dopperpuhl et al., "A 200-MHz 64-b Dual-Issue CMOS Microprocessor," *IEEE Journal of Solid State Circuits*, vol. 27, no. 11, pp. 1555–1567, Nov. 1992.

[Elmasry91] M. Elmasry, Ed., *Digital MOS Integrated Circuits II*, IEEE Press, 1991.

[Friedman84] V. Friedman and S. Liu, "Dynamic Logic CMOS Circuits," *IEEE Journal of Solid State Circuits*, vol. SC-19, no. 2, pp. 263–266, April 1984.

[Glasser85] L. Glasser and D. Dopperpuhl, *The Design and Analysis of VLSI Circuits*, Addison-Wesley, 1985.

[Goncalvez83] N. Goncalvez and H. De Man, "NORA: A Racefree Dynamic CMOS Technique for Pipelined Logic Structures," *IEEE Journal of Solid State Circuits*, vol. SC-18, no. 3, pp. 261–266, June 1983.

[Heller84] L. Heller et al., "Cascade Voltage Switch Logic: A Differential CMOS Logic Family," *Proc. IEEE ISSCC Conference*, pp. 16–17, February 1984.

[Hodges88] D. Hodges and H. Jackson, *Analysis and Design of Digital Integrated Circuits*, McGraw-Hill, 1988.

[Krambeck82] R. Krambeck et al., "High-Speed Compact Circuits with CMOS," *IEEE Journal of Solid State Circuits*, vol. SC-17, no. 3, pp. 614–619, June 1982.

[Lee86] C. M. Lee and E. Szeto, "Zipper CMOS," *IEEE Circuits and Systems Magazine*, pp. 10–16, May 1986.

[Liu93] D. Liu and C. Svensson, "Trading Speed for Low Power by Choice of Supply and Threshold Voltages," *IEEE Journal of Solid State Circuits*, vol. SC-28, no 1, pp. 10–17, January 1993.

[Murphy81] B. Murphy and R. Edwards, "A CMOS 32b Single Chip Microprocessor," *Proc. ISCC 81*, pp. 230–231, 1981.

[Rabaey95] J. Rabaey and M. Pedram, *Low Power Design Methodolgies*, Kluwer, 1995.

[Radhakrishnan85] D. Radhakrishnan, S. Whittaker, and G. Maki, "Formal Design Procedures for Pass-Transistor Switching Circuits," *IEEE Journal of Solid State Circuits*, vol. SC-20, no. 2, pp. 531–536, April 1985.

[Shimohigashi93] K. Shimohigashi and K. Seki, "Low-Voltage ULSI Design," *IEEE Journal of Solid State Circuits*, vol. 28, no. 4, April 1993.

[Shoji88] M. Shoji, *CMOS Digital Circuit Technology*, Prentice Hall, 1988.

[Uyemura88] J. Uyemura, *Fundamentals of MOS Digital Integrated Circuits*, Addison-Wesley, 1988.

[Uyemura92] J. Uyemura, *Circuit Design for CMOS VLSI*, Kluwer, 1992.

[Veendrick84] H. Veendrick, "Short-Circuit Dissipation of Static CMOS Circuitry and Its Impact on the Design of Buffer Circuits," *IEEE Journal of Solid State Circuits*, vol. SC-19, no 4, pp. 468–473, August 1984.

[Weste93] N. Weste and K. Eshragian, *Principles of CMOS VLSI Design: A Systems Perspective*, Addison-Wesley, 1993.

[Yano90] K. Yano et al., "A 3.8 ns CMOS 16 × 16 b Multiplier Using Complimentary Pass-Transistor Logic," *IEEE Journal of Solid State Circuits*, vol. SC-25, no 2, pp. 388–395, April 1990.

4.8 Exercises and Design Problems

1. [E, None, 4.2] Implement the equation $X = ((\overline{A} + \overline{B})(\overline{C} + \overline{D} + \overline{E}) + \overline{F})\overline{G}$ using complementary CMOS. Size the devices so that the output resistance is the same as that of an inverter with an NMOS $W/L = 1$ and PMOS $W/L = 3$.
2. [M, None, 4.2] CMOS LOGIC
   - **a.** Do the following two circuits (Figure 4.61) implement the same logic function? If yes, what is that logic function? If no, give Boolean expressions for both circuits.
   - **b.** Will these two circuits' output resistances always be equal to each other?
   - **c.** Will these two circuits' rise and fall times always be equal to each other? Why or why not?
3. [E, None, 4.2] The transistors in the circuits of the preceding problem have been sized to give an output resistance of 19.5 kΩ for the worst-case input pattern. This output resistance can vary, however, if other patterns are applied.
   - **a.** What input pattern (*A–E*) gives the lowest output resistance when the output is low? What is the value of that resistance?
   - **b.** What inputs (*A–E*) give the lowest output resistance when the output is high? What is the value of that resistance?

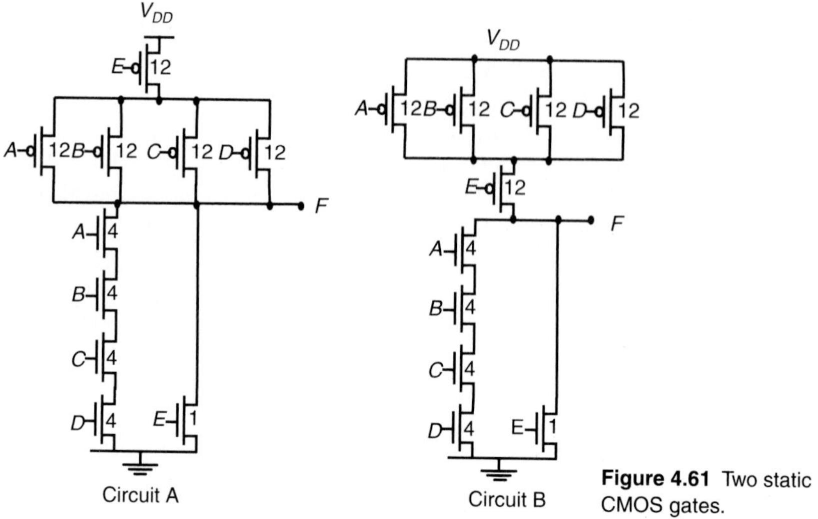

Figure 4.61 Two static CMOS gates.

4. [C, None, 4.2] For analysis purposes, it is often useful to collapse series and parallel connections of transistors into a single transistor equivalent with an *effective* $(W/L)_{eff}$. For simplicity, assume that all inputs are tied to the same voltage.

   a. For the series transistors of Figure 4.62, derive an approximate expression for $(W/L)_s$ in terms of $(W/L)_{1,2}$. Extend this formula to a series connection of N transistors.

   b. Repeat (a), finding $(W/L)_p$ for the parallel connection of transistors, shown also in Figure 4.62.

   c. Discuss the limitations of these approximations.

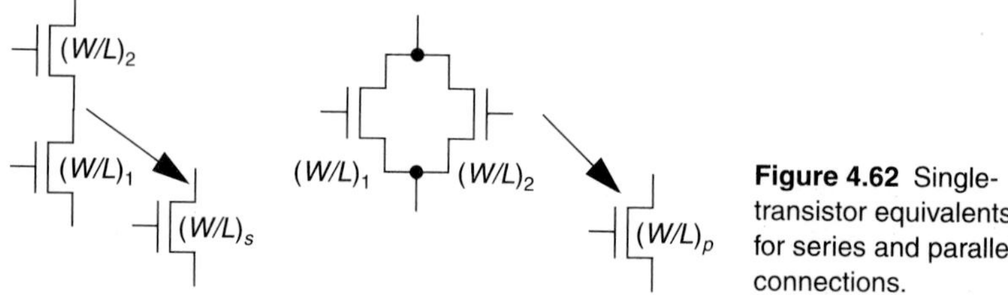

Figure 4.62 Single-transistor equivalents for series and parallel connections.

5. [E, None, 4.2] What is the logic function of circuits A and B in Figure 4.63? Which one is a dual network and which one is not? Is the nondual network still a valid static logic gate? Explain. List any advantages of one configuration over the other.

6. [E, None, 4.2] Compute the following data for the pseudo-NMOS inverter shown in Figure 4.64:

   a. V_{OL} and V_{OH}

   b. NM_L and NM_H

   c. The average power dissipation: (1) for V_{in} low, and (2) for V_{in} high.

7. [E, None, 4.2] For the pseudo-NMOS inverter of problem 6 and an output load of 3 pF, calculate t_{plH}, t_{pHL}, and t_p. Are the rising and falling delays equal? Why or why not?

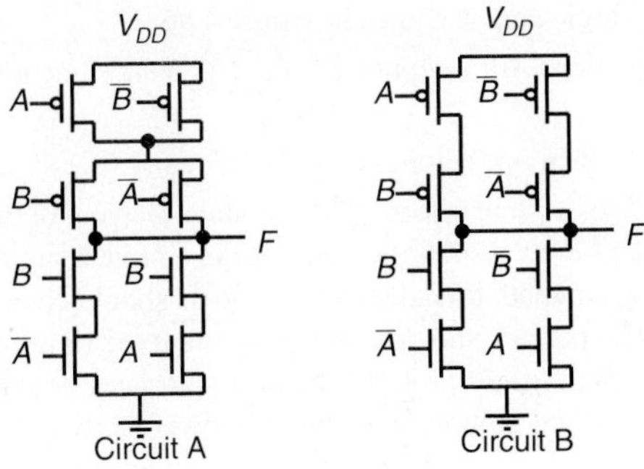

Figure 4.63 Two logic functions.

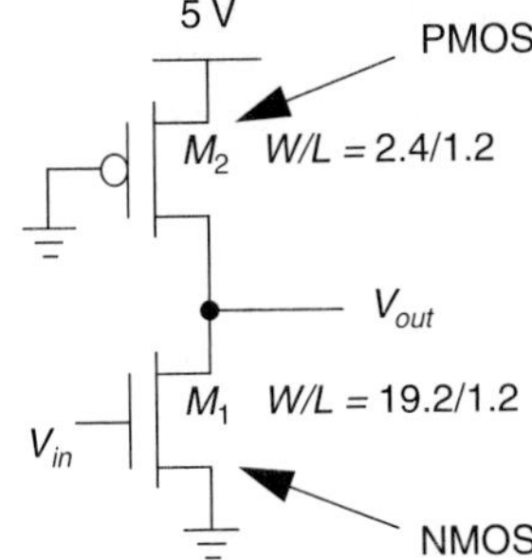

Figure 4.64 Pseudo-NMOS inverter.

8. [M, None, 4.2] Design a three-input ratioed NOR gate using W/L_{eff} = 1.8/0.9 NMOS pull-down transistors. The total load at the output is 50 fF.

a. Set V_{OL} = 0.1 V and V_{DD} = 5 V. Give the necessary size of the pull-up device, and calculate t_{pLH} and t_{pHL} if the pull-up device is: (1) a resistor; (2) a depletion NMOS device (V_T = –2.0); (3) a PMOS device.

b. Repeat for V_{OL} = 0.4 V.

9. [M, SPICE, 4.2] Consider the circuit of Figure 4.65.

a. What is the output voltage if only one input is high? If all four inputs are high?

b. What is the average static power consumption if at any time each input turns on with an (independent) probability of 0.5? 0.1?

c. Compare the obtained results with SPICE.

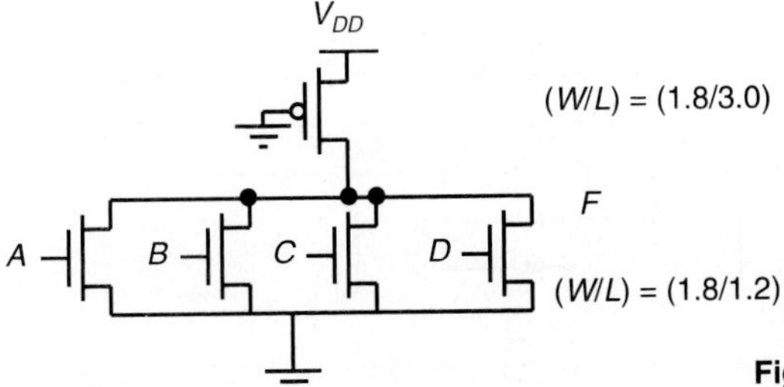

Figure 4.65 Pseudo-NMOS gate.

10. [M, None, 4.2] Implement $F = A\overline{BC} + \overline{A}CD$ (and $\overline{F}$) in DCVSL. Assume A, B, C, D, and their complements are available as inputs. Make sure to use the minimum number of transistors.

11. [E, Layout, 4.2] A complex logic gate is shown in Figure 4.66.

a. Write the Boolean equations for outputs F and G. What function does this circuit implement?

b. What logic family does this circuit belong to?

c. Assuming $W/L = 4.8/1.2$ for *all* transistors, either produce a layout of the gate using an editor such as Magic, or hand-draw a stick diagram for the circuit. Your stick diagram should conform to the following datapath layout style: (1) Inputs should enter the layout from the left in polysilicon; (2) The outputs should exit the layout at the right in polysilicon (this is reasonable, since the outputs would probably be driving transistor gate inputs of the next cell to the right). (3) Power and ground lines should run vertically in metal 1.

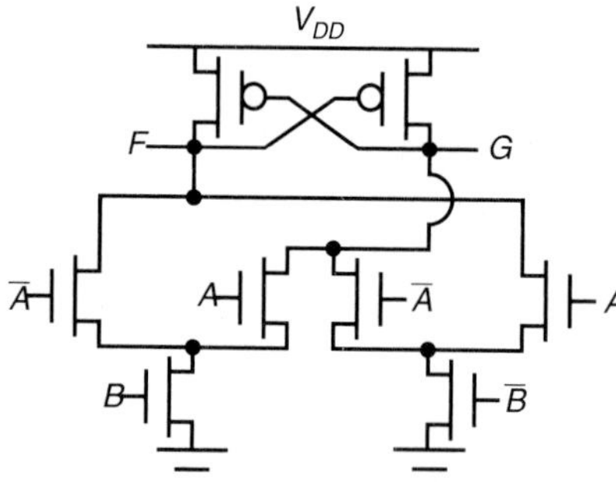

Figure 4.66 Two-input complex logic gate.

12. [E, SPICE, 4.2] An extremely simple bit-error detection scheme consists of appending a parity bit to a group of data bits. In an even parity scheme, the parity bit is assigned a value such that the total number of 1's in the new word is even (including the parity bit). After transmission of the word, single-bit errors can be detected by checking the parity of the received word.

As a trivial example of this scheme, consider an even-parity checking circuit for a single data bit and its associated parity bit. The Boolean function for the error flag is given by $Error = P \oplus D$, and the DCVSL implementation is shown in Figure 4.67. The error signal evaluates high when the input word is not of even parity.

Use SPICE (level 2) to simulate the parity checker circuit shown in Figure 4.67. Demonstrate correct circuit operation and find the worst-case t_{PHL} and t_{PLH} (using the error signal as the output node). Assume 1 ns rise and fall times for the inputs. Assume a W/L of 4.8/1.2 for NMOS devices and 3.6/1.2 for PMOS. In computing diffusion areas and perimeters, you may use the approximations: $AD = AS = W \times 3\ \mu m$ and $PD = PS = W + 6\ \mu m$.

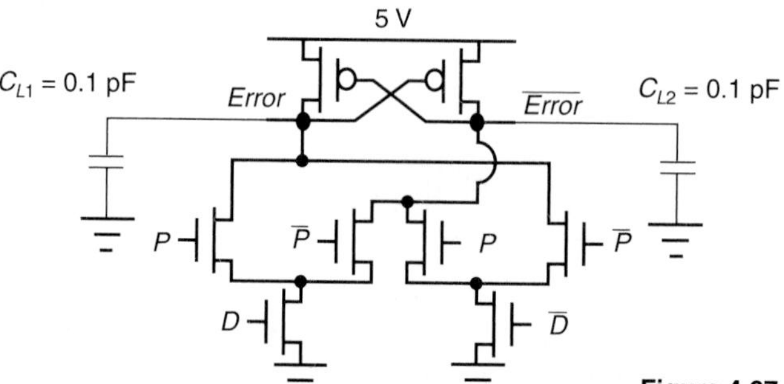

Figure 4.67 DCVSL parity checker.

13. [M, None. 4.2] Figure 4.68 contains a pass-gate logic network.

a. Determine the truth table for the circuit. What logic function does it implement?

b. Assuming 0 and 5 V inputs, size the PMOS transistor to achieve a $V_{OL} = 0.3$ V.

c. If the PMOS were removed, would the circuit still function correctly? Does the PMOS transistor serve any useful purpose?

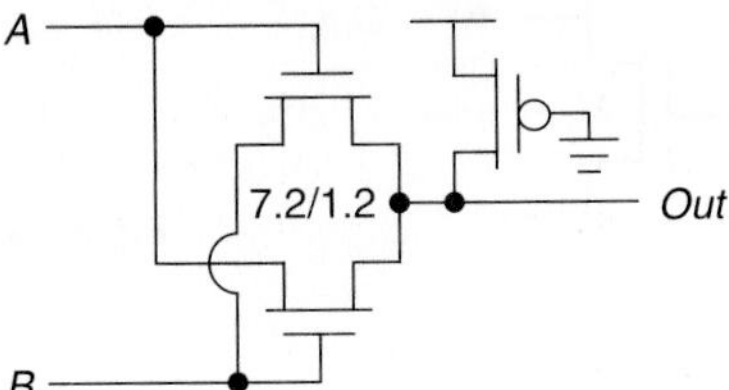

Figure 4.68 Pass-gate network.

14. [E, None, 4.2] An inverter is driven by a single NMOS pass gate (Figure 4.69). Assume that the total capacitance at node x, C_x, equals 25 fF.

a. What are t_{pLH} and t_{pHL} at node x?

b. If node x is high 50% of the time, what is the static power consumption in the inverter?

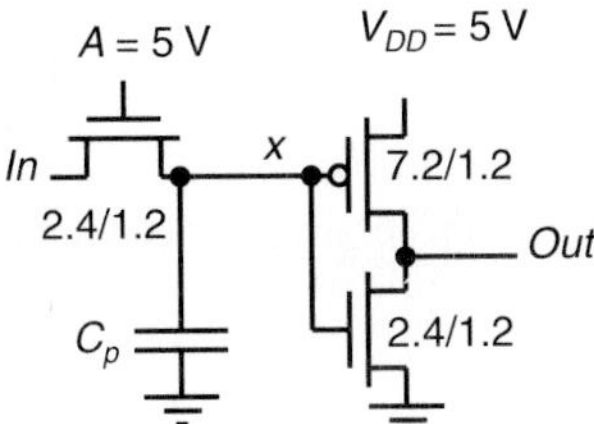

Figure 4.69 Pass-gate logic.

15. [M, None, 4.2] Effects of scaling on pass-gate logic.

a. If a process has a t_{buf} of 0.4 nsec, R_{eq} of 8k ohms, and C of 12 fF, what is the optimal number of stages between buffers in a pass-gate chain?

b. Suppose that if the dimension of this process are shrunk by a factor S, R_{eq} scales as $1/S^2$, C scales as $1/S$, and t_{buf} scales as $1/S^2$. What is the expression for the optimal number of buffers as a function of S? What is this number if $S = 2$?

16. [C, None, 4.2] Consider the circuit of Figure 4.70. Let $C_x = 50$ fF, M_r has $W/L_{eff} = 1.8/1.5$, M_n has $W/L_{eff} = 1.8/0.9$. Assume the output inverter doesn't switch until its input equals $V_{DD}/2$.

a. How long will it take M_n to pull down node x from 5 V to 2.5 V if In is at 0 V and B is at 5 V?

b. How long will it take M_n to pull up node x from 0 V to 2.5 V if V_{In} is 5 V and V_B is 5 V?

c. What is the value of V_B necessary to pull down V_x to 2.5 V when $V_{In} = 5$ V?

17. [M, None, 4.3] Sketch the waveforms at x, y, and z for the given inputs (Figure 4.71). You may approximate the time scale, but be sure to compute the voltage levels. Assume that $V_T = 1.0$ V when body effect is a factor.

18. [E, None, 4.3] Consider the circuit of Figure 4.72.

a. Give the logic function of x and y in terms of A, B, and C. Sketch the waveforms at x and y for the given inputs. Do x and y evaluate to the values you expected from their logic functions? Explain.

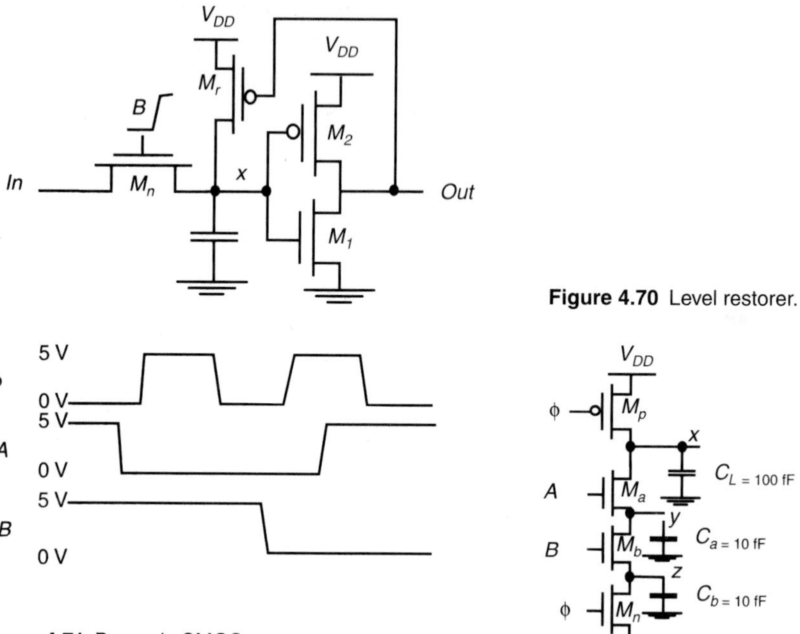

Figure 4.70 Level restorer.

Figure 4.71 Dynamic CMOS.

b. Redesign the gates using *np*-CMOS to eliminate any race conditions. Sketch the waveforms at x and y.

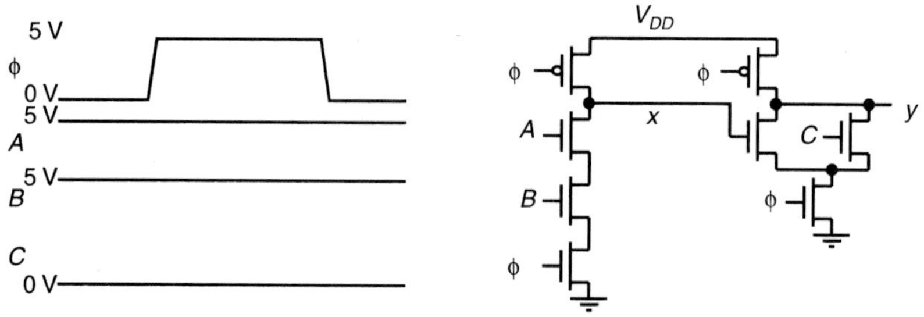

Figure 4.72 Cascaded dynamic gates.

19. [M, None, 4.3] Suppose we want to implement the two logic functions given by $F = A + B + C$ and $G = A + B + C + D$. Assume both true and complementary signals are available.

a. Implement these functions in dynamic CMOS as cascaded ϕ stages so as to minimize the total transistor count.

b. Discuss any conditions under which this implementation would fail to operate properly.

c. Design an *np*-CMOS implementation of the same logic functions. Does this design display any of the difficulties of part (b)?

20. [C, Spice, 4.3] Figure 4.73 shows a dynamic CMOS circuit in DOMINO logic. In determining source and drain areas and perimeters, you may use the following approximations: $AD = AS = W \times 3\ \mu m$ and $PD = PS = W + 6\ \mu m$. Also, assume 1 nsec rise/fall times for all inputs, including the clock. Furthermore, you may assume that all the inputs and their complements are available. Finally, assume that all inputs change during the precharge phase of the clock cycle.

a. What Boolean functions are implemented at outputs F and G? If A and B are interpreted as two-bit binary words, $A = A_1A_0$ and $B = B_1B_0$, then what interpretation can be applied to output G?

b. Which gate (1 or 2) has the highest potential for harmful charge sharing and why? What sequence of inputs (spanning two clock cycles) results in the worst-case charge-sharing scenario? Using SPICE, determine the extent to which charge sharing affects the circuit for this worst case.

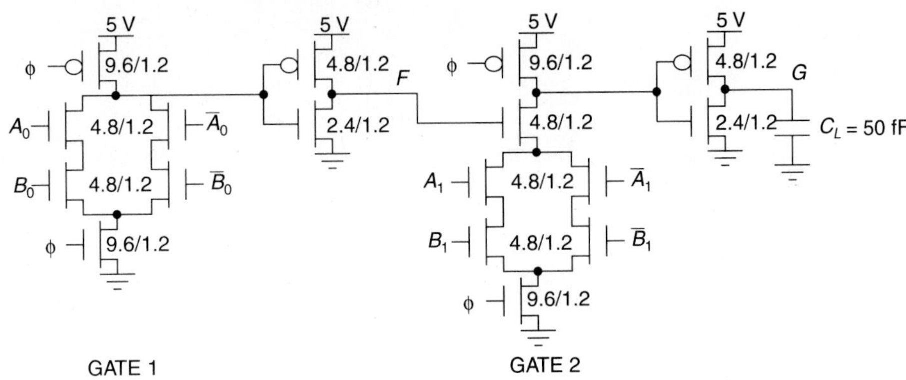

Figure 4.73 DOMINO logic circuit.

21. [M, Spice, 4.3] In this problem you will consider methods for eliminating charge sharing in the circuit of Figure 4.73. You will then determine the performance of the resulting circuit.

a. In problem 20 you determined which gate (1 or 2) suffers the most from charge sharing. Add a single 9.6/1.2 PMOS precharge transistor (with its gate driven by the clock ϕ and its source connected to V_{DD}) to one of the nodes in that gate to maximally reduce the charge-sharing effect. What effect (if any) will this addition have on the gate delay? Use SPICE to demonstrate that the additional transistor has eliminated charge sharing for the previously determined worst-case sequence of inputs.

b. For the new circuit (including additional precharge transistor), find the sequence of inputs (spanning two clock cycles) that results in the worst-case delay through the circuit. Remember that precharging is another factor that limits the maximum clocking frequency of the circuit, so your input sequence should address the worst-case precharging delay.

c. Using SPICE on the new circuit and applying the sequence of inputs found in part (b), find the maximum clock frequency for correct operation of the circuit. Remember that the pre-charge cycle must be long enough to allow all precharged nodes to reach ~90% of their final values before evaluation begins. Also, recall that the inputs (A, B and their complements) should not begin changing until the clock signal has reached 0 V (precharge phase), and they should reach their final values before the circuit enters the evaluation phase.

22. [C, None, 4.2–3] For this problem, refer to the layout of Figure 4.74.

a. Draw the schematic corresponding to the layout. Include transistor sizes.

b. What logic function does the circuit implement? To which logic family does the circuit belong?

c. Does the circuit have any advantages over fully complementary CMOS?

d. Calculate the worst-case V_{OL} and V_{OH}.

e. Calculate the worst-case t_{pHL} as well as the parasitic capacitances necessary for this computation. Neglect wiring capacitances, and assume that the gate is driving an identical gate.

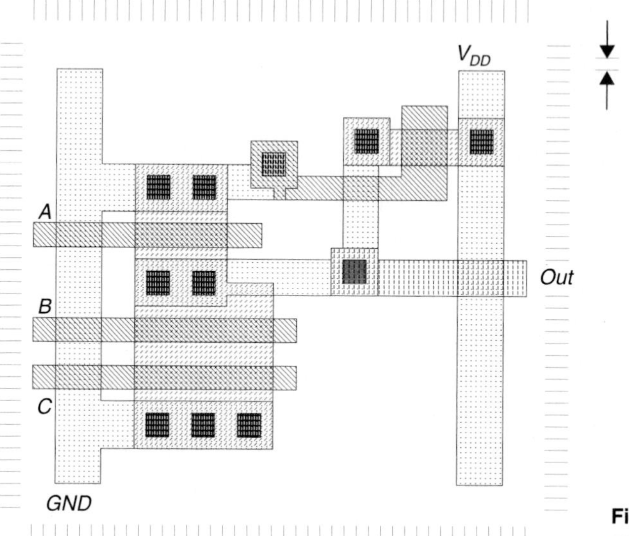

Figure 4.74 Layout of complex gate.

23. [E, None, 4.4] Derive the truth table, state transition graph, and output transition probabilities for a three-input exclusive-or gate with independent, identically distributed, uniform white-noise inputs.

24. [C, None, 4.4] Figure 4.75 shows a two-input multiplexer. For this problem, assume independent identically-distributed uniform white noise inputs.

a. Does this schematic contain reconvergent fan-out? Explain your answer.

b. Ascertain the exact signal (P_1) and transition ($P_{0 \to 1}$) formulas for nodes X, Y, and Z for: (1) a static, fully complementary CMOS implementation, and (2) a dynamic CMOS implementation.

25. [M, None, 4.4] Compute the switching power consumed by the multiplexer of Figure 4.75, assuming that all significant capacitances have been lumped into the three capacitors shown in the figure, where $C = 0.3$ pF. Also, assume that $V_{DD} = 5$ V and that input events occur at a frequency of 50 MHz. Perform this calculation for the following:

a. A static fully-complementary CMOS implementation

b. A dynamic CMOS implementation

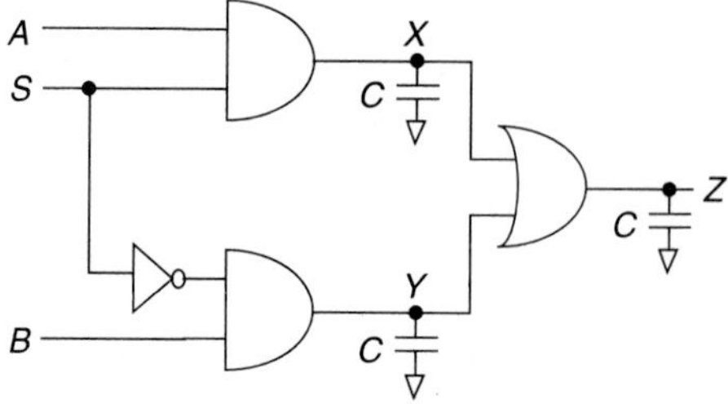

Figure 4.75 Two-input multiplexer

DESIGN PROJECT

Design, lay out, and simulate a CMOS four-input XOR gate in the standard 1.2 micron CMOS process. You can choose any logic circuit style, and you are free to choose how many stages of logic to use: you could use one large logic gate or a combination of smaller logic gates. The supply voltage is set at 3.0 V! Your circuit must drive an external 20 fF load in addition to whatever internal parasitics are present in your circuit.

The primary design objective is to minimize the propagation delay of the worst-case transition for your circuit. The secondary objective is to minimize the area of the layout. At the very worst, your design must have a propagation delay of no more than 1.5 nsec and occupy an area of no more than 2000 square microns, but the faster and smaller your circuit, the better. Be aware that, when using dynamic logic, the precharge time should be made part of the delay.

The design will be graded on the magnitude of $A \times t_p^2$, the product of the area of your design and the square of the delay for the worst-case transition.

APPENDIX

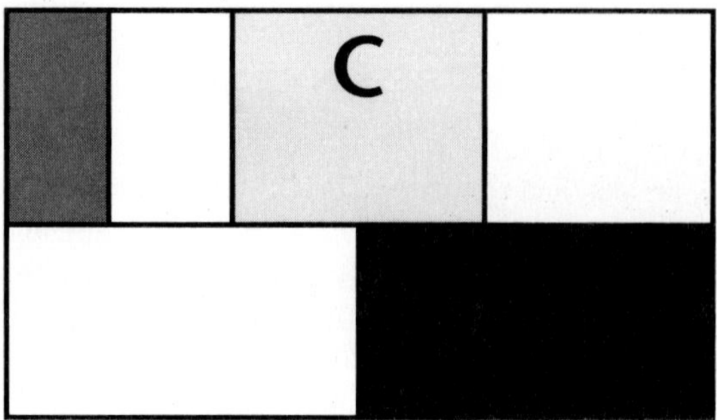

LAYOUT TECHNIQUES FOR COMPLEX GATES

Weinberger and standard-cell layout techniques

Euler graph approach

In Chapter 4, we have discussed in detail how to construct the schematics of complex gates and how to size the transistors. The last step in the design process is to derive a layout for the gate (also called cell) or, in other words, to determine the exact shape of the various polygons composing the gate layout. The composition of a layout is strongly influenced by the *interconnect approach*. How does the cell fit in the overall chip layout and how does it communicate with neighboring cells? Keeping these issues in mind from the start results in denser designs with less parasitic capacitance.

Two approaches are worth pointing out (although many others can be envisioned). In a first technique, called the *Weinberger approach* [Weinberger67], the data wires (inputs and outputs) are routed (in metal) parallel to the supply rails and perpendicular to the diffusion areas, as illustrated in Figure C.1. Transistors are formed at the cross-points of the polysilicon signal wires (connected to the horizontal metal wires) and the diffusion zones. The "over-the-cell" wiring approach makes the Weinberger technique particularly suited for bit-sliced data paths, which are the topic of Chapter 7. An example of a cell implemented using this approach is the dynamic NAND gate, shown in Figure 4.34.

In the *standard-cell technique*, signals are routed in polysilicon perpendicular to the power distribution (Figure C.2). This approach tends to result in a dense layout for static

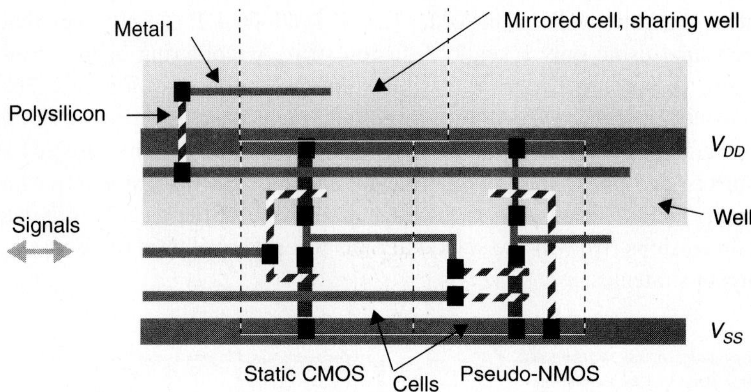

Figure C.1 The Weinberger approach for complex gate layout (using a single metal layer).

CMOS gates, as the vertical polysilicon wire can serve as the input to both the NMOS and the PMOS transistor. Examples of cells implemented using the standard-cell approach are shown in Figure 4.10, Figure 4.16, and Figure 4.32. Interconnections between cells are generally established in so-called routing channels, as demonstrated in Figure C.2. The standard-cell approach is very popular at present, due to its high degree of automation. An example of this technique is the gate-matrix approach ([Wing85, Huang89]). For a detailed description of the design automation tools, supporting the standard-cell approach, we refer to Chapter 11.

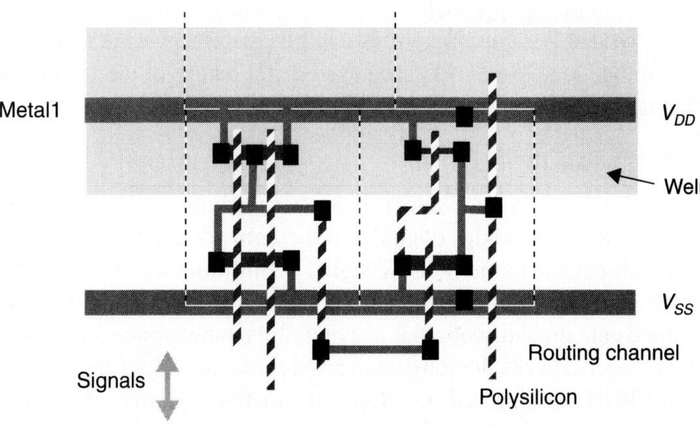

Figure C.2 The standard-cell approach for complex gate layout.

The common use of this layout strategy makes it worth analyzing how a complex Boolean function can be mapped efficiently onto such a structure. For density reasons, it is desirable to realize the NMOS and PMOS transistors as an unbroken row of devices with abutting source/drain connections and with the gate connections of the corresponding NMOS and PMOS transistors aligned. This approach requires only a single strip of diffusion in both wells. To achieve this goal, a careful ordering of the input terminals is necessary. This is illustrated in Figure C.3, where the logical function $\overline{x} = (a + b) \cdot c$ is implemented. In the first version, the order {a c b} is adopted. It is easily seen that no solution can be found using only a single diffusion strip. A reordering of the terminals (for instance using {*a b c*}), generates a feasible solution, as shown in Figure C.3b. Observe that the "layouts" in Figure C.3 do not represent actual mask geometries, but are rather symbolic diagrams of the gate topologies. Wires and transistors are represented as dimensionless objects, and positioning is relative, not absolute. Such representations are called *stick diagrams* and are often used at the conception time of the gate, before determining the actual dimensions. We will use stick diagrams whenever we want to discuss gate topologies or layout strategies.

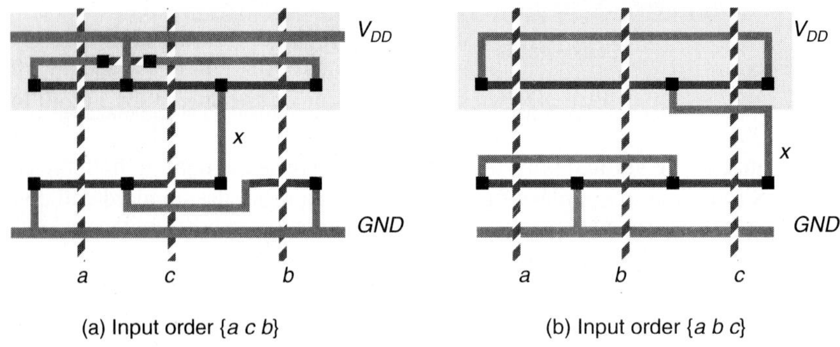

Figure C.3 Stick Diagram for $\overline{x} = (a + b) \cdot c$.

Fortunately, a systematic approach has been developed to derive the permutation of the input terminals, so that complex functions can be realized by uninterrupted diffusion strips that minimize the area [Uehara81]. The systematic nature of the technique also has the advantage that it is easily automated. It consists of two steps:

1. **Construction of *logic graph*.**

   The logic graph of a transistor network (or a switching function) is the graph of which the vertices are the nodes (signals) of the network and the edges represent the transistors. Each edge is named for the signal controlling the corresponding transistor. Since the PUN and PDN networks of a static CMOS gate are dual, their corresponding graphs are dual as well; that is, a parallel connection is replaced by a series one and vice versa. This is demonstrated in Figure C.4, where the logic graphs for the PDN and PUN networks of the Boolean function $\overline{x} = (a + b) \cdot c$ are overlaid (notice that this approach can be used to derive dual networks).

2. **Identification of Euler paths.**

   An Euler path in a graph is defined as a path through all nodes in the graph, such that each edge in the graph is only visited once. Identification of such a path is important as a result of the following finding:

   An ordering of the inputs leading to an uninterrupted diffusion strip of NMOS (PMOS) transistors is possible if there exists an Euler path in the logic graph of the PDN (PUN) network.

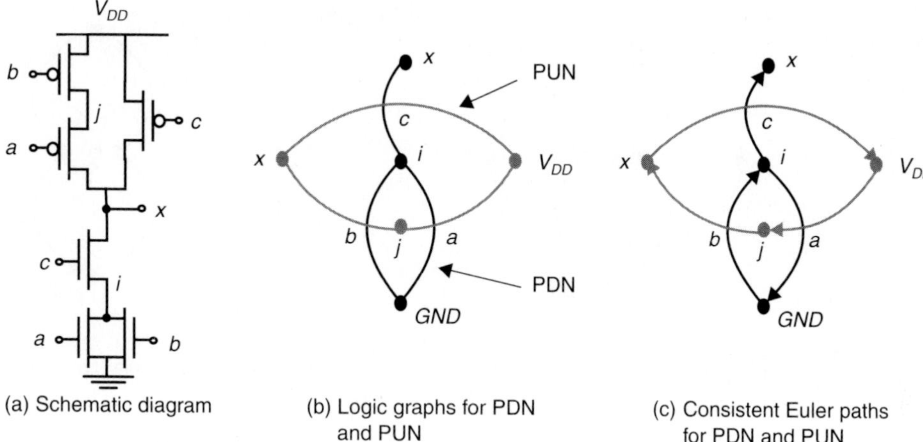

Figure C.4 Schematic diagram, logic graph, and Euler paths for $x = (a + b) \cdot c$

The sequence of edges in the Euler path equals the ordering of the inputs in the gate layout. To obtain the same ordering in both the PUN and PDN networks (as is necessary if we want to use a single poly strip for every input signal), the Euler paths must be *consistent*, that is, they must have the same sequence. The reasoning behind this finding is easily clarified: to form an interrupted strip of diffusion, all transistors must be visited in sequence; that is, the drain of one device is the source of the next one. This is equivalent to traversing the logic graph along an Euler path. Be aware that Euler paths are not unique: many different solutions might exist.

Consistent Euler paths for the example of Figure C.4a are shown in Figure C.4c. The layout associated with this solution is shown in Figure C.3b. An inspection of the logic diagram of the function shows that $\{a\ c\ b\}$ is an Euler path for the PUN, but not for the PDN. A single-diffusion-strip solution is, hence, nonexistent (Figure C.3a).

Example C.1 Derivation of Layout Topology of Complex Logic Gate

As an example, let us derive the layout topology of the following logical function:

$$x = \overline{ab + cd}.$$

The logical function and one consistent Euler path are shown in Figure C.5a and Figure C.5b. The corresponding layout is shown in Figure C.5c.

One should be aware that the existence of consistent Euler paths depends upon the way the Boolean expressions (and the corresponding logic graphs) are constructed. For instance, no consistent Euler paths can be found for $\overline{x} = a + b \cdot c + d \cdot e$, but the function $\overline{x} = b \cdot c + a + d \cdot e$ has a simple solution (confirm that this is true—preserve the ordering of the function when constructing the logic graphs). A restructuring of the function is sometimes necessary before a set of consistent paths can be identified. This could lead to an exhaustive search over all possible path combinations. Fortunately, a simple algorithm to avoid this plight has been proposed in [Uehara81]. A discussion of this is beyond our scope, however, and we refer the interested reader to the Uehara-Van Cleemput text.

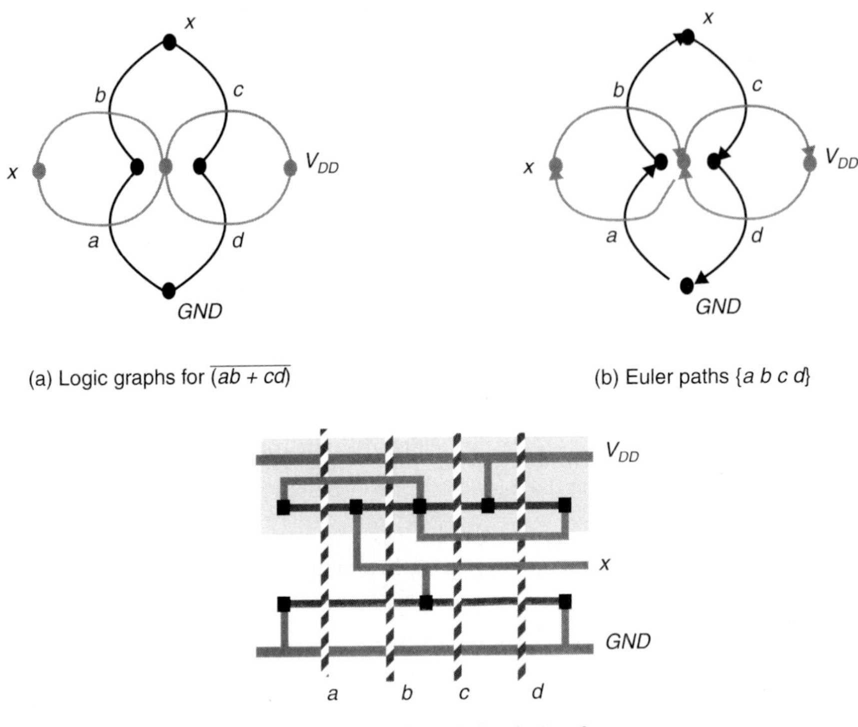

Figure C.5 Deriving the layout topology for $x = \overline{(ab + cd)}$.

Finally, it is worth mentioning that the layout strategies presented are not the only possibilities. Sometimes, for instance, it might be more effective to provide multiple diffusion strips, stacked vertically. In this case, a single polysilicon input wire can serve as the input for multiple transistors. This might be beneficial for certain gate structures, such as the NXOR gate. A case-by-case analysis is, therefore, recommended.

To Probe Further

A good overview of cell-generation techniques can be found in [Rubin87, pp. 116–128]. Some of the landmark papers in this area are listed below.

[Huang89] S. Huang and O. Wing, "Improved Gate Matrix Layout," *IEEE Trans. on CAD*, vol. 8, no. 8, pp. 875–889, 1989.

[Rubin87] S. Rubin, *Computer Aids for VLSI Design*, Addison-Wesley, 1987.

[Uehara81] T. Uehara and W. Van Cleemput, "Optimal Layout of CMOS Functional Arrays," *IEEE Trans. on Computers*, vol. C-30, no. 5, pp. 305–311, May 1981.

[Weinberger67] A. Weinberger, "Large Scale Integration of MOS Complex Logic: A Layout Method," *IEEE Journal of Solid State Circuits*, vol. 2, no. 4, pp. 182–190, 1967.

[Wing85] O. Wing, S. Huang, and R. Wang, "Gate Matrix Layout," *IEEE Trans. on CAD*, vol. 4, no. 3, pp. 220–231, 1985.

VERY HIGH PERFORMANCE DIGITAL CIRCUITS

Digital design in bipolar and BiCMOS technologies

Designing high speed logic in GaAs

Extreme performance using cryogenic techniques

5.1 Introduction

This chapter deals with the design of digital integrated circuits with operating speeds in the multiple hundreds of MHz, and even the GHz range. These speeds are desirable in the core processors of super- and mainframe computers, and even high-end workstations. High performance is also a prime requirement in the domain of high-speed signal-acquisition apparatus, such as digital sampling oscilloscopes. Measurement equipment must always be faster than the circuits it observes; hence the need for high-speed logic. With the advent of optical fiber, digital communication systems have been extended into the Gbits/sec area and need extremely high-speed front-end circuitry. Finally, the availability of high-speed technologies, such as bipolar silicon or the even more advanced GaAs, simplifies the task of automating the design process for high-performance circuits.

This chapter commences with a detailed overview of the bipolar design approaches for complex combinational logic. In earlier chapters we have seen that bipolar digital design is capable of achieving propagation delays that are between 2 and 5 times faster than what can be achieved in a comparable CMOS technology. After a discussion of ECL and related approaches, we briefly analyze alternative bipolar gates. The analysis of bipolar circuitry is concluded with a discussion of the BiCMOS technology that combines the speed and driving capabilities of bipolar with the density and compactness of CMOS. The basic BiCMOS gate is analyzed in detail, followed by a discussion of its applications and scaling prospects.

In our quest for ever higher speeds, even bipolar designs eventually reach a maximum. When extreme performance is a necessity, room-temperature silicon is replaced by other semiconductor materials such as gallium arsenide (GaAs). The lure of GaAs and other compound semiconductors is a substantial increase in carrier mobility and, hence, performance. An alternative solution is to opt for operation at a reduced temperature. Lowering the temperature reduces the delay of traditional semiconductor components. Some materials even have the property of becoming superconductive when operated below a certain temperature, which eliminates resistivity altogether. Circuits with mind-boggling performance can be conceived using these technologies. The second half of this chapter is devoted to a discussion of these nonmainstream design approaches.

5.2 Bipolar Gate Design

In the subsequent sections, we first discuss how the generic ECL gate, introduced in Chapter 3, can be extended to implement complex logic functions. A number of modifications to the generic gate are introduced to increase its performance or to improve its scalability. The section concludes with a short overview of some alternative bipolar logic styles.

5.2.1 Logic Design in ECL

To recapitulate, the basic structure of the ECL gate is repeated in Figure 5.1. It consists of a bias network generating a reference voltage V_{ref}, a differential pair Q_1-Q_2, and a pair of emitter-follower output drivers. From a logic point of view, one of the useful properties of

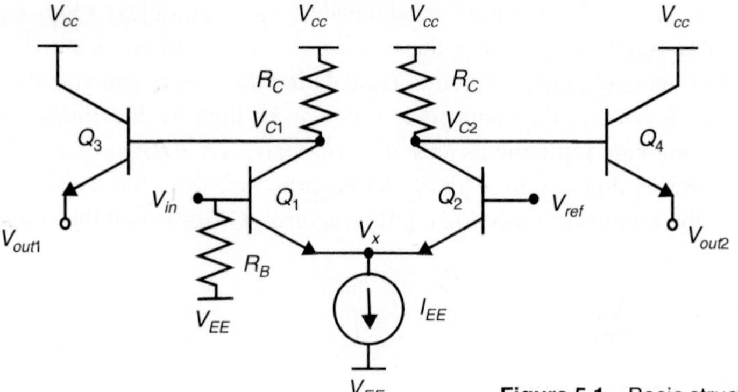

Figure 5.1 Basic structure of the ECL gate.

the basic ECL gate is that it uses a *complementary logic style*, which means that the complement of each logic signal is available. This is similar to the CMOS CPL design technique discussed earlier. This approach avoids the need for extra inverters, as is the case in inverting logic families such as complementary CMOS. It has been observed that for a given logic function, static CMOS takes approximately twice as many gates as ECL.

Example 5.1 Advantages of Complementary Logic

A 4-bit ALU in CMOS requires 98 gates using NAND logic, while an ECL implementation can be conceived with only 57 gates. Even more important, the average number of cascaded logic stages in a typical signal path equals 6.1 for CMOS and 2.9 for ECL. This reduction of the critical path by a factor of 2.1 has a significant impact on performance [Masaki92].

One of the simplest gates to implement in ECL is the OR/NOR function, which can be realized by adding extra input transistors in parallel, as shown in Figure 5.2. Turning one of the two input transistors on is sufficient to divert the bias current to the left branch of the current switch and to pull V_{C1} down. This is equivalent to realizing a NOR gate. The complementary node V_{C2} goes high at the same time, implementing the OR function. The static and dynamic characteristics of the standard ECL gate are only marginally affected by this modification.

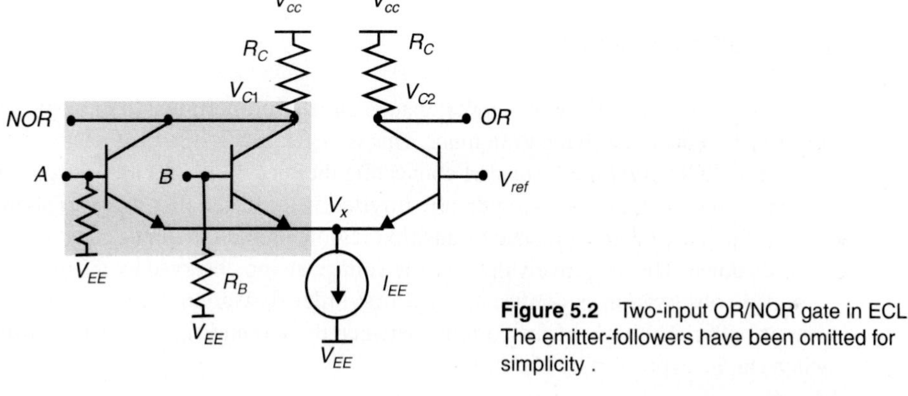

Figure 5.2 Two-input OR/NOR gate in ECL. The emitter-followers have been omitted for simplicity .

Realizing an (N)AND operation is somewhat more complex. One approach to implement complex functions is to use the *wired-OR* configuration, as shown in Figure 5.3. Simply wiring the gate outputs together (also called *dotting*) is equivalent to an "OR-ing." It is sufficient for one of the connected outputs to be high for the combined result to be high. The resulting gate implements $\overline{(A+B)} + \overline{(C+D)} = \overline{(A+B) \cdot (C+D)}$. The advantage of this approach is that complex gates can be constructed on the fly by simply connecting or wiring the outputs of basic logic OR structures. Observe that the combined gate

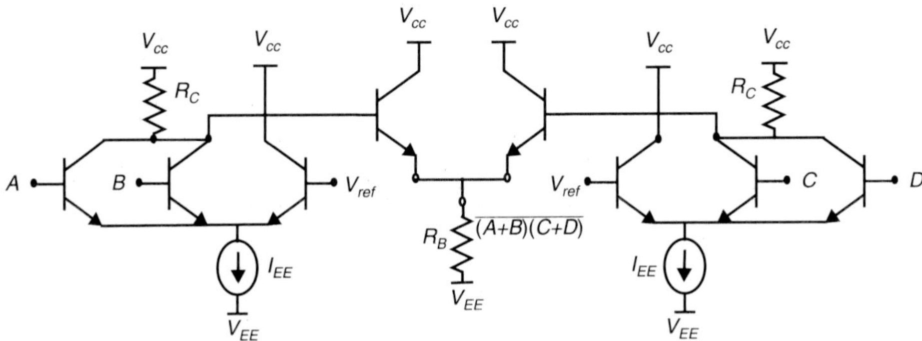

Figure 5.3 Wired-OR configuration, realizing $\overline{(A + B) \cdot (C + D)}$.

is not complementary anymore, since only $\overline{(A+B) \cdot (C+D)}$ is available, not its complement. Figure 5.3 also shows that in case only one output of an ECL gate is used, it is perfectly reasonable to omit the collector resistance at the other side.

The ECL circuits discussed so far are of the so called *single-ended* nature. The central component of the gate is a current switch, with one side connected to the input(s) and the other side to a reference voltage centered in the middle of the logic swing. To ensure sufficient noise margins, it is essential that the low and high input levels differ a number of ϕ_T from the reference voltage. Remember from Chapter 3 that the transition region of the ECL inverter approximately equals 120 mV to 240 mV. For that reason, the logic swing of a single-ended family is normally chosen to range between 500 and 800 mV. The extra safety margin is necessary to accommodate variations in V_{ref}, supply voltage, process parameters, or temperature.

5.2.2 Differential ECL

Since the propagation delay of a digital gate is directly proportional to the logic swing, it is attractive to reduce the swing even more. This is made possible by a simple modification of the generic ECL structure. Instead of connecting the second input of the current switch to a reference voltage, it can be driven by the inverted value of the first input, as shown in Figure 5.4. Such a gate has a reduced transition region. While one input goes up, the second one goes down. This effectively doubles the voltage swing observed by the differential pair.

This observation is confirmed by a first-order derivation of the V_{IH} and V_{IL} of the inverter/buffer. The ratio of the currents between the left and right branches of the current switch can be expressed by Eq. (5.1).

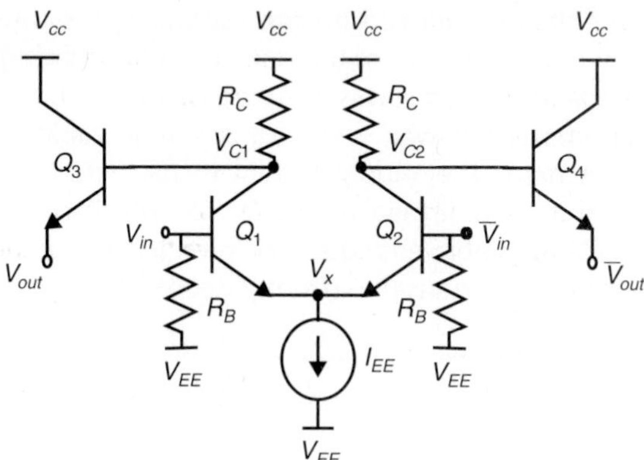

Figure 5.4 Differential ECL buffer.

$$\frac{I_{C1}}{I_{C2}} = \frac{I_S e^{V_{in}/\phi_T}}{I_S e^{\overline{V}_{in}/\phi_T}} = e^{\frac{V_{in} - \overline{V}_{in}}{\phi_T}} \tag{5.1}$$

The current ratio is an exponential function of the difference between the signal and its inverse, which is twice as large as the difference between the signal and a fixed reference voltage, assuming that the inverse signal changes at the same rate. Using the alternative definition of V_{IL} and V_{IH} (I_C = 1% or 99% of I_{EE}, respectively), the width of the transient region can be computed as follows

$$\frac{I_{C1}}{I_{EE}} = \frac{e^{\frac{V_{in} - \overline{V}_{in}}{\phi_T}}}{1 + e^{\frac{V_{in} - \overline{V}_{in}}{\phi_T}}} = \alpha = 0.99$$

$$V_{IH} - V_{IL} = \phi_T \ln\left(\frac{\alpha}{1 - \alpha}\right) = 120 \text{ mV (at room temperature)} \tag{5.2}$$

As expected, a reduction by a factor of 2 with respect to the single-ended structure is obtained. This, in turn, allows for a reduction in voltage swing. Values as small as 200 mV are not uncommon!

Design Considerations

The differential approach offers several important advantages:

- The *sensitivity to supply noise* is reduced. Local drops in the supply voltages affect both V_{in} and $\overline{V}_{in}$ in a similar way. This has no impact on the operation of the gate, since only the difference between input signals is important. This is not true in the single-ended case, where the input signal is compared to a fixed reference voltage that varies as a function of temperature, process parameters, and noise levels.

- In addition, the *switching noise* introduced on the supply lines is substantially reduced in the differential case as a result of the reduced voltage swing and the balanced load. For the single-ended case, it is assumed that only one side of the gate is connected to an emitter-follower, which causes large variations in the supply current during switching. The supply current in the differential case is approximately constant. This is confirmed in Figure 5.5, which compares the supply current (or power) of the single-ended and differential logic structures for both a positive and negative signal transition (from [Greub91]).

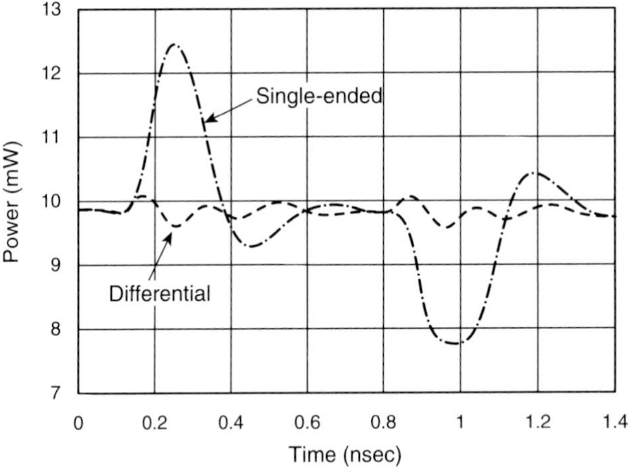

Figure 5.5 Instantaneous power consumption (or supply current) for differential and single-ended buffers (from [Greub91]). Parameters: $V_{CC} - V_{EE} = 5$ V, $I_{EE} = 400$ μA, $I_{emitter\text{-}follower} = 800$ μA, V_{swing}(differential) = V_{swing}(single-ended)/2 = 250 mV.

- Because differential logic circuits tend to require a lower number of cascaded gates for a given function, they are generally faster than their single-ended counterparts.

The differential approach also has some potential deficiencies. The implementation of a fully differential style requires complementary logic networks similar to the DCVSL CMOS logic discussed in Chapter 4. This translates to a larger number of transistors, as demonstrated in Figure 5.6, which shows the schematic diagrams of a three-input OR and AND gate. The OR/NOR structure requires six transistors in contrast to the four needed in the single-ended case.

Differential logic structures are based on the same fundamental concept as the simple ECL inverter, namely, *current steering*. Based on the value of the inputs, the current provided by the current source is guided along either the left or the right branch of the differential pair, causing one output to go low and the other to remain high. This is easily verified for the gates presented in Figure 5.6. Be aware that a current path has to be available for all possible operation conditions!

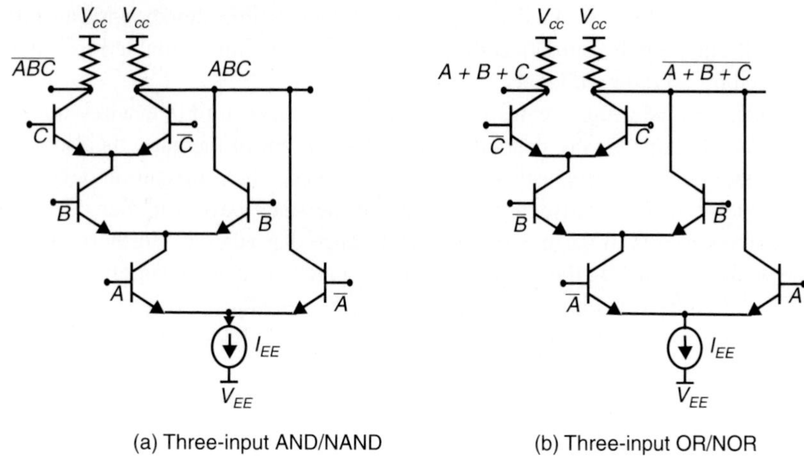

(a) Three-input AND/NAND (b) Three-input OR/NOR

Figure 5.6 AND and OR gates in differential ECL (emitter-followers are omitted).

While the idea seems simple enough, there is a catch, as illustrated in Figure 5.7a. For performance reasons, Q_3 should not go into saturation (see Section 3.4.1). Consequently when Q_3 is on (or V_{in3} is high), it holds that $V_{C3} \geq V_{B3}$. To turn Q_2 on, V_{B2} must be at $V_{C3} + V_{BE(on)} \geq V_{B3} + V_{BE(on)}$ or, $V_{in2} \geq V_{in3} + V_{BE(on)}$. This means that all inputs connecting to transistors at *level 2* (Figure 5.7a) need a dc offset of at least $V_{BE(on)}$ with respect to inputs to *level 3* devices.[1] Similar offsets are needed when connecting to devices placed higher in the input network. Achieving these offsets is accomplished by adjusting the output stage of the (preceding) gate, as demonstrated in Figure 5.7a. Connecting the base and

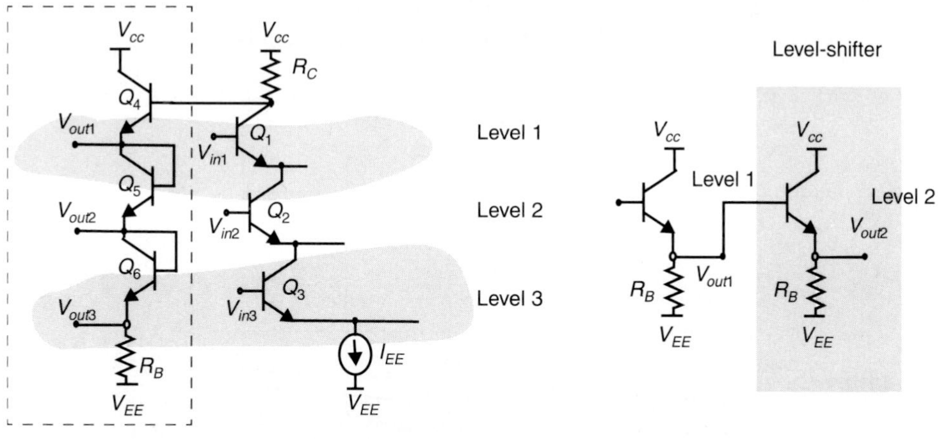

(a) Multilevel emitter-follower output stage (b) Level 1-to-2 converter

Figure 5.7 Adjusting voltage levels in differential ECL using level shifters.

[1] Because ECL signals are normally referenced with respect to V_{CC}, it is common to refer to level 1 signals as those that are one V_{BE} below V_{CC}, level 2 as 2 $V_{BE\text{'s}}$ below V_{CC}, and so on.

collector of transistors Q_4 and Q_5 turns those devices into diodes with an on-voltage of $V_{BE(on)}$. By tapping the outputs at the various diode positions, suitable input signals can be generated for all required levels.

This level-shifting creates extra complexity, since multiple wires may need to be routed for the same signal, depending upon the fan-out of the gate. Its also puts a restriction on the number of transistors that can be stacked. The propagation delay is a function of the output level, because signals lower on the stack have a higher delay. Adding too many layers results in an intolerable performance degradation. Furthermore, the number of layers is restricted by the available voltage range between the supply rails. For instance, the *level 3* signals in Figure 5.7a range from $V_{CC} - 3V_{BE(on)}$ (high) to $V_{CC} - 3V_{BE(on)} - V_{swing}$ (low). A typical differential ECL library (e.g. [Tektronix93]) uses at most three layers.

Another way to address noncompatible signal levels, while avoiding the complex, multilevel output emitter-follower, is to insert level-shifting circuits whenever needed. A *level 1-to-level 2* converter is shown in Figure 5.7b. In this way, all standard logic gates can be designed with a sole level 1 output, and level-shifting buffers are introduced only when connecting to multilevel gates.

One additional property of the differential logic style is worth mentioning. The single-ended ECL gate tends to exhibit radically different values for its output rise and fall times, because these are set by different circuit elements. This results in different values of t_{pLH} and t_{pHL}, because the switching point of the gate is determined by comparing the input signal with the fixed reference voltage. In the differential circuit, the switching is determined by the crossing of the input signal and its inverse. Although these signals may have very different slopes, this does not affect the delay of the gate, which is thus independent of the direction of the transition. Figure 5.8 compares the delay of a single-ended versus a differential buffer as a function of the capacitance. While the delay of the single-ended version is a strong function of the transition direction, only one delay exists for the differ-

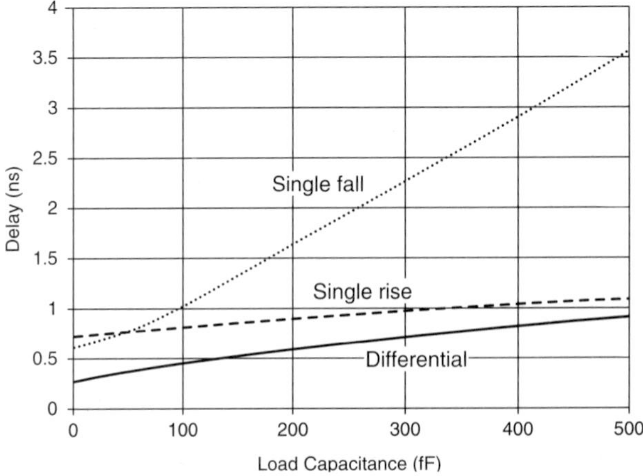

Figure 5.8 Performance comparison of a differential and a single-ended buffer (from [Tektronix93]). Process parameters: f_T = 12 GHz, $A_{E,min}$ = 0.6 μm × 2.4 μm, worst-case modeling. Both cells use 100 μA tree and 100 μA emitter-follower currents.

ential case. Notice also the smaller delay of the differential buffer caused by the reduced swing. Finally, observe that the value of the load capacitance has only a minor impact on the delay in contrast to CMOS.

Example 5.2 Differential ECL Buffer

Figure 5.9 shows the schematics and layout of a differential ECL buffer ([Tektronix93]). Its unloaded delay is 180 psec and increases by 330 psec for every pF of load capacitance for a total current of 0.6 mA. The input capacitance of the gate equals 50 fF. For a 5 V supply voltage, this translates into a PDP of 600 fJ (unloaded).

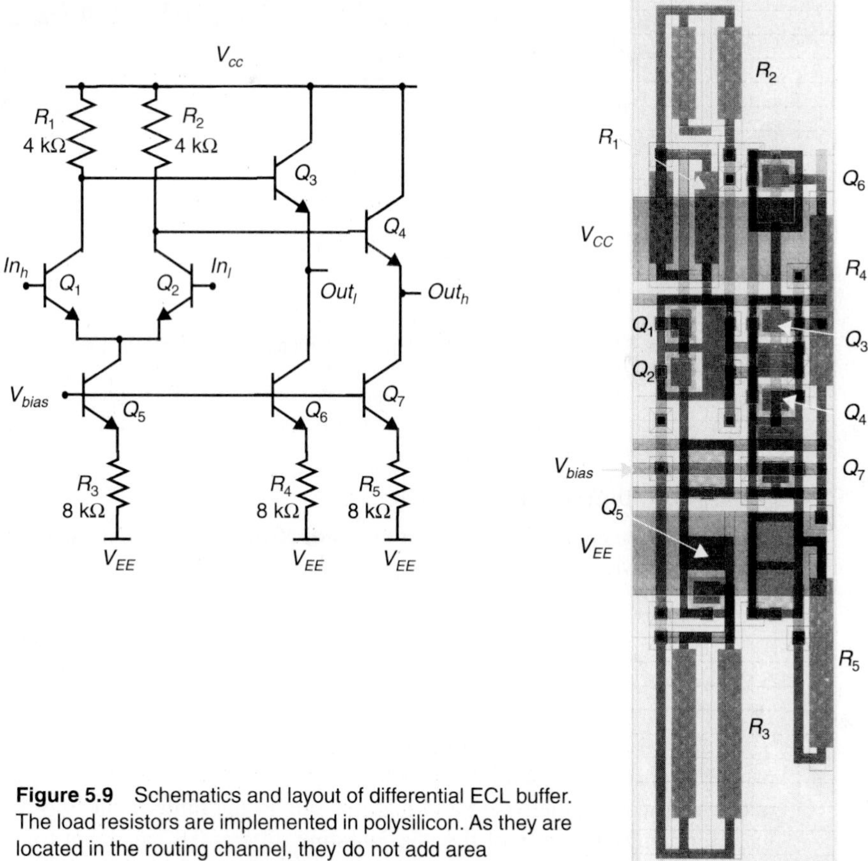

Figure 5.9 Schematics and layout of differential ECL buffer. The load resistors are implemented in polysilicon. As they are located in the routing channel, they do not add area overhead. *Courtesy of Tektronix.*

Example 5.3 Differential ECL Gate

The circuit schematics of a four-input multiplexer is shown in Figure 5.10. Verify its functionality by checking how the current flows for each of the combinations of the signals A and B. The logic swing at the output equals 400 μA × 625 Ω = 250 mV. To function correctly, the input signals i_{0-3} must be at level 1 (or $V_{CC} - V_{BE(on)}$), while the B and A signals should be at levels 2 and 3, respectively. The static power consumption of the differential switch is easily computed and equals 400 μA × 5 V = 2 mW.

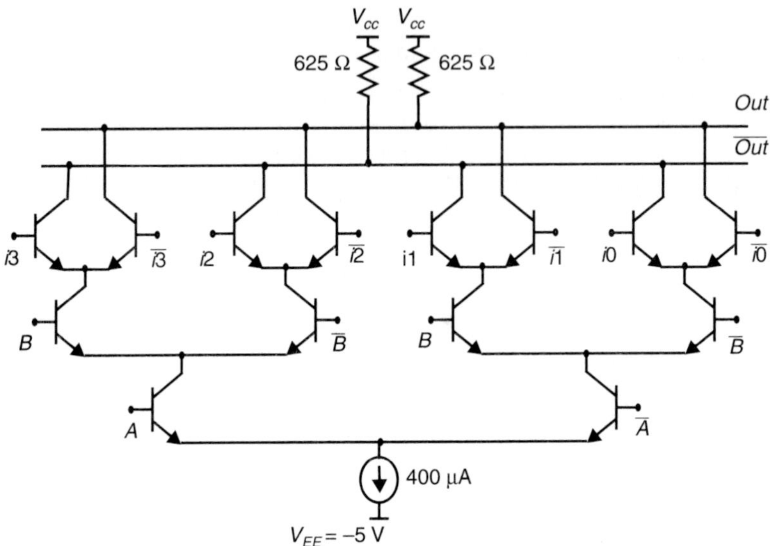

Figure 5.10 Differential ECL four-input multiplexer (emitter-followers have been omitted); V_{CC} = 0 V.

Problem 5.1 Differential ECL Gate

Derive the circuit topology for a two-input XOR gate in both differential and single-ended design styles.

5.2.3 Current Mode Logic

The emitter-follower output structure of the ECL gate makes it possible to drive large capacitive loads with small delay. Additionally, the output structure supports a large fan-out, which would otherwise be impossible due to the finite input impedance of the ECL gate, formed by the base of the nonsaturating differential pair input transistor in parallel with the pull-down resistor. On the negative side, the output stage adds a considerable area overhead, as is apparent in Figure 5.9. It also consumes a substantial amount of power, which prevents the realization of complex integrated circuits. For instance, a 300 MHz 32-bit microprocessor implemented in the single-ended ECL logic style [Jouppi93] has been reported. The power dissipation of the 486k transistor device equals 115 W! To disperse the excess heat generated at these power levels, special (and expensive) cooling approaches are an absolute necessity.

The capacitive loads on the internal nodes of a complex logic block such as an adder tend to be small, and the fan-out is generally restricted to 1 or 2. For these small loads, it is reasonable to eliminate the emitter-follower. When a bipolar gate only consists of a differential pair, the logic style is called *current mode logic* (CML). A differential version of a CML gate is obtained by eliminating transistors Q_3 and Q_4, as well as the pull-down resistors in Figure 5.4. Eliminating the follower results in a drastic reduction in power consumption, while keeping the propagation delay reasonable for small loads. Figure 5.11 plots the propagation delay of a differential ECL and a CML buffer as a function of the logic swing for a fan-out of 1 (with and without the extra load of an interconnect wire of 0.5 mm long). CML is actually faster for small logic swings and small loads. Its delay increases rapidly with the logic swing, as the transistors start to saturate during the switching. On the other hand, the static power dissipation of the CML gate is only 2 mW versus the 10 mW of the ECL gate. CML is, therefore, only used within cells where the fan-out is small, the interconnect length is short, and power dissipation and cell area is an issue.

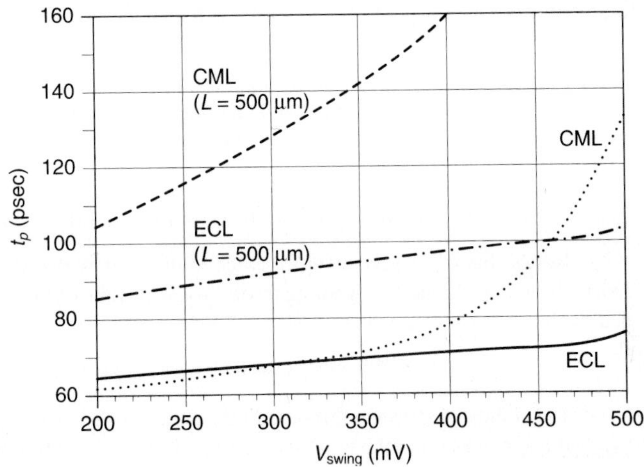

Figure 5.11 Propagation delays of ECL and CML buffers versus logic swing for different loads (unloaded and loaded with a 500 μm interconnect wire). Design parameters: fan-out = 1, I_{EE} = 400 μA, $I_{follower}$ = 800 μA (from [Greub91]).

Example 5.4 CML Gate Characteristics

A differential CML buffer (obtained by removing the emitter-followers in Figure 5.4) has been designed using our generic bipolar technology. The value of the current source is set to 0.4 mA and R_C = 625 Ω. The supply voltages V_{CC} and V_{EE} are set to 0 V and -1.7 V, respectively. This translates into a voltage swing of 0.25V with V_{OH} = 0 V and V_{OL} = –0.25 V. **Observe that the differential nature of the gate eliminates the need for output level-shifting, necessary in single-ended gates.**

The VTC of the gate is plotted in Figure 5.12a. The simulated values ofthe logic levels are consistent with the manual analysis. The values of V_{IH} and V_{IL} (using the unity gain definition) evaluate to –0.085 V and –0.165 V, which results in both a low and a high noise margin of 0.085 V. Compared to the ECL gate of Example 3.13, all dc parameters (width of transition region, logic swing, and noise margins) have been halved. It should be mentioned that the dc analysis was performed with perfectly complementary inputs (V_{in2} = –0.25 V – V_{in1}).

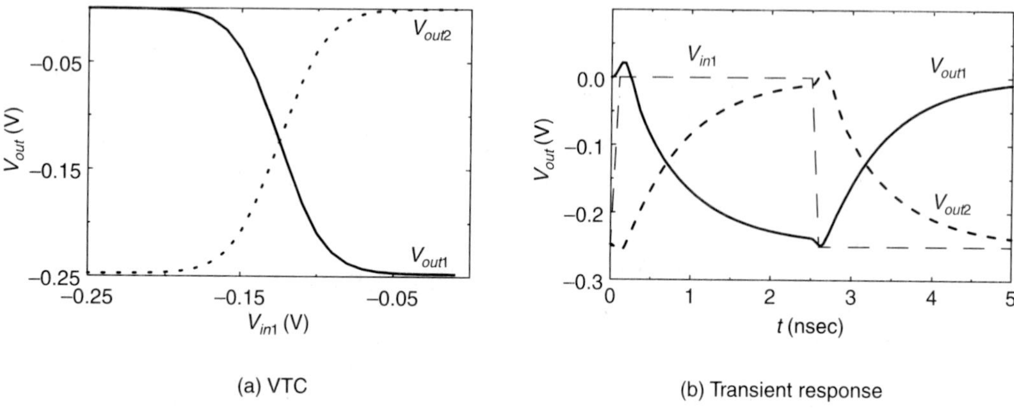

Figure 5.12 VTC and transient response of CML gate ($V_{in2} = -0.25 - V_{in1}$).

For our standard bipolar technology, the load capacitance (for a single fan-out) can be computed by the same techniques used in Example 3.15 for the ECL gate. Assuming a fan-out capacitance of 60 fF, we can approximate the capacitance at the output node as follows:

$$C_L = C_{fan\text{-}out} + C_C = C_{fan\text{-}out} + (C_{cs1} + 2\,C_{bc1}) =$$
$$60 \text{ fF} + 0.67 \times 47 \text{ fF} + 2 \times 1.01 \times 22 \text{ fF} = 136 \text{ fF}$$

Notice that C_{bc} is accounted for twice to incorporate the Miller effect. The weighting functions are the K_{eq} factors. Ignoring the turn-on (off) time of the transistors, the propagation delay can now be approximated:

$$t_p = 0.69\,R_C\,C_L = 0.69 \times 625 \times 136 = 59 \text{ psec}$$

The simulated transient response of the circuit is shown in Figure 5.12b. The values of t_{pLH} and t_{pHL} evaluate to 69 and 63 psec respectively, or t_p = 66 psec. [2] This number rapidly degrades for larger fan-out values. For instance, for a fan-out of 3, the delay climbs to 102 psec!

The static power consumption of the gate is independent of the logic state and equals 0.4 mA × 1.7 V = 0.68 mW. When switching the gate at the fastest possible speed, the dynamic power consumption evaluates to

$$P_{dyn} = C_L\,(V_{CC} - V_{EE})\,V_{swing} / t_p = 0.98 \text{ mW}$$

Observe that this expression takes into account that at least one collector gets charged for every transition. The differential nature of the structure effectively doubles the dynamic consumption.

Combining the obtained numbers results in a PDP of 102 fJ (for a fan-out of 1), which compares very favorably to the 750 fJ obtained for both the CMOS and ECL inverters discussed in Chapter 3.

---

[2] These numbers are obtained using the traditional definitions of t_{pHL} and t_{pLH}. As mentioned earlier, these definitions do not make much sense for a differential logic family, where the propagation delay should be determined by measuring the time between the cross-over points of the differential input and output signals.

5.2.4 ECL with Active Pull-Downs

The realization of a fast, low-power ECL-type gate is hampered by the following constraints or restrictions.

- In order to reduce t_{pHL}, the value of R_B has to be kept small; t_{pHL} normally dominates the transient performance.
- Small values of R_B result in high values of the static power consumption.

  Scaling the technology to smaller lithographic dimensions has only a minor impact on this picture.
- The supply voltage of the traditional, single-ended ECL gate cannot be dropped much below 3 V_{BE} ($\approx$ 2.5 V). V_{OH} equals $V_{CC} - V_{BE}$; the voltage drop between base and emitter of the input transistor adds another V_{BE}, and an additional V_{BE} is needed to ensure the proper operation of the current source of the differential pair. Since V_{BE} is a built-in voltage that is only marginally affected by technology parameters, this voltage is not reduced in scaled technologies. The power consumption/gate can hence only be reduced by reducing the current levels. This is in contrast to CMOS, where the supply voltage is expected to keep dropping as the technology feature sizes are reduced.
- R_B is not a function of technology parameters and is mainly set by the allowable power-consumption level. Fitting more gates on a die requires a reduction of the consumption/gate, which means a higher value for R_B. This adversely affects the performance (or keeps it constant at best), as $t_{pHL} \sim R_B C_L$. C_L might decrease proportional to the scaling factor, although this depends upon what portion of the load is composed of wiring capacitance.

Both of the above arguments, nonscalability of the supply voltage and the pull-down resistance, represent major obstacles to a continued performance improvement of the ECL logic family unless some substantial changes are made to the structure. Short of completely eliminating the emitter-follower (as is the case in CML), one approach is to replace the pull-down resistor by an active network [Chuang92]. Such an active pull-down network provides ample current when switching, but operates at reduced current levels in standby.

An example of such a circuit is shown in Figure 5.13. This *ac-coupled active pull-down* circuit utilizes a capacitor to strongly turn on the pull-down *npn*-transistor during a negative-going output transition, and to turn it off during a positive one. The steady-state current of the device is set by the dc-bias network. This approach reduces the stand-by current while substantially improving the performance. The capacitor completely blocks dc signals. Extra biasing devices are needed, however, to establish the steady-state current in the output stage. This approach has been used to realize the ECL buffers with a propagation delay of 23 psec, which is among the fastest achieved in room-temperature silicon [Toh89].

Another approach is to present a variable load to the emitter-follower. The circuit in Figure 5.14 uses an NMOS device to achieve that goal. This approach can be seen as an introduction to the BiCMOS technology, which merges bipolar and MOS transistors in the same process and is treated in Section 5.3. The NMOS device is turned on more strongly during a high-to-low transition, providing more pull-down current and hence increased

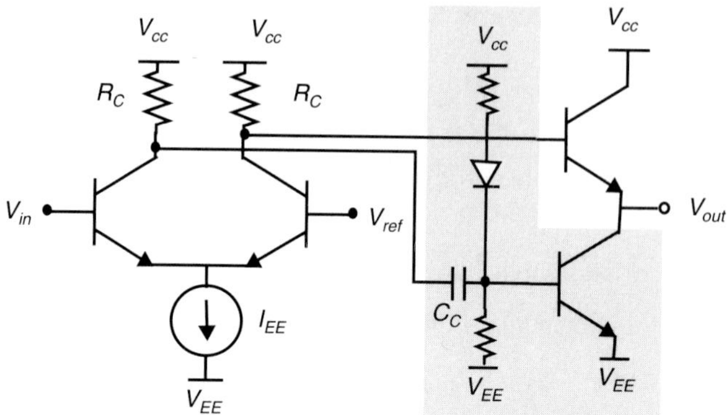

Figure 5.13 Ac-coupled, active pull-down ECL buffer.

performance. In the high-output state, the gate-source voltage of the transistor and the static power consumption are reduced. It has been reported that for the same performance (of 200 psec), the structure presented consumes four times less power than the ECL circuit with the resistive pull-down [Chen92].

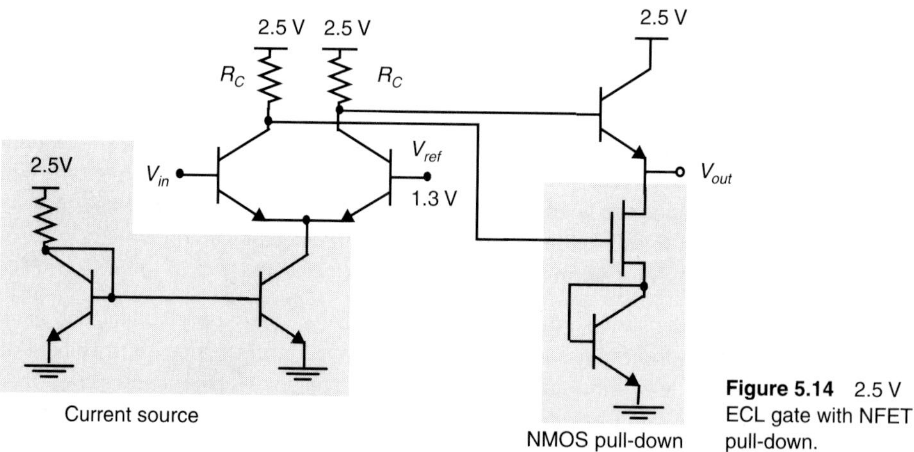

Figure 5.14 2.5 V ECL gate with NFET pull-down.

Observe how the circuit of Figure 5.14 is designed to operate at the *minimum possible supply voltage* (approximately three times V_{BE}). The signal swing is approximately equal to V_{BE} and is centered around 1.3 V (1.5 V_{BE} below the supply voltage V_{CC}). The diode in the output stage reduces the dc standby current. The current source of the input stage is implemented by a bipolar current mirror, which reduces the required voltage drop over the source. The reference voltage of 1.3 V is chosen to avoid the saturation of the current-source transistor.

Problem 5.2 Active Emitter-Follower Load

Instead of using a resistive load for the emitter-follower circuit, a current source can also be employed, as shown in Figure 5.15. The same circuit was shown earlier in the differential buffer of Figure 5.9. Discuss the impact of this option on performance and power consumption.

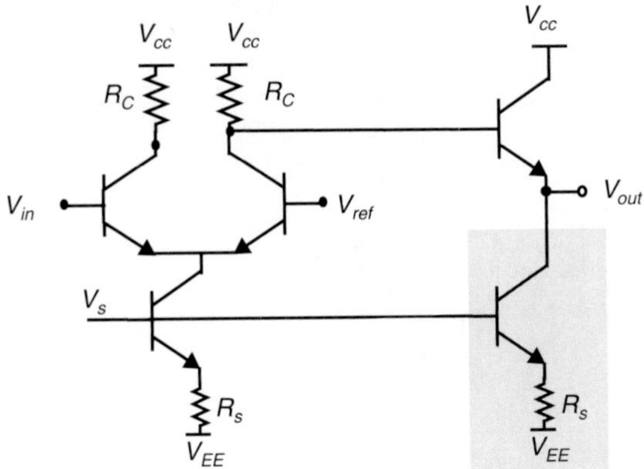

Figure 5.15 Current-source pull-down.

Various other approaches have been proposed to improve the performance of the ECL structure or to reduce its power consumption [Chuang92]. While these techniques have resulted in spectacular performances at reasonable power levels, whether these approaches will help ECL to compete with deep submicron CMOS in times to come is an open question [Masaki92].

5.2.5 Alternative Bipolar Logic Styles

A large variety of bipolar digital gates have been proposed over time, covering a wide span of speed and power requirements. The most popular among them has been the *transistor-transistor-logic* (TTL) gate. TTL dominated the *discrete logic component* market, which supports individually packaged NAND and NOR gates, multiplexers or bus-drivers, for more than two decades. This picture has changed in the 1980s due to two major factors:

1. Discrete logic gates in static CMOS became competitive in speed at a lower power cost.
2. The advent of programmable logic components such as PLDs and FPGAs, discussed in Chapter 11, made it possible to *program* complex random logic functions (equivalent to hundreds of TTL gates) on a single component. This results in a large reduction in board real-estate cost, while adding flexibility.

The last factor in particular has influenced the demise of TTL. Due to its historical impact and for the sake of completeness, a brief examination of the basic TTL structure is worth the effort however. It consists of three units, as shown in Figure 5.16:

- *The input stage*, which consists of a multi-emitter bipolar transistor and performs an AND-ing of the inputs: both inputs of the transistor have to be high for its collector to be high.
- The *phase splitter*, which generates two signals with opposite phases. These signals are used to drive the output stage.
- The totem-pole (push-pull) *output structure*. Only one of the bipolar transistors of the structure is ON in steady-state mode, since the controlling signals have opposite

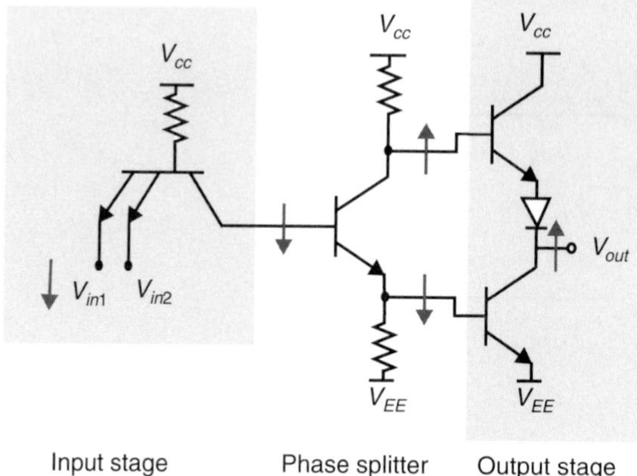

Figure 5.16 Generic TTL NAND structure. The arrows denote the direction of the signal transitions for one of the input signals going low.

phases. In contrast to the emitter-follower output stage, this structure has the advantage that no static power is consumed, while ample current drive is still available. The disadvantage is the need for a phase splitter. Saturation of the output transistors also degrades the performance. Elaborate techniques (using, for instance, Schottky Barrier diodes—see Appendix D) have been devised to avoid saturation.

Because the TTL gate is an intricate composition, its actual operation is complex, but not particularly fast. A detailed analysis of its behavior would lead us astray and would contribute little to the understanding of contemporary digital design approaches. The interested reader should refer to the numerous available textbooks and reference works that treat TTL design in extensive detail (e.g., [Hodges88]).

One legacy of the TTL era has endured. The TTL logic levels have become a de facto standard due to their widespread usage. Input and output signals of integrated circuits must often still comply with this standard. An overview of these requirements is given in Table 5.1.

Table 5.1 TTL signal requirements, valid for popular TTL families such as low-power Schottky (LS) and fast (F) TTL (from [Buchanan90]).

V_{IL}(max), V	V_{IH}(min), V	V_{OL}(max), V	V_{OH}(min), V
0.8	2.0	0.5	2.5

Besides ECL and TTL, a number of other bipolar logic families have emerged and vanished over the years. Three examples are integrated injection logic (I^2L or MTL), Schottky transistor logic, and integrated Schottky logic ([Hodges88]). A detailed discussion is not warranted since none are actively used in current integrated circuit designs.

One gate structure, called *nonthreshold logic* (NTL), is gradually gaining some acceptance in high-performance designs, due to its very low power-delay product ([Ichino87]), and is therefore worth analyzing. The basic structure of an NTL gate is shown in Figure 5.17. It consists of an input (logic) stage and an emitter-follower output structure. The input structure resembles the RTL gate, briefly discussed in Chapter 3, where it was concluded that its performance suffers from the saturation of the pull-down

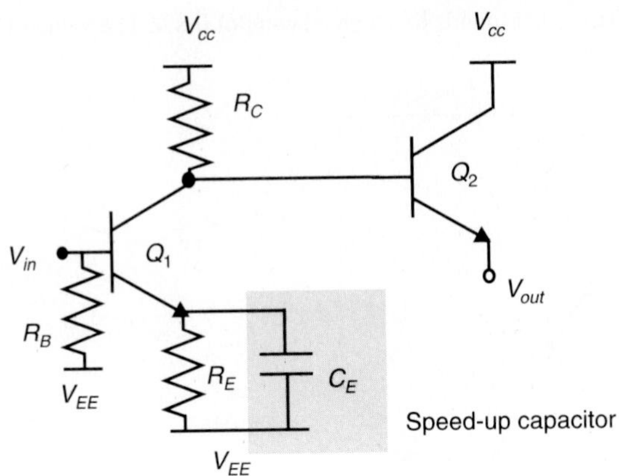

Figure 5.17 Nonthreshold logic (NTL) gate structure.

device. This is avoided in the NTL gate by adding the emitter-degenerating resistor R_E, which ensures that Q_1 stays in the forward-active mode for the voltage input range of interest. The emitter-follower is added to provide additional fan-out drive capability, eliminating another deficiency of the RTL gate. The dc parameters can be derived by ignoring the base current of the emitter-follower.

$$
\begin{aligned}
V_{OH} &= V_{CC} - V_{BE(on)} \\
V_{OL} &= V_{CC} - V_{BE(on)} - \frac{R_C}{R_E}(V_{OH} - V_{BE(on)}) \\
&= \left(\frac{R_E - R_C}{R_E}\right)V_{CC} + \left(\frac{2R_C - R_E}{R_E}\right)V_{BE(on)}
\end{aligned}
\tag{5.3}
$$

The gain of the gate in the transient region is approximately proportional to $(-R_C/R_E)$, as can be derived using the small-signal approach. A large value of the ratio enhances the voltage gain and the noise margin, but increases the gate delay. $R_C = 2R_E$ seems to be a good compromise. Under those conditions, V_{OL} evaluates to $3V_{BE(on)} - V_{CC}$, and the total signal swing equals $2V_{CC} - 4V_{BE(on)}$. As V_{OL} has to be larger than 0, this translates to the constraint that V_{CC} has to be kept smaller than 3 $V_{BE(on)}$. Ensuring that Q_1 does not saturate over the input range of interest puts a lower bound on V_{CC} of approximately 2 $V_{BE(on)}$.

A remarkable feature of this gate is that the voltage-transfer characteristic displays only a single strong nonlinearity, corresponding to the turning on of Q_1. For the rest of the operation range Q_1 stays in the forward-active mode, because saturation is avoided during normal operation. Hence the name *nonthreshold* logic. The lack of a well-defined low output level makes this approach particularly sensitive to noise. The attentive reader will have observed already that the gate retains the regenerative property as long as $R_C > R_E$.

Example 5.5 VTC of an NTL Gate

The simulated VTC of an NTL gate is shown in Figure 5.18 for the following parameters: $V_{CC} = 1.9$ V, $R_C = 2\ R_E = 0.5$ kΩ. Assume that $V_{be(on)} = 0.75$ V. The manually derived values for V_{OL} and V_{OH} (0.35 V and 1.15 V, respectively) closely correspond to the simulated values marked on the simulation. This corresponds to a signal swing of 0.8 V. The value of V_{IL} equals 0.8 V, while V_{IH} cannot be defined.

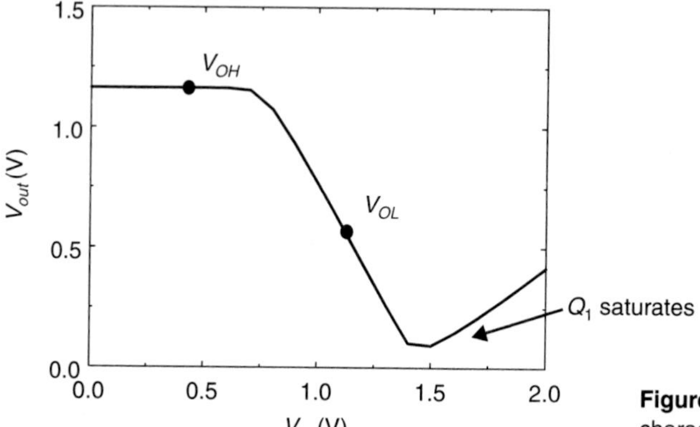

Figure 5.18 Voltage-transfer characteristic of an NTL gate.

While the emitter degeneration acts as a negative feedback and helps create the non-threshold characteristic, its effects are not desirable from a transient point of view. Raising the voltage at the base of Q_1 also raises its emitter voltage, reducing the current available for discharging the collector capacitance. The effects of this negative feedback can be reduced by adding an extra decoupling capacitor C_E at the emitter node, as shown in Figure 5.17. The larger the capacitance, the larger the decoupling effect and the speed-up. This capacitor can be implemented with a reverse-biased diode.

Example 5.6 Transient Response of the NTL Gate

The simulated transient response of the NTL gate is shown in Figure 5.19 (for a fan-out of 1 and a pull-down resistance of 2 kΩ). The response is shown without and with a decoupling capacitance of 0.3 pF. It can be observed that the decoupling capacitance only affects the high-to-low transition. The simulated value of the propagation delay with decoupler equals 41 psec.

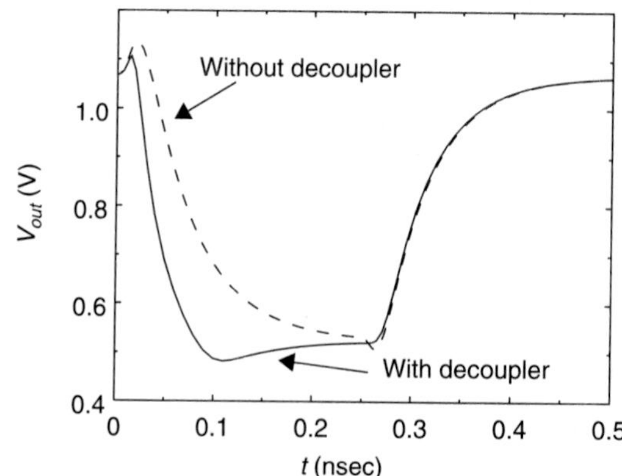

Figure 5.19 Transient response of an NTL gate (without and with an emitter-decoupling capacitance of 0.3pF).

The static power consumption of the gate can be derived. For a low input, only the emitter-follower consumes static power, while both the logic and the output structures are consuming power for a high input. Adding all factors results in a static consumption of 2.28 mW. The dynamic consumption, when clocked at the maximum possible rate, equals 2.3 mW. Reduction of the consumption is possible by reducing the voltage swing, the supply voltage, and the circuit performance. For this particular version of the gate, the PDP evaluates to 376 pJ.

The NOR operation is the logic function most easily implemented in NTL, and is constructed by a parallel connection of the input transistors. More complex functions can be realized by transistor stacking. This requires extra level shifters, while saturation should be carefully avoided.

5.3 The BiCMOS Approach

Complementary MOS offers an inverter with near-perfect characteristics such as high, symmetrical noise margins, high input and low output impedance, high gain in the transition region, high packing density, and low power dissipation. Speed is the only restricting factor, especially when large capacitors must be driven. In contrast, the ECL gate has a high current drive per unit area, high switching speed, and low I/O noise. For similar fan-outs and a comparable technology, the propagation delay is about two to five times smaller than for the CMOS gate. However, this is achieved at a price. The high power consumption makes very large scale integration difficult. A 100k-gate ECL circuit, for instance, consumes 60 W (for a signal swing of 0.4 V and a power supply of 4 V). The typical ECL gate also has inferior dc characteristics compared to the CMOS gate—lower input impedance and smaller noise margins.

In recent years, improved technology has made it possible to combine complimentary MOS transistors and bipolar devices in a single process at a reasonable cost. A cross-section of a typical BiCMOS process is shown in Figure 5.20. A single *n*-epitaxial layer is used to implement both the PMOS transistors and bipolar *npn* transistors. Its resistivity is chosen so that it can support both devices. An n^+-buried layer is deposited below the epitaxial layer to reduce the collector resistance of the bipolar device, which simultaneously increases the immunity to latchup. The *p*-buried layer improves the packing density,

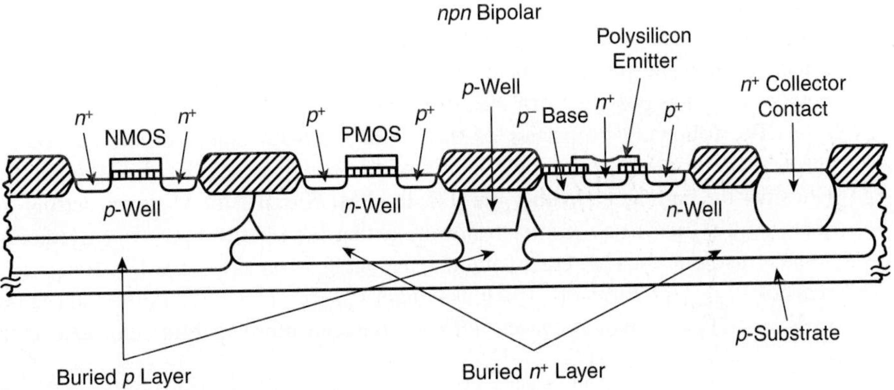

Figure 5.20 Cross-section of BiCMOS process (from [Haken89]).

because the collector-collector spacing of the bipolar devices can be reduced. It comes at the expense of an increased collector-substrate capacitance.

This technology opens a wealth of new opportunities, because it is now possible to combine the high-density integration of MOS logic with the current-driving capabilities of bipolar transistors. A *BiCMOS inverter*, which achieves just that, is discussed in the following section. We first discuss the gate in general and then provide a more detailed discussion of the steady-state and transient characteristics, and the power consumption. The section concludes with a discussion of the usage of BiCMOS and the future outlook. Most of the techniques used in this section are similar to those used for CMOS and ECL gates (Chapter 3), so we will keep the analysis short and leave the detailed derivations as an exercise.

5.3.1 The BiCMOS Gate at a Glance

As was the case for the ECL and CMOS gates, there are numerous versions of the BiCMOS inverter, each of them with slightly different characteristics. Discussing one is sufficient to illustrate the basic concept and properties of the gate. A template BiCMOS gate is shown in Figure 5.21a. When the input is high, the NMOS transistor M_1 is on, causing Q_1 to conduct, while M_2 and Q_2 are off. The result is a low output voltage (Figure 5.21b). A low V_{in}, on the other hand, causes M_2 and Q_2 to turn on, while M_1 and Q_1 are in the off-state, resulting in a high output level (Figure 5.21c). In steady-state operation, Q_1 and Q_2 are never on simultaneously, keeping the power consumption low. An attentive reader may notice the similarity between this structure and the TTL gate of Figure 5.16. Both use a bipolar *push-pull* output stage. In the BiCMOS structure, the input stage and the phase-splitter are implemented in MOS, which results in a better performance and higher input impedance.

The impedances Z_1 and Z_2 are necessary to remove the base charge of the bipolar transistors when they are being turned off. For instance, during a high-to-low transition on the input, M_1 turns off first. To turn off Q_1, its base charge has to be removed. This happens through Z_1.. Adding these resistors not only reduces the transition times, but also has a positive effect on the power consumption. There exists a short period during the transition when both Q_1 and Q_2 are on simultaneously, thus creating a temporary current path between V_{DD} and *GND*. The resulting current spike can be large and has a detrimental effect on both the power consumption and the supply noise. Therefore, turning off the devices as fast as possible is of utmost importance.

The following properties of the voltage-transfer characteristic can be derived by inspection. First of all, the logic swing of the circuit is smaller than the supply voltage. Consider the high level. With V_{in} at 0 C, the PMOS transistor M_2 is on, setting the base of Q_2 to V_{DD}. Q_2 acts as an emitter-follower, so that V_{out} rises to $V_{DD} - V_{BE(on)}$ maximally. The same is also true for V_{OL}. For V_{in} high, M_1 is on. Q_1 is on as long as $V_{out} > V_{BE(on)}$. Once V_{out} reaches $V_{BE(on)}$, Q_1 turns off. V_{OL} thus equals $V_{BE(on)}$.[3] This reduces the total voltage swing to $V_{DD} - 2V_{BE(on)}$, which causes not only reduced noise margins, but also increases the

[3] Given enough time, the output voltage will eventually reach the ground rail. Once Q_1 is turned off, a resistive path to ground still exists through M_1-Z_1. Due to the high resistance of this path, this takes a substantial amount of time. It is therefore reasonable to assume that $V_{OL} = V_{BE(on)}$.

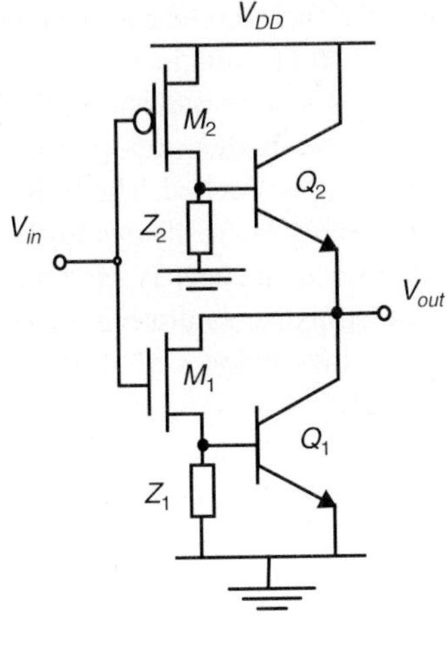

(a) A generic BiCMOS gate

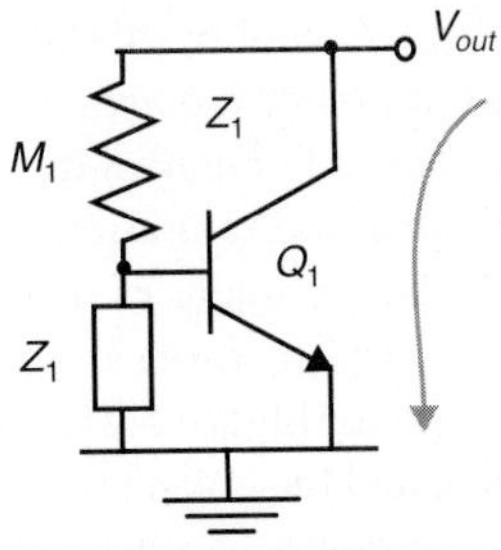

(b) Equivalent circuit for high-input signal

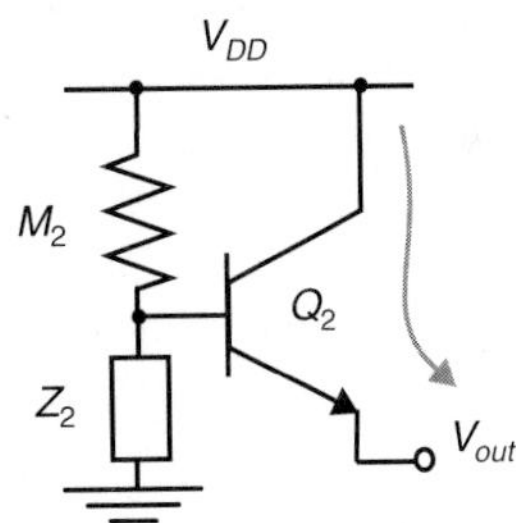

(c) Equivalent circuit for low-input signal

Figure 5.21 The BiCMOS gate.

power dissipation. Consider for instance the circuit of Figure 5.22, where the BiCMOS gate is shown with a single fan-out for $V_{in} = 0$. The output voltage of $V_{DD} - V_{BE(on)}$ fails to turn the PMOS transistor of the subsequent gate completely off, since $V_{BE(on)}$ is approximately equal to the PMOS threshold. This leads to a steady-state leakage current and power consumption. Various schemes have been proposed to get around this problem, resulting in gates with logic swings equal to the supply voltage at the expense of increased complexity. Some of these schemes will be discussed later. Aside from this difference, the VTC of the BiCMOS inverter is remarkably similar to that of CMOS.

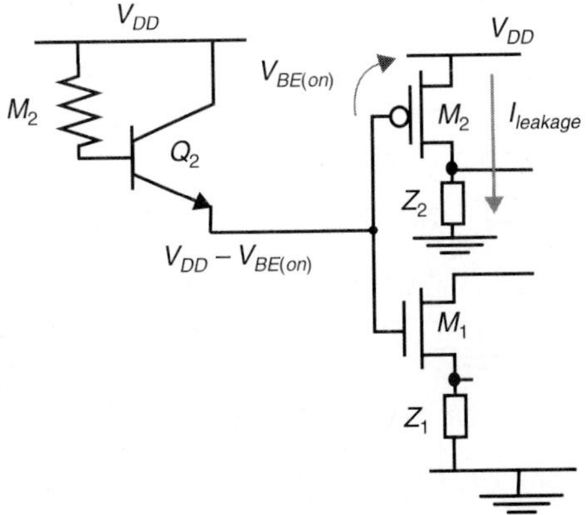

Figure 5.22 Increased power consumption due to reduced voltage swing.

The propagation delay of the BiCMOS inverter consists of two components: (1) turning the bipolar transistors on (off), and (2) (dis)charging the load capacitor. From our discussion of the RTL gate (Chapter 3), we learned how important it was to keep the bipolar transistors out of the saturation region. Building and removing the base charge of a saturated transistor requires a considerable amount of time and results in a slow gate. One of the attractive features of the BiCMOS inverter is that the structure prevents both Q_1 and Q_2 from going into saturation. They are either in forward-active mode or off. For the high output level, Q_2 remains in the forward-active mode when V_{OH} is reached. The PMOS transistor M_2 acts a resistor, ensuring that the collector voltage of M_2 is always higher than its base voltage (Figure 5.21c). Similarly, at the low-output end, M_1 acts as a resistor between the base and the collector of Q_1, preventing the device from ever saturating (Figure 5.21b). The base charge is, therefore, kept to a minimum, and the devices are turned on and off quickly.

Consequently, it is reasonable to assume that for typical capacitive loads, the delay is dominated by the capacitor (dis)charge times. To analyze the transient behavior of the inverter, assume that the load capacitance C_L is the dominating capacitance. Consider first the low-to-high transition. In this case, the equivalent circuit of Figure 5.23a is valid. Q_1 is switched off fast, as its base charge is removed through Z_1. The load capacitor C_L is charged by the current multiplier M_2-Q_2. The source current of M_2 is fed into the base of Q_2 and multiplied with the β_F of Q_2 (assuming that Q_2 operates in the forward-active region). This produces a large charging current of $(\beta_F + 1)\,(V_{DD} - V_{BE(on)} - V_{out}) / R_{on}$ (with R_{on} the equivalent on-resistance of the PMOS transistor). During the high-to-low transition, the equivalent circuit of Figure 5.23b is valid. Q_2 is turned off through Z_2. Once again, the combination M_1-Q_1 acts as a β_F current multiplier. Assuming that the resistance of M_2 in the forward-active mode equals R_{on}, the discharge current equals $(\beta_F + 1)\,(V_{out} - V_{be(on)}) / R_{on}$ (assuming that $R_{on} << Z_1$). The current multiplication factor makes the BiCMOS gate more effective than the CMOS inverter for large capacitive loads.

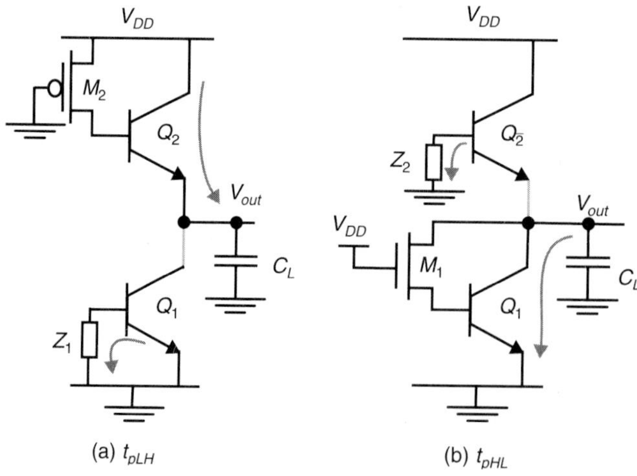

Figure 5.23 Transient behavior of BiCMOS inverter.

In summary, the BiCMOS inverter exhibits most of the properties of the static CMOS inverter. In addition, it displays excellent capacitance-driving capabilities as a result of the push-pull bipolar output stage. The price is a slightly more complex gate and a more complex and expensive fabrication process.

5.3.2 The Static Behavior and Robustness Issues

The use of resistive elements makes the BiCMOS gate of Figure 5.21 unattractive for real designs. A number of slightly modified, more popular circuits are shown in Figure 5.24. In the first circuit (a), the impedances Z_1 and Z_2 are replaced by active impedances (or transistors) that are only turned on when needed. It still has the unfortunate property that a diode voltage drop is lost at the high end of the output range and to a lesser degree also at the low end. Circuit (b) has similar properties. The main difference between the two topologies resides in the transient behavior. Circuit (c) remedies the voltage-drop problem.

Deriving the other parameters of the VTC of the BiCMOS inverter manually is truly complex due to the large number of devices and their interplay. We restrict ourselves to SPICE simulations.

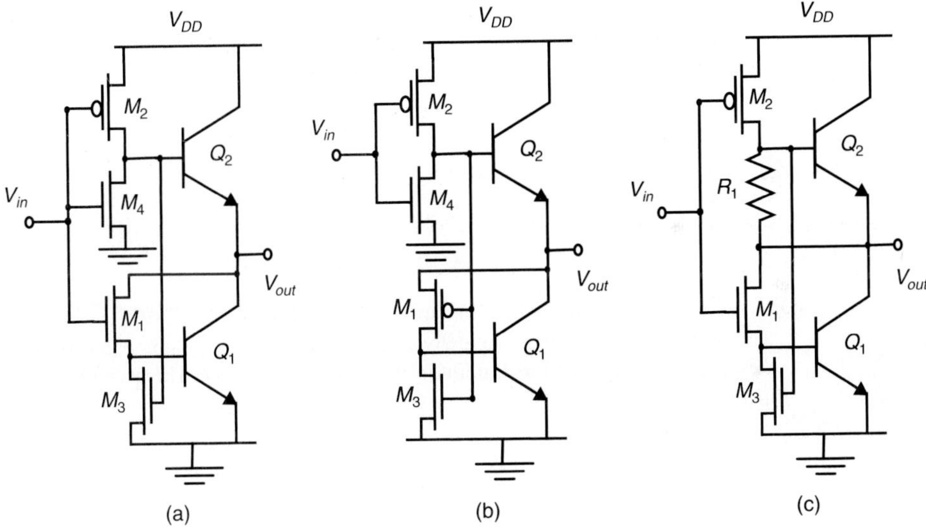

Figure 5.24 Alternative topologies for BICMOS inverters.

Example 5.7 VTC of a BiCMOS Inverter

The voltage-transfer characteristic of the inverter of Figure 5.24b is simulated using SPICE. A contrived BiCMOS process is employed that merges the MOS devices and bipolar transistors described by the models of Chapter 2. The NMOS and bipolar transistors are minimum size, while the PMOS transistors are made twice as wide as the NMOS devices. The supply voltage V_{DD} is set at 5 V.

The resulting VTC is shown in Figure 5.25. The complex shape of the curve is caused by the complex interactions among the large number of active devices present in the circuit. To clarify the behavior, we have also plotted the dc transfer characteristics for the base voltages of transistors Q_1 and Q_2. In the transient region between 2 V and 3.5 V, none of the bipolar transistors are really on. Also, the PMOS device M_1 only turns on after M_3 turns off and when V_{base2} is sufficiently below V_{out}. This causes Q_1 to turn on and creates an additional drop

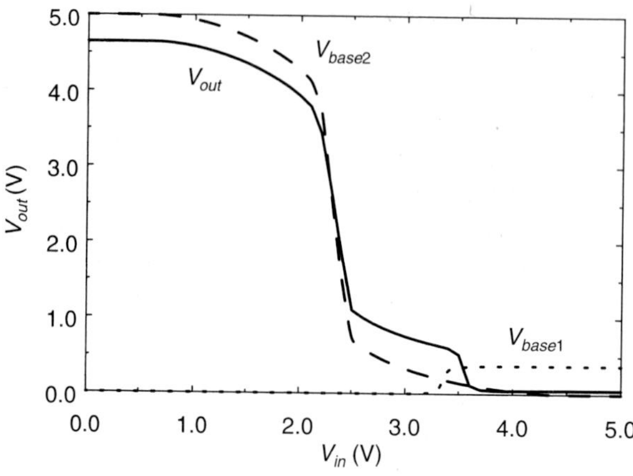

Figure 5.25 Simulated voltage-transfer characteristic for BiCMOS inverter of Figure 5.24b.

in the output voltage around $V_{in} \approx 3.5$ V. Notice, furthermore, that V_{OH} is higher than expected. This results from the fact that Q_2 still carries some emitter current when the base-emitter voltage is smaller than $V_{BE(on)}$. The following dc parameters can be extracted:

$$V_{OH} = 4.64 \text{ V}; V_{OL} = 0.05 \text{ V}$$

$$V_{IL} = 1.89 \text{ V}; V_{IH} = 3.6 \text{ V}$$

$$V_M = 2.34 \text{ V}$$

$$NM_L = 1.84 \text{ V}; NM_H = 1.04 \text{ V}$$

Although the noise margins are not as good as for the CMOS inverter, they are still within the acceptable range. Actually, the projected value of V_{IH} is open for discussion. We could also pick the first break-point in the VTC ($V_{in} \approx 2.5$ V), which yields even better noise margins.

An example of a BiCMOS inverter that does not suffer from a reduced voltage swing is shown in Figure 5.24c. The resistor R_1 (in combination with M_2) provides a resistive path between V_{DD} and V_{out} and slowly pulls the output to V_{DD} once Q_2 is turned off, as demonstrated in Figure 5.26. Full-rail BiCMOS circuits are the subject of active research.

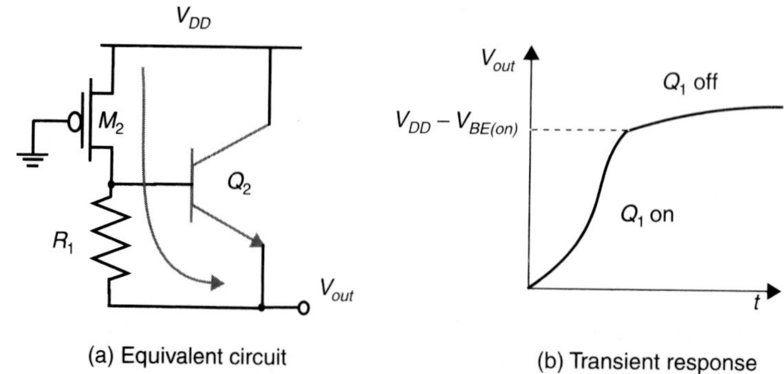

Figure 5.26 Low-to-high transition in full-swing BiCMOS circuit.

5.3.3 Performance of the BiCMOS Inverter

The BiCMOS inverter exhibits a substantial speed advantage over CMOS gates when driving large capacitive loads. This results from the current-multiplying effect of the bipolar output transistors. As in the ECL case, deriving accurate expressions for the propagation delay is nontrivial. The gate consists of a large number of active devices (up to six) and contains a number of internal nodes, each of which could have a dominant effect on the transient response. Although detailed studies have been presented in the literature (e.g., [Rosseel88]), we restrict ourselves to a simplified analysis. This establishes a first-order model for the delay. SPICE simulations can then be used to establish a more quantitative result.

Consider first the low-to-high transition in the circuit of Figure 5.21a. Assume that the input signal is switching very fast and that its rise/fall times can be ignored. After turning off M_1, the impedance Z_1 allows the base charge of Q_1 to drain to ground. Since the transistor was operating in forward-active mode, the stored charge is small, and Q_1 turns off fast. To a first order, we can therefore assume that this has no impact on the propagation delay and that Q_1 is turned off instantaneously. Under those conditions, the transient behavior can be modeled by the equivalent circuit of Figure 5.27a.

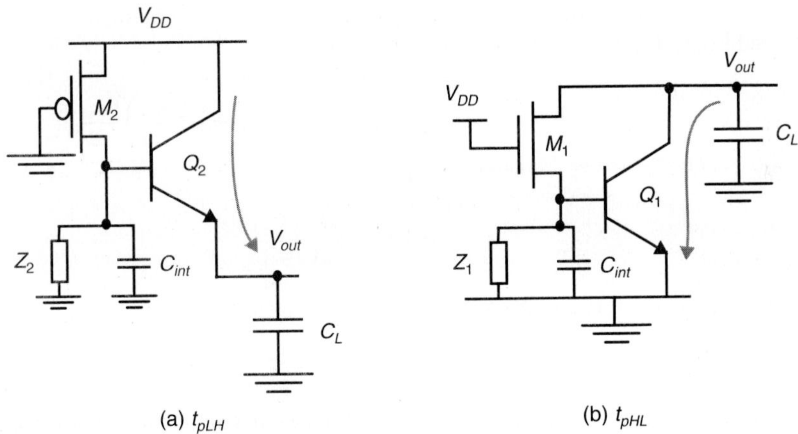

Figure 5.27 Equivalent circuits for transient analysis.

The propagation delay consists of two components. First, the capacitor C_{int} has to be charged to $V_{BE(on)}$ through M_2 to turn on Q_2. Once this point is reached, Q_2 acts as an emitter-follower, and C_L gets charged. Approximative expressions can be derived for both time intervals:

$$t_{turn\text{-}on} = \frac{C_{int}V_{BE(on)}}{I_{charge1}} \tag{5.4}$$

with $I_{charge1}$ the average charging current.

$$I_{charge1} = \frac{I_{M2}(V_{int} = 0) + I_{M2}(V_{int} = V_{BE(on)}) - V_{BE(on)}/Z_2}{2} \tag{5.5}$$

As Z_2 is normally a large resistor, the latter component of the charging current can be ignored. The PMOS device operates in saturation mode in this time interval, providing ample current; therefore, $t_{turn\text{-}on}$ is small.

To compute the second component of the propagation delay, where Q_2 acts as an emitter-follower, we can use the reflection rule (similar to the analysis of the ECL gate) to merge the internal and external circuit nodes into a single node. C_L now appears in parallel with C_{int}, but its value is divided by $(\beta_F + 1)$. This is equivalent to stating that the base current of Q_2 is multiplied by that factor. The corresponding delay is now readily computed:

$$I_{charge} = \frac{\left(C_{int} + \frac{C_L}{\beta_F + 1}\right)\frac{V_{swing}}{2}}{I_{charge2}} \tag{5.6}$$

$I_{charge2}$ equals the average charging current during that interval. This consists primarily of the current through M_2 (ignoring the current loss through Z_2). The value of V_{swing} is determined by gate topology, but normally equals $V_{DD} - 2\,V_{BE(on)}$, as was apparent from the dc analysis. The value of $I_{charge2}$ is comparable to the average PMOS charging current, as observed in a CMOS inverter with similar-size devices.

The overall value of the low-to-high propagation delay is obtained by combining Eq. (5.5) and Eq. (5.6).

$$\begin{aligned} t_{pLH} = t_{turn\text{-}on} + t_{charge} &= \frac{C_{int} V_{BE(on)}}{I_{charge1}} + \frac{\left(C_{int} + \frac{C_L}{\beta_F + 1}\right)\frac{V_{swing}}{2}}{I_{charge2}} \\ &= a \times C_{int} + \frac{b \times C_L}{\beta_F + 1} \end{aligned} \tag{5.7}$$

This delay consists of two components:

1. A fixed component that is proportional to C_{int} and is normally small. C_{int} is a lumped capacitance, composed of contributions of the PMOS device (diffusion capacitance) and the bipolar transistor (*be*- and *bc*-junction capacitance and base-charge capacitance).

2. The second component is proportional to the load capacitance C_L. The loading effect is substantially reduced by the $(\beta_F + 1)$ current multiplier introduced by the bipolar transistor.

It is interesting to compare this result with the delay of a CMOS inverter, assuming similar-size MOS transistors. The following linear approximation of the delay of the CMOS inverter is valid.

$$t_{pLH(CMOS)} = c \times C_{int} + d \times C_L \tag{5.8}$$

In comparing Eqs. (5.7) and (5.8), we realize that the values of the coefficients are approximately equal ($a \approx c$ and $b \approx d$), as determined by the current through the PMOS and the voltage swing, which are of the same order in both designs. C_{int} is substantially larger in the BiCMOS case due to the contributions of the bipolar device. These observations allow us to draw an approximate plot of t_{pLH} versus the load capacitance C_L for both the CMOS and BiCMOS gates (Figure 5.28).

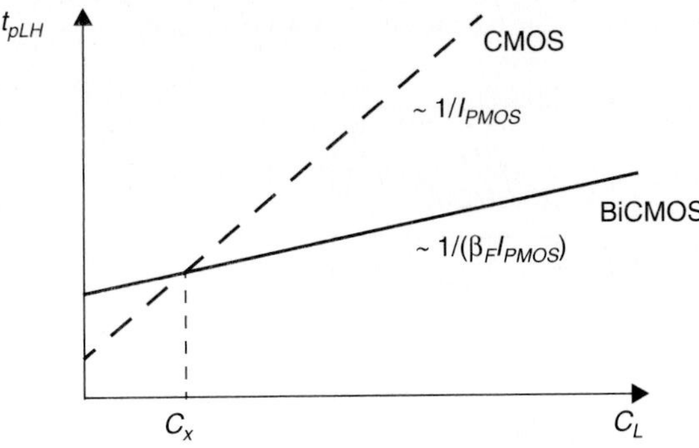

Figure 5.28 Propagation delay of BiCMOS and CMOS gates as a function of C_L.

For very low values of C_L, the CMOS gate is faster than its BiCMOS counterpart due to the smaller value of C_{int}. For larger values of C_L, the bipolar output transistors easily provide the extra drive current, and the BiCMOS gate becomes superior. Although the cross-over point C_x is technology-dependent, it typically ranges from $C_L \approx 50$ to 250 fF. As a result, BiCMOS inverters are normally used as buffers to drive large capacitances. They are not very effective for the implementation of the internal gates of a logic structure (such as an ALU), where the associated load capacitances are small. One must also remember that the complexity of the BiCMOS gate incurs an important area overhead. Consider carefully when and where to use BiCMOS structures.

A similar analysis holds for the high-to-low transition. It is assumed that Q_2 turns off instantaneously, as its base charge is quickly removed through Z_2. The resulting equivalent circuit is shown in Figure 5.27b. Once again, the delay consists of two contributions:

1. Turning on Q_1. This requires the charging of the internal capacitance C_{int} through the NMOS device.

2. Discharging C_L through the combined network of NMOS and bipolar transistor. Ignoring the current loss through Z_1, all the drain current of M_1 sinks into the base of Q_1. Assuming forward-active operation, this results in a collector current β_F times larger. The total discharge current equals $(\beta_F + 1)\, I_{NMOS}$.

Hence, the following approximative expression is valid

$$t_{pHL} = t_{turn\text{-}on} + t_{discharge} = \frac{C_{int} V_{BE(on)}}{I_{charge3}} + \frac{\left(C_{int} + \frac{C_L}{\beta_F + 1}\right)\frac{V_{swing}}{2}}{I_{charge4}} \tag{5.9}$$

Eq. (5.9) closely resembles the one derived for the t_{pLH}. It is worth mentioning that C_{int} is not constant and changes between turn-on and discharge modes. An average value over the complete operation range produces acceptable first-order results.

Example 5.8 Propagation Delay of a BiCMOS Inverter

The propagation delay of the BiCMOS buffer of Example 5.7 is simulated using SPICE for a load of 1 pF. The result is plotted in Figure 5.29 and compared with the performance of the CMOS inverter of Example 3.7 (for a similar load). The propagation delay of 0.86 nsec for the BiCMOS gate compares favorably to the 6.0 nsec of the CMOS inverter.

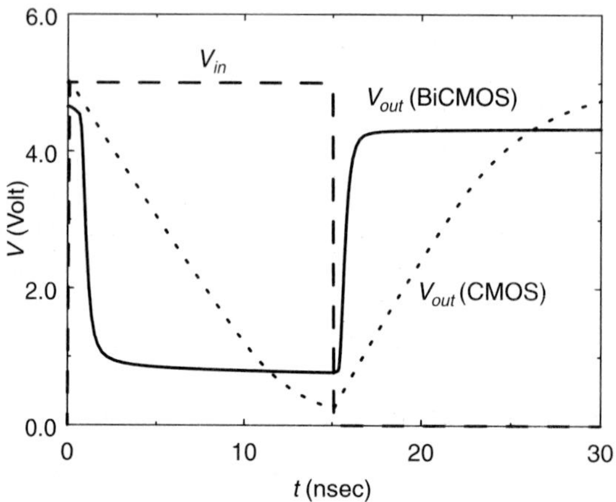

Figure 5.29 Transient response of BiCMOS and CMOS inverters for a load of 1 pF.

Notice the reduced voltage swing of the BiCMOS gate. The loss at both the high and low levels is, however, substantially smaller than the 0.7 V ($V_{BE(on)}$) suggested by the first-order model and is approximately equal to 0.4 V. For very low capacitive loads, the CMOS gate is approximately 5.5 times faster than its BiCMOS counterpart. This is illustrated in Figure 5.30, where the propagation delays of the CMOS and BiCMOS gates are plotted as a function of C_L. The cross-over point, where BiCMOS becomes faster than CMOS, is situated around 100 fF. Notice that for C_L values below 1 pF the propagation delay of the BiCMOS gate is virtually independent of the load capacitance. For those load values, the capacitance of the internal nodes (C_{int}) dominates the performance, and the factor attributable to C_L is negli-

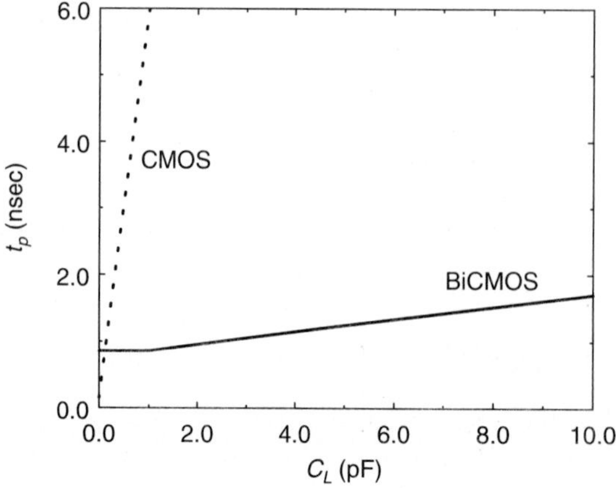

Figure 5.30 Simulated propagation delays of CMOS and BiCMOS gates as a function of C_L.

gible. The measured slope of the CMOS curve is approximately 64 times steeper, which is somewhat lower than the expected value of $\beta_F + 1$ (or 101). The discrepancy is due to a number of inefficiencies in the BiCMOS gates, such as the V_{BE} losses.

The analysis derived above is correct as long as the current flowing through the bipolar transistors is limited. Large currents might adversely affect the speed of the gate due to the second-order effects listed below.

- **Collector Resistance R_c**—The equivalent circuits of Figure 5.27 ignore the presence of the collector resistance r_c between the extrinsic collector contact and the intrinsic collector-base junction. The voltage drop over r_c causes the transistor to saturate even though the extrinsic V_{CE} is larger than 0.7 V, as is guaranteed by the BICMOS buffer design. For instance, a collector resistance of 100 ohms conducting a transient current of 1mA causes a voltage drop of 0.1 V. When driving large capacitive loads, currents in excess of 5 mA are regularly observed. The transistor accordingly saturates, causing a deterioration of the propagation delay; t_p is then composed of the time to get the transistor into saturation, followed by a discharging of the load capacitance with a time constant $r_C C_L$ (Eq. (5.10)). This problem can be avoided by increasing the size of the transistor, decreasing r_C.

$$t_{p(HL,LH)} = t_{turn\text{-}on} + t_{sat} + \alpha r_C C_L \tag{5.10}$$

- **High-Level Injection—**This effect occurs when the density of electrons transported across the collector-base space is comparable to the doping of the collector. The resulting *base push-out* effectively increases the width of the base and degrades the switching performance of the transistor. A typical parameter to quantify the onset of high-level injection is the *knee current* I_k, which is (in practice) the value of the collector current at which the forward current gain β_F is reduced to 50% of its value. Similar to the degradation caused by the collector resistance, high-level injection effects can be avoided by increasing the emitter area of the transistor.

5.3.4 Power Consumption

The BiCMOS gate performs in the same manner as the CMOS inverter in terms of power consumption. Both gates display almost no static power consumption, while the dynamic dissipation is dominated by the (dis)charging of the capacitors. When driving small loads, the latter factor is slightly larger for the BiCMOS gate, due to the increased complexity of the gate. On the other hand, when driving very large capacitors, BiCMOS becomes favorable. To achieve comparable performance, CMOS drivers consist of a cascade of gradually increasing inverters (discussed in Chapter 8). The power dissipated in charging the internal capacitances becomes an important fraction of the overall consumption. This cascading is, generally, not needed in BiCMOS.

The short-circuit currents during switching might be smaller or larger for BiCMOS, depending upon the level of circuit optimization. The superior current-driving capabilities of the bipolar transistors produce steeper signal slopes and, consequently, a faster transition through the transition region. This potential advantage is, however, easily annihilated by intrinsic RC delays in the gate. A small differential delay might cause the bipolar transistors to be on simultaneously for a longer time, causing a large direct current to flow (remember the high transconductance of the bipolar transistors). All in all, only precise simulations that include parasitic capacitances and resistances can tell exactly which gate is more power efficient.

5.3.5 Technology Scaling

Because the BiCMOS technology is a merger of CMOS and bipolar technologies, the scaling behavior of the BiCMOS gate is determined by the combined properties of both technologies. Scaling down the dimensions generally results in improved performance.

Unfortunately, the BiCMOS gate inherits one of the most important deficiencies of the bipolar technology: **built-in voltages such as $V_{BE(on)}$ do NOT scale.** The performance of the BiCMOS gate suffers in an important way when the supply voltage is reduced. Consider the equivalent circuit of Figure 5.27b. The current through M_1 during the discharging of C_L is proportional to $(V_{GS} - V_T) = (V_{in} - V_{BE(on)} - V_T)$. V_{in} itself suffers from a $V_{BE(on)}$ loss for most BiCMOS gate topologies. Therefore, $I_{NMOS} \sim (V_{DD} - 2V_{BE(on)} - V_T) \approx (V_{DD} - 2.2\text{ V})$. This leaves ample current drive for $V_{DD} = 5$ V. For lower supply voltages, a substantial degradation in performance can be observed that eventually causes the CMOS gate to be faster even for large capacitive loads. This is illustrated in Figure 5.31, where the propagation delays of CMOS and BiCMOS gates are plotted as a function of V_{DD} (for two different technologies; see [Raje91]). We can see that using BiCMOS does not make much sense for supply voltages below 3 V.

This deficiency ultimately hampers the future usefulness of BiCMOS, because voltage scaling is an absolute necessity for submicron devices. The conception of a low-voltage BiCMOS structure is currently a hot research topic [Embabi93].

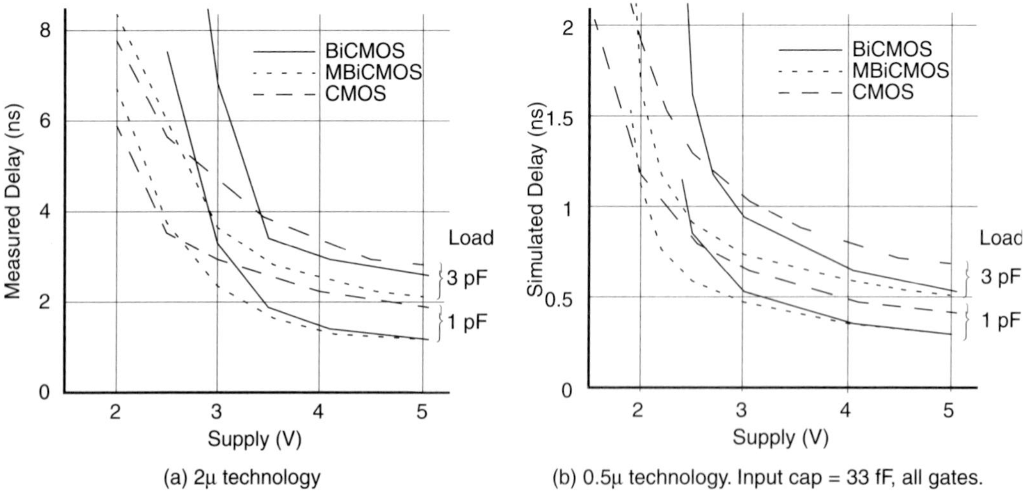

Figure 5.31 Propagation delays of CMOS and BiCMOS gates as a function of V_{DD} for 2.0 and 0.5 μm technologies (from [Raje91]). The BiCMOS gates analyzed are the structures of Figure 5.24a (called BiCMOS) and 5.24b (called MBiCMOS).

5.3.6 Designing BiCMOS Digital Gates

The analysis of a number of industrial BiCMOS designs rapidly reveals that the BiCMOS gate is almost uniquely used for *buffering or driving purposes*. When driving large fan-out, high-capacitive busses, and off-chip signals, the bipolar output stage helps to provide large currents, using only a small area and consuming less power compared to the CMOS buffer. Therefore, the BiCMOS design approach has its major impact in circuits such as memories and gate arrays where large capacitive loads are common. The topic of driving large capacitances is discussed in detail in Chapter 8, while memories and gate arrays are treated in Chapters 10 and 11.

The limited usage of the bipolar device seems to be a waste of a valuable resource. Once the step is made to the more expensive BiCMOS technology, it is worth exploiting its capabilities to a maximal degree. This requires some rethinking of traditional design approaches, which might explain the reluctance of the designer to freely mix MOS and bipolar transistors in a design. The proliferation of BiCMOS circuit into the design of functions such as ALUs is also hampered by the reduced packing density. MOS devices of the same type can be placed in the same well, which means that the distances between the devices are short. On the other hand, bipolar transistors must be placed in separate n-wells, which significantly reduces the device density. This constraint can be somewhat relaxed by merging *npn* transistors and PMOS devices in the same well.

Extending the basic BiCMOS inverter gate to more complex logic operations is straightforward. The logic function only affects the CMOS part of the gate, while the bipolar output circuitry remains unchanged. An example of a two-input NAND gate is shown in Figure 5.32. Both pull-up and pull-down networks are implemented using the traditional CMOS approach. Extension to other gates is trivial.

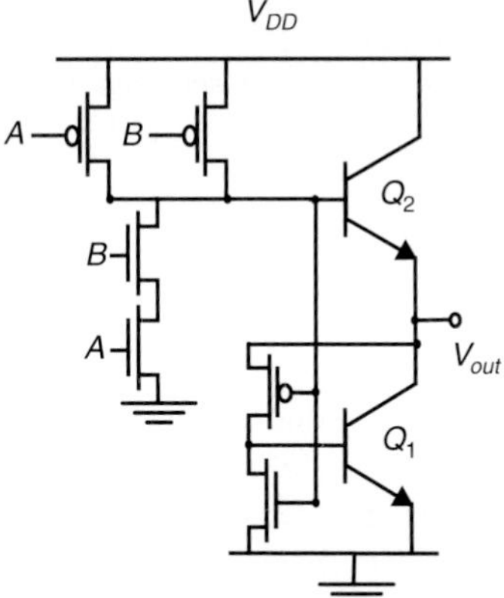

Figure 5.32 Two-input BiCMOS NAND gate.

The most important issue is to determine when it is beneficial to employ such a gate in a combinational circuit. As established above, the BiCMOS approach is advantageous for a large load capacitance (or, equivalently, fan-out). An additional advantage is that the bipolar output stage isolates the internal, logic nodes of the gate from the load. This allows for an optimization of the MOS logic transistors based on the gate topology *only* (for instance, using a progressive sizing), regardless of the load, because the latter is taken care of by the bipolar output buffer. This addresses a major disadvantage of the complex CMOS gate, as pointed out in Chapter 4. For a BiCMOS structure to be the gate of choice requires either a large fan-out, or a complex gate. For instance, it has been shown [Rosseel88] that the BiCMOS two-input NAND gates becomes superior over its CMOS equivalent for a fan-out of four gates. This crosspoint comes even earlier for more complex gates.

Some caution is advisable. Due to the lower packing density of the bipolar transistor and the more complex nature of the gate, an area penalty is involved. Only when speed is an issue (for instance on the critical path of a design) and the load capacitance is substantial should a BiCMOS gate be considered.

More innovative structures are possible as well. One possible application of a mixed bipolar-MOS design has already been shown in Figure 5.14, where a MOS transistor serves as an active pull-down in an ECL gate. Another possible option is to use bipolar amplifiers to speed up the performance of traditional CMOS circuits. For instance, a faster circuit can be conceived by reducing the signal swing in the critical path and propagating that reduced swing to the next stage through a bipolar amplifier [Rosseel89]. The larger transconductance of the bipolar device makes it more sensitive to smaller signal swings. Getting a similar gain requires substantially larger MOS devices.

Example 5.9 BiCMOS Inverter Layout

Figure 5.33 shows the layout of a BiCMOS inverter/driver using the circuit topology of Figure 5.24a (from [Embabi93]). Observe how the PMOS transistor M_2 and the bipolar device Q_2 are merged in the same n-well. This saves the isolation area that would be needed if two separate wells were used. Still, the bipolar output stage represents a substantial area overhead. From inspection of the layout, the device sizes can be derived: $M_1 = (32/2)$, $M_2 = (56/2)$, $M_3 = M_4 = (6/2)$ (all sizes in λ). The emitter areas for Q_1 and Q_2 equal $3 \times 10 \times 2\ \lambda^2$. Notice how M_1 and M_2 are implemented as two parallel transistors, while the emitters of the bipolar devices are implemented as three separate strips. This helps to adapt the aspect ration of the cell.

It is left as an exercise for the reader to simulate the steady-state and transient performance of the cell.

5.4 Digital Gallium Arsenide Design *

The combination of the latest manufacturing technology and advanced circuit design makes it possible to realize inverters with propagation delays of around 20 psec in silicon bipolar. Once again, treat this value with caution, as it is obtained in an ideal structure such as a ring oscillator (with a fan-out of 1). The actual gate delay that can be achieved in actual designs is at least twice as large, and more often many times higher. When faster switching speeds are required such as in the next generation of super-computers or in the front-end of advanced radio-telecommunication devices, silicon-based designs run out of steam.

This does not mean that going faster is out of the question. Other semiconductor materials, such as gallium arsenide (GaAs) and silicon-germanium (SiGe), have switching properties that exceed the performance of silicon. In the next sections, some attention is

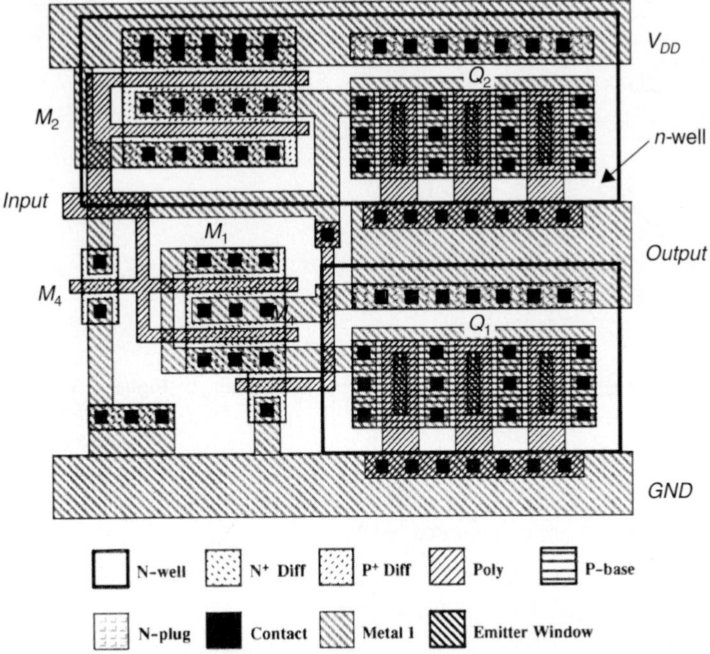

Figure 5.33 Layout of BiCMOS inverter of Figure 5.24a (from [Embabi93]).

devoted to the design of digital gates in these technologies. Although these approaches represent only a small fraction of the digital design market, it is valuable to have an impression of how digital design for very high speed is conducted. We will limit ourselves to a discussion of GaAS-based design.

5.4.1 GaAs Devices and Their Properties

GaAs Material Properties

The performance of submicron silicon MOS devices is constrained by the maximum electron-drift velocity υ_{sat}, which approximately equals 10^7 cm/sec. As this is an intrinsic property of the material, faster switching speeds can only be achieved by a scaling of the technology or by exploring other semiconductor materials. An example of the latter is gallium-arsenide, which is a compound semiconductor material that has the intrinsic capability of being approximately twice as fast as silicon. Figure 5.34 plots the carrier velocity of electrons and holes in both GaAs and Si as a function of the electrical field. For lower field strengths, the velocity is proportional to the field for all carriers. For higher field values, the carrier velocity saturates to approximately 10^7 cm/sec, independent of the material or the carrier. The most important lesson to be learned from this graph is that at lower values of the electrical field, GaAs electrons display a higher velocity, peaking at 2×10^7 cm/sec before dropping to the saturation value. The velocity increase is due to the lower *effective mass* m_e of the GaAs electrons compared to Si. When operated at low electrical fields,

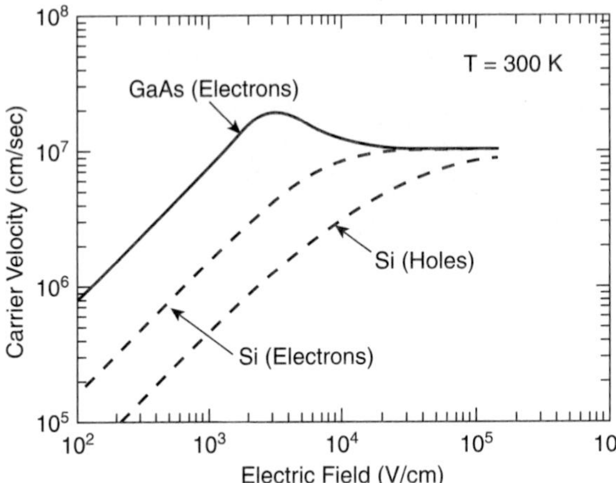

Figure 5.34 Measured carrier velocity versus electric field for Si and GaAs [Sze69].

GaAs has the capability of being substantially faster than Si. A number of the important properties of GaAs are enumerated below.

- When operated at low electrical fields, *n*-type GaAs circuits can be *twice as fast as silicon circuits*. To exploit this feature requires operation at low voltages (around 1 V).
- This difference becomes even more significant for *light doping levels*, where the electron mobility can reach 8000 to 9000 cm^2/Vsec at room temperature. This is approximately 10–20 times higher than silicon. Special devices such as the HEMT (*high electron mobility transistor*) have been developed to exploit this feature. These structures produce some of the fastest logic available, especially at lower temperatures.
- Unfortunately, holes in GaAs do not exhibit equally desirable properties. The hole velocity is approximately 15–20 times lower compared to the GaAs electron. This means that the *complimentary structures are not as desirable* in GaAs as they were in Si.
- Due to very high levels of surface-state charge, structures like *metal-oxide-semiconductor transistors are not possible*. Most GaAs designs, therefore, make use of MESFET (metal-semiconductor field-effect transistor) devices, which are introduced in the next section.
- Pure GaAs is *semi-insulating* with a resistivity between 10^7 and 10^9 Ω·cm at room temperature. This means that devices made of doped GaAs can be isolated from each other by the insertion of undoped material, although additional isolation can be provided by selective ion implantation. This is more area-effective than the field-oxide approach in CMOS. It also has the advantage of reducing the parasitic capacitance.
- Due to a larger band-gap and the semi-insulating substrate, GaAs has the advantage of being more *immune to radiation effects*. It is therefore attractive for space and

military applications where it is in direct competition with silicon-on-insulator technologies.

- Finally, but most importantly, GaAs is an extremely brittle and fragile material. For this particular reason, GaAs wafers tend to be no larger than three inches, whereas six-inch and larger silicon wafers are common. This results in a *reduced manufacturing efficiency.* Moreover, getting a reasonable yield has been challenging due to the high defect density in the basic material and the tight device requirements.

The MESFET Device

As mentioned previously, the lack of a MOS-style device has made the MESFET (metal-semiconductor FET) the device of choice in GaAs design. A cross-section of an *n*-MESFET is shown in Figure 5.35. It consists of a conductive *n*-type surface channel with

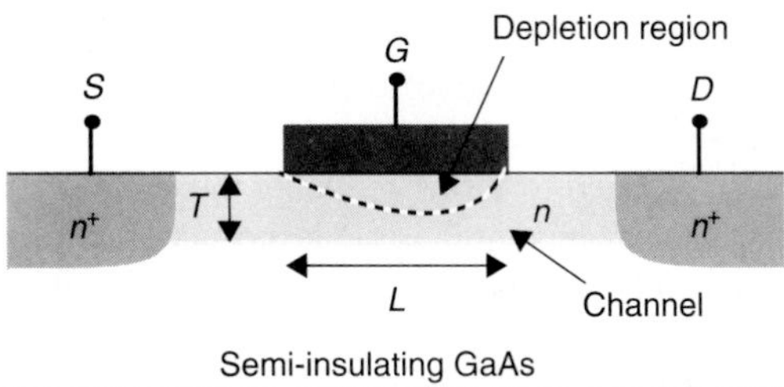

Figure 5.35 Cross-section of a GaAs MESFET.

a thickness *T*, located between two n^+ ohmic contacts that act as source and drain. Semi-insulating GaAs is used as the substrate material. The device control terminal (the gate) is implemented by depositing a metal (typically Ti/Pd/Au, although Al, W, and Pt alloys work as well) on a section of the channel, so that a *Schottky-barrier diode is created.* A Schottky diode is a metal-semiconductor junction formed by depositing a small metal contact onto a lightly doped *n*-type semiconductor. The resulting diode relies on single-carrier conduction, what results in fast switching times. Appendix D offers a more detailed discussion of this device.

The principles of operation and the basic relationships governing the device are similar to those of the silicon-junction FETs (JFET) and MOSFETs. The MESFET is a unipolar transistor, which means that conduction is dominated by one type of carrier, in this particular case the electron. Its operation is can be understood from the following qualitative analysis:

- Under zero-bias conditions ($V_{GS} = 0$), a depletion region is created under the gate due to the built-in voltage of the metal-semiconductor junction, as shown in Figure 5.35. This reduces the number of mobile carriers present in the channel. Typically, the thickness of the depletion layer is smaller than *T*, which means the channel is still conducting. The nominal MESFET transistor is therefore a *depletion* device.

- Applying a positive voltage to the gate reduces the width of the depletion layer. This increases the conductivity of the channel. However, once the gate-channel voltage is sufficiently high, the Schottky diode becomes forward-biased, and current starts flowing into the gate. As no further increase in conductivity occurs from that point on, this condition is to be avoided in general. The voltage at which forward conduction of the gate occurs depends upon the gate metal and ranges around 0.5–0.7 V.

- When the gate voltage is reduced, the depletion region extends, and the channel conductivity drops owing to the smaller cross-sectional area of the conducting channel. This trend continues until the depletion region extends through the conductive layer (or $W_{depl} > T$) and pinches off the channel. At that point, the transistor is turned off as the channel conductivity drops to 0. The gate-source voltage needed to make this happen is called the *pinch-off voltage* V_P. The pinch-off voltage is analogous to the threshold voltage in a MOS transistor and is a function of the thickness of the channel, the doping level, and the built-in voltage of the Schottky diode. Typical values of V_P for the depletion devices range from –0.7 V to –2.5 V.

As the depletion device is the generic MESFET GaAs transistor, it is tempting to build logic gates using only these devices, as was the case in the earlier days of GaAs digital design. Unfortunately, such a design approach requires the availability of multiple supply voltages—the gate-source voltage has to be negative to turn the device off, while the drain-source voltage has to be positive for conduction.

An enhancement device can be realized by either reducing the channel thickness or by ion implantation, ensuring that channel pinch-off is achieved under zero-bias conditions. The logic swing for an enhancement-based logic gate is limited to the difference between the pinch-off voltage, which normally ranges between 0 and 0.2 V, and the voltage at which the Schottky-barrier diode begins to conduct (around 0.7 V). Under these low swing conditions, small variations of the pinch-off voltage can have a dramatic impact on the circuit's functionality or performance. The realization of complex circuits, therefore, requires an accurate control of the channel thickness over the complete wafer to minimize pinch-off voltage variations.

The device model for the GaAs MESFET is remarkably similar to the MOSFET one. An extra factor is included to incorporate the effects of velocity saturation.

$$I_D = \begin{cases} 0 \text{ for } (V_{GS} < V_P) \\ \beta(V_{GS} - V_P)^2(1 + \lambda V_{DS})\tanh(\alpha V_{DS}) \text{ for } (V_{GS} > V_P) \end{cases} \tag{5.11}$$

This model, called the *Curtice model* after its inventor [Curtice80], includes both the linear and saturation regions and is an empirical fit using the hyperbolic tangent function. The gate diode is modeled by the traditional diode equation

$$I_D = I_S(e^{V_G/n\phi_T} - 1) \tag{5.12}$$

A later model, named Raytheon, improved the Curtice model on two fronts: (1) improved I_D versus V_{GS}, and (2) better capacitance models. The parameters for some state-of-the-art GaAs devices are given in Table 5.2. For the same process, the threshold voltages for the enhancement and depletion devices can vary between (0.18 V … 0.3 V) and (–0.735 V … –0.92 V), respectively. To obtain the actual values for a particular transistor, the β and I_S values have to be multiplied by the device ratio (W_{eff} / L_{eff}) and the effective gate area ($W_{eff} \times L_{eff}$), respectively. W_{eff} and L_{eff} stand for the effective transistor width and length.

$$\begin{aligned} W_{eff} &= W - \Delta W \\ L_{eff} &= L - \Delta L \end{aligned} \tag{5.13}$$

For the process presented here, ΔL and ΔW equal 0.4 μm and 0.15 μm, respectively.

Table 5.2 Typical transistor parameters for a 1.0 μm GaAs process.

	β (A/V²)	V_{P0}(V)	λ (1/V)	α (1/V)	I_S (A)	n
Enhancement	250×10^{-6}	0.23	0.2	6.5	0.5×10^{-3}	1.16
Depletion	190×10^{-6}	–0.825	0.0625	3.5	10^{-2}	1.18

Example 5.10 GaAs MESFET Current-Voltage Characteristics

The simulated voltage-current characteristics of a (4 μm/1 μm) GaAs MESFET enhancement transistor, implemented in the GaAs process of Table 5.2, are plotted in Figure 5.36. The most important feature differentiating the device from the MOSFET transistor is the presence of the Schottky diode between gate and channel. From the V_{GS}-I_D curve, we can see that this diode becomes forward-biased once V_{GS} approximately equals 0.75 V. This means that for low values of V_{DS}, the drain current actually becomes negative, as observed in the I_D-V_{DS} curves (encircled). Notice also that the device is velocity-saturated for most of the voltage range of interest.

We can see a close correspondence between the simulated results and the model of Eq. (5.11). For instance, for $V_{GS} = 0.5$ V and $V_{DS} = 2$ V, the Curtice model Eq. (5.11) yields the following current value:

$$\begin{aligned} I_D &= \left(\frac{4-0.15}{1-0.4}\right) \times 250 \times 10^{-6}(0.5-0.23)^2(1+0.2\times 2)\tanh(6.5\times 2) \\ &= 163.7\ \mu\text{A} \end{aligned} \tag{5.14}$$

which is close to the simulated value.

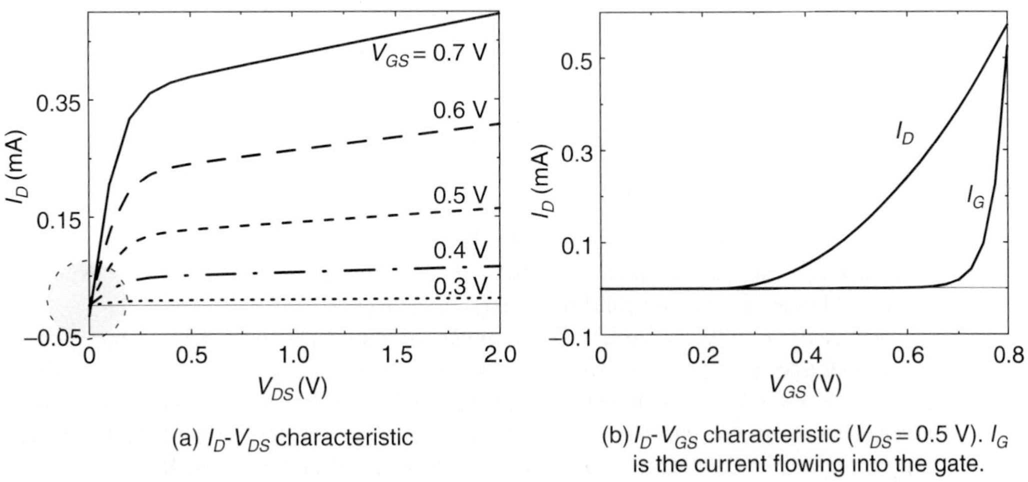

Figure 5.36 Current-voltage characteristics of GaAs enhancement transistor (W= 4 μm, L = 1 μm).

The Curtice model is by no means the ultimate in the modeling of GaAs MESFETs. More advanced models include effects such as drain-induced threshold variations, as well as more complex curve-fitting techniques.

The HEMT Device

While the MESFET is used in the majority of the GaAs digital designs, the HEMT (High Electron Mobility Transistor) is the device of choice when extreme performance is required. The cross-section of such a device is shown in Figure 5.37. Its operation relies on

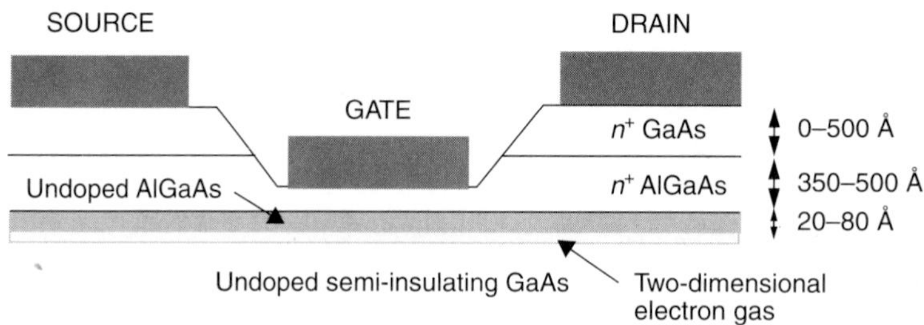

Figure 5.37 Cross-section of AlGaAs/GaAs high electron mobility transistor (HEMT) (from [Dingle78]).

the fact that mobility of the carriers is much higher in an undoped region than in a doped material. The HEMT structure separates the donor regions (n^+ AlGaAs) that produce the electrons, but impede high mobilities, from the conducting channel (undoped GaAs) with its very high mobility. AlGaAs is selected as donor material because it has a wider bandgap (1.8 eV) than GaAs (1.4 eV). This causes free electrons from the ionized donors to diffuse to the undoped material due to the electron's inherent affinity to move to the lower bandgap region. Electron mobilities of 8500 cm^2/Vsec can be achieved in HEMT transistors, compared to the channel mobilities of 4500 cm^2/Vsec in GaAs MESFETs (at 300 K). The situation is even more extreme at lower temperatures (e.g., 77 K, the temperature of liquid nitrogen), where impurity scattering is the dominant mechanism limiting carrier velocity. Mobilities of up to 50,000 cm^2/Vsec have been obtained for HEMT devices operating in this temperature range.

From an operation point of view, the device of Figure 5.37 belongs to the class of the MESFETs with the gate Schottky diode formed by the junction of the gate metal and the n^+AlGaAs layer. This diode has the advantage of having a larger turn-on voltage (~ 1V) than its GaAs counterpart, which provides larger noise margins. Depletion and enhancement devices can be constructed. Consequentially, the GaAs MESFET gate structures described below are just as applicable to HEMT devices.

In addition to the MESFET HEMT, other structures have been devised that demonstrate extreme performance, such as the *heterojunction bipolar transistor* (HBT) [Asbec84]. A discussion of these devices would lead us too far astray. It suffices to say that heterojunction devices provide the highest performance at present (barring superconducting gates) and are intensively used in the most demanding applications, such as radio front-ends operating in the high GHz range.

5.4.2 GaAs Digital Circuit Design

Buffered FET Logic

The design of reliable digital circuits in the GaAs MESFET technology has proven to be quite a challenge. A first approach is to use *depletion devices only*. An example of such a gate, implemented in the so-called buffered FET logic (BFL) logic style, is shown in Figure 5.38. Two supply voltages, V_{DD} and V_{SS} (4 V and –2.5 V, respectively) are needed.

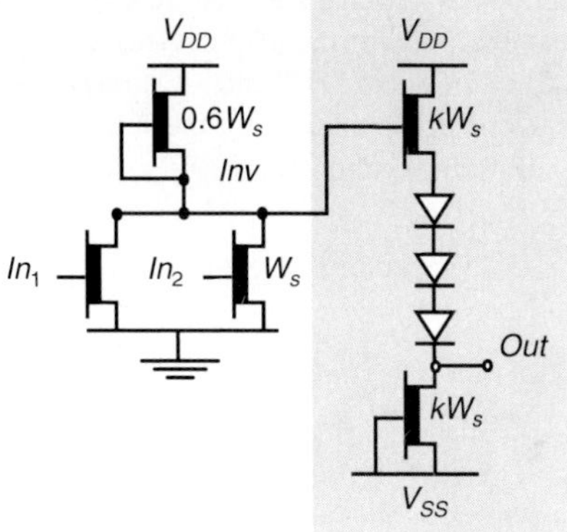

Figure 5.38 Two-input NOR gate in buffered FET logic (BFL). The italic numbers indicate the relative transistor sizes. Notice also the symbols used for the depletion MESFET devices.

The first stage implements the logic function, in this case a two-input NOR gate, and is similar to a traditional depletion-load NMOS gate (see Chapter 4). The main difference is that the pull-down devices are depletion transistors as well, requiring negative input levels to turn them off. The low input level has to be lower than V_P. On the other hand, V_{OH} cannot be higher than $V_{D(on)}$, which is the turn-on voltage of the Schottky diode. The output of the depletion-load inverter is located between *GND* and V_{DD}. A source-follower output stage with level-shifting diodes is inserted to adjust these levels so that all stated requirements are met. This output stage has the additional advantage of making the performance of the gate relatively insensitive to fan-out loading or capacitive loads.

This structure has proven to be relatively insensitive to processing and power-supply variations, which is useful when the processing is not well controlled.

Example 5.11 Parameter Variations in BFL

The impact of variations in the pinch-off voltage of the MESFET devices on the dc parameters of a nine-input BFL NOR gate was examined in [Milutinovic90]. Changing the threshold from –1.25 V to –2.0 V causes only a 0.6 V shift in the switching threshold V_M, while V_{OH} and V_{OL} change by at most 0.2 V. A further change in the pinch-off voltage to –2.2 V causes the gate to fail.

The structure also suffers from three major disadvantages:

1. It is based on ratioed logic, which means rise and fall times can be very different.
2. The power consumption is high. The dissipation per gate is typically between 5 and 10 mW, most of which can be attributed to the output stage. This prevents its use in large-scale designs (> 2000 gates).
3. It uses two supply voltages, which is not attractive from a system perspective.

Example 5.12 DC Characteristics of the BFL Inverter

A BFL inverter is designed using the depletion devices characterized in Table 5.2. All devices have a (*W/L*) ratio of (4 μm/1 μm) with the exception of the load device, which is made 0.6 times smaller. The supply voltages V_{DD} and V_{SS} are set at 3.5 V and –2 V, respectively.

The current through the inverter stage is approximated by the following expression, assuming that both devices are on and that all Schottky diodes are off:

$$\left(\frac{3.85}{0.6}\right)(V_{in} + 0.825)^2(1 + 0.0625V_0)\tanh(3.5V_o)$$

$$= \left(\frac{2.25}{0.6}\right)(0.825)^2(1 + 0.0625(3.5 - V_o))\tanh(3.5(3.5 - V_o))$$

Solving this equation for various values of V_{in} yields the voltage-transfer characteristic of the input stage. Finding this solution is made complex by the presence of the transcendental functions. This can be addressed by using recursive equations solvers. For instance, for V_{in} = 0V, a V_{out} of 0.25 V is found, which is extremely close to the simulated value. The results of a dc analysis are found in Figure 5.39 for a fan-out of one identical gate. The simulation plots the output of the inverter as well as the buffer stages. Two issues are worth raising:

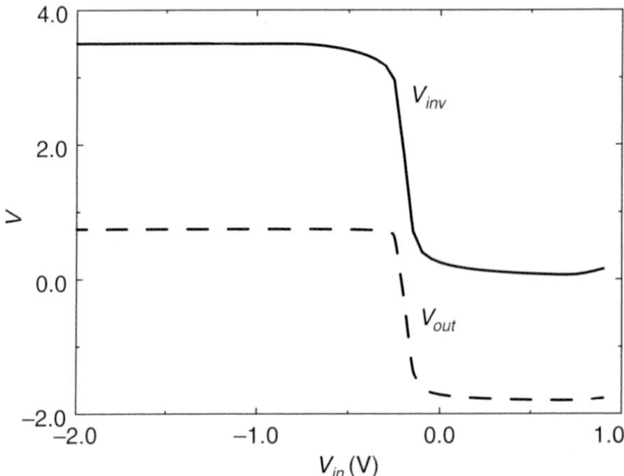

Figure 5.39 Voltage-transfer characteristic of BFL inverter (V_{DD} = 3.5 V, V_{SS} = –2 V).

1. For input values lower than the threshold of the input transistor, the output of the buffer stage (V_{inv}) is high and equals 3.5 V. Once the input device is turned on, the output starts to drop until a value of 70 mV is reached for an input of 0.6 V. Raising the input even more turns on the input diode, and the output starts to rise.

2. V_{out} tracks V_{in} fairly accurately. The high value of V_{out} is set by the input diode of the fan-out gate, which turns on around 0.75 V.

The following gate characteristics can be derived from the simulation:

$$V_{OH} = 0.75\ \text{V},\ V_{OL} = -1.8\ \text{V},\ V_{IH} = -0.06\ \text{V},\ V_{IL} = -0.25\ \text{V},\ V_M = -0.2\ \text{V}$$
$$NM_H = 0.81\ \text{V},\ NM_L = 1.55\ \text{V}.$$

Direct-Coupled FET Logic

One way to avoid the dual supply voltages is to use a combination of depletion and enhancement devices. The latter are used for the implementation of the logic function, while the depletion transistors serve as loads. The resulting structure is, not surprisingly, the depletion-load inverter, which is well known from the early MOS digital design era. In GaAs jargon, such a structure is called a direct-coupled FET logic (DCFL) gate, an example of which is shown in Figure 5.40. Some major differences from its MOS counterpart are worth mentioning.

- The value of the high input voltage is limited by the onset of the gate conduction in the pull-down transistor, which ranges between 0.6 and 0.7 V.
- The low level is dangerously close to the threshold voltage of ± 0.1 V. This requires a strict control of the pinch-off voltage.
- Furthermore, the gate inherits all the bad properties of the depletion-load gate, such as asymmetrical transient response and static power consumption.

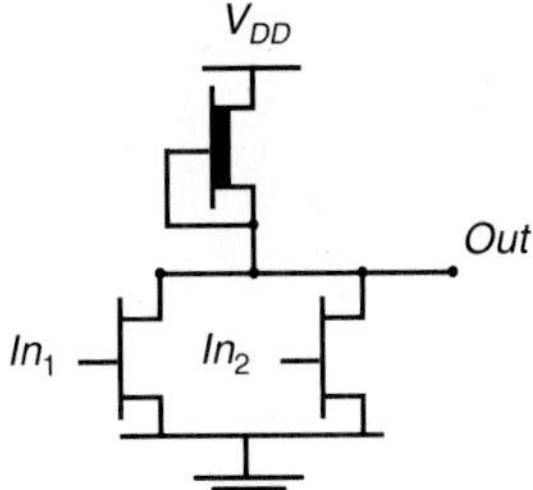

Figure 5.40 Two-input NOR gate in direct-coupled FET logic (DCFL).

While achieving propagation delays similar to BFL, the DCFL structure consumes substantially less static and dynamic power. This is illustrated in Figure 5.41, which plots the propagation delay of DCFL and BFL gates as a function of the power consumption, all designed in a 1 micron technology. An order of magnitude difference in power dissipation is observed for similar performance, if one manages to keep the threshold under control. The BFL gate, on the other hand, has superior fan-out driving capabilities.

Source-Coupled FET Logic

The concerns about the limitations of FET threshold control in DCFL have prompted the development of another logic family with a wide allowable threshold range. The inspira-

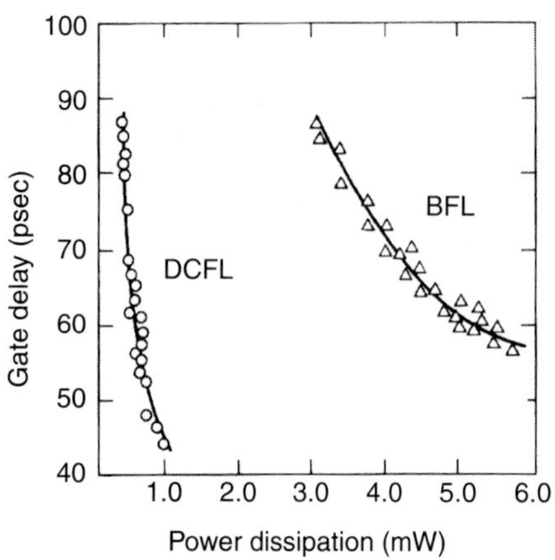

Figure 5.41 Comparing the power consumption and gate delay of DCFL and BFL gates (from [Singh86]).

tion for this family, called source-coupled FET logic (SCFL), can be directly traced to the bipolar ECL structure. It consists of a differential pair and two source-follower output buffers with diode level-shifters (Figure 5.42). Proper operation of the gate requires only that the input transistors of the differential pair be well matched. As with ECL, the power supply noise is reduced, making it possible to operate with small noise margins. All other considerations raised with respect to ECL gates, such as differential versus single-ended, are valid here as well. While SCFL overcomes the tight threshold control associated with DCFL and is intrinsically faster, its power dissipation is higher than DCFL but less than BFL.

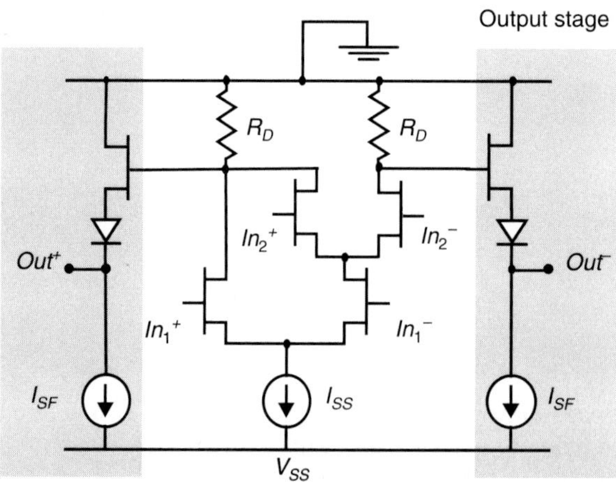

Figure 5.42 Two-input NOR gate in differential source-coupled FET logic (SCFL).

Example 5.13 MESFET Source-Follower

Consider an inverter in SCFL with $V_{D(on)} = 0.7$ V, $V_{OH} = -1.3$ V, $V_{OL} = -1.7$ V. Using the transistor parameters of Table 5.2, determine the value of the I_{SF} such that the voltage drop between the gate and source of the source-follower equals 0.5 V in the midpoint of the voltage transition (assuming that the W_{eff} / L_{eff} of the source-follower devices equals 10).

In the midpoint of the voltage swing, it holds that $V_{out} = -1.5$ V. From the input data, we derive the following data for the source-follower: $V_{DS} = 0.8$ V and $V_{GS} = 0.5$ V. Plugging these numbers into Eq. (5.11) yields a required drain-source current of 0.21 mA. For this current level, the voltage drop over the source-follower is virtually constant over the complete voltage range of interest. Because the transistor operates in the saturation region, a variation of only 9 mV can be observed between the high and the low output levels.

It is left as an exercise for the reader to determine the value of R_D and the sizes of the transistors in the current switch. You may assume that $I_{SS} = I_{SF}$.

GaAs Performance: A Comparison

To put the gates presented here in perspective, Table 5.3 presents the measured performance of a number of GaAs logic families (from [Hodges88]). The table presents the delay for a fan-out of 1 (t_{p0}), the sensitivity to fan-out ($\Delta t_p/FO$), and capacitance ($\Delta tp/C_L$) and the power consumption per gate P.

Table 5.3 Typical performance of GaAs logic families.

Logic Family	t_{p0} (psec)	$\Delta t_p/FO$ (psec/FO)	t_p/C_L (psec/fF)	P (mW/gate)
BFL (1 μm)	90	20	0.67	10
BFL (0.5 μm)	54	12	0.67	10
DCFL (1 μm)	54	35	1.84	0.25
SCFL	BFL range	low	low	~5
DCFL HEMT (0.5 μm - 77 K)	11	7	0.32	1.3

GaAs Design Space

Digital GaAs excels in the domain of extremely fast, small-scale integration components—frequency dividers, counters, (de) multiplexers—where multi-GHz operation has been achieved. For instance, an 8-bit multiplexer implemented in BFL has been demonstrated to run at 3 Gbits/sec. These circuits can be of interest in very high speed communication systems.

Yield issues, as well as power dissipation limitations rapidly come into play when attempting large-scale integration. To demonstrate what can be achieved, consider first the case of the digital multiplier. The average gate/delay as a function of the power dissipation for a number of multipliers is plotted in Figure 5.43. Actual gate propagation delays of 60 psec and 170 psec have been achieved for the HEMT and MESFET technologies, respectively. This translates into multiplication delays of 4 nsec for a 16 × 16 multiplier (at room temperature) with the power dissipation in the multiple (1–6) watt range.

Other larger-scale modules have been implemented, such as SRAM memories (4.1 nsec access time for a 16K memory) and gate arrays (up 3000 gates operating at

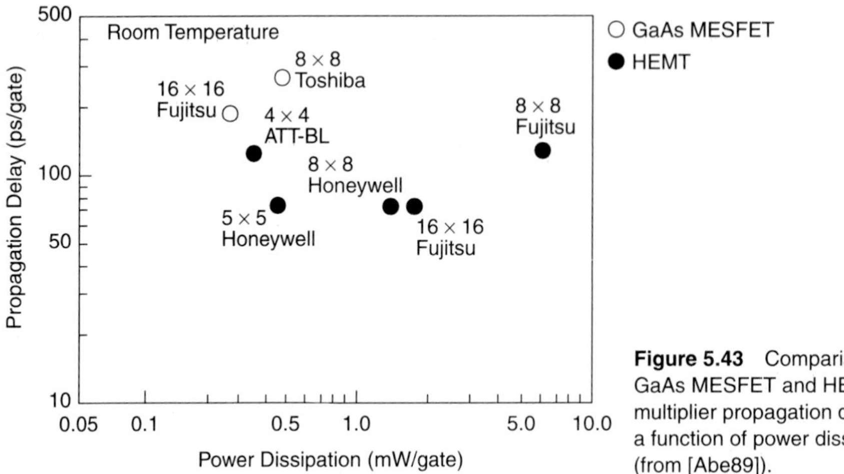

Figure 5.43 Comparison of GaAs MESFET and HEMT multiplier propagation delays as a function of power dissipation (from [Abe89]).

700 MHz). Multiple attempts have been made to use GaAs technology for the implementation of supercomputer and mainframe processors as well as microprocessors, but these efforts have largely proven unsuccessful. Although working prototypes have been demonstrated, manufacturing and economic constraints have prevented these components from reaching the market.

5.5 Low-Temperature Digital Circuits *

An alternative approach to higher performance is to operate the devices at lower temperatures. The carrier mobility in most devices increases dramatically when the temperature is reduced. Besides the increased mobility, cooling further enhances the performance and reliability of digital integrated circuits, improving the subthreshold slope, the junction leakage current and capacitance, and the interconnection resistance. Some non-scalable parameters such as the thermal voltage, are also reduced when the temperature is lowered.

While this sounds attractive, cooling comes at substantial cost. High-quality coolers are expensive, bulky, and consume extra power. The most popular cooling media in use are the inert gases, nitrogen and helium, which have boiling temperatures of 77 K and 4.2 K, respectively. While liquid nitrogen is inexpensive, and cooling costs are moderate, operating at liquid helium temperatures allows for superconductive operation.

In this section, we briefly discuss the operation of silicon at lower temperatures as well as the nature and potential of superconducting digital circuitry.

5.5.1 Low-Temperature Silicon Digital Circuits

Cooling results in an increase of both saturation velocity and carrier mobility for MOS devices. Simultaneously, the junction capacitance is reduced due to the *freeze-out* effect, which means that the dopant atoms hold on to their extra electrons and holes at low temperatures. This results in wider depletion regions, and consequently smaller capacitances. All the above helps to reduce the intrinsic delay of the MOS gate. The impact of cooling on some of the MOS device parameters is shown in Table 5.4.

Table 5.4 Measured device parameters as a function of temperature. The numbers quoted are for an NMOS transistor, with the corresponding values for a PMOS device between brackets (from [Ghoshal93a]).

Parameter	300 K	77 K	4 K
V_T (V) (@I_D = 0.1 μA)	0.12 (0.08)	0.3 (–0.18)	0.35 (–0.29)
μ_{fe}(cm²/V·sec)	490 (220)	2300 (1000)	4400 (3500)
I_{Dsat} (mA/mm)	31 (16)	57 (29)	61 (30)
Subthreshold slope (mV/decade)	74 (81)	21 (28)	5.7 (9.4)

Combining the increased current drive with the reduced capacitance results in a performance increase by a factor of two-to-three for liquid nitrogen operation, and even more when operating at 4 K.

At the same time, *leakage currents* are substantially reduced, because the leakage current of a junction (I_S) is a strong function of the temperature ($\sim e^{qV_j/kT}$). The reduced sub-threshold slope of the device further reduces leakage and makes it possible to operate at lower threshold voltages. At 4 K, a dynamic gate behaves as a static structure, and refresh is no longer necessary.

Finally, reducing the temperature also decreases the interconnect *resistivity*, because the carriers have less thermal energy, and the scattering rate is subsequentially reduced. At liquid nitrogen temperatures, the resistance of aluminum wires improves by a factor five to six [Bakoglu90].

Besides the cost and difficulty of the providing high-quality cooling environments, operation at a reduced temperature has some disadvantages.

- The mentioned carrier freeze-out increases the resistance of the source and drain regions, since fewer mobile carriers are available. It also causes the threshold voltage to increase, as Table 5.4 shows. Due to the freeze-out, less of the ion-implanted impurities in the channel are being ionized.
- Threshold voltages in cooled MOS devices tend to drift with time due to hot-electron trapping effects, as carriers injected into the gate are more likely to be trapped. This effect can be remedied by operating at lower voltages.
- The current gain of *bipolar devices* degrades at lower temperatures due to bandgap narrowing and reduced emitter-base injection. While this helps to suppress parasitic effects such as latchup and subthreshold conduction in MOS transistors, it precludes the use of bipolar gates at temperatures lower than 77 K.

Cooling has been frequently used in the design of high-performance mainframe and supercomputing systems. For instance, the ETA supercomputer (1987) uses liquid nitrogen cooling to reduce its cycle time from 14 nsec at room temperature to 7 nsec. Another emerging approach is to combine MOS silicon structures with superconducting logic. This exploits the extreme performance of the superconducting circuitry with the high density of MOS. It is worth noting that dynamic circuits exhibit a better behavior at liquid helium temperatures, as leakage is eliminated and noise signals are reduced [Ghoshal93b].

5.5.2 Superconducting Logic Circuits

The use of superconductivity in digital circuits dates back to the 1950s. The development of the Josephson junction at IBM [Josephson62] spurred the quest for a superconducting computer. While this effort faltered in the early 1980s, the 1990s witnessed a renewed interest in this technology for two reasons: (1) the discovery of high-temperature, superconducting alloys, and (2) the introduction of niobium junctions, which provide increased reliability and performance compared to the earlier lead-alloy-based junctions. Before discussing the Josephson junction, which is the prime switching element in superconducting logic, we first describe superconductivity.

Superconductivity

A number of materials have the property to *conduct current without resistance* when cooled below a critical temperature T_c. Until recently, most of the known superconducting materials exhibited this desirable property only when cooled close to absolute zero. In the late 1980s, a new class of superconducting ceramic materials was discovered with critical temperatures around and above 100 K. This discovery is important, because it substantially reduces the cooling cost, using liquid nitrogen as a coolant. New composites with ever higher critical temperatures are still being discovered, raising hopes for the availability of room-temperature superconductivity in the near future. One warning with respect to those materials should be heeded: the onset of superconductivity is not only a function of the temperature, but also of the density of the current flowing through the material (J) and the magnetic field present (Φ).

$$T_c = f(J, \Phi) \tag{5.15}$$

Raising either the current density or the magnetic field above a critical value causes the material to revert to the resistive state. For instance, the compound material yttrium-barium-copper-oxide (or YBCO) has a nominal critical temperature of 95 K, which is substantially above the 77 K of liquid nitrogen. Unfortunately, the maximum current density allowed at 77 K equals 4 $\mu A/\mu^2$, which is too low to be useful in digital circuit design.

The potential impact of superconductivity on circuit design is quite large. It enables the transmission of signals over long wires without any resistive loss. This decreases the propagation delay while lowering the power dissipation. Currents can be stored in inductive loops for an almost infinite time, providing for a simple memory structure. In contrast to most digital circuits that can be modeled as *RC*-networks, the first-order model for a superconducting component is closer to an *LC*-network.

The most obvious application of superconductivity in the digital arena is to use traditional devices such as MOS transistors, interconnected by superconducting wires. While this approach helps to address some of the interconnect issues raised in Chapter 8, its impact on overall circuit performance is limited, affecting only the delay of the *RC*-dominated wires. A potential application is the distribution of clocks with minimal skew.

More impressive performance benefits are obtained when employing superconducting switching devices as well. Using this approach, switching delays in the range of picoseconds can be obtained, which is almost an order of magnitude faster than what can be obtained with semiconductor devices. The most popular of these devices is the Josephson junction.

The Josephson Junction

The Josephson junction (abbreviated JJ) was discovered in the early 1960s at the IBM Watson center [Josephson62]. It consists of two layers of superconducting material separated by a very thin insulator (between 1 and 5 nm), as shown in Figure 5.44. The material of choice in current superconducting design is niobium, which has a critical temperature of 9 K. The niobium process has the advantage of being substantially more reliable than the lead-alloy processes used in the earlier JJ implementations.

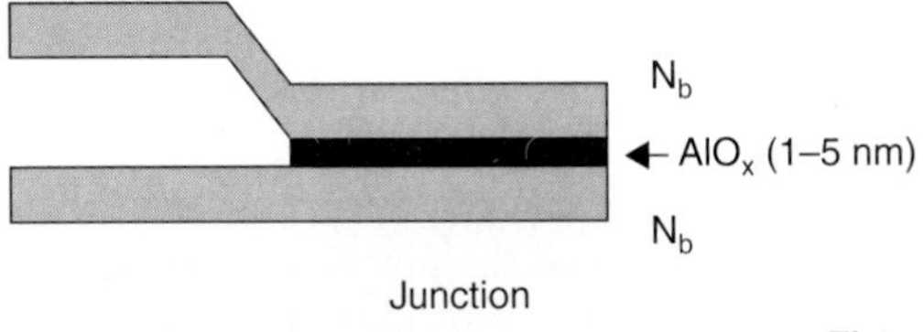

Figure 5.44 Nb-AlO_x-Nb Josephson junction.

The Josephson junction is a tunneling device. In the superconductive mode, electrons tunnel from one electrode to the other without voltage loss. The oxide barrier acts as a superconductor. Raising the current level (or adjusting the magnetic field) causes the device to revert to the resistive state, which results in a fixed voltage drop over the junction. For the Nb/AlO_x/Nb junction, the voltage drop over the junction V_G equals 2.8 mV (!).[4]

Consider a JJ junction connected to a shunt resistor R_L and a current source I_S, as shown in Figure 5.45b. The behavior of the circuit can be understood by combining the *I-V* characteristic of the junction, represented as a cross in the schematic, with the load-line of the resistor (Figure 5.45a). Assume that the junction is initially in the superconducting state. The voltage over the junction is zero, independent of the current level (interval A-B in Figure 5.45a). Raising the current keeps the junction in the superconducting state as long as it does not exceed a critical level (I_{cr}). Larger currents (point B) cause the junction to switch to the voltage state represented by the black curve. The voltage over the junction is constant for most of the current range of interest and equal to the energy-gap voltage V_G. A linear resistive-like behavior is observed for higher current levels. The operation point C is determined by the cross-section between the *I-V* characteristics of the junction and the resistor ($= I_S - V_{JJ}/R_L$). Most of the current is transferred to the load resistor. The critical

[4] The difference between the superconducting and resistive operation modes of the junction is that in the former, the electrons tunnel through the junction in pairs. For a more detailed description of the underlying concepts, please refer to [VanDuzer89].

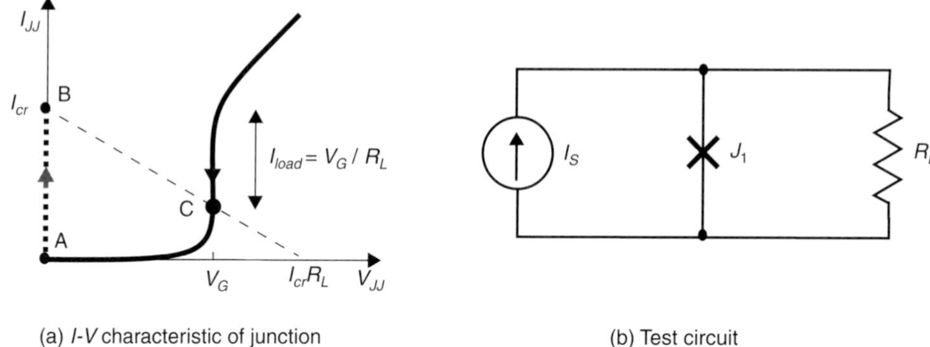

(a) *I-V* characteristic of junction (b) Test circuit

Figure 5.45 *I-V* characteristic of JJ junction when shunted with load resistor.

current I_{cr}, where the switching occurs, is a function of the junction area as well as the applied magnetic field.

As is apparent in Figure 5.45a, the Josephson junction displays a hysteresis-like behavior. The junction remains in the voltage stage, even when the current level is dropped below I_{cr}. For the junction to revert to the superconductive state, it is necessary to reduce the current level to zero, as illustrated by the arrow on the curve.

From the above discussion, it becomes clear that the junction can be modeled as a two-terminal device with two operation modes: the superconducting, zero-voltage and the resistive, fixed-voltage state. In a typical operation, the junction is biased with a current I_{bias}, slightly smaller than the critical current I_{cr}. If a switching action is required, the current is slightly raised, causing the junction to revert to the resistive state. Another approach is to apply a magnetic field so that the critical current I_{cr} is lowered below I_{bias}, which has a similar effect (Eq. (5.15)). Lowering the biasing current to zero resets the device to the superconducting state, after which the next operation cycle can commence.

The Josephson junction has the disadvantage of being a two-terminal device. This property makes it less favorable for digital operations, as no isolation exists between inputs and outputs. A control terminal can be added by overlaying an insulated thin (superconductive) wire on top of the junction. Suppose now that the junction is biased with a bias current somewhat below I_{cr}. Routing a current through the control wire causes a magnetic field to pass through the plane of the junction, what reduces the critical current. When the critical current drops below the bias current, the junction becomes resistive. The resulting structure has perfectly isolated input and output terminals.

In general, it is rare to use a single junction in a digital circuit. It is more advantageous to use two or more junctions connected in a superconducting loop or an assembly of loops. Such a circuit is called an *interferometer* or *superconducting quantum interference device* (SQUID). An example of a two-junction SQUID is shown in Figure 5.46. A magnetically-coupled control terminal has been added. The magnetic coupling is captured by the mutual inductance *M*. The *I-V* characteristic of the SQUID structure is similar to the single junction, but tends to offer larger noise margins.[5]

[5] The actual characteristics of the SQUID are somewhat more complex. For a full discussion of its operation (which is beyond the scope of this text), please refer to [VanDuzer89].

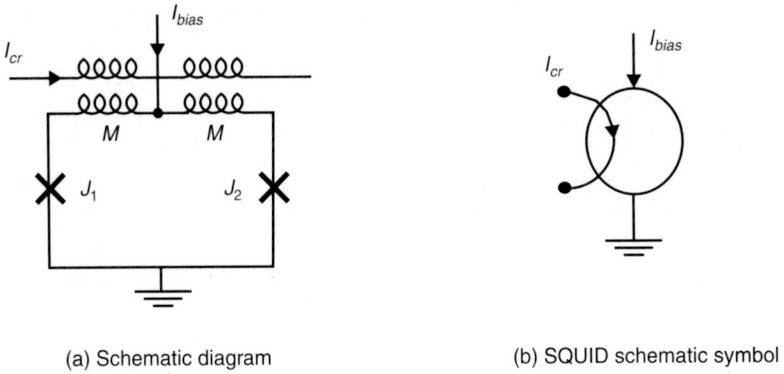

(a) Schematic diagram (b) SQUID schematic symbol

Figure 5.46 SQUID with magnetically-coupled control wire.

The main attraction of the Josephson junction is its extremely fast switching time. Gate delays in the range of picoseconds have been recorded, which is substantially below what can be achieved in semiconductor technologies, and opens the door for multi-GHz digital circuits. The switching speed is mostly limited by parasitic circuit effects, not by intrinsic constraints. One word of caution: while switching from the superconductive to the resistive state proceeds with incredible speed, the reverse operation (the *resetting* of the junction) is comparatively slow and can take up to 20 psec. The reset phase can be compared to the precharging operation in dynamic MOS circuits. As in the dynamic approach, the impact of the "dead time" on the overall performance can be minimized by adopting the correct system architecture. For instance, it is typical for JJ circuits to operate in a pipelined mode with multiple clocks, where one stage is evaluating while the others are being reset.

Superconducting Digital Circuits

On the basis of the type of control mechanism employed, we can divide Josephson digital circuits into two classes. In the first class, switching between the two states is accomplished by current overdrive or *current injection*, while the second class uses *magnetic coupling* [Hasuo89]. The concepts behind both approaches are illustrated in Figure 5.47, where simplified implementations of a two-input OR gate are shown.

Consider first the current-injection approach (Figure 5.47a). The SQUIDs are powered by a pulsed current source, which delivers a current I_{bias}, smaller than I_{cr}. If none of the inputs is high, the junctions in the SQUID stay in the superconducting mode, and $V_{out} = 0$ V. If either input A or B is high, an extra current flows into the loop through the resistors R_L. The combination of the bias and the injected currents exceeds the critical current and causes the junctions in the loop to become resistive. The output of the gate moves from 0 V in the superconducting state to the gap voltage of 2.8 mV. The bias current is diverted from the loop into the connecting fan-out gates, assuming that the on-resistance of the junctions is higher than R_L. Since the output current flows into the SQUID loop of the connecting gates (which is equivalent to stating that the input-impedance of the gate is small), fan-out gates must be connected in parallel.

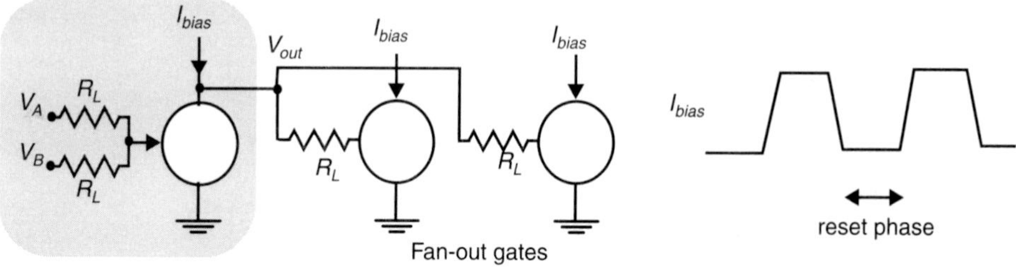

(a) Current-injection gate with fan-out

(c) Bias current waveform

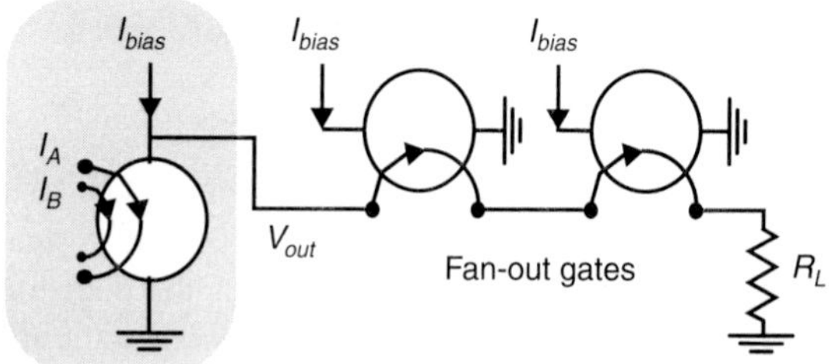

(b) Magnetically coupled gate with fan-out

Figure 5.47 Josephson junction logic families.

The magnetically coupled approach (Figure 5.47b) relies on a similar idea. If both inputs are low, the SQUID operates in the superconducting mode ($V_{out} = 0$ V). Applying an input current to one (or both) of the inputs generates a magnetic field that reduces the critical current below the applied bias current. The SQUID switches to the resistive state, and the output switches to high ($V_{out} = 2.8$ mV). As the input of the gate is physically isolated from the output due to the magnetic coupling, the output signal can be serially connected to multiple cascaded gates.

To initiate the next logic operation, the bias current is lowered to zero (Figure 5.47c), and the junctions are reset to the superconducting state.

While these two circuits give an impression of how a Josephson junction logic family can be constructed, the picture is by no means complete. Multiple variants of those logical families have been devised over the years, each of them with varying fan-out, noise-margin and switching-speed properties [Hasuo89]. In fact, a third class of logic styles has emerged called the *hybrid* style. Logic circuits of this class combine current injection and magnetic coupling to achieve better noise margins and faster switching speed. A member of the hybrid gate class is the popular MVTL gate (*modified variable threshold logic*), which is the logic style of choice in most of the larger-scale superconducting designs (see [Kotani90]), and is pictured in Figure 5.48. Assume that all junctions are initially in the superconducting mode and that an input current I_{in} is applied, which could be the OR-ing of two input currents. This current is coupled magnetically to the SQUID loop consisting of junctions J_1 and J_2. At the same time, I_{in} is also injected into the loop through junction J_3. The combination of both current injection and magnetic coupling accelerates the switching of the junctions J_1 and J_2 to the resistive state. If R_i is chosen to be smaller than the load resistance R_L, the bias current is diverted towards R_i instead of the fan-out gate. This causes J_3 to change state and to become resistive, which routes the input current I_{in} into R_i and

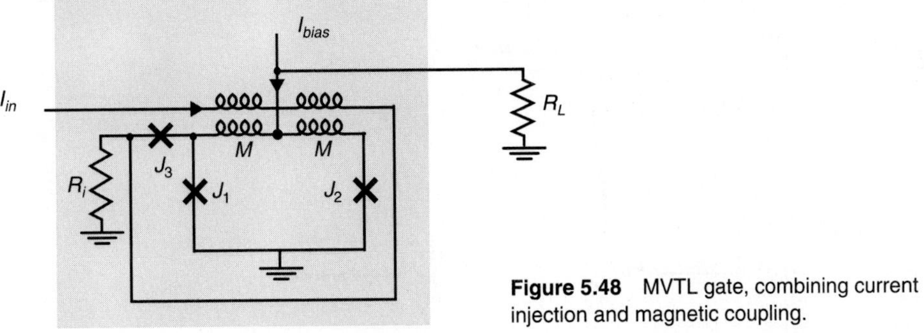

Figure 5.48 MVTL gate, combining current injection and magnetic coupling.

deflects the bias current to the fan-out gate. The purpose of J_3 is to provide isolation between input and output, a desirable property for digital gates that is typically not present in the current-injection logic families. The hybrid nature of the structure that combines injection and coupling results in extra-fast operation speeds. In fact, propagation delays of 1.5 psec (!) have been measured for a two-input MVTL OR-gate with a single fan-out.

Example 5.14 An MVTL Gate

The layout of an MVTL two-input OR gate is shown in Figure 5.49. The input voltages In_1 and In_2 are converted into a current with the aid of the input resistors R_{in1} and R_{in2}. The wire carrying this current is routed on top of the SQUID loop and provides the required magnetic coupling. The bias current is delivered through the resistor R_{bias}, connected to the pulsed supply voltage V_{bias}. The resistor R_D is added to dampen parasitic oscillations in the superconducting LC loop. The gate is implemented in a Nb/AlO$_x$/Nb technology with a 3 μm × 3 μm minimum junction area.

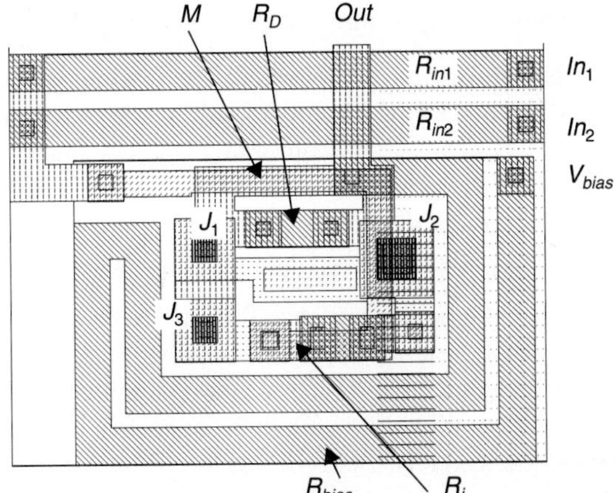

Figure 5.49 Layout of a two-input MVTL NOR gate (from [Mehra94]).

The simulated transient response of the gate is plotted in Figure 5.50. The observed gate delay approximately equals 20 psec. The small oscillations on the output signal are due to inductive effects. The hysteresis effect of the Josephson junction is apparent. It is necessary to lower the bias current to 0 to reset the output signal.

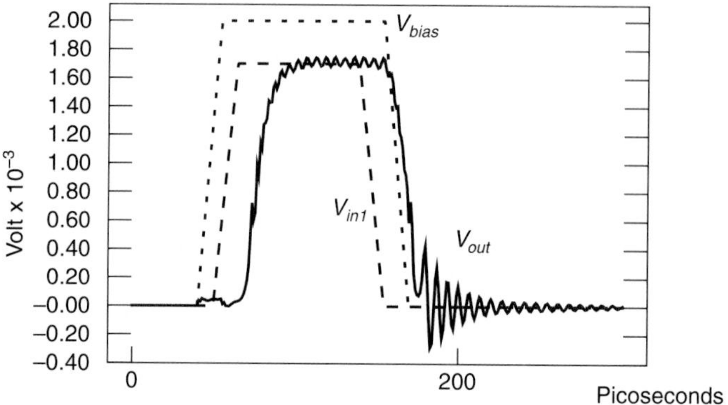

Figure 5.50 Transient response of a two-input MVTL NOR gate (V_{in2} = 0). The bias voltage is scaled to fit on the same scale as the input and output signals. Notice how lowering V_{in1} does not cause V_{out} to revert to the superconductive state. This is only accomplished by resetting the bias current.

Even though the gates shown above seem simple, Josephson junction digital design is far from trivial for a number of reasons.

- The gates, in general, are *noninverting*. Implementing an inverter requires a complex clocking scheme. This deficiency can be addressed by using differential logic, and by providing both signal polarities simultaneously, as is customary in the CPL and ECL logic styles discussed earlier.
- The circuits are powered by an *ac-power supply* (or clock). Distributing such a clock at high speeds is complicated. Be aware that a minimum dead time is necessary to ensure resetting of the junctions in between logic operations. To address this issue, complex clocking schemes with up to three clocks are commonly used.
- *Interfacing with the external world* is complicated. The internal signals in a Josephson junction design have a logic swing of only 2.8 mV, while the external world typically requires much larger swings. The conversion process introduces additional delay that hampers the overall performance. Additionally, every connection to the outside world has to pass through the cooling dewar and introduces heat leaks. The number of connections should therefore be kept to an absolute minimum.
- In general, design at this extreme performance level is exceedingly difficult, since we must address an array of second-order parasitic effects. Signals start to behave as electromagnetic waves, and inductive effects become significant (as discussed in Chapter 8). To keep the delays associated with those parasitics from becoming dominant, a careful sizing of the load resistors is necessary.

Example 5.15 A Josephson Signal Processor

Notwithstanding these difficulties, a number of high-density, high-performance circuits have been realized in the Josephson technology. One of the most complex implementations up to the date of writing is a 1 GOPS Digital Signal Processor [Kotani90]. The circuit consists of 6,300 MVTL gates and counts 23,000 Josephson Junctions. The average delay per gate equals only 5.3 psec/gate. An 8 × 8 multiplication takes a mere 240 psec! The total power consumption of the circuitry, when clocked at 1 GHz, totals 12 mW. This very low dissipation can be attributed to the small logic swing of 2.8 mV. Unfortunately, this benefit is dwarfed by the large amount of power dissipated by the cooling dewar.

5.6 Perspective: When to Use High-Performance Technologies

In this chapter, we have examined non-MOS digital circuit design styles. One of the important lessons to be learned is that the nature and the properties of the semiconductor device have a tremendous impact on the preferred gate topology. Structures that are performing very well in a CMOS technology do not necessarily do well in a silicon-bipolar or GaAs environment. For instance, the high transconductance of the bipolar transistor makes the differential pair an attractive solution. Using the same approach in CMOS requires either very large transistors, or larger voltage swings, and results in a suboptimal solution.

Exploiting the specific properties of a technology can yield a substantial performance improvement. Propagation delays for unloaded inverters can go as low as 25 psec for silicon bipolar, around 10 psec for GaAs HEMT devices, and in the single psec range for superconducting Josephson junction designs. The record speed at present is held by the superconducting single flux-quantum devices, which can switch in 1 psec [Likharev91]. All these numbers demonstrate that multi-GHz digital data processing is a definite possibility.

Besides a pure digital performance advantage, other properties might make a given technology or a device more attractive for a given set of applications.

- Even though it is doubtful that bipolar silicon will maintain its speed advantage over MOS in the deep-submicron region, one must remember that the bipolar transistor also has superior analog properties. This makes this technology attractive for high-performance *mixed-signal applications* like those in radio and networking circuits. These applications require a close-knit mixture of high-speed digital and analog capability, which is inherently available in bipolar silicon.
- The BiCMOS technology has found widespread use in designs where *the capacitive load is huge or unpredictable*. Examples of such structures are memories and gate arrays, where BiCMOS has resulted in important speed-ups. One can argue that pure CMOS designs can achieve similar driving capabilities in approximately the same area. Accomplishing this requires a careful optimization (using a chain of inverters with gradually increasing transistor sizes, as will be discussed in Chapter 8) and typically consumes more power. The memory applications furthermore exploit the analog superiority of the bipolar transistor to implement *high-performance sense amplifiers*.
- GaAs has proven to be effective in the production of *semiconductor lasers*. This makes the material desirable for optical applications, such as fiber-optic networks. Another property of GaAs is its radiation hardness, which makes it attractive for space and defense applications.

- Although the cost of cooled circuit-operation is high, this is not an issue for a number of applications, where extreme performance is an absolute must. The area of *instrumentation* is an excellent example in that perspective. There will always be a need for measurement devices that operate at a higher speed than the device under test.

This analysis concludes with a philosophical consideration. The chapter demonstrates that achieving extreme switching speeds comes at a substantial cost in design effort. Traditional design methodologies and design automation techniques become useless. Interconnections become a significant part of the circuit schematic at these high frequencies and introduce noise and delay (this is extensively discussed in Chapter 8). The design of a reliable high-performance circuit typically turns into a lengthy analysis and optimization process. It is furthermore not obvious that scaling technologies into the deep submicron regions will result in sustained performance improvements. Power considerations, for instance, might provide an upper limit on the switching frequencies that are attainable.

Before opting for one of the higher performing, but less designer-friendly and expensive design technologies, we should consider if the performance gain cannot be obtained by other means, for instance by using *concurrent processing*. Too often the clock speed is used as the dominant performance metric. Frequently, the same system performance can be obtained by running multiple slower elements in parallel. This might come at some expense in area but with greatly reduced design effort. This tendency is becoming prevalent in the high-performance computer arena, where supermainframe computers with their extremely high switching speeds are gradually losing out against parallel implementations.

5.7 Summary

The following concepts were introduced in this chapter:

- ECL is certainly the most popular bipolar logic style at present. Complex gates in traditional ECL are based on a combination of the OR/NOR structure and the *wired-or* approach. Variants of the generic gate such as *current mode logic* and *differential logic* help to trade off between density, performance, reliability, driving capacity, and power consumption.

- A continued scaling of the performance of bipolar gates requires further modifications, such as active pull-down networks. One of the main problems with the bipolar design technique is the *nonscalability of the supply voltage* below approximately 2 V, as dictated by the built-in diode voltage drop. Variations of the structure or different bipolar logic styles, such as NTL, might become more appealing at that point.

- The BiCMOS gate combines *the density of CMOS with the current-drive capabilities of bipolar.* This results in a fast gate structure that outperforms CMOS, especially when driving large capacitances. The performance gain is obtained by using a push-pull bipolar output stage, which delivers a β_F current gain with respect to similar CMOS structure. However, this requires a more expensive technology and a more complex gate structure. One of the main challenges facing BiCMOS design is to maintain that performance gain at lower voltage levels.

- GaAs is a semiconductor material that has the potential to outperform silicon by a factor of 2 (or even higher when heterojunction devices are employed). Achieving this performance boost is complicated by the limited set of device options. Most GaAs designs at present use a MESFET device as the main building block. The main challenge in the design of a high-performance MESFET device is to deal simultaneously with low supply voltages, small logic swings, and variations in the device parameters. A number of logic families have been devised that address these issues. The most popular ones are BFL, DFL, and SCFL. GaAs designs are attractive for the implementation of small building blocks with very high performance, such as those needed in networking and communication systems.
- Heterojunction devices are gaining rapid recognition as one of the promising techniques for future very high performance design. They exploit the high carrier velocity obtained at low doping levels in GaAs and in other compound semiconductors such as silicon-germanium (SiGe).
- Reducing the ambient operating temperature of a digital circuit results in a significant performance improvement. Cooling a silicon MOS design to the liquid nitrogen range boosts the performance by a factor 2 to 3.
- The fastest digital devices at present use the superconducting technology and achieve switching speeds in the picosecond range. The fundamental building block for most of these designs is the Josephson junction, a current/flux-controlled device with a hysteresis-like behavior. The high performance does not come for free. Providing the necessary cooling medium requires an expensive, bulky dewar. Design at these high speeds is also anything but easy. The main application domain of these devices has therefore been in areas where this extreme performance is essential, such as instrumentation.
- A number of exciting developments, such as the emergence of the high-temperature superconductors, hybrid silicon-superconductor, and other new devices, such as flux quantum transistors, might change this picture in the coming decades.

5.8 To Probe Further

A number of specialized textbooks have recently been published on both bipolar and GaAs digital design, a number of which are listed below. Excellent overviews of the state-of-the-art techniques can be found in [Elmasry94], [Embabi93], and [Long90]. Once again, the *IEEE Journal of Solid-State Circuits* and the ISSCC conference proceedings are the common source to consult regarding the latest developments in each of these technologies.

REFERENCES

[Abe89] M. Abe et al., "Ultrahigh-Speed HEMT LSI Circuits", in *Submicron Integrated Circuits*, ed. R. Watts, Wiley, pp. 176-203, 1989.

[Alvarez89] A. Alvarez, *BiCMOS Technology and Applications*, Kluwer Academic Publishers, Boston, 1989.

[Asbec84] P. Asbec et al., "Application of Heterojunction Bipolar Transitsors to High-Speed, Small Scale Digital Integrated Circuits," *IEEE GaAs IC Symposium*, pp. 133–136, 1984.

[Bakoglu90] H. Bakoglu, *Circuits, Interconnections and Packaging for VLSI*, Addison-Wesley, 1990.

[Buchanan90] J. Buchanan, *CMOS/TTL Digital Systems Design*, McGraw-Hill, 1990.

[Chen92] C. Chen, "2.5 V Bipolar/CMOS Circuits for 0.25 μm BICMOS Technology," *IEEE Journal of Solid-State Circuits*, vol. 27, no. 4, April 1992.

[Chuang92] C.T. Chuang, "Advanced Bipolar Circuits," *IEEE Circuits and Systems Magazine*, pp. 32–36, November 1992.

[Curtice80] W. Curtice, "A MESFET Model for Use in the Design of GaAs Integrated Circuits," *IEEE Trans. Microwave Theory and Tech.*, vol. MTT-28, pp. 448–456, 1980.

[Dingle78] R. Dingle et al., "Electron Mobilities in Modulation-Doped Semiconductor Hetero-junction Superlattices," *Appl. Phys. Letters*, vol. 33, no. 7, pp. 665–667, October 1987.

[Emasry94] M. Elmasry, ed., *BiCMOS Integrated Circuit Design*, IEEE Press, 1994.

[Embabi93] S. Embabi, A. Bellaouar, and M. Elmasry, *Digital BiCMOS Integrated Circuit Design*, Kluwer Academic Publishers, Boston, 1993.

[Ghoshal93a] U. Ghoshal, L. Huynh, T., Van Duzer, and S. Kam, "Low-Voltage, Nonhysteretic Operation of CMOS Transistors at 4K," *IEEE Electron Device Letters*, 1993.

[Ghoshal93b] U. Ghoshal, D. Hebert, and T. Van Duzer, "Josephson-CMOS Memories," *1993 ISSCC Conference*, vol. 36, pp. 54–55, 1993.

[Greub91] H. Greub et al., "High Performance Standard Cell Library and Modeling Technique for Differential Advanced Bipolar Current Tree Logic," *IEEE Journal of Solid-State Circuits*, vol. 26, no. 5, pp 749–762, May 1991.

[Haken89] R. Haken et al., "BiCMOS Process Technology," in [Alvarez89], pp. 63–124, 1989.

[Hasuo89] S. Hasuo and T. Imamura, "Digital Logic Circuits," *Proc. of the IEEE*, vol. 77, no. 8, pp. 1177–1193.

[Hodges88] D. Hodges and H. Jackson, *Analysis and Design of Digital Integrated Circuits*, McGraw-Hill, 1988.

[Ichino87] H. Ichino, "A 50-psec 7K-gate Masterslice Using Mixed Cells Consisting of an NTL Gate and LCML Macrocell," *IEEE Journal of Solid-State Circuits*, SC-22, pp. 202–207, 1987.

[Josephson62] B. Josephson, "Possible New Effects in Superconductive Tunneling," *Phys. Letters*, vol. 1, pp. 251, 1962.

[Jouppi93] N. Jouppi et al., "A 300 MHz 115 W 32b Bipolar ECL Microprocessor," *Dig. Technical Papers ISSCC Conf.*, pp 84–85, 1993.

[Kanopoulos89] N. Kanopoulos, *Gallium Arsenide Integrated Circuits: A Systems Perspective*, Prentice Hall, 1989.

[Kotani90] S. Kotani et al., "A 1 GOPS 8b Josephson Signal Processor," *Dig. Tech. Papers ISSCC Conf.*, pp. 148–149, February 1990.

[Likharev91] K. Likharev and V. Semenov, "RSFQ Logic/Memory Family: A New Josephson-Junction Technology for Sub-Terahertz-Clock-Frequency Digital Systems," *IEEE Trans. Applied Supercond.*, vol. 1, pp. 3–28, March 1991.

[Long90] S. Long and S. Butner, *Gallium Arsenide Digital Integrated Circuit Design*, McGraw-Hill, 1990.

[Masaki92] A. Masaki, "Deep-Submicron CMOS Warms Up to High-Speed Logic," *IEEE Circuits and Devices Magazine*, pp. 18–24, November 1992.

[Mehra94] R. Mehra, "Digital Filter Design with High Performance Superconducting Technology," Masters thesis, University of California, Berkeley, 1994.

[Milutinovic90] V. Milutinovic, ed., *Microprocessor Design for GaAs Technology*, Prentice Hall, 1990.

[Raje91] P. Raje et al., "MBiCMOS: A Device and Curcuit Technique Scalable to the Sub-micron, Sub-2V Regime," *Digest of Technical Papers ISSCC Conf.*, vol. 34, pp. 150–151, February 1991.

[Rocchi90] M. Rocchi, *High Speed Digital IC Technologies*, Artech House, 1990.

[Rosseel88] G. Rosseel et al., "Delay Analysis for BiCMOS Drivers," *BCTM*, pp. 220–222, 1988.

[Rosseel89] G. Rosseel et al., "A single-ended BiCMOS Sense Circuit for Digital Circuits," *Proceedings ISSCC Conference*, pp. 114–115, February 1989.

[Singh86] H. Singh et al., "A Comparative Study of GaAs Logic Families Using Universal Shift Resistors and Self-Aligned Gate Technology," *IEEE GaAs IC Symposium*, pp. 11–15, 1986.

[Sze69] S. Sze, *Physics of Semiconductor Devices*, Wiley Interscience, 1969.

[Tektronix93] *GST-1 Standard Cell IC User Documentation*, Tektronix, Portland, 1993.

[Toh89] K. Toh et al., "A 23 psec/2.1 mW ECL Gate with an ac-coupled Active Pull-down Emitter-follower Stage," *IEEE Journal of Solid State Circuits*, SC-24, no. 5, pp 1301–1305, 1989.

[Van Duzer89] T. Van Duzer, "Superconductor Digital IC's," in *VLSI Handbook*, Ed. J. De Giacomo, McGraw-Hill, pp. 16.1–16.21, 1989.

5.9 Exercises and Design Problems

For all problems, use the device parameters provided in Chapter 2 (as well as the book cover), unless otherwise mentioned.

1. [E, None, 5.2] For each of the following, state whether CMOS or bipolar design styles are preferred and why.
   - **a.** Lower power dissipation
   - **b.** Lower delay times (speed)
   - **c.** Driving large capacitances
   - **d.** Large noise margins
   - **e.** Which style generally has a higher input impedance? Is a high input impedance beneficial or detrimental to digital design? Why?
2. [E, None, 5.2] A differential CML gate is shown in Figure 5.51.
   - **a.** Determine the important points of the VTC (V_{OH}, V_{OL}, V_{IH}, V_{IL}, and V_M). For the computation of V_{IH} and V_{IL}, you can use the simplified definition (V_{IL} is defined as the point where transistor Q_1 is carrying 1% of current through the current source, while at V_{IH}, transistor Q_1 carries 99% of the total current).
   - **b.** Assume that the output *Out* connects to the input terminals of five identical gates (or has a fan-out of 5). Recompute the values of V_{OH} and V_{OL} under these conditions.
3. [M, None, 5.2] Consider the circuit of Figure 5.52.
   - **a.** Describe the intended logic function,
   - **b.** The circuit has one major problem. Describe **precisely** that problem.
   - **c.** Present a technique for solving the problem.

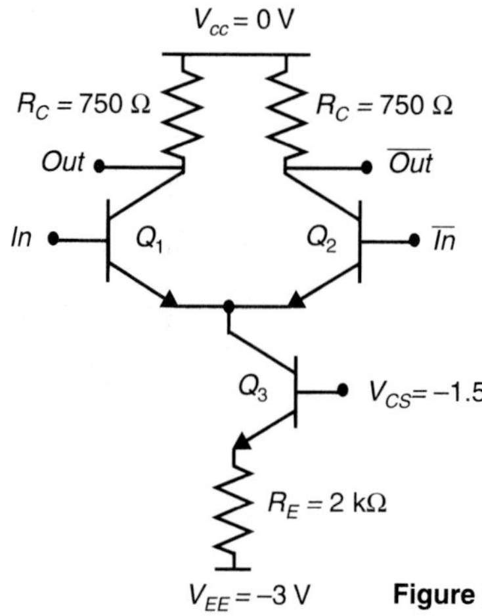

Figure 5.51 Differential CML gate.

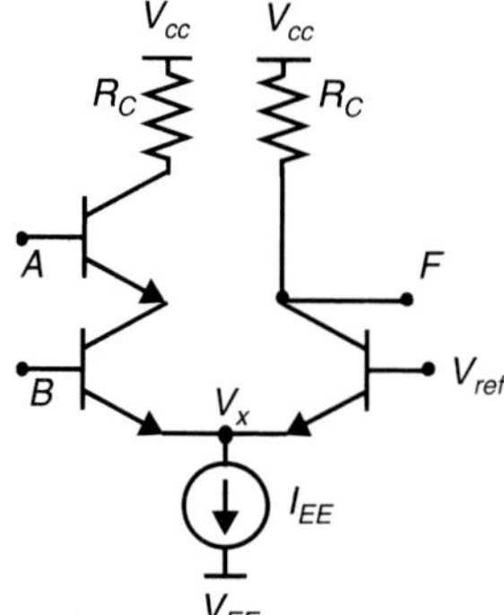

Figure 5.52 Bipolar digital gate with serious problem.

4. [M, None, 5.2] Implement the logic for a 1-bit ripple-carry adder cell using both the single-ended ECL and the differential CML logic styles. What performance/cost trade-offs exist between the two implementations?

5. [C, None, 5.2] Figure 5.55 shows the schematic of an ECL inverter. Assume output loading C_L = 40 fF and a lumped internal capacitance C_C = 120 fF. Parameters: r_B = 100 Ω, $C_{BC} = C_{BE}$ = 20 fF, V_{CEsat} = 0.1 V, $I_S = 10^{-14}$ A, β = 100, τ_F = 10 psec, τ_S = 1 nsec.
   a. Compute t_{pHL}.
   b. Compute the static power consumption of the gate.
   c. Express the dynamic power consumption as a function of switching frequency.
   d. What is the frequency at which the dynamic power equals the static power?
   e. Explain why dynamic power consumption is almost never an important factor in this gate. (Does it make sense to operate the inverter at this frequency?)

6. [M&D, SPICE, 5.3] Determine the transistor sizes for the BiCMOS driver of Figure 5.24a so that a $t_{pHL} = t_{pLH}$ = 5 nsec is achieved for load capacitance C_L = 40 pF, while minimizing area. V_{DD} = 5 V. Refine your design with SPICE. Explain any deviations.

7. [M, None, 5.3] Consider the circuit configuration of Figure 5.54.
   a. Determine the values of V_{OL} and V_{OH}.

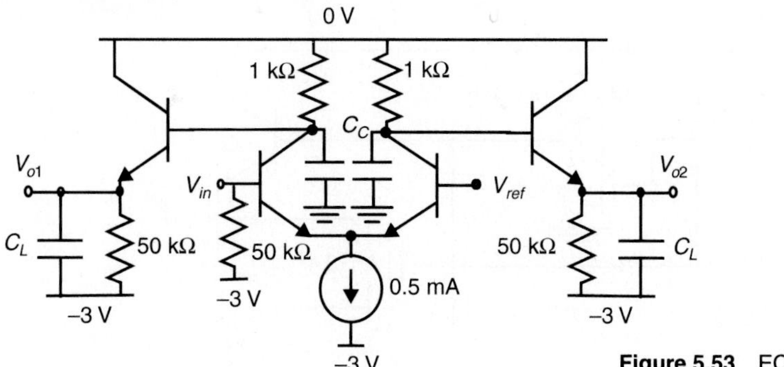

Figure 5.53 ECL inverter.

b. Explain why Q_1 normally operates in **forward-active** mode during the low-to-high transition.

c. In reality, the bipolar transistor can saturate due to the parasitic collector r_c. Explain why.

d. For a supply voltage of 3.3 V, determine the maximum value of r_c so that Q_1 never saturates during the low-to-high transition.

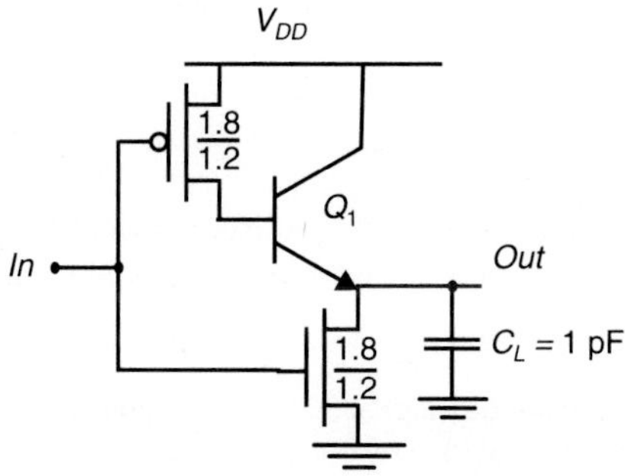

Figure 5.54 Bipolar-MOS circuit.

8. [E, None, 5.3] Consider again Figure 5.54.

a. Derive an expression for t_{pLH} as a function of V_{DD}. Assume that $\lambda = 0$.

b. Compare the t_{pLH} of this gate to an equivalent CMOS inverter, where the PMOS-bipolar combination is replaced by a single PMOS device with $(W/L) = (1.8/1.2)$. Determine which one is faster (for $V_{DD} >> V_T$) and derive an approximate expression for the speed ratio. Perform only a **qualitative** analysis.

9. [C, SPICE, 5.3] A BiCMOS gate is given in Figure 5.55.

a. What is the function of the gate?

b. Hand calculate V_{OH}, V_{OL}, V_{IL}, and NM_L. V_{IL} is defined as the point where M_3 and Q_2 turn on. Draw the VTC of the circuit. Assume a sharp transition at V_{IL} on the VTC curve.

c. Use SPICE to plot the VTC. Determine V_{OH}, V_{OL}, V_{IH}, V_{IL}, V_M, NM_H, and NM_L from the plot.

d. Compare the results of parts (b) and (c). Do hand-calculated V_{OH} and V_{OL} agree with SPICE? If not, give two reasons for the difference.

10. [E, SPICE, 5.3] For the circuit of Figure 5.55, plot t_p as a function of output loading for values of C_L between 0 and 10 pF. (Use SPICE to find t_{pHL} and t_{pLH} for several data points.) Compute

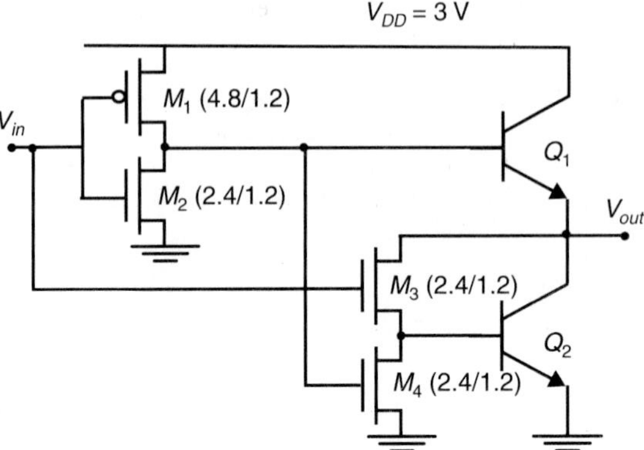

Figure 5.55 BiCMOS gate.

the slope of the curve. If SPICE experiences convergence errors, try: .option method=gear maxord=3.

11. [E, None, 5.4] List the main benefits of using GaAs for digital design. What are the main drawbacks of GaAs circuits?

12. [E, HSPICE, 5.4] Draw the VTC of the GaAs inverter circuit of Figure 5.56. Sweep the input signal between 0 and 0.7 V. Assume (a) a depletion and (b) an enhancement device. Compare the manual results with the output of HSPICE. Use the following models for the MESFET transistors.

```
.model enh njf
+ vto=0.23 beta=250u lambda=0.2 alpha=6.5 ucrit=0 gamds=0 ldel=-0.4u wdel=-0.15u
+ rsh=210 n=1.16 is=0.5m level=3 sat=0 acm=1 capop=1 gcap = 1.2 m crat = 0.666

.model dp njf
+ vto=-0.825 beta=190u lambda=0.065 alpha=3.5 ucrit=0 gamds=0 ldel=-0.4u wdel=-0.15u
+ rsh=210 n=1.18 is=10m level=3 sat=0 acm=1 capop=1 gcap = 1.2 m crat = 0.666
```

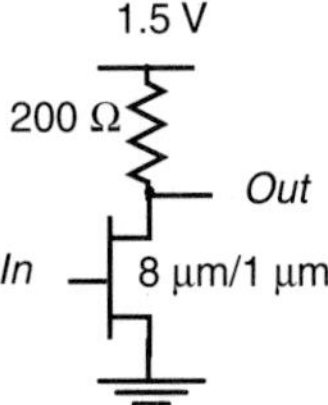

Figure 5.56 GaAs inverter.

13. [M, HSPICE, 5.4] Determine the propagation delay of the buffered FET NOR of Example 5.12 as a function of the load capacitance (using HSPICE). Use the models given in Problem 12. For the Schottky barrier diode, use a MESFET with the drain shorted to the source. Discuss the obtained results.

14. [C, HSPICE, 5.4]
    a. Sketch the schematic of a two-input buffered-FET NAND gate. Include the level-shifting output stage(s). Explain the obtained results.
    b. Simulate the VTC of the obtained gate (for both inputs).
    c. What are the power consumption and propagation delay of the circuit of part (b).

15. [E, None, 5.5] Using the parameters of Table 5.4, determine the speed-up obtained when cooling a CMOS inverter from 300 K to 4 K.

16. [M, None, 5.5] Consider the Josephson junction circuit shown in Figure 5.57. Assume that $I_{bias} < I_{c1} < I_{c2}$.

a. Determine the logic function of the circuit. Describe its basic operation.

b. Explain why this circuit exhibits input-output isolation. In doing so, assume that a current I_x is injected into the gate from the output.

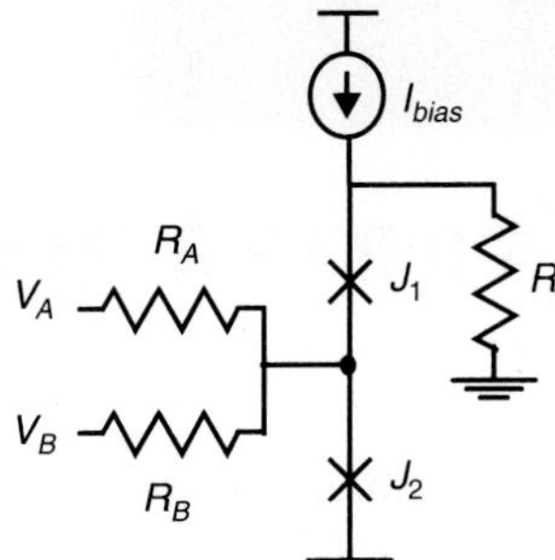

Figure 5.57 Josephson junction circuit.

17. [C, None, 5.5] Discuss the operation of the circuit of Figure 5.58. What is its function? What are the output levels?

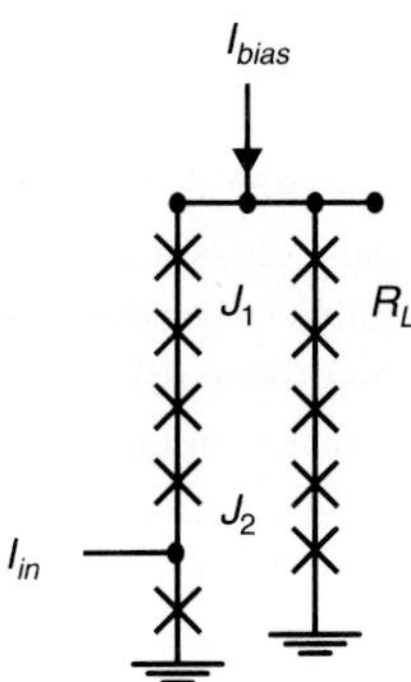

Figure 5.58 Josephson junction circuit.

APPENDIX

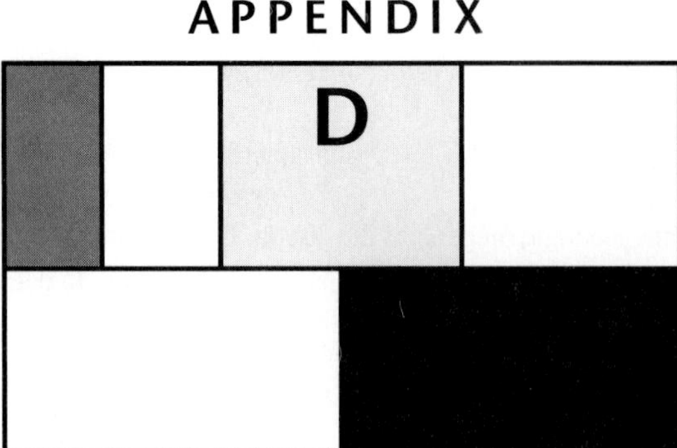

THE SCHOTTKY-BARRIER DIODE

Single-carrier diode device with fast switching characteristics

In the discussion of the diode switching transients in Chapter 2, it became clear that a major part of the switching time is spent in (re)moving the excess minority carrier charge. Reducing the amount of charge stored (and the associated capacitance) is therefore of prime importance when fast-switching diodes are required. For this reason, *pn*-junction diodes are often replaced by Schottky-barrier diodes in fast bipolar digital circuits. The advantage of the Schottky diode is that all conduction is via electrons flowing in metal or *n*-type material. In other words, only majority carriers contribute to the current flow, in contrast to the *pn*-junction diode, where the current is carried by minority carriers. Hence, the minority-charge storage effects do not affect the switching speed, and the transient response is almost uniquely determined by the depletion-region capacitance, which is substantially smaller.

The Schottky-barrier diode is a metal-semiconductor junction formed by depositing a small metal contact onto a lightly doped *n*-type semiconductor ($N_D \leq 10^{-16}$ cm^{-3}). A cross section of such a diode is shown in Figure D.1, which also shows its schematic symbol.

A detailed discussion of the operation of the Schottky diode is beyond the scope of this text. It suffices to understand that the conducting electrons in the semiconductor are at a higher energy state than those in the metal. Upon contact, these electrons tend to migrate to the metal, creating a depletion region at the contact boundary. The immobile charge of the depletion region creates a potential barrier, which prevents the flow of electrons from semiconductor to metal. Applying a positive voltage to the metal reduces the potential bar-

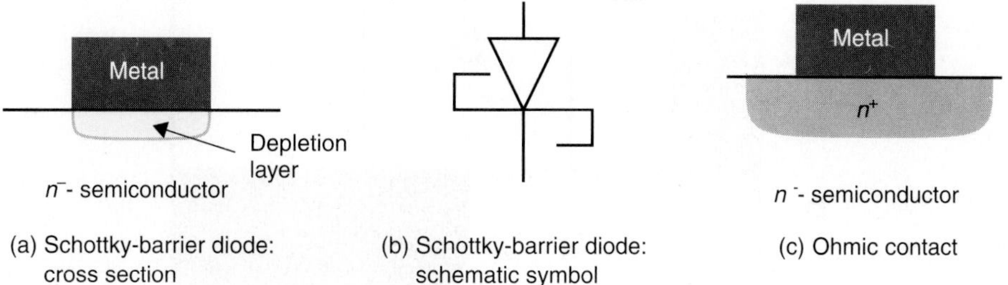

Figure D.1 Schottky-barrier diode and ohmic contact.

rier and causes a current flow. Reducing the voltage to the metal blocks the current flow and results in a reverse-bias condition. For more information on the Schottky diode operation, see the device textbooks, such as [Muller86].

From the intuitive analysis above, it comes as no surprise that the behavior of the Schottky-barrier diode is similar to that of the *pn*-junction. The current obeys the ideal diode equation, although the expression for the saturation current is different, as shown in Eq. (D.1). K_{SB} is an empirical constant, and ϕ_B is the barrier height, which depends uniquely upon the two materials composing the diode.

$$
\begin{aligned}
I_D &= I_S(e^{V_D/\phi_T} - 1) \\
&\text{with} \\
I_S &= K_{SB}T^2 e^{-\phi_B/\phi_T}
\end{aligned}
\tag{D.1}
$$

The on-voltage of the diode depends upon the materials selected. For example, using aluminum and titanium-silicide for the metal yields on-voltages of 0.69 V and 0.59 V, respectively.

If every semiconductor-metal contact would form a diode, building a functional integrated circuit would be hard. Fortunately, when the semiconductor is heavily doped, the depletion layer becomes so narrow that electrons can travel in either direction through the potential barrier using a mechanism called *tunneling,* and current flows equally well in both directions. Such a contact is called an *ohmic contact* and is used extensively in integrated circuits. A cross section of an ohmic contact is shown in Figure D.1c.

REFERENCES

[Muller86] R. Muller and T. Kamins, *Device Electronics for Integrated Circuits,* 2nd ed., John Wiley and Sons, 1986.

CHAPTER

DESIGNING SEQUENTIAL LOGIC CIRCUITS

Implementation techniques for flip-flops, latches, oscillators, pulse generators, and Schmitt triggers

Static versus dynamic realization

Avoiding races in sequential circuits

6.1 Introduction

The logic circuits described in Chapters 4 and 5 all have one property in common—the outputs are a logic combination of the **current** input signals. This chapter presents another class of logic circuits called *sequential logic* circuits. In these circuits, the inputs not only depend on the current values of the inputs, but also on **preceding** input values. In other words, a sequential circuit remembers some of the past history of the system—it has memory.

We can implement this memory function in a circuit in two ways. The first approach uses *positive feedback*, or regeneration. Here, one or more output signals are intentionally connected back to the inputs. This results in a class of elements called *multivibrator circuits*. The bistable element, or flip-flop, is its most popular representative, but other elements such as monostable and astable circuits are also frequently used.

A second approach is to use *charge storage* as a means to store signal values. This approach, which is very popular in the MOS world, requires regular refreshing, as charge tends to leak away with time. It is therefore called *dynamic*, in contrast with the regenerative approach, with which a signal value can be held indefinitely and is therefore called *static*.

The bistable elements, both static and dynamic, add an invaluable class of hardware modules to the library, namely the register. This element, which allows for the storage of previous signal values, is an essential component in the design of any digital processor. An elaborate discussion of the design and analysis of bistable sequential circuits is therefore appropriate and constitutes the core of this chapter. Other multivibrators, although less common, are often useful as well. The astable multivibrator acts as an oscillator and can be used as a clock generator, while the monostable multivibrator serves as a pulse generator. A discussion of both of those elements is included at the end of the chapter.

6.2 Static Sequential Circuits

6.2.1 Bistability

Two inverters connected in cascade are shown in Figure 6.1a, along with a voltage-transfer characteristic typical of such a circuit. Shown plotted are the VTCs of the first inverter, that is, V_{o1} versus V_{i1}, and the second inverter (V_{o2} versus V_{o1}). The latter plot is rotated to accentuate that $V_{i2} = V_{o1}$. Assume now that the output of the second inverter V_{o2} is connected to the input of the first V_{i1}, as shown by the dotted lines in Figure 6.1a. The resulting circuit has only three possible operation points (*A*, *B*, and *C*), as demonstrated on the combined VTC. The following important conjecture is easily proven to be valid:

> Under the condition that the gain of the inverter in the transient region is larger than 1, only *A* and *B* are stable operation points, and *C* is a metastable operation point.

This condition holds for every inverter we have discussed in previous chapters. Suppose that the cross-coupled inverter pair is biased at point *C*. A small deviation from this bias

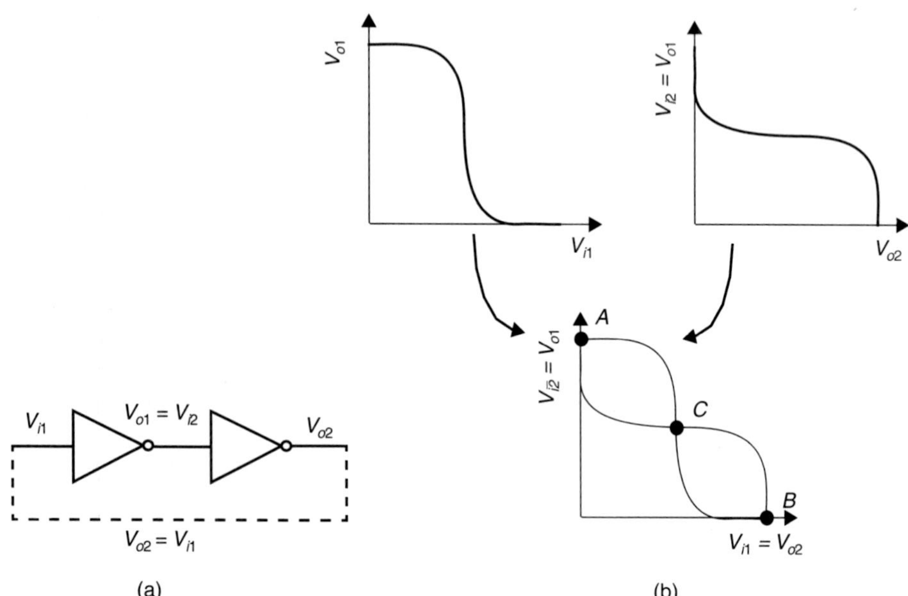

Figure 6.1 Two cascaded inverters (a) and their VTCs (b).

point, possibly caused by noise, is amplified and *regenerated* around the circuit loop. This is a consequence of the gain around the loop being larger than 1. The effect is demonstrated in Figure 6.2a. A small deviation δ is applied to V_{i1} (biased in C). This deviation is amplified by the gain of the inverter. The enlarged divergence is applied to the second inverter and amplified once more. The bias point moves away from C until one of the operation points A or B is reached. In conclusion, C is an unstable operation point. Every deviation (even the smallest one) causes the operation point to run away from its original bias. The chance is indeed very small that the cross-coupled inverter pair is biased at C and stays there. Operation points with this property are termed *metastable*.

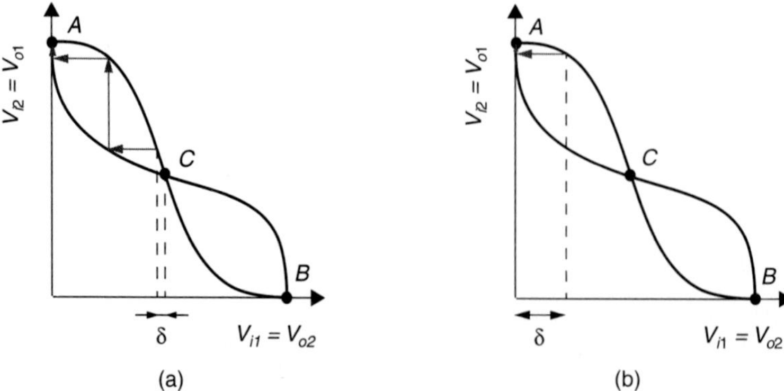

Figure 6.2 Metastability.

On the other hand, A and B are stable operation points, as demonstrated in Figure 6.2b. In these points, the **loop gain is much smaller than unity.** Even a rather large deviation from the operation point is reduced in size and disappears.

Hence the cross-coupling of two inverters results in a *bistable* circuit, that is, a circuit with two stable states, each corresponding to a logic state. The circuit serves as a memory, storing either a 1 or a 0 (corresponding to positions A and B).

In order to change the stored value, we must be able to bring the circuit from state A to B and vice-versa. Since the precondition for stability is that the loop gain G is smaller than unity, we can achieve this by making A (or B) temporarily unstable by increasing G to a value larger than 1. This is generally done by applying a trigger pulse at V_{i1} or V_{i2}. For instance, assume that the system is in position A ($V_{i1} = 0$, $V_{i2} = 1$). Forcing V_{i1} to 1 causes both inverters to be on simultaneously for a short time and the loop gain G to be larger than 1. The positive feedback regenerates the effect of the trigger pulse, and the circuit moves to the other state (B in this case). The width of the trigger pulse need be only a little larger than the total propagation delay around the circuit loop, which is twice the average propagation delay of the inverters.

In summary, a bistable circuit has two stable states. In absence of any triggering, the circuit remains in a single state (assuming that the power supply remains applied to the circuit), and hence remembers a value. A trigger pulse must be applied to change the state of the circuit. Another common name for a bistable circuit is *flip-flop*.

6.2.2 Flip-Flop Classification

In the logic design world, a number of different flip-flop (FF) types are known. The simplest one is the *SR flip-flop* (or set-reset flip-flop). One possible implementation, using only NOR gates, is shown in Figure 6.3a. The logic symbol for this circuit is given in Figure 6.3b. This circuit is similar to the cross-coupled inverter pair with the inverters replaced by NOR gates. The second input of the NOR gates is connected to the trigger inputs (S and R), which makes it possible to force the outputs Q and $\overline{Q}$ to a given state. These outputs are complimentary. When both S and R are 0, the flip-flop is in a quiescent state and both outputs retain their value. If a positive (or 1) pulse is applied to the S input, the Q output is forced into the 1 state (with $\overline{Q}$ going to 0). Vice versa, a 1 pulse on R resets the flip-flop and the Q output goes to 0. The length of the trigger pulse has to be larger than the loop delay of the cross-coupled pair, as we have already noted.

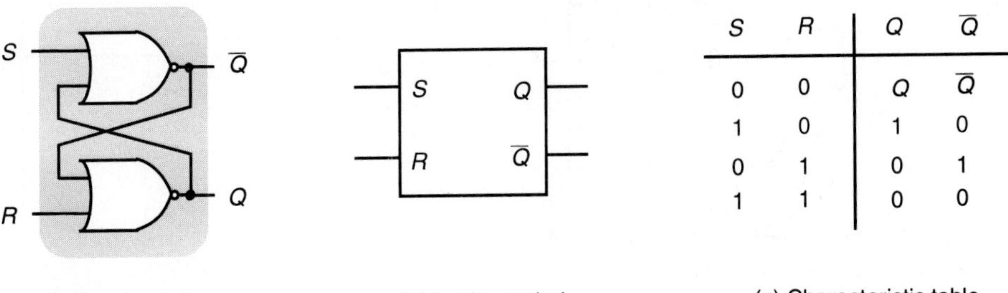

S	R	Q	$\overline{Q}$
0	0	Q	$\overline{Q}$
1	0	1	0
0	1	0	1
1	1	0	0

(a) Schematic diagram (b) Logic symbol (c) Characteristic table

Figure 6.3 NOR-based *SR* flip-flop.

These results are summarized in the *characteristic table* of the flip-flop, shown in Figure 6.3c. The characteristic table is the truth table of the gate and lists the output states as functions of all possible input conditions. The first three input combinations of the table have already been discussed. When both S and R are high, both Q and $\overline{Q}$ are forced to zero. Since this does not correspond with our constraint that Q and $\overline{Q}$ must be complementary, this input mode is considered to be forbidden. The problem with this operation condition is that when the input triggers return to their zero levels, the resulting state of the latch is unpredictable and depends on whatever input is last to go low.

The *SR* flip-flop can also be implemented with NAND gates, as shown in Figure 6.4. As we can deduce from the circuit diagram and the characteristic table, the quiescent state of the latch corresponds to both S and R high. A negative-going (or 0) pulse on S or R respectively sets or resets the flip-flop. Having both S and R equal to zero is forbidden. The small circles at the inputs of the NAND-gate *SR* flip-flop indicate that the gate is triggered by a negative-going pulse, or operates on so-called *negative logic*. This is in contrast with the NOR-based FF, which triggers on positive-going pulses, and is therefore said to operate on *positive logic*.

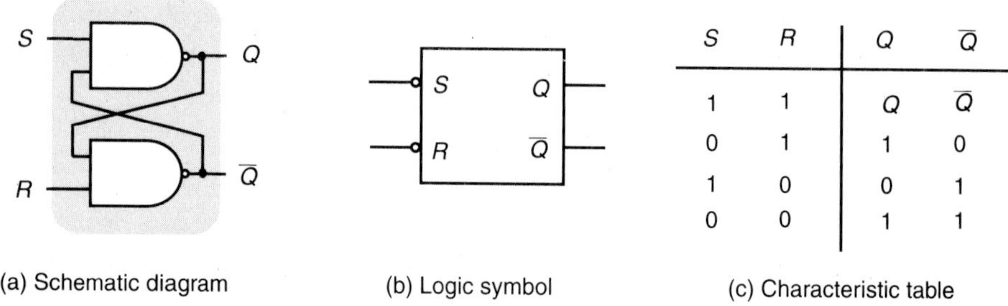

S	R	Q	$\overline{Q}$
1	1	Q	$\overline{Q}$
0	1	1	0
1	0	0	1
0	0	1	1

(a) Schematic diagram (b) Logic symbol (c) Characteristic table

Figure 6.4 NAND-based *SR* flip-flop

The ambiguity of having a nonallowed mode caused by trigger pulses going active simultaneously can be circumvented by adding two feedback lines to the circuit. The resulting device is called the *JK flip-flop*. An all-NAND version is shown in Figure 6.5a.

An important addition is the *clock input* ϕ. This ensures that changes in the output logic states of the flip-flops in a design are synchronized with each other. A circuit in which all changes in state are related to a change in a clock signal (or a number of clock

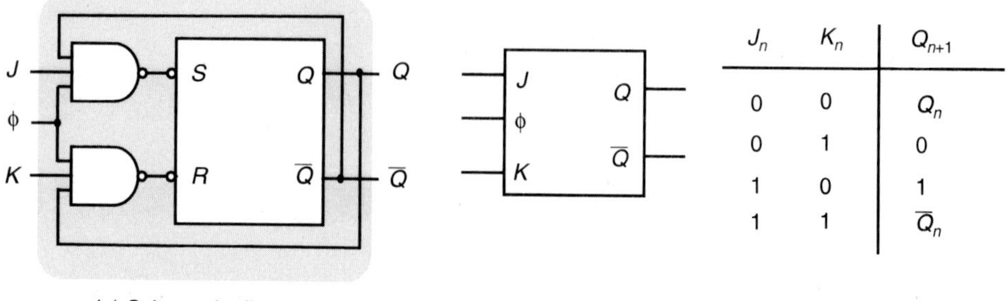

J_n	K_n	Q_{n+1}
0	0	Q_n
0	1	0
1	0	1
1	1	$\overline{Q}_n$

(a) Schematic diagram (b) Logic symbol (c) Characteristic table

Figure 6.5 *JK* flip-flop.

signals) is called a *synchronous* circuit. The majority of the currently designed circuits belong to this class.

The logic symbol and characteristic table of the *JK* flip-flop are given in Figure 6.5b and c. The setting and resetting of the flip-flop is performed as for the *SR* FF, except that these events are now synchronized with respect to the clock by way of the two NAND gates at the inputs of the FF. As long as the clock ϕ is low, the *S* and *R* inputs of the FF are both at 1, and the state of the flip-flop remains unchanged regardless of the values of the *J* and *K* inputs. Input-signal events can only propagate to the output when the clock is high. Another major difference between the *SR* and *JK* flip-flops is that the forbidden state is eliminated. When both *J* and *K* inputs are high and the clock ϕ goes high, the feedback of the output values Q and $\overline{Q}$ causes the flip-flop to toggle its state. For instance, when the flip-flop is in the set state ($\overline{Q} = 0$, $Q = 1$), only *R* goes low as a result of the feedback, and the flip-flop moves to the reset state.

There is a catch, however. In the above example, the feedback disables the *K* and enables the *J* input after toggling. If the clock is still high, this causes the flip-flop to change state again. This repeated and unwanted toggling puts some stringent constraints on the clock pulse width. To function correctly, the pulse width of the clock has to be less than the propagation delay of the FF. This restriction is removed in other FF structures.

The characteristic table of the *JK* FF is similar to the one of the *SR* FF, but the forbidden mode is now replaced by the toggle mode. This is expressed in the characteristic table by stating that the next value of Q (denoted as Q_{n+1}) is the reverse of the current value of Q (indicated as Q_n). Note that input changes can only propagate when the clock ϕ is high, or in other words, when the inputs are synchronized.

Some derived forms of the *JK* FF are commonly used and have received proper names. One is the *toggle* or *T flip-flop*. This is a *JK* FF where the only allowed operation mode is the toggle operation, which can be achieved by tying both *J* and *K* permanently together, as shown in Figure 6.6a. Another special FF is the *delay* or *D flip-flop,* which is used copiously in digital systems for the temporary storage of data. Here, the Q output of the gate is a replica of the *D* input. An inverter between the *S* and *R* inputs of the FF ensures that the inputs are always complementary Figure 6.6(b). Therefore, a 1 input results in a setting of the FF, while a 0 resets the device. As suggested by its name, the D-FF simply generates a delayed version of the input signal that is synchronized with the clock ϕ.

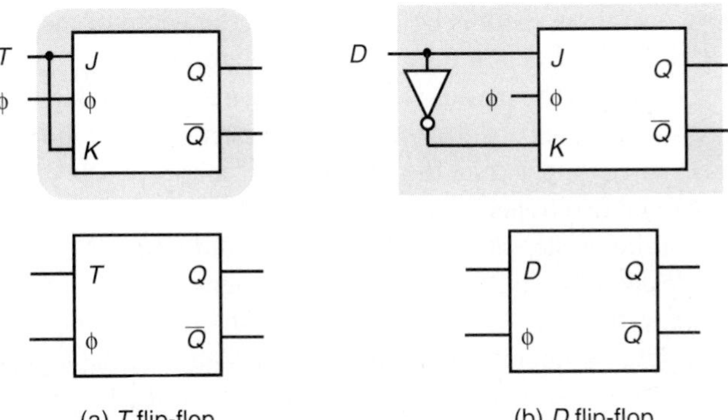

Figure 6.6 Derived flip-flops.

6.2.3 Master-Slave and Edge-Triggered FFs

The *JK* flip-flop presented above is also called a *latch*. A flip-flop is a latch if the gate is transparent while the clock is high (low) [Hill74]. Any change at the input is reflected at the output after a nominal delay. The latch is said to open with the rising of the clock. Data is accepted continuously until the clock goes down and the latch closes.

The transparent nature of the latch can cause some severe problems. Consider the simple circuit of Figure 6.7. As long as the clock is high, the output of the flip-flop oscillates back and forth between the 0 and the 1 states. This phenomenon is called a *race* (or *race-around)* condition and can only be avoided by making the pulse width of ϕ smaller than the propagation delay of the loop. Since the loop delay in the example is small, and probably smaller than the pulse width, this situation has a major chance of occurring. The result of this repetitive toggling is that the output is undetermined when the clock goes low. Observe that a *JK* flip-flop has an intrinsic race problem when *J* and *K* are high, as discussed earlier.

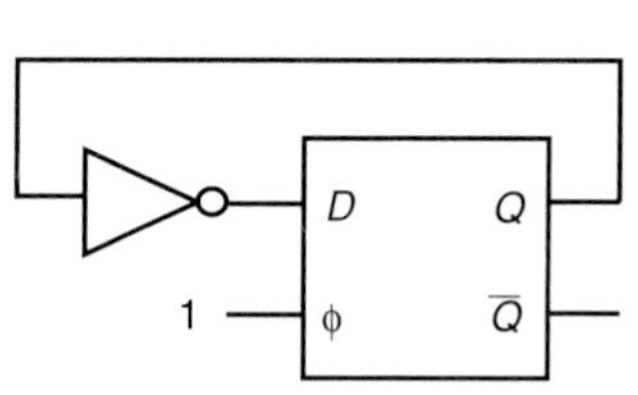

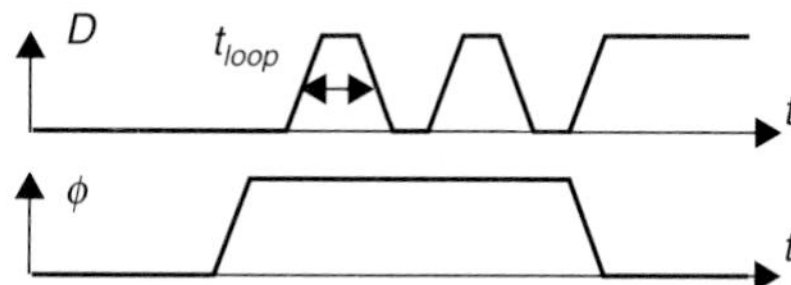

Figure 6.7 Race (-around) problem in latch-based designs (during $\phi = 1$).

One way to avoid a race is to use the *master-slave* approach. Master-slave flip-flops are built by cascading two basic flip-flops with opposite clock phases, as illustrated in Figure 6.8. The first flip-flop, called the master, becomes operational when the clock ϕ goes high. During this period, inputs *J* and *K* are enabled, and the intermediate signals *SI* and *RI* can be changed. The feedback of the Q and $\overline{Q}$ signals ensures that the flip-flop acts as a *JK* flip-flop that toggles when both *J* and *K* are high. The $\overline{\phi}$ clock is low in this time period, putting the second FF, called the slave, in the *hold* mode.This prevents the changes in *SI* and *RI* from propagating to the Q and $\overline{Q}$ outputs. On the falling edge of the clock, ϕ goes down, freezing the state of the master latch. A very short time later, the NAND input gates of the slave latch are enabled, and the changes in *SI* and *RI* are propagated to the outputs. Due to this master-slave operation, there is no limitation on the width of the clock pulse, since the master latch is disabled at the time the outputs are changing, and no racing around can occur. In other words, the master-slave principle makes sure that the feedback path of the flip-flop is always interrupted either at the master or slave side. The pulse lengths of ϕ and $\overline{\phi}$ have to be larger than the propagation delays of the master (slave) latches for this structure to be functional.

The logic symbol of the master-slave flip-flop is given in Figure 6.8b. The little circle on the ϕ input indicates that the state of the flip-flop changes at the falling edge of the clock. Very often, the basic flip-flop circuit is extended with some additional inputs, as indicated by gray lines on the symbol of Figure 6.8b (not included in the logic diagram). These are the *asynchronous* inputs, which can change the state of the flip-flop regardless

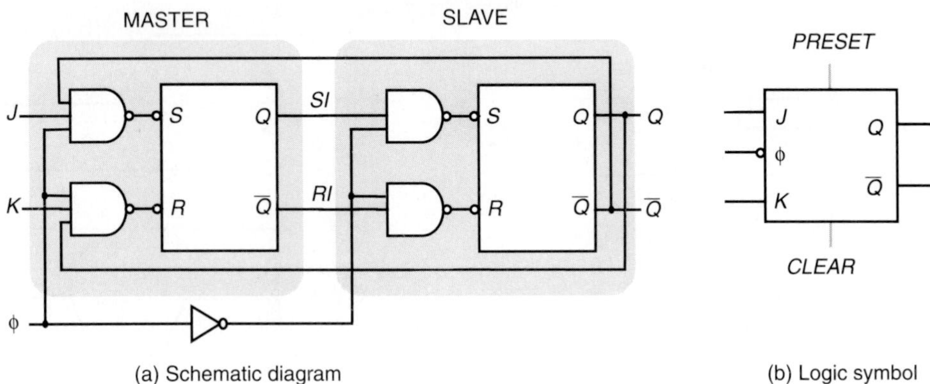

Figure 6.8 Master-slave flip-flop.

of the clock state. Typical direct inputs are the *PRESET* and *CLEAR* inputs that initialize the flip-flop state to either 1 or 0.

A problem with the master-slave approach is that the circuit is sensitive to changes in the input signals as long as ϕ is high. In other words, the input signals must stay constant when the clock is active. Assume that the flip-flop is reset ($\overline{Q} = 1$). This means that the *J*-input NAND gate of the master FF is enabled. Any *spike* or *glitch* on this input, possibly caused by the switching of a neighboring signal, might cause the master latch to be set. Bringing the *J* input back to zero does not change the state of the latch, because it can only be reset by a pulse on the *K* input, which is currently disabled! Thus one can say that the *J* input has "caught" a 1 that is subsequently transferred to the slave when ϕ goes low. This problem is therefore known as *one-catching,* or *level-sensitive*, and it can be avoided by keeping the length of the ϕ pulse as small as possible. This might not always be possible due to constraints imposed by the rest of the system. One way to circumvent this problem is to make use of *edge-triggered* devices.

The idea behind the edge-triggered approach is to allow the state of the flip-flop to change only at the rising (or falling) edge of the clock. Input events at other points in time are ignored, so that one-catching is avoided. This is generally achieved by ensuring that the *S* and *R* pulses, as presented to the actual latch, are of a controlled and narrow width and occur synchronously with a clock transition. The width of the pulses can be manipulated in many different ways, all of which require a timing element as a reference. The delay of an *RC* network or the propagation delay of a logic gate are commonly used timing references.

Consider, for instance, the circuit of Figure 6.9a. When ϕ is high, the output of gate N_2 is always 1 (independent of *In*), which means that the gate is disabled. During the same period, the input *In* can propagate through N_1. As shown on the waveform diagram of Figure 6.9b, a high input causes *X* to go low. On the falling edge of ϕ, N_1 is disabled and N_2 enabled. *X* is forced to 1. This happens only after a period of time equal to the propagation delay of N_1. During a short time interval, N_2 sees both of its inputs at 0 and goes low. *X* eventually reaches the 1 value, and *Out* goes back to 1. Consequently, a short, low-going pulse appears at the output of N_2 with a length approximately equal to the propagation delay of N_1.

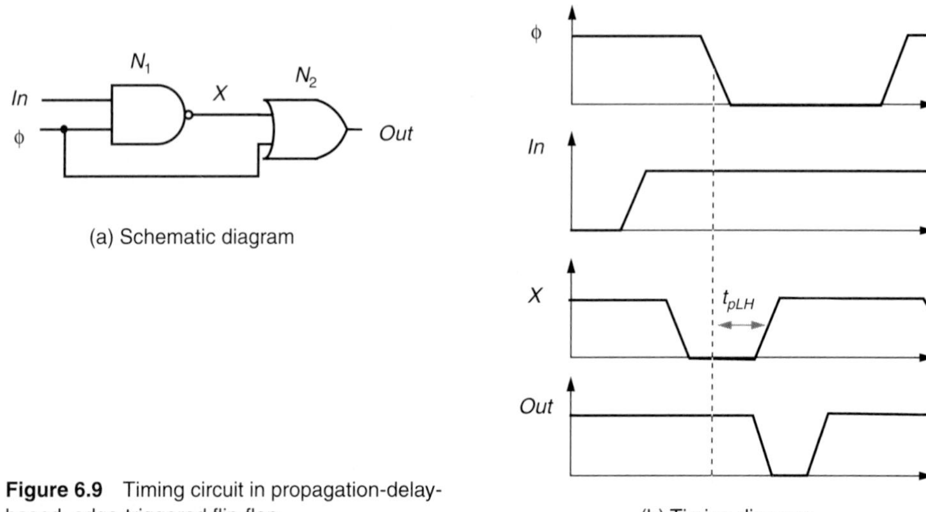

Figure 6.9 Timing circuit in propagation-delay-based, edge-triggered flip-flop.

This idea is applied in the edge-triggered *JK* flip-flop of Figure 6.10. The values of *J* and *K* are sampled at the low-going edge of ϕ and generate short pulses at the *S* or *R* inputs of the latch. The state of the flip-flop thus only changes when the clock goes low, and the output state is determined by the values of the *J* and/or *K* inputs just prior to the clock going low.

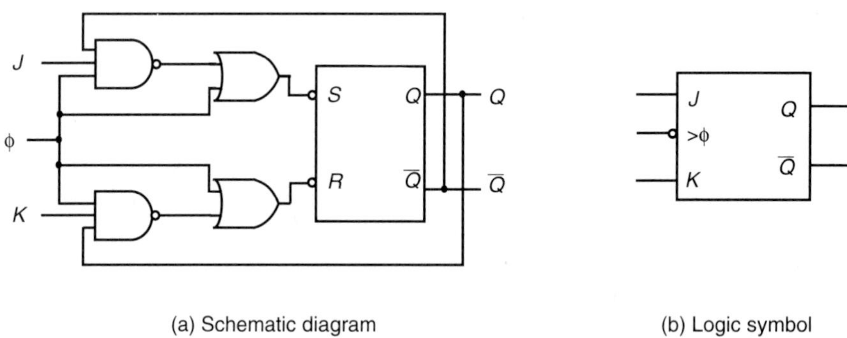

Figure 6.10 Negative edge-triggered *JK* flip-flop.

In order to function correctly, the edge-triggered flip-flop requires the inputs to be stable some time before the clock goes low. This period is called the *set-up time* of the FF and in this case is approximately equal to the propagation delay of the input gate. Other flip-flop designs also require the state of the *J* and *K* lines to be held for some time after the clock edge, known as the *hold* time. The definitions of set-up and hold times are illustrated in the timing diagram of Figure 6.11. The performance of an FF is furthermore qualified by its *propagation delay* t_{pFF}, which equals the time it takes for the output to change after the occurrence of a clock edge.

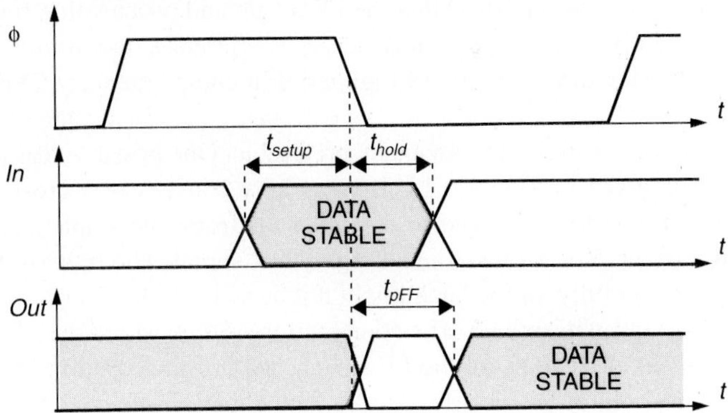

Figure 6.11 Definition of set-up time, hold time, and propagation delay of a flip-flop.

This collection of timing parameters helps to determine the speed of a sequential machine, which consists of a block of combinational logic and a number of (edge-triggered) flip-flops storing the state, as in Figure 6.12. With $t_{p,comb}$ the longest propagation delay through the combinational network, the following relation must hold for the circuit to operate correctly:

$$t_{pFF} + t_{p,\,comb} + t_{setup} < T \tag{6.1}$$

where T is the period of the clock-signal. Eq. (6.1) determines the maximum clock speed for which the circuit still performs correctly.

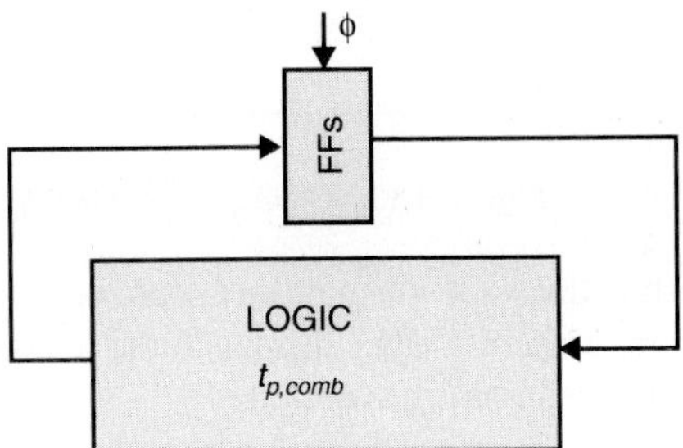

Figure 6.12 Maximum speed of sequential machine using edge-triggered FFs.

The logic symbol of the edge-triggered flip-flop is shown in Figure 6.10b. The symbol is differentiated from the master-slave by the small > sign at the φ input. The presence or absence of a small circle at the clock input indicates whether the outputs change on the falling or the rising edge of the clock. The former case is called *negative edge-triggered* (with dot), while the latter is called *positive edge-triggered.*

The subsequent sections discuss the implementation of the static flip-flop in both the CMOS and bipolar technologies.

6.2.4 CMOS Static Flip-Flops

The simplest way to implement the static flip-flop structures introduced in the previous sections would be to replace the gates by their static CMOS implementations. This produces complex compositions that use an excessive numbers of transistors, and are in general large and slow. Observe from Eq. (6.1) that the FF set-up and propagation times have a direct impact on the global circuit performance. As a reference, the straightforward implementation of the master-slave flip-flop of Figure 6.8 in complementary CMOS consumes 38 transistors.

More efficient implementations are obviously preferable. One possible realization of a clocked SR flip-flop is given in Figure 6.13. It essentially consists of a cross-coupled inverter pair. The extra transistors are used to drive the FF from one stable state to the other upon application of the S or R pulses and a high clock signal. The transistors of the FF must be dimensioned carefully, or the latch might not be switchable. Consider the case where Q is high and an R pulse is applied. The combination of transistors M_4, M_7, and M_8 forms a ratioed inverter. In order to make the FF switch, we must succeed in bringing Q below the switching threshold of the inverter M_1-M_2. Once this is achieved, the positive feedback causes the flip-flop to invert states. This requirement forces us to increase the sizes of transistors M_5, M_6, M_7, and M_8.

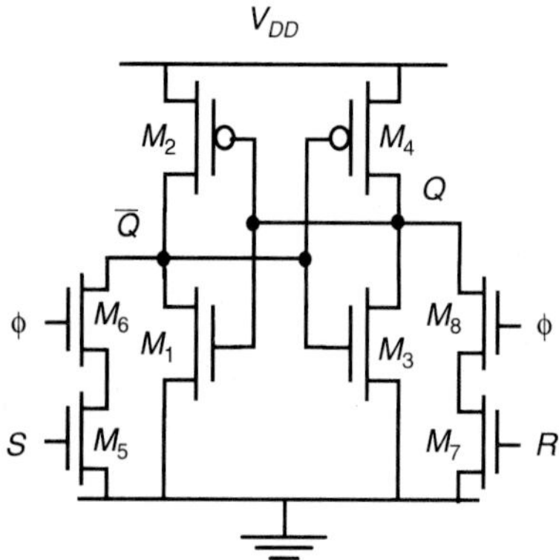

Figure 6.13 CMOS clocked *SR* flip-flop.

It is important to realize that this flip-flop does not consume any static power. Due to the feedback, one inverter is in the high state, while the other resides in the low state. No static paths between V_{DD} and GND can exist except during switching.

Example 6.1 Transistor Sizing of Clocked SR Flip-Flop

Assume that the cross-coupled inverter pair is designed with minimum-size transistors, and that those transistors are scaled so that the inverter threshold V_M is located at $V_{DD}/2$. For the 1.2 μm CMOS technology, the following transistor sizes were selected: $(W/L)_{M1} = (W/L)_{M3} = (1.8/1.2)$, and $(W/L)_{M2} = (W/L)_{M4} = (5.4/1.2)$.

Assuming $Q = 1$, determine the minimum sizes of M_5, M_6, M_7, and M_8 to make the device switchable.

To switch the FF from the $Q = 1$ to the $Q = 0$ state, it is essential that the low level of the ratioed, pseudo-NMOS inverter $(M_7$-$M_8)$-M_4 be below the switching threshold of the inverter M_1-M_2 that equals $V_{DD}/2$. It is reasonable to assume that as long as $V_Q > V_M$, $V_{\overline{Q}}$ equals 0, and the gate of transistor M_4 is grounded. The boundary conditions on the transistor sizes can be derived by equating the currents in the inverter for $V_Q = V_{DD} / 2$, as given in Eq. (6.2). We assume that M_7 and M_8 have identical sizes. Under that condition, the pull-down network can be modeled by a single transistor M_{78}, whose length is twice the length of the individual devices.

$$k_{n,\,M78}\left((V_{DD}-V_{Tn})\frac{V_{DD}}{2}-\frac{V_{DD}^2}{8}\right) = k_{p,\,M4}\left((V_{DD}-|V_{Tp}|)\frac{V_{DD}}{2}-\frac{V_{DD}^2}{8}\right) \tag{6.2}$$

For $V_{Tn} \approx |V_{Tp}|$, this results in the condition that $k_{n,M78} \geq k_{p,M4}$, which translates to the following size constraint:

$$(W/L)_{M78} \geq (\mu_p/\mu_n)\,(W/L)_{M4} = (W/L)_{M3}$$

The latter part of the expression stems from the requirement that in order to set V_M of the inverter M_3-M_4 to $V_{DD}/2$, it is necessary to make the PMOS device M_4 (μ_n/μ_p) times wider than the NMOS transistor M_3. This finally results into minimum size requirements for M_7 and M_8 (and by considering symmetry, M_5 and M_6).

$$(W/L)_{M7} = 2(W/L)_{M78} \geq 2(W/L)_{M3}$$

or $(W/L)_{M7} \geq (3.6/1.2)$.

Figure 6.14 plots the simulated voltage-transfer characteristic (V_Q versus V_R) of the FF for various values of the (W/L) ratios of transistors M_7-M_8. Observe that the flip-flop barely switches for $(W/L) = (3.6/1.2)$. However, once the switching point is reached ($V_R \approx 4.7$ V), the device switches very abruptly. This is due to the positive feedback that comes into effect even when the inverter M_1-M_2 is turned on ever so lightly. This positive feedback effect was ignored in our manual analysis, which explains the small deviation between the predicted minimum size (3.6/1.2) versus the simulated one (3.3/1.2). The simulation further demonstrates that a device size of (1.8/1.2) fails to switch the gate, while a larger device size of (7.2/1.2) places the switching point somewhere in the middle of the voltage range. This is clearly the desirable solution, as it maximizes the noise margins and increases the switching speed.

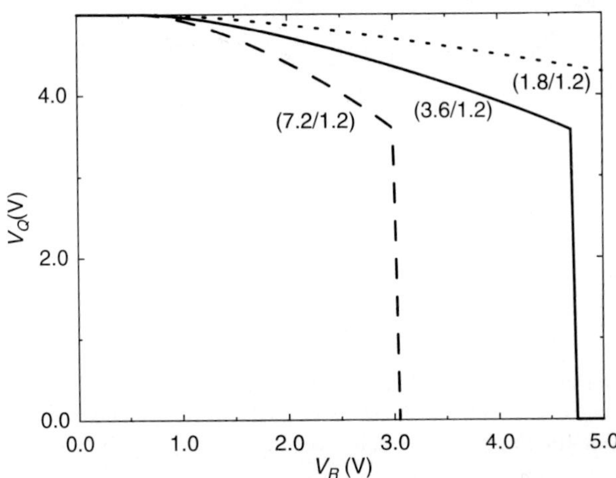

Figure 6.14 Simulated voltage-transfer characteristic of *SR* FF (V_Q versus V_R).

The positive feedback effect makes a manual derivation of propagation delay of the *SR* FF (and other FFs) nontrivial. Some simplifications are therefore necessary. Consider, for instance, the flip-flop of Figure 6.13, where Q and $\overline{Q}$ are set to 0 and 1, respectively. A pulse is applied at node S, causing the flip-flop to toggle. In the first phase of the transient, node $\overline{Q}$ is being pulled down by transistors M_5 and M_6. Since node Q is initially low, the PMOS device M_2 is on while M_1 is off. The transient response is hence determined by the pseudo-NMOS inverter formed by (M_5-M_6) and M_2. Once $\overline{Q}$ reaches the switching threshold of the CMOS inverter M_3-M_4, this inverter reacts and the positive feedback comes into action, turning M_2 off and M_1 on. This accelerates the pulling down of node $\overline{Q}$. From this analysis, we can derive that the propagation delay of node Q is approximately equal to the delay of the pseudo-NMOS inverter formed by (M_5-M_6) and M_2. To obtain the delay for node Q, it is sufficient to add the delay of the complementary CMOS inverter M_3-M_4.

Example 6.2 Propagation Delay of Static SR Flip-Flop

The transient response of the flip-flop of Figure 6.13, as obtained from simulation, is plotted in Figure 6.15. The devices are sized as described in Example 6.1, and a load of a single inverter is assumed for each FF output. The flip-flop is initially in the reset state, and an S-pulse is applied. As we can observe, this results first in a discharging of the $\overline{Q}$ output while Q stays at 0. Once the switching threshold of the inverter M_3-M_4 is reached, the Q output starts to rise. The delay of this transient is solely determined by the M_3-M_4 inverter, which is hampered by the slow rise time at its input. From the simulation results, we can derive that $t_{p\overline{Q}}$ and t_{pQ} equal 270 psec and 710 psec, respectively.

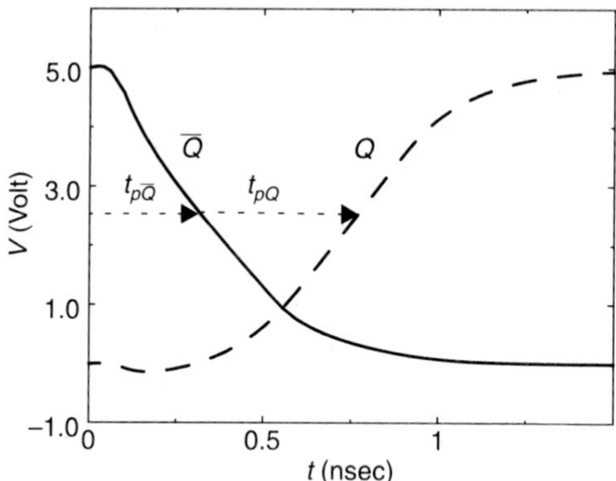

Figure 6.15 Transient response of *SR* flip-flop.

Problem 6.1 Complimentary CMOS *SR* FF

Instead of using the modified *SR* FF of Figure 6.13, it is also possible to use a fully complementary implementation. Derive the transistor schematic (which consists of 12 transistors). This circuit is more complex, but switches faster and consumes less switching power. Explain why.

The circuit of Figure 6.13 forms the basic building block for both master-slave and edge-triggered flip-flops. For instance, a master-slave D-FF implementation based on the latch of Figure 6.13 requires only 22 (10 + 10 + 2) transistors compared to the 38 for the fully complementary solution.

A related approach uses transmission gates (or pass-transistors) under control of the clock signals to connect the triggering pulses to the internal nodes of the cross-coupled inverter pair. The corresponding *SR* latch is shown in Figure 6.16 and requires only 6 transistors. Due to its simplicity, this circuit topology is very popular in the design of static memories, as is discussed in Chapter 10. The structure should be treated with care when used in a datapath—the pass-transistor is a bidirectional device and does not isolate input from output. Consider the circuit of Figure 6.17, which contains two cascaded flip-flops. The feedback inverter of the second FF couples back through the pass-transistor and can prevent the first gate from switching. At best, we can expect a serious deterioration of the transient response. The only way to avoid this feedback is to introduce a unidirectional gate such as a buffer between the two FFs.

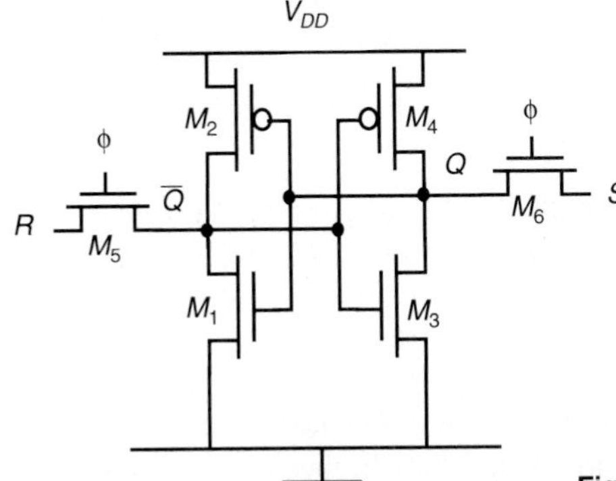

Figure 6.16 Clocked CMOS 6-transistor *SR* flip-flop.

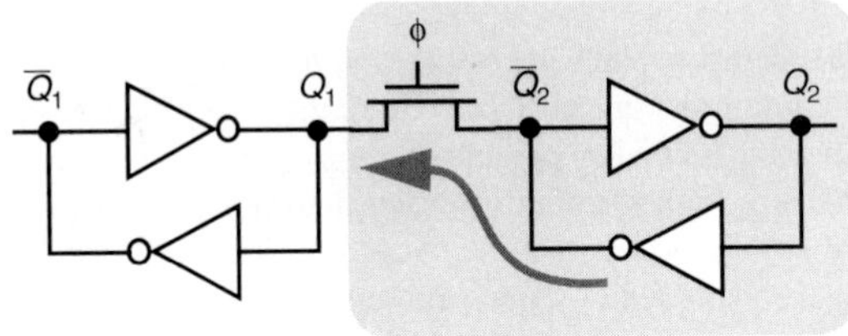

Figure 6.17 Cascade of SR FFs with pass-transistor access devices. The arrow shows how the feedback inverter of the second FF affects the output of the first FF (for $\phi = 1$).

Problem 6.2 SR Flip-Flop with Pass-Transistor Access Devices

Discuss the implications of using an NMOS-only pass-transistor in the circuit of Figure 6.16 instead of full transmission gates.

6.2.5 Bipolar Static Flip-Flops

The standard ECL gate implements the OR/NOR function (Section 3.4). A simple NOR-type *SR* latch is thus readily implemented in ECL, as shown in Figure 6.18. Notice that the circuit is inherently differential, and no voltage references are required. Furthermore, the current source has been simplified to a single resistor. Be aware that more elaborate cur-

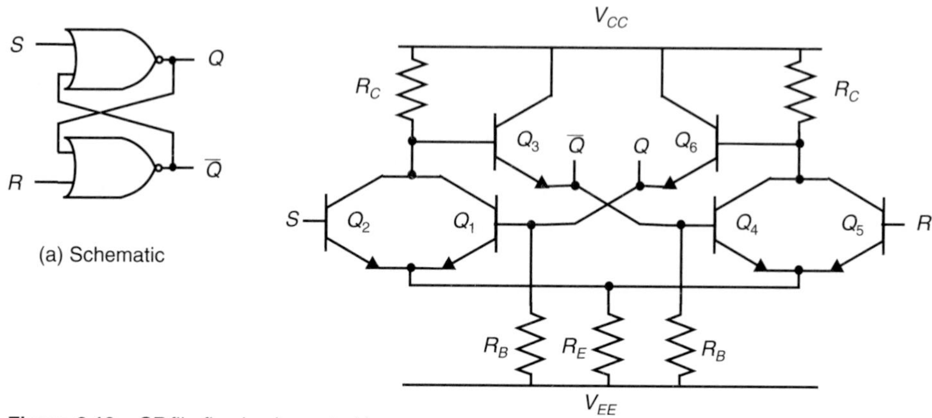

Figure 6.18 *SR* flip-flop implemented in ECL technology.

rent sources can easily be used. When both S and R are low, the flip-flop resides in one of its stable states. Depending upon the stored value, either Q_1 or Q_2 is on, and the current flows through the left or right branches of the differential pair, respectively. Raising S (or R) diverts the current from the right to the left side (or vice-versa) and causes the flip-flop to change states. The proposed circuit topology is only one out of many possible configurations. For instance, the emitter-followers can be eliminated from the feedback path and used only to drive the output load. If this load is small, they can be avoided all together, similar to the approach taken in current mode logic (CML).

Example 6.3 ECL *SR* Flip-Flop

For the circuit of Figure 6.18, determine the values of R_E and R_C so that the voltage swing at the output equals 0.5 V. Use a first-order analysis, but ensure that all devices are operating in forward-active mode. Assume the following parameters: $V_{BE(on)} = 0.7$ V, $V_{BE(sat)} = 0.8$ V, $V_{CE(sat)} = 0.1$ V, $\beta_F = 100$ (or very high for the first-order analysis). The supplies are set to $V_{CC} = GND$ and $V_{EE} = -3.5$ V.

For the first part of the analysis (steady-state operation), we may assume that both the S and R signals are low. We position the flip-flop in the 1 state, which means $Q = 1$ or, equivalently, that Q_1 is on and Q_4 is off. Both Q_2 and Q_5 are off as well. Assuming that the transistors are in forward-active mode and that β_F is high enough to ignore the base currents, the following equations can be easily derived with V_{Cn}, V_{Bn}, and V_{En} respectively the collector, base, and emitter voltages of transistor Q_n.

$$
\begin{gathered}
V_{B1} = V_{CC} - V_{BE(on)} \qquad V_{E1} = V_{B1} - V_{BE(on)} = V_{CC} - 2V_{BE(on)} \\
V_{C1} = V_{CC} - I_{C1}R_C \qquad V_{\overline{Q}} = V_{C1} - V_{BE(on)} \\
I_{C1} \approx I_{E1} = (V_{E1} - V_{EE})/R_E \\
V_Q = V_{C2} - V_{BE(on)} \qquad V_{C2} = V_{CC}
\end{gathered} \tag{6.3}
$$

This produces the following equations for V_Q and $V_{\overline{Q}}$:

$$V_Q = V_{C2} - V_{BE(on)}$$
$$V_{\overline{Q}} = V_{CC} - \frac{R_C}{R_E}(V_{CC} - 2V_{BE(on)} - V_{EE}) - V_{BE(on)} \tag{6.4}$$

Solving Eq. (6.4) yields a relationship between R_C and R_E: $R_E = 4.2\ R_C$. The absolute values of R_C and R_E, as well as R_B, are determined by the desired current levels and the required transient response time. To make the flip-flop switch, it is sufficient to bring either S or R to the V_{OH} level (= –0.7 V).

This is confirmed in the transient simulation of Figure 6.19, where the FF is brought from the 0 state to the 1 state by applying a pulse on the S input. The observed voltage swing (0.44 V) is somewhat lower than the predicted 0.5 V due to the finite base currents. We can also see that the S pulse must be just wide enough to bring the flip-flop over the switching point. Once this is accomplished, bringing the S pulse low has no effect, and the gate safely evolves towards the 1 state.

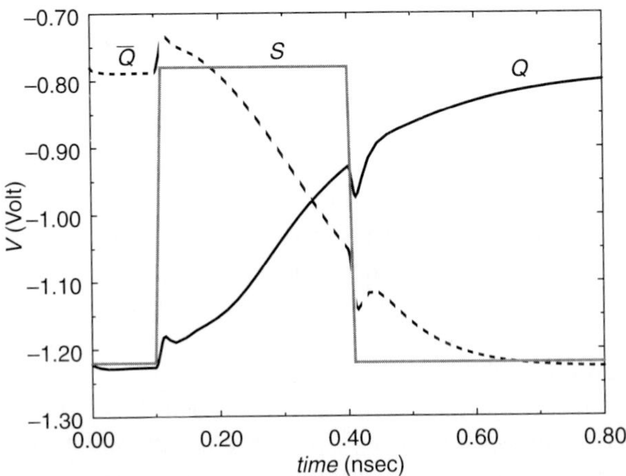

Figure 6.19 Simulated transient response of the *SR* flip-flop of Figure 6.18. The current level is set to 0.5 mA (or R_E = 4.2 kΩ and R_C = 1 kΩ).

While the SR flip-flop structure of Figure 6.18 is rather straightforward, complete ECL flip-flops, such as clocked, edge-triggered, or master-slave FFs, require complex circuit topologies, that are tailored and tuned to obtain the maximum performance. The analysis of those structures is rather involved and is beyond the scope of this text.

6.3 Dynamic Sequential Circuits

Storage in a static sequential circuit relies on the concept that a cross-coupled inverter pair produces a bistable element and can thus be used to memorize binary values. This approach has the useful property that a stored value remains valid as long as the supply voltage is applied to the circuit, hence the name *static*. The static FF is, furthermore, robust and not too sensitive to disturbances on the signal lines. In order to change the state of the FF, the trigger pulse must have a minimum length that is approximately equal to the propagation delay of the inverter chain. The major disadvantage of the static gate is its complexity. Especially in computational structures such as pipelined datapaths, the large size of the static register cell becomes a dominant and restrictive feature.

Positive feedback is not the only means to implement a memory function. A capacitor can act as a memory element as well. The absence of charge denotes a 0, while its presence stands for a stored 1. No capacitor is ideal, unfortunately, and some charge leakage is always present. A stored value can only be kept for a limited amount of time, which is typically in the range of milliseconds. If one wants to preserve signal integrity, a periodic *refresh* of its value is necessary. Hence the name *dynamic* storage. Reading the value of the stored signal from a capacitor without disrupting the charge requires the availability of a device with a high input impedance. Such a device is readily available in a MOS technology, but harder to come by in the bipolar world. Therefore, *bipolar sequential circuits tend to be uniquely static*, while in MOS both approaches are commonly used.

6.3.1 The Pseudostatic Latch

Dynamic CMOS latches can be implemented in many different ways, some of which are presented here. A first approach, called the *pseudostatic* latch, is shown in Figure 6.20a. Once again, the FF core consists of two inverters. The feedback loop is closed as long as ϕ is low or $\overline{\phi}$ high, as shown in Figure 6.20b. In this mode, the loop gain is high and the circuit operates as a bistable element. On the other hand, when the clock ϕ goes high, the feedback loop opens, the loop gain becomes zero, and the input value is sampled and stored on the internal capacitors.[1] In a sense, the flip-flop of Figure 6.20a is still a static circuit, since it retains the stored value as long as ϕ is kept low and the supply voltage is applied. It is worth mentioning that the circuit presented is of the class of the *level-sensitive* circuits. The circuit is in a transparent mode as long as ϕ is high, and its input appears directly at the output, delayed only by the propagation delay of the inverters.

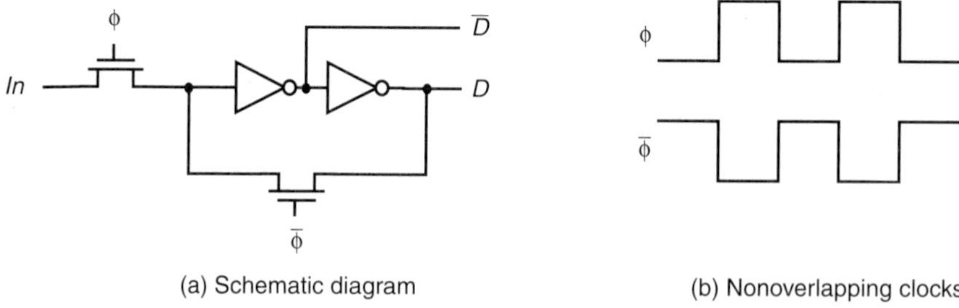

(a) Schematic diagram (b) Nonoverlapping clocks

Figure 6.20 Pseudo-static latch

[1] For the sake of simplicity, we represent a switch as a single NMOS pass-transistor in this chapter. When one wants to avoid the resulting threshold loss, either full transmission gates or level-restoring devices must be used.

A master-slave D FF is constructed by cascading two of the above latches, while reversing the clocks for the second unit (Figure 6.21a). The problem with this circuit is that ϕ and $\overline{\phi}$ might overlap (Figure 6.21b), which may cause two types of failures:

- Node A can become undefined as it is driven by both In and B when ϕ and $\overline{\phi}$ are simultaneously high.
- The input signal In can propagate through both the master and the slave FF if ϕ and $\overline{\phi}$ are simultaneously high for a long enough period. This may cause a *race* condition to occur, which can destroy the state of the FF.

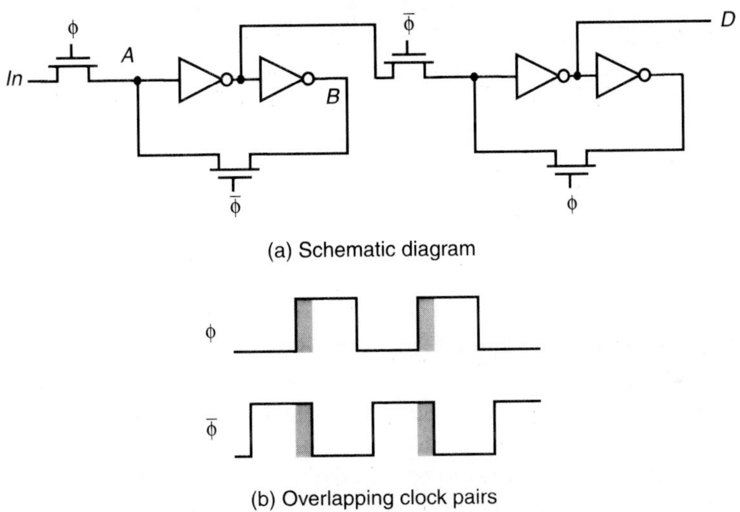

Figure 6.21 Master-slave pseudostatic *D* flip-flop.

Clock overlap can be the result of the clock-generation process, when, for instance, $\overline{\phi}$ is derived from ϕ with the aid of an inverter with its associated propagation delay. Another reason could be the clock-routing network. One clock can experience more propagation delay than the other due to the RC-delay of the wires or due to different loading factors. This effect is called *clock skew* and has become a dominant problem in current high-performance designs. A more detailed discussion of clock skew, its associated effects, and how to cure them follows in Chapter 9, which presents an elaborate discussion on clocking issues.

Those problems can be avoided by using two *nonoverlapping clocks* ϕ_1 and ϕ_2 instead (Figure 6.22), and by keeping the nonoverlap time $t_{\phi 12}$ between the clocks large enough such that no overlap occurs even in the presence of clock-routing delays. During the nonoverlap time, the FF is in the high-impedance state—the feedback loop is open, the loop gain is zero, and the input is disconnected. Leakage will destroy the state if this condition holds for too long a time. Therefore $t_{\phi 12}$ must be kept smaller than 1 to 2 msec to avoid malfunction (this is never a real constraint in actual designs). Hence the name *pseudostatic*: the FF employs a combination of static and dynamic storage approaches depending upon the state of the clock.

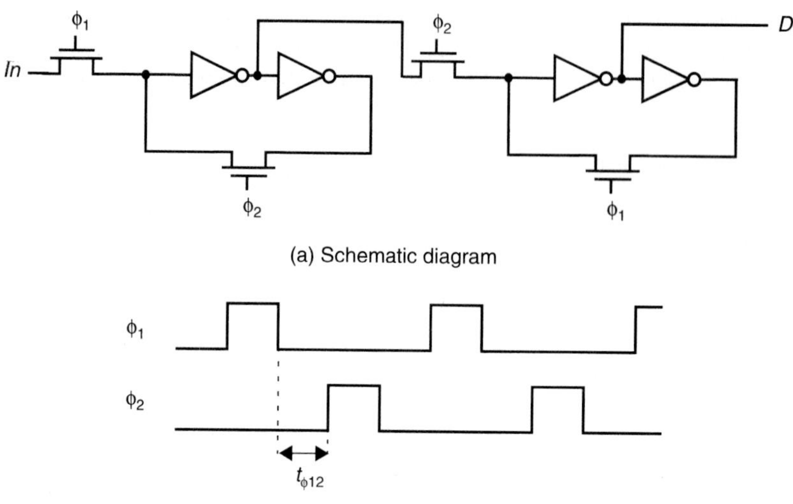

Figure 6.22 Pseudostatic two-phase *D* FF.

6.3.2 The Dynamic Two-Phase Flip-Flop

We can reduce the complexity of the flip-flop even more by using a fully dynamic approach, as shown in the master-slave *D* FF of Figure 6.23. The gate presented requires only six transistors, but the state has to be refreshed at periodic intervals to prevent a loss due to charge leakage. This circuit is often used in pipelined datapaths and register files for micro- or signal processors. In those modules, data storage in the latches is overwritten on a periodic base anyway, and an explicit refresh is not necessary.

A disadvantage of the circuit is that correct operation requires the availability of two nonoverlapping clocks, or four if complementary transmission gates are used. Similar to

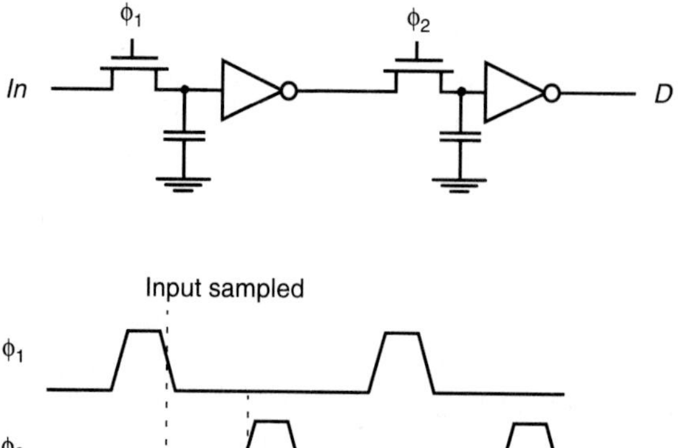

Figure 6.23 Dynamic master-slave *D* FF and corresponding timing diagram.

the pseudostatic gate, the dynamic FF is sensitive to overlap between the ϕ_1 and ϕ_2 clocks. If the clocks are simultaneously high for long enough, both switches are on simultaneously, which creates a race condition. It is therefore important to ensure that the nonoverlapping condition is valid over the complete chip by making $t_{\phi 12}$ sufficiently large. Unfortunately, this has a negative impact on the circuit performance, as will be demonstrated in Chapter 9. Furthermore, distributing a pair (or two pairs) of nonoverlapping clocks over a large die and ensuring that they remain nonoverlapping is cumbersome and counterproductive. A number of approaches to circumvent this constraint have therefore been devised, as will become apparent in the following sections.

Problem 6.3 Dynamic Master-Slave *D* FF

Given the dynamic master-slave *D* FF of Figure 6.23, determine the values of the set-up time as well as the propagation delay of the register.

6.3.3 The C²MOS Latch

The pseudodynamic or dynamic *D* FFs malfunction when the clocks (ϕ-$\overline{\phi}$ or ϕ_1-ϕ_2) overlap for a length of time. Figure 6.24 shows an ingenious *D* FF, which is insensitive to overlap. This circuit is called the *C²MOS register* [Suzuki73].

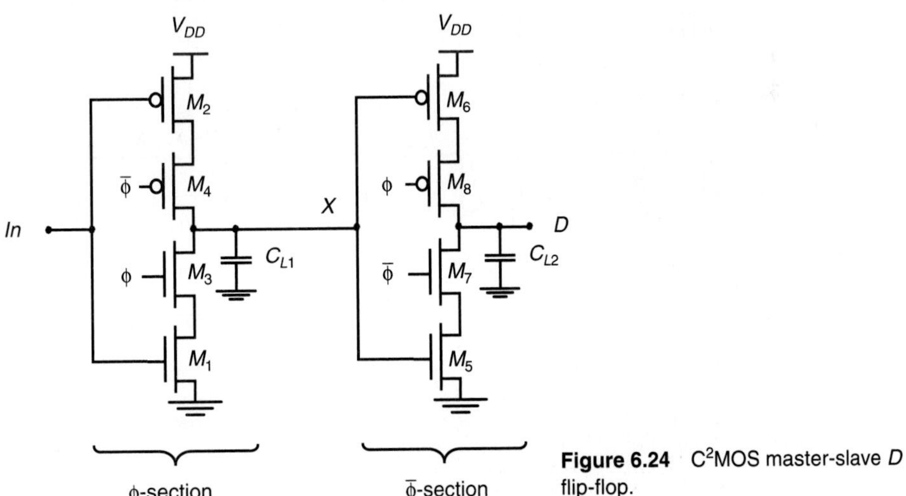

Figure 6.24 C²MOS master-slave *D* flip-flop.

The register operates in two phases.

1. $\phi = 1$ ($\overline{\phi} = 0$): The ϕ-section acts as an inverter, because M_3 and M_4 are on. It is said to be in the *evaluation mode*. Meanwhile, the $\overline{\phi}$ section is in a high-impedance mode, or in a *hold mode*. Both transistors M_7 and M_8 are off, decoupling the output from the input. The output *D* retains its previous value stored on the output capacitor C_{L2}.

2. The roles are reversed during $\phi = 0$: The first section is in hold mode (M_3-M_4 off), while the second section evaluates (M_7-M_8 on). The value stored on C_{L1} propagates to the output node.

The overall circuit operates as a negative edge-triggered master-slave *D* FF. During $\phi = 1$, the input is sampled and stored on C_{L1}. When $\phi = 0$, this value is transferred to C_{L2} and appears at the output of the gate. Most important is the following observation:

> A C^2MOS register with (ϕ-$\overline{\phi}$) clocking is insensitive to overlap, as long as the rise and fall times of the clock edges are sufficiently small.

In order to establish the truth of this statement, we have to examine both the (0-0) and (1-1) overlap cases. In the (1-1) overlap case, the circuit simplifies to the network shown in Figure 6.25a. Inspection of this circuit reveals that a race through this network is simply not possible. An input signal cannot propagate to the output, since only the pull-down networks are enabled. As the circuit consists of a cascade of inverters, signal propagation requires one pull-up followed by a pull-down, or vice-versa, which is not feasible in the situation presented. This effectively breaks any cycle that might exist in the circuit. A similar situation occurs in the case of a (0-0) overlap (Figure 6.25b). Here, only the pull-up networks are enabled, and signal propagation from input to output is impossible.

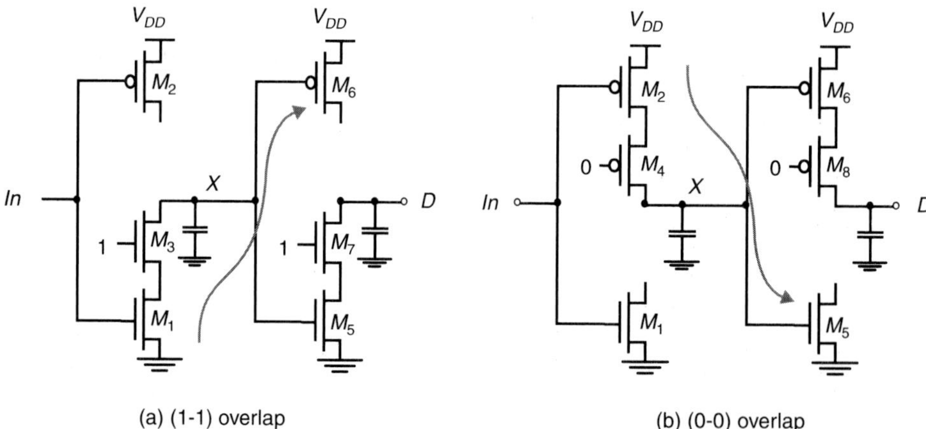

Figure 6.25 C^2MOS *D* FF during overlap periods. No feasible signal path can exist between *In* and *D*, as illustrated by the arrows.

In summary, it can be stated that the C^2MOS latch is insensitive to clock overlaps because those overlaps activate either the pull-up or the pull-down networks of the latches, but never both of them simultaneously. If the *rise and fall times of the clock* ($t_{r\phi}$, $t_{f\phi}$) are sufficiently slow, however, there exists a time slot where both the NMOS and PMOS transistors are conducting. This creates a path between input and output that can destroy the state of the circuit. Simulations have shown that the circuit operates correctly as long as the clock rise time $t_{r\phi}$ (or fall time $t_{f\phi}$) is smaller than approximately five times the propagation delay of the latch (t_{pC2MOS}). This criterion is not too stringent and is easily met in practical designs. The impact of the rise and fall times is illustrated in Figure 6.26, which plots the simulated transient response of a C²MOS D FF for clock slopes of respectively 0.1 and 3 nsec (for an initial propagation delay of approximately 0.3 nsec). For slow clocks, the potential for a race condition exists.

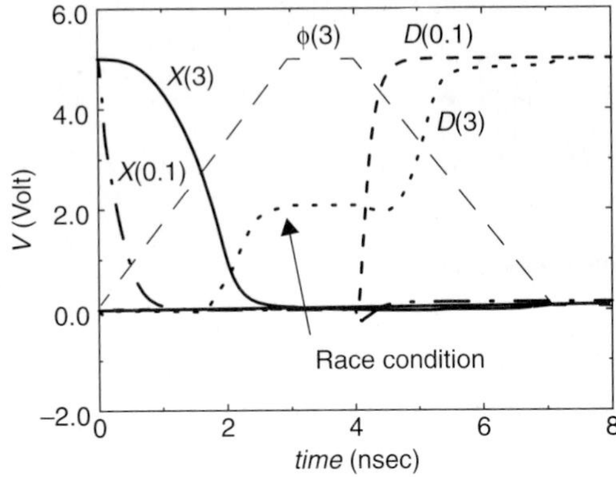

Figure 6.26 Transient response of C²MOS FF for 0.1 nsec and 3 nsec clock rise (fall) times assuming $In = 1$.

The C²MOS latch is especially useful for very high speed operation. In that case, avoiding clock overlap becomes hard. Since the cell requires fewer contacts than the pass-transistor-based cell, its layout is more compact. It is easily verified that this circuit also operates under control of a two-phase clock pair (ϕ_1-ϕ_2).

Sideline: A Brief Introduction to Pipelining

The need for fast, yet small registers especially arises in pipelined datapath structures. *Pipelining* is a popular design technique often used to accelerate the operation of the datapaths of micro- and signal processors. The idea is easily explained with the example of Figure 6.27a. The goal of the presented circuit is to compute $\log(|a - b|)$, where both a and b represent streams of numbers, that is, the computation must be performed on a large set of input values. Using Eq. (6.1), we can determine the minimal clock period T_{min} necessary to ensure a correct evaluation of the results.

$$T_{min} = t_{p,\,reg} + t_{p,\,logic} + t_{setup,\,reg} \tag{6.5}$$

with $t_{p,reg}$ and $t_{setup,reg}$ the propagation delay and the setup times of the registers, respectively. For the sake of simplicity, we assume that the registers are edge-triggered D FFs. The term $t_{p,logic}$ stands for the worst-case delay path through the combinatorial network, which consists of the adder, absolute value, and logarithm functions. The latter delay is generally much larger than the delays associated with the registers and dominates the circuit performance. *Pipelining* is a means to break this performance bottleneck. Assume that we introduce registers between the logic blocks, as shown in Figure 6.27b. This causes the computation for one set of input data to spread over a number of clock periods, as shown in Table 6.1. The result for the data set (a_1, b_1) only appears at the output after three clock-periods. At that time, the circuit has already performed parts of the computations for the next data sets, (a_2, b_2) and (a_3, b_3). The computation is performed in an assembly-line fashion, hence the name pipeline.

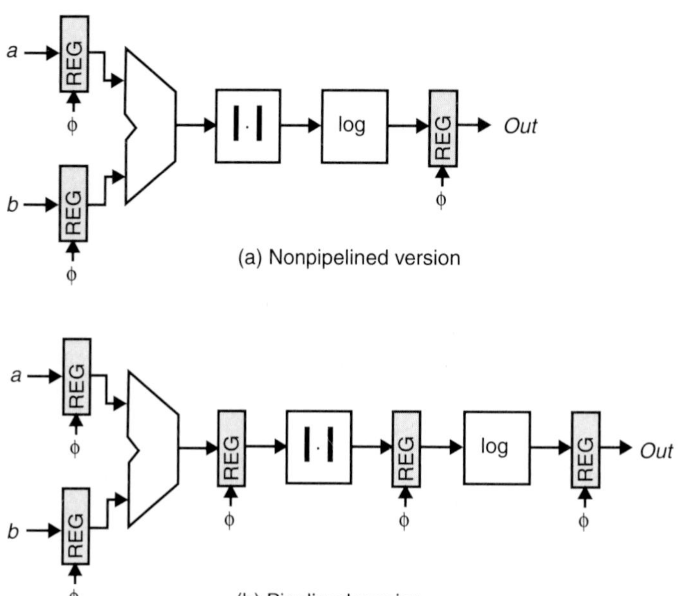

Figure 6.27 Datapath for the computation of log(|*a* + *b*|).

Table 6.1 Example of pipelined computations.

Clock Period	Adder	Absolute Value	Logarithm
1	$a_1 + b_1$		
2	$a_2 + b_2$	$\|a_1 + b_1\|$	
3	$a_3 + b_3$	$\|a_2 + b_2\|$	$\log(\|a_1 + b_1\|)$
4	$a_4 + b_4$	$\|a_3 + b_3\|$	$\log(\|a_2 + b_2\|)$
5	$a_5 + b_5$	$\|a_4 + b_4\|$	$\log(\|a_3 + b_3\|)$

The advantage of the pipelined operation becomes apparent when examining the maximum clock speed (or the minimum clock period) of the modified circuit. The combinational circuit block has been partitioned into three sections, each of which has a smaller propagation delay than the original function. This effectively reduces the value of the minimum allowable clock period:

$$T_{min,\,pipe} = t_{p,\,reg} + max(t_{p,\,add}, t_{p,\,abs}, t_{p,\,\log}) + t_{setup,\,reg} \tag{6.6}$$

Suppose that all logic blocks have approximately the same propagation delay, and that the latch delays are still ignorable with respect to the logic delays. The pipelined network outperforms the original circuit by a factor of three under these assumptions, or $T_{pipe,min} = T_{min}/3$. The increased performance comes at the relatively small cost of two additional registers and an increased latency.[2] This explains why pipelining is popular in the implementation of very high performance datapaths. Be aware, however, that adding extra pipeline stages only makes sense up to a certain point. When the delays of the registers become comparable to the logic delays, no extra performance is gained. Adding extra stages only increases the hardware overhead.

6.3.4 NORA-CMOS—A Logic Style for Pipelined Structures

The concept of the C^2MOS latch can now be extended to support the effective implementation of pipelined circuits. In the following discussion, we use without loss of generality the ϕ-$\overline{\phi}$ notation to denote a two-phase clock system. All the results presented are equally valid for a two-phase ϕ_1-ϕ_2 clocking approach.

Consider first the pipelined circuit of Figure 6.28. The pipeline registers are implemented using pass-transistor-based D latches. When the clocks ϕ and $\overline{\phi}$ are nonoverlapping, correct pipeline operation is obtained. Input data is sampled on C_1 at the negative

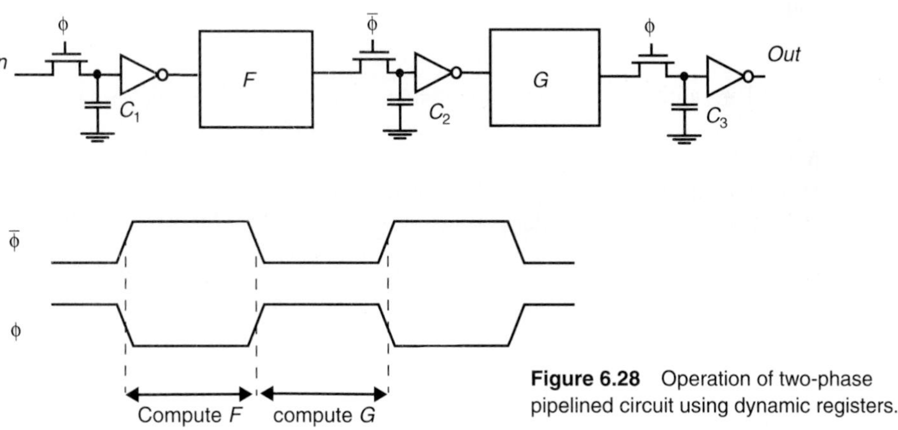

Figure 6.28 Operation of two-phase pipelined circuit using dynamic registers.

edge of ϕ; the result of the first logic block is stored on C_2 when $\overline{\phi}$ goes low, and so on. The nonoverlapping of the clocks ensures correct operation. When the clocks do overlap, however, a race condition can occur. Under correct conditions, the value stored on C_2 is the result of passing the previous input (stored on the falling edge of ϕ on C_1) through the logic function *F*. When ϕ overlaps $\overline{\phi}$, the next input is already being applied to *F*, and its effect might propagate to C_2 before $\overline{\phi}$ goes low. In other words, a *race* develops between the previous input and the current one. Which value wins depends upon the logic function

[2] Latency is defined here as the number of clock cycles it takes for the data to propagate from the input to the output. For the example at hand, pipelining increases the latency from 1 to 3. An increased latency is in general acceptable, but can cause a global performance degradation if not treated with care.

F, the overlap time, and the value of the inputs since the propagation delay is often a function of the applied inputs. The latter factor makes the detection and elimination of race conditions nontrivial.

The same circuit can also be implemented using C^2MOS latches, as shown in Figure 6.29. The operation is similar to the one discussed above. This topology has one additional, important property:

> A C^2MOS-based pipelined circuit is race-free as long as all the logic functions F (implemented using static logic) between the latches are noninverting.

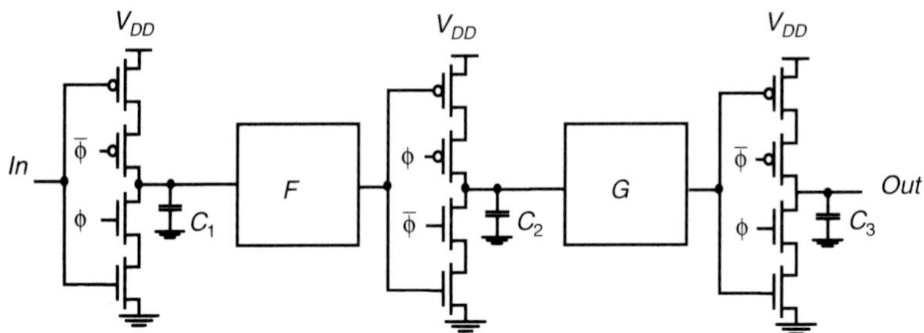

Figure 6.29 Pipelined datapath using C^2MOS latches.

The validity of the above argument can be explained as follows. During a (1-1) overlap between ϕ and $\overline{\phi}$, all C^2MOS latches, simplify to pure pull-down networks (see Figure 6.25). The only way a signal can race from stage to stage under this condition is when the logic function F is inverting, as illustrated in Figure 6.30, where F is replaced by a single, static CMOS inverter. Similar considerations are valid for the (0-0) overlap.

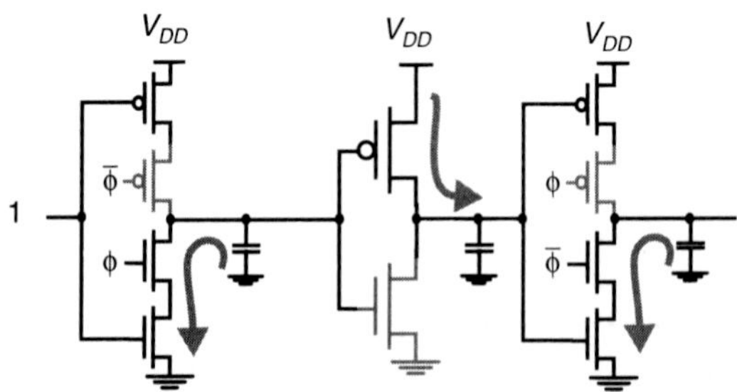

Figure 6.30 Potential race condition during (1-1) overlap in C^2MOS-based design.

Based on this concept, a logic circuit style called *NORA-CMOS* was conceived [Goncalves83]. NORA stands for NO-RAce logic and targets the implementation of fast, pipelined datapaths using dynamic logic. It combines *C^2MOS* pipeline registers and *np-CMOS* dynamic logic function blocks (see Chapter 4). Each module consists of a block of combinational logic that can be a mixture of static and dynamic logic, followed by a C^2MOS latch. Logic and latch are clocked in such a way that both are simultaneously in either evaluation, or hold (precharge) mode. A block that is in evaluation during $\phi = 1$ is called a *ϕ-module*, while the inverse is called a *$\overline{\phi}$-module*. Examples of both classes are shown in Figure 6.31 a and b, respectively. The operation modes of the modules are summarized in Table 6.2.

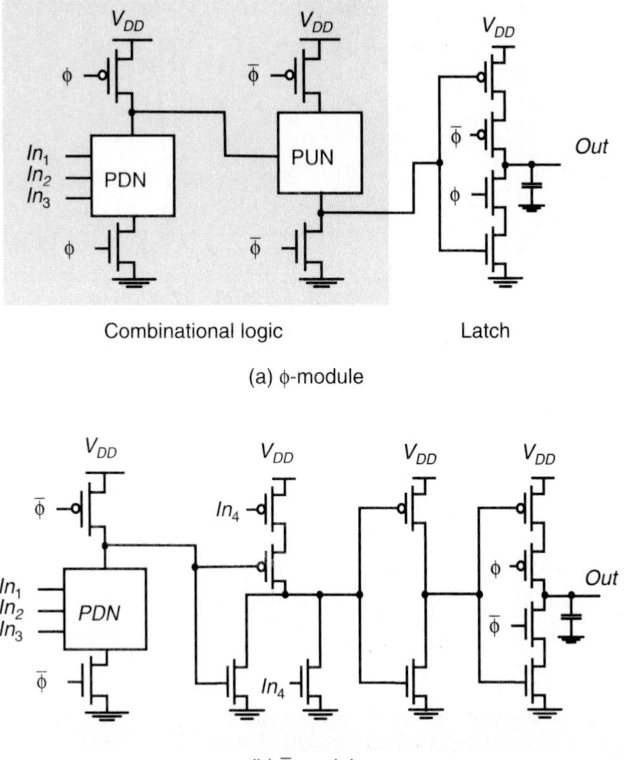

Figure 6.31 Examples of NORA CMOS Modules.

Table 6.2 Operation modes for NORA logic modules.

	ϕ **block**		$\overline{\phi}$ **block**	
	Logic	**Latch**	**Logic**	**Latch**
$\phi = 0$	Precharge	Hold	Evaluate	Evaluate
$\phi = 1$	Evaluate	Evaluate	Precharge	Hold

A NORA datapath consists of a chain of alternating ϕ and $\overline{\phi}$ modules. While one class of modules is precharging with its output latch in hold mode, preserving the previous output value, the other class is evaluating. Data is passed in a pipelined fashion from module to module. The resulting datapath combines high performance with high layout density. To illustrate this concept, a pipelined chain of two inverters and its simulated transient response are shown in Figure 6.32. The various operation modes of both logic and register are easily discerned. Other examples of NORA logic appear in Chapter 7.

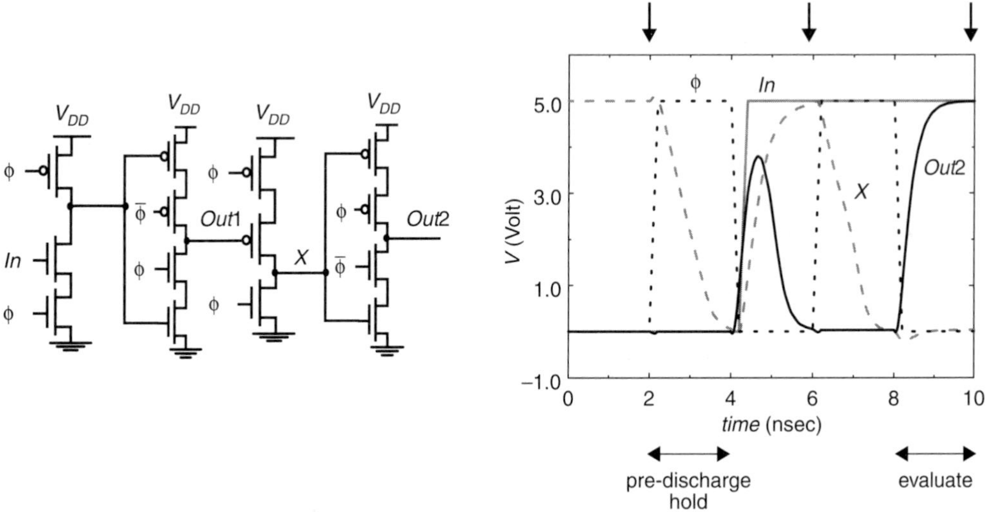

Figure 6.32 Simulated response of pipelined, cascaded inverters. The plots show an internal, dynamic node and the output node of the second C^2MOS register. Observe how the input takes two clock ticks to traverse to the output. The arrows on the top point out when the output is valid.

NORA offers designers a wide range of design choices. Dynamic and static logic can be mixed freely, and both ϕp and ϕn dynamic blocks can be used in cascaded or in pipelined form. With this freedom of design, extra inverter stages, as required in DOMINO-CMOS, are most often avoided.

Design Rules

In order to ensure correct operation, two important rules should always be followed:

- **The dynamic-logic rule:** Inputs to a dynamic ϕn (ϕp) block are only allowed to make a single $0 \rightarrow 1$ ($1 \rightarrow 0$) transition during the evaluation period (Chapter 4).
- **The C^2MOS rule:** In order to avoid races, the number of static inversions between C^2MOS latches should be even.

The presence of dynamic logic circuits (with precharge and evaluation) requires the introduction of some extensions to the latter rule. Consider the situation pictured in Figure 6.33a. During precharge ($\phi = 0$), the output register of the module has to be in hold mode, isolating the output node from the internal events in the module. Assume now that a (0-0) overlap occurs. Node A gets precharged to V_{DD}, while the latch simplifies to a pull-up network (Figure 6.33b). It can be observed that under those circumstances the output node charges to V_{DD}, and the stored value is erased! This malfunctioning is caused by the fact that the number of static inversions between the last dynamic node in the module and the latch is odd, which creates an active path between the precharged node and the output.

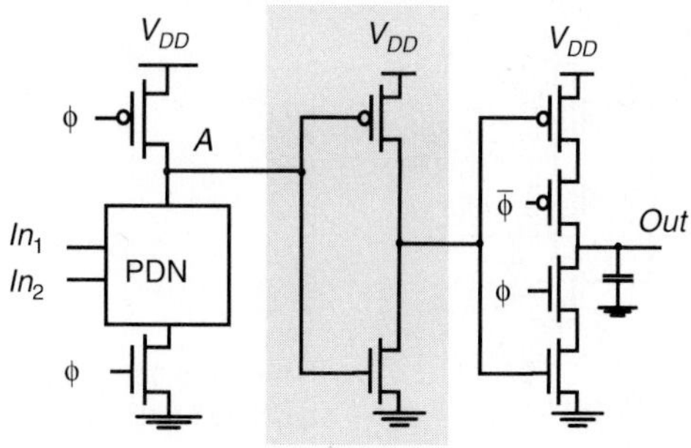

(a) Circuit with odd number of static inversions between dynamic logic stage and register

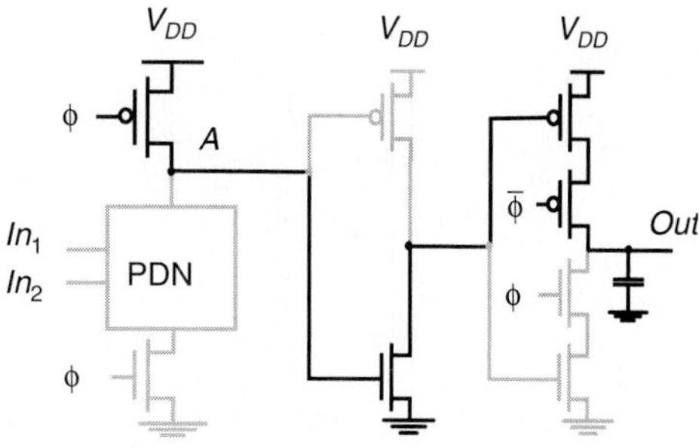

(b) Same circuit during (0-0) clock overlap.

Figure 6.33 Extended C^2MOS rules.

This translates into the following rule: The number of static inversions between the last dynamic block in a logic function and the C^2MOS latch should be *even*. This and similar considerations lead to a reformulated C^2MOS rule [Goncalvez83].

Revised C^2MOS Rule

- The number of static inversions between C^2MOS latches should be even (in the absence of dynamic nodes); if dynamic nodes are present, the number of static inverters between a latch and a dynamic gate in the logic block should be even. The number of static inversions between the last dynamic gate in a logic block and the latch should be even as well.

Adhering to the above rules is not always trivial and requires a careful analysis of the logic equations to be implemented. This often makes the design of an operational NORA-CMOS structure cumbersome. Furthermore, NORA is a dynamic logic style and is thus sensitive to problems such as charge sharing and leakage. Its use should only be considered when maximum circuit performance is a must.

6.3.5 True Single-Phase Clocked Logic (TSPCL)

It now turns out that the NORA design style can easily be simplified, so that a **single clock** (without an inverse clock) is sufficient to correctly operate a dynamic-sequential CMOS circuit [Yuan89]. This requires a redesign of the C^2MOS latch. The modified latches are shown in Figure 6.34 and are called the *doubled n-C^2MOS* and the *doubled p-C^2MOS* latches (replacing the ϕ- and $\overline{\phi}$-C^2MOS latches, respectively). Consider the doubled *n*-C^2MOS latch. When ϕ is high, the latch is in the transparent *evaluate mode* and corresponds to two cascaded inverters; hence it is noninverting. On the other hand, when $\phi = 0$, both inverters are disabled, and the latch is in *hold-mode*. Only the pull-up networks are

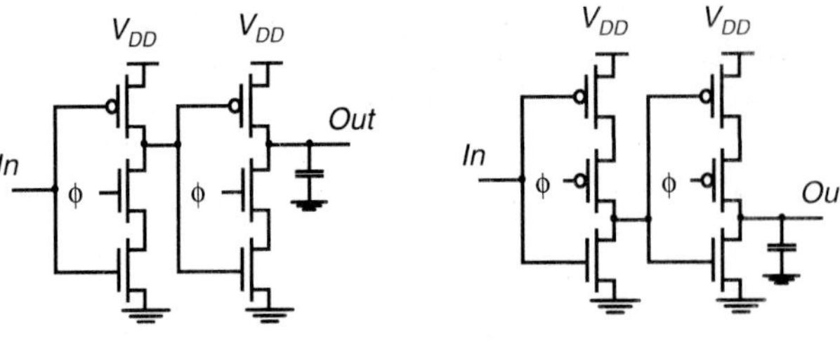

Figure 6.34 Doubled C^2MOS latches.

still active, while the pull-down circuits are deactivated. As a result of the dual- stage approach, no signal can ever propagate from the input of the latch to the output in this operation mode. Races are thus eliminated.

Compared to the C^2MOS approach, this design has the advantage that virtually all constraints are removed. There are no even-inversion constraints between two latches or between a latch and a dynamic block. Dynamic and static circuits can be mixed freely. The logic functions can be included in the n-C^2MOS or p-C^2MOS latches, or placed between them, as shown in Figure 6.35. The only disadvantage of this approach is an increased number of transistors per latch (six instead of four).

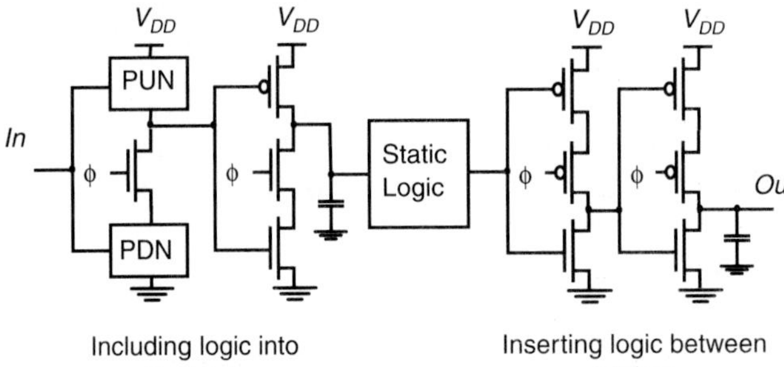

Figure 6.35 Adding logic to the TSPC approach.

The latches can further be simplified into the circuits of Figure 6.36, where only the first inverter is controlled by the clock. Besides the reduced number of transistors, these circuits have the advantage that the clock load is reduced by half. On the other hand, not all node voltages in the latch experience the full logic swing. For instance, the voltage at node A (for $V_{in} = 0$ V) maximally equals $V_{DD} - V_{Tn}$, which results in a reduced drive for the output NMOS transistor and a loss in performance.

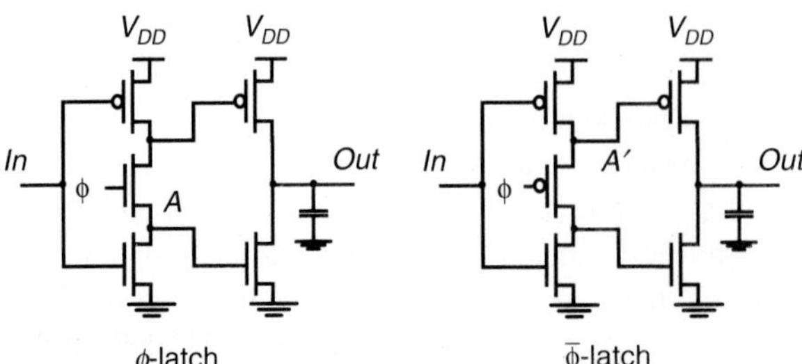

Figure 6.36 Simplified TSPC latch (also called split-output).

The resulting design methodology is called *true single-phase clock logic (TSPC)*, because it allows for the implementation of dynamic sequential circuits with a single clock. The basic TSPC latches can be combined in many different ways to implement all essential sequential components. For instance, a number of implementations of an edge-triggered *D* FF using TSPC are shown in Figure 6.37. While the first two circuits use the doubled latch, the last version employs the five-transistor split-output latch. The resulting gate achieves almost the same speed as the first two, but reduces the clock load substantially (two clock connections instead of four). The latter benefit turns out to be of major importance, especially for circuits employing many registers such as large shift-registers.

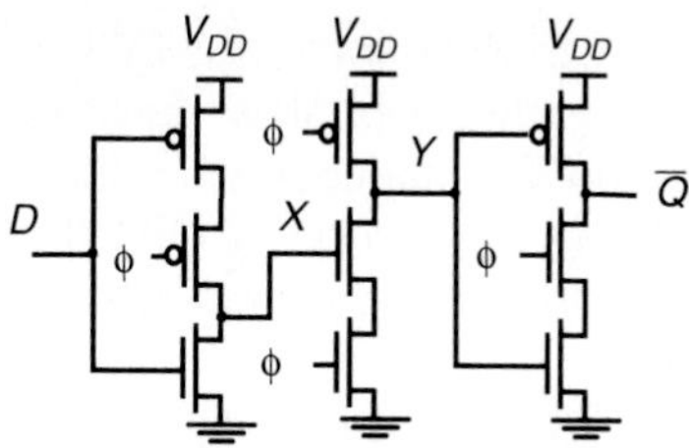

(a) Positive edge-triggered *D* flip-flop

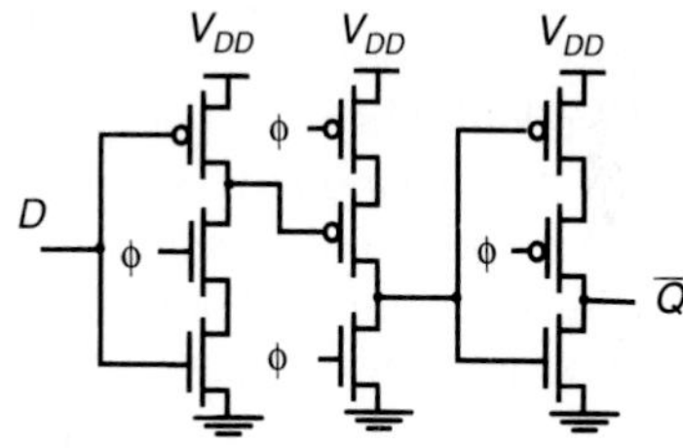

(b) Negative edge-triggered *D* flip-flop

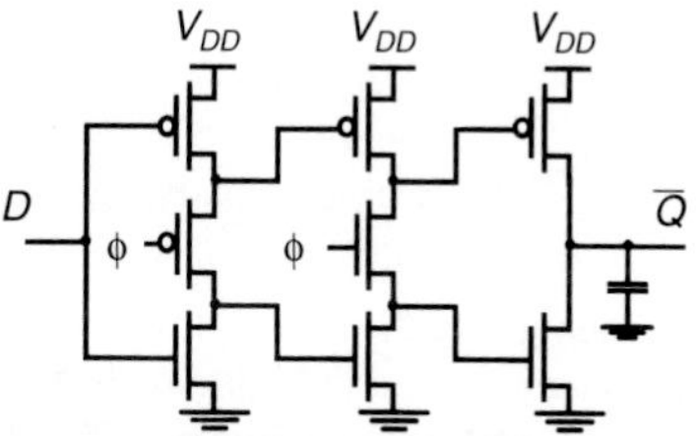

(c) Positive edge-triggered *D* flip-flop using split-output latches

Figure 6.37 *D*-FFs in TSPC.

Depending upon the surrounding logic, further simplifications can be introduced in the TSPC register structure. For instance, a NORA ϕ-section where all combinational logic is implemented with dynamic gates, can be reduced to the structure of Figure 6.38. Since the upper PMOS transistor of the C^2MOS latch is always off during precharge, we

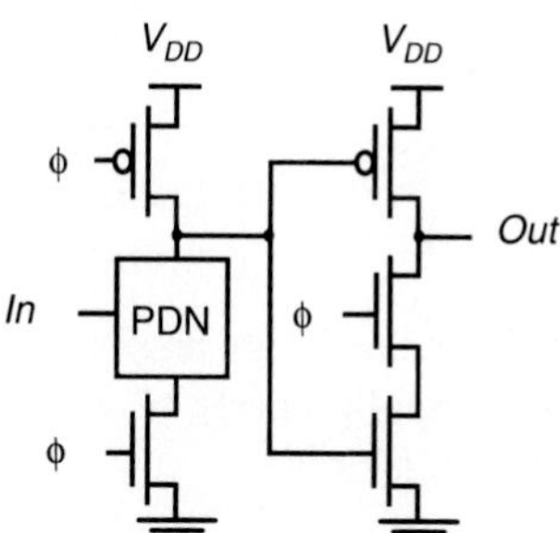

Figure 6.38 A NORA ϕ-section in TSPC.

can eliminate the $\overline{\phi}$-clocked enabling transistor of the C^2MOS latch without further consequences.

TSPC has been used successfully for the implementation of a number of very high speed CMOS circuits. For instance, TSPC was the circuit style of choice for the implementation of the datapath registers in the alpha microprocessor of Digital Equipment Corporation [Dopperpuhl92], which achieves a clock speed of 200 MHz in a 0.75 μm CMOS technology.

Design Considerations

One word of caution is necessary: Similar to the C^2MOS latch, the TSPC latch malfunctions when the *slope of the clock* is not sufficiently steep. Slow clocks cause both the NMOS and PMOS clocked transistors to be on simultaneously, resulting in undefined values of the states and race conditions. The clock slopes should therefore be carefully controlled. If necessary, local buffers must be introduced to ensure the quality of the clock signals.

Furthermore we must not forget that the TSPC register is dynamic. The high impedance of the storage nodes make the circuit very sensitive to noise and leakage. Often, a feedback transistor is added to the structure (in the style of DOMINO CMOS) to make the gate pseudostatic.

Example 6.4 TSPC Edge-Triggered Flip-Flop

To illustrate the operation of a TSPC flip-flop, we simulated the transient response of the positive edge-triggered FF of Figure 6.37a. The results are plotted in Figure 6.39. The signals in the bottom plot represent the internal signals *X* and *Y*, as marked in Figure 6.37a. This circuit clearly exhibits a positive edge-triggered behavior. We can also discern the operation modes of the two composing latches. The response displays all the typical artifacts of a dynamic circuit. For example, the output signal rises substantially above the supply voltage for a negative-going edge of the clock. This reflects the effect of the precharging of node Y, being capacitively coupled to the output node.

From the top-level plot, the propagation delay of the FF for a low-to-high transition of the output can be derived as the distance between the 50% crosspoints of the positive edge of the clock and the output signal. For a fan-out of 1, a delay of 520 psec is observed. No special transistor optimizations were applied in the design of this flip-flop, and all devices are minimum-size.

6.4 Non-Bistable Sequential Circuits

In the preceding sections, we have focused our attention on the design and control of a single sequential element (register or latch). The most important property of such a circuit is that it has two stable states. Circuits of this class are called *bistable*.

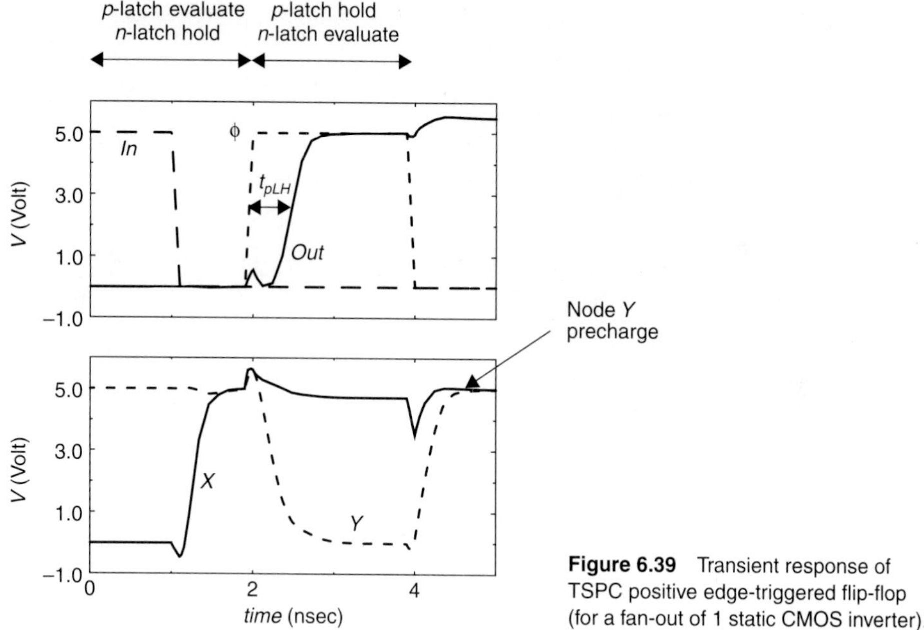

Figure 6.39 Transient response of TSPC positive edge-triggered flip-flop (for a fan-out of 1 static CMOS inverter).

The bistable element is not the only sequential circuit of interest. Other regenerative circuits can be cataloged as *astable* or *monostable*. The former ones act as oscillators and can, for instance, be used for on-chip clock generation. The latter ones serve as pulse generators, also called *one-shot circuits*. Another interesting regenerative circuit is the Schmitt trigger. This component has the useful property of showing hysteresis in its dc characteristics, because its switching threshold is variable and depends upon the direction of the transition (low-to-high or high-to-low). This peculiar feature can come in handy in noisy environments.

6.4.1 The Schmitt Trigger

A Schmitt trigger [Schmitt38] is a device with two important properties:

1. It responds to a slowly changing input waveform with a *fast transition time at the output*.

2. The voltage-transfer characteristic of the device displays *different switching thresholds for positive- and negative-going input signals*. This is demonstrated in Figure 6.40, where a typical voltage-transfer characteristic of the Schmitt trigger is shown (and its schematics symbol). The switching thresholds for the low-to-high and high-to-low transitions are called V_{M+} and V_{M-}, respectively. The *hysteresis voltage* is defined as the difference between the two.

One of the main uses of the Schmitt trigger is to turn a noisy or slowly varying input signal into a clean digital output signal. This is illustrated in Figure 6.41. Notice how the

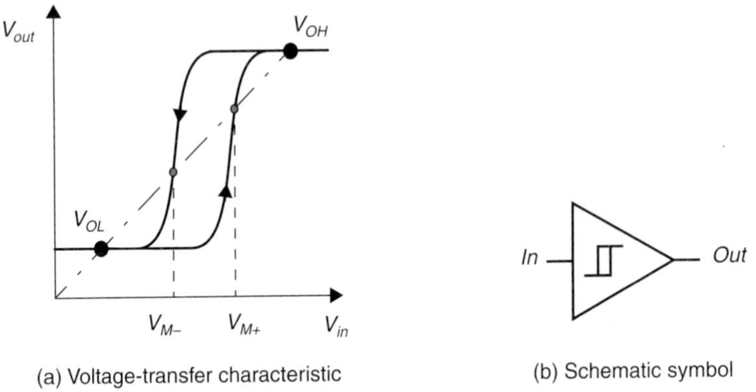

(a) Voltage-transfer characteristic (b) Schematic symbol

Figure 6.40 Noninverting Schmitt trigger.

hysteresis suppresses the ringing on the signal. At the same time, the fast low-to-high (and high-to-low) transitions of the output signal should be observed. For instance, steep signal slopes are beneficial in reducing power consumption by suppressing direct-path currents. The "secret" behind the Schmitt trigger concept is the use of positive feedback.

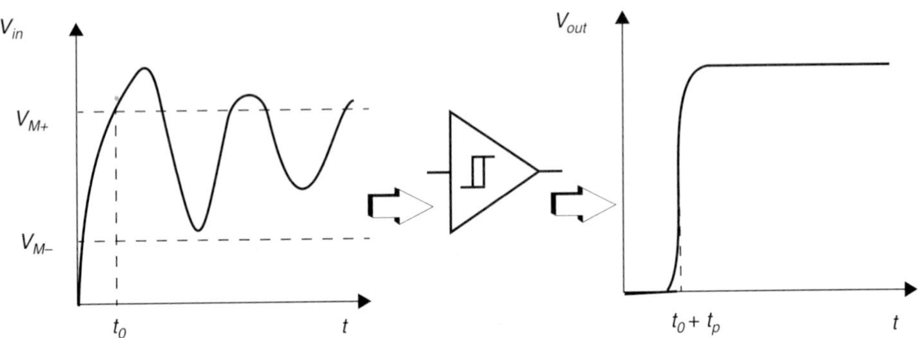

Figure 6.41 Noise suppression using a Schmitt trigger.

Emitter-Coupled Schmitt Trigger

The idea of using positive feedback to create a transfer characteristic with hysteresis is best explained with the aid of the simple bipolar circuit of Figure 6.42a. The dc currents through the transistors Q_1 and Q_2 are set to different values by making $R_l > R_2$. Combining this with the feedback caused by the coupling of the transistor emitters leads to hysteresis as different input-signal values are required to turn the transistors on or off.

Assume that the input voltage is initially low, so that transistor Q_1 is turned off. Q_2 is on and saturated. The latter condition is fulfilled as long as R_1 and R_2 have comparable values. For Q_2 to be in saturation, its base current I_{B2} must be larger than the collector current divided by β_F.

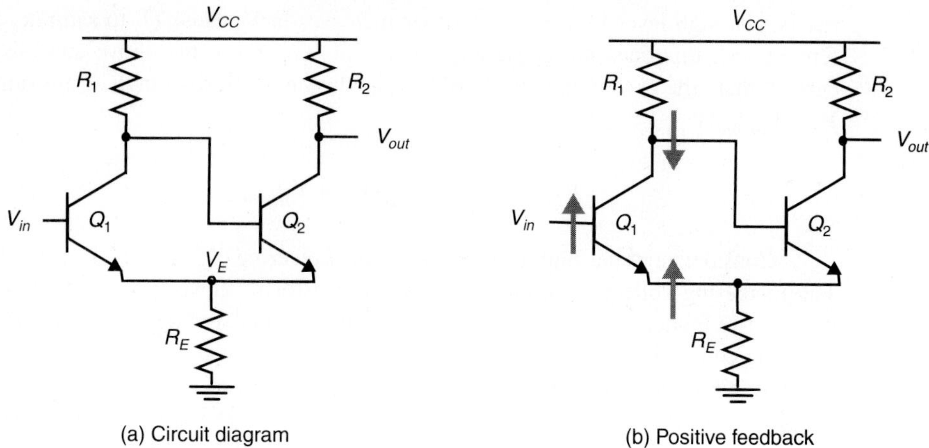

Figure 6.42 Noninverting, emitter-coupled Schmitt trigger.

$$I_{B2} > I_{C2}/\beta_F$$

or

$$\frac{V_{CC} - V_E - V_{BE(sat)}}{R_1} > \frac{V_{CC} - V_E - V_{CE(sat)}}{R_2\beta_F} \tag{6.7}$$

or

$$R_1 < R_2\beta_F\left(\frac{V_{CC} - V_E - V_{CE(sat)}}{V_{CC} - V_E - V_{BE(sat)}}\right) \approx R_2\beta_F$$

where V_E is the voltage at the emitters of Q_1 and Q_2. Under this condition, the output voltage V_{out} $(= V_{OL})$ is computed by equating the currents through transistor Q_2.

$$I_{E2} = \frac{V_E}{R_E} = I_{B2} + I_{C2} = \frac{V_{CC} - V_E - V_{BE(sat)}}{R_1} + \frac{V_{CC} - V_E - V_{CE(sat)}}{R_2} \tag{6.8}$$

Assuming that V_{cc} is substantially larger than $V_{BE(sat)}$ and $V_{CE(sat)}$, the following approximating values for V_E and V_{OL} can be derived:

$$V_E \approx \frac{R_E}{R_E + R_1 \| R_2} V_{cc}$$

and

$$V_{OL} = V_E + V_{CE(sat)} \tag{6.9}$$

We now increase the input voltage V_{in} steadily until it reaches the value $V_E + V_{BE(on)}$. Any further raising of V_{in} causes Q_1 to conduct and the voltage at its collector V_{C1} to fall as a result of the collector current of Q_1. Simultaneously, V_E increases as it equals $V_{in} - V_{BE(on)}$ (Figure 6.42b). The base-emitter voltage of Q_2 hence diminishes rapidly as the base voltage is dropping while its emitter voltage is rising. This regenerative action turns Q_2 off very rapidly, and a swift transition is observed at the output that ultimately reaches its high level $V_{OH} = V_{CC}$. Further increases in V_{in} cause Q_1 to saturate. Since Q_2 remains off, this does not cause any change in V_{OH}. From the above analysis, we can derive that the switching threshold V_{M+} during a low-to-high transition equals $V_E + V_{BE(on)}$:[3]

$$V_{M+} \approx \frac{R_E}{R_E + R_1 \parallel R_2} V_{cc} + V_{BE(on)} \tag{6.10}$$

Consider now the high-to-low transition. Lowering V_{in} reduces V_E as well as V_{C1}. The increasing collector current causes the transistor to move from saturation into forward-active mode, and raises V_{CE1} and, consequently, V_{BE2} (which are identical). Q_2 turns on at the point where $V_{CE1} = V_{BE2} = V_{BE(on)}$. The regenerative action now operates in the opposite direction. The voltage at the base of Q_2 increases as I_{C1} decreases due to the diversion of current from Q_1 to Q_2, while its emitter voltage is dropping as V_{in} is lowered. Q_1 is turned off, Q_2 saturates, and the output makes a swift transition to V_{OL}. As stated, the switching point for the negative-going transition is situated at the point where $V_{CE1} = V_{BE(on)}$. Since Q_1 and Q_2 are in forward-active mode and at the edge of conduction, respectively, at that point, the following expression is valid,

$$V_E = (V_{CC} - V_{BE(on)})\frac{R_E}{R_1 + R_E} \approx V_{CC}\frac{R_E}{R_1 + R_E} \tag{6.11}$$

which translates into

$$V_{M-} = V_E + V_{BE(on)} \approx V_{CC}\frac{R_E}{R_1 + R_E} + V_{BE(on)} \tag{6.12}$$

Comparing Eq. (6.12) and Eq. (6.10) shows that the hysteresis voltage (the difference between V_{M+} and V_{M-}) is a function of the discrepancy between R_1 and R_2. The larger R_1, the larger the difference.

Example 6.5 Emitter-Coupled Schmitt Trigger

Determine the values of V_{M-} and V_{M+} for the emitter-coupled Schmitt trigger of Figure 6.42. The values of R_1, R_2, and R_E are set to 4 kΩ, 2.5 kΩ and 1 kΩ, respectively, with $V_{CC} = 5$ V. Assume $\beta_F = 100$, $V_{BE(on)} = 0.7$ V, $V_{BE(sat)} = 0.8$ V, and $V_{CE(sat)} = 0.1$V.

Working out the above equations yields the following values: $V_{OH} = 5$ V, $V_{OL} = 1.92$ V, $V_{M+} = 2.52$ V, and $V_{M-} = 1.56$ V. This results in a hysteresis value of 0.96 V. Simulation produced similar values: $V_{OH} = 5$ V, $V_{OL} = 1.94$ V, $V_{M+} = 2.62$ V, and $V_{M-} = 1.73$ V. The deviations are mainly due to discrepancies between the actual values of V_{BE} and V_{CE} and the ones projected above.

CMOS Schmitt Trigger

A similar type of operation can be observed in the CMOS Schmitt trigger of Figure 6.43. The reasoning behind this circuit is that the switching threshold of a CMOS inverter is

[3] This is an approximation based on the assumption that the transition is very swift.

determined by the (k_n/k_p) ratio between the NMOS and PMOS transistors. Increasing the ratio results in a reduction of the threshold, while decreasing it results in an increase in V_M. Adapting the ratio depending upon the direction of the transition results in a shift in the switching thresholds and a hysteresis effect. This adaptation is achieved with the aid of feedback.

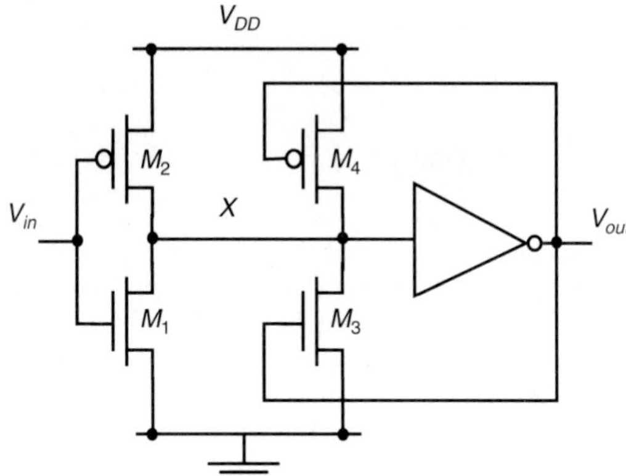

Figure 6.43 CMOS Schmitt trigger.

Suppose that V_{in} is initially equal to 0, so that $V_{out} = 0$ as well. The feedback loop biases the PMOS transistor M_4 in the conductive mode while M_3 is off. The input signal effectively connects to an inverter consisting of two PMOS transistors in parallel (M_2 and M_4) as a pull-up network, and a single NMOS transistor (M_1) in the pull-down chain. This modifies the effective transistor ratio of the inverter to $k_{M1}/(k_{M2}+k_{M4})$, which moves the switching threshold upwards.

Once the inverter switches, the feedback loop turns off M_4, and the NMOS device M_3 is activated. This extra pull-down device speeds up the transition and produces a clean output signal with steep slopes.

A similar behavior can be observed for the high-to-low transition. In this case, the pull-down network originally consists of M_1 and M_3 in parallel, while the pull-up network is formed by M_2. This reduces the value of the switching threshold to V_{M-}.

Example 6.6 CMOS Schmitt Trigger

Determine the sizes of transistors M_3 and M_4, so that the switching thresholds for the Schmitt trigger of Figure 6.43 are set to 1.5 and 3.5 V, respectively. You may assume that all other devices are minimum-size (1.8 μm/1.2 μm for NMOS and 5.4 μm/1.2 μm for PMOS transistors). The supply voltage is set to 5 V.

Consider first V_{M+}. To determine the threshold of the low-to-high transition, we analyze the simplified circuit of Figure 6.44, where the PMOS device M_4 is put in parallel with M_2 as a result of the feedback. By equating the current through the devices, we can derive an expression that relates V_{M+} to the size of transistor M_4:

$$\frac{k_1}{2}(V_{M+} - V_{Tn})^2 =$$
$$\frac{k_2}{2}(V_{DD} - V_{M+} - |V_{Tp}|) + k_4\left((V_{DD} - |V_{Tp}|)(V_{DD} - V_{M+}) - \frac{(V_{DD} - V_{M+})^2}{2}\right) \tag{6.13}$$

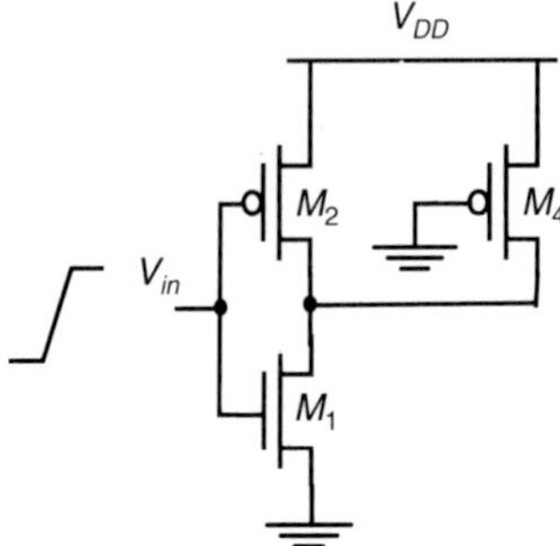

Figure 6.44 Equivalent circuit to determine V_{M+}.

Inserting the transistor parameters and setting the value of V_{M+} to 3.5 V yields a required value for k_4 of 107 mA/V^2. This corresponds to a (W/L) ratio of approximately 4. In a similar way, a ratio of approximately 1.3 can be derived for the NMOS transistor M_3.

The circuit was simulated using SPICE. The dimensions of M_3 were set to (1.2 μm/1.2 μm) for M_3 and (3.6 μm/1.2 μm) for M_4. This corresponds to the derived ratios, while also taking into account the effects of the lateral diffusion. The results are plotted in Figure 6.45. Observe the fast transitions at the output node as a result of the positive feedback. The simulated values of 1.53 V for V_{M-} and 3.7 V for V_{M+} are in close correspondence with the predicted ones.

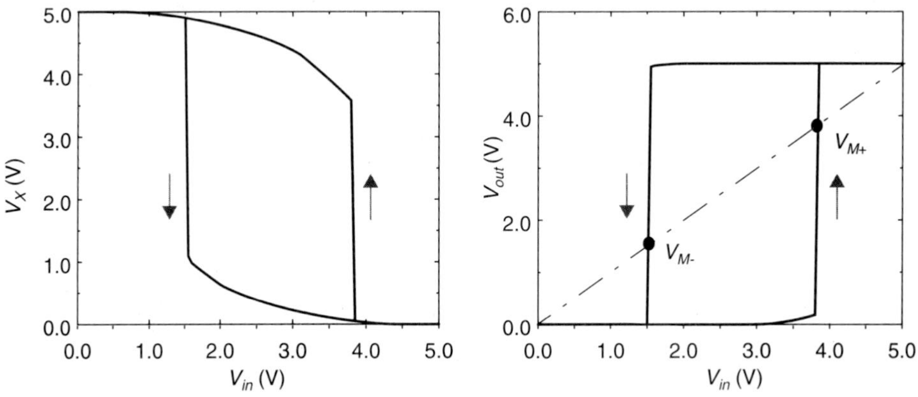

Figure 6.45 Voltage-transfer characteristic for CMOS Schmitt trigger of Figure 6.43 (nodes *X* and *Out*). The arrows denote the direction of the input transition (rising or falling).

Problem 6.4 An Alternative CMOS Schmitt Trigger

Another CMOS Schmitt trigger is shown in Figure 6.46. Discuss the operation of the gate, and derive expressions for V_{M-} and V_{M+}.

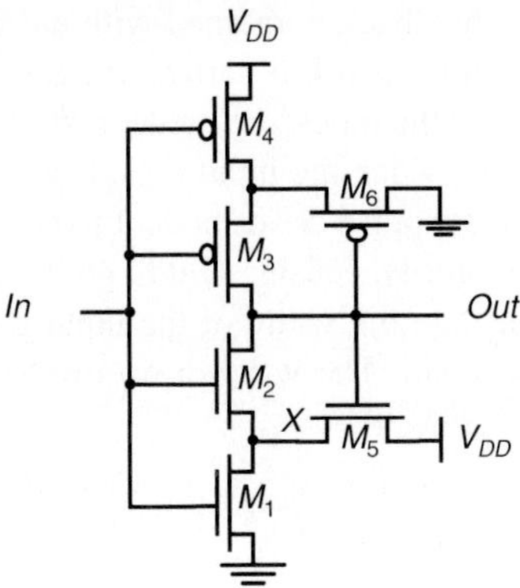

Figure 6.46 Alternate CMOS Schmitt trigger.

6.4.2 Monostable Sequential Circuits

A monostable element is a circuit that generates *a pulse of a predetermined width* every time the quiescent circuit is triggered by a pulse or transition event. It is called *monostable* because it has only one stable state (the quiescent one). A trigger event, which is either a signal transition or a pulse, causes the circuit to go temporarily into another quasi-stable state. This means that it eventually returns to its original state after a time period determined by the circuit parameters. This circuit, also called a *one-shot*, is useful in generating pulses of a known length. This functionality is required in a wide range of applications. A notorious example is the *address transition detection* (ATD) circuit, used for the timing generation in static memories. This circuit detects a change in a signal, or group of signals, such as the address or data bus, and produces a pulse to initialize the subsequent circuitry.

One class of one-shots uses a simple delay element to control the duration of the pulse. The concept is illustrated in Figure 6.47. In the quiescent state, both inputs to the EXOR are identical, and the output is low. A transition on the input causes the EXOR inputs to differ temporarily and the output to go high. After a delay t_d (of the delay element), this disruption is removed, and the output goes low again. A pulse of length t_d is created. The delay circuit can be realized in many different ways, such as an *RC*-network or a chain of basic gates. An example of such a monostable circuit was given earlier when discussing edge-triggered flip-flops (Section 6.2.3, Figure 6.9). The circuit presented uses a simple gate as the delay element. The resulting pulse has a width equal to the propagation delay of that gate.

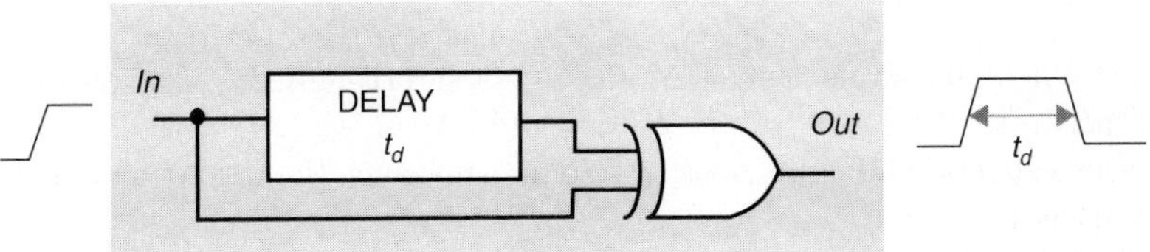

Figure 6.47 Transition-triggered one-shot.

A second class of monostable circuits uses feedback combined with an RC timing network to produce an output pulse of a fixed width, given a narrow trigger pulse. An example of such a circuit is shown in Figure 6.48. In the quiescent mode, node B is pulled high by the resistor, node Out is low, and A is high, since the input signal In is low. The capacitor C is discharged. Triggering In with a narrow pulse drives node A to GND. Due to the capacitive coupling through C, B follows temporarily, causing Out to go high. Assuming that the gate delays are negligible (or smaller than the width of the input pulse), Out reaches the high level before the input goes low again. The waveforms are displayed in Figure 6.48b.

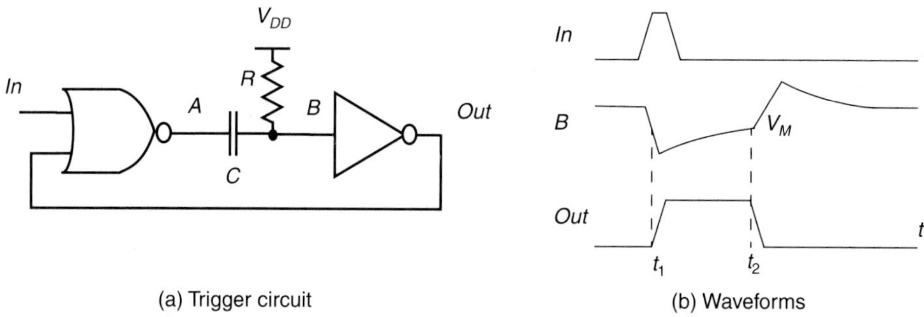

Figure 6.48 Monostable trigger.

This situation is not a stable one. The voltage at B starts to rise because the resistor R pulls up the node with a time constant RC. At time t_2, the voltage at B reaches the switching threshold of the output gate, which toggles. Out goes low, causing A to go high again. The capacitive coupling results in an overshoot at node B that gradually disappears. The circuit reverts to the quiescent state. The width of the output pulse t_d is determined primarily by the time-constant RC and the switching threshold of the output gate V_M.

We can derive an analytical expression for t_d by evaluating the voltage waveform at node B during the transition:

$$V_B(t) = V_{DD}\left(1 - e^{-\frac{t - t_1}{RC}}\right) \tag{6.14}$$

Taking into account that $V_B(t_2) = V_M$ yields

$$t_d = t_2 - t_1 = RC \times \ln\left(\frac{V_{DD}}{V_{DD} - V_M}\right) \tag{6.15}$$

For $V_M = V_{DD}/2$, this evaluates to 0.69 RC — not surprisingly the 50% transition point of the RC network.

The network of Figure 6.48 has some important limitations and problems that restrict its usefulness.

1. The RC time-constant has to be substantially larger than the propagation delays of the composing gates.

2. The switching threshold V_M is relatively sensitive to spreads in the production process. This translates directly into changes in the pulse width, causing fluctuations over different production runs. Circuits that reduce this sensitivity have been proposed, but a discussion of those is beyond the scope of this text.

On the other hand, the circuit has the positive property that R, C, and V_M are relatively insensitive to variations in the operating temperature. The generated pulse width is thus reasonably stable over a wide range of temperatures.

6.4.3 Astable Circuits

An astable circuit has no stable states. The output oscillates back and forth between two quasi-stable states with a a period determined by the circuit parameters. One of the main applications of oscillators is the on-chip generation of clock signals.

The ring oscillator (Chapter 3) is a simple, but important, example of an astable circuit. It consists of an odd number of inverters connected in a circular chain. Due to the odd number of inversions, no stable operation point exists, and the circuit oscillates with a period equal to $2 \times t_p \times N$, with N the number of inverters in the chain and t_p the propagation delay of the composing gates.

The simulated response of a ring oscillator with five stages is shown in Figure 6.49 (all gates use minimum-size devices). The observed oscillation period equals 2.57 nsec, which corresponds to a gate propagation delay of 257 psec. By tapping the chain at various points, different phases of the oscillating waveform are obtained (phases 1, 3, and 5 are displayed in the plot). A wide range of clock signals with different duty-cycles and phases can be derived from those elementary signals using simple logic operations.

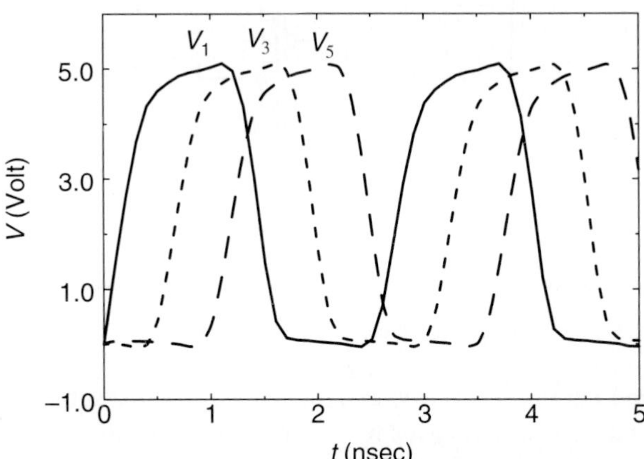

Figure 6.49 Simulated five-stage ring oscillator waveforms. The outputs of stages 1, 3, and 5 are shown.

The generic ring oscillator produces a waveform with a fixed oscillating frequency. Often it is necessary that the frequency be controllable or programmable. An example of such a circuit is the *voltage-controlled oscillator (VCO)*, whose oscillation frequency is proportional to the value of a control voltage. The ring oscillator is easily adapted to that purpose. To vary the oscillating period, we have to manipulate the propagation delay of the composing inverters. We can achieve this, for instance, by controlling the current available to (dis)charge the load capacitance of the gate.

Such a circuit, called a *current-starved inverter*, is presented in Figure 6.50 [Jeong87]. In this gate, the maximal (dis)charge current of the inverter is limited by adding some extra devices. The NMOS transistor M_3, controlled by the input voltage V_{contr}, acts as a current source with value I_{ref} and sets an upper limit on the discharge current. Lowering V_{contr} reduces the discharge current and, hence, increases t_{pHL}. Similarly, the charging current is controlled with the aid of M_5, which also acts as a current source with value I_{ref}. The current through M_5 is translated into a charge current with the aid of the current mirror formed by the PMOS devices M_6 and M_4. M_6 acts as a diode and sets a bias voltage V_{GS6} that is controlled by I_{ref}. With $V_{GS4} = V_{GS6}$ and assuming that both devices operate in the saturation region, it follows that $I_{DS4} = I_{DS6} = I_{ref}$. Since both M_3 and M_5 operate in the saturation region, a quadratic relation exists between V_{contr} and I_{ref} (and t_p). As such, the oscillating frequency of the VCO can be controlled over quite a large range.

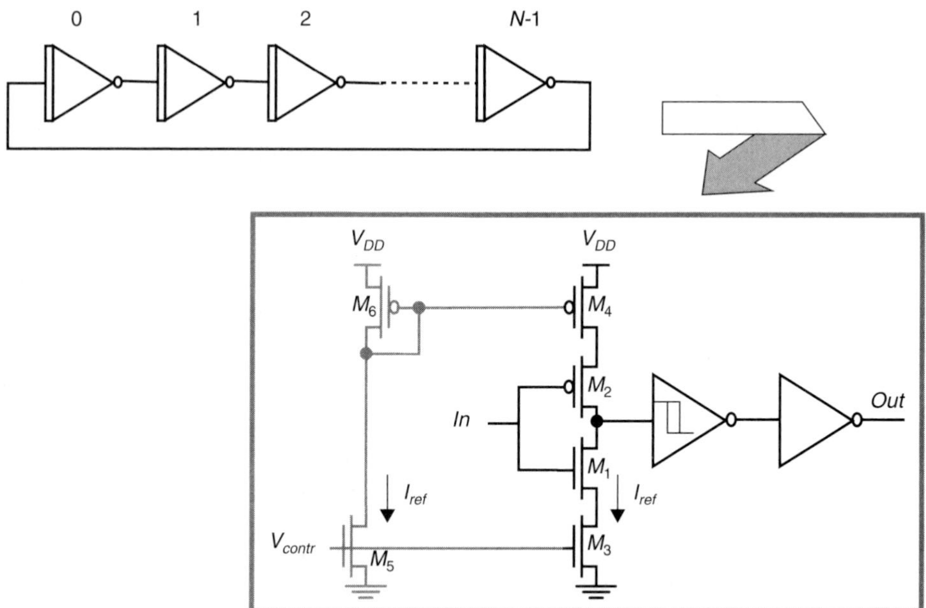

Figure 6.50 Voltage-controlled oscillator based on current-starved inverters.

Under low operating current levels, the current-starved inverter suffers from slow rise and fall times at its output. This is resolved by feeding its output into a Schmitt trigger. An extra inverter is necessary to correct the signal polarity. Observe that transistors M_5 and M_6 can be shared over all inverters in the chain.

Example 6.7 Current-Starved Inverter

A current-starved inverter uses minimum-size devices for M_1 and M_2. M_3 and M_4 are made five times wider. Determine the operation range of the control voltage, assuming a supply voltage of 5 V.

We restrict this analysis to t_{pHL} (similar considerations hold for the low-to-high transition). The maximum current of the minimum-size inverter is easily computed:

$$I_{sat} = 0.5(1.8/0.9)19.6 \times 10^{-6}(5 - 0.74)^2 = 357\ \mu\text{A}$$

To get a similar current in the control transistor M_3, we need the following control voltage:

$$I_{sat} = 0.5(9/0.9)19.6 \times 10^{-6}(V_{contr} - 0.74)^2$$

Solving for V_{contr} yields 1.17 V. This means that the most dramatic impact on the propagation delay is obtained between 0.74 V and 1.17 V. In reality, the range is larger, mostly because the control transistor comes out of saturation earlier for larger values of the control signal. Figure 6.51 plots the simulated t_{pHL} as a function of V_{contr}. Only a minor change in the delay is observed for $V_{contr} > 1.5$ V. A virtually quadratic dependence between control voltage and delay is apparent from the simulation, as projected.

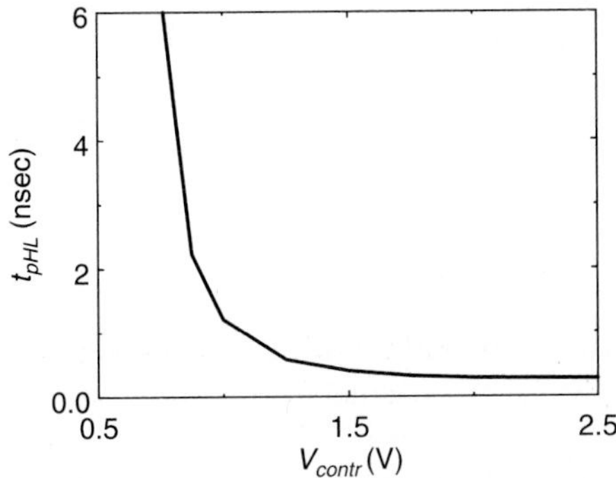

Figure 6.51 t_{pHL} of current-starved inverter as a function of the control voltage.

By analogy to the monostable element, it is also possible to conceive an astable circuit composed of an *RC* network combined with feedback. The circuit of Figure 6.52 presents a convenient way to generate a clock signal for a CMOS circuit under the condition that the frequency-stability requirements are not too stringent. Assume that the inverters have a switching threshold $V_M = V_{DD}/2$, the gate delays are negligibly small with respect to the *RC* time-constant, and the output voltage of the gates changes instantaneously when the input voltage crosses V_M.

At time $t = 0$, we assume V_{int} to be rising. When crossing V_M, it causes gate 1 and gate 2 to toggle. As a result, V_{out1} and V_{out2} go low and high, respectively. The abrupt change in V_{out2} is capacitively coupled to node *Int*, which jumps from V_M to $V_M + V_{DD} = 3V_{DD}/2$, all of which occurs at time $t = 0$. The voltage at node *Int* now starts to decay exponentially towards GND with a time-constant *RC*. After some time, it crosses V_M again, this time in the falling direction, toggling gates 1 and 2. V_{out1} and V_{out2} go to 1 and 0, respectively, and V_{int} jumps to $-V_{DD}/2$ due to the capacitor *C*. V_{int} starts to decay towards V_{DD} with a time-constant *RC* until it reaches V_M once more, and the complete cycle is repeated.

The circuit, called *a relaxation oscillator*, has no stable state and oscillates with a period $T = 2\,(\log 3)RC$. This value is only valid for T much larger than the gate delay, and R larger than the output resistance of the inverter. We can derive more precise expressions by taking the gate delays and the finite output resistance into account. Finally, a word of

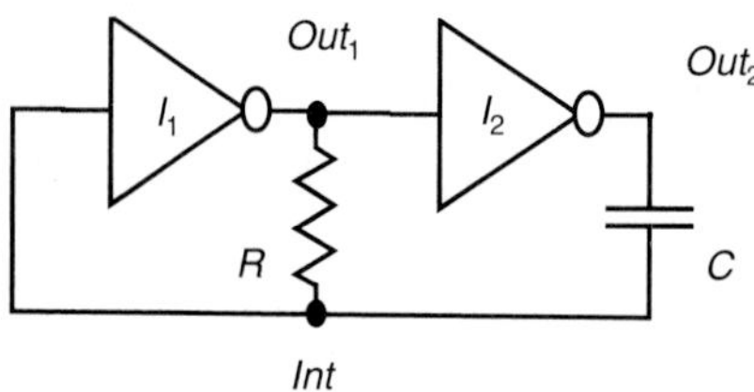

Figure 6.52 Relaxation oscillator.

caution: the voltage waveform at node 1 surpasses the range between V_{DD} and GND. This might cause problems in integrated circuits, since the inverse-biased diffusion junctions might be turned on. Extreme caution is hence required when designing an oscillator of this class.

Problem 6.5 Relaxation Oscillator

Draw the voltage waveforms for the circuit of Figure 6.52. Derive the expression for the oscillation period given above.

6.5 Perspective: Choosing a Clocking Strategy

A crucial decision that must be made in the earliest phases of a chip design is how to clock the chip. The synchronization of the various operations occurring in a complex circuit so that it operates reliably under all circumstances is one of the most intriguing challenges facing the digital designer of the next decade. Choosing the right clocking scheme affects both the functionality and the performance of a circuit.

A number of widely used clocking schemes were introduced in this chapter. One scheme uses a pair of two-phase nonoverlapping clocks, while the other employs a single clock in an edge-triggered approach. Designers actually used to adopt even more complex clocking schemes, distributing up to eight clock signals. The advantage of this multi-phase approach is that it allows for a more efficient utilization of the time, which typically results in higher-performance solutions. Distributing these clocks so as to avoid race conditions has become more and more nightmarish with increases in chip complexity and size. Loads on clock lines are huge (> 1 nF), while the delay in propagating the clock signal from the source to its destinations over the long clock wires is substantial (in the range of nanoseconds).

The general trend in high-performance CMOS VLSI design is therefore to *use simple clocking schemes*, even at the expense of performance. Most automated design methodologies, such as standard cells, employ a single-phase, edge-triggered approach, based on static flip-flops. But the tendency towards simpler clocking approaches is also apparent in high-performance designs such as microprocessors. An example is the Alpha processor of Digital Equipment Corporation [Dopperpuhl92], which uses the single-phase, dynamic TSPC approach.

This issue is so important that a more detailed discussion of the timing and synchronization of digital circuits forms the topic of Chapter 9.

6.6 Summary

This chapter has explored the subject of sequential digital circuits. The following topics were discussed:

- The cross-coupling of two inverters creates a *bistable* circuit, called a *flip-flop*. A third potential operation point turns out to be metastable; that is, any diversion from this bias point causes the flip-flop to converge to one of the stable states.
- From a transient perspective, a flip-flop is characterized by three parameters: *the setup time, the hold time, and the propagation delay.*
- A major challenge in the design of sequential circuits is to avoid *race conditions*. Master-slave and edge-triggered flip-flops were conceived especially with this purpose in mind. Variants of these structures include *JK*, *D* and *T* flip-flops.
- The static flip-flop uses positive feedback as the storage mechanism. The resulting structure is robust and retains its charge as long as the supply voltage is applied. Static flip-flops are slow and require a large area, however.
- The dynamic flip-flop approach relies on charge storage. It has the advantage of yielding simpler and faster flip-flops, but it is unfortunately more susceptible to error due to leakage, dynamic artifacts, and synchronization discrepancies.
- The problem of race conditions can be addressed either by using multiple clock phases, or by employing special register structures. The C^2MOS and the TSPC latches are examples of the latter.
- The combination of dynamic logic with dynamic latches can produce extremely fast computational structures. An example of such an approach, the NORA logic style, is very effective in pipelined datapaths.
- Monostable structures have only one stable state. They are useful as pulse generators.
- Astable multivibrators, or oscillators, possess no stable state. The ring oscillator is the best-known example of a circuit of this class.
- Schmitt triggers display hysteresis in their dc characteristic and fast transitions in their transient response. They are mainly used to suppress noise.

6.7 To Probe Further

The basic concepts of sequential gates can be found in many logic design textbooks (e.g., [Mano82] and [Hill74]). The design of sequential circuits in both CMOS and bipolar is amply documented in most of the traditional digital circuit handbooks.

References

[Dopperpuhl92] D. Dopperpuhl et al., "A 200 MHz 64-b Dual Issue CMOS Microprocessor," *IEEE JSSC*, vol. 27, no. 11, Nov. 1992, pp. 1555–1567.

[Goncalves83] N. Goncalves and H. De Man, "NORA: a racefree dynamic CMOS technique for pipelined logic structures," *IEEE JSSC*, vol. SC-18, no. 3, June 1983, pp. 261–266.

[Haznedar91] H. Haznedar, *Digital Microelectronics*, Benjamin/Cummings, 1991.

[Hill74] F. Hill and G. Peterson, *Introduction to Switching Theory and Logical Design*, Wiley, 1974.

[Hodges88] D. Hodges and H. Jackson, *Analysis and Design of Digital Integrated Circuits*, McGraw-Hill, 1988.

[Jeong87] D. Jeong et al., "Design of PLL-based clock generation circuits," *IEEE JSSC*, vol. SC-22, no. 2, April 1987, pp. 255–261.

[Mano82] M. Mano, *Computer System Architecture*, Prentice Hall, 1982.

[Schmitt38] O. H. Schmitt, "A Thermionic Trigger," *Journal of Scientific Instruments*, vol. 15, January 1938, pp. 24–26.

[Shoji88] M. Shoji, *CMOS Digital Circuit Technology*, Prentice Hall, 1988.

[Suzuki73] Y. Suzuki, K. Odagawa, and T. Abe, "Clocked CMOS calculator circuitry," *IEEE Journal of Solid State Circuits*, vol. SC-8, December 1973, pp. 462–469.

[Veendrick92] H. Veendrick, *MOS ICs: From Basics to ASICs*, VCH, Weinheim, 1992.

[Yuan89] J. Yuan and Svensson C., "High-Speed CMOS Circuit Technique," *IEEE JSSC*, vol. 24, no. 1, February 1989, pp. 62–70.

6.8 Exercises and Design Problems

1. [E, None, 6.2] The indicated waveforms are applied to the *JK* master-slave flip-flop of Figure 6.53. For this problem, assume that gate delays are short compared to the input signal time scale.
   a. Sketch the waveforms that appear at the Q_M and Q_S outputs of the master and slave latches. Assume that the flip-flop is initially in the reset state.
   b. Do the waveforms exhibit any 1's-catching behavior?

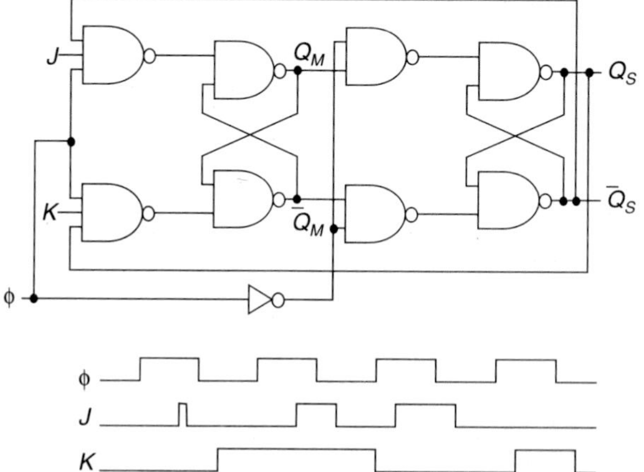

Figure 6.53 *JK* master-slave flip-flop.

2. [E, None, 6.2] Figure 6.18 proposes an ECL implementation of an *SR* flip-flop. For $R = S = 0$, what values of R_B, R_E, and R_C are required to establish quiescent currents of 0.75 mA in the differential pair devices and of no more than 0.1 mA in the emitter-follower devices? Assume $V_{CC} = 0$, $V_{EE} = -3$ V, $V_{swing} = 0.75$ V, $V_{BE(on)} = 0.75$ V, $V_{BE(sat)} = 0.85$ V, $V_{CE(sat)} = 0.1$ V. Neglect base currents for first-order analysis.
3. [M&D, SPICE, 6.2] Design the *JK* flip-flop of Figure 6.5 using static CMOS minimum-size devices. $V_{DD} = 5$ V.

a. Estimate the set-up and hold times and the propagation delay in terms of gate delays. Find these gate delays using SPICE under the appropriate loading conditions. Compare your total estimated delay with a SPICE simulation.

b. Use progressive transistor sizing and appropriate ordering to improve the speed. Find the new set-up and hold times and the propagation delays using SPICE.

4. [M, None, 6.2] Figure 6.54 contains an ECL sequential circuit.

a. What function does this circuit perform? Where might a circuit like this be used?

b. In this logic family is it reasonable to assume that both true and complementary inputs are available? Explain.

c. What is the purpose of the *pnp* transistors Q_9 and Q_{10}? If they were removed and their inputs were connected directly to Q_7 and Q_8, what would be the effect on the circuit functionality and performance?

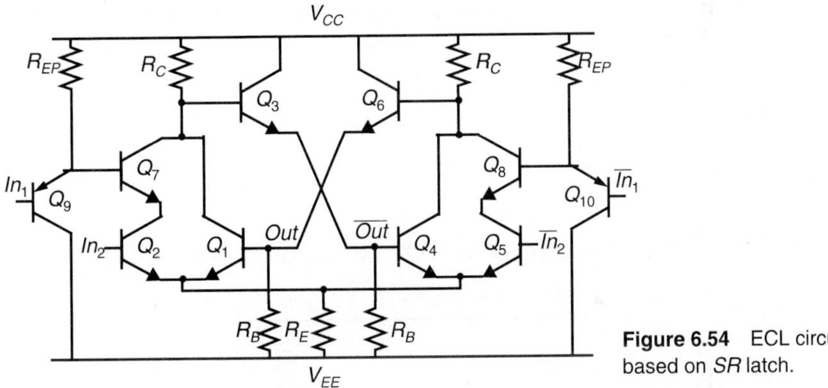

Figure 6.54 ECL circuit based on *SR* latch.

5. [E, SPICE, 6.3] A dynamic master-slave *D* flip-flop was introduced in Figure 6.23. Assume that there is no charge leakage and that a nonoverlapping clock pair ϕ-ϕ with 50% duty cycle is available. The sizes of all NMOS and PMOS transistors are 1.8 μm/1.2 μm and 3.6 μm/1.2 μm, respectively. Assume V_{DD} = 5 V, and C_L = 20 fF at the *D*-output.

a. Provide a first-order estimate of the parasitic capacitances at the inverter inputs.

b. Determine the set-up time and the propagation delay of the register, and compare with SPICE.

6. [E, None, 6.3] Consider dynamic implementations of the *D* flip-flop using either the C^2MOS approach of Figure 6.24 or the TSPC approach of Figure 6.37a. The flip-flops are driven by clocks ϕ and ϕ with a 50% duty cycle.

a. Discuss the impact of clock overlap on the behavior of both flip-flops.

b. Discuss the impact of finite clock rise and fall times on the behavior of both flip-flops.

7. [M, SPICE, 6.3] For the circuits of Problem 6,

a. Use SPICE to determine the worst-case clock slopes that can be allowed before the circuit starts to fail. Assume (1.8/1.2) NMOS and (3.6/1.2) PMOS devices. The supply voltage is set at 5 V.

b. Determine the maximum clock frequency of both structures. Perform a first-order manual analysis before confirming the result using SPICE. A circuit fails if any of it signals does not come within 90% of its final value before the next clock event is initiated.

8. [M&D, SPICE, 6.3] Design a pipelined circuit of the function $F = (AB + BC) + D$. Use C^2MOS latches for the implementation.

   a. Draw the schematics of the circuit assuming you are allowed to introduce two pipeline stages between input and output. Try both static and dynamic implementations of the logic. Place the latches so that a maximum clock speed is obtained. What limits clock speeds in both cases.

   b. Using SPICE, determine the maximum clocking frequency for part a. How long (in nsec) does it take for an input to appear at output node? Assume $V_{DD} = 5$ V and (1.8/1.2) NMOS and (3.6/1.2) PMOS devices.

9. [M&D, None, 6.3] Pipeline the function $F_{n+1} = (\overline{AB} + \overline{BF_n}) + C$ with as many stages as possible. Include the input and output latches.

   a. Design it at the gate level. Estimate the minimum clock period in terms of gate delays. Under what conditions can a race occur, assuming fast clock rise and fall times.

   b. Sketch the transistor schematics using basic TSPC latches.

   c. Sketch the schematics using TSPC latches with logic integrated into the latches.

   d. Discuss the pros and cons of the latter approach.

10. [M, None, 6.4] A designer has come up with a novel Schmitt trigger circuit (Figure 6.55).

   a. Determine the values of V_{out} for $V_{in} = 0$ V and $V_{in} = 3$ V.

   b. Determine the values of V_M+ (switching point for a low-to-high input transition) and V_{M-} (high-to-low transition).

   c. Draw the resulting VTC. Mark V_{OH}, V_{OL}, V_{M-} and V_{M+}.

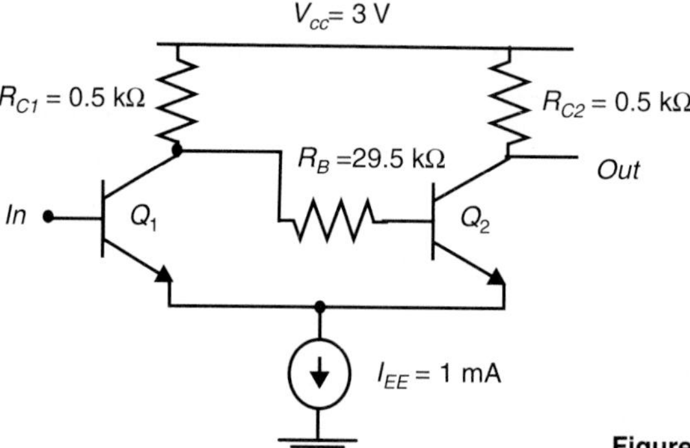

Figure 6.55 Bipolar Schmitt trigger.

11. [C, None, 6.4] A pseudo-NMOS Schmitt trigger is shown in Figure 6.56. To ease hand calculations, you may assume that the PMOS transistor remains in saturation over the entire range of operation. $(W/L)_1 = 1.8/1.2$ and $(W/L)_{2,3,4} = 18/1.2$.

   a. Compute V_{OH} and V_{OL}.

   b. Compute the switching points V_{M+} (for the input making a low-to-high transition) and V_{M-} (for the input making a high-to-low transition).

12. [M, None, 6.4] Figure 6.57 shows an alternative CMOS Schmitt trigger circuit. Sketch the VTC, including approximate values for V_{M+} and V_{M-}.

13. [E, None, 6.4] Figure 6.58 shows an alternative monostable multivibrator. Assume that gate delays are negligible relative to the RC time-constant. Further assume that the gates have input-protection diodes that limit the input voltage to the range $-0.8 < V_{in} < V_{DD} + 0.8$.

   a. Sketch waveforms for the voltages at X, Y, and Out using an appropriate trigger waveform. Label your sketches with critical voltages and time-constants.

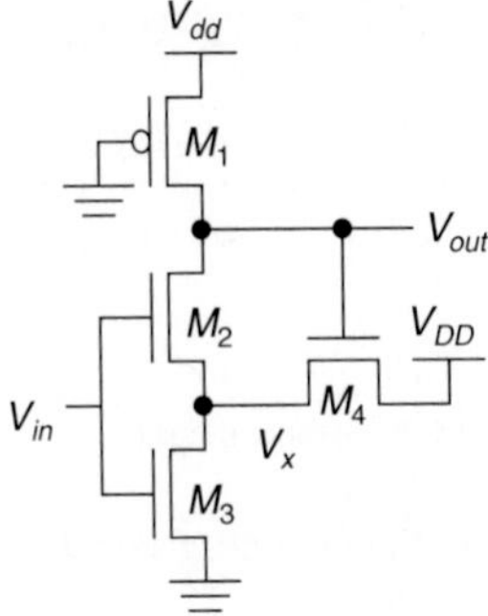

Figure 6.56 Pseudo-NMOS Schmitt trigger.

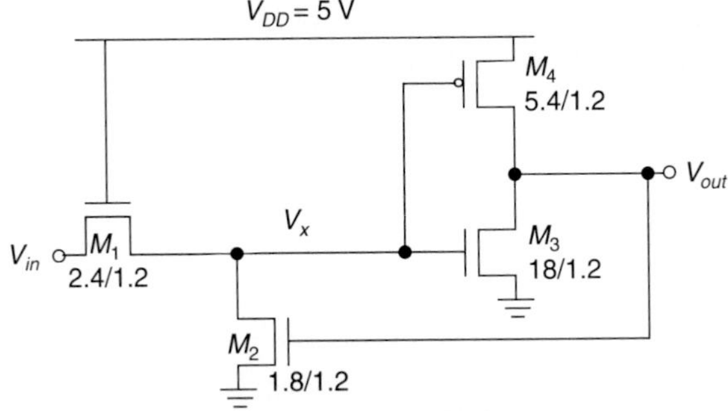

Figure 6.57 Alternative CMOS Schmitt trigger.

b. Derive equations for $V_Y(t)$ and the output pulse width, t_{pw}. The switching threshold of the inverter is V_M.

c. Find the value of C required for t_{pw} = 50 μs when R = 20 kΩ and $V_M = V_{DD}/2$.

d. Find the time required after the pulse for node Y to recover to within 0.1 V of its final value using the R and C from part c.

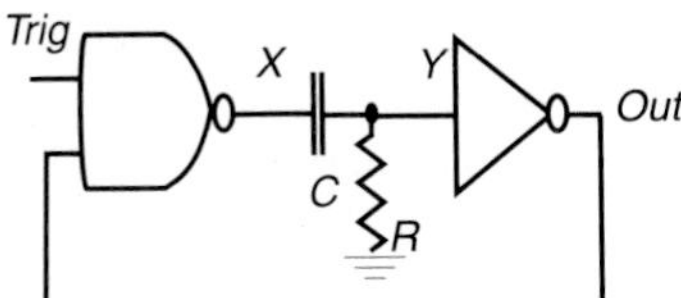

Figure 6.58 Alternative monostable multivibrator.

14. [E, None, 6.4] A multivibrator circuit is shown in Figure 6.59. The Schmitt trigger is inverting and has a rail-to-rail swing. You may ignore the propagation delay of the Schmitt trigger (or assume that $RC >> t_{p,schmitt}$).

a. Draw the waveforms at the nodes V_i and V_o, and determine the function of the circuit (bistable, monostable, or astable).

b. Derive an expression for the most important timing parameter of the above circuit as a function of the supply voltage, the Schmitt-trigger thresholds V_{M-} and V_{M+}, and the circuit parameters R and C. The nature of the timing parameter depends upon the type of multivibrator (e.g., pulse width for a monostable, propagation delay for a bistable, and oscillation period for an astable).

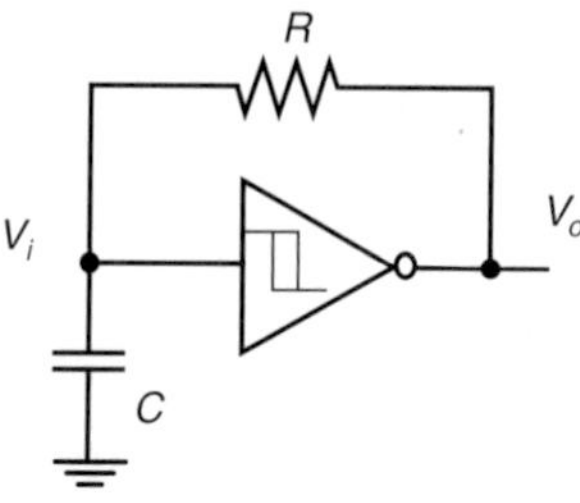

Figure 6.59 Multivibrator circuit.

15. [M, None, 6.4] An alternative oscillator is shown in Figure 6.60. Draw the signal waveforms for this innovative network. Determine the *oscillation frequency*. Discuss the advantage of this circuit. You may assume that the delay of the inverters, the resistances of the MOS transistors, and all internal capacitors can be ignored. The inverter switch point is set at 1.65 V. Assume that nodes *Y* and *Z* are initially at 0 V and 3.3 V, respectively.

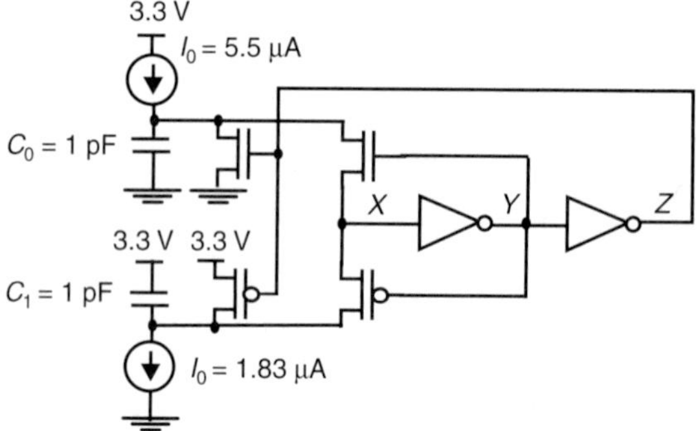

Figure 6.60 Oscillator.

16. [E, None, 6.4] A five-stage ring oscillator is selected for the implementation of a VCO. The oscillation frequency should be 5 MHz for a supply voltage of 3.3 V. The basic inverter circuit used in the oscillator is shown in Figure 6.61. Due to their large sizes, you may consider transistors M_2 and M_3 to be ideal switches (with *zero on-resistance*).

a. Determine the dc voltages needed at nodes *H* and *L* needed to get the required oscillation frequency. Consider only the capacitance shown in the figure. Ignored the channel-length modulation factor. Also make sure that rise and fall times are identical.

b. Confirm the result of part *a* using SPICE.

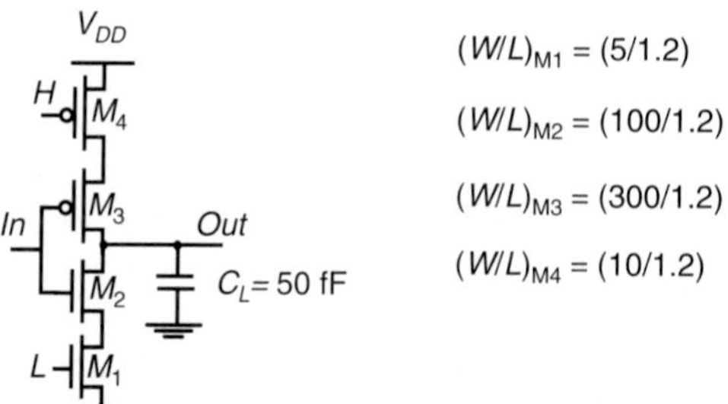

$(W/L)_{M1} = (5/1.2)$

$(W/L)_{M2} = (100/1.2)$

$(W/L)_{M3} = (300/1.2)$

$(W/L)_{M4} = (10/1.2)$

Figure 6.61 Current-starved inverter.

c. The output waveforms of the inverter shown will have very slow rise and fall times (which leads to high short-circuit currents and power consumption). Discuss briefly how you would modify the circuit to eliminate this problem (while maintaining the oscillation frequency).

DESIGN PROBLEM

Design a 4-bit random number generator, to be used for test pattern generation. The random number generator uses the linear feedback shift Register approach, that is described in Chapter 11. The block diagram of the generator is shown in Figure 6.62.

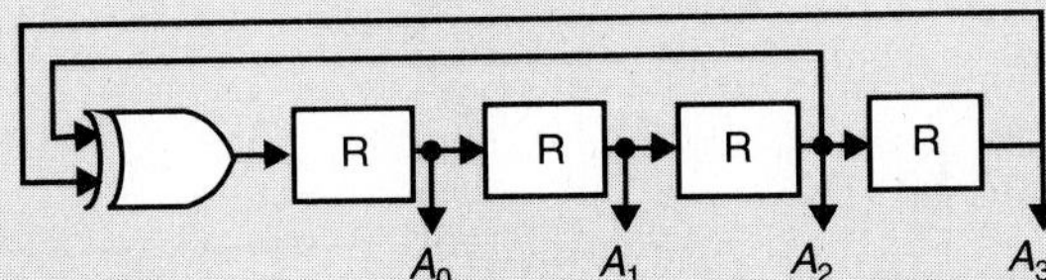

Figure 6.62 Pseudo-random number generator.

The goal of the project is *ultimate speed* and nothing else. This means that you should minimize the clock period to the maximal extent. Area and power are not in issue in this project. You are free to choose any CMOS implementation style for the design, including but not limited to: complementary CMOS, ratioed logic, DCVSL, pass-transistor logic, CPL, and dynamic logic. Feel free to mix the logic families in your design. All complimentary signals must be internally generated, and any number of levels of logic may be used. Registers can be dynamic or static. You are free to use the clocking strategy of your choice (single phase, two phase, four phase, ...). *Make sure, however, that races do not occur.*

A number of constraints have to be taken into account: (1) A power supply of 3.3 V should be used. (2) The output signals should settle to within 10% of their final value before the next clock event can be introduced. (3) The noise margins should be at least 10% of the voltage swing. (4) Each output bit of the generator should have a 50 fF load. (5) You are given a primary clock signal with a rise and fall time of 200 psec and a duty cycle of 50%. All other clock signals should be derived from this primary signal using actual logic (e.g., complementary clocks, nonoverlapping clocks, clocks with a faster rise and fall time, etc.).

PART

II

A SYSTEMS PERSPECTIVE

"Art, it seems to me,
should simplify.
That, indeed, is very nearly
the whole
of the higher
artistic process;
finding what conventions of form
and what of detail
one can do without
and yet preserve
the spirit of the whole."

Willa Sibert Cather
On the Art of Fiction, 1920

CHAPTER

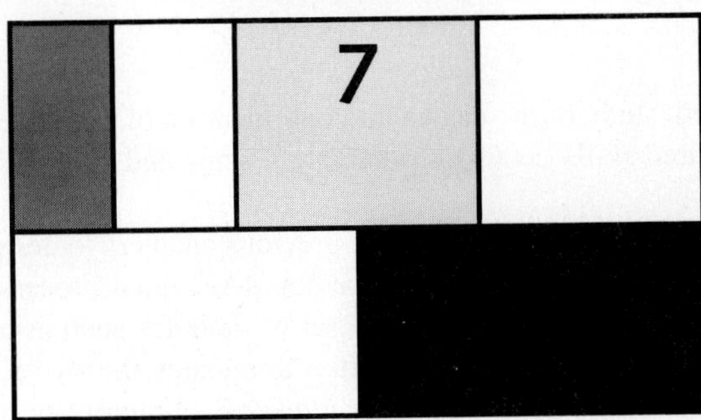

DESIGNING ARITHMETIC BUILDING BLOCKS

Designing adders, multipliers, and shifters for performance, area, or power

Logic and system optimizations for datapath modules

Higher-level models of arithmetic modules

7.1 Introduction

After the in-depth study of the design and optimization of the basic digital gates, it is time to test our acquired skills on a somewhat larger scale and put them in a more system-oriented perspective.

We will apply the techniques of the previous chapters to design a number of circuits often used in the datapaths of microprocessors and signal processors. More specifically, we discuss the design of a representative set of modules such as adders, multipliers, and shifters. The speed of these elements often dominates the overall system performance. Hence, a careful design optimization is required. It rapidly becomes obvious that the design task is not straightforward. For each module, there exist multiple equivalent logic and circuit topologies, each of which has its own benefits and disadvantages in terms of area, speed, or power.

Although far from complete, the analysis presented helps focus on the essential trade-offs that must be made in the course of the digital design process. You will see that optimization at only one design level—for instance, through transistor sizing only—leads to inferior designs. A global picture is of crucial importance. A good digital designer focuses his attention on the gates, circuits, or transistors that have the largest impact on his goal function. The noncritical parts of the circuit can be developed routinely. We will develop first-order performance models that foster understanding of the fundamental mechanics of a module.The discussion also clarifies which computer aids can help to simplify and automate this phase of the design process.

Before analyzing the design of the arithmetic modules, a short discussion of the role of the datapath in the digital-processor picture is useful. This not only helps highlight the specific design requirements for the datapath, but also puts the rest of this book in perspective. Other processor modules, such as the input/output, controller, and memory modules, have different requirements and are discussed in subsequent chapters. After an analysis of the area-time trade-offs in the design of adders, multipliers, and shifters, we will use the same structures to illustrate some of the power-minimization approaches introduced in Chapter 4. The chapter concludes with a short perspective on datapath design and its trade-offs. Appendix E, following this chapter, discusses a number of the layout approaches used in high-speed datapaths.

7.2 Datapaths in Digital Processor Architectures

An analysis of the components of a simple processor puts the different classes of digital circuits and their usage in perspective. Such a processor could be the brain of a personal computer (PC) or the heart of a compact disc player. A typical block diagram is shown in Figure 7.1 and is composed of a number of building blocks that occur in one form or another in almost every digital processor.

- **The datapath** is the core of the processor; it is where all computations are performed. The other blocks in the processor are support units that store either the results produced by the datapath or help to determine what will happen in the next cycle. A typical datapath consists of an interconnection of basic combinational func-

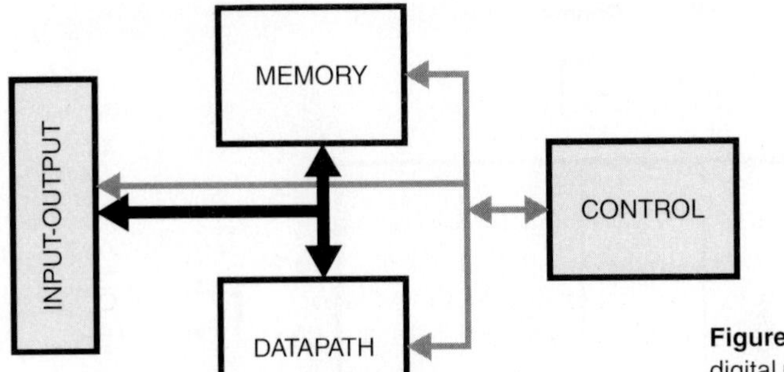

Figure 7.1 Composition of a generic digital processor. The arrows represent the possible interconnections.

tions, such as logic (AND, OR, EXOR) or arithmetic operators (addition, multiplication, comparison, shift). Intermediate results are stored in registers. The design of the arithmetic operators is the topic of this chapter.

- **The control module** determines what actions happen in the processor at any given point in time. A controller can be viewed as a finite state machine (FSM). It consists of registers and logic, and is hence a sequential circuit. The logic can be implemented in different ways—either as an interconnection of basic logic gates, often called *random logic*, or in a more structured fashion using programmable logic arrays (PLAs) and instruction memories.
- **The memory module** serves as centralized data storage. A broad range of different memory classes exist. The main difference between those classes is in the way data can be accessed, such as read-only versus read-write, sequential versus random access, or single-ported versus multiported access. Another way of differentiating between memories is related to their data-retention capabilities. Dynamic memory structures must be refreshed periodically to keep their data, while static memories keep their data as long as the power source is turned on. Finally, memory structures such as flash memories conserve the stored data even when the supply voltage is removed. A single processor might combine different memory classes. For instance, random access memory can be used to store data and read-only memory to store instructions.
- **The interconnect and input-output circuitry**—the interconnect network joins the different processor modules to one another as well as to the outside world. Unfortunately, the wires composing the interconnect network, are nonideal and present a capacitive, resistive, and inductive load to the driving circuitry. With growing die-sizes, the length of the interconnect wires tends to grow, resulting in increasing values for these parasitics.

Datapaths are often arranged in a *bit-sliced* organization, as shown in Figure 7.2. Instead of operating on single-bit digital signals, the data in a processor is arranged in a *word*-based fashion. For instance, a 32-bit processor operates on data words that are 32 bits wide. This is reflected in the organization of the datapath. Since the same operation has to be performed on each bit of the data word, the datapath consists of 32 identical slices, each of them operating on a single bit. Hence the word *bit-sliced*. The datapath designer can concentrate on the design of a single slice that is repeated 32 times.

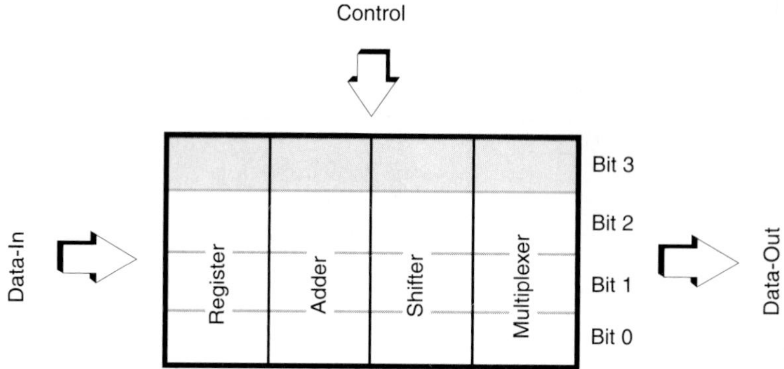

Figure 7.2 Bit-sliced datapath organization.

7.3 The Adder

Addition is probably the most commonly used arithmetic operation. Often it is also the speed-limiting element. Therefore, careful optimization of the adder is of utmost importance. This optimization can proceed either at the logic or circuit level. Typical logic-level optimizations try to rearrange the Boolean equations so that a faster or smaller circuit is obtained. An example of such a logic optimization is the *carry look-ahead adder* discussed later in the chapter. Circuit optimizations, on the other hand, manipulate transistor sizes and circuit topology to optimize the speed. Before considering both optimization processes, we provide a short summary of the basic definitions of an adder circuit (as defined in any book on logic design [e.g., Davio83]).

7.3.1 The Binary Adder: Definitions

The truth table of a binary full adder is given in Table 7.1. A and B are the adder inputs, C_i is the carry input, S is the sum output, and C_o is the carry output. The Boolean expressions for S and C_o are given in Eq. (7.1).

Table 7.1 Truth table for full adder.

A	B	C_i	S	C_o	Carry status
0	0	0	0	0	delete
0	0	1	1	0	delete
0	1	0	1	0	propagate
0	1	1	0	1	propagate
1	0	0	1	0	propagate
1	0	1	0	1	propagate
1	1	0	0	1	generate
1	1	1	1	1	generate

$$\begin{aligned} S &= A \oplus B \oplus C_i \\ &= A\overline{B}\overline{C}_i + \overline{A}B\overline{C}_i + \overline{A}\overline{B}C_i + ABC_i \\ C_o &= AB + BC_i + AC_i \end{aligned} \tag{7.1}$$

It is often useful from an implementation perspective to define S and C_O as functions of some intermediate signals G (Generate), D (Delete), and P (Propagate). $G = 1$ $(D = 1)$ ensures that a carry bit will be *generated* (*deleted*) at C_O independent of C_i, while $P = 1$ guarantees that an incoming carry will propagate to C_O. Expressions for these signals can be derived from inspection of the truth table.

$$\begin{aligned} G &= AB \\ D &= \overline{A}\overline{B} \\ P &= A + B \text{ (or } P = A \oplus B) \end{aligned} \tag{7.2}$$

S and C_o can be rewritten as functions of P and G (or D)

$$\begin{aligned} C_o(G, P) &= G + PC_i \\ S(G, P) &= P \oplus C_i \end{aligned} \tag{7.3}$$

Notice that G and P are only functions of A and B and are not dependent upon C_i. In a similar way, we can also derive expressions for $S(D,P)$ and $C_o(D,P)$.

An N-bit adder can be constructed by cascading N full-adder circuits in series, connecting $C_{o,k-1}$ to $C_{i,k}$ for $k = 1$ to $N-1$, and the first carry-in $C_{i,0}$ to 0 (Figure 7.3). This configuration is called a *ripple-carry adder* since the carry bit "ripples" from one stage to the other. The delay through the circuit depends upon the number of logic stages that must be traversed and is a function of the applied input signals. For some input signals, no rippling effect occurs at all, while for others the carry has to ripple all the way from the *least-significant (lsb)* to the *most-significant bit (msb)*. The propagation delay of such a structure (also called the *critical path*) is defined as the *worst-case delay over all possible input patterns*.

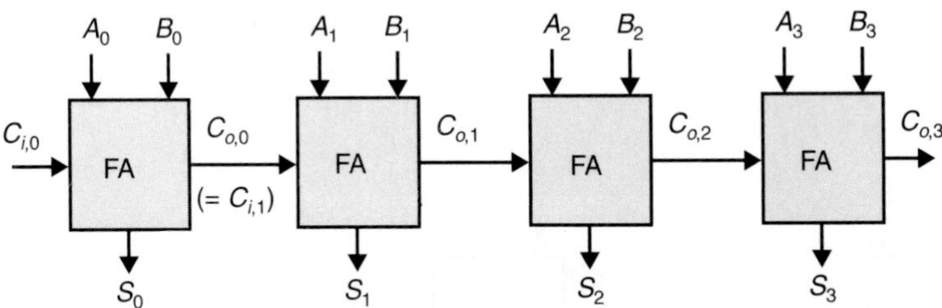

Figure 7.3 Four-bit ripple-carry adder: topology.

In the case of the ripple-carry adder, the worst-case delay happens when a carry generated at the least significant bit position propagates all the way to the most significant bit. The delay is then proportional to the number of bits in the input words N and is approximated by Eq. (7.4).

$$t_{adder} \approx (N-1)t_{carry} + t_{sum} \tag{7.4}$$

where t_{carry} and t_{sum} equal the propagation delays from C_i to C_o and S, respectively.[1]

Example 7.1 Propagation Delay of Ripple-Carry Adder

Derive the values of A_k and B_k ($k = 0...N-1$) so that the worst-case delay is obtained for the ripple-carry adder.

The worst-case condition requires that a carry be generated at the lsb position. Since the input carry of the first full adder C_{i0} is always 0, this means that both A_0 and B_0 must equal 1. All the other stages must be in propagate mode. Hence, either A_i or B_i must be high, but not both at the same time. Finally, we would like to physically measure the delay as a transition on the msb sum-bit. Assuming an initial value of 0 for S_{N-1}, we must arrange a $0 \rightarrow 1$ transition. This is achieved by setting both A_{N-1} and B_{N-1} to 0 (or 1), which yields a high sum-bit given the incoming carry of 1.

For example, the following values for A and B trigger the worst-case delay for an 8-bit addition. The rightmost bit represents the msb in this binary representation. Observe that this is only one of the many worst-case patterns. Derive some others as an exercise.

A: 0000001; B: 01111111

Two important conclusions can be drawn from Eq. (7.4).

- The propagation delay of the ripple-carry adder is *linearly* proportional to N. This property becomes increasingly important when designing adders for the wide data-paths ($N = 16...128$) that are desirable in current and future computers.
- When designing the full adder cell for a fast ripple-carry adder, it is far more important to optimize t_{carry} than t_{sum}, since the latter has only a minor influence on the total value of t_{adder}.

Before starting an in-depth discussion on the circuit design of full adder cells, an additional logic property of the full adder is worth mentioning.

Inverting all inputs to a full adder results in inverted values for all outputs.

This property, also called *the inverting property*, is expressed in Eq. (7.5) and will be extremely useful when optimizing the speed of the ripple-carry adder. It states that the circuits of Figure 7.4 are identical.

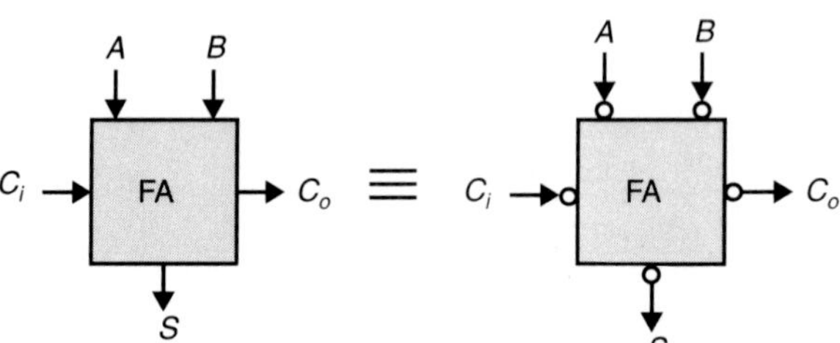

Figure 7.4 Inverting property of the full adder. The circles in the schematics represent inverters.

[1] Eq. (7.4) assumes that for the lsb the delay from the input signals A_0 (or B_0) to $C_{o,0}$ is equal to t_{carry}. Although not completely correct, this approximation is acceptable and helps to simplify the expression.

$$\begin{aligned} \bar{S}(A, B, C_i) &= S(\bar{A}, \bar{B}, \bar{C_i}) \\ \overline{C_o}(A, B, C_i) &= C_o(\bar{A}, \bar{B}, \bar{C_i}) \end{aligned} \quad (7.5)$$

7.3.2 The Full Adder: Circuit Design Considerations

Static Adder Circuit

One way to implement the full adder circuit is to take the logic equations of Eq. (7.1) and translate them directly into complementary CMOS circuitry. Some logic manipulations can help to reduce the transistor count. For instance, it is advantageous to share some logic between the sum- and carry-generation subcircuits, as long as this does not slow down the carry generation, which is the most critical part, as stated previously. An example of such a reorganized equation set is given in Eq. (7.6). The equivalence with the original equation set is easily verified.

$$\begin{aligned} C_o &= AB + BC_i + AC_i \\ S &= ABC_i + \overline{C_o}(A + B + C_i) \end{aligned} \quad (7.6)$$

The corresponding adder design, using complementary static CMOS, is shown in Figure 7.5 and requires 28 transistors. Besides consuming a large area, this circuit is also slow:

- Long chains of series PMOS transistors are present in both *carry*- and *sum* generation circuits.
- The intrinsic load capacitance of the C_o signal is large and consists of two diffusion and six gate capacitances plus the wiring capacitance.

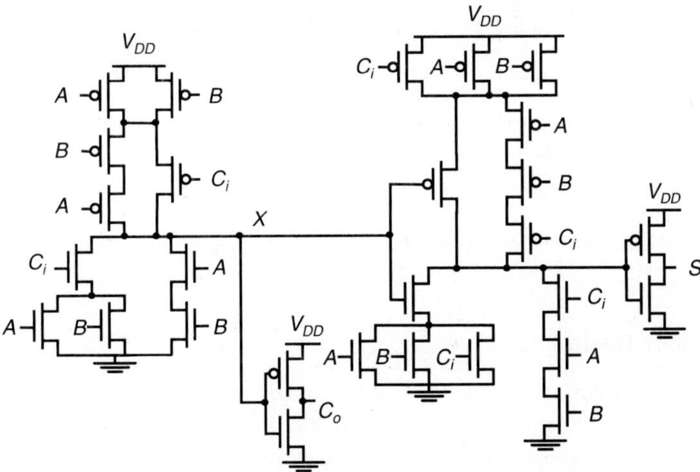

Figure 7.5 Complementary static CMOS implementation of full adder.

- The carry-generation circuit requires two inverting stages per bit. As mentioned above, minimizing the carry-path delay is the prime goal of the designer of high-speed adder circuits.
- The sum generation requires one extra logic stage, but this is not that important, since this factor appears only once in the propagation delay Eq. (7.4).

Although slow, the circuit includes some smart design tricks. Notice that in the first gate of the carry-generation circuit, the NMOS and PMOS transistors connected to C_i are placed as close as possible to the output of the gate. This is a direct application of a circuit-optimization technique discussed in Section 4.2—transistors on the critical path should be placed as close as possible to the output of the gate. For instance, in stage k of the adder, signals A_k and B_k are available and stable long before $C_{i,k}$ ($= C_{o,k-1}$) arrives after rippling though the previous stages. In this way the capacitances of the internal nodes in the transistor chain are precharged or discharged in advance. On arrival of $C_{i,k}$, only the capacitance of node X has to be (dis)charged. Putting the $C_{i,k}$ transistors closer to V_{DD} and GND would require not only the (dis)charging of the capacitance of node X but also of the internal capacitances.

The speed of this circuit can now be improved gradually by using some of the adder properties discussed in the previous section. First of all, the number of inverting stages in the carry path can be reduced by exploiting the inverting property—inverting all the inputs of a full adder cell also inverts all the outputs. This rule allows us to eliminate an inverting gate in a carry chain, as demonstrated in Figure 7.6. The only disadvantage is that this design needs different cells for the even and odd slices of the adder chain.

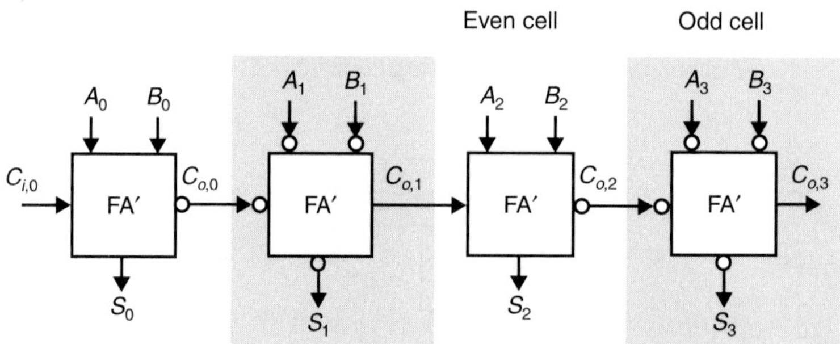

Figure 7.6 Inverter elimination in carry path. FA′ stands for a full adder FA without the inverter in the carry path.

Improved Adder Design

An improved adder circuit, also called the *symmetrical, or mirror adder*, is shown in Figure 7.7 [Weste85]. Its operation is based on Eq. (7.3). The carry generation circuitry is worth analyzing. First, the carry-inverting gate is eliminated, as dictated by the previous section. Secondly, the PDN and PUN networks of the gate are not dual. Instead, they form a smart implementation of the propagate/generate/delete function: when either D or G is high, $\overline{C}_o$ is set to V_{DD} or GND, respectively. When the conditions for a *Propagate* are valid

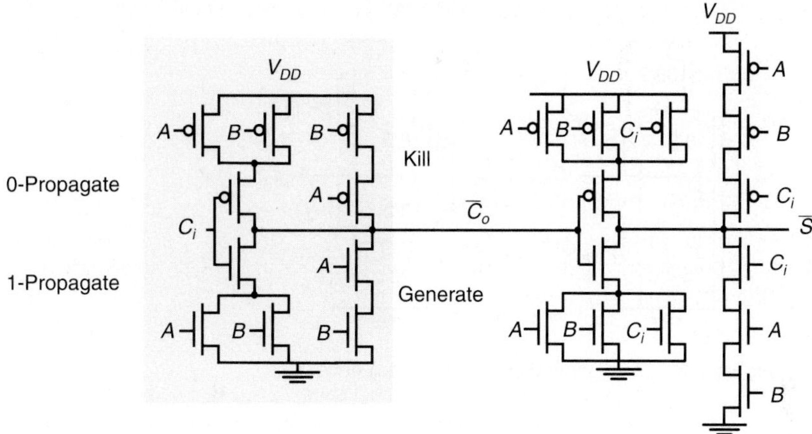

Figure 7.7 Mirror adder—circuit schematics.

(or P is 1), the incoming carry is propagated (in inverted format) to $\overline{C}_o$. This results in a considerable reduction in both area and speed. The analysis of the output circuitry is left to the reader. Some other observations are worth considering.

- This full adder cell requires only 24 transistors.
- The NMOS and PMOS chains are completely symmetrical. This guarantees identical rising and falling transitions if the NMOS and PMOS devices are properly sized. A maximum of two series transistors can be observed in the carry-generation circuitry.
- When laying out the cell, the most critical issue is the minimization of the capacitance at node $\overline{C}_o$. The reduction of the diffusion capacitances is particularly important.
- The capacitance at node $\overline{C}_o$ is composed of four diffusion capacitances, two internal gate capacitances, and six gate capacitances in the connecting adder cell, or a total of ± 12 gate capacitances assuming that the diffusion capacitance is approximately equal to a gate capacitance (see Chapter 3). This is identical to the fully complementary implementation of Figure 7.6.
- The transistors connected to C_i are placed closest to the output of the gate.
- Only the transistors in the carry stage have to be optimized for speed. All transistors in the sum stage can be minimum-size.

Example 7.2 Static Adder Design

Consider a slight modification of the static ripple-carry cell of Figure 7.5. A differential approach is used; that is, every full adder cell generates both C_o and $\overline{C}_o$. The crucial gates of the cell are depicted in Figure 7.8. It is left as an exercise for the reader to fill in the rest of the cell (use transmission-gate EXORs as much as possible). The transistor sizes for our 1.2 μm CMOS process are annotated on the schematics (in λ). Observe how a progressive sizing is used. Explain why the PMOS transistor connected to P is smaller than the one connected to C_{in} in the first gate.

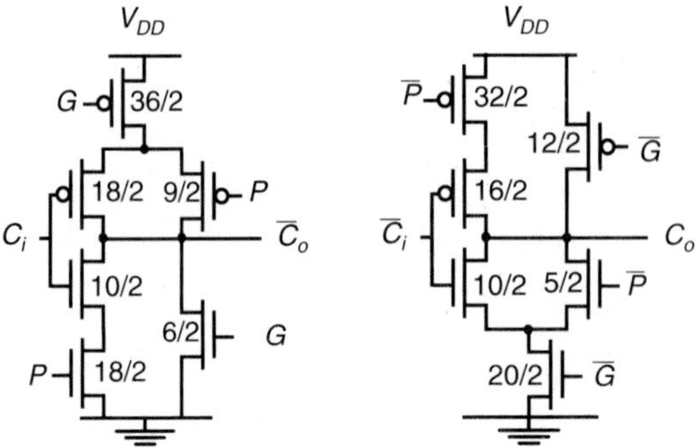

Figure 7.8 Carry-generation circuitry of full-adder cell.

From simulations, we can derive a delay model in the style of Eq. (7.4).

$$t_{add} = t_{A,B \to co} + (N-2)t_{ci \to co} + t_{ci \to s}$$

with $t_{A,B \to co} = 1.63$ nsec, $t_{ci \to co} = 0.32$ nsec, and $t_{ci \to s} = 1$ nsec, for a rise time of 2 nsec at the inputs. This yields the following expression for t_{add}

$$t_{add} = 1.99 + 0.32N \text{ nsec}$$

A 32-bit addition thus takes 12.23 nsec.

Dynamic Adder Design

Most of the adder circuits discussed above can also be designed using dynamic circuit styles such as DOMINO CMOS or *np*-CMOS. An implementation of an *np*-CMOS adder is shown in Figure 7.9. The basic cell requires only 17 transistors, ignoring the extra inverters required for the input or output signals. The alternating even and odd carry stages are realized using NMOS and PMOS networks respectively. The sum-generation networks are also implemented in alternating device types. While it is a direct implementation of Eq. (7.3), the reduced capacitance of the dynamic circuitry results in a substantial speed-up over the static implementation. The load capacitance on the carry bit approximately equals seven equivalent gate capacitances—three diffusion and four gate capacitances.

Example 7.3 Dynamic Adder Design

A layout of the dynamic adder design of Figure 7.9 is plotted in Figure 7.10. Observe the bit-sliced organization. Close to minimum-size transistors are used virtually everywhere except for the carry-generation circuitry. The large devices used there (up to 48/1.2) are easily recognizable in the layout. In the 1.2 μm CMOS technology, the total area per bit equals 120 μm × 31 μm, for a very small carry delay of 200 psec/bit. This delay number ignores the precharge time, which is at least of a similar length.

In addition, the dynamic approach allows for the conception of alternative circuit diagrams that are hard to realize in a static way. An example of such a circuit is shown in Figure 7.11. This adder is called a *Manchester Carry-Chain Adder,* and uses a cascade of pass-transistors to implement the carry chain. *Propagate* and *Generate* signals are gener-

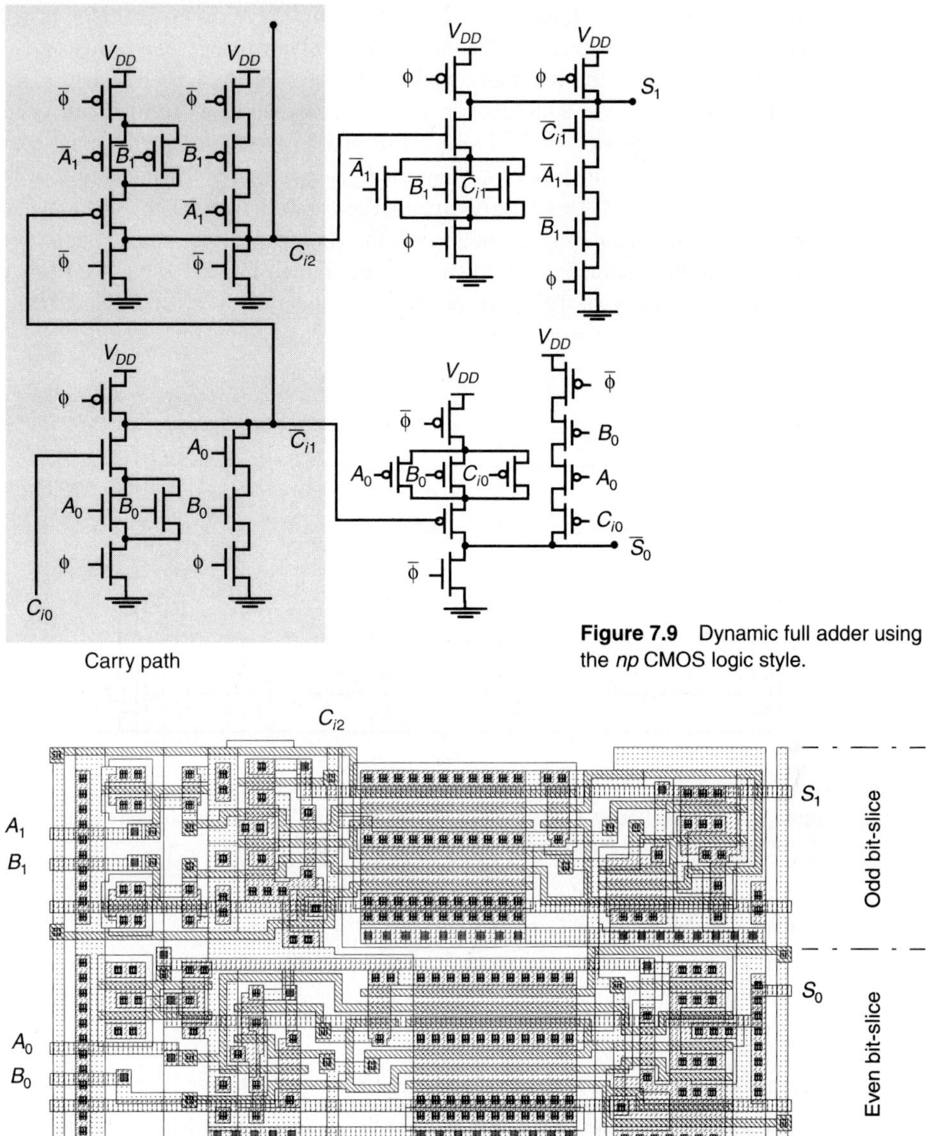

Figure 7.9 Dynamic full adder using the *np* CMOS logic style.

Figure 7.10 Layout of dynamic full adder using the *np*-CMOS logic style (from [Kleinfelder91]).

ated in the traditional way using, for instance, pass-transistor logic. During the precharge phase ($\phi = 0$), all intermediate nodes of the pass-transistor chain are precharged to V_{DD}. During the evaluation phase, the A_k node is discharged when there is an incoming carry and the *Propagate* signal P_k is high, or when the *Generate* signal for stage k (G_k) is high.

The capacitance per node on the carry chain is very small and equals only four diffusion capacitances. Unfortunately, the distributed *RC*-nature of the carry chain results in a propagation delay that is quadratic in the number of bits N. To avoid this, it is necessary to insert signal-buffering inverters. The optimum number of stages per buffer depends on the buffer delay t_{pbuf} and the resistance and capacitance of the pass-transistors, as was discussed in Section 4.2. An additional speed-up is obtained by a careful sizing of the transistors in the pass-transistor chain. During a discharge of the complete chain, transistor M_0 has to sink the largest amount of current, or $I_{M0} > I_{M1} > \ldots > I_{M4}$. Therefore, it is advantageous to size the transistors progressively. The same is true for the discharge transistors connected to the *Generate* signals and the *Evaluate* transistors. Short propagation delays/bit can be obtained using this technique. For instance, for our 1.2 μm technology, we measured a delay of 192 psec per bit. This delay is very sensitive to any parasitic capacitances at the precharged nodes.

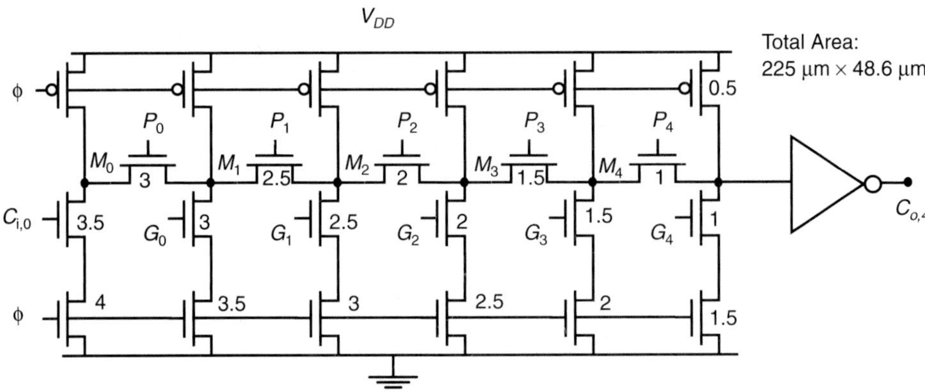

Figure 7.11 Manchester carry-chain adder (5-bit section). The annotated numbers indicate the relative transistor sizes. A unit-size transistor measures (6/1.2).

Problem 7.1 Manchester Carry Chain

The carry chain uses only NMOS transistors in the pass-transistor network (instead of full transmission gates). Discuss why this is an acceptable approach in this configuration.

Example 7.4 Transistor Sizing in the Manchester Carry Chain

The worst-case delay of the carry chain of the adder in Figure 7.11 can be modeled by the linearized *RC* network of Figure 7.12. The linearized on-resistance of the minimum-size transistor is assumed to equal 20 kΩ, and the linearized diffusion capacitance contributed by each minimum-size device is set to 5 fF. The diffusion capacitance at each node in the chain is the sum of the capacitive contributions of two pass-transistors, a pull-down and a precharge device.

The propagation delay of this *RC* network is expressed by the following equation:

$$t_p = 0.69 \sum_{i=1}^{N} C_i \left(\sum_{j=1}^{i} R_j \right) \tag{7.7}$$

where N is the number of nodes in the network and C_i the capacitance of node i to ground. The last term in the equation (the resistive sum) represents the total resistance between node i and

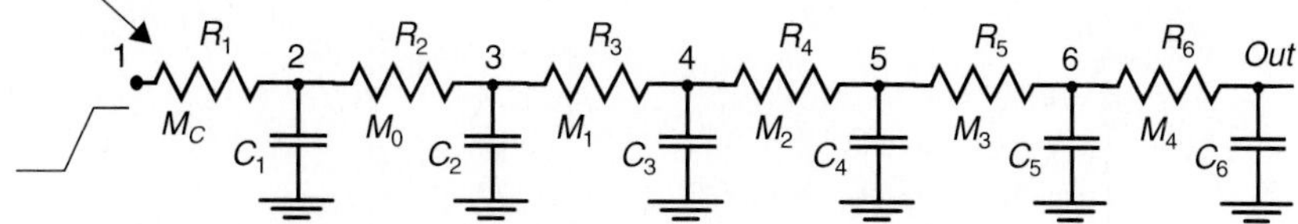

Figure 7.12 Equivalent network to determine propagation delay of carry chain.

the source node (node 1). This expression, called the *Elmore delay*, is discussed further in Chapter 8. The delay of the *RC* network of Figure 7.12 can now be determined

$$t_p = 0.69(C_1R_1 + C_2(R_1 + R_2) + C_3(R_1 + R_2 + R_3) + C_4(R_1 + R_2 + R_3 + R_4) + C_5(R_1 + R_2 + R_3 + R_4 + R_5) + C_6(R_1 + R_2 + R_3 + R_4 + R_5 + R_6))$$

Since R_1 occurs six times in the expression, it makes sense to minimize this contribution by making the first transistor larger than the other ones, or to use progressive scaling.

First consider the case where all transistors are minimum-size. The capacitance at each node is estimated to equal 20 fF, and all resistances are set to 20 kΩ.

$$t_p = 0.69C(6R + 5R + 4R + 3R + 2R + R) = 0.69 \times 21 \times RC = 5.8 \text{ nsec}$$

Assume now that the stages are made progressively larger, starting from a minimum-size transistor at the output. The (W/L) of the next-to-last transistor is scaled by a factor k (> 1), which means that its resistance is divided by k, while its associated capacitances are approximately increased by a factor k. The following expressions hold

$$C_i = kC_{i+1};\ R_i = R_{i+1}/k$$

$$C_6 \approx 20 \text{ fF};\ R_6 = 20 \text{ k}\Omega$$

This yields the following expression for t_p

$$t_p = 0.69CR(1 + 2k + 3k^2 + 4k^3 + 5k^4 + 6k^5)/k^5$$

Figure 7.13 plots the propagation delay (normalized with respect to *0.69RC*) and the area of the transistor chain (normalized with respect to a minimum-size transistor) as a function of k. Observe that the area increases dramatically with k, which effectively excludes the use of large scaling factors. The delay starts from $21RC$ for the nonscaled version and decreases sharply for k between 1 and 1.5. For instance, a k-factor of 1.5 reduces the delay by 40% (to $0.69 \times 12 \times RC$) at the expense of a 3.5-fold increase in area. A smaller scaling factor (e.g., k = 1.2) does not effect the area much (only 1.65 times larger), but still yields a reasonable speed-up of 20%. Notice that the minimum possible delay for very large values of k equals six time-constants. This means that a linear dependence on the number of *RC*-stages is achieved in contrast to the quadratic dependence of the nonscaled implementation.

WARNING: Be aware that the above analysis represents only a first-order model. Extensive simulation on extracted layouts is necessary to fine-tune the transistor sizes.

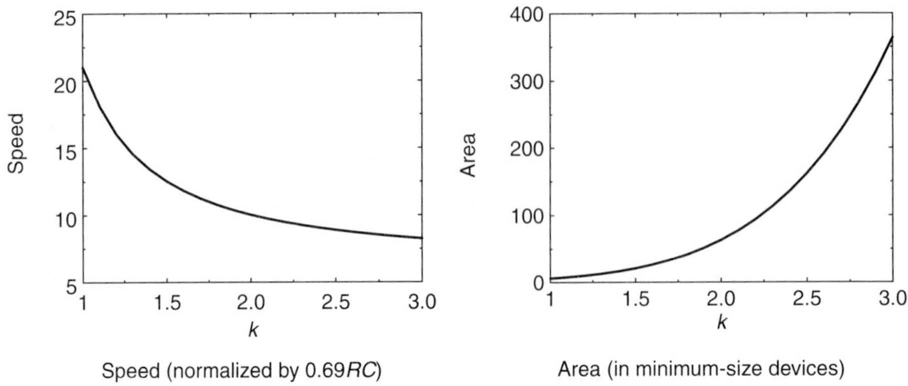

Figure 7.13 Speed and area of Manchester carry chain as a function of scaling factor *k*.

Pipelined Adder

One way to break the dependency between the addition time and the number of bits is to employ the pipelining technique introduced in Chapter 6. Pipeline registers are inserted on the carry path of the adder so that the subsequent sum bits are produced in different time intervals. The pipelining reduces the critical path and the clock period of the adder to a single carry- and sum-generation stage. An N-bit addition now takes N clock cycles, but a result is produced every clock cycle because N additions are performed simultaneously. This extreme usage of pipelining is called *bit-level pipelining* and has been used effectively in signal-processing applications.

An example of a pipelined adder based on the NORA-CMOS dynamic circuit approach is shown in Figure 7.14. The even and odd bits are implemented as ϕ-blocks and $\overline{\phi}$-blocks, respectively. One stage is evaluating, while the next one is in the precharging mode. Observe also how within a single stage, the *np*-CMOS approach is used to cascade gates. Further pipelining can be achieved by inserting an extra register between the carry- and sum-generating circuitry, reducing the critical path to a single carry generation. This incurs an extra cost, because all inputs to the sum-generating circuitry (A_i, B_i, and C_i) must be delayed as well by inserting extra registers.

7.3.3 The Binary Adder: Logic Design Considerations

The ripple-carry adder is only practical for the implementation of additions with a relatively small word length. Fast computers such as mainframes or supercomputers require additions with a word length up to 128 bits. The linear dependence of the adder speed on the number of bits makes the usage of ripple adders rather impractical. Therefore, logic optimizations are necessary, resulting in adders with $t_p < O(N)$. A number of those are discussed briefly below. We concentrate on the circuit design implications, since most of these structures are well-known from traditional logic design.

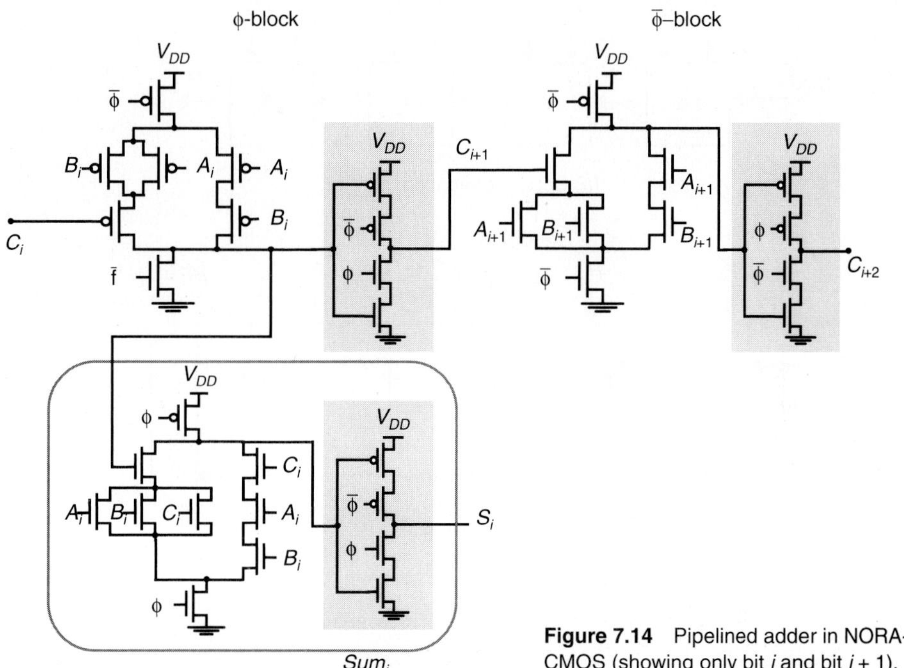

Figure 7.14 Pipelined adder in NORA-CMOS (showing only bit *i* and bit *i* + 1).

The Carry-Bypass Adder

Consider the four-bit adder block of Figure 7.15a and suppose that the values of A_k and B_k ($k = 0 \ldots 3$) are such that all *Propagate* signals P_k ($k = 0 \ldots 3$) are high. An incoming carry $C_{i,0} = 1$ propagates under those conditions through the complete adder chain and causes an outgoing carry $C_{o,3} = 1$. In other words,

$$\text{if } (P_0 P_1 P_2 P_3 = 1) \text{ then } C_{o,3} = C_{i,0} \qquad \text{else either DELETE or GENERATE occurred.} \tag{7.8}$$

This information can be used to speed up the operation of the adder, as shown in Figure 7.15b. When $BP = P_0P_1P_2P_3 = 1$, the incoming carry is forwarded immediately to the next block through the bypass transistor M_b. Hence the name *carry-bypass adder* [Turrini89]. If this is not the case, the carry is obtained via the normal route.

Figure 7.16 shows the possible carry-propagation paths when the full adder circuit is implemented in Manchester-carry style. This picture demonstrates how the bypass speeds up addition: either the carry propagates through the bypass path, or a carry is generated somewhere in the chain. In both cases, the delay is smaller than the normal ripple configuration. The area overhead incurred by adding the bypass path is small and typically ranges between 10 and 20%.

Let us now compute the delay of an *N*-bit adder. At first we assume that the total adder is divided in (*N/M*) equal-length bypass stages, each of which contains *M* bits. An

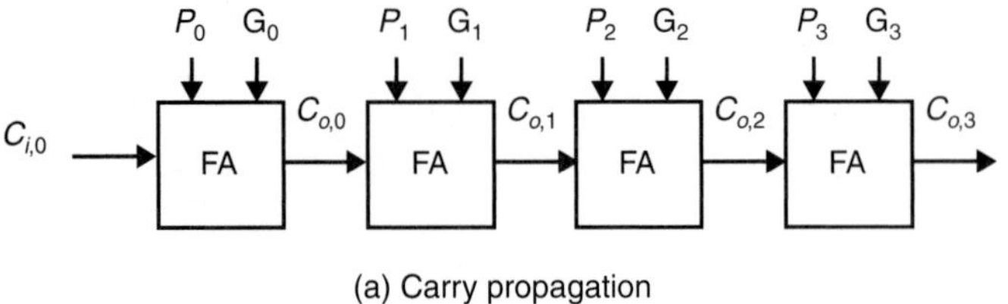

(a) Carry propagation

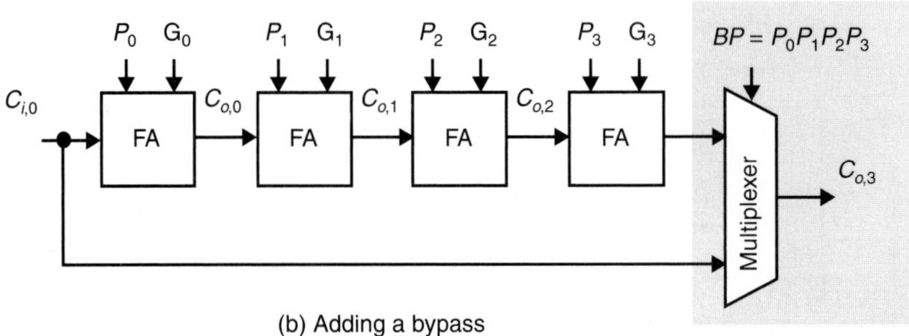

(b) Adding a bypass

Figure 7.15 Carry-bypass structure—basic concept.

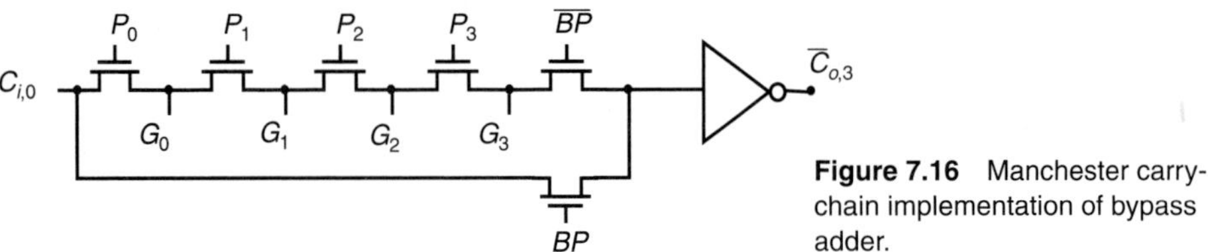

Figure 7.16 Manchester carry-chain implementation of bypass adder.

approximating expression for the total propagation time can be derived from Figure 7.17a and is given in Eq. (7.9).

$$t_p = t_{setup} + Mt_{carry} + \left(\frac{N}{M} - 1\right)t_{bypass} + Mt_{carry} + t_{sum} \tag{7.9}$$

with the composing parameters defined as follows:

- t_{setup}—the fixed overhead time to create the *Generate* and *Propagate* signals.
- t_{carry}—propagation delay through a single bit. The worst-case carry-propagation delay through a single stage is then approximately M times larger.
- t_{bypass}—propagation delay through the bypass multiplexer of a single stage.
- t_{sum}—the time to generate the sum of the final stage.

The critical path is marked on the block diagram of Figure 7.17 in shaded gray. From Eq. (7.9), it follows that t_p is still linear in the number of bits N, since in the worst case the carry has to propagate through (N/M – 1) bypass stages.The optimal number of bits per bypass block is determined by technological parameters such as the extra delay of the

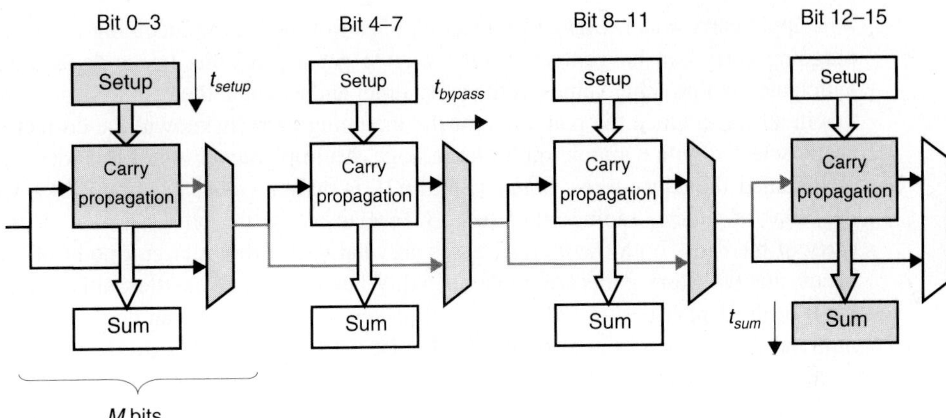

Figure 7.17 (N = 16) carry-bypass adder—composition. The worst-case delay path is shaded in gray.

bypass-selecting multiplexer, the buffering requirements in the carry chain, and the ratio of the delay through the ripple and the bypass paths.

Although still linear, the slope of the delay function increases in a more gradual fashion than for the ripple-carry adder, as pictured in Figure 7.18. This difference is substantial for larger adders. Notice that the ripple adder is actually faster for small values of N. The overhead of the extra bypass multiplexer makes the bypass structure not interesting for small adders. The crossover point depends upon technology considerations and is normally situated between 4 and 8 bits.

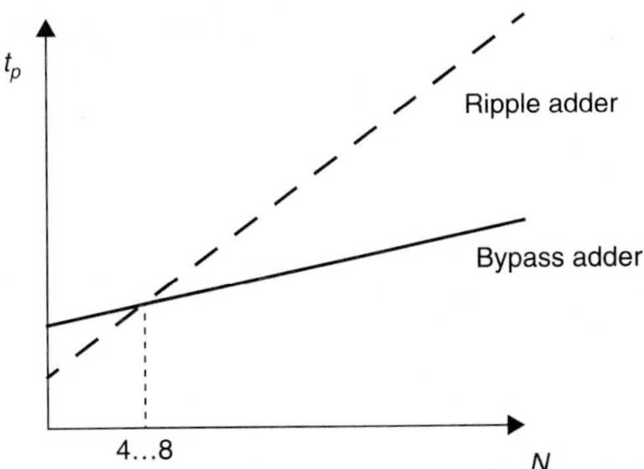

Figure 7.18 Propagation delay of ripple-carry versus carry-bypass adder.

Problem 7.2 Delay of Carry-Bypass Adder

Determine an input pattern that triggers the worst-case delay in a 16-bit (4 × 4) carry-bypass adder. Assuming that $t_{carry} = t_{setup} = t_{bypass} = t_{sum} = 1$, determine that delay and compare it to that of a normal ripple adder.

The Linear Carry-Select Adder

In a ripple-carry adder, every full-adder cell has to wait for the incoming carry before an outgoing carry can be generated. One way to get around this linear dependency is to anticipate both possible values of the carry input and evaluate the result for both possibilities in advance. Once the real value of the incoming carry is known, the correct result is easily selected with a simple multiplexer stage. An implementation of this idea, appropriately called the *carry-select adder* [Bedrij62], is demonstrated in Figure 7.19. Consider the block of adders, adding bits k to $k+3$. Instead of waiting on the arrival of the output carry of bit $k - 1$, both the 0 and 1 cases are analyzed. From a circuit point of view, this means that two carry paths are implemented. When $C_{o,k-1}$ finally settles, either the result of the 0 or the 1 path is selected in the multiplexer section, which can be performed with a minimal delay time. As is evident from Figure 7.19, the hardware overhead of the carry-select adder is restricted to an additional carry path and a multiplexer, and equals about 30% more than a ripple-carry structure.

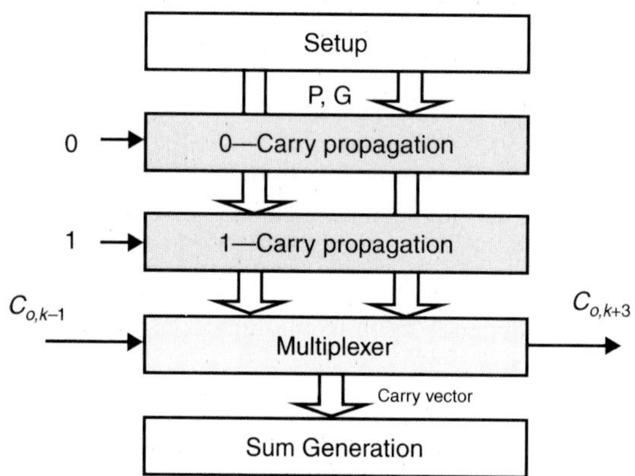

Figure 7.19 Four-bit carry-select module—topology.

A full carry-select adder is now constructed by chaining a number of equal-length adder stages, as done in the carry-bypass approach (Figure 7.20). From inspection of the circuit, we can derive a first-order model of the worst-case propagation delay of the module. The critical path is shaded in gray in Figure 7.20.

$$t_{add} = t_{setup} + Mt_{carry} + \left(\frac{N}{M}\right)t_{mux} + t_{sum} \tag{7.10}$$

where t_{setup}, t_{sum}, and t_{mux} are fixed delays. N and M represent the total number of bits, and the number of bits per stage, respectively. t_{carry} is the delay of the carry through a single full-adder cell. The carry delay through a single block is proportional to the length of that stage, or equals Mt_{carry}.

The propagation delay of the adder is once again linearly proportional to N (Eq. (7.10)). The reason for this linear behavior is that the *block-select* signal that selects between the 0 and 1 solutions still has to ripple through all stages in the worst case.

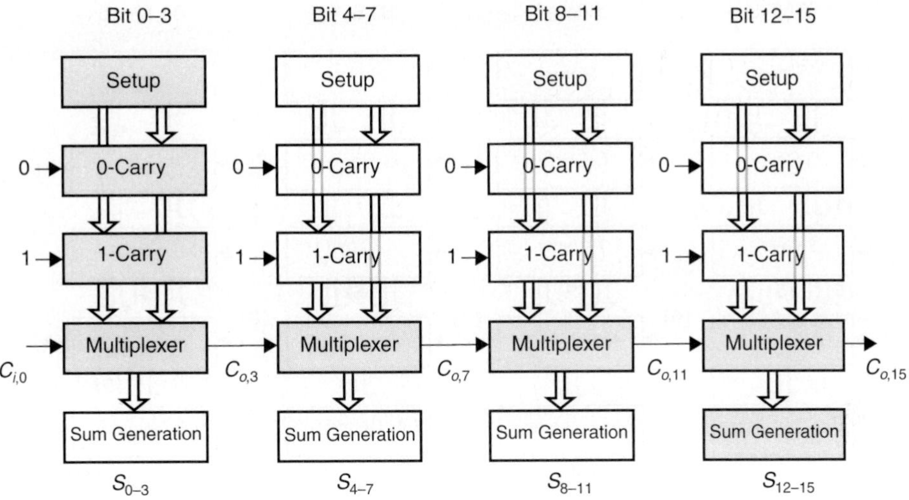

Figure 7.20 16-bit, linear carry-select adder. The critical path is shaded in gray.

Problem 7.3 Linear Carry-Select Delay

Determine the delay of a 16-bit linear carry-select adder using unit delays for all cells. Compare with the results of Problem 7.2. Compare various block configurations.

The Square Root Carry-Select Adder

The next structure illustrates how an alert designer can make a major impact. To optimize a design, it is essential to locate the critical timing path first. Consider the case of a 16-bit linear carry-select adder. To simplify the discussion, assume that the full-adder and multiplexer cells have an identical propagation delay equal to a normalized value of 1. The worst-case arrival times of the signals with respect to the time the input is applied at the different network nodes are marked and annotated in Figure 7.21a. This analysis demonstrates that the critical path of the adder ripples through the multiplexer networks of the subsequent stages.

One striking opportunity comes to light. Consider the multiplexer gate of the last adder stage. The inputs to this multiplexer are the results of the two carry chains and the block-multiplexer signal from the previous stage. A major mismatch between the signal arrival times can be observed. The results of the carry chains are stable long before multiplexer signal arrives. It makes sense to equalize the delay through both paths. This can be achieved by progressively adding more bits to the subsequent stages in the adder, requiring more time for the generation of the carry signals. For instance, the first stage can add 2 bits, the second contains 3, the third has 4, and so forth, as demonstrated in Figure 7.21b. The annotated arrival times show that this adder topology is faster than the linear organization, even though an extra stage is needed. In fact, the same propagation delay is also valid for a 20-bit adder. Observe that the discrepancy in arrival times at the multiplexer nodes has been eliminated.

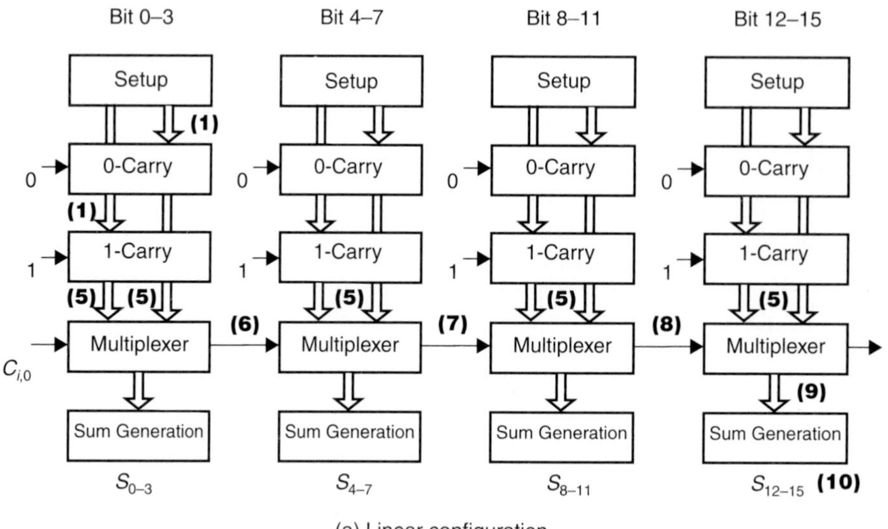

(a) Linear configuration

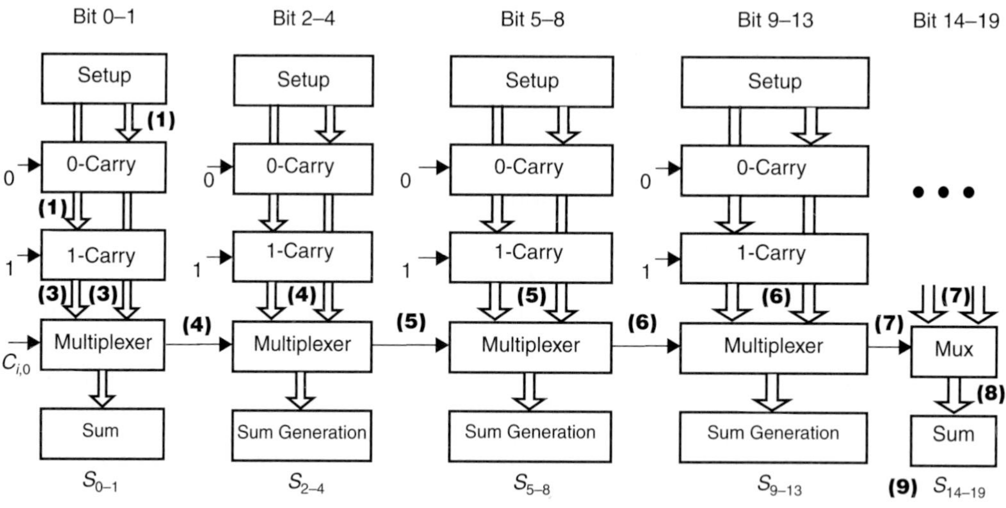

(b) Square root configuration

Figure 7.21 Worst-case signal arrival times in carry-select adders. The signal arrival times are marked in **bold**.

In effect, the simple trick of making the adder stages progressively longer results in an adder structure with sublinear delay characteristics. This is illustrated by the following analysis. Assume that an N-bit adder contains P stages, and the first stage adds M bits. An additional bit is added to each subsequent stage. The following relation then holds:

$$N = M + (M+1) + (M+2) + (M+3) + \ldots + (M+P-1)$$
$$= MP + \frac{P(P-1)}{2} = \frac{P^2}{2} + P\left(M - \frac{1}{2}\right) \tag{7.11}$$

If $M << N$ (e.g., $M = 2$, and $N = 64$), the first term dominates and Eq. (7.11) can be simplified.

$$N \approx \frac{P^2}{2} \tag{7.12}$$

and

$$P \approx \sqrt{2N} \tag{7.13}$$

Eq. (7.13) can be used to express t_{add} as a function of N by rewriting Eq. (7.10):

$$t_{add} = t_{setup} + Mt_{carry} + (\sqrt{2N})t_{mux} + t_{sum} \tag{7.14}$$

The delay is proportional to $\sqrt{N}$ for large adders ($N >> M$), or $t_{add} = O(\sqrt{N})$. This square-root relation has a major impact. This is illustrated in Figure 7.22, where the delays of both the linear and square-root select adders are plotted as a function of N. It can be observed that for large values of N, t_{add} almost becomes a constant.

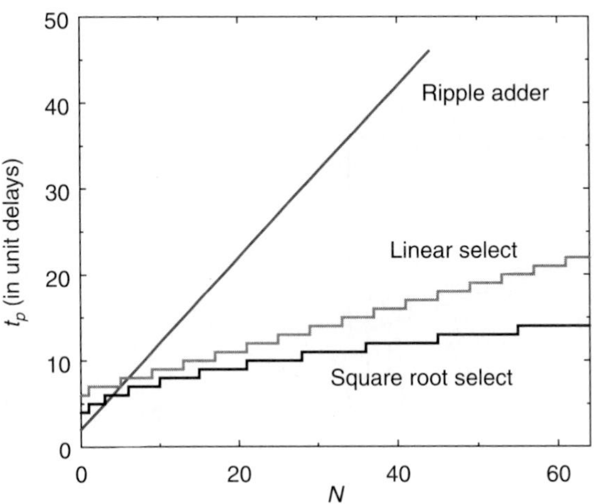

Figure 7.22 Propagation delay of square-root carry-select adder versus linear ripple and select adders. The unit delay model is used to model the cell delays.

Problem 7.4 Square Root Bypass Adder

The idea of making subsequent stages longer in the presence of linear ripple effects is a general technique and can also be used for other structures besides carry-select. For instance, we can also construct a square root carry-bypass adder. Determine the topology of such a structure and derive an expression for the propagation delay. Hint: be aware that the sum-generation delay is a function of the length of the stage, in contrast to the select adder. This factor would start to dominate the overall delay and requires some adjustments to the topology of the select adder.

The Carry-Lookahead Adder

When designing even faster adders, it is essential to get around the rippling effect of the carry that is still present in one form or another in both the carry-bypass and carry-select adders. The *carry-lookahead* principle offers a possible way to do so [McSorley61]. As stated before, the following relation holds for each bit position in an N-bit adder:

$$C_{o,k} = f(A_k, B_k, C_{o,k-1}) = G_k + P_k C_{o,k-1} \tag{7.15}$$

The dependency between $C_{o,k}$ and $C_{o,k-1}$ can be eliminated by expanding $C_{o,k-1}$:

$$C_{o,k} = G_k + P_k(G_{k-1} + P_{k-1} C_{o,k-2}) \tag{7.16}$$

or in a fully expanded form,

$$C_{o,k} = G_k + P_k(G_{k-1} + P_{k-1}(\ldots + P_1(G_0 + P_0 C_{i,0}))) \tag{7.17}$$

Observe that $C_{i,0} = 0$. This expanded relationship can be used to implement an N-bit adder. For every bit, the carry and sum outputs are independent of the previous bits. The ripple effect has thus been effectively eliminated, and therefore the addition time should be independent of the number of bits. A block diagram of the overall composition of a carry-lookahead adder is shown in Figure 7.23.

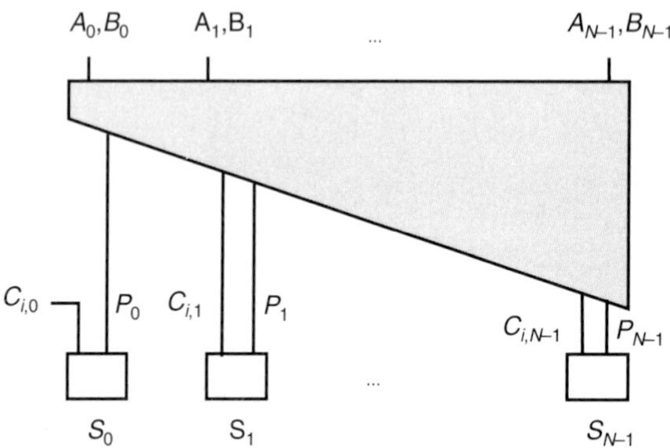

Figure 7.23 Conceptual diagram of a carry-lookahead adder.

This high-level model contains some hidden dependencies. When we study the detailed schematics of the adder, it becomes obvious that the constant addition time is wishful thinking and that the real delay is at least linear with the number of bits. This is illustrated by Figure 7.24, where a possible circuit implementation of Eq. (7.17) is shown for $N = 4$. The large fan-in of the circuit makes it prohibitively slow for larger values of N. Implementing it with simpler gates requires multiple logic levels. In both cases, the propagation delay increases. Furthermore, the fan-out on some of the signals tends to grow excessively, slowing down the adder even more, since the propagation delay of a gate is proportional to its load. For instance, the signals G_0 and P_0 appear in the expression for every one of the subsequent bits. Hence, the capacitance on these lines is substantial. Finally, the area of the implementation grows progressively with N. Therefore, the lookahead structure as suggested by Eq. (7.16) is only useful for small values of N (≤ 4).

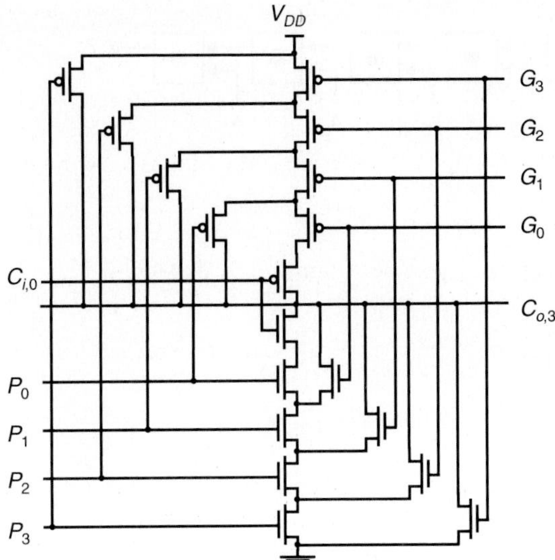

Figure 7.24 Schematic diagram of 4-bit lookahead adder (from [Weste85]).

The Logarithmic Lookahead Adder

Although a constant addition time is clearly not feasible, it is possible to restructure the lookahead adder in such a way that an $O(\log_2 N)$ delay order is obtained [Brent82]. This restructuring is based on a basic property of associative operators.

> Consider a generic associative operator, called the *dot* operation (•). The associativity property implies that the following statement is valid: $(a \bullet b) \bullet c = a \bullet (b \bullet c)$. Under those conditions, combining N arguments using the • operator can be executed with a critical path equal to $(\log_2 N)\, t_\bullet$, where $t_\bullet$ is the propagation delay of the dot-operator. Implementing the same computation in a linear, ripple-like fashion results in a propagation delay of $(N-1)\, t_\bullet$.

The property is illustrated in Figure 7.25 for the case of $N = 8$. The linear (or ripple) topology is illustrated in Figure 7.25a and has a critical path of 7 $t_\bullet$. The logarithmic one (shown in Figure 7.25b) combines its operands in a tree-like fashion and has a delay of only 3 $t_\bullet$. The number of operators stays exactly the same. How the associative property helps to go from one structure to the other is made clear in Eq. (7.18).

$$\begin{aligned}&(((((A_1 + A_2) + A_3) + A_4) + \ldots) + A_8) \\ &= (((A_1 + A_2) + (A_3 + A_4)) + (\ldots + (\ldots + A_8)))\end{aligned} \tag{7.18}$$

This property can be applied to the case of an N-bit adder. This requires the definition of a • operator that establishes the following relationship between two tuples (g, p). A *tuple* is an ordered set of values, in this particular case containing two values.

$$(g, p) \bullet (g', p') = (g + pg', pp') \tag{7.19}$$

In other words, the • operator is a function that takes in two sets of two inputs ((g,p) and (g', p')) and produces a set of two outputs $(g+pg', pp')$. We can verify that this dot-

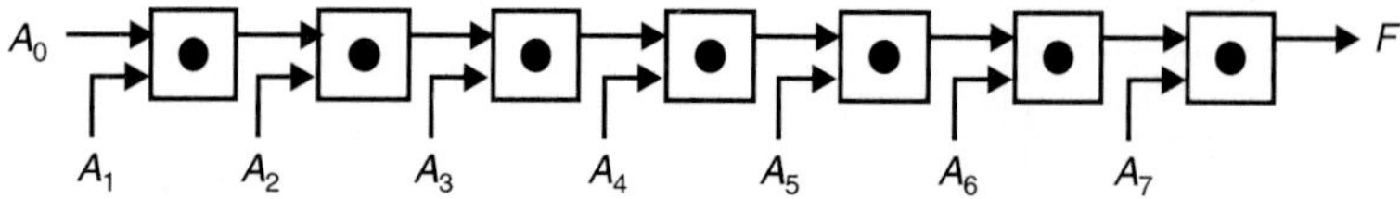

(a) Linear

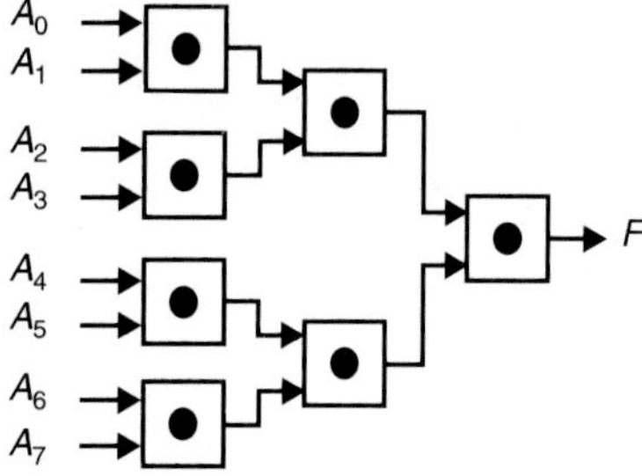

(b) Logarithmic

Figure 7.25 Configurations of associative operators.

operation is associative, but not commutative. We must introduce two extra functions α and β to access the elements of the tuple.

$$\begin{aligned} g &= \alpha(g, p) \\ p &= \beta(g, p) \end{aligned} \tag{7.20}$$

Eq. (7.17) can now be rewritten with the aid of the dot-operator:

$$\begin{aligned} C_{o,0} &= G_0 + PC_{i,0} = G_0 = \alpha(G_0, P_0) \\ C_{o,1} &= G_1 + G_0P_1 = \alpha((G_1, P_1) \bullet (G_0, P_0)) \\ &\ldots \\ C_{o,k} &= \alpha((G_k, P_k) \bullet (G_{k-1}, P_{k-1}) \bullet \ldots \bullet (G_0, P_0)) \end{aligned} \tag{7.21}$$

where G_i and P_i are the *Generate* and *Propagate* signals of the ith bit of the adder.

Suppose that a hardware cell implementing the dot-operator is available. While multiple possible implementations of Eq. (7.21) exist, the binary tree organization that is attainable due to the associativity property, clearly yields the highest performance. The complete organization of an 8-bit adder is shown in Figure 7.26.

The forward binary tree realizes the following carry signals:

$$\begin{aligned} C_{o,0} &= \alpha(G_0, P_0) \\ C_{o,1} &= \alpha((G_1, P_1) \bullet (G_0, P_0)) \\ C_{o,3} &= \alpha(((G_3, P_3) \bullet (G_2, P_2)) \bullet ((G_1, P_1) \bullet (G_0, P_0))) \\ C_{o,7} &= \alpha((C_{4-7}, P_{4-7}) \bullet (C_{o,3}, P_{o,3})) \end{aligned} \tag{7.22}$$

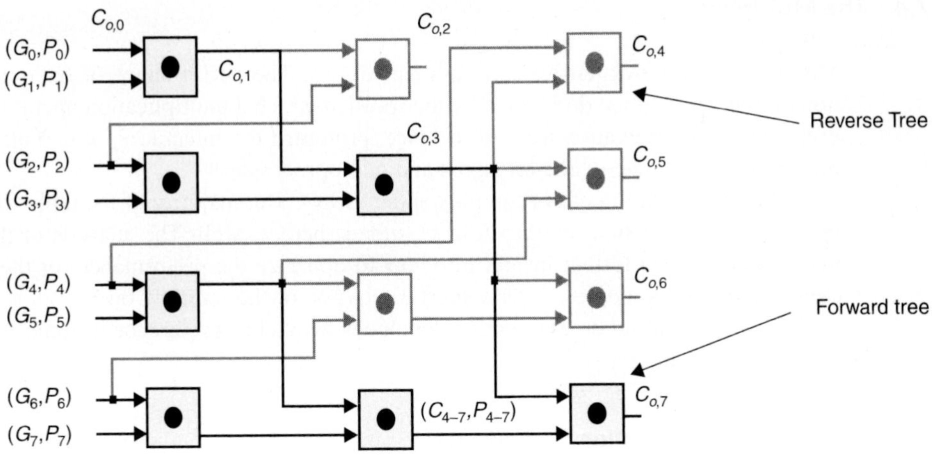

Figure 7.26 An 8-bit Brent-Kung structure. The forward binary tree is colored gray.

In order to generate the complete set of carry bits, this forward binary-tree structure is not sufficient, since it only produces $C_{o,0}$, $C_{o,1}$, $C_{o,3}$, and $C_{o,7}$. An *inverse binary tree* is needed to realize the other carry bits (as shown in gray lines in Figure 7.26).This structure combines intermediate results to produce the remaining carry bits. It is left for the reader to verify that this structure produces the correct expressions for all carry bits.

The resulting structure, commonly called the Brent-Kung adder, has a number of worthwhile properties:

- The number of logic levels is proportional to $\log_2 N$. The same is valid for the propagation delay.
- Gate fan-in is limited. The fan-out can be large, but this can be handled by careful buffering.
- The layout is very compact. As Figure 7.26 shows, the cells can be fitted together in a perfect jigsaw puzzle.
- Once the carry bits are available, the sum bits are easily derived in constant time.

On the average, a lookahead adder is more than 100% larger than a ripple adder, but has dramatic speed advantages for larger adders. The logarithmic behavior makes it preferable over bypass or select adders for larger values of N. The exact value of the crosspoint depends heavily on technology and circuit design factors.

The discussion of adders is by no means complete. Due to its impact on the performance of computational structures, the design of fast adder circuits has been the subject of many publications. It is even possible to construct adder structures with a propagation delay *independent of the number of bits*. Examples of those are the carry-save and redundant binary arithmetic structures [Swartzlander90]. These adders require number encoding and decoding steps whose delay is a function of N. Therefore, they are only interesting when embedded in larger structures such as multipliers or high-speed signal processors.

7.4 The Multiplier

Multiplications are expensive and slow operations. The performance of many computational problems is often dominated by the speed at which a multiplication operation can be executed. This observation has, for instance, prompted the integration of complete multiplication units in state-of-the-art signal and microprocessors.

Multipliers are in effect complex adder arrays. The majority of the topics discussed in the preceding section are therefore of interest here as well. The analysis of the multiplier gives us some further insight into how to optimize the performance (or the area) of complex circuit topologies. After a short discussion of the multiply operation, we discuss the basic array multiplier. The section concludes with a brief reference to some alternative or improved multiplier structures.

7.4.1 The Multiplier: Definitions

Consider two *unsigned* binary numbers X and Y that are M and N bits wide respectively. To introduce the multiplication operation, it is useful to express X and Y in a binary representation.

$$X = \sum_{i=0}^{M-1} X_i 2^i \qquad Y = \sum_{j=0}^{N-1} Y_j 2^j \tag{7.23}$$

with $X_i, Y_j \in \{0, 1\}$. The multiplication operation is then defined as follows:

$$\begin{aligned} Z = X \times Y &= \sum_{k=0}^{M+N-1} Z_k 2^k \\ &= \left(\sum_{i=0}^{M-1} X_i 2^i\right)\left(\sum_{j=0}^{N-1} Y_j 2^j\right) = \sum_{i=0}^{M-1}\left(\sum_{j=0}^{N-1} X_i Y_j 2^{i+j}\right) \end{aligned} \tag{7.24}$$

A common way to implement this operation is to resort to an approach similar to manually computing a multiplication. The multiplicand is consecutively multiplied with every bit of the multiplier, resulting in a number of *partial products*. These intermediate results are added after the proper shifting has been applied. The approach is illustrated in Figure 7.27. This set of operations can be mapped directly into hardware. The resulting structure is called an *array multiplier*.

7.4.2 The Array Multiplier

The composition of an array multiplier is shown in Figure 7.28. There is a one-to-one correspondence between this hardware structure and the manual multiplication of Figure 7.27. The generation of a partial product requires a multiplication by 1 or 0. In logic terms, this is nothing more than an AND operation. Generating the N partial products requires N M-bit AND gates. Most of the area of the multiplier is devoted to the adding of the N par-

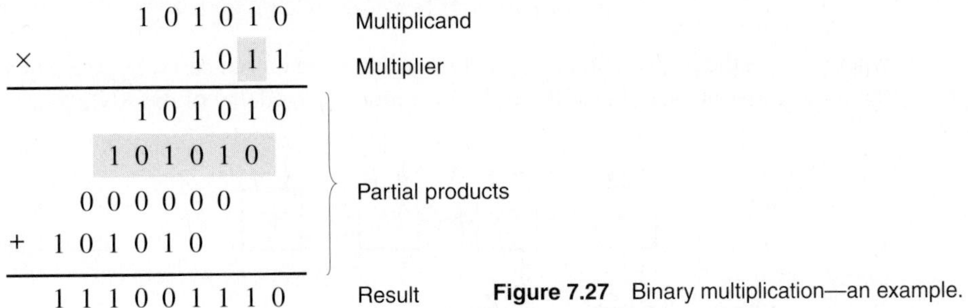

Figure 7.27 Binary multiplication—an example.

tial products, which requires $N - 1$ M-bit adders. The shifting of the partial products is performed by simple routing and does not require any active logic. The overall structure can easily be compacted into a rectangle, resulting in a very efficient layout.

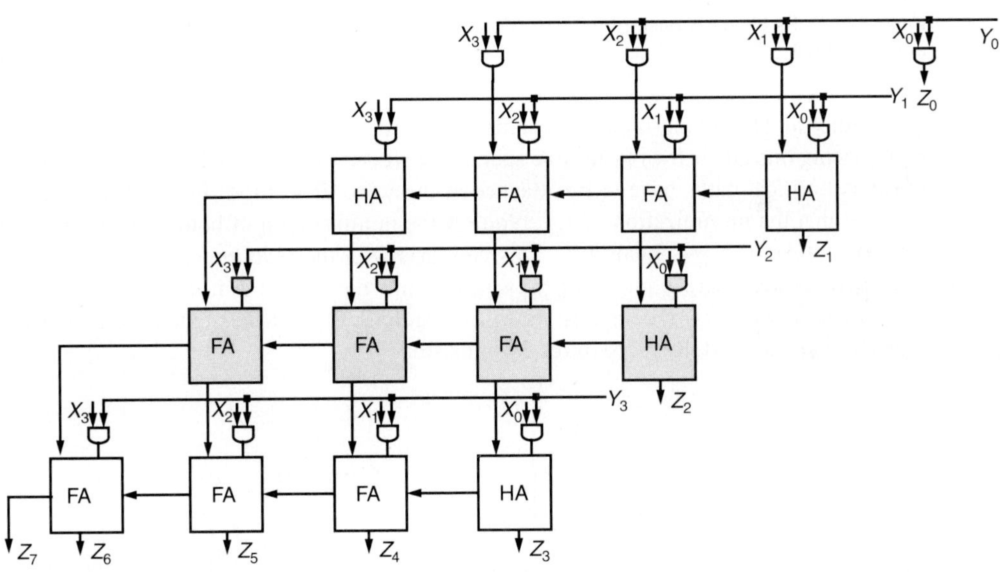

Figure 7.28 A 4 × 4 bit-array multiplier for unsigned numbers—composition. HA stands for a half adder, or an adder cell with only two inputs. The hardware for the generation and addition of one partial product is shaded in gray.

Due to the array organization, determining the propagation delay of this circuit is not straightforward. Consider the implementation of Figure 7.28. The partial sum adders are implemented as ripple-carry structures. Performance optimization requires that the critical timing path be identified first. This turns out to be nontrivial. In fact, a large number of almost identical-length paths can be identified. Two of those are highlighted in Figure 7.29. A closer look at those critical paths yields an expression for the propagation delay of the multiplier.

$$t_{mult} \approx [(M-1)+(N-2)]t_{carry} + (N-1)t_{sum} + t_{and} \quad (7.25)$$

where t_{carry} is the propagation delay between input and output carry, t_{sum} the delay between the input carry and sum bit of the full adder, and t_{and} the delay of the AND gate.

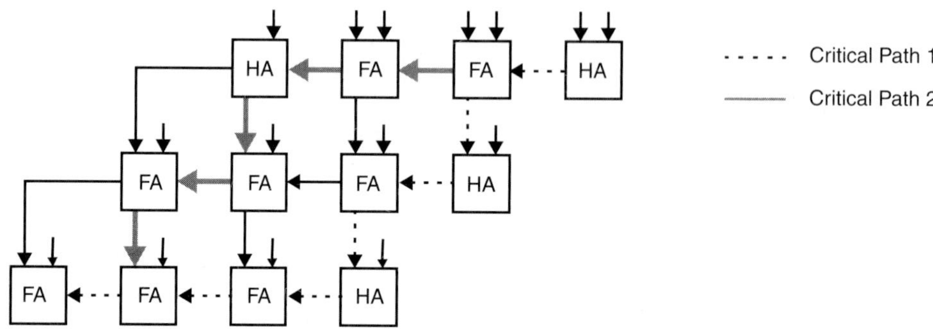

Figure 7.29 Ripple-carry based 4 × 4 multiplier (simplified diagram). Two of the possible critical paths are highlighted.

Since all critical paths have the same length, speeding up one of them—for instance by replacing one adder by a faster one such as a carry-select adder—does not make much sense. All critical paths have to be attacked at the same time. From Eq. (7.25), it can be deduced that the minimization of t_{mult} requires the minimization of both t_{carry} and t_{sum}. In the ideal situation, t_{carry} equals t_{sum}. This contrasts with the requirements for adder cells discussed before, where a minimal t_{carry} was of prime importance. A full-adder circuit with comparable t_{sum} and t_{carry} delays is shown in Figure 7.30. It makes extensive use of the transmission gate EXOR and requires 24 transistors.

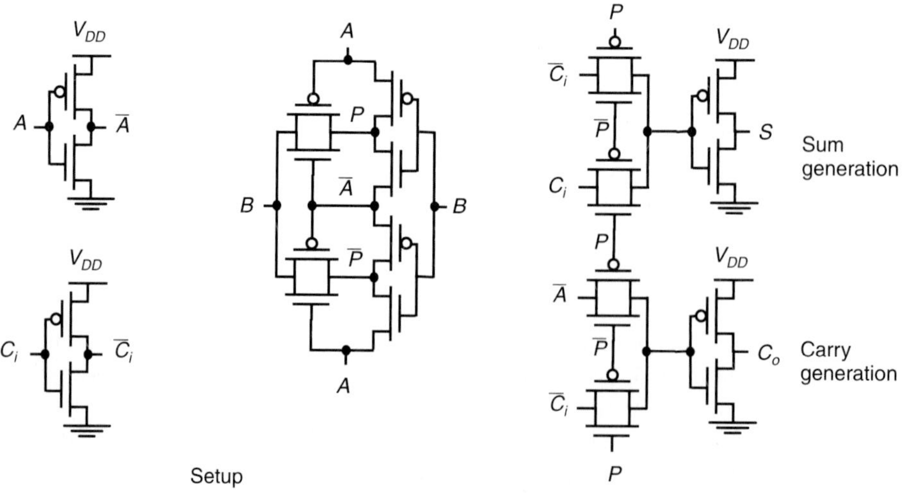

Figure 7.30 Transmission-gate-based full-adder cell with sum and carry delays of similar value (after [Weste93]).

Problem 7.5 Signed-Binary Multiplier

The multiplier presented in Figure 7.28 only handles unsigned numbers. Adjust the structure so that two's-complement numbers are also accepted.

Due to the large number of almost identical critical paths, increasing the performance of the structure of Figure 7.29 can only be achieved with careful transistor sizing. The benefits of this time-consuming effort are marginal. A more efficient multiplier structure is obtained by noticing that the multiplication result does not change when the output carry bits are passed diagonally downwards instead of to the right (Figure 7.31). An extra adder is added (called a *vector-merging* adder) to generate the final result. Such a multiplier is called a *carry-save multiplier* [Denyer81], because the carry bits are not immediately added, but are rather "saved" for the next adder stage. While this structure has a slightly increased area cost (one extra adder), it has the advantage that its worst-case critical path is uniquely defined, as highlighted in Figure 7.31 and expressed in Eq. (7.26).

$$t_{mult} = (N-1)t_{carry} + t_{and} + t_{merge} \tag{7.26}$$

still assuming that $t_{add} = t_{carry}$. A simple way to reduce the propagation delay of this structure is to minimize t_{merge}. This is achieved by using a fast adder implementation such as a carry-select or a lookahead structure for the merging adder.

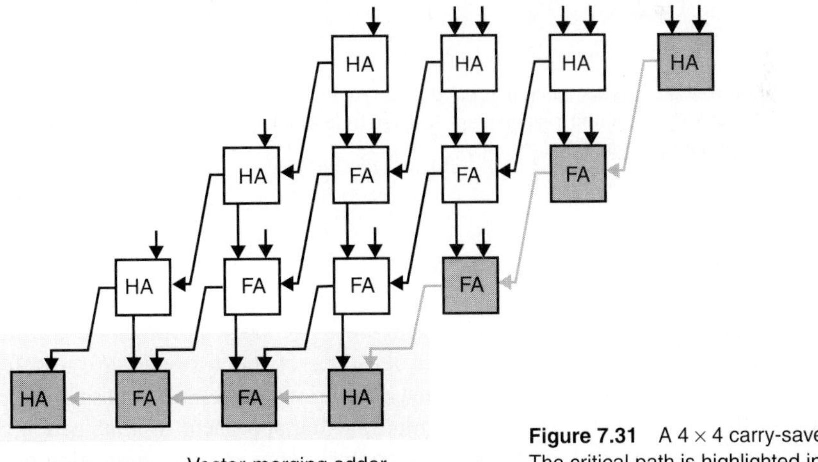

Figure 7.31 A 4 × 4 carry-save multiplier. The critical path is highlighted in gray.

Design Consideration

The attentive reader observes that the left-most half-adder on each row has only one input and hence can be trivially reduced to a wire. The resulting floorplan then consists of 4 rows of 3 adder cells each.

Example 7.5 Carry-Save Multiplier

When mapping the carry-save multiplier of Figure 7.31 onto silicon, one has to take into account some other topological considerations. To ease the integration of the multiplier into the rest of the chip, it is advisable to make the outline of the module approximately rectangular. A floorplan for the carry-save multiplier that achieves this goal is shown in Figure 7.32. Observe the regularity of the topology. This makes the generation of the structure amenable to automation.

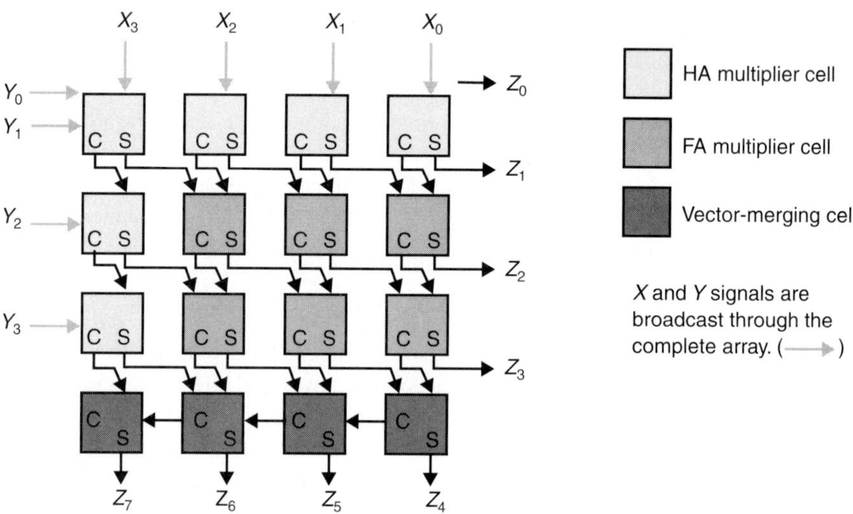

Figure 7.32 Rectangular floorplan of carry-save multiplier. Different cells are differentiated by shades of gray. *X* and *Y* signals are AND-ed before being added. The left-most column of cells is redundant and can be eliminated.

7.4.3 Other Multiplier Structures

A multitude of alternative multiplier circuits have been proposed in the literature, a detailed discussion of which is beyond the scope of this textbook. We therefore restrict ourselves to a short overview. When area is of prime concern, it is possible to reduce the cost of the multiplier by using a time-multiplexing approach. A combination of a single adder and a storage element is used to iteratively compute the summation of the partial products. In the extreme case, even the sum of two *N*-bit numbers can be time-multiplexed on a single full-adder cell. This approach, called *bit-serial multiplication*, results in a very small implementation area [Smith88]. Unfortunately, re-using the same cell in an iterative fashion has a detrimental impact on the performance. A first-order approximation of the propagation delay of the bit-serial multiplier predicts a quadratic complexity: $t_{mult} = M \times N \times t_{carry} = O(M \cdot N)$. This makes this approach useless for large values of M and N.

On the other hand, the speed of the multiplication can be substantially increased by using a special encoding of the multiplier word (*Booth encoding*) that reduces the number of required addition stages [Booth51]. Instead of the traditional binary encoding, the multiplier word is recoded into a radix-4 scheme.

$$Y = \sum_{j=0}^{(N-1)/2} Y_j 4^j \text{ with } (Y_j \in \{-2, -1, 0, 1, 2\}) \quad (7.27)$$

The advantage of the recoding is that the number of partial products and hence the number of additions is halved, which results in a speed-up as well as area reduction. The only expense is a somewhat more involved multiplier cell. While the multiplication with {0, 1} is equivalent to an AND operation, multiplying with {–2, –1, 0, 1, 2} requires a combination of inversion and shift logic. The encoding can be performed on the fly and requires some simple logic gates. Virtually every multiplier currently in use employs the Booth scheme (see [Weste93] for a detailed description of a Booth multiplier).

The partial-sum adders can also be rearranged in a tree-like fashion, similar to the structure used in the Brent-Kung lookahead adder (Figure 7.33). In Figure 7.33a, a vertical slice is extracted from a generic carry-save multiplier (as shown in Figure 7.31). This diagram demonstrates that data ripples from top to bottom, similar to what occurs in the ripple-carry adder. The number of stages equals the number of bits in the multiplier word minus 2. Using a procedure similar to the one followed for the logarithmic carry-lookahead adder, this linear chain can be translated into a tree structure (Figure 7.33b). This topology, which has an $O(\log_2 N)$ multiplication time, is called the *Wallace multiplier* [Wallace64]. While substantially faster than the carry-save structure for large multiplier word lengths, it has the disadvantage of being very irregular. This complicates the task of coming up with a dense and efficient layout. Consequently, Wallace multipliers are only used in designs where performance is critical and design time is only a secondary consideration.

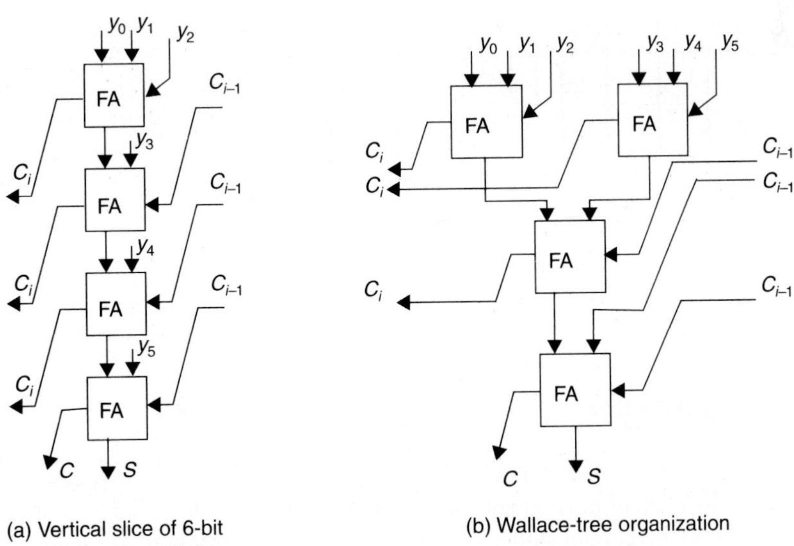

Figure 7.33 Wallace-tree multiplier—basic concept. The *y*-signals represent the partial products.

All these techniques can be combined to yield a multiplier structure with extremely high performance. For instance, a 54 × 54 multiplier can achieve a propagation delay of only 10 nsec in a 0.5 μm CMOS technology by combining Booth encoding with a Wallace tree and by using a mixed carry-select, carry-lookahead, vector-merging adder [Mori91]. More information on these multipliers (and others) can be found in the references [e.g., Swartzlander90].

7.5 The Shifter

The shift operation is another essential arithmetic operation that requires adequate hardware support. It is used extensively in floating-point units, scalers, and multiplications by constant numbers. The latter can be implemented as a combination of add and shift operations. Shifting a data-word left or right over a constant amount is a trivial hardware operation and is implemented by the appropriate signal wiring. A programmable shifter, on the other hand, is more complex and requires active circuitry. In essence, such a shifter is nothing less than an intricate multiplexer circuit. A simple one-bit left-right shifter is shown in Figure 7.34. Depending on the control signals, the input word is either shifted left or right or remains unchanged. Multi-bit shifters can be built by cascading a number of these units. This approach rapidly becomes complex, unwieldy, and slow for larger shift values. Therefore, a more structured approach is advisable. Two commonly used shift structures, the barrel shifter and the logarithmic shifter, are discussed.

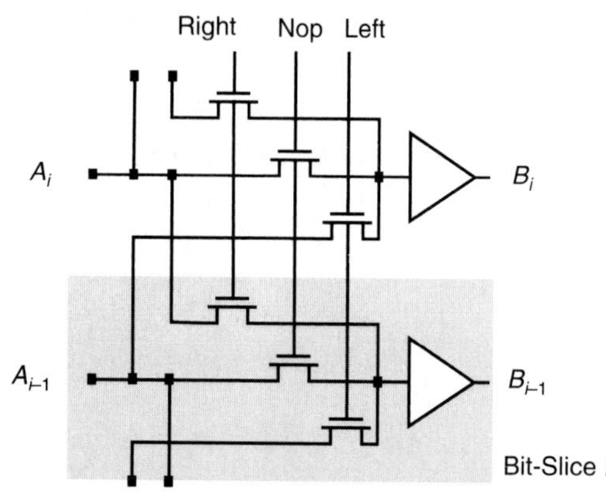

Figure 7.34 One–bit (left/right) programmable shifter. The data passes undisturbed under the Nop condition.

7.5.1 Barrel Shifter

The structure of a barrel shifter is shown in Figure 7.35. It consists of an array of transistors, where the number of rows equals the word length of the data, and the number of columns equals the maximum shift width. In this particular case, both are set equal to four. The control wires are routed diagonally through the array. A major advantage of this shifter is that the signal has to pass through at most one transmission gate. In other words, the propagation delay is theoretically constant and independent of the shift value or shifter size. This is not true in reality because the capacitance at the input of the buffers rises linearly with the maximum shift-width.

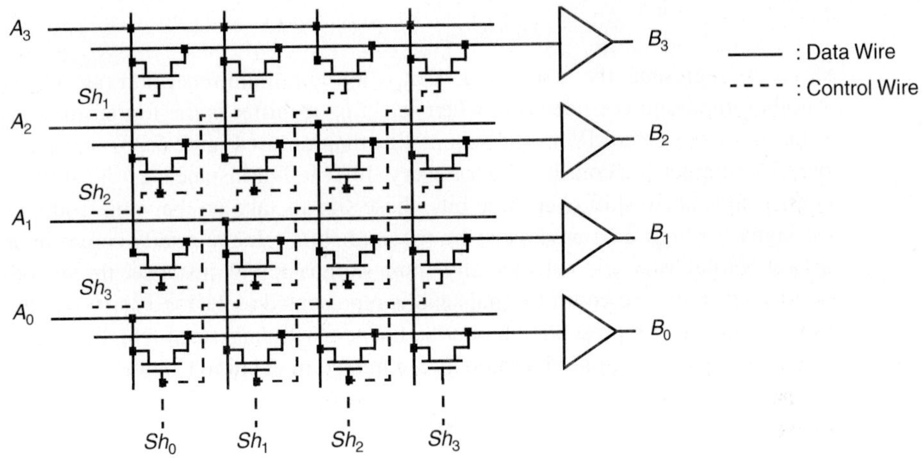

Figure 7.35 Barrel shifter with a programmable shift-width from 0 to 3 bit to the right. The structure supports automatic repetition of the sign bit (A_3), also called sign-bit extension.

An important property of this circuit is that the layout size is not dominated by the active transistors as was the case in all other arithmetic circuits, but by the number of wires running though the cell. More specifically, the size of the cell is bounded by the pitch of the metal wires! Consider the layout of Figure 7.36. It is clear that the width of the cell is constrained by the two vertical metal wires. This observation allows us to estimate the width of the shifter cell as a function of the maximum shift-width M.

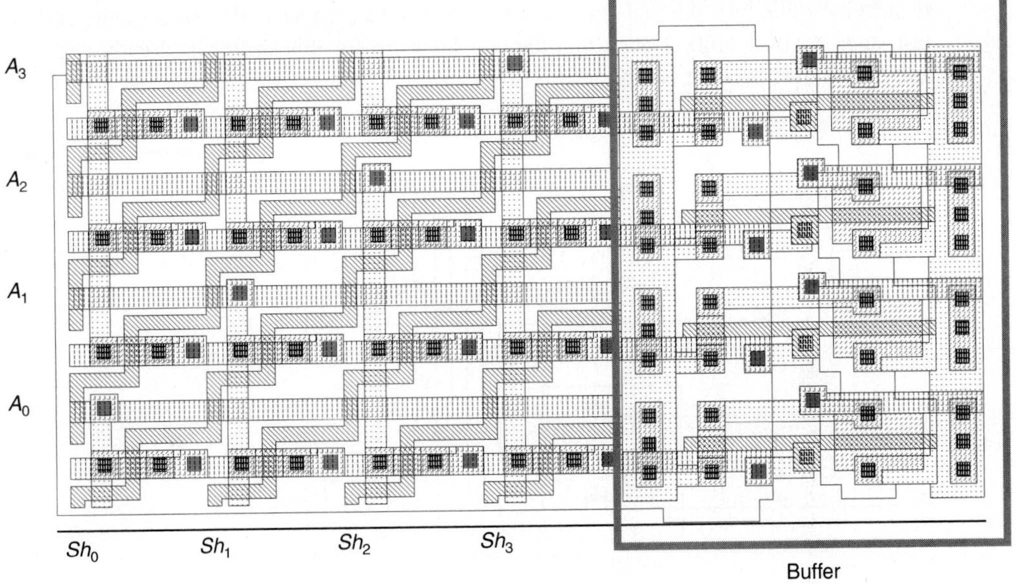

Figure 7.36 4 bit (by 4) barrel shifter—layout. See also Colorplate 11.

$$width_{barrel} \approx 2 \times p_m \times M \tag{7.28}$$

where p_m represents the metal pitch, that is the minimum repetition rate for metal wires. Another important consideration when selecting a shifter is the format in which the shift value must presented. From the schematics diagram of Figure 7.35, we can see that the barrel shifter needs a control wire for every shift bit. For instance, a 4-bit shifter needs four control signals. To shift over three bits, these signals take on the value 1000. Only one of the signals is high. In a processor, the required shift value normally comes in an encoded, binary format which is substantially more compact. For instance, the encoded control word needs only two control signals and is represented as 11 for a shift over three bits. To translate the latter representation into the former (with only one bit high), an extra module, called a *decoder*, is required. Decoders are treated in detail in Chapter 10.

Problem 7.6 Two's Complement Shifter

Explain why the shifter shown in Figure 7.35 implements a two's complement shift.

7.5.2 Logarithmic Shifter

While the barrel shifter implements the whole shifter as a single array of pass-transistors, the logarithmic shifter uses a staged approach. The total shift value is decomposed into shifts over powers-of-two. A shifter with a maximum shift width of M consists of a $\log_2 M$ stages, where the ith stage either shifts over 2^i or passes the data unchanged. An example of a shifter with a maximum shift value of seven bits is shown in Figure 7.37. For instance, to shift over five bits, the first stage is set to shift mode, the second to pass mode, and the last stage again to shift. Notice that the control word for this shifter is already encoded, and no separate decoder is required.

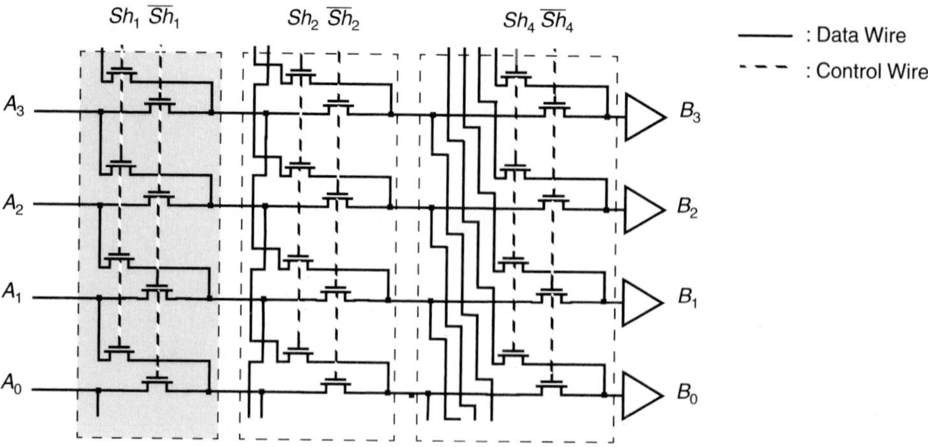

Figure 7.37 Logarithmic shifter with maximal shift width of 7 bits to the right (only the 4 least significant bits are shown).

Once again, the size of this shifter is determined by the metal pitch of the vertical wires. The number of wires actually varies per stage, as is apparent from Figure 7.37 and the layout of Figure 7.38. The width of the cell can be estimated using Eq. (7.29):

$$width_{\log} \approx p_m(2K + (1 + 2 + \ldots + 2^{K-1})) = p_m(2^K + 2K - 1) \tag{7.29}$$

where $K = \log_2 M$.

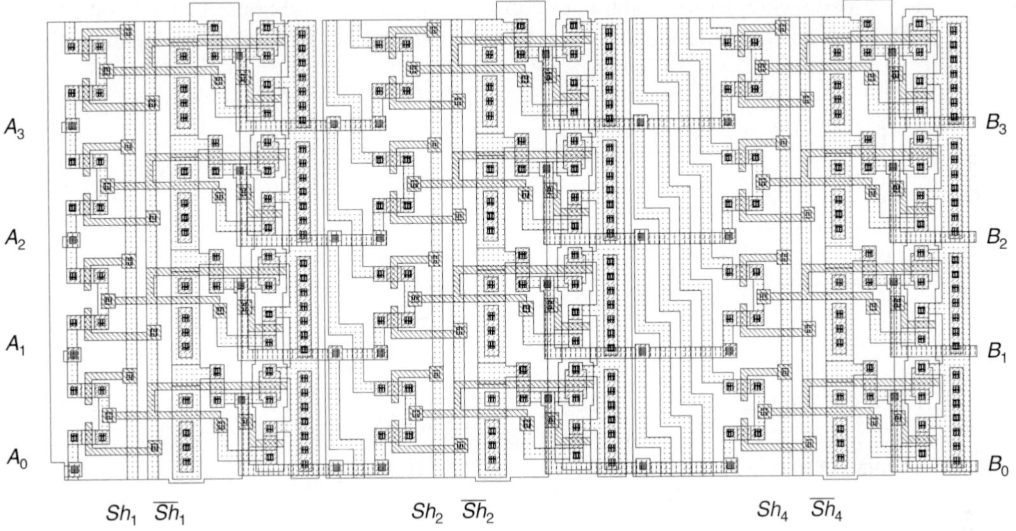

Figure 7.38 Logarithmic shifter—layout. See also Colorplate 12.

Comparing this result to the size of the barrel shifter given by Eq. (7.28) shows that the logarithmic shifter is always smaller than the barrel shifter when only the wiring is considered. For larger values of M, it is definitely the structure of choice.

The speed of the logarithmic shifter depends upon the shift width in a logarithmic way, since an M-bit shifter requires $\log_2 M$ stages. Furthermore, the series connection of pass-transistors slows the shifter down for larger shift values. A careful introduction of intermediate buffers is therefore necessary, as discussed in Chapter 4.

In general, it can be concluded that a barrel shifter is appropriate for smaller shifters. For larger shift values, the logarithmic shifter becomes more effective, both in terms of area and speed. Furthermore, the logarithmic shifter is easily parameterized, allowing for automatic generation. The most important message of this section is that the exploitation of regularity in an arithmetic operator can lead to dense and high-speed circuit implementations.

7.6 Other Arithmetic Operators

In the previous sections, we only discussed a subset of the large number of arithmetic circuits required in the design of micro- and signal processors. Besides adders, multipliers, and shifters, others operators such as comparators, dividers, counters, and goniometric operators (sine, cosine, tangent) are often needed. A full analysis of all these circuits would lead us too far astray and is beyond the scope of this book. We refer the interested reader to some of the excellent references available on this topic, [Swartzlander90] [Davio83] [Weste93].

The reader should be aware that most of the design ideas introduced in this chapter apply to these other operators as well. For instance, comparators can be devised with a linear, square root, and logarithmic dependence on the number of bits. In fact, some operators are simple derivatives of the adder or multiplier structures. For example, Figure 7.39 shows how a two's complement subtracter can be realized by combining a two's complement adder with an extra inversion stage, or how a subtracter can be used to implement $A \geq B$.

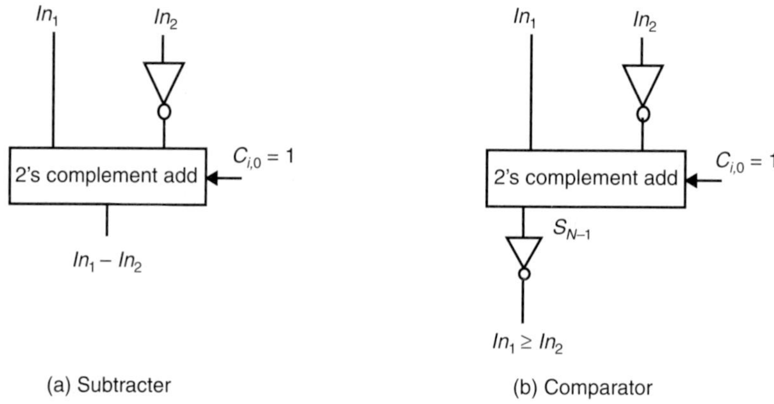

Figure 7.39 Arithmetic structures derived from a full adder.

Problem 7.7 Comparator

Derive a logic diagram for a comparator that implements the following logic functions: ≥, =, ≤.

7.7 Power Considerations in Datapath Structures

In the preceding discussion on adders, multipliers, and shifters, we have mainly explored the speed-area trade-off and ignored power considerations. In this section, we briefly analyze the third dimension of the design exploration space. Since most of the approaches to minimize power have already been introduced in Chapter 4, the following discussion serves mostly as an illustration of the concepts advanced there. Similar to the approach followed in Section 4.4.5, the power minimization approaches are subdivided into two classes: reducing the supply voltage and decreasing the effective capacitance.

7.7.1 Reducing the Supply Voltage

A reduction in supply voltage results in quadratic power savings, but a negative side effect is a loss in performance. At the datapath level, this loss of performance can be compensated for by other means, such as logical and architectural optimizations. For instance, a ripple-carry adder can be replaced by a faster structure such as a lookahead adder. The latter implementation is larger and more complex, which translates into a larger value for the physical capacitance. This is more than offset by fact that the faster adder can run at a lower supply voltage for the same performance.

While the ripple-versus-lookahead trade-off operates at the logical level, Example 7.6 illustrates how architectural optimizations can be employed to compensate for the effect of a reduced V_{DD}.

Example 7.6 Minimizing the Power Consumption Using Parallelism

To illustrate how architectural techniques can be used to compensate for reduced speed, a simple 8-bit datapath consisting of an adder and a comparator is analyzed, assuming a 2 µm CMOS technology [Chandrakasan92]. As shown in Figure 7.40, inputs A and B are added, and the result is compared to input C. Assume that the worst-case delay through the adder, comparator, and latch is approximately 25 nsec at a supply voltage of 5 V. At best, the system can be clocked with a clock period of $T = 25$ nsec. When required to run at this maximum possible throughput, it is clear that the operating voltage cannot be reduced any further since no extra delay can be tolerated. We use this topology as the reference datapath for our study and present power-improvement numbers with respect to this reference. The average power consumed by the reference datapath is given by Eq. (7.30).

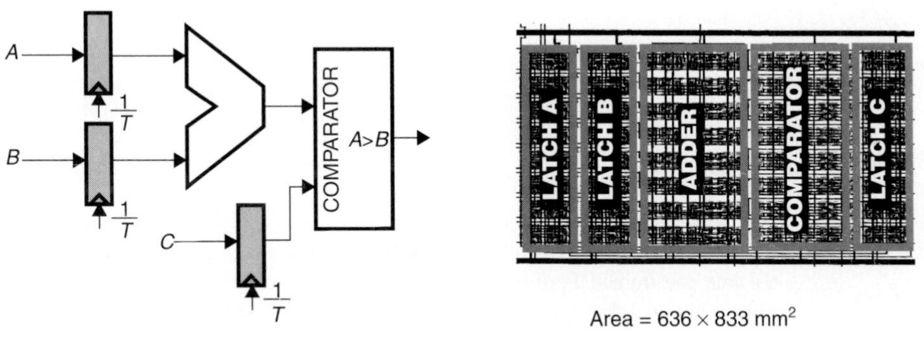

Figure 7.40 A simple datapath with corresponding layout.

$$P_{ref} = C_{ref} V_{ref}^2 f_{ref} \tag{7.30}$$

where C_{ref} is the total effective capacitance being switched per clock cycle. The effective capacitance can be determined by averaging the energy over a sequence of random input patterns with a uniform distribution.

One way to maintain throughput while reducing the supply voltage is to utilize a parallel architecture. As shown in Figure 7.41, two identical adder-comparator datapaths connect in parallel, allowing each unit to work at half the original rate while maintaining the original throughput. Since the speed requirements for the adder, comparator, and latch have decreased from 25 nsec to 50 nsec, the voltage can be dropped from 5 V to 2.9 V (the voltage that doubles the delay can be deduced from Figure 4.56). While the datapath capacitance has increased by a factor of two, the operating frequency has correspondingly decreased by a factor of two. Unfortunately, there is also a slight increase in the total effective capacitance due to the extra routing and data multiplexing. This results in an increased capacitance by a factor of 2.15. The power for the parallel datapath is thus given by

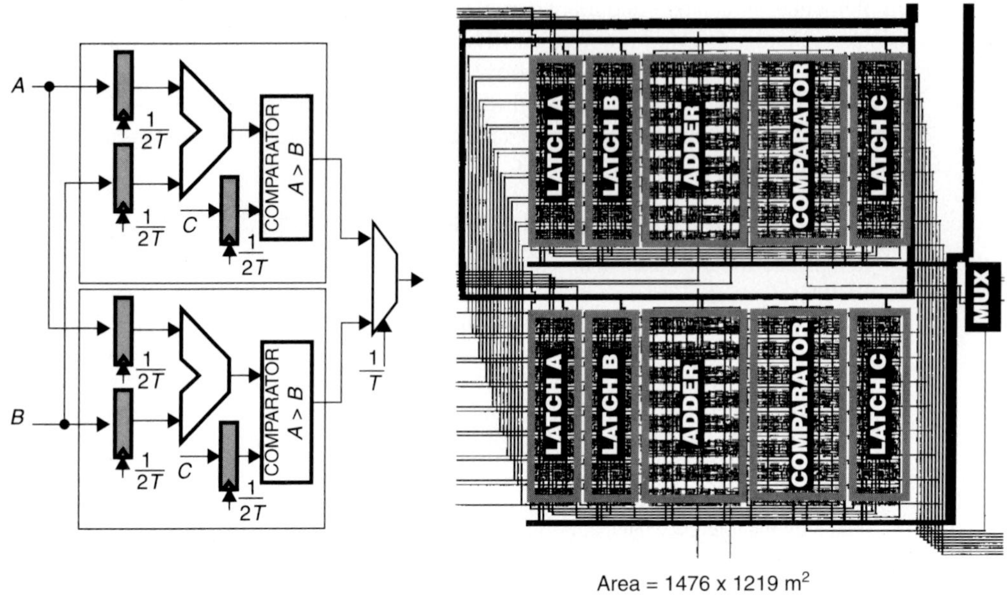

Figure 7.41 Parallel implementation of the simple datapath.

$$P_{par} = C_{par}V^2_{par}f_{par} = (2.15C_{ref})(0.58V_{ref})^2\left(\frac{f_{ref}}{2}\right) \approx 0.36P_{ref} \tag{7.31}$$

The approach presented *trades off area for power*, as the resulting area is approximately 3.4 times larger than the original design. This technique is only applicable when the design is not area-constrained. Furthermore, parallelism introduces extra routing overhead, which might cause additional dissipation. Careful optimization is needed to minimize this overhead.

Parallelism is not the only way to compensate for the loss in performance. Other architectural approaches, such as the use of pipelining, can accomplish the same goal. The most important message in the above analysis is that if power dissipation is the prime concern, dropping the supply voltage is the most effective means to achieve that goal. The subsequent loss in performance can be compensated for, if necessary, by an increase in area. Within certain bounds, this is acceptable, since area is no longer the compelling issue it used to be due to the dramatic increase in integration levels of the last decade.

Problem 7.8 Reducing the Supply Voltage Using Pipelining

A pipeline stage is introduced between the adder and the comparator of the reference datapath of Figure 7.40. You may assume that this roughly halves the propagation delay of the logic, while it increases the capacitance by 15%. Obviously an extra pipeline register is needed on input C as well. Determine how much power can be saved by this approach, given that the throughput has to remain constant compared to the reference datapath. Comment on the area overhead.

7.7.2 Reducing the Effective Capacitance

As the effective capacitance is the product of the physical capacitance and the switching activity, minimization of both factors is recommended.

Circuit and Logic Styles

One approach is to select a circuit style with low capacitance and/or switching activity. Figure 7.42 plots the power-delay product of an 8-bit adder, implemented in a variety of circuit and logic styles, as a function of the propagation delay [Chandrakasan92]. The very small internal capacitances of the complementary pass transistor logic style (CPL), combined with the reduced voltage swing, results in a very power-efficient structure. On the other end of the spectrum, we can locate the carry-select adder and the DCVSL structure. Both approaches suffer from their differential nature, which tends to increase the switching activity. [2]

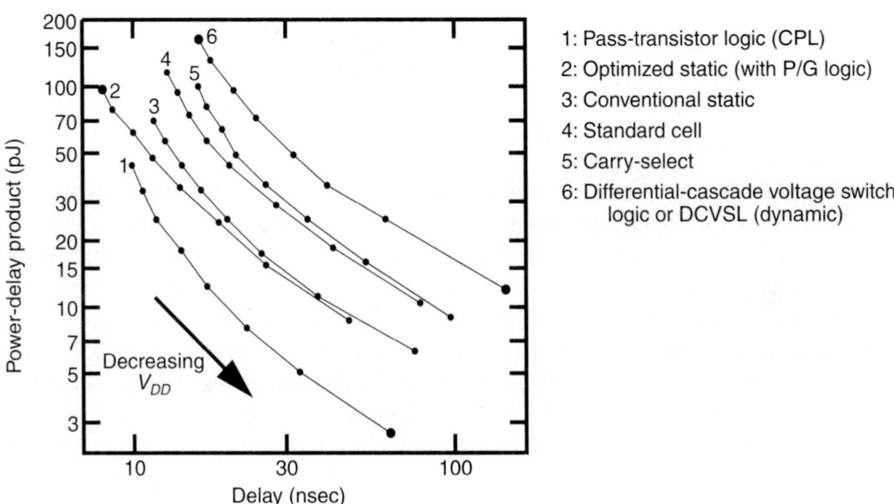

Figure 7.42 Power-delay product of 8-bit adders as a function of the propagation delay with the supply voltage as parameter (implemented in a 2 μm CMOS technology).

Glitching

Dynamic hazards, or *glitching,* may be a major contribution to the dissipation in complex structures such as adders and multipliers. The wide discrepancy in the lengths of the signal paths in some of those structures can be the cause of spurious transients. This is demonstrated in Figure 7.43, which displays the simulated response of a 16-bit ripple adder for all inputs going simultaneously from 0 to 1. A number of the sum bits are shown as a function of time. The sum signals should be 0 for all bits. Unfortunately, a 1 appears briefly at all of

[2] Be aware that the results presented are for a particular implementation in a particular technology. The conclusions might differ for other technologies.

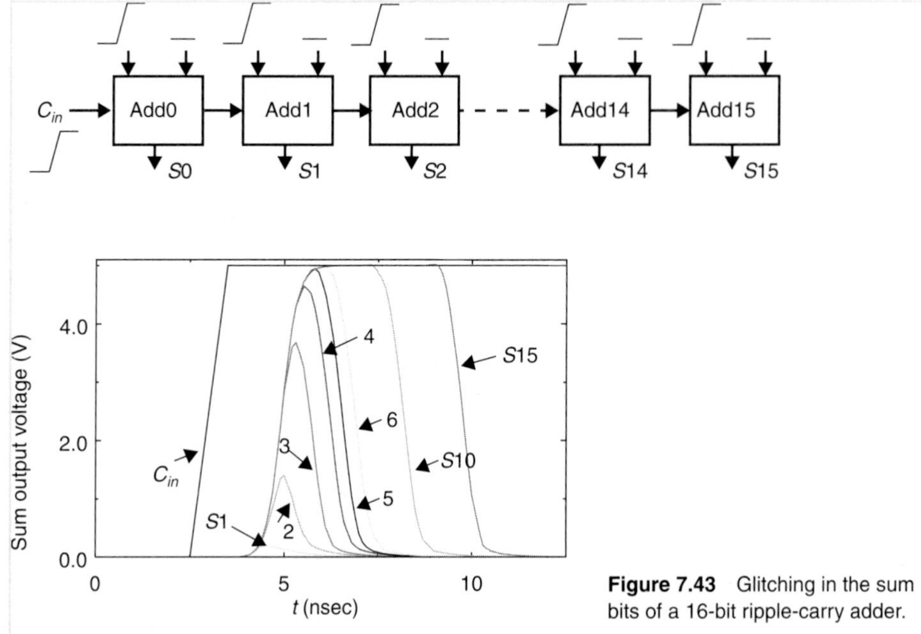

Figure 7.43 Glitching in the sum bits of a 16-bit ripple-carry adder.

the outputs, since the carry takes a significant amount of time to propagate from the first bit to the last. Notice how the glitch becomes more pronounced as it travels down the chain.

A dramatic reduction in glitching activity can be obtained by selecting structures with balanced signal paths. The Brent-Kung lookahead structure and the Wallace-tree multiplier both have that property and should, therefore, be more attractive from a power point of view. In fact, an inspection of the lookahead structure of Figure 7.26 reveals that the timing paths to the inputs of dot operators are of a similar length. Some small deviations might occur due to differences in loading and fan-out.

This speculation is confirmed by experimental results, as shown in Table 7.2, which tabulates the (simulated) power-delay product for a variety of adder and multiplier structures [Callaway92].[3] Simulations indicate that in a straightforward design of a 32-bit carry-save multiplier, on the average 99,000 signal transitions occur per multiplication. For the balanced Wallace-tree multiplier, this number is reduced to 19,500. Ignoring extra parameters, such as capacitance values, this means a reduction in power dissipation by a factor of 5!

Reduced Switching Activity

As mentioned before, multiplexing multiple operations on a single hardware unit can have a detrimental effect on the power consumption. Besides increasing the physical capacitance, it can also increase the switching activity, as demonstrated by the example of the shared bus in Chapter 4. This is illustrated with a simple experiment in Figure 7.44 that

[3] The Dadda multiplier is another multiplier structure with logarithmic behavior. Although somewhat superior to the Wallace multiplier in theoretical performance, its irregular structure makes it less amenable to efficient implementation.

Table 7.2 Simulated, normalized power-delay product for a number of adder and multiplier structures (from [Callaway92]).

64-bit adder		32-bit multiplier	
Ripple carry	1	Carry save	1
Carry bypass	0.27	Wallace	0.05
Carry select	0.27	Dadda	0.04
Carry lookahead	0.15		

compares the power consumption of two counters running simultaneously. In the first case, both counters run on separate hardware, while in the second case they are multiplexed on the same unit. Figure 7.44b plots the number of switching events as a function of the skew between the two counters. The nonmultiplexed case is always superior, except when both counters run in a completely synchronous fashion. The multiplexing tends to randomize the data signals presented to the operational unit, which results in increased switching activity. When power consumption is a concern, it is often beneficial to avoid the excessive reuse of resources. Observe that CMOS hardware units consume only negligible amounts of power when idle. Providing dedicated, specialized operators only presents an extra cost in area, while being generally beneficial in terms of speed and power.

Taking all the aforementioned factors together, power savings of one or more orders of magnitude are possible. This can be complemented by system-level techniques such as the power-down and the clock-speed reduction approaches that are commonly used in portable applications such as laptop computers and video cameras.

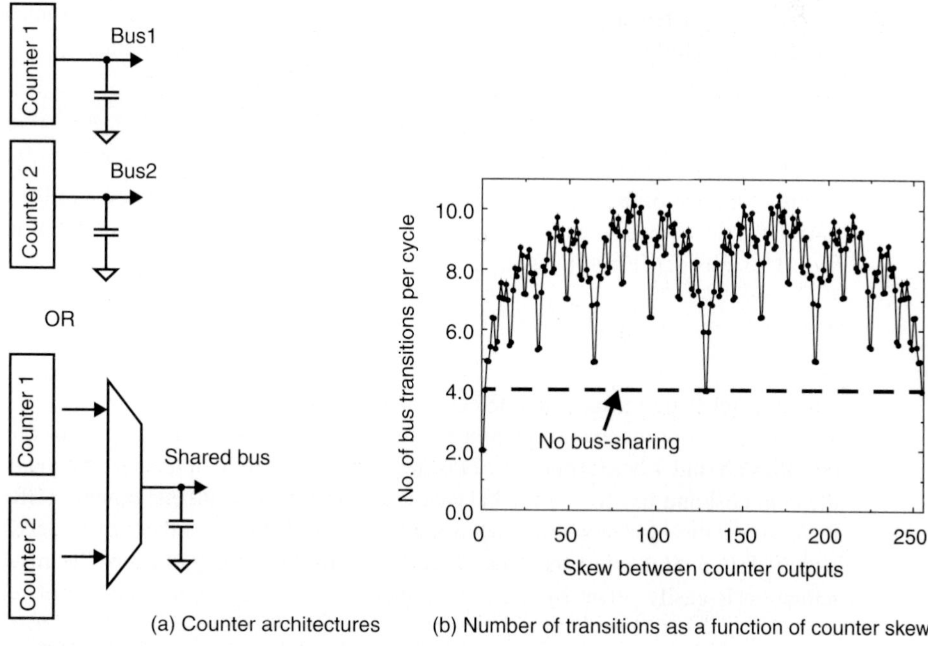

Figure 7.44 Multiplexing increases the switching activity.

7.8 Perspective: Design as a Trade-off

The analysis of the adder and multiplier circuits makes it clear again that digital circuit design is a trade-off game between area, speed, and power requirements. This is demonstrated in Figure 7.45, which plots the area and speed for the adders discussed above as a function of the number of bits (for a 3 μm CMOS technology).[4] The overall project goals and constraints determine which factor is dominant.

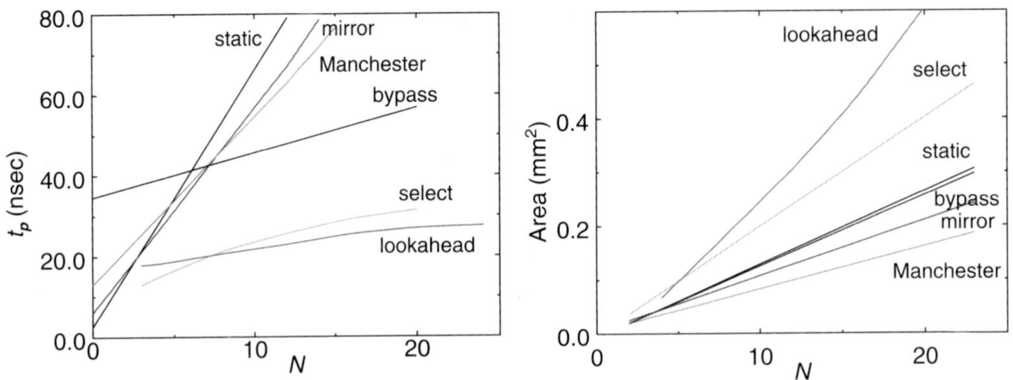

Figure 7.45 Area and propagation delay of various adder structures as a function of the number of bits *N* (for a 3 μm CMOS technology). Based on results from [Vermassen86].

The die area has a strong impact on the cost of an integrated circuit. A larger chip size means that fewer parts fit on a single wafer. The die area also has an impact on the *yield of the manufacturing process*. The yield *Y* is defined as a percentage:

$$Y = \frac{\text{Number of good chips on wafer}}{\text{Number of chips on wafer}} 100 \tag{7.32}$$

and is a function of the die area and the defect density. Over the years, a number of models have been developed to express this dependency. A common model for small-to-medium size chips was developed by Murphy [Murphy64]:

$$Y = \left[\frac{1 - e^{-AD}}{AD}\right]^2 \tag{7.33}$$

where *A* and *D* are the area and defect density, respectively.

Due to improvements in process technology, modern processes can boast defect densities around 1 defect per cm^2. This trend has reduced the impact of the yield factor on the chip cost and has tipped the balance towards the nonrecurring expense (NRE) part of the cost equation for moderate-size designs. The NRE includes the cost of developing the circuit design. Reducing the area can help to improve the yield, but the resulting cost reduction is easily offset by a dramatic increase in design time. The use of regular and

[4] Be aware that these results are for a particular implementation in a particular technology. Extrapolation to other technologies should be done with care.

modular design techniques and design automation typically incurs an area penalty, but this is more than offset by the reduction in design time and time-to-market. Area is therefore losing some of its importance as a design parameter.

Performance and power dissipation on the other hand can have a dramatic impact on the viability of a product. Ultimate performance is what makes the newest microprocessor sell, and the lowest possible power consumption is a great marketing argument for a cellular phone. Understanding the market of a product is therefore essential when deciding on how to play the trade-off game.

One should also be aware that area and speed can also act as design constraints. Excessive area or power requirements easily can make a design nonfeasible.

In this context it is worth summarizing some of the important design concepts that have been repeated in the course of this chapter.

1. The most important message is to select the *right structure* before starting an elaborate circuit optimization. Going for the optimal performance of a complex structure by rigorously optimizing transistor sizes and topologies might not always be the best choice. Optimizations at higher levels of abstraction, such as the logic level, can often generate more dramatic results. Simple first-order calculations can help give a global picture on the pros and cons of a proposed structure.

2. Determine the *critical timing path* through the circuit and focus most of your optimization efforts on that part of the circuit. Computer-aided design tools are available to help determine the critical paths and size the transistors appropriately.

3. Circuit size is not always determined by the number of transistors, but also by other factors such as *wiring and the number of vias and contacts*. These factors are becoming even more important with shrinking dimensions or when extreme performance is a goal, as will become obvious in the next chapter.

4. Although an obscure optimization can sometimes help to get a better result, be wary if this results in an irregular and convoluted topology. *Regularity and modularity* are a designer's best friend.

7.9 Summary

In this chapter, we have studied the implementation of arithmetic data path operators from an area, performance, and power perspective. Special attention was devoted to the development of *first-order performance models* that allow for a fast analysis and comparison of various logic structures before diving into the time-consuming transistor-level optimizations.

- A datapath is best implemented in a *bit-sliced* fashion. A single layout slice is used repetitively for every bit in data word. This regular approach eases the design effort and results in fast and dense layouts.

- A *ripple-carry* adder has a performance that is linearly proportional to the number of bits. Circuit optimizations concentrate on reducing the delay of the carry path. A number of circuit topologies were examined, showing how careful optimization of the circuit topology and the transistor sizes helps to reduce the capacitance on the carry bit.

- Other adder structures use logic optimizations to increase the performance. The performance of *carry-select and carry-lookahead* adders depends on the number of bits in square root and logarithmic fashion, respectively. This increase in performance comes at a penalty in area.
- A *multiplier* is nothing more than a collection of cascaded adders. Its critical path is far more complex, and performance optimizations proceed along vastly different routes. The carry-save technique relies on a logic manipulation to turn the adder array into a regular structure with a well-defined critical timing path that can easily be optimized.
- The performance and area of a programmable shifter are dominated by the wiring. The exploitation of regularity can help to minimize the impact of the interconnect wires. This is exemplified in the barrel and the logarithmic shifter structures.
- Finally, *power consumption* can be reduced substantially by the proper choice of circuit, logical, or architectural structure. This might come at the expense of area, but area might not be that critical in the age of submicron devices.
- The use of parallelism and pipelining can help to *reduce the supply voltage*, while maintaining the same performance. The *effective capacitance* can be reduced by avoiding waste, as introduced for instance by excessive multiplexing.

7.10 To Probe Further

The literature on arithmetic and computer elements is vast. Important sources for newer developments are the *Proceedings* of the workshop on computer arithmetic, the *IEEE Journal on Computers* and the *IEEE Journal of Solid-State Circuits* (for integrated circuit implementation). An excellent collection of the most significant papers in the area can be found in some IEEE Press reprint volumes [Swartzlander90]. A number of other references are provided for further reading.

REFERENCES

[Bedrij62] O. Bedrij, "Carry Select Adder," *IRE Trans. on Electronic Computers*, vol. EC-11, pp. 340–346, 1962.

[Booth51] A. Booth, "A Signed Binary Multiplication Technique," *Quart. J. Mech. Appl. Math.*, vol. 4., part 2, 1951.

[Brent82] R. Brent and H.T. Kung, "A Regular Layout for Parallel Adders," *IEEE Trans. on Computers,* vol. C-31, no. 3, pp. 260–264, March 1982.

[Callaway92] T. Callaway and E. Swartzlander, "Optimizing Arithmetic Elements for Signal Processing," *Proc. IEEE VLSI Signal Processing Workshop V,* IEEE Press, pp. 91–100, October 1992.

[Chandrakasan92] A. Chandrakasan, S. Sheng, and R. Brodersen, "Low Power CMOS Digital Design," *IEEE Journal of Solid State Circuits,* vol. SC-27, no. 4, pp. 1082–1087, April 1992.

[Dadda65] L. Dadda, "Some Schemes for Parallel Multipliers," *Alta Frequenza,* vol. 34, pp. 349–356, May 1965.

[Davio83] M. Davio, J. Deschamps, and A. Thayse, *Digital Systems with Algorithm Implementation*, John Wiley and Sons, 1983.

[Denyer 81] P. Denyer and D. Myers, "Carry-save Arrays for VLSI Signal Processing," in *VLSI 81*, John Gray, ed., pp. 151–160, 1981.

[Elmasry91] Elmasry, M. ed., *Digital MOS Integrated Circuits II*, IEEE Press, 1991.

[Kleinfelder91] S. Kleinfelder and J. Judy, "Two Fast Low Power Domino Carry Chain Adders," EE 241 Project Reports, J. Rabaey, instructor, Berkeley, 1991.

[Kuninobu87] S. Kuninobu, "Design of High Speed MOS Multiplier and Divider Using Redundant Binary Representation," *Proc. Symp. on Computer Arithmetic*, pp. 80–85, 1987.

[McSorley61] O. McSorley, "High Speed Arithmetic in Binary Computers," *IRE Proceedings*, vol. 49, pp. 67–91, 1961.

[Mead80] C. Mead and L. Conway, *Introduction to VLSI Systems*, Addison-Wesley, 1980.

[Mori91] L. Mori et al., "A 10 ns 54 ¥ 54b Parallel Structured Full Array Multipler with 0.5 mm CMOS Technology," *IEEE Journal of Solid State Cicruits*, vol. 26, no. 4, pp. 600–605, April 1991.

[Murphy64] B. Murphy, "Cost-Size Optima of Monolithic Integrated Circuits," *IEEE Proceedings*, vol. 52, pp. 1537–1545, Dec. 1964.

[Smith88] S. Smith and P. Denyer, *Serial-Data Computation*, Kluwer Academic Publishers, 1988.

[Swartzlander90] E. Swartzlander, ed., *Computer Arithmetic—Part I and II*, IEEE Computer Society Press, 1990.

[Turrini89] S. Turrini, "Optimal Group Distribution in Carry-Skip Adders," *Proc.of the 9th Symposium on Computer Arithmetic*, pp. 96–103, 1989.

[Vermassen86] H. Vermassen, "Mathematical Models for the Complexity of VLSI" (in dutch), Master's thesis, Katholieke Universiteit Leuven, Belgium.

[Wallace64] C. Wallace, "A Suggestion for a Fast Multiplier," *IEEE Trans. on Electronic Computers*, EC-13, pp. 14–17, 1964.

[Weste85&93] N. Weste and K. Eshragian, *Principles of CMOS VLSI Design: A Systems Perspective*, 2nd ed., Addison-Wesley, 1985–1993.

[Yano90] K. Yano et al., "A 3.8 nsec CMOS 16× 16b Multiplier Using Complimentary Pass-Transistor Logic," *IEEE Journal of Solid State Circuits*, vol. 25, no. 2, pp. 388–395, April 1990.

7.11 Exercises and Design Problems

1. [E, None, 7.6] For this problem you are given a cell library consisting of full adders and two-input Boolean logic gates (i.e. AND, OR, INVERT, etc.).
   a. Design an N-bit two's complement subtracter using a minimal number of Boolean logic gates. The result of this process should be a diagram in the spirit of Figure 7.6. Specify the value of any required control signals (e.g., C_{in}).
   b. Express the delay of your design as a function of N, t_{carry}, t_{sum}, and the Boolean gate delays (t_{and}, t_{or}, t_{inv}, etc.).
2. [M, None, 7.6] A magnitude comparator for unsigned numbers can be constructed using full adders and Boolean logic gates as building blocks. For this problem you are given a cell library consisting of full adders and arbitrary fan-in logic gates (i.e., AND, OR, INVERTER, etc.).
   a. Design an N-bit magnitude comparator with outputs $A \geq$ B and $A = B$ using a minimal number of Boolean logic gates. The result of this process should be a diagram in the spirit of Figure 7.6. Specify the value of any required control signals (e.g., C_{in}).

b. Express the delay of your design in computing the two outputs as a function of N, t_{carry}, t_{sum}, and the Boolean gate delays (t_{and}, t_{or}, t_{inv}, etc.).

3. [E, None, 7.6] Determine the logic function of the circuit of Figure 7.46 (nodes c and d. Derive an expression for its propagation delay as a function of the number of bits.

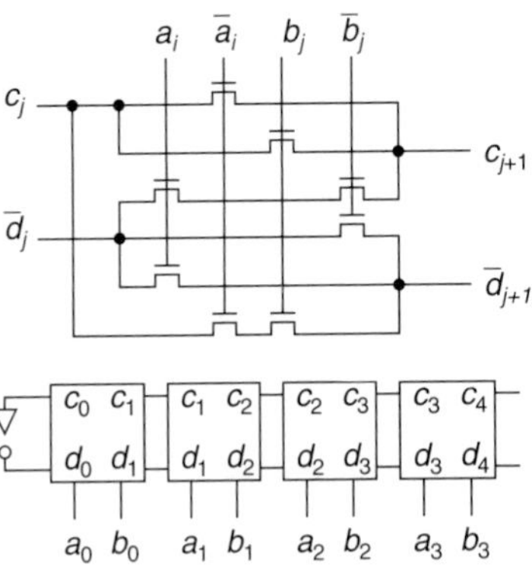

Figure 7.46 Arithmetic module.

4. [E, None, 7.6] The circuit of Figure 7.47 implements a 1-bit datapath function in dynamic (precharge/evaluate) logic.

a. Write down the Boolean expressions for outputs F and G. On which clock phases are outputs F and G valid?

b. To what datapath function could this unit be most directly applied (e.g., addition, subtraction, comparison, shifting)?

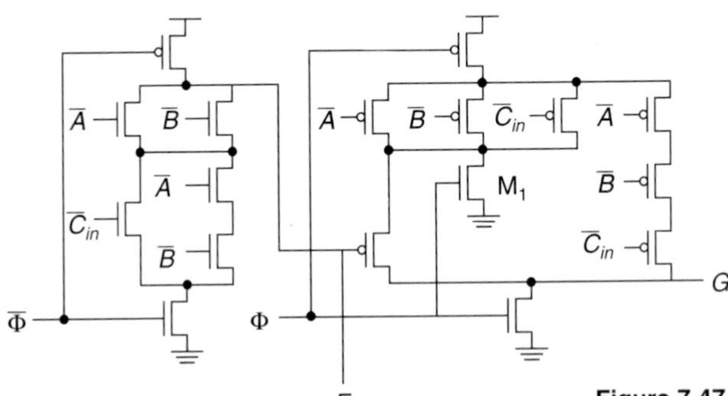

Figure 7.47 Datapath module bit-slice.

5. [M, None, 7] Consider the dynamic logic circuit of Figure 7.47.

a. What is the purpose of transistor M_1? Is there another way to achieve this end that would reduce capacitive loading on the Φ?

b. How can the evaluation phase of F be sped up by rearranging transistors? No transistors should be added, deleted, or resized.

c. Can evaluation of G be sped up in the same manner? Why or why not?

6. [E&D, SPICE, 7.3] The adder circuit of Figure 7.48 makes extensive use of the transmission gate EXOR. $V_{DD} = 5$ V.

a. Explain how this gate operates. Derive the logic expression for the various circuit nodes. Why is this a good adder circuit?

b. Derive a first-order approximation of the capacitance on the C_o-node in equivalent gate-capacitances. Assume that gate and diffusion capacitances are approximately identical. Compare your result with the circuit of Figure 7.7.

c. Assume that all transistors with the exception of those on the carry path are minimum-size. Use 10/1.2 NMOS and 30/1.2 PMOS devices on the carry-path. Using SPICE simulation, derive a value for all important delays (input-to-carry, carry-to-carry, carry-to-sum).

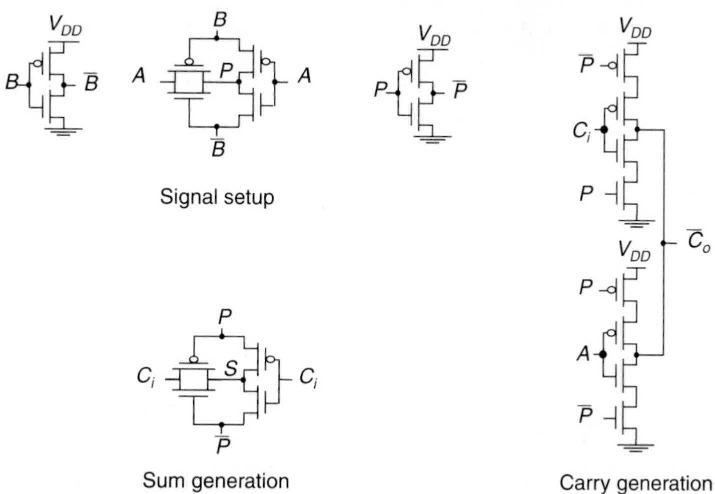

Figure 7.48 Quasi-clocked adder circuit.

7. [M&D, None, 7.3] Dynamic implementation of the 4-bit carry-lookahead circuitry of Figure 7.24 can significantly reduce the required transistor count.

a. Design a DOMINO-logic implementation of Eq. (7.17). Compare the transistor counts of the two implementations.

b. What is the worst-case propagation delay path through this new circuit?

c. Are there any charge-sharing problems associated with your design? If so, modify your design to alleviate these effects.

8. [C, None, 7.3] Figure 7.49 shows a popular adder structure called the conditional-sum adder. Figure 7.49a shows a four-bit instance of the adder, while Figure 7.49b gives the schematics of the basic adder cell. Notice that only pass-transistors are used in this implementation.

a. Derive Boolean descriptions for the four outputs of the one-bit conditional adder cell.

b. Based on the results of *a* describe how the schematic of Figure 7.49a results in an addition.

c. Derive an expression for the propagation delay of the adder as a function of the number of bits N. You may assume that a switch has a constant resistance R_{on} when active and that each switch is identical in size.

9. [M, None, 7.3] Consider replacing all of the NMOS evaluate transistors in a dynamic Manchester carry chain with a single common pull-down as shown in Figure 7.50a. Assume that each NMOS transistor has $(W/L)_N = 2.4/1.2$ and each PMOS has $(W/L)_P = 7.2/1.2$. Further assume that parasitic capacitances can be modeled by a 20 fF capacitor on each of the internal

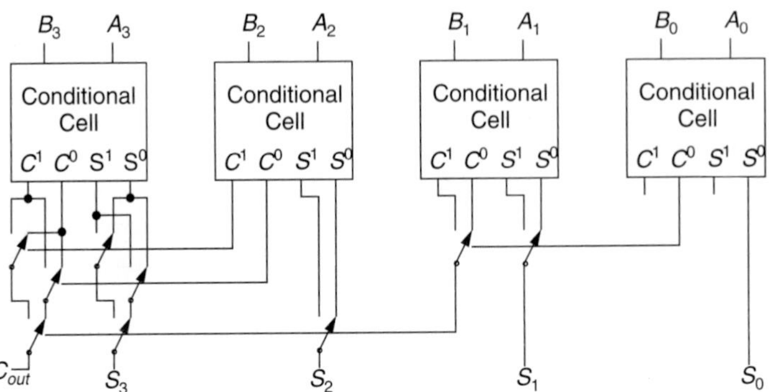

(a) Four-bit conditional-sum adder

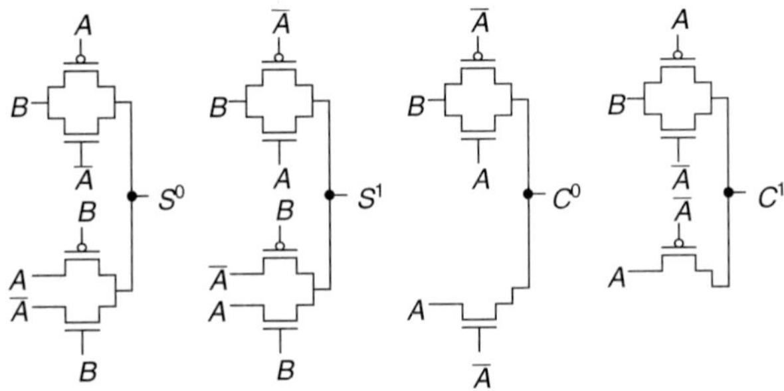

(b) Conditional adder cell

Figure 7.49 Conditional-sum adder.

nodes: A, B, C, D, E, and F. Assume all transistors can be modeled as linear resistors with an on-resistance, $R_{on} = 10\ \text{k}\Omega$.

a. Does this variation perform the same function as the original Manchester carry chain? Explain why or why not.

b. Assuming that all inputs are allowed only a single zero-to-one transition during evaluation, will this design involve charge-sharing difficulties? Justify your answer.

c. Complete the waveforms of Figure 7.50b for $P_0 = P_1 = P_2 = P_3 = 5$ V and $G_0 = G_1 = G_2 = G_3 = 0$ V. Compute and indicate t_{pHL} values for nodes A, E, and F. Compute and indicate when the 90% precharge levels are obtained.

10. [M, None, 7.4] An array multiplier consists of rows of adders, each producing partial sums that are subsequently fed to the next adder row. In this problem, we consider the effects of pipelining such a multiplier by inserting registers between the adder rows.

a. Redraw Figure 7.29 inserting word-level pipeline registers as required to achieve maximal benefit to throughput for the 4 × 4 multiplier. *Hint: you must use additional registers to keep the input bits synchronized to the appropriate partial sums.*

b. Repeat *a* for a carry-save, as opposed to ripple-carry, architecture.

c. For each of the two multiplier architectures, compare the critical path, throughput, and latency of the pipelined and nonpipelined versions.

d. Which architecture is better suited to pipelining, and how does the choice of a vector-merging adder affect this decision?

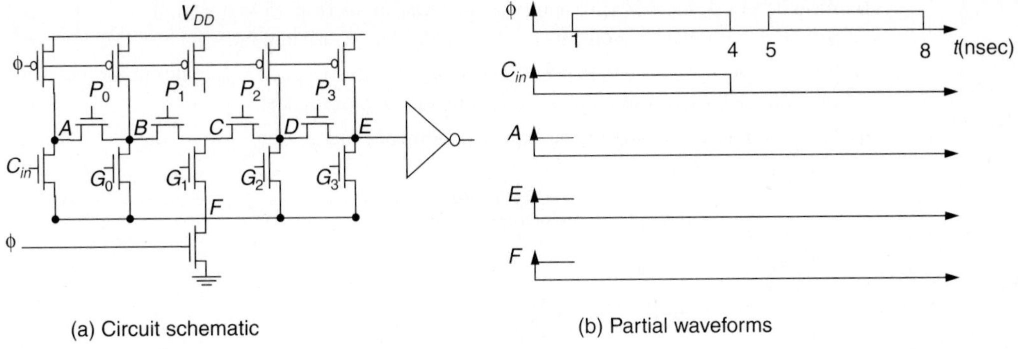

(a) Circuit schematic (b) Partial waveforms

Figure 7.50 Alternative dynamic Manchester carry-chain adder.

11. [E, None, 7.4] A chip consists of several modules: controllers, memories, datapaths, and so on. *Floorplanning* is the process of assembling the various units so that the overall chip area is minimized. In order to minimize wasted silicon area, modules are typically designed to be more or less rectangular, as odd shapes often will not fit well with other modules.

a. Derive a rectangular floorplan for the unsigned 4×4 array multiplier of Figure 7.29. Show all input and output lines, as well as all intercell communications.

b. Suppose that the multiplier is used in a manner that guarantees: $0 \leq X \times Y \leq 15$. Which bits of the output Z are still utilized? Without altering your basic floorplan, redraw the diagram including only those cells still required to produce the necessary output bits. What is the basic shape of your new floorplan?

c. Comment on how the reduced design of part *b* compares to the original multiplier in terms of area, delay, and energy (ignore glitching).

d. On a chip composed of otherwise rectangular modules, how useful is the area reduction offered by the floorplan of part *b*?

e. If the chip contains two reduced-output-width multipliers, however, can you think of a way to effectively exploit the area savings achieved in part *b*?

12. [M, None, 7.5] The cell widths derived for the barrel shifter and the logarithmic shifter and summarized in Eqs. (7.28) and (7.29) account only for metal-wire-pitch constraints. Obviously, transistors and overhead will occupy some area as well and will contribute to the overall shifter widths. For the barrel shifter, assume a width overhead of 2.4 μm per cell and for the log shifter assume 30.6 μm per cell. Assume $p_m = 4.8$ μm.

a. Plot W_{barrel} and W_{log} as M varies from 1 to 32 bits using Eqs. (7.28) and (7.29) (i.e., ignoring the overhead component).

b. Write new equations for W_{barrel} and W_{log} that account for the new overhead terms.

c. Plot the revised equations for W_{barrel} and W_{log} as M varies from 1 to 32 bits.

d. Compare the plots of parts *a* and *c* explaining their differences.

e. About when does the log shifter become more attractive than the barrel shifter in terms of area, according to the revised equations?

13. [C, None, 7.5] Figure 7.51 shows the schematic and layout for an inverting, two-bit (right) barrel shifter. For this problem, assume $V_{DD} = 5$ V and $C_L = 0.1$ pF. The layout is to scale, and the grid markings bounding it are spaced at $2\lambda = 1.2$ μm.

a. Does this implementation perform sign extension? If so, explain how; if not, describe the sign-bit behavior.

b. Assume $(A_1, A_0) = (5\text{ V}, 0\text{ V})$ and (Sh_0, Sh_1) transitions from (5 V, 0 V) to (0 V, 5 V) at $t = 0$. Sketch approximate waveforms for A_1, A_0, Sh_0, Sh_1, X, B_1, and B_0.

c. For the inputs of b calculate the parasitic capacitances C_{P1} and C_{P2} required for the analysis of the delay t_{pHL} from Sh_1 to B_0. Ignore wiring capacitance.

d. For the inputs of b compute the delay t_{pHL} from Sh_I to B_0.

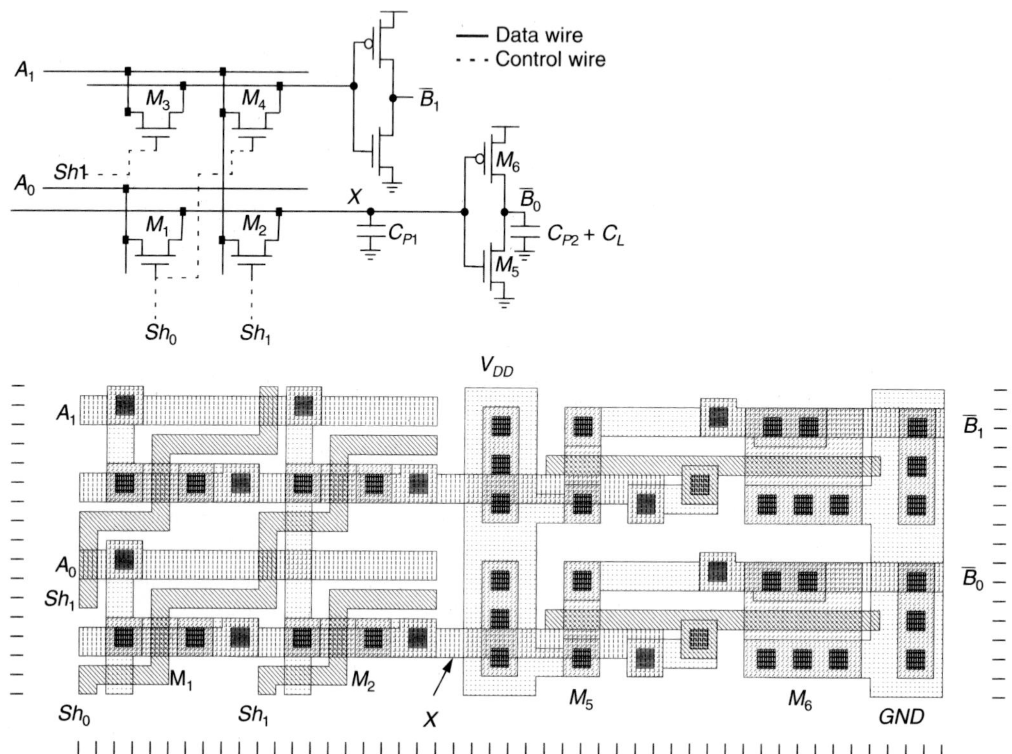

Figure 7.51 Two-bit barrel shifter.

14. [E, None,7.7] Consider the circuit of Figure 7.52. Modules A and B have a delay of 20 nsec and 65 nsec at 5 V, and switch 30 pF and 112 pF respectively. The register has a delay of 4 nsec and switches 0.2 pF. Adding a pipeline register allows for reduction of the supply voltage while maintaining throughput. How much power can be saved this way? Delay with respect to V_{DD} can be approximated from Figure 4.56.

15. [E,None,7.7] Repeat Problem 14, using parallelism instead of pipelining. Assume that a 2-to-1 multiplexer has a delay of 8 nsec at 5 V and switches 0.6 pF. Try parallelism by 2 and by 4. Which one is preferable?

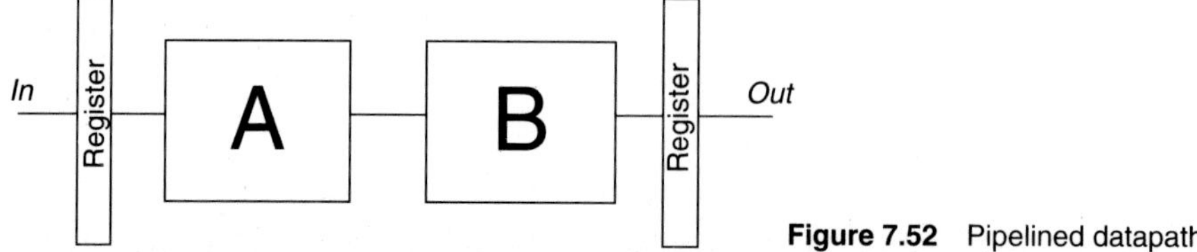

Figure 7.52 Pipelined datapath.

16. [C& D, None, 7.7–7.8] The circuit in Figure 7.53 performs some simple arithmetic. Table 7.3 enumerates area and the capacitance switched per operation. Delay with respect to V_{DD} can be approximated from Figure 4.56. The multiplier has a worst-case delay of 30 nsec at 5 V, and the adder has a worst-case delay of 35 nsec at 1.5 V. Assume negligible delays through muxes and registers. The circuit is part of a video processing system that requires a new sample to be processed every 100 nsec.

a. What is the minimal voltage supply so that the circuit meets the throughput constraints? What is the power dissipated by the circuit? What is the total area?

b. Use pipelining to enable voltage scaling in order to minimize power consumption. How many pipeline stages are optimal? What are the supply voltage, power, and area of this new circuit?

c. Use parallelism to enable voltage scaling. What are the supply voltage, power, and area of this new circuit?

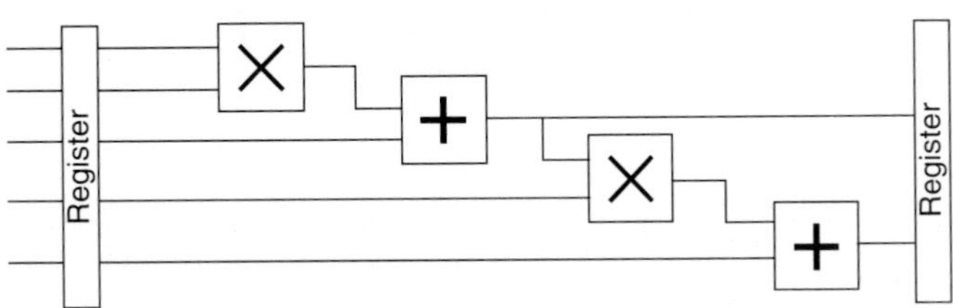

Figure 7.53 Arithmetic circuit.

Table 7.3 Module library.

Unit	Width × height (microns)	Capacitance/operation
Multiplier	1600 × 1500	25 pF
Adder	690 × 210	1.4 pF
Register	580 × 64	0.2 pF
Multiplexer	690 × 64	0.3 pF

DESIGN PROJECT

Design a 16-bit adder circuit that has a worst-case delay of 50 nsec at a supply voltage of 2 V. Minimize the power consumption. Pick one particular circuit style (possible choices (but not limited to): static CMOS, *np*-CMOS, Manchester, pass-transistor EXOR-based). Area is not a constraint.

APPENDIX

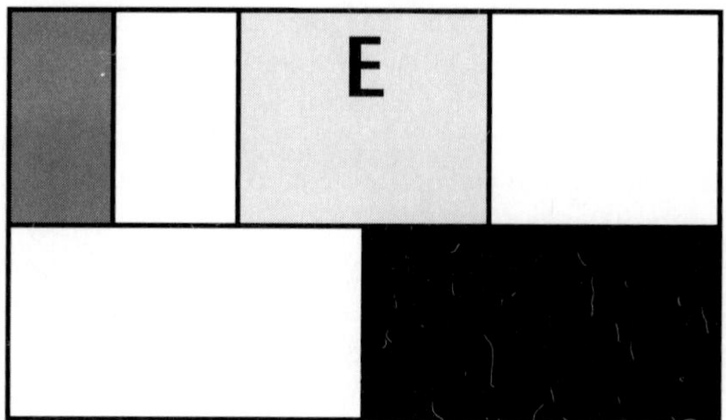

FROM DATAPATH SCHEMATICS TO LAYOUT

Bit-sliced datapath layout

The stringent performance requirements often imposed on datapath structures affect the way the circuit topology is translated into silicon. Extreme care is needed to minimize the parasitic capacitance on critical signal nets. The extra capacitance due to wiring is of particular concern. Keeping the structure dense and compact while avoiding long interconnections is a prime goal. This explains why the design of the datapath is still a manual, labor-intensive process in high-performance microprocessors, while the conception of other units, such as a controller, is heavily automated.

Fortunately, the regular, bit-sliced structure of the datapath simplifies the daunting task of the design engineer. In a two-metal layer process, a number of layout strategies are commonly used when implementing bit-sliced modules. The basic topology and the choice of routing layers for two of them are illustrated in Figure E.1.

In approach I, the wells are oriented horizontally and are shared between neighboring slices. The V_{DD} rail is shared as well. This requires the mirroring of even and odd slices around the horizontal axis. In the second layout strategy, the well and the supply lines run from top to bottom and can be shared between neighboring cells, for example, between an adder cell and a multiplexer. The latter requires those cells to be connected by abutment, which precludes the insertion of vertical routing channels. This approach is therefore better for pure manual design.

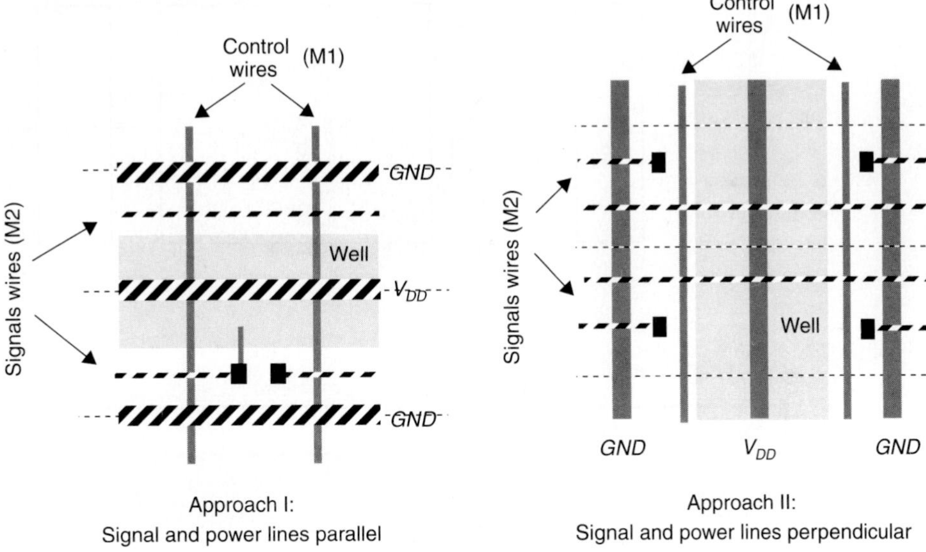

Figure E.1 Layout strategies for bit-sliced datapaths, assuming an *n*-well CMOS technology. The oriented horizontally, which means that data is routed from left to right (or vice versa) and control fr bottom.

The impact of the layout strategy is illustrated in Figure E.2 with the example of a four-bit datapath. The structure is automatically generated using a *datapath compiler* [Brodersen92] and employs Strategy II. Some features of this design are worth highlighting.

- While power and ground are distributed vertically (in Metal1), global supply rails are routed on top and bottom in Metal2, distributing the supply voltages to the cells. These wires should be dimensioned so that they can carry the peak current required by the connecting cells.
- Vertical routing channels are provided to enable the connection of the terminals of neighboring cells. These prevent the sharing of the wells between the cells. The traditional Strategy II approach is only found within the adder cell, where the logic depth requires cascaded gates. Wells and supply lines are shared between these cells.
- The datapath contains four regular slices and a so-called control slice. The latter buffers the incoming control signals and provides enough driving capacity to handle the potentially large fan-out. Some logic functionality can also be provided, such as providing inverse control signals. This is, for instance, the case for the transmission gates in the multiplexer cell. In the example of Figure E.2, only the multiplexers and the tri-state buffer need control signals. In addition, clock signals must be distributed to the register cells.

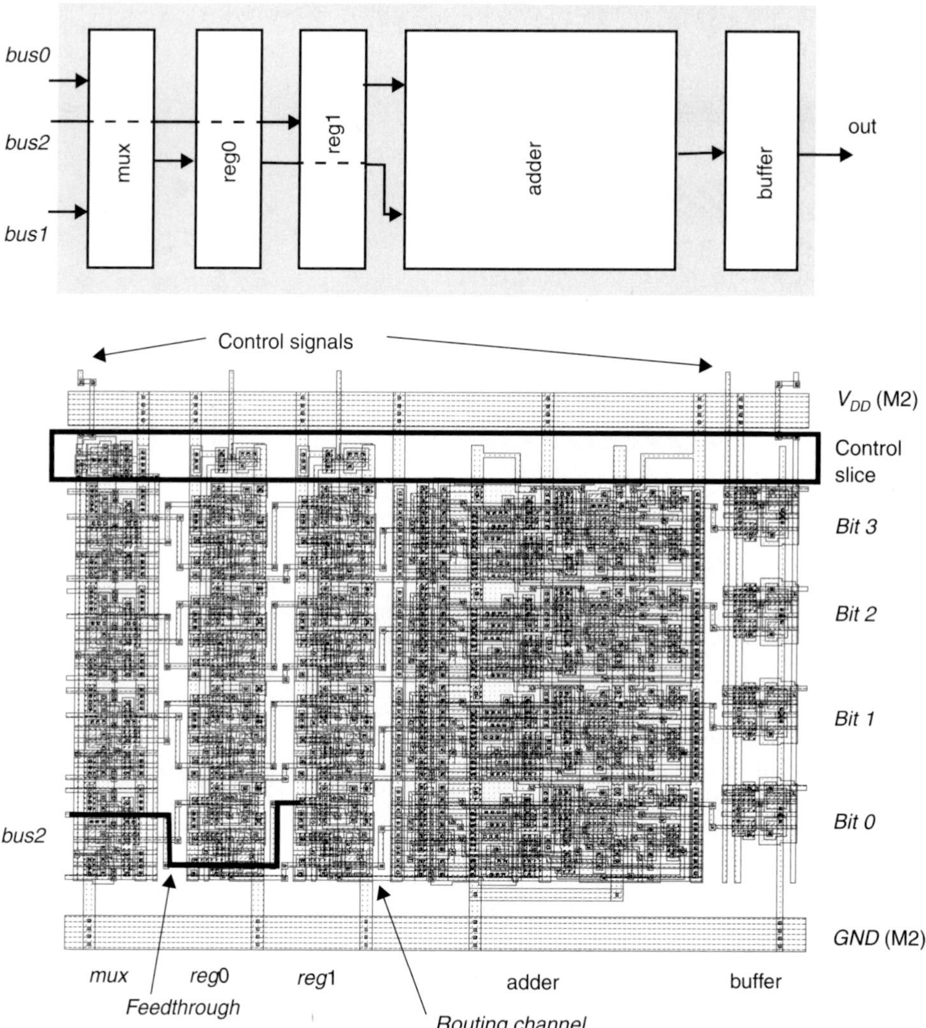

Figure E.2 Layout of 4-bit data path using layout strategy II.

- The bit-sliced datapath approach uses a linear placement paradigm; that is, all cells are placed on a one-dimensional axis. As a result, some connections must traverse between non-neighboring cells. To keep the wire length and the associated parasitic capacitance small, *feedthroughs* are used that exploit unused area in a cell to provide a wiring path between the left and right sides. An example of a feedthrough is shown in Figure E.2, where *bus*2 is routed to *reg*1 over the *mux* and *reg*0. It is often advantageous to make cells somewhat larger than necessary to enable a minimum number of feedthroughs. When running out of feedthroughs, extra wiring space must be provided between the slices. This not only increases the overall size of the datapath, but results also in longer wire lengths for both the internal wires in the cell and the interconnect wires.

Problem E.1 Placement in Datapath

The linear placement of Figure E.2 is not a unique solution. Explain why the proposed solution is probably the better one.

To Probe Further

An excellent description of automatic layout generation for datapaths appears in [Brodersen92].

[Brodersen92] R. Brodersen, ed., "Anatomy of a Silicon Compiler," Kluwer Academic Publishers, 1992.

CHAPTER

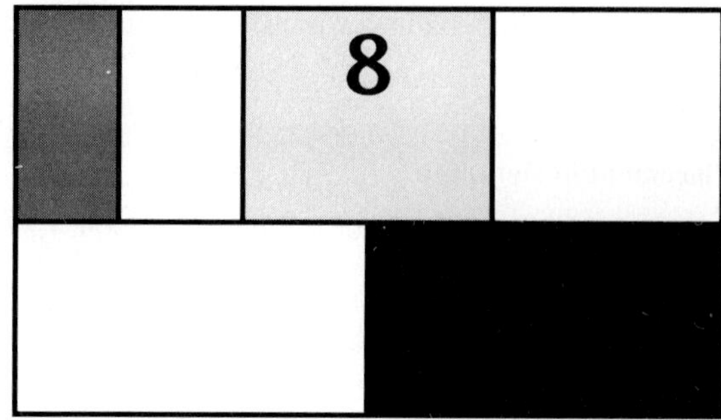

COPING WITH INTERCONNECT

Modeling of capacitive, resistive, and inductive parasitics of interconnect

Effects of parasitics on performance and how to cope with them

Noise induced by interconnect causes reliability problems

Comments on packaging of integrated circuits

8.1 Introduction

The previous chapters have established that the speed of a logic gate is proportional to the load capacitance. Until now, this load was assumed to be a composition of the intrinsic capacitances of the gate and its fan-out. It was demonstrated that under these circumstances a scaling of the device dimensions leads to decreasing gate delays.

This analysis ignores the contribution of the interconnect wires that introduce a range of parasitic effects. While the ideal wire does not affect circuit performance at all, a real wire in an integrated circuit or system introduces capacitive, resistive, and inductive parasitics that can have a dominant influence on the circuit operation. Even more unfortunate, these parasitic effects display a scaling behavior that is different from the active devices discussed so far. Their effects increase as device dimensions are reduced and dominate the performance in submicron technologies. This situation is aggravated by the fact that improvements in technology make the production of ever-larger die sizes economically feasible, which results in an increase in the average length of an interconnect wire and in the associated parasitic effects. It is therefore worthwhile to analyze the effects of interconnect on performance and to discuss various ways to cope with them.

As mentioned, interconnect introduces three types of parasitic effects: capacitive, resistive, and inductive. All three have a dual effect on the circuit behavior.

1. An induction of noise, which affects the reliability of the circuit.
2. An increase in propagation delay.

In this chapter, we discuss each of these effects in sequence.

8.2 Capacitive Parasitics

At the current CMOS switching speeds, the capacitance of the interconnect wire is definitely the most important parasitic. It is usually the only one that the designer of a standard, nonaggressive CMOS circuit has to consider. Increasing the load capacitance of a gate raises the propagation delay, as earlier discussions have made clear. Furthermore, interwire capacitance introduces crosstalk and hence affects the robustness of a design.

First, we analyze the different components of the interconnect capacitance and determine how they behave with a scaling of the technology. This is followed by a discussion of a number of circuit solutions that can help reduce the influence of the capacitance on performance and reliability.

8.2.1 Modeling Interconnect Capacitance

The capacitance of an interconnect wire can be modeled by the parallel-plate capacitor model of Figure 8.1. If the width of the wire is substantially larger than the thickness of the insulating material, it may be assumed that the electrical-field lines are orthogonal to the capacitor plates. Under those circumstances, and assuming SiO_2 as the insulating material, the total capacitance of the wire can be modeled as

$$C_{int} = \frac{\varepsilon_{ox}}{t_{ox}} WL \tag{8.1}$$

where W and L are respectively the width and length of the wire, and $\varepsilon_{ox} = 3.97\ \varepsilon_0 = 3.5 \times 10^{-11}$ F/m and represents the permittivity of the oxide. The wires are normally routed on top of the field oxide, which is substantially thicker than the gate oxide, resulting in a smaller capacitance per unit area. A set of typical wire capacitances is given in Table 8.1. This table collects the values of the capacitance per unit area for the routing layers of a 1 μm CMOS technology.

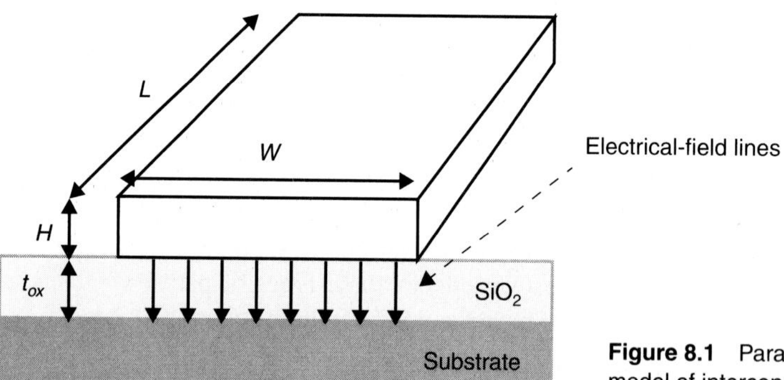

Figure 8.1 Parallel-plate capacitance model of interconnect capacitance.

Table 8.1 Capacitance/unit area for typical interconnect layers (1 μm CMOS).

Interconnect Layer	fF/μm²
Polysilicon to substrate	0.058 ± 0.004
Metal1 to substrate	0.031 ± 0.001
Metal2 to substrate	0.015 ± 0.001
Metal3 to substrate	0.010 ± 0.001
n^+ Diffusion to substrate (@ 0 V)	0.36 ± 0.02
p^+ Diffusion to substrate (@ 0 V)	0.46 ± 0.06

Scaling the technology by a factor S reduces most interconnect parameters, such as the width W and thickness t_{ox} along the same lines. Unfortunately, the same is not true for the wire length L. The scaling behavior of the wire length depends on the wire locality. Interconnect wires can typically be divided into two major classes: local, intramodule interconnections, and global wires that provide communication between large modules. This is illustrated in Figure 8.2, which plots a typical distribution of the interconnect-wire lengths on a chip [Kang87]. The distribution has two peaks, one around 0.1 L_D that represents the local interconnect, and another around 0.5 L_D representing the global wires. L_D stands for the die size, which is approximately equal to $\sqrt{A_D}$, with A_D the chip area. Empirical formulas have been derived that relate the average length of the global wire to the chip area. An example of such an expression, proposed by Sorkin [Sorkin87], is given in Eq. (8.2).

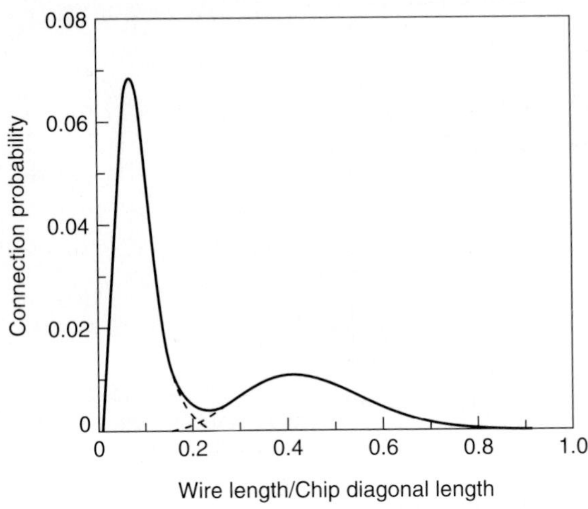

Figure 8.2 Typical distribution of wire lengths in an integrated circuit.

$$L_{av} \approx \frac{\sqrt{A_D}}{3} \tag{8.2}$$

Hence, the overall scaling behavior of the wiring capacitance can be expressed through a parameter $S_{C,wire}$ that is a function of the technology scaling factor S (Chapter 3) and the wire-length scaling factor S_L:

$$S_{C,\,wire} = \frac{S \times S_L}{S} = S_L \tag{8.3}$$

or, the capacitance scales in proportion to the wire length. This equation assumes that all interconnect dimensions besides the length (t_{ox}, W, H) scale in the same way as the transistor dimensions.

The length of a local interconnect wire approximately scales with the technology scaling factor S, as the increased packing density of the devices makes it possible to similarly reduce the wire lengths. Consequently, its capacitance is reduced by the same factor as the device dimensions and the gate speed. On the other hand, the length of the global interconnect wires grows proportionally to the die size, as expressed by Eq. (8.2). Improvements in material quality and manufacturing techniques have made it possible to reliably fabricate ever-larger die sizes. This is demonstrated in Figure 8.3, where the die sizes of some typical static and dynamic memory circuits are plotted as a function of the number of memory bits/chip. The latter tends to increase by a factor of four every three years. An approximately linear increase in area can be observed. A similar trend has been reported for logic circuits such as microprocessors. The scaling factors S_L and $S_{C,wire}$ are thus smaller than 1. This means that the global wiring capacitance tends to grow with technology scaling. A growing disparity between the decreasing intramodule gate delays and the delays of driving the large capacitances of the intermodule wires can be observed. Addressing this disparity requires effective circuit techniques, as discussed later in this section.

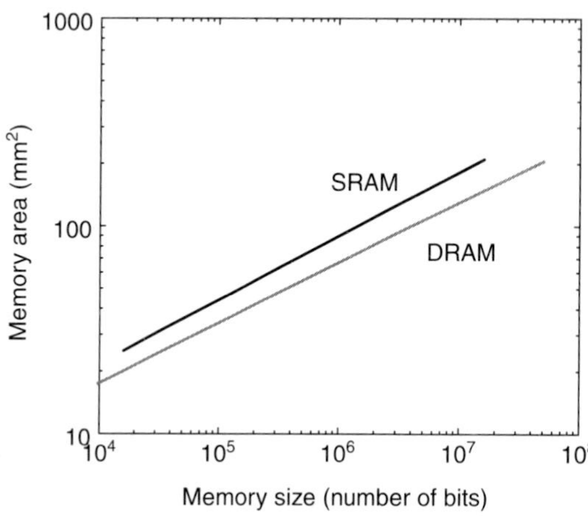

Figure 8.3 Average die size of SRAM and DRAM memories as a function of memory capacity.

Example 8.1 Intra- and Intermodule Wiring Capacitance

This disparity is best illustrated with some numbers. The estimated average intra-module wiring capacitance for a 0.2 μm CMOS technology and a die size of 2 cm will equal 50 fF, versus 1.9 pF for the average intermodule wire [Masaki92].

The actual situation is even worse. Scaling the vertical dimension H of the wire with the technology scaling factor S results in an increase in wiring resistance, which is discussed in Section 8.3. Therefore, it has been a common tendency over the years to scale the thickness H by a different and smaller ratio than the other device dimensions. This reduces the *W/H-ratio*, which approaches unity in advanced processes. Under those circumstances, the parallel-plate model assumed above becomes inaccurate. The capacitance between the side-walls of the wires and the substrate, called the *fringing capacitance*, can no longer be ignored and contributes to the overall capacitance. This effect is illustrated in Figure 8.4a.

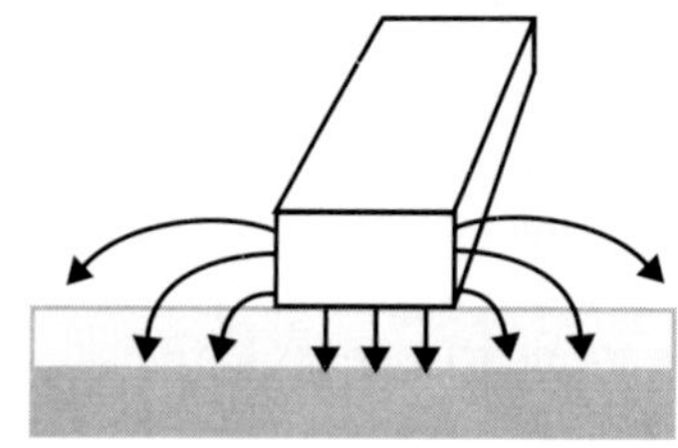

(a) Fringing fields

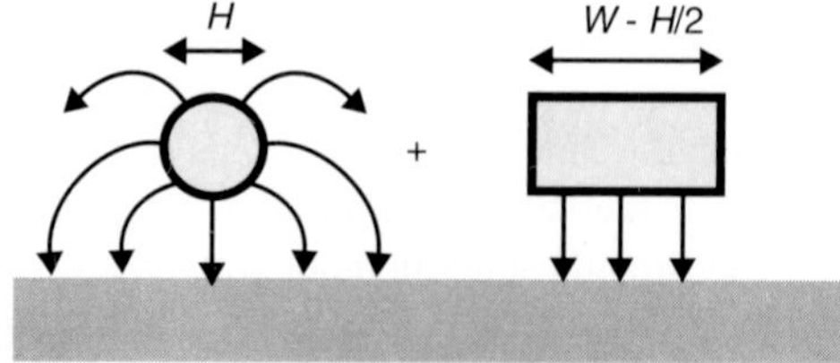

(b) Model of fringing field capacitance.

Figure 8.4 The fringing-field capacitance. The model decomposes the capacitance into two contributions: a parallel-plate capacitance, and a fringing capacitance, modeled by a cylindrical wire with a diameter equal to the thickness of the wire.

The total capacitance is approximated as the sum of two components (Figure 8.4b), a parallel-plate capacitance determined by the orthogonal field between a wire of width $W - H/2$ and the ground plane, and a fringing capacitance modeled by a cylindrical wire with a dimension equal to the interconnect thickness H. The resulting analytical model for C_{fringe} is rather complex. An empirical formula for C_{in} that is relatively accurate and efficient to compute is given in [Vdmeijs84].

$$C_{int} = \varepsilon_{ox} L\left[\left(\frac{W}{t_{ox}}\right) + 0.77 + 1.06\left(\frac{W}{t_{ox}}\right)^{025} + 1.06\left(\frac{H}{t_{ox}}\right)^{0.5}\right] \tag{8.4}$$

Even so, it is customary to use tables to determine the contribution of the fringing capacitance. Typical values of the fringing capacitance per unit length for a 1 μm CMOS technology are shown in Table 8.2. Figure 8.5 plots the value of the wiring capacitance as a function of (W/H). For larger values of (W/H) the total capacitance approaches the parallel-plate model. For (W/H) smaller than 1.5, the fringing component actually becomes the dominant component. The fringing capacitance can increase the overall capacitance by a factor between 1.5 and 3 for small line widths.

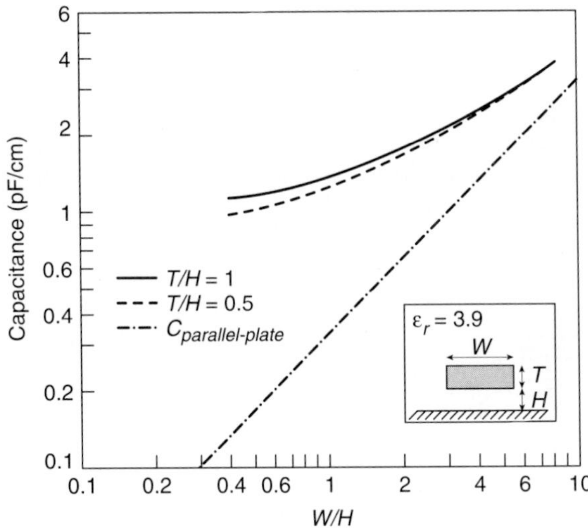

Figure 8.5 Capacitance of interconnect wire as a function of (W/H), including fringing-field effects (from [Schaper83]).

Table 8.2 Fringing capacitance per unit length (for a 1 μm CMOS process).

	Fringing capacitance (fF/μm)
Polysilicon to substrate	0.043 ± 0.004
Metal1 to substrate	0.044 ±0.001
Metal2 to substrate	0.035 ±0.001
Metal3 to substrate	0.033 ±0.001

Example 8.2 Capacitance of Metal Wire

Some global signals, such as clocks, are distributed all over the chip. The length of those wires can be substantial. For die sizes between 1 and 2 cm, wires can reach a length of 10 cm and have associated wire capacitances of substantial value. Consider an aluminum wire 10 cm long and 4 μm wide. We can compute the value of the total capacitance using the data presented in Table 8.1 and Table 8.2.

$$\text{Area (parallel-plate) capacitance: } 0.1 \times 10^6 \times 4 \times 0.031 = 12.4 \text{ pF}$$

$$\text{Fringing capacitance: } 0.1 \times 10^6 \times 2 \times 0.044 = 8.8 \text{ pF}$$

$$\text{Total capacitance: } 21.2 \text{ pF.}$$

Notice the factor 2 in the computation of the fringing capacitance, which takes the two sides of the wire into account.

Finally, the nonproportional scaling of the horizontal and vertical dimensions of the wiring also results in an increase in *interwire capacitance*. The thickness H of the wire remains rather constant, while the minimum distance D between the wires on the same layer or different layers is reduced (Figure 8.6). Typical values of the interwire capacitances for a 1μm CMOS process are shown in Table 8.3. The increasing contribution of the interwire capacitance to the total capacitance with decreasing feature sizes is best illustrated by Figure 8.7. In this graph, it is assumed that oxide and wire thickness are held constant while scaling all other dimensions. When W becomes smaller than 1.75 H, the interwire capacitance starts to dominate. It is especially this interwire capacitance that causes a reliability problem in the form of cross talk.

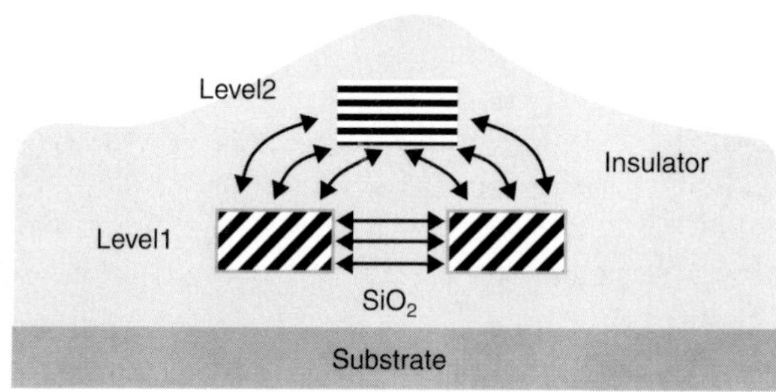

Figure 8.6 Interwire capacitance.

Table 8.3 Inter-wire capacitance for 1 μm CMOS process.

	Area capacitance (fF/mm2)	Fringing capacitance (fF/mm)
Metal1 to polysilicon	0.055	0.049
Metal2 to polysilicon	0.022	0.040
Metal2 to metal1	0.035	0.046

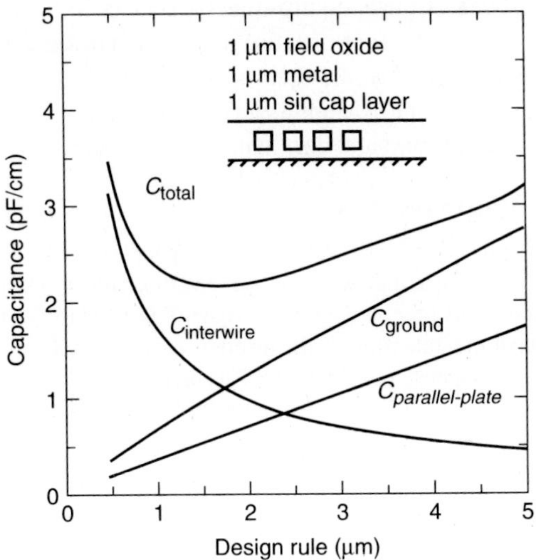

Figure 8.7 Interconnect capacitance as a function of design rules. It consists of a capacitance to ground and an inter-wire capacitance (from [Schaper83]).

8.2.2 Capacitance and Reliability—Cross Talk

An unwanted coupling from a neighboring signal wire to a network node introduces an interference that is generally called *cross talk*. The resulting disturbance acts as a noise source and can lead to hard-to-trace intermittent errors, since the injected noise depends upon the transient value of the other signals routed in the neighborhood. In integrated circuits, this intersignal coupling can be both capacitive and inductive, as shown earlier in Figure 3.1. Capacitive cross talk is the dominant effect at current switching speeds.

It is obviously not a good idea to allow the capacitance between two signal wires to grow too large if one wants to keep cross talk at a minimum. It is, for instance, bad practice to have two wires run parallel for a long distance. One is often tempted to do just that when distributing the two clocks in a two-phase system or when routing a bus. Capacitive crosstalk between signals on the same layer can be reduced by keeping the distance between the wires large enough or by inserting a *shielding* wire—*GND* or V_{DD}—between the two signals (Figure 8.8).

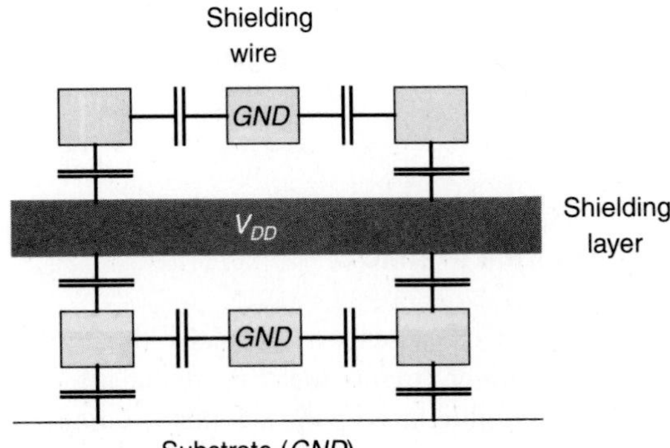

Figure 8.8 Cross section of routing layers, illustrating the use of shielding to reduce capacitive cross talk.

The interwire capacitance between signals on different layers can be further reduced by the addition of extra routing layers. When four or more routing layers are available, we can fall back to an approach often used in printed circuit board design. Every signal layer is interleaved with a *GND* or V_{DD} metal plane (Figure 8.8). This effectively turns the interwire capacitance into a capacitance-to-ground and eliminates interference.

Example 8.3 interwire Capacitance and Cross Talk

Consider the dynamic logic circuit of Figure 8.9. The storage capacitance C_X of the dynamic node X is composed of the diffusion capacitances of the pre- and discharge transistors and the gate capacitance of the connecting inverter. A nonrelated signal Y is routed as a metal1 wire over the polysilicon gate of one of the transistors in the inverter. This creates a parasitic capacitance C_{XY} with respect to node X. Suppose now that node X is precharged to 5 V and that signal Y undergoes a transition from 5 V to 0 V. The charge redistribution causes a voltage drop ΔV_X on node X.

$$\Delta V_X = \frac{C_X}{C_X + C_{XY}} 5V \tag{8.5}$$

Assume that C_X equals 25 fF. An overlap of $5 \times 5\ \mu m^2$ between metal1 and polysilicon results in a parasitic capacitance of 1.9 fF ($5 \times 5 \times 0.055 + 2 \times 5 \times 0.049$), as obtained from Table 8.3. The computation of the fringing effect in the case of overlapping wires is complex and depends upon the relative orientation of the wires. We assumed in this analysis that two sides of the cross section contribute. The impact of the fringing capacitance is very small anyhow in this particular case. A 5 V transition on Y thus causes a voltage disturbance of 0.35 V on the dynamic node. Combined with other parasitic effects, such as charge redistribution and clock feedthrough, this might lead to circuit malfunctioning.

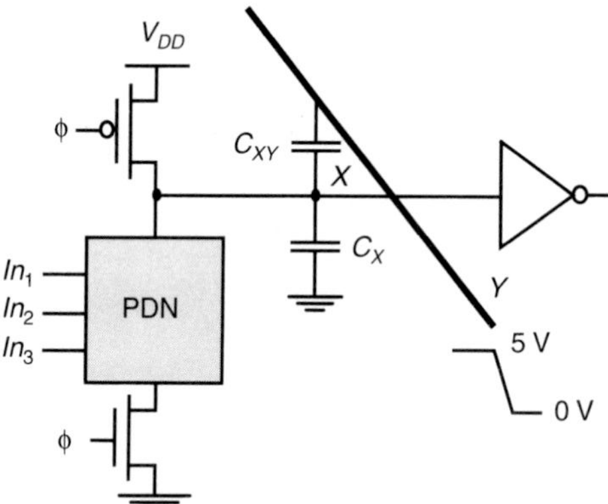

Figure 8.9 Cross talk in dynamic circuits.

8.2.3 Capacitance and Performance in CMOS

The increasing values of the interconnect capacitances, especially those of the global wires, emphasize the need for effective driver circuits that can (dis)charge capacitances with sufficient speed. This need is further highlighted by the fact that in complex designs a single gate often has to drive a large fan-out and hence has a large capacitive load. Typical examples of large on-chip loads are busses, clock networks, and control wires. The latter include, for instance, reset and set signals. These signals control the operation of a large number of gates, so fan-out is normally high. Other examples of large fan-outs are encountered in memories where a large number of storage cells is connected to a small set of control and data wires. The capacitance of these nodes is easily in the multipicofarad range. The worst case occurs when signals go off-chip. In this case, the load consists of the package wiring, the printed circuit board wiring, and the input capacitance of the connected ICs or components. Typical off-chip loads range from 20 to 50 pF, which is a thousand times higher than a standard on-chip load. Driving those nodes with sufficient speed becomes one of the most crucial design problems.

In Chapter 3, an approximate expression was derived for the propagation delay of a CMOS inverter:

$$\begin{aligned} t_p &= \frac{t_{pLH} + t_{pHL}}{2} \\ &\approx \frac{C_L V_{DD}}{2}\left(\frac{1}{k_n(V_{DD} - V_{Tn})^2} + \frac{1}{k_p(V_{DD} - |V_{Tp}|)^2}\right) \\ &\approx \frac{C_L}{2V_{DD}}\left(\frac{1}{k_n} + \frac{1}{k_p}\right) \text{ for } (|V_T| \ll V_{DD}) \end{aligned} \tag{8.6}$$

Note that t_p is directly proportional to C_L. The delay for driving an off-chip capacitance would, therefore, be a thousand times larger than the delay of an internal gate with small fan-out if the same driver circuit were used. This is clearly unacceptable. Increasing the device sizes and providing more (dis)charge current helps only to a certain extent. The input capacitance of the larger devices loads the previous logic stage, which partially annihilates the gain in performance. Therefore, a buffer circuit must be placed between the logic stage and load capacitance. The design and optimization of such a buffer gate is the topic of the remaining sections, where the following cases are examined.

1. Buffering with a single inverter
2. Buffering with multiple inverting stages
3. Driving off-chip capacitors (case study)
4. High-impedance, or tri-state buffers
5. Alternative circuit techniques for driving large capacitances

Before starting this discussion, we revisit the transistor-sizing problem. When cascading multiple inverter stages, the optimal ratio between NMOS and PMOS transistors varies from what was derived for the generic case in Chapter 3.

Cascading Inverter Stages

Eq. (8.6) shows that equalizing k_p and k_n produces identical values for the t_{pLH} and t_{pLH} of a static CMOS inverter. This requires the PMOS transistor to be made $\varepsilon = \mu_n/\mu_p$ times larger than the NMOS device and produces an optimal t_p for a single gate.

The above statement is not valid when cascading multiple inverters. Consider the cascaded (and identical) gates of Figure 8.10. The load capacitance of the first inverter consists of the sum of its own diffusion capacitances, the gate capacitances of the fan-out gate, and the wiring capacitance.

$$C_L = (C_{dp1} + C_{dn1}) + (C_{gp2} + C_{gn2}) + C_W \tag{8.7}$$

where C_{dp1} and C_{dn1} are respectively the equivalent drain-diffusion capacitors of the PMOS and NMOS transistors of the first inverter, and C_{gp2} and C_{gn2} the gate capacitances of the transistors of the second gate. C_W represents the wiring capacitance.

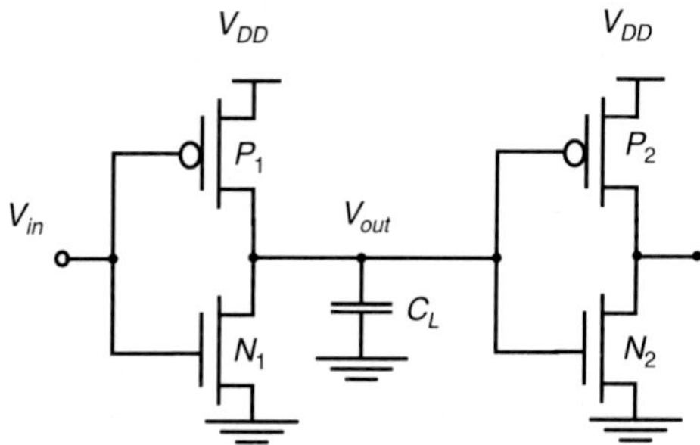

Figure 8.10 Set of cascaded inverters.

Assume that the PMOS transistors are designed α times wider than the NMOS devices ($\alpha = (W/L)_p/ (W/L)_n$). The transistor capacitors scale in approximately the same way, or $C_{dp} = \alpha C_{dn}$, and $C_{gp} = \alpha C_{gn}$. Rewriting Eq. (8.7) yields

$$C_L = (1+\alpha)(C_{dn} + C_{gn}) + C_W = (1+\alpha)C_n + C_W \tag{8.8}$$

with

$$C_n = C_{dn} + C_{gn}$$

The propagation delay of the first gate can be expressed as a function of α.

$$t_p = \frac{(1+\alpha)C_n + C_W}{2V_{DD}}\left(\frac{1}{k_n} + \frac{1}{k_p}\right) = \frac{(1+\alpha)C_n + C_W}{2V_{DD}k_n}\left(1 + \frac{\varepsilon}{\alpha}\right) \tag{8.9}$$

where ε represents the ratio between the device transconductance parameters. The optimal value of α is determined by setting $\partial t_p/\partial\alpha$ to 0.

$$\alpha_{opt} = \sqrt{\varepsilon\left(1 + \frac{C_W}{C_n}\right)} \tag{8.10}$$

This means that $\alpha_{opt} \approx \sqrt{\varepsilon}$ when the wiring capacitance is negligible ($C_W << C_n$), as is typically the case in a chain of inverters. For $\varepsilon = 2.5$, α_{opt} now evaluates to 1.6. This value trades off a larger charge current (improving t_{pLH}) versus a smaller load capacitance (improving t_{pHL}). If the wiring capacitance dominates, larger values of α_{opt} should be used. Observe that we had previously derived an optimal value of ε for the noncascaded case.

From the above analysis, we can conclude that it is advantageous to make the PMOS transistor somewhat smaller than the ε ratio when chaining inverters. This results in an asymmetrical response ($t_{pLH} > t_{pHL}$), but tends to optimize the global response of the system, while also being desirable from an area perspective.

Using a Single Inverter as Buffer

Consider the following problem: a minimum-size inverter that is part of the logic circuitry of an ALU of a microprocessor must drive a load capacitance C_L that is x times larger than the input capacitance C_i of the inverter. To drive this large load, a single inverter-buffer is inserted between the logic gate and load capacitance. The PMOS and NMOS transistors of the second gate are u times wider than those of the first one. We assume that the ratio α between PMOS and NMOS devices is identical for both gates. The question is what value of u minimizes the propagation delay of the combined inverter set, $t_p = t_{p,inv} + t_{p,buf}$?

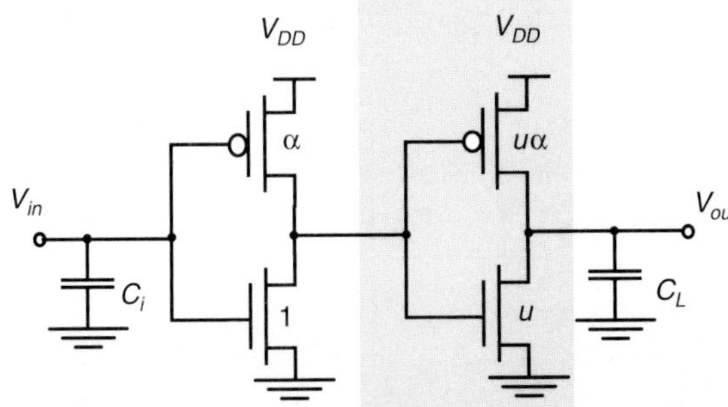

Figure 8.11 Single-inverter buffer.

Define as t_{p0} the propagation delay of a minimum-size gate with a single minimum-size inverter as fan-out. The total propagation delay of the inverter chain of Figure 8.11 can then be approximated.

$$t_p = ut_{p0} + \frac{x}{u}t_{p0} = \left(u + \frac{x}{u}\right)t_{p0} \tag{8.11}$$

The first inverter is slowed u times due to the larger load. The second inverter-buffer benefits from its u-times-larger transistor sizes to drive the x-times-larger load C_L.

The optimum value of u is derived by setting the partial derivative of t_p with respect to u to zero. This yields the following results:

$$u_{opt} = \sqrt{x} \tag{8.12}$$

and

$$t_{p,opt} = 2t_{p0}\sqrt{x} \tag{8.13}$$

Design Consideration

From the above, we can determine when it makes sense to introduce a buffer by comparing the delay of a single inverter driving C_L to an inverter-buffer combination.

$$x t_{p0} > 2 t_{p0} \sqrt{x} \tag{8.14}$$

From Eq. (8.14), we can deduce that inserting a buffer only makes sense when $x > 4$, which is equivalent to a fan-out of four similar gates.

Multiple-Stage Buffer

When driving very large capacitors, a single buffer stage might not be sufficient. For instance, when driving an off-chip capacitance that is a thousand times larger than the on-chip capacitances, the best propagation delay achievable with a single buffer stage equals $2\sqrt{1000}\ t_{p0} = 64\ t_{p0}$. Although much better than the unbuffered delay of 1000 t_{p0}, this delay is still unacceptable. A more powerful approach is to use a chain of N inverters, as shown in Figure 8.12.

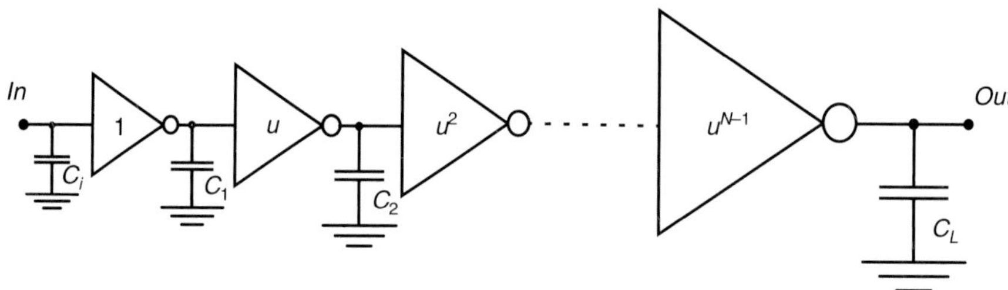

Figure 8.12 *N*-stage buffer.

From the results of the two-stage buffer, one can surmise that the minimum delay is obtained by dividing the delay equally over all N stages. This is achieved by gradually scaling up all stages with a constant factor u, which produces an identical delay $t_{p,stage} = u.t_{p0}$ for each stage, since u is the ratio between input and output capacitance of each stage. To achieve the same delay for the last stage, it is necessary that $C_L = x\ C_{in} = u^N C_{in}$, which yields the following expressions:

$$t_p = N \times u \times t_{p0}$$

and (8.15)

$$N = \frac{\ln(x)}{\ln(u)}$$

Equating the derivative of t_p with respect to u to 0 produces the optimal scaling factor [Mead80].

$$u_{opt} = e = 2.7182$$

and (8.16)

$$t_{p,opt} = e\ \ln(x) t_{p0} = e\ \ln\left(\frac{C_L}{C_i}\right) t_{p0}$$

The optimal buffer design scales consecutive stages in an exponential fashion. Figure 8.13 plots $t_p/t_{p0} = \ln(x)\ (u/\ln(u))$ as a function of u for various values of x. Observe that the curves are rather flat around the optimum value of e. Values of u equal to 3 or 4 are almost equally acceptable and help reduce the required number of buffer stages and the area. Table 8.4 enumerates the values of $t_{p,opt}/t_{p0}$ for the unbuffered design, the single inverter buffer, and the cascaded inverter buffer for a variety of values for x. Observe the impressive speed-up obtained with cascaded buffers when driving very large capacitive loads.

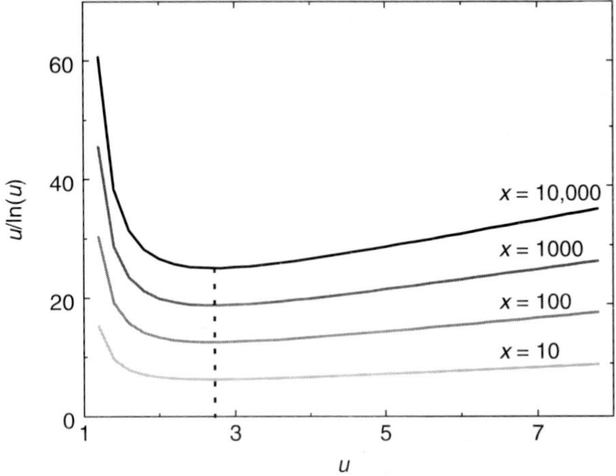

Figure 8.13 t_p/t_{p0} versus u for various values of x.

Table 8.4 t_{opt}/t_{p0} versus x for various driver configurations.

x	Unbuffered	Single Buffer	Cascaded Buffers
10	10	6.3	6.3
100	100	20	12.5
1000	1000	63	18.8
10,000	10,000	200	25.0

Cascaded Inverter Drivers—Revisited

The above computations assume that the load capacitance of each inverter scales linearly with the size of the connecting inverter. This is not completely correct, since the load capacitance consists of three factors: diffusion capacitance of the driving gate, wiring capacitance, and gate capacitance of the connecting gate. Only the latter component scales with the size of the fan-out. Consequently, the results presented tend to be pessimistic. Assume that the output diffusion capacitance of a gate is equal to α times its input gate capacitance, and the wiring capacitance is ignorable. A gate connecting to a u-times-larger inverter is then loaded with an output capacitance $(u + a)\ C_i$. Under those circumstances, we can prove that the optimal scaling factor can be derived from the following implicit equation:[1]

[1] An in-depth study of the sizing and optimization of CMOS buffers can be found in [Hedensteirna87].

$$u_{opt} = e^{(u_{opt} + \alpha)/u_{opt}} \tag{8.17}$$

Notice that setting α equal to 0 yields $u_{opt} = e$, which is consistent with the previously derived results. The value of the optimum tapering factor is plotted as a function of α in Figure 8.14. Increasing values of α result in a larger scaling factor. When $\alpha = 1$, as is approximately the case for most current technologies, u_{opt} evaluates to 3.6. This larger value translates in a more area-efficient solution than the value of 2.78 derived above.

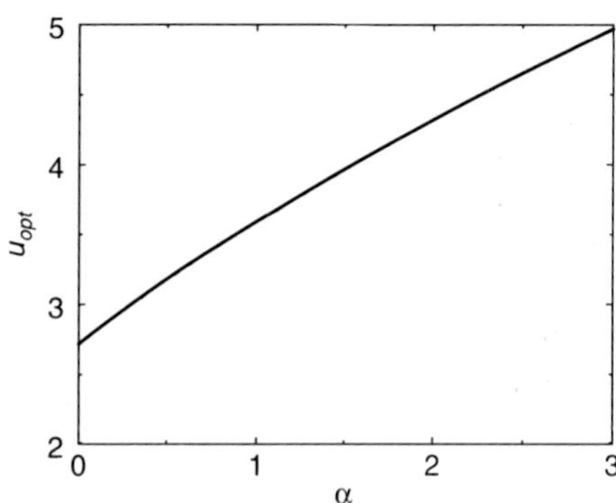

Figure 8.14 Optimum tapering factor *u* as a function of α, the ratio between the intrinsic output capacitance and the input gate capacitance of the inverter.

Problem 8.1 Cascaded CMOS Buffers

Derive Eq. (8.17). The approach is similar to the one used in deriving the ideal equation, Eq. (8.16), which ignores the intrinsic capacitance. Determine the value of the propagation delay as a function of α.

Case Study: Driving Off-Chip Capacitances

With shrinking physical dimensions and increasing circuit speeds, communication with the outside world becomes one of the toughest design problems to solve. The number of connections going off-chip is roughly proportional to the complexity of the circuitry on the chip. This relationship was first observed by E. Rent of IBM (published in [Landman71]), who translated it into an empirical formula that is appropriately called *Rent's rule*. This formula relates the number of input/output pins to the complexity of the circuit, as measured by the number of gates.

$$P = K \times G^{\beta} \tag{8.18}$$

where K is the average number of I/Os per gate, G the number of gates, β the Rent exponent, and P the number of I/O pins to the chip. β varies between 0.1 and 0.7. Its value depends strongly upon the application area, architecture, and organization of the circuit, as demonstrated in Table 8.5. Clearly, microprocessors display a very different input/output behavior than memories.

In current and future technologies, where the number of gates on a chip approaches 1 million, chips with more than 200 pins are common. Packages with over 350 pins are becoming a necessity. This puts tough requirements on the bonding-pad design in terms of

Table 8.5 Rent's constant for various classes of systems ([Bakoglu90])

Application	β	*K*
Static memory	0.12	6
Microprocessor	0.45	0.82
Gate array	0.5	1.9
High-speed computer (chip)	0.63	1.4
High-speed computer (board)	0.25	82

noise immunity. The simultaneous switching of a lot of pads, each driving a large capacitor, causes large transient currents and creates voltage fluctuations on the power and ground lines. This tends to reduce the noise margins, as we discuss later in this chapter.

Furthermore, technology scaling reduces the internal capacitances on the chip, while off-chip capacitances remain approximately constant—typically between 20 and 50 pF. This results in a net increase in the required driving ratio x. In a consistent system-design-approach, we expect the off-chip propagation delays to scale in the same way as the on-chip delays. This puts an even tougher burden on the bonding-pad buffer design.

Example 8.4 Output Buffer Design

The problems associated with a bonding-pad driver design are illustrated with a simple design example. Consider the case where an on-chip minimum-size inverter has to drive an off-chip capacitor C_L of 20 pF. C_i approximately equals 10 fF for a standard gate in a 1.2 μm CMOS process. This corresponds to a t_{p0} of approximately 0.2 nsec. The ratio between C_L and C_i equals 2000, which definitely calls for a multistage buffer design. From Eq. (8.16), we learn that seven stages result in a near-optimal design with a scaling factor u of 2.96 and an overall propagation delay of 4.1 nsec. From a PMOS / NMOS ratio of $\sqrt{\varepsilon}$ —which equals 1.58 for our standard process parameters—and a minimum dimension of 1.2 μm, we can derive the widths for the NMOS and PMOS transistors in the consecutive inverter stages, as shown in Table 8.6.

Table 8.6 Transistor sizes for optimally-sized cascaded buffers.

Stage	1	2	3	4	5	6	7
W_n (μm)	1.8	5.3	15.8	47.7	138.2	409.0	1210.7
W_p (μm)	2.8	8.4	24.9	73.8	218.3	646.2	1912.8

This solution obviously requires some extremely large transistors with gate widths of almost 2 mm! The overall size of this buffer, which achieves the optimal off-chip delay, equals the size of several thousand minimum-size inverters. This is clearly prohibitive, since a complex chip requires a large number of those drivers (see Rent's Rule).

Fortunately, it is not necessary to achieve the optimal buffer delay in most cases. Off-chip communications can often be performed at a fraction of the on-chip clocking speeds. Relaxing the delay from 4.1 nsec to 10 nsec still allows for off-chip clock speeds of over 25 Mhz, while substantially reducing the buffering requirements. This redefines the buffer design problem.

Given a maximum propagation delay time $t_{p,max}$, determine the number of buffer stages N and the required scaling factor u, such that the overall area is minimized. This is equivalent to finding a solution that sets t_p as close as possible to $t_{p,max}$.

The optimization problem is now reformulated into the finding of the minimum integer value of N that obeys the constraints of Eq. (8.19).

$$\frac{t_{p,\,max}}{t_{p0}} \geq \ln(x)\frac{u}{\ln(u)} = N \times x^{1/N} \tag{8.19}$$

This transcendental optimization problem can be solved using a small computer program or using mathematical packages such as MATLAB [Etter93] or MATHEMATICA. Figure 8.15 plots the right-hand side of this equation as a function of N for some values of x. For a given value of capacitor ratio x and a maximum value of the delay (t_{pmax}/t_{po}), the minimum number of buffers N is derived from inspection of the appropriate curve.

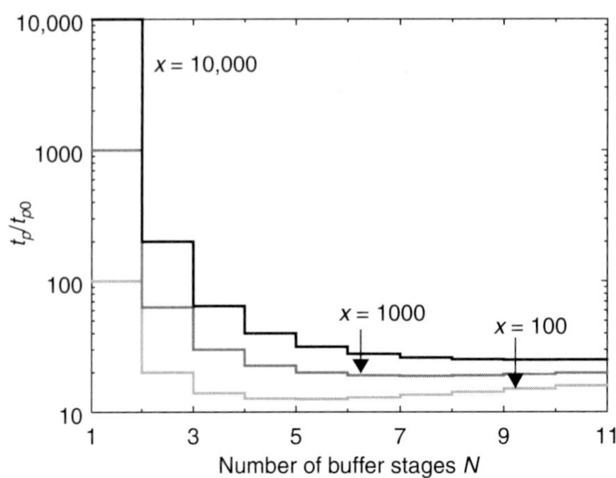

Figure 8.15 t_p/t_{p0} as a function of the number of buffer stages for various values of x.

Example 8.5 Output Driver Design—Revisited

Applying these results to the bonding-pad driver problem and setting $t_{p,max}$ to 10 nsec results in the following solution: $N = 3$, $u = 12.5$, and $t_p = 7.56$ nsec. The required transistor sizes are summarized in Table 8.7.

Table 8.7 Transistor sizes of redesigned cascaded buffer.

Stage	1	2	3
W_n (μm)	1.8	22.7	286.0
W_p (μm)	2.8	35.3	444.5

The overall area of this solution is approximately 20 times smaller than the optimum solution, while its speed is reduced by less than a factor of 2. This clearly shows that *designing a circuit for the right speed, not for the maximum speed,* pays off!

Design Consideration

Even this redesigned buffer requires wide transistors. One must be careful when designing those devices, since the large value of W translates into very long gate connections. Long polysilicon wires tend to be highly resistive, which degrades the switching performance. This problem can be addressed in a number of ways. For instance, a wide transistor can be constructed by connecting many smaller transistors in parallel (Figure 8.16a) or by using ring- or even spiral-shaped devices (Figure 8.16b). In each approach, the resistance of the gate is reduced with the aid of a low-resistance metal bypass connecting the shorter polysilicon sections. An example of a pad driver designed using these techniques is shown in Figure 8.17.

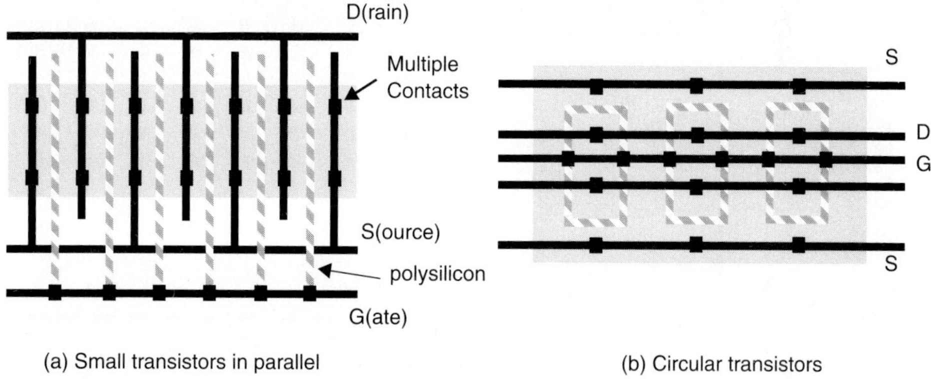

Figure 8.16 Approaches to implementing very wide transistors.

The design of bonding pad-drivers is obviously a critical and nontrivial task that is further complicated by noise and reliability considerations. For instance, the large transient currents resulting from the switching of the huge output capacitance can cause latchup to occur. Multiple well and substrate contacts supplemented with guard rings help avoid the onset of this destructive effect.

Fortunately, a number of novel packaging technologies that reduce off-chip capacitance are currently emerging. An overview of the evolution of the packaging technology is given later in this chapter.

Tri-State Buffers

Busses are essential in most digital systems. A bus is a bundle of wires that connect a set of sender and receiver devices such as processors, memories, disks, and input/output devices. When one device is sending on the bus, all other sending devices should be disconnected. This can be achieved by putting the output buffers of those devices in a high-impedance state Z that effectively disconnects the gate from the output wire. Such a buffer has three possible states—0, 1, and Z—and is therefore called a *tri-state* buffer.

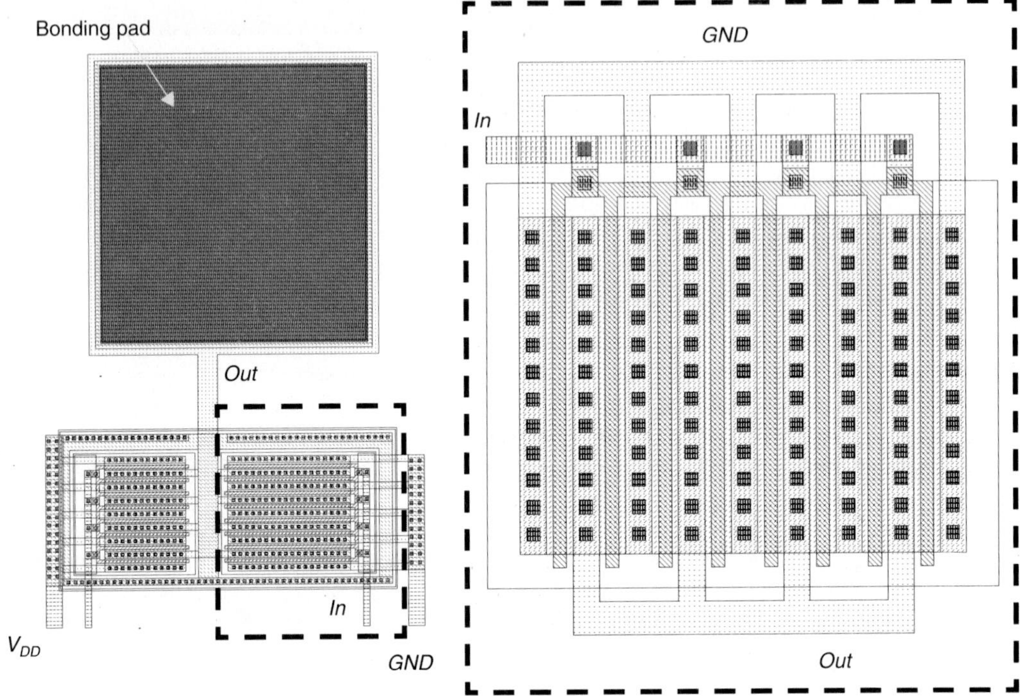

Figure 8.17 Layout of final stage of bonding-pad driver. The plot on the right side is a magnification of the NMOS transistor connected between *GND* and *Out*. See also Colorplate 13.

Implementing a tri-state inverter is straightforward in CMOS. Simultaneously turning off the NMOS and PMOS transistor produces a floating output node that is disconnected from its input. Two possible implementations of a tri-state buffer are shown in Figure 8.18. While the first one is simple, the second is more appropriate for driving large capacitances because of the reduced number of series transistors.

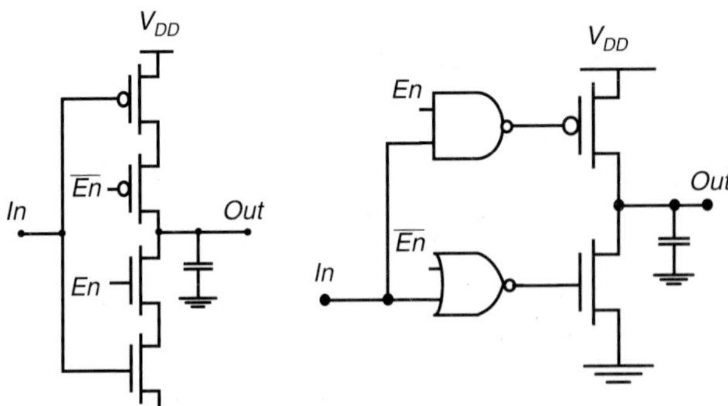

Figure 8.18 Two possible implementations of a tri-state buffer. *En* = 1 enables the buffer.

Alternative Techniques for Driving Large Capacitances

Increasing the size of the driver transistor, and thus increasing the average current I_{av} during switching, is only one way of coping with the delay caused by the load capacitance. An analysis of the propagation delay of a gate, as given by Eq. (8.20), reveals that for a given load capacitance, the delay can also be reduced by decreasing the voltage swing V_{swing}.

$$t_p = \frac{C_L V_{swing}}{I_{av}} \tag{8.20}$$

Besides an increased performance, lower signal swings also reduce the dynamic power consumption, which can be substantial when the load capacitance is large. On the negative side of the balance, reduced signal swings result in smaller noise margins. Furthermore, CMOS gates are not particularly effective in detecting and reacting to small signal changes, because of the relatively small transconductance of the MOS device. In order to work properly and to achieve high performance, reduced-swing circuits normally require amplifier circuits. The overhead of those extra amplifiers is only justifiable for network nodes with a large fan-in, where the amplifier can be shared over many input gates. Typical examples of such nodes are the data or address buses of a microprocessor or the data lines in a memory array. In the latter case, the amplifier is called a *sense amplifier.* Sense amplifiers in memories most often use *differential* or *double-ended* detection techniques, where the signal is compared to a reference signal or a signal moving in an opposite direction. In this section, we only discuss *single-ended* techniques, where we try to detect an absolute change in voltage. Double-ended sensing is discussed in Chapter 10. From the many known approaches we have chosen two: precharging and charge redistribution.

1. Precharging

A first approach to speeding up the response of large fan-in circuits such as buses is to make use of precharging, an example of which is shown in Figure 8.19. During $\phi = 0$, the bus wire is precharged to V_{DD} through transistor M_2. Because this device is shared by all input gates, it can be made large enough to ensure a fast precharging time. During $\phi = 1$, the bus capacitance is conditionally discharged by one of the pull-down transistors. This operation is slow because the large capacitance C_{bus} must be discharged through the small pull-down device M_1.

A speed-up at the expense of noise margin can be obtained by observing that all transitions on the bus are from high-to-low during evaluation. A faster response can be obtained by moving the switching threshold of the subsequent inverter upwards. This results in an asymmetrical gate. In a traditional inverter design, M_3 and M_4 are sized so that t_{pHL} and t_{pLH} are identical, and the switching threshold (V_M) of the inverter is situated around 0.5 V_{DD}. This means that the bus voltage V_{bus} has to drop over $V_{DD}/2$ before the output inverter switches. This can be avoided by making the PMOS device larger, moving V_M upwards. This causes the output buffer to start switching earlier.

The precharged approach can result in a substantial speed-up for the driving of large capacitive lines. However, it also suffers from all the disadvantages of dynamic circuit techniques—charge-sharing, leakage, and inadvertent charge loss as a result of cross talk and noise. Cross talk between neighboring wires is especially an issue in densely wired bus networks. The reduced noise margin NM_H of the asymmetrical read-out inverter makes this circuit particularly sensitive to these parasitic effects. Extreme caution and extensive simulation is required when designing large precharged networks. However, the speed benefit is sometimes worth the effort.

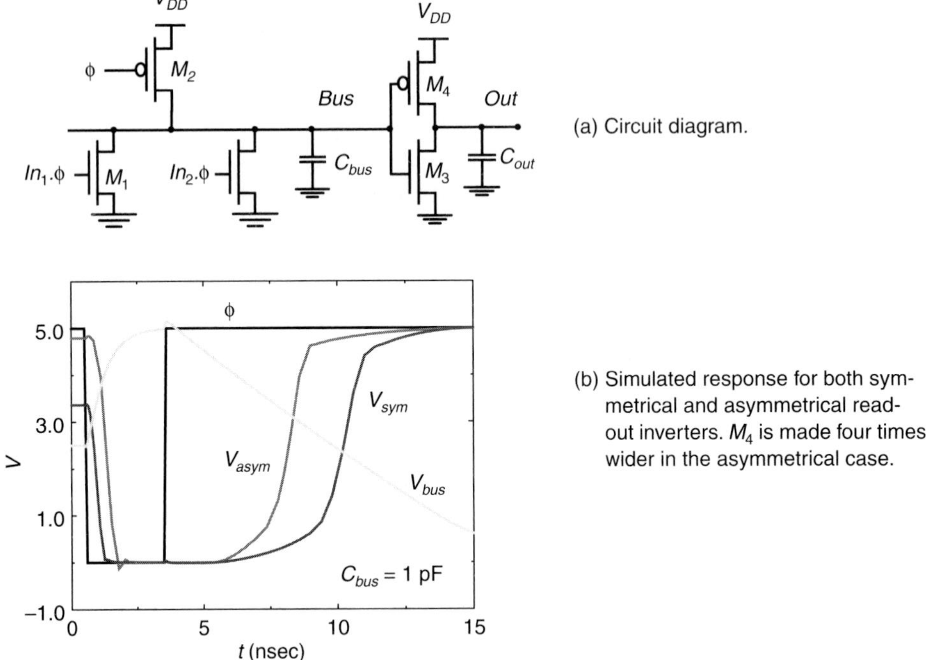

(a) Circuit diagram.

(b) Simulated response for both symmetrical and asymmetrical readout inverters. M_4 is made four times wider in the asymmetrical case.

Figure 8.19 Precharged bus.

The simulated response of a precharged bus network is shown in Figure 8.19b. The output signal is plotted for both symmetrical and asymmetrical output inverters.

An even faster response can be obtained at the expense of power consumption by precharging the interconnection to the most sensitive point of the output inverter, its switching threshold V_M. During the evaluation phase, a small deviation from this center point causes the output inverter to switch to either the high or low output level. Unfortunately, biasing an inverter around V_M means that both PMOS and NMOS devices are simultaneously on, which results in static power consumption.

An example of a $V_{DD}/2$ precharged configuration is shown in Figure 8.20a. During ϕ_{pre}, the input of the inverter M_3-M_4 is short-circuited to its output or $V_{in} = V_{out} = V_M = V_{DD}/2$ when M_3 and M_4 are appropriately sized. The output inverter M_5-M_6 is thus biased in the middle of the transition region. During the evaluation phase ($\phi_{eval} = 1$), the short-circuit is disabled, and the bus capacitance is pulled either high or low by the input pull-up or pull-down transistors. A small variation of the bus voltage rapidly results in a large output voltage swing, as the output inverter actually operates as an amplifier. When the switching frequency is high enough, the large bus capacitance only swings over small voltage differences, as can be seen from the approximated voltage waveforms shown in Figure 8.20b. Notice the small swing of V_{bus}.

The performance improvement comes at the expense of large static currents during precharging. Circuits have been proposed to reduce the static component by turning off the precharging transistors after the equilibrium voltage $V_{DD}/2$ is reached [Bakoglu90].

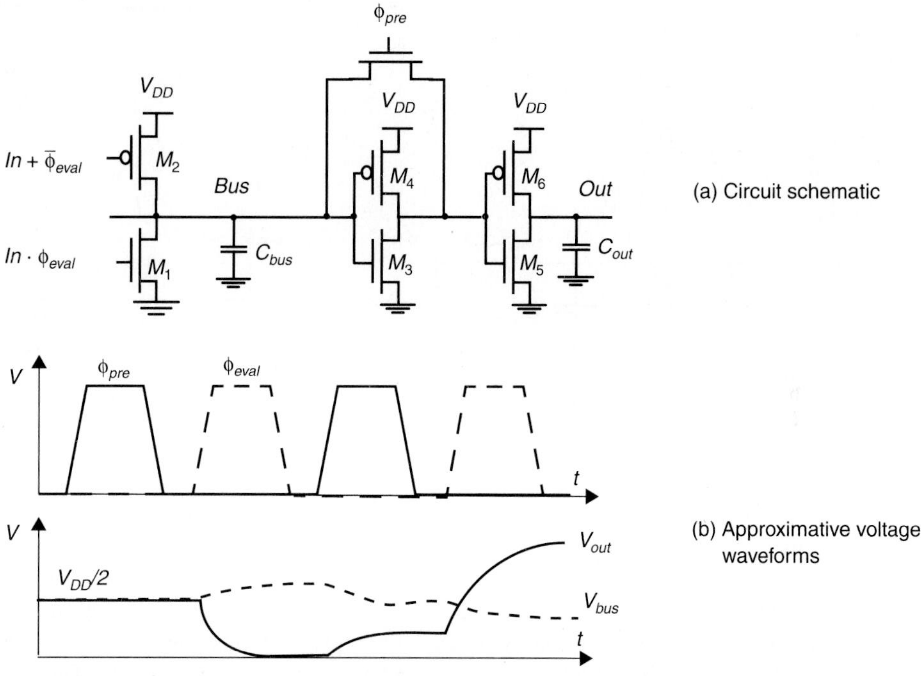

Figure 8.20 Interconnect precharged to $V_{DD}/2$

2. Charge-Redistribution Amplifiers

The precharged bus can yield even higher performances by combining it with a charge-sharing operation, as illustrated in Figure 8.21 [Heller75]. The basic idea is to exploit the imbalance between a large capacitance C_{large} and a much smaller component C_{small}. The two capacitors are isolated by the pass-transistor M_1.

The initial voltages on nodes L and S (V_{L0} and V_{S0}) are precharged to $V_{ref} - V_{Tn}$ and V_{DD} by connecting node S to the supply voltage. Because of the voltage drop over M_1, V_L only precharges to $V_{ref} - V_{Tn}$.

When one of the pull-down devices (e.g., M_2) turns on, node L with its large capacitance slowly discharges. As long as $V_L \geq V_{ref} - V_{Tn}$ transistor M_1 is off, and V_S remains constant. Once V_L drops below the trigger voltage $(V_{ref} - V_{Tn})$, M_1 turns on. A charge redistribution is initiated, and nodes L and S equalize. This can happen very fast due to the small capacitance on the latter node. A small voltage variation on node L translates into a large voltage drop on node S, as is illustrated in the simulated transient response of Figure 8.21b. The circuit hence acts as an amplifier. The resulting signal can be fed into an inverter with a switching threshold larger than $V_{ref} - V_{Tn}$ to produce a rail-to-rail swing.

The disadvantage of the charge-redistribution amplifier is that it operates with a very small noise margin. A small variation of node L due to noise or leakage may cause an erroneous discharge of S. Careful design and analysis is therefore necessary.

This concept has served as the inspiration for a variety of circuits, one of which is presented here. A modified version of the pseudo-NMOS inverter is shown in Figure 8.22.

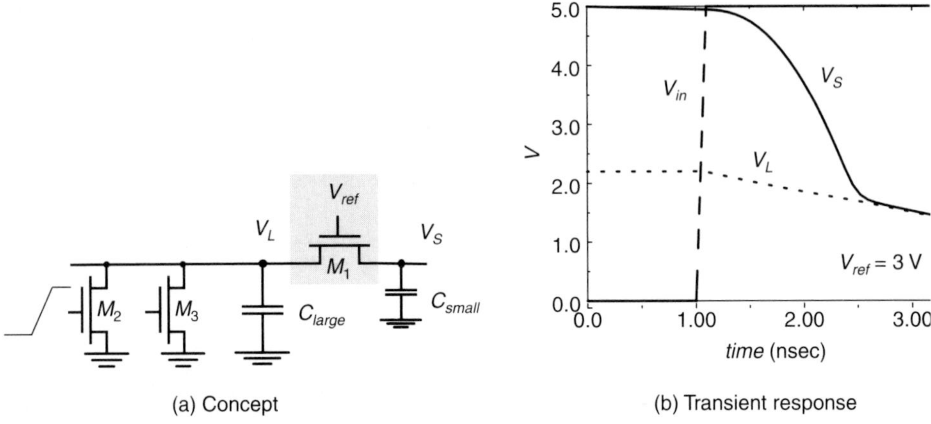

Figure 8.21 Charge-redistribution amplifier.

A cascode device[2] with its gate connected to a fixed reference voltage V_{ref} is inserted between load and driver transistor. The voltage swing at the large-capacitive node *Bus* is restricted to $V_{ref} - V_{Tn}$ at the high end and V_{OL} at the low end. The latter value is determined by the ratio between the load and driver impedances, as is typical in ratioed logic. The output node *Out* with the small capacitive load exhibits a much larger voltage difference between high and low states, V_{DD} and $V_{OL} + V_{cascode}$ respectively. The voltage drop over the cascode transistor is kept small by making transistor M_2 wide. All pull-down transistors can be kept minimum-size, which is an important advantage in a large fan-in gate. The effect of charge redistribution results in a fast transient response at the output node. Analog designers recognize this circuit as a cascode amplifier.

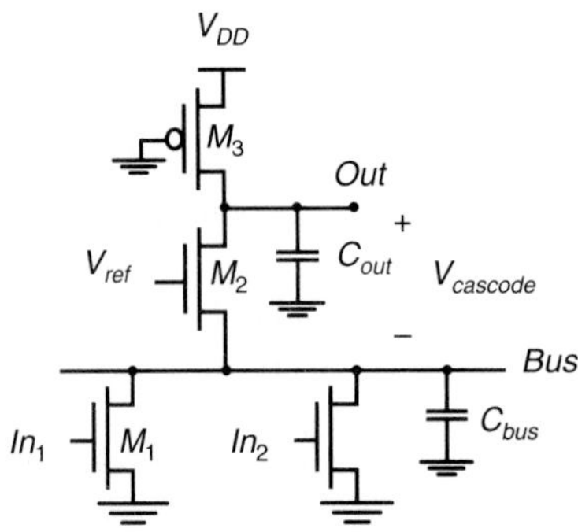

Figure 8.22 Charge-redistribution bus amplifier.

Example 8.6 Cascode Charge-Redistribution Amplifier

Consider the circuit in Figure 8.22. Assume that both M_1 and M_3 are minimum-size devices (1.8/1.2), while M_2 is chosen to be substantially wider (27/1.2). Determine the logic levels for both the *Bus* and the *Out* nodes. Assume the generic 1.2 μm CMOS parameters. The supply voltage is set to 5 V and V_{ref} to 3 V.

[2] A cascode device is a transitor with its gate connected to a fixed bias voltage. It typically serves as an impedance converter.

The high voltage levels are determined to be equal to V_{DD} and $V_{ref} - V_{Tn}(V_{ref})$ for V_{out} and V_{bus} respectively. The latter must be solved iteratively due to the body-effect factor, and equals 2.03 V.

The low voltage levels are obtained by setting $I_{M1} = I_{M2} = I_{M3}$. This produces a very complex set of equations that are hard to solve manually. An approximated solution can be obtained by assuming that M_3 saturates. **Do not forget to check the validity of these assumptions at the end of the analysis.** I_{M3} then evaluates to

$$I_{M3} = \frac{1}{2}\frac{1.8}{(1.2-0.3)}(5.4 \times 10^{-6})(5-0.74))^2 = 98 \times 10^{-6}\text{A}$$

Equating I_{M3} to I_{M1}, assuming that M_1 operates in linear operation, and solving for V_{bus} yields $V_{bus,low} = 0.63$ V. The logic swing on the bus thus equals 1.4 V, which is substantially smaller than the rail-to-rail swing. The voltage drop over the cascode device can be derived by setting $I_{M2} = I_{M3}$. It is safe to ignore the body effect and the second-order term in the current equation—M_2 obviously operates in the linear mode. The resulting voltage drop of 0.1 V over M_2 means $V_{out,low} = 0.63\text{ V} + 0.1\text{ V} = 0.73$ V. Hence, the output node experiences a logic swing that is close to rail-to-rail. Observe that the saturation assumption for M_3 is not really correct, however, the resulting deviation is only minor.

These results are confirmed by SPICE simulations, which predict 2.2 V and 5 V for the high voltages and 0.44 V and 0.5 V for the low voltages for the *Bus* and *Out* nodes respectively.

Problem 8.2 DSL Gate

The cascode principle has been exploited for other purposes, such as reducing the logic swing at the internal nodes of a complex logic gate (Figure 8.23). The DSL (differential split-level logic) is a variant of the DCVSL logic family introduced in Chapter 4.

Determine how the extra cascode devices help by analyzing the logic levels. Under what circumstances is this gate faster than a traditional DCVSL gate? Finally, determine if this gate exhibits static power consumption.

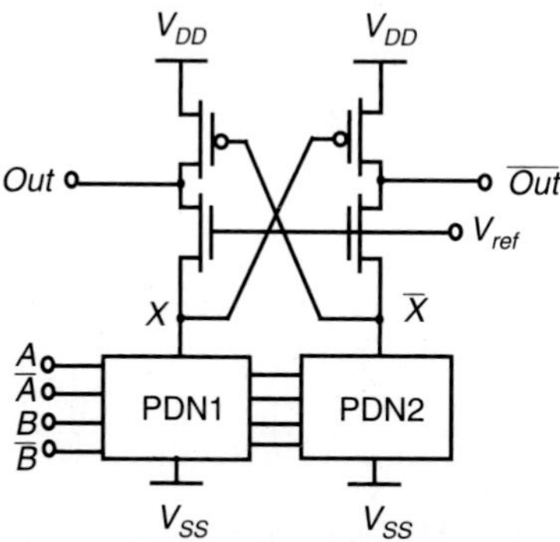

Figure 8.23 Differential split-level logic.

8.2.4 Capacitance and Performance in Bipolar Design

When driving large capacitive loads, bipolar devices are at a clear advantage compared to MOS. Providing large (dis)charging currents for a given driving voltage requires area, and the area penalty is much larger for MOS devices. This difference was stressed in Chapter 2, and is summarized briefly below in Table 8.8. For comparable technologies, the transit time of a charge carrier through the MOS channel τ_T and the bipolar base-transit time τ_F tend to be of comparable values (row 1). For instance, for a 0.8 µm CMOS process and a channel length of 0.8 µm, $\tau_T \approx 8$ psec. For a bipolar process with a base width of 0.07 µm, a τ_F of approximately 6 to 10 psec is obtained. The real difference between the two devices is the available charge per unit channel or emitter area, which is much higher for the bipolar device and can be developed with a far smaller driving voltage (row 2). Therefore, bipolar transistors are at a distinct area advantage over MOS devices when driving large currents is required, and this advantage holds even when scaling to submicron dimensions (row 3) [Solomon82].

Table 8.8 Analysis of driving capabilities of CMOS and bipolar transistors (see Chapter 2.)

	CMOS	**Bipolar**
Transit time	$\tau_T = L/\nu_{sat}$	$\tau_F = W_b^2/2D_b$
Channel or base charge	$Q_I \approx A_G(V_{GS} - V_T)$	$Q_F \approx A_E e^{Vbe/\phi T}$
Current	$I_{DS} = Q_I/\tau_T$	$I_C = Q_F/\tau_F$

Large capacitive loads can thus be driven rather easily in bipolar designs and do not require the special circuit attention and optimization needed in CMOS. This difference is amply illustrated with the example of Figure 8.24, where a 10 pF capacitance is charged using a source-follower and an emitter-follower configuration (for comparable device parameters and with $A_G = A_E$).

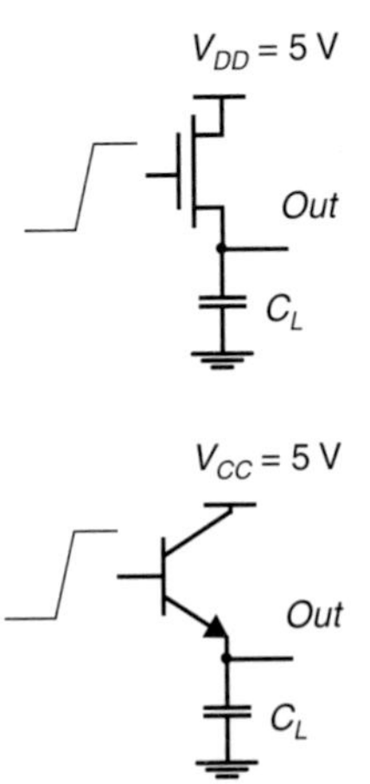

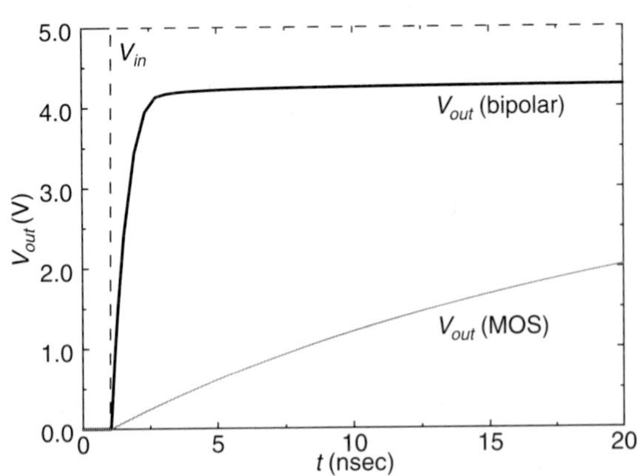

Figure 8.24 Charging a capacitor using an MOS source-follower and a bipolar emitter-follower.

Consequently, bipolar gates are relatively insensitive to capacitive fan-out loads, and transistor sizing is rarely necessary. This is further exemplified by the results shown in Figure 8.25, which plots the propagation delay (t_{pLH}) of an ECL gate as a function of the capacitive load and for different device sizes of the emitter-follower transistors. For smaller capacitance values, the delay is rather independent of the load and is dominated by the intrinsic delay of the gate. The gate with the smaller emitter-follower is somewhat faster due to the smaller intrinsic capacitances. At the other end of the spectrum, the delay is dominated by the charging of the load capacitor through the emitter-follower. For large capacitor values (>1 pF), it helps to increase the size of the emitter-follower since this reduces the parasitic resistances r_c and r_b, which keeps the transistor out of saturation for a longer period of time for large switching currents. However, the difference is minor. Notice that for a load capacitance of 100 pF, the propagation delay is still only 2.4 nsec! A 1000-fold increase in capacitive load increased the delay by only a factor of 24, which demonstrates the low-sensitivity argument.

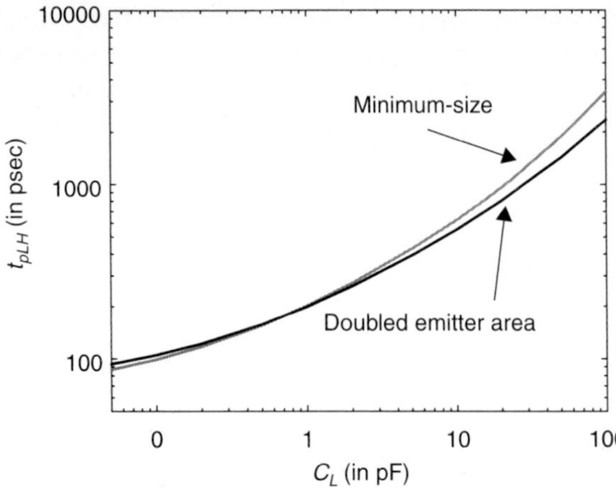

Figure 8.25 Delay of ECL gate as a function of load capacitance.

On the other hand, the *high-to-low transition* of an ECL gate is dominated by the R_BC_L time-constant and increases linearly as a function of C_L. Driving large loads fast requires small values of R_B, which means increased power consumption. A load resistance of 50 Ω discharges a 100 pF capacitive load with a delay of 3.35 nsec, which is similar to the charging time derived above. This comes at the expense of approximately 0.1 mA in standby current. More advanced pull-down structures might therefore be necessary, as discussed in Chapter 5.

The small voltage swing of ECL circuits offers another performance advantage when driving large capacitance values. This is especially valuable when going off-chip, since this is where the largest capacitive loads are present. Some high-performance memories combine MOS memory arrays with ECL input/output drivers for precisely this reason. This approach combines the best of both worlds—high density in the array as a result of the MOS device properties and fast input/output, resulting from the superior driving capabilities of the ECL gates.

Similar observations hold for the BiCMOS gate as well. Its main attraction is the driving capability and the relative insensitivity to capacitive loading. For instance, BiCMOS has been used extensively in gate arrays, where the length of the interconnection wire is unknown beforehand. A careful optimization of the the multi-stage CMOS driver is difficult. The propagation delay of the BiCMOS driver, in contrast, is rather insensitive to variations in capacitive load. For instance, a BiCMOS gate array was used to implement a digital filter circuit with a clock rate of 100 Mhz [Yoshino90]. Achieving such clock rates with a CMOS-only approach requires extensive optimization and transistor sizing. BiCMOS drivers are also used extensively in the design of fast memories, where large fan-in gates are very common.

Design Consideration

In conclusion, when driving large capacitances we must carefully weigh a number of often contradictory factors such as the speed requirements, power constraints, design methodology, and economics. CMOS structures can be optimized to yield relatively fast driver structures at the expense of design time, area, and power consumption. The latter is caused by the internal capacitance of the inverter chain, which almost equals the load capacitance and contributes substantially to the dynamic power consumption. Bipolar and BiCMOS approaches, on the other hand, rely on more expensive technologies that require the designer to deal with level conversion and interfacing issues as well as more complex gate topologies.

Example 8.7 Power Dissipation in CMOS and BiCMOS Drivers

Let us first consider the dynamic power consumption of a cascaded CMOS buffer. When driving a capacitance C_L with an N-stage buffer using a scaling factor u, the total internal capacitance of the buffer equals

$$C_{int} = C_i(1 + u + u^2 + \dots + u^{N-1}) = C_i\left(\frac{u^N - 1}{u - 1}\right) \tag{8.21}$$

This is only marginally smaller than the load capacitance itself, which equals $C_i u^N$. For example, the internal capacitance of the buffer designed in Example 8.4 equals 10.4 pF, compared to the load capacitance of 20 pf. This means that the dynamic power consumed by the switching of the internal capacitances is almost as large as the power needed to switch the load. This is not the case in the BiCMOS buffer, where the internal capacitance is small.

This result was confirmed in [Abo92], where it was demonstrated that the *energy-delay product* needed for driving a 20 pF load in the 1.2 mm CMOS technology was reduced by a factor of 3 when going from the best CMOS to the best BiCMOS design. The large gain is due to the reduced delay of the buffer and the reduced internal capacitance.

8.3 Resistive Parasitics

In this section, we discuss the impact of wiring resistance on reliability and performance. Before doing so, we first analyze the effects of scaling on the wiring resistance and introduce a number of approximative models for resistive wires.

8.3.1 Modeling and Scaling of Interconnect Resistance

The resistance of a uniform slab of conducting material can be expressed as

$$R = \frac{\rho}{H}\frac{L}{W} \tag{8.22}$$

where ρ is the resistivity of the material. Since H is a constant for a given technology, Eq. (8.22) can be rewritten as follows:

$$R = R_{\square}\frac{L}{W} \tag{8.23}$$

with

$$R_{\square} = \frac{\rho}{H} \tag{8.24}$$

the *sheet resistance* of the material, having units of Ω/□. To obtain the resistance of a wire, simply multiply the sheet resistance by the length of the wire and divide by the width. Typical values of the sheet resistance of various interconnect materials are given in Table 8.9.

Table 8.9 Sheet resistance values for a typical 1.0 μm CMOS process.

Material	Sheet Resistance (Ω/□)
n^+ diffusion	10 ± 2
p^+ diffusion	10 ± 2
n-well (under field)	1150 ± 250
polysilicon (H = 0.33 μm)	10 ± 2
metal1, metal2	0.07 ± 0.006

From this table, we conclude that metal is the preferred material for the wiring of long interconnections. Polysilicon should only be used for local interconnect. Although the sheet resistance of the diffusion layer (n^+, p^+) is comparable to that of polysilicon, the use of diffusion wires should be avoided due to its large capacitance and the associated RC delay.

Transitions between routing layers add extra resistance to a wire, called the *contact resistance*. The preferred routing strategy is thus to keep signal wires on a single layer whenever possible and to avoid excess contacts or vias. It is possible to reduce the contact resistance by making the contact holes larger. Unfortunately, current tends to concentrate around the perimeter in a larger contact hole. This effect, called *current crowding*, puts a practical upper limit on the size of the contact. The following contact resistances (for minimum-size contacts) are measured for the 1.0 μm process: 21 Ω for metal1 to n^+, p^+, or polysilicon; 2 Ω for metal2 to metal1.

A full scaling of the technology increases the sheet resistance due to the reduction of the height factor H. Assuming that the width and height of the wire scale with the technology scaling factor S, the following scaling behavior of the interconnect resistance can be derived:

$$S_R = \frac{S_L}{S^2} \tag{8.25}$$

where S_L is the scaling factor for the wire length. The resistance grows linearly with the scaling factor S for *local interconnections,* where $S_L = S$ due to the reduced cross-section of the wire. The situation is worse for *global interconnections*, whose length tends to grow in proportion to the die size. The global-wire resistance grows in an almost cubic fashion, if we assume that the die size (and the length of the global wire) scales as $S_L = 1/S$. The increasing resistance affects the circuit operation and performance in a dual way: *increased propagation delay* as a result of *RC* delay and *resistive* (or *ohmic*) *voltage drop*.

The scaling factor of the *RC* delay can be determined by combining Eq. (8.3) and Eq. (8.25).

$$S_{RC} = \frac{S_L^2}{S^2} \tag{8.26}$$

The *RC* delay of the wire remains constant for local interconnect ($S = S_L$),while the intrinsic gate delay is decreasing. The *RC* delay of the interconnect thus becomes increasingly important for submicron technologies. The situation is far worse for global interconnect, where the time-constants grow in a fourth-order fashion (for $S_L = 1/S$) when scaling. Therefore, either technological or circuit innovations and precautions are necessary to avoid a dramatic performance degradation.

The resistive nature of real interconnect wires also causes a degradation in the signal levels due to the ohmic voltage drop over the wire. The scaling behavior of the voltage drop can be determined by observing that the average current in the network decreases with S for a full scaling where all the voltage levels are scaled with S (Table 3.6 in Chapter 3).

$$\Delta V = R \times I \tag{8.27}$$

such that

$$S_{\Delta V} = \frac{S_L}{S} \tag{8.28}$$

ΔV remains approximately constant for local interconnections ($S = S_L$), but increases substantially for global interconnections such as system busses and power- and clock-distribution networks. The ohmic effects are even more pronounced in the constant-voltage scaling model that is closer to reality.

In the following paragraphs, we discuss how to cope with the above effects at the circuit level. However, trying to cope with these effects at the circuit level alone is hard and expensive in terms of area and/or performance. Substantial efforts are therefore undertaken to solve at least part of the problem at the technology level. These approaches can be summarized in three categories.

1. **Selective Scaling**—While scaling all horizontal dimensions (W and L), the vertical dimension of the wires (H) can be kept approximately constant. This freezes the sheet resistance at the expense of a more complex processing technology, where the vertical process dimensions become comparable to the horizontal ones. Besides placing an extra burden on the process, the selective scaling also has a negative effect on the interconnect capacitance, because fringing capacitance and interwire capacitances gain in importance, as discussed in Section 8.2.

2. **Interconnect Materials with Lower Sheet Resistance**—Examples of such materials are the various *silicides* that are now available in a wide range of processes. A

silicide is a compound formed using silicon and a refractory metal. This creates a highly conductive material that can withstand high-temperature process steps without melting. Examples of silicides are WSi_2, $TiSi_2$, $PtSi_2$, and TaSi. WSi_2, for instance, has a resistivity ρ of 130 μΩ-cm, which is approximately eight times lower than polysilicon. The silicides are most often used in a configuration called a *polycide*, which is a simple layered combination of polysilicon and a silicide. A typical polycide consists of a lower level of polysilicon with an upper coating of silicide and combines the best properties of both materials—good adherence and coverage (from the poly) and high conductance (from the silicide). A MOSFET fabricated with a polycide gate is shown in Figure 8.26. This structure is especially effective in structures such as memories, where the gates of a whole array of transistors must be driven by a single polysilicon wire to keep the cell area to a minimum. The *RC* delay along that wire dominates the performance.

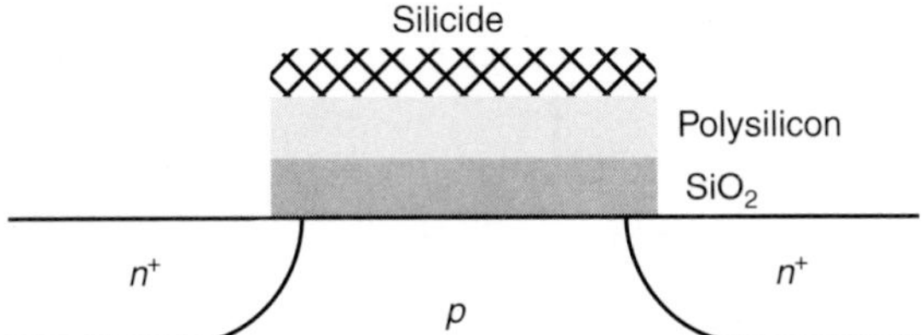

Figure 8.26 A polycide-gate MOSFET.

In high-performance circuits, even the low resistivity of aluminum wires tends to create an intolerable speed degradation. Therefore, materials with a lower resistivity such as gold (Au) are often used as the metal interconnect layer in technologies such as ECL and GaAs. Gold has a resistivity of 2.2 μΩ-cm at room temperature, which is slightly lower than the 2.8 μΩ-cm of Al.

3. **Multiple Interconnect Layers**—Having more interconnect layers tends to reduce the average wire length for two reasons: (1) a straight connection between two points becomes feasible as fewer obstacles are present, and (2) the overall area is reduced because interconnect wires can be placed on top of each other as well as on top of logic components. The local connections can use the lower and denser metallization levels, while the upper layers with wider and thicker lines are used by the global interconnections, such as power and clock nets. Two metal layers are available in almost all state-of-the-art processes at present, while three and more layers are more common.

Interconnect technology is an active research area, and substantial innovation is expected in the years to come. Some promising approaches include high-temperature superconducting wires (discussed in Chapter 5), and optical interconnect [Wada92, Tewksbury94].

8.3.2 Resistance and Reliability—Ohmic Voltage Drop

Current flowing through a resistive wire results in an ohmic voltage drop that degrades the signal levels. This is especially important in the power distribution network, where current levels can easily reach 100 mA, if not amperes. Consider a 15 mm long V_{DD} or *GND* wire with a current of 1 mA per μm width. This current is about the maximum that can be sustained by an aluminum wire due to *electromigration*, which is discussed in the subsequent section. Assuming a sheet resistance of 0.07 Ω/□, the resistance of this wire (per μm width) equals 1 kΩ. A current of 1 mA/μm results in a voltage drop of 1 V. The altered value of the voltage supply reduces noise margins and changes the logic levels as a function of the distance from the supply terminals. This is demonstrated by the circuit in Figure 8.27, where an inverter placed far from the power and ground pins connects to a device closer to the supply. The difference in logic levels caused by the *IR* voltage drop over the supply rails might partially turn on transistor M_1. This can result in an accidental discharging of the precharged, dynamic node *X*, or cause static power consumption if the connecting gate is static.

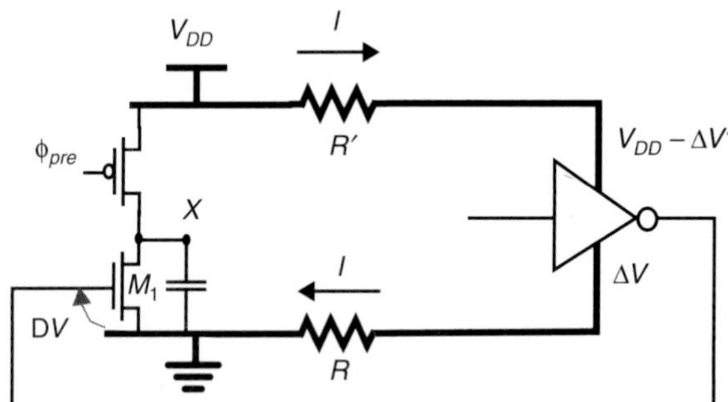

Figure 8.27 Ohmic voltage drop on the supply rails reduces the noise margins.

The most obvious solution to this problem is to reduce the maximum distance between the supply pins and the circuit supply connections. This is achieved by using a *finger-shaped* power distribution network, as shown in Figure 8.28a. This network reduces the maximal wire length between source and destination. It also avoids multiple cross-over between V_{DD} and *GND*. The latter should be avoided as much as possible, since contacts and vias provide extra resistance and are reliability hazards. When a contact between different layers is necessary, make sure to maximize the contact perimeter (not the area), since the current tends to concentrate around the perimeter (*current crowding*). This can be achieved by providing multiple smaller contacts.

Another solution is to provide multiple power and ground pins and use the superior board-level metallization, implemented in low-resistance copper, to connect the different power and ground networks (Figure 8.28b).

Designing and laying out the power distribution network must be done in the early phases of the design process before beginning the actual design of the logic modules. It is an essential part of the *floorplanning* process that determines the overall layout of a complex chip based on early estimates of sizes and current requirements of the individual modules. The distribution network is then gradually refined during the circuit and physical design of the chip. Note that the design of the power network requires knowledge of *peak currents (or power),* in contrast to the analysis and minimization of power dissipation, where the average current is of more importance.

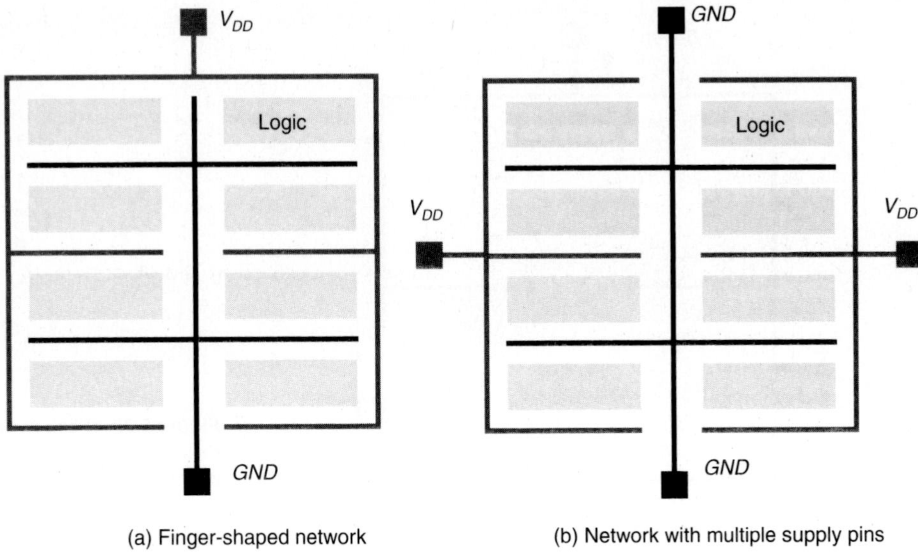

(a) Finger-shaped network (b) Network with multiple supply pins

Figure 8.28 Power-distribution networks.

Example 8.8 Sizing of the Power Distribution Network

Assume that each module of the circuit of Figure 8.28a draws a peak supply current of 100 mA. The die size measures 1 cm. The power network is designed in metal1 and has been appropriately sized so that the peak current in the wire does not exceed 1 mA/μm as shown by the wire dimensions in Figure 8.29. In the worst case, this could result in a voltage drop of 1.4 V at the point the farthest away from the supply terminal, assuming that the resistance of the metal wire equals 0.07 Ω/□. This is unacceptable, and a resizing of the wires seems appropriate. Fortunately, the peak currents of the different modules are generally staggered in time, resulting in less stringent requirements.

Another option besides resizing the wire width is to provide more supply terminals. For instance, the worst-case voltage drop for the configuration of Figure 8.28b under similar conditions equals 0.7 V.

8.3.3 Electromigration

The current density (current per unit area) in a metal wire is limited due to an effect called *electromigration*. A *direct* current in a metal wire running over a substantial time period, causes a transport of the metal ions. Eventually, this causes the wire to break or to short-circuit to another wire. This type of failure will only occur after the device has been in use for some time. Some examples of failure caused by migration are shown in Figure 8.30. Notice how the first photo clearly shows hillock formation in the direction of the electron current flow.

The rate of the electromigration depends upon the temperature, the crystal structure, and the current density. The latter is the only factor that can be effectively controlled by

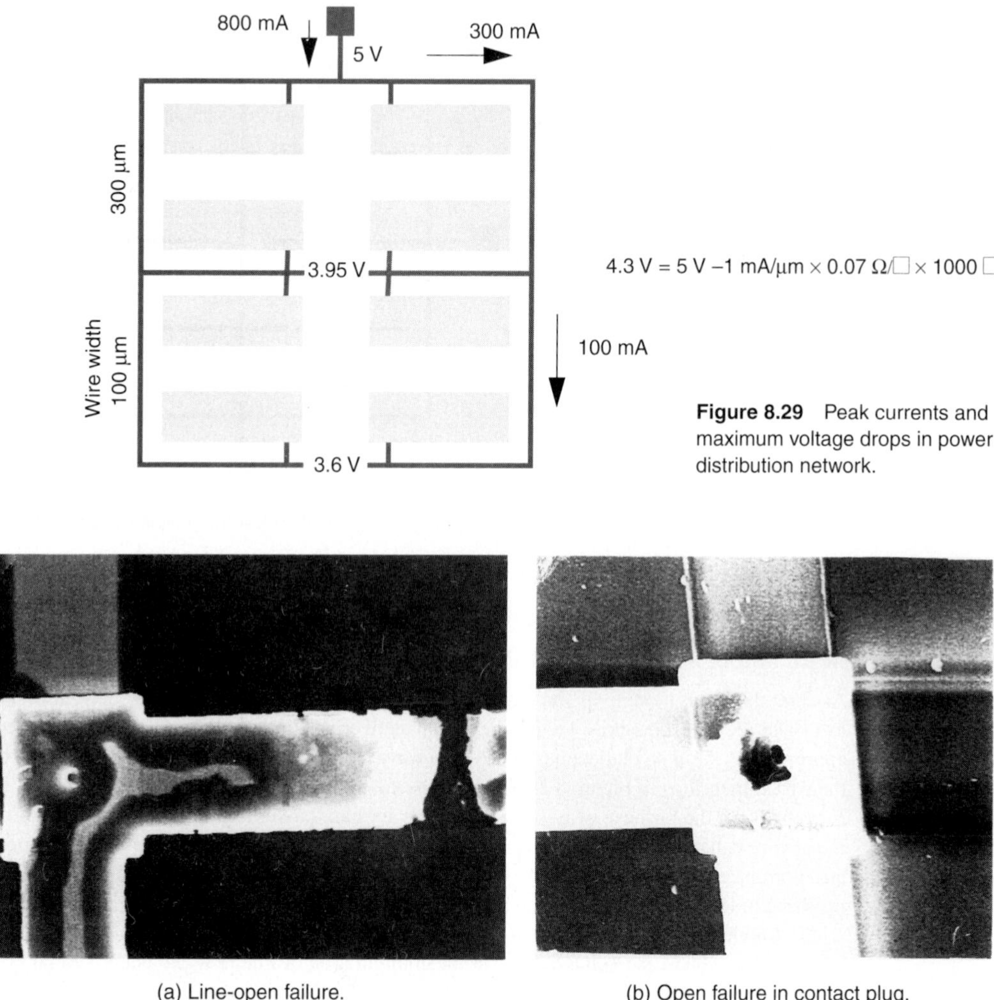

Figure 8.29 Peak currents and maximum voltage drops in power distribution network.

(a) Line-open failure.

(b) Open failure in contact plug.

Figure 8.30 Electromigration-related failure modes (*Courtesy of N. Cheung and A. Tao, U.C. Berkeley*).

the circuit designer. Keeping the current below 0.5 to 1 mA/µm normally prevents migration. This parameter can be used to determine the minimal wire width of the power and ground network. Signal wires normally carry an ac-current and are less susceptible to migration. The bidirectional flow of the electrons tends to anneal any damage done to the crystal structure. Most companies impose a number of strict wire-sizing guidelines on their designers, based on measurements and past experience. Recent research results have shown that many of these rules tend to be overly conservative [Tao94].

At the technology level, a number of precautions can be taken to reduce the migration risk. One option is to add alloying elements (such as Cu or Tu) to the aluminum to prevent the movement of the Al ions. Another approach is to control the granularity of the ions.

8.3.4 Resistance and Performance—RC Delay

Modeling *RC* Delay

Several options exist when modeling the delay behavior of a resistive wire. A first approach lumps the total resistance into one single R and similarly combines the global capacitance into a single capacitor C. This simple model, called the *lumped RC model* (Figure 8.31a), is pessimistic and inaccurate for long interconnect wires. For those wires, a distributed rc model (Figure 8.31b) is more appropriate. L represents the total length of the wire, while r and c stand for the resistance and capacitance per unit length. A symbolic representation of the distributed rc line is given in Figure 8.31c.

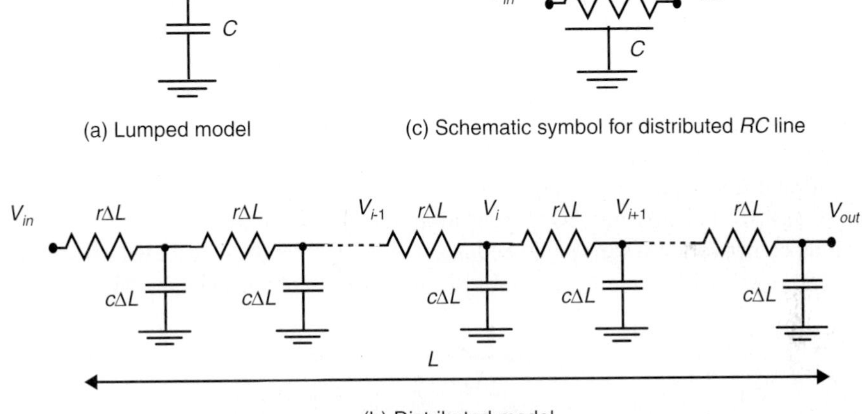

Figure 8.31 *RC* delay of interconnect wire.

The voltage in node i of this network can be determined by solving the following differential network equation:

$$c\Delta L\frac{\partial V_i}{\partial t} = \frac{(V_{i+1} - V_i) + (V_i - V_{i-1})}{r\Delta L} \tag{8.29}$$

The correct behavior of the distributed rc line is then obtained by reducing ΔL asymptotically to 0. For $\Delta L \to 0$, Eq. (8.29) becomes the well-known *diffusion equation*:

$$rc\frac{\partial V}{\partial t} = \frac{\partial^2 V}{\partial x^2} \tag{8.30}$$

where V is the voltage at a particular point in the wire, and x is the distance between this point and the signal source. No closed-form solution exists for this equation. It is common practice to use approximative solutions [Bakoglu90].

The dominant time constant at the output node of the discrete network of Figure 8.31b can be approximated by Eq. (8.31). A technique to generate that expression is presented in a later section.

$$\tau(V_{out}) = rc(\Delta L)^2\left(\frac{N(N+1)}{2}\right) \tag{8.31}$$

where $N = L/\Delta L$ represents the number of sections in the approximation. For $N \to \infty$, this equation reduces to Eq. (8.32).

$$\tau(V_{out}) = \frac{rc\ L^2}{2} \tag{8.32}$$

The time-constant grows quadratically with the length of the wire. Doubling the length of a wire increases its delay by a factor of four! The signal delay of long wires therefore tends to be dominated by the *RC* effect. Notice that the lumped model would have predicted a time-constant of $RC = rc\ L^2$, which is twice as large as the actual value. This demonstrates that the lumped model is inappropriate for modeling long *RC*-lines.

This is further illustrated in Figure 8.32, where the simulated step responses of the distributed and lumped *RC*-network models are compared. Lumping the resistance and the capacitance of a long interconnection wire results in a pessimistic estimation of the propagation delay. Some of the important reference points in the step response are tabulated in Table 8.10. For instance, the propagation delay (defined at 50% of the final value) of the lumped network not surprisingly equals 0.69 *RC*, while the distributed network has a delay of 0.38 *RC*, with *R* and *C* the total resistance and capacitance of the wire.

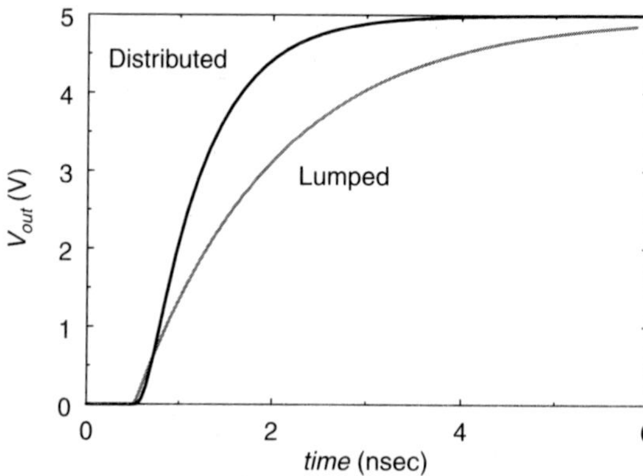

Figure 8.32 Simulated step response of lumped and distributed *RC* wires.

Table 8.10 Step response of lumped and distributed *RC* networks—points of Interest.

Voltage range	Lumped *RC* network	Distributed *RC* network
$0 \to 50\%$ (t_p)	0.69 *RC*	0.38 *RC*
$0 \to 63\%$ (τ)	*RC*	0.5 *RC*
$10\% \to 90\%$ (t_r)	2.2 *RC*	0.9 *RC*

When simulating long wires using a circuit simulator such as SPICE, an accurate model of the wire is essential. Some circuit simulators support a distributed *RC* model as a primitive element. HSPICE, for instance, uses a generalized resistor model called *wire*, where the capacitance to ground can be defined.

Such a model is regrettably not available in most versions of SPICE. Fortunately, a simple yet accurate model can be constructed by approximating the distributed *RC* by a lumped *RC* network. Figure 8.33 shows some of these approximations ordered along increasing precision and complexity. The accuracy of the model is determined by the number of stages. For instance, the error of the π3 model is less than 3%, which is generally sufficient.

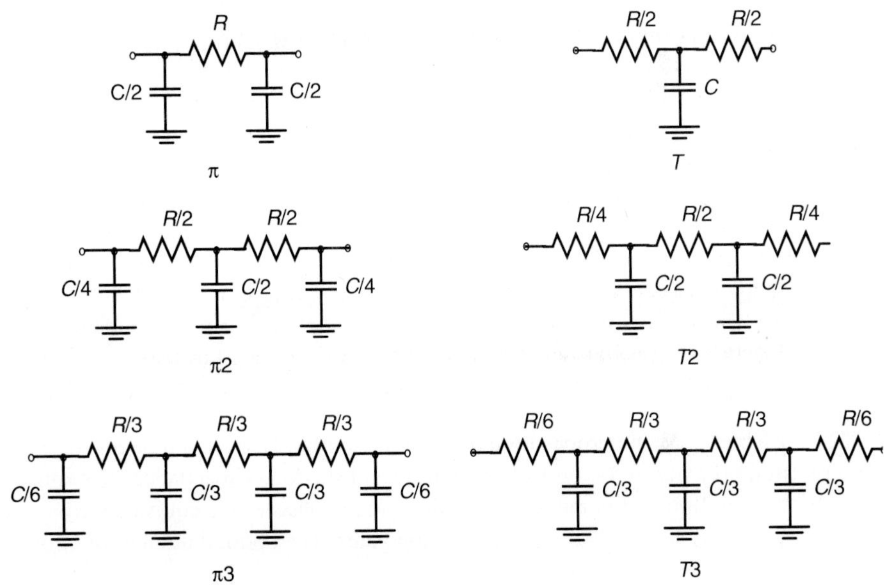

Figure 8.33 Simulation models for distributed *RC* line.

Example 8.9 *RC* delay of Polysilicon Wire

Consider the propagation delay of a polysilicon wire with a width of 1 μm. From the data of Table 8.1 and Table 8.9, we can compute the propagation delay of a 1 mm wire.

$$t_p = 0.38\,(1000)^2\,(10)(0.058 + 0.043\,{*}2) = 0.54 \text{ nsec}$$

This approximately equals two to three times the basic gate delay. The total resistance and capacitance of the wire are 10 kΩ and 0.15 pF respectively. Making the wire five times longer increases the delay by a factor of 25 to 13.6 nsec, which is unacceptable.

As a base for comparison, a similar wire implemented in the n^+ diffusion layer would yield a delay of 1.2 nsec for 1 mm and 29.3 nsec for 5 mm length. Be aware that the capacitance of a diffusion wire is nonlinear and a function of the applied voltage.

How to Reduce *RC* Delays

A first option for reducing *RC* delays is to use better interconnect materials when they are available and appropriate. For instance, a 5 mm metal1 wire has a delay of only 0.08 nsec, which can be ignored. Long polysilicon wires are sometimes hard to avoid though. Examples are address lines in memories. Keeping them in polysilicon contributes substantially to the memory density. In those circumstances, it is possible to reduce the propagation delay by introducing intermediate buffers, also called *repeaters*, in the interconnect line (Figure 8.34). Making an interconnect line *M* times shorter reduces its propagation delay quadratically, which is sufficient to offset the extra delay of the repeaters when the wire is sufficiently long.

$$t_p = 0.38rc\left(\frac{L}{M}\right)^2 M + (M-1)t_{pbuf} \tag{8.33}$$

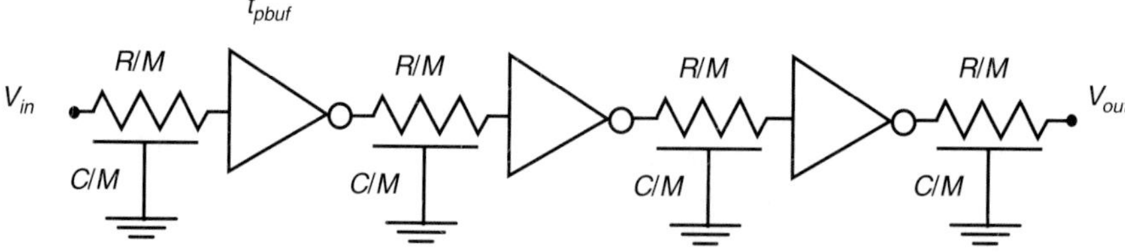

Figure 8.34 Reducing *RC* interconnect delay by introducing repeaters.

where t_{pbuf} is the propagation delay of the buffer. One should be aware that t_{pbuf} is a function not only of the buffer size, but also of the load capacitance. For simplicity, the latter factor is ignored in the following derivation. Observe the equivalence between this design problem and the chain of transmission gates. The optimal number of buffers can be found by setting $\partial t_p/\partial M = 0$.

$$M = L\sqrt{\frac{0.38rc}{t_{pbuf}}} \tag{8.34}$$

Example 8.10 Optimized Delay of *RC* Chain

Eq. (8.34) indicates that 5 sections would be optimal for the 5 mm long polysilicon wire, assuming a t_{pbuf} of 0.5 nsec. This results in an overall delay time of 4.74 nsec, an important improvement compared to the 13.6 nsec for the unbuffered line.

Delay of Complex RC Networks

Until now, we have discussed the *RC* delay of a single wire. Very often wires have multiple branches, and the delay analysis of such an *RC* network becomes far more involved. Fortunately, a number of useful theorems have been developed that can help estimate the delay of complex *RC* networks. A short discussion of these theorems is appropriate.

1. The Elmore Delay

Before considering a branched RC network, let us first analyze the simple, non-branched RC chain shown in Figure 8.35. Assume that all capacitances are initially pre-charged to V_{DD}. The objective is to find a simple close-form formula that gives the time it takes for the voltage at a node i to drop from V_{DD} to 0.5 V_{DD} after a zero-going step is applied at the source. This is equivalent to deriving the equivalent first-order time-constant (also called the first moment of the impulse response). Elmore [Elmore48] showed that the following expression is valid for a cascaded N-stage RC chain:

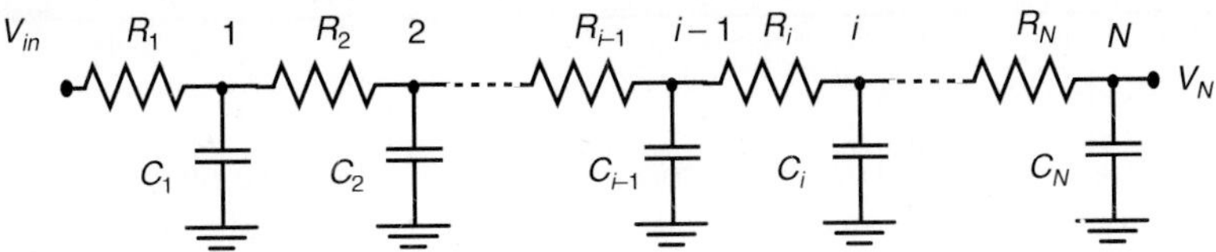

Figure 8.35 *RC* chain.

$$\tau_N = \sum_{i=1}^{N} R_i \sum_{j=i}^{N} C_j = \sum_{i=1}^{N} C_i \sum_{j=1}^{i} R_j \tag{8.35}$$

This is equivalent to stating that the first-order time-constant τ_i at a node i is found as the sum of a number of components. This can be derived by determining the path between i and the source node. Since this is a chain, there exists only one such path. Each node j on the path contributes a fraction of τ. This fraction is determined as the product of the capacitance (to ground) at node j and the total resistance between j and the source.

As an example, consider node 2 in the RC chain of Figure 8.35. Its time-constant consists of two components contributed by nodes 1 and 2. The component of node 1 consists of C_1R_1 with R_1 the total resistance between the node and the source, while the contribution of node 2 equals $C_2(R_1 + R_2)$. The equivalent time constant at node 2 equals $C_1R_1 + C_2(R_1 + R_2)$. τ_i of node i can be derived in a similar way.

$$\tau_i = C_1R_1 + C_2(R_1 + R_2) + \ldots + C_i(R_1 + R_2 + \ldots + R_i)$$

Example 8.11 Time-Constant of Distributed *RC* Line

The Elmore formula can be used to derive the time-constant of a distributed RC line as shown in Figure 8.31b. This differs from the circuit of Figure 8.35 in that all of the resistors and capacitors are identical. For an N-stage network, Eq. (8.35) evaluates to the following expression:

$$\tau_N = \Delta L^2(cr + c(2r) + \ldots + c(Nr)) = \Delta L^2 cr(1 + 2 + \ldots + N) = \Delta L^2 cr\, N(N+1)/2$$

This is identical to Eq. (8.31), which was provided without proof or derivation.

The Elmore delay formula has proven to be extremely useful. Besides making it possible to approximate simple RC chains, the formula can also be used to approximate the propagation delay of complex transmission-gate networks and pull-down (up) transistor networks. In the switch model, transistors are replaced by their equivalent, linearized on-resistance. The evaluation of the propagation delay is then reduced to the analysis of the resulting *RC* network.

Unfortunately, most of those networks cannot be modeled as *RC* chains and the Elmore expression cannot be used. An extension is offered by the Penfield-Rubinstein-Horowitz theorem.

WARNING: Be aware that an *RC*-chain is characterized by a number of time-constants. The Elmore expression determines the volume of only the dominant one, and presents thus a first-order approximation.

2. Analyzing *RC* Trees

Given a resistor network in the form of a tree, in which the resistor at the root r of the tree is connected to ground by a voltage source that switches from 1 V to 0 V at $t = 0$ (Figure 8.36). Every node k of the network ($k \neq s$) has a capacitance to ground C_k and is set to an initial voltage $V_k(0)$. Besides nodes s and ground, the network contains N nodes.

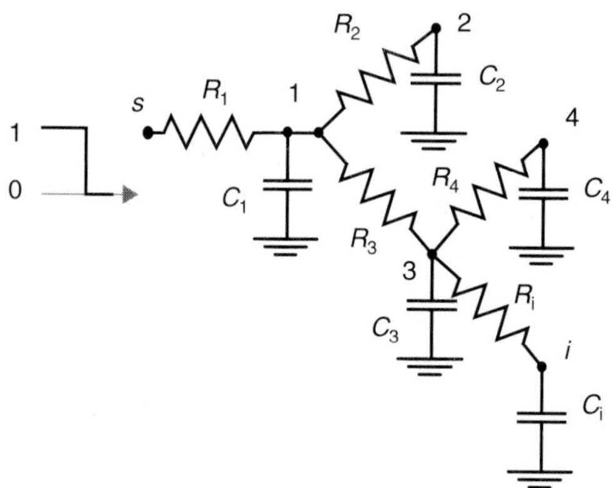

Figure 8.36 Tree-structured *RC* network.

Under those conditions, the following always holds for node i ($i \neq s$):

$$\int_0^{\infty} v_i(t)dt = \sum_{k=1}^{N} C_k V_k(0) R_{i,k} \tag{8.36}$$

with

$$R_{i,k} = \sum R_j \Rightarrow (R_j \in [path(i \to s) \cap path(k \to s)])$$

$R_{i,k}$ is the resistance shared among the paths from the root node s to nodes k and i. For the circuit in Figure 8.36, $R_{i4} = R_1 + R_3$ while $R_{i2} = R_1$. The integral of Eq. (8.36) evaluates to τ_i, when one assumes that the waveform at node i can be approximated by e^{-t/τ_i}, or that a single dominant time-constant exists. This theorem covers the Elmore formula as a special case.

The above results are part of a publication by Rubinstein, Penfield, and Horowitz that establishes precise minimum and maximum bounds on the voltage waveforms in an *RC* tree [Rubinstein83]. These bounds provide a convenient tool for the manual analysis of complex RC or transistor-capacitance networks. They have also found ample application in computer-aided timing analyzers at the switch and functional level [Horowitz83].

Example 8.12 *RC* Delay of a Tree-Structured Network

Using Eq. (8.36), a first-order time-constant can be computed for node i in the network of Figure 8.36.

$$\tau_i = R_1C_1 + R_1C_2 + (R_1 + R_3)C_3 + (R_1 + R_3)C_4 + (R_1 + R_3 + R_i)C_i$$

8.4 Inductive Parasitics

Besides having a parasitic resistance and capacitance, interconnect wires also exhibit an inductive parasitic. An important source of parasitic inductance is introduced by the bonding wires and chip packages. Even for intermediate-speed CMOS designs, the current through the input-output connections can experience fast transitions that cause voltage drops as well as ringing and overshooting, phenomena not found in RC circuits. At higher switching speeds, wave propagation and transmission line effects can come into the picture. Both effects are analyzed in this section. First, we discuss the source of these inductive parasitics and their quantitative values.

8.4.1 Sources of Parasitic Inductances

The inductive model of a wire depends upon the type of wire and its environment. Some of the environments encountered in the design of electrical systems are shown in Figure 8.37. Coaxial cables and triplate strip line are mostly found in printed-circuit (pc) boards or interconnections between pc boards. Interconnect lines on a chip normally act as microstrip lines, while the chip bonding wires can be modeled as wires above the ground plane. The inductance of a bonding wire is typically about 1 nH/mm, while the inductance per package pin ranges between 7 and 40 nH per pins depending on the type of package as well as the positioning of the pin on the package boundary [Steidel83]. On the other hand, an on-chip wire typically has an inductance of about 1–2 nH/mm. Typical values of the parasitic inductances and capacitances for a number of commonly used packages are summarized in Table 8.11. When considering CMOS design, we are concerned only with the inductance of the package pins. For faster switching technologies such as ECL, GaAs, and superconducting technologies, inductance of wires on the board and chip come into play.

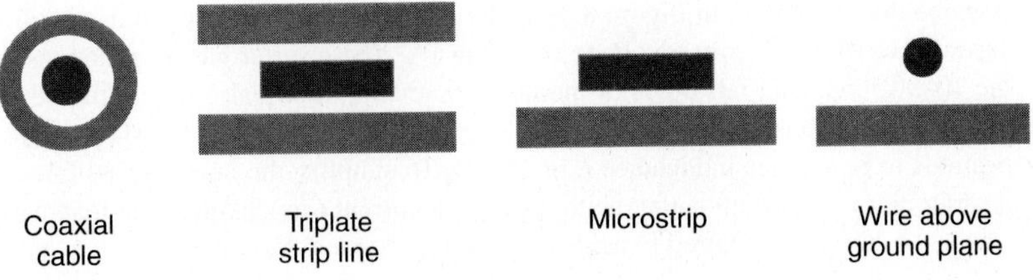

Figure 8.37 Cross-sections of typical wire types and environments found in the design of electronic

Table 8.11 Typical capacitance and inductance values of package and bonding styles (from [Steidel83] and [Franzon93]).

Package Type	Capacitance (pF)	Inductance (nH)
68-pin plastic DIP	4	35
68-pin ceramic DIP	7	20
256-pin grid array	1–5	2–15
Wire bond	0.5–1	1–2
Solder bump	0.1–0.5	0.01–0.1

8.4.2 Inductance and Reliability—$L\frac{di}{dt}$ Voltage Drop

Inductors resist a change in the current flowing through them by raising a voltage drop over the inductor (Figure 8.38). The value of the voltage drop equals

$$v_L(t) = L\frac{di_L}{dt} \tag{8.37}$$

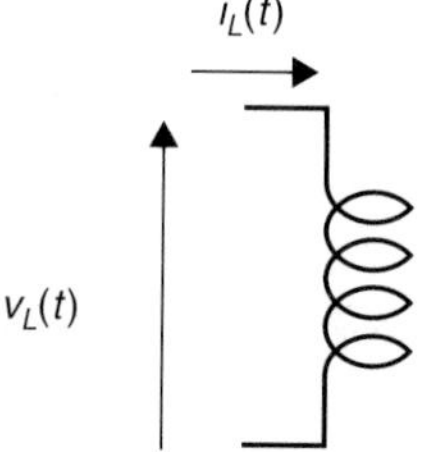

Figure 8.38 Voltage over inductor as a function of change in current.

During each switching action, a transient current is sourced from (or sunk into) the supply rails to charge (or discharge) the circuit capacitances, as modeled in Figure 8.39. Both V_{DD} and V_{SS} connections are routed to the external supplies through bonding wires and package pins and possess a nonignorable series inductance. Hence, a change in the transient current creates a voltage difference between the external and internal supply voltages. This situation is especially severe at the output pads, where the driving of the large external capacitances generates large current surges. The deviations on the internal supply voltages affect the logic levels and result in reduced noise margins.

Example 8.13 Noise Induced by Inductive Bonding Wires and Package Pins

Assume that the circuit in Figure 8.39 is the last stage of an output pad driver, driving a load capacitance of 20 pF over a voltage swing of 5 V. The inverter has been dimensioned so that the 10–90% rise and fall times of the output signal (t_r, t_f) equal 4 nsec. Since the power and ground connections are connected to the external supplies through the supply pins, both connections have a series inductance L of 10 nH. To simplify the analysis, assume also that the inverter acts as a current source with a constant current (dis)charging the load capacitance. A current of 20 mA is required to achieve the 4 nsec rise and fall times.

$$I_{avg} = (20 \text{ pF} \times (0.9 - 0.1) \times 5 \text{ V}) / 4 \text{ nsec} = 20 \text{ mA}$$

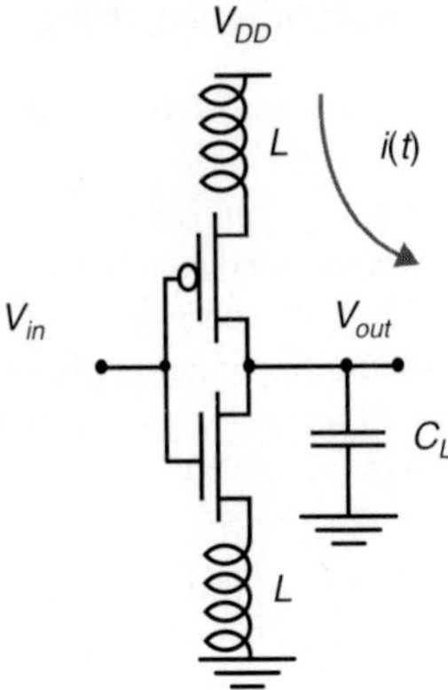

Figure 8.39 Inductive coupling between external and internal supply voltages.

The left side of Figure 8.40 plots the evolution of output voltage, inductor current, and inductor voltage over time. The abrupt, discontinuous current change causes sharp voltage spikes over the inductor. In reality, the charging current is not a constant. A better approximation assumes that the current rises linearly to a maximum after which it drops back to zero, again in a linear fashion. This is shown by the dotted lines in Figure 8.40. Under this model, the voltage drop over the inductors equals

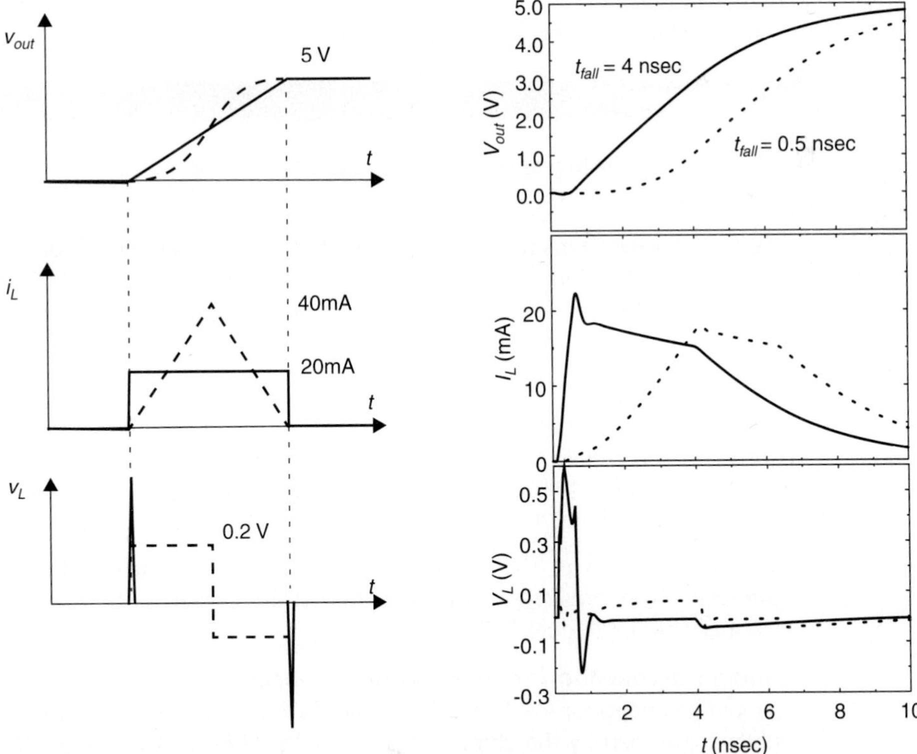

Figure 8.40 Signal waveforms for output driver connected to bonding pads. The results of an actual simulation are shown on the right side.

$$v_L = L\, di_L/dt = (10\text{ nH} \times 40\text{ mA}) / 2\text{ nsec} = 0.2\text{ V}.$$

The peak current of 40 mA results from the fact that the total delivered charge—or the integrated area under I_L—is fixed and equals 20 pF × 5V.

An actual simulation is shown on the right side of Figure 8.40. The observed voltage drop over the inductor and the shape of the voltage waveform depend upon the fall time of the input signal. For a steep input signal (t_f = 0.5 nsec), the voltage spike on the supply equals 0.6 V, and the waveform resembles the first scenario. The noise on the supply reduces to 0.1 V for the input with the slower fall time at the expense of an increased propagation delay. Its waveform is much closer to the second case.

In an actual circuit, a single supply pin serves a large number of gates or output drivers. A simultaneous switching of those drivers causes even worse current transients and voltage drops. As a result, the internal supply voltages deviate in a substantial way from the external ones. For instance, the simultaneous switching of the 16 output drivers of an output bus would cause a voltage drop of at least 1.5 V if the supply connections of the buffers were connected to the same pin on the package.

Improvements in packaging technologies are leading to ever-increasing numbers of pins per package. Packages with more than 350 pins are currently available. Simultaneous switching of a substantial number of those pins results in huge spikes on the supply rails that are bound to disturb the operation of the internal circuits as well as other external components connected to the same supplies.

Design Techniques

Fortunately, the following approaches are available to the designer to address the $L(di/dt)$ problem.

1. **Separate power pins for I/O pads and chip core**—Since the I/O drivers require the largest switching currents, they also cause the largest current changes. It is wise to isolate the center of the chip, where most of the logic action occurs, from the drivers by providing different power and ground pins.

2. **Multiple power and ground pins**—In order to reduce the *di/dt* per supply pin, we can restrict the number of I/O drivers connected to a single supply pin. Typical numbers are five to ten drivers per supply pin. Be aware that this number depends heavily upon the switching characteristics of the drivers such as the number of simultaneously switching gates and the rise and fall times.

3. **Careful selection of the positions of the power and ground pins on the package**—The inductance of pins located at the corners of the package is substantially higher (Figure 8.41).

4. **Adding decoupling capacitances on the board**—These capacitances, which should be added for every supply pin, act as local supplies and stabilize the supply voltage seen by the chip. They separate the bonding-wire inductance from the inductance of the board interconnect (Figure 8.42). The bypass capacitor, combined with the inductance, actually acts as a low-pass network that filters away the high-frequency components of the transient voltage spikes on the sup-

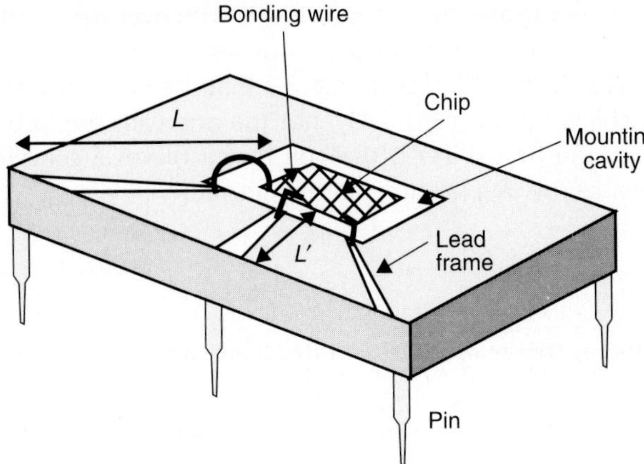

Figure 8.41 The inductance of a bonding-wire/pin combination depends upon the pin position.

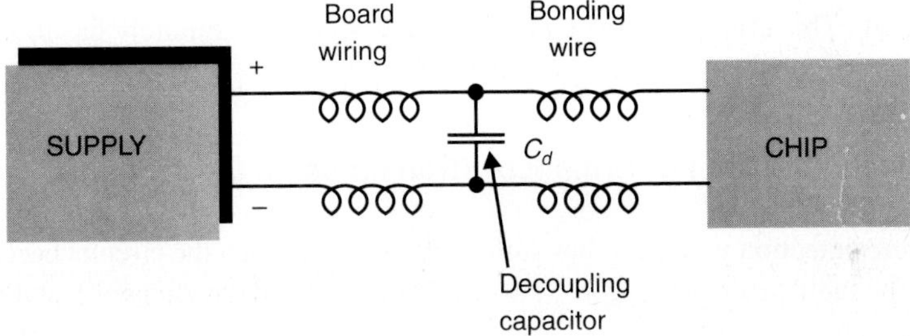

Figure 8.42 Decoupling capacitors isolate the board inductance from the bonding wire and pin inductance.

ply lines.[3] In high-performance circuits with high switching speeds and steep signal transitions, it is becoming common practice to integrate decoupling capacitances on the chip, which ensures cleaner supply voltages. For example, a total of 128 nF (!) of decoupling capacitance is integrated on the DEC Alpha microprocessor chip [Dopperpuhl92]. This capacitance is implemented using gate oxide and distributed all over the chip, especially under the data busses.

5. **Increase the rise and fall times** of the off-chip signals to the maximum extent allowable.

6. **Use advanced packaging technologies** such as surface-mount or hybrids that come with a substantially reduced capacitance and inductance per pin. For instance, we can see from Table 8.11 that the bonding inductance of a chip mounted in flip-chip style on a substrate using the solder-bump techniques is reduced to 0.1nH, which is 50 to 100 times smaller than for standard packages. A short discussion of the different packaging techniques and their properties follows at the end of this chapter.

[3] A word of caution. The combination of inductor-capacitance might act as an underdamped resonator, making the noise problem actually worse. A careful insertion of damping resistance is often required.

The above analysis demonstrates that output drivers with over-designed rise and fall times are not only expensive in terms of area, but also might affect the circuit operation and reliability. We concluded in Section 8.2.3 that the better driver in terms of area is the one that achieves a specified delay, not the one with the fastest delay. When noise is considered, the best driver is the one that achieves a specified delay with the maximum allowable rise and fall times at the output.

Problem 8.3 Design of Output Driver with Reduced Rise/Fall Times

Given a cascaded output driver designed for a given delay that produces excessive noise on the supply lines, define the best approach to address the noise problem: (a) Scale down all stages of the buffer; (b) Scale down the last stage only; (c) Scale down all stages except the last one.

Finally, be aware that the mutual inductance between neighboring wires also introduces cross talk. This effect is not yet a concern in CMOS but definitely has to be considered for higher switching speeds ([Johnson93]).

8.4.3 Inductance and Performance—Transmission Line Effects

When an interconnection wire becomes sufficiently long or when the circuits become sufficiently fast, the inductance of the wire starts to dominate the delay behavior, and transmission line effects must be considered. This is more precisely the case when the rise and fall times of the signal become comparable to the time of flight of the signal waveform across the line as determined by the speed of light. Until recently, this was only the case for the fastest digital designs implemented in the GaAs or ECL technologies. As advancing technology increases line lenghts and switching speeds, this situation will become more common. Transmission line effects are soon to be considered in the fastest CMOS designs.

In this section, we first analyze the transmission line model. Next, we apply it to the current semiconductor technology and determine when those effects should be actively considered in the design process. Techniques to minimize the impact of the transmission line behavior are discussed.

Transmission Line Model

Similar to the resistance and capacitance of an interconnect line, the inductance is distributed over the wire. A distributed *RLC* model of a wire, known as the transmission line model, becomes the most accurate approximation of the actual behavior. The transmission line has the prime property that a signal propagates over the interconnection medium as a *wave*. This is in contrast to the distributed *RC* model, where the signal *diffuses* from the source to the destination governed by the diffusion equation, Eq. (8.30). In the wave mode, a signal propagates by alternatively transferring energy from the electric to the magnetic fields, or equivalently from the capacitive to the inductive modes.

Consider the point x along the transmission line of Figure 8.43 at time t. The following set of equations holds:

$$\frac{\partial v}{\partial x} = -ri - l\frac{\partial i}{\partial t}$$
$$\frac{\partial i}{\partial x} = -gv - c\frac{\partial v}{\partial t} \tag{8.38}$$

Assuming that the leakage conductance g equals 0, which is true for most insulating materials, and eliminating the current i yields the *wave propagation equation*, Eq. (8.39).

$$\frac{\partial^2 v}{\partial x^2} = rc\frac{\partial v}{\partial t} + lc\frac{\partial^2 v}{\partial t^2} \tag{8.39}$$

where r, c, and l are the resistance, capacitance, and inductance per unit length, respectively.

At present, the longer wire lengths cause most of the transmission line problems to occur at the printed-circuit board level. Not many problems are found at the integrated-circuit level yet. Due to the high conductivity of the interconnect material used at the board level (copper), the resistance of the transmission line can be ignored, and a simplified capacitive/inductive model, called the *lossless transmission line*, is appropriate. The lossless model is assumed in most of the subsequent discussion. The lossy model is only discussed briefly at the end. For the lossless line, Eq. (8.39) simplifies to the *ideal wave* equation:

$$\frac{\partial^2 v}{\partial x^2} = lc\frac{\partial^2 v}{\partial t^2} = \frac{1}{\nu^2}\frac{\partial^2 v}{\partial t^2} \tag{8.40}$$

A step input applied to a lossless transmission line propagates along the line with a speed ν, given by Eq. (8.41).

$$\nu = \frac{1}{\sqrt{lc}} \tag{8.41}$$

Even though the values of both l and c depend on the geometric shape of the wire, their product is a constant and is only a function of the surrounding media. An alternative formulation of ν can be derived from Maxwell's laws.

$$\nu = \frac{1}{\sqrt{\varepsilon\mu}} = \frac{c_0}{\sqrt{\varepsilon_r\mu_r}} \tag{8.42}$$

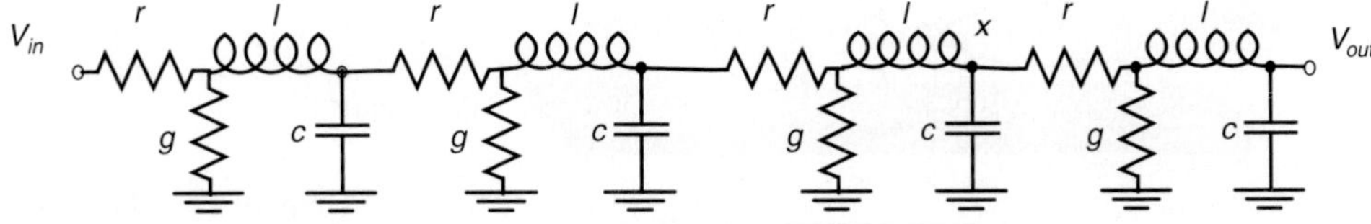

Figure 8.43 Lossy transmission line.

where ε and μ are respectively the dielectric constant and magnetic permeability of the surrounding medium. ε_r and μ_r are defined as the relative values of both dielectric constant and permeability of the medium with respect to vacuum and c_0 equals the speed of light (30 cm/nsec) in a vacuum. The propagation speeds for a number of materials used in the fabrication of electronic circuits are tabulated in Table 8.12. The propagation speed for

Table 8.12 Dielectric constants and wave-propagation speeds for various materials used in electronic circuits (from [Bakoglu90]).

Dielectric	ε_r	Propagation speed (cm/nsec)
Vacuum	1	30
SiO_2	3.9	15
PC board (epoxy glass)	5.0	13
Alumina (ceramic package)	9.5	10

SiO_2 is two times slower than in a vacuum, which means that transmission line effects are more pronounced in the former medium. The propagation delay per unit wire length (t_p) of a transmission line is the inverse of the speed:

$$t_p = \sqrt{lc} \tag{8.43}$$

Let us now analyze how a wave propagates along a lossless transmission line. Suppose that a voltage step V has been applied at the input and has propagated to point x of the line (Figure 8.44). All currents are equal to 0 at the right side of x, while the voltage over the line equals V at the left side. An additional capacitance cdx must be charged for the wave to propagate over an additional distance dx. This requires the following current:

$$I = \frac{dQ}{dt} = c\frac{dx}{dt}V = c\nu V = \sqrt{\frac{c}{l}}V \tag{8.44}$$

since the propagation speed of the signal dx/dt equals ν. This means that the signal sees the remainder of the line as an impedance,

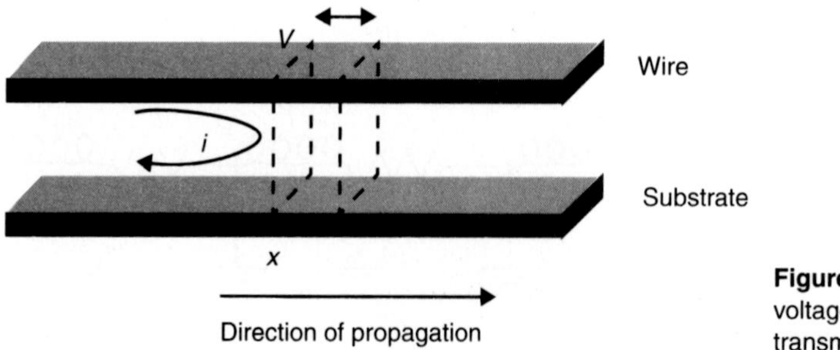

Figure 8.44 Propagation of voltage step along a lossless transmission line.

$$Z_0 = \frac{V}{I} = \sqrt{\frac{l}{c}}.$$

This impedance, called the *characteristic impedance* of the line, is a function of the dielectric medium and the geometry of the conducting wire and isolator (Eq. (8.45)). A typical value of Z_0 for IC interconnect ranges from 50 to 100 Ω.

$$Z_0 = \sqrt{\frac{l}{c}} = l\nu = \frac{1}{c\nu} \qquad (8.45)$$

Example 8.14 Propagation Speeds of Signal Waveforms

The information of Table 8.12 shows that it takes 1.5 nsec for a signal wave to propagate from source-to-destination on a 20 cm wire deposited on an epoxy printed-circuit board. If transmission line effects were an issue on silicon integrated circuits, it would take 0.13 nsec for the signal to reach the end of a 2 cm wire.

Termination

The behavior of the transmission line is strongly influenced by the termination of the line. The termination determines how much of the wave is reflected upon arrival at the wire end. This is expressed by the *reflection coefficient* ρ that determines the relationship between the voltages and currents of the incident and reflected waveforms.

$$\rho = \frac{V_{refl}}{V_{inc}} = \frac{I_{refl}}{I_{inc}} = \frac{R - Z_0}{R + Z_o} \qquad (8.46)$$

where R is the value of the termination resistance. The total voltages and currents at the termination end are the sum of incident and reflected waveforms.

$$\begin{aligned} V &= V_{inc}(1 + \rho) \\ I &= I_{inc}(1 - \rho) \end{aligned} \qquad (8.47)$$

Three interesting cases can be distinguished, as illustrated in Figure 8.45. In case (a) the terminating resistance is equal to the characteristic impedance of the line. The termination appears as an infinite extension of the line, and no waveform is reflected. This is also demonstrated by the value of ρ, which equals 0. In case (b), the line termination is an open circuit ($R = \infty$), and $\rho = 1$. The total voltage waveform after reflection is twice the incident one as predicted by Eq. (8.47). Finally, in case (c) where the line termination is a short circuit, $R = 0$, and $\rho = -1$. The total voltage waveform after reflection equals zero.

The transient behavior of a complete transmission line can now be examined. It is influenced by the characteristic impedance of the line, the series impedance of the source Z_S, and the loading impedance Z_L at the destination end, as shown in Figure 8.46.

Consider first the case where the wire is open at the destination end, or $Z_L = \infty$, and $\rho_L = 1$. An incoming wave is completely reflected without phase reversal. Under the assumption that the source impedance is resistive, three possible scenarios are sketched in Figure 8.47: $R_S = 5\ Z_0$, $R_S = Z_0$, and $R_S = 1/5\ Z_0$.

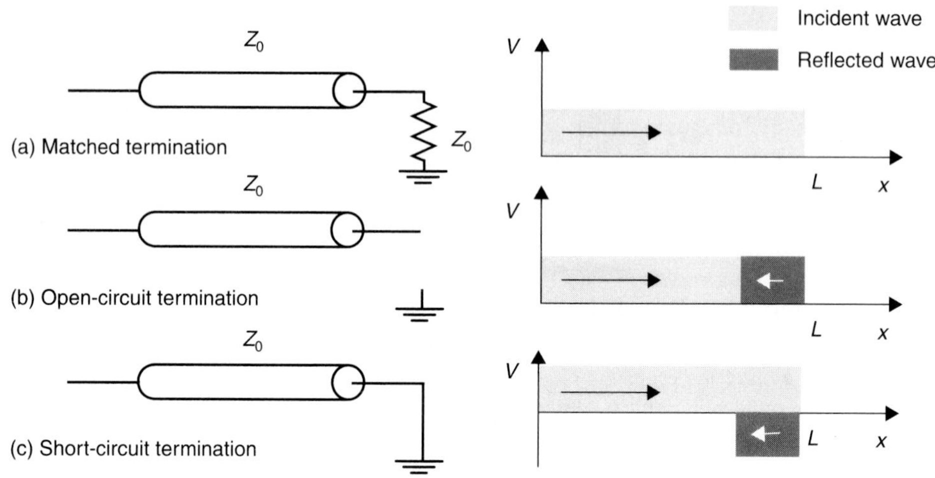

Figure 8.45 Behavior of various transmission line terminations.

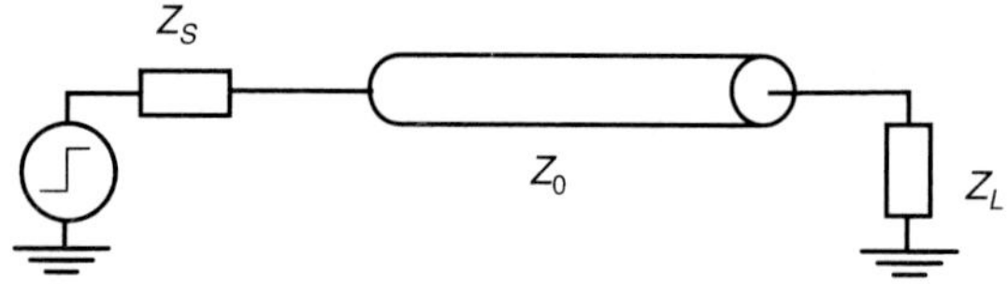

Figure 8.46 Transmission line with terminating impedances.

1. Large source resistance—$R_S = 5\ Z_0$ (Figure 8.47a)

Only a small fraction of the incoming signal V_{in} is injected into the transmission line. The amount injected is determined by the resistive divider formed by the source resistance and the characteristic impedance Z_0.

$$V_{source} = (Z_0 / (Z_0 + R_S))\ V_{in} = 1/6 \times 5\ \text{V} = 0.83\ \text{V} \tag{8.48}$$

This signal reaches the end of the line after L/v sec, where L stands for the length of the wire and is fully reflected, which effectively doubles the amplitude of the wave ($V_{dest} = 1.67$ V). The time it takes for the wave to propagate from one end of the wire to the other is called the *time-of-flight*, $t_{flight} = L/v$. Approximately the same happens when the wave reaches the source node again. The incident waveform is reflected with an amplitude determined by the source reflection coefficient, which equals 2/3 for this particular case.

$$\rho_S = \frac{5Z_0 - Z_0}{5Z_0 + Z_0} = \frac{2}{3} \tag{8.49}$$

The voltage amplitude at source and destination nodes gradually reaches its final value of V_{in}. The overall rise time is, however, many times L/v.

When multiple reflections are present, as in the above case, keeping track of waves on the line and total voltage levels rapidly becomes cumbersome. Therefore a graphical

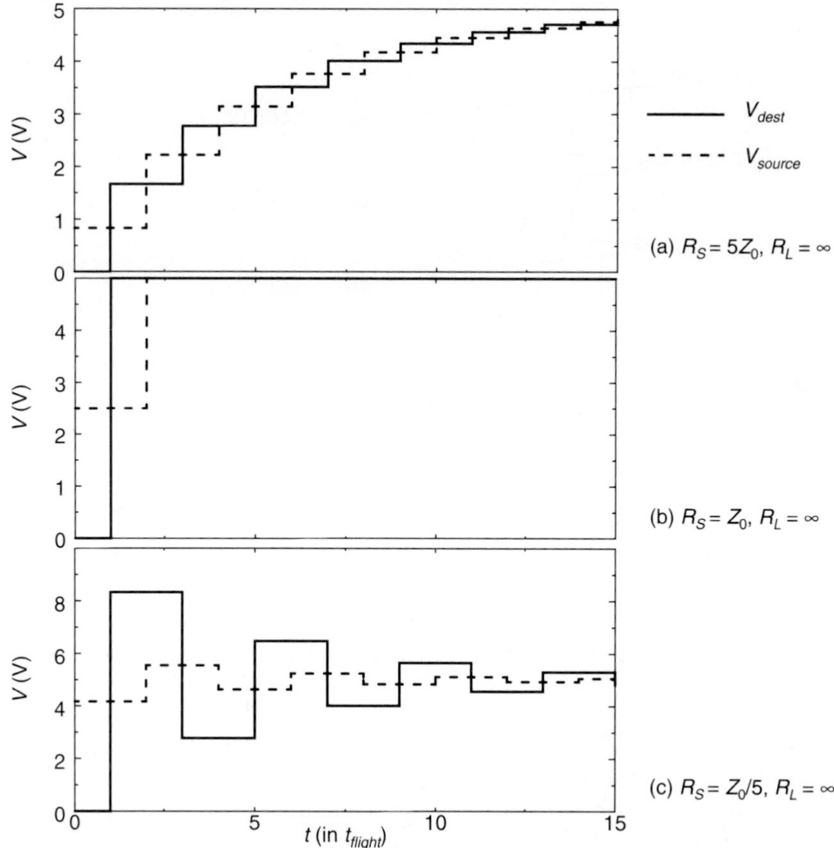

Figure 8.47 Transient response of transmission line.

construction called the *lattice diagram* is often used to keep track of the data (Figure 8.48). The diagram contains the values of the voltages at the source and destination ends, as well as the values of the incident and reflected wave forms. The line voltage at a termination point equals the sum of the previous voltage, the incident, and reflected waves.

2. Small source resistance—$R_S = Z_0/5$ (Figure 8.47c)

A large portion of the input is injected in the line. Its value is doubled at the destination end, which causes a severe overshoot. At the source end, the phase of the signal is reversed ($\rho_S = -2/3$). The signal bounces back and forth and exhibits severe ringing. It takes multiple L/v before it settles.

3. Matched source resistance—$R_S = Z_0$ (Figure 8.47b)

Half of the input signal is injected at the source. The reflection at the destination end doubles the signal, so that the final value is reached immediately. It is obvious that this is the most effective case. Matching the line impedance at the source end is called *series termination*. Note that the above analysis is an ideal one, as it is assumed that the input signal

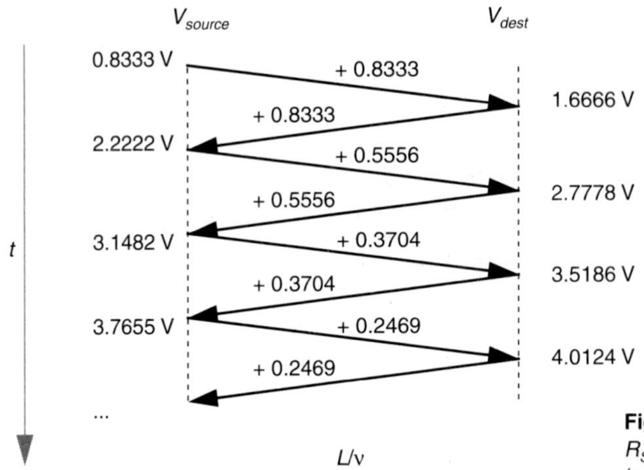

Figure 8.48 Lattice diagram for $R_S = 5\ Z_0$ and $R_L = \infty$. $V_{step} = 5$ V, (as in Figure 8.47a).

has a zero rise time. In real conditions the signals are substantially smoother, as demonstrated in the simulated response of Figure 8.49 (for $R_S = Z_0/5$ and $t_r = t_{flight}$).

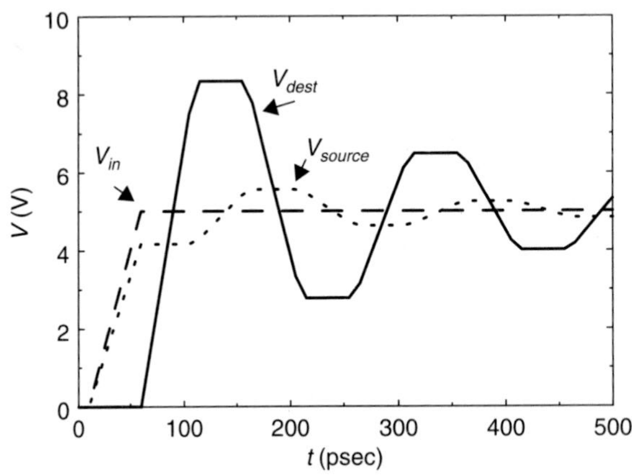

Figure 8.49 Simulated transient response of lossless transmission line for finite input rise times ($R_S = Z_0/5$, $t_r = t_{flight}$).

Problem 8.4 Transmission Line Response

Derive the lattice diagram of the above transmission line for $R_S = Z_0/5$, $R_L = \infty$, and $V_{step} = 5$ V. Also try the reverse picture—assume that the series resistance of the source equals zero, and consider different load impedances.

Similar considerations are valid when the termination is provided at the destination end, called *parallel termination*. Matching the load impedance to the characteristic impedance of the line once again results in the fastest response. This leads to the following conclusion.

To avoid potentially disastrous transmission line effects such as ringing or slow propagation delays, the transmission line should be terminated, either at the

source (series termination), or at the destination (parallel termination) with a resistance equal to the characteristic impedance Z_0 of the transmission line.

Example 8.15 Termination of ECL Circuits

A typical case of parallel termination occurs in the design of high-speed ECL circuits, where the pull-down resistance at the input of the gate is used as the terminating impedance. Figure 8.50a shows the example of two cascaded ECL gates connected by a 100 Ω transmission line that is 2 cm long. The simulated transient response is plotted in Figure 8.50b for parallel termination resistances of 10 kΩ and 100 Ω. Notice how the change in termination results in a shift in the voltage levels. This is to be expected, because larger load currents for the emitter-follower translate to larger base currents and hence a dc-offset. The circuit with the matched termination obviously displays a faster and cleaner response.

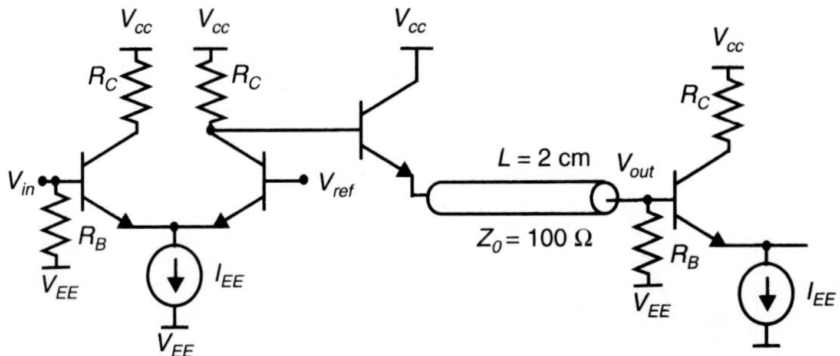

(a) Cascaded ECL gates

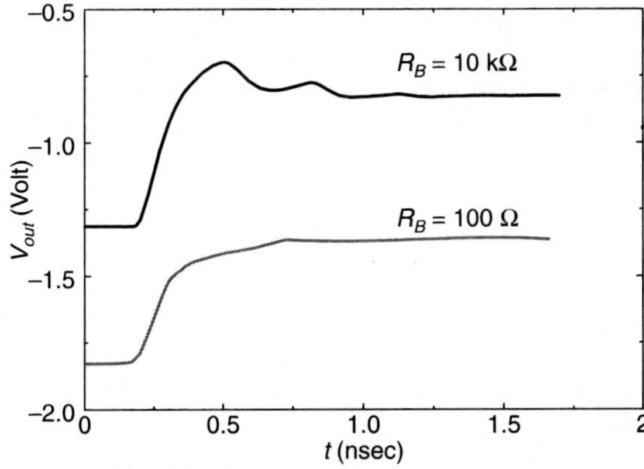

(b) Simulated response

Figure 8.50 Transmission line effects in ECL circuits.

Example 8.16 Capacitive Termination

Loads in MOS digital circuits tend to be of a capacitive nature. One might wonder how this influences the transmission line behavior and when the load capacitance should be taken into account.

The characteristic impedance of the transmission line determines the current that can be supplied to charge capacitive load C_L. From the load's point of the view, the line behaves as a resistance with value Z_0. The transient response at the capacitor node, therefore, displays a time constant Z_0C_L. This is illustrated in Figure 8.51, which shows the simulated transient response of a series-terminated transmission line with a characteristic impedance of 50 Ω loaded by a capacitance of 2 pF. The response shows how the output rises to its final value with a time-constant of 100 psec (= 50 Ω × 2 pF) after a delay equal to the time-of-flight of the line.

This asymptotic response causes some interesting artifacts. After $2\,t_{flight}$, an unexpected voltage dip occurs at the source node that can be explained as follows. Upon reaching the destination node, the incident wave is reflected. This reflected wave also approaches its final value asymptotically. Since V_{dest} equals 0 initially instead of the expected jump to 5 V, the reflection equals −2.5 V rather than the expected 2.5 V. This forces the transmission line temporarily to 0 V, as shown in the simulation. This effect gradually disappears as the output node converges to its final value.

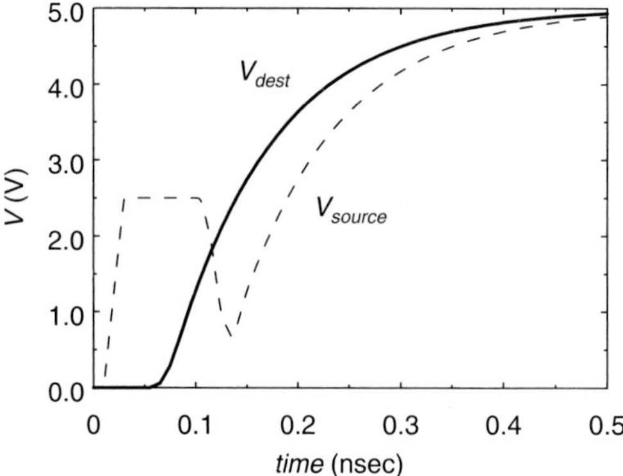

Figure 8.51 Capacitively terminated transmission line: R_S = 50 Ω, $R_L = \infty$, C_L = 2 pF, Z_0 = 50 Ω, t_{flight} = 50 psec.

The propagation delay of the line equals the sum of the time-of-flight of the line (= 50 psec) and the time it takes to charge the capacitance (= $0.69\,Z_0\,C_L$ = 69 psec). This is exactly what the simulation yields. In general, we can say that the capacitive load should only be considered in the analysis when its value is comparable to or larger than the total capacitance of the transmission line [Bakoglu90].

Lossy Transmission Line

While board and module wires are thick and wide enough to be treated as lossless transmission lines, the same is not entirely true for on-chip interconnect where the resistance of the wire is an important factor. The lossy transmission line model should be applied instead. Going into detail about the behavior of a lossy line would lead us to far astray. We therefore only discuss the effects of resistive loss on the transmission line behavior in a qualitative fashion.

Series resistance has a dual effect on the response of the line. These effects are demonstrated in Figure 8.52 for the case of a step input. First, the size of the step is attenuated as the wave propagates over the line. Secondly, the wave reaches its final value in an *RC*-like way. The farther it is from the source, the more the response resembles the behavior of a distributed *RC* line. In fact, the resistive effect becomes dominant, and the line behaves as a distributed *RC* line when *R* (the total resistance of the line) > $2\,Z_0$.

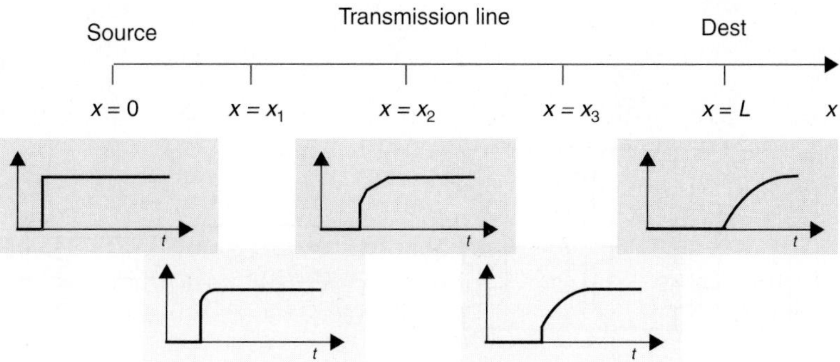

Figure 8.52 Step response of lossy transmission line.

Be aware that the actual wires on chips, boards, or substrates behave in a far more complex way than predicted by the above analysis. For instance, branches on wires, often called *transmission line taps,* cause extra reflections and can affect both signal shape and delay. Since the analysis of these effects is very involved, the only meaningful approach is to use computer analysis and simulation techniques.

SPICE Model of Transmission Line

SPICE supports a lossless transmission line model that was used in some of the simulations discussed previously. The line characteristics are defined by the characteristic impedance Z_0, while the length of the line can be defined in either of two forms. A first approach is to directly define the *transmission delay TD*, which is equivalent to the time-of-flight. Alternatively, a frequency F may be given together with NL, the dimensionless, normalized electrical length of the transmission line, which is measured with respect to the wavelength in the line at the frequency F. The following relation is valid.

$$NL = F \cdot TD \tag{8.50}$$

No lossy transmission line model is currently provided. When necessary, loss can be added by breaking up a long transmission line into shorter sections and adding a small series resistance in each section to model the transmission line loss. Be careful when using this approximation. First of all, the accuracy is still limited. Secondly, the simulation speed might be severely effected, since SPICE chooses a time step that is less than or equal to half of the value of TD. For small transmission lines, this time step might be much smaller than what is needed for transistor analysis.

Example 8.17 Design of an Output Driver—Revisited

As a conclusion to this section, we simulate an output driver that includes the majority of the parasitic effects introduced in this chapter. The schematics of the driver are shown in Figure 8.53a.

The inductance of the power, ground, and signal pins as well as the transmission line behavior of the board have been included (assuming a wire length of 45cm, which is on the large side, but not uncommon in complex systems). The first simulation of Figure 8.53b shows the output signals for a parallel termination of 10 kΩ. Severe ringing is observed. In an actual

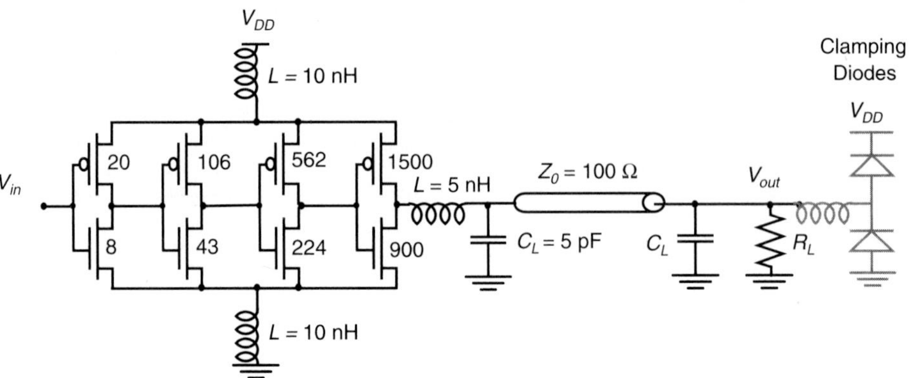

(a) Output driver schematic.

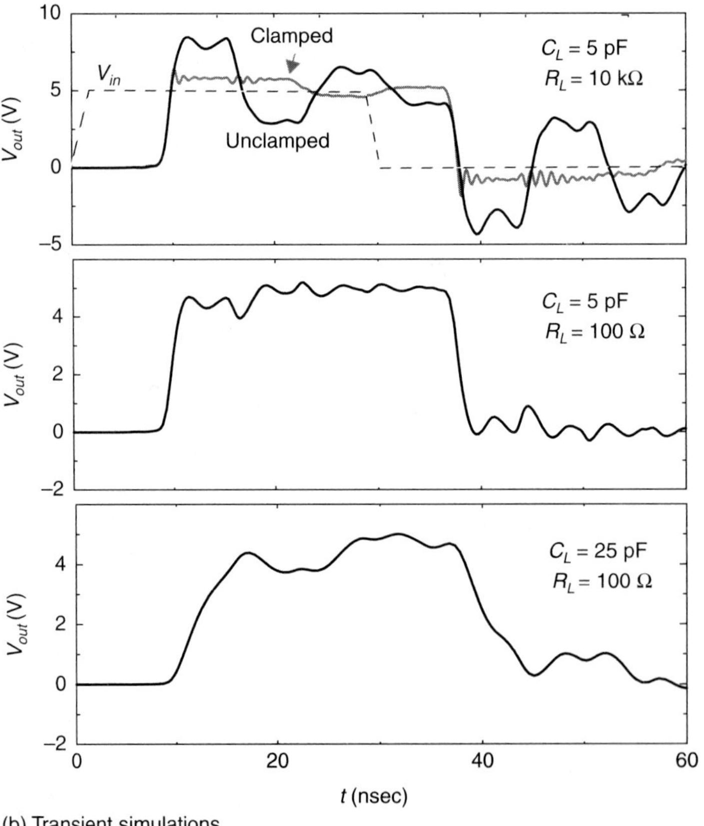

(b) Transient simulations

Figure 8.53 Simulation of output driver for various terminations and loads.

design, this ringing is not as prominent. Large over- and under-shoots are eliminated by the input-protection diodes of the fan-out device, as demonstrated by the shaded curve. This clamping action injects large current spikes in the substrate which may trigger latchup effects and is therefore to be avoided.

Next the line is terminated with a series resistance of 100 Ω, equal to the characteristic impedance of the line. Settling occurs almost instantaneously. In a final simulation, the load capacitance is increased from 5 pF to 25 pF. This does not influence the settling behavior, but increases the propagation delay of the driver as well as the rise and fall times of the output signal.

8.5 Comments on Packaging Technology

The package plays a fundamental role in the operation and performance of a component. Besides providing a means of bringing signal and supply wires in and out of the circuit, it also removes the heat generated by the circuit and provides mechanical support. Finally, its also protects the die against environmental conditions such as humidity.

The preceding discussions have demonstrated that the packaging technology can have a tremendous impact on the performance of a high-speed microprocessor or signal processor. This influence is further accentuated by the reduction in internal signal delays as a result of technology scaling. Up to 50% of the delay of a high-performance computer is currently due to packaging delays, and this number is expected to rise. The search for higher-performance packages with fewer inductive or capacitive parasitics has accelerated in recent years. Traditional dual-in-line, through-hole mounted packages are rapidly being replaced by other approaches such as surface-mount and multichip module techniques. It is useful for the circuit designer to be aware of the available options and their pros and cons.

Due to its multifunctionality, a good package must comply with a large variety of requirements.

- **Electrical requirements**—Pins should exhibit low capacitance (both interwire and to the substrate), resistance, and inductance. A large Z_0 is desirable to optimize transmission line behavior. Observe that intrinsic integrated-circuit impedances are high.
- **Mechanical and thermal properties**—The heat-removal rate should be as high as possible. Mechanical reliability requires a good matching between the thermal properties of the die and the chip carrier. Long-term reliability requires a strong connection from die to package as well as from package to board.
- **Low Cost**—Cost is always one of the more important properties. While ceramics have a superior performance over plastic packages, they are also substantially more expensive. Packing density is a major factor in reducing board cost. Finally, there is an increasing demand for a high input/output count as predicted by Rent's Rule in Eq. (8.18). Packages with over 300 pins are no exception anymore. Increasing the pin count either requires an increase in the package size or a reduction in the pitch between the pins. Both have a profound effect on the packaging economics.

Packages can be classified in many different ways —by their main material, the number of interconnection levels, and the means used to remove heat. In this short section, we can only glance briefly at each of those issues.

8.5.1 Package Materials

The most common materials used for the package body are ceramic and plastics. The latter have the advantage of being substantially cheaper, but suffer from inferior thermal properties. For instance, the ceramic Al_2O_3 (Alumina) conducts heat better than SiO_2 and the Polyimide plastic, by factors of 30 and 100 respectively. Furthermore, its thermal expansion coefficient is substantially closer to the typical interconnect metals. The disadvantage of alumina and other ceramics is their high dielectric constant, which results in large interconnect capacitances.

8.5.2 Interconnect Levels

The traditional packaging approach uses a two-level interconnection strategy. The die is first attached to an individual chip carrier or substrate. The package body contains an internal cavity where the chip is mounted. These cavities provide ample room for many connections to the chip leads (or pins). The leads compose the second interconnect level and connect the chip to the global interconnect medium, which is normally a PC board. Complex systems contain even more interconnect levels, since boards are connected together using backplanes or ribbon cables. The first two layers of the interconnect hierarchy are illustrated in the drawing of Figure 8.41.

This deep hierarchy of interconnect levels is becoming unacceptable in today's complex designs with their higher levels of integration, large signal counts, and increased performance requirements. The trend is toward reducing the number of levels. In the future, improved manufacturing, design, and testing capabilities will make it possible to integrate a complex computer system with all its peripherals on a single piece of semiconductor. For the time being, attention is focused on the elimination of the first level in the packaging hierarchy. Instead of housing dies in individual packages, they are mounted directly on the interconnect medium or board. This packaging approach is called the multichip module technique (or MCM) and results in a substantial increase in packing density as well as improved performance. The following sections provide a brief overview of the interconnect techniques used at levels one and two of the interconnect hierarchy, followed by a short discussion of the MCM technology.

Interconnect Level 1 —Die-to-Package-Substrate

For a long time, *wire bonding* was the technique of choice to provide an electrical connection between die and package. In this approach, the backside of the die is attached to the substrate using glue with a good thermal conductance. Next, the chip pads are individually connected to the lead frame with aluminum or gold wires. The wire-bonding machine use for this purpose operates much like a sewing machine. An example of wire bonding is shown in Figure 8.54. Although the wire-bonding process is automated to a large degree, it has some major disadvantages.

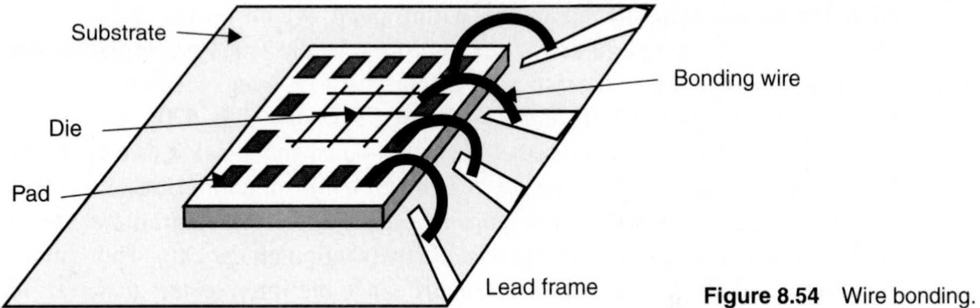

Figure 8.54 Wire bonding.

1. Wires must be attached serially, one after the other. This leads to longer manufacturing times with increasing pin counts.
2. Larger pin counts make it substantially more challenging to find bonding patterns that avoid shorts between the wires.
3. Bonding wires have inferior electrical properties, such as a high individual inductance (5 nH or more) and mutual inductance with neighboring signals.
4. The exact value of the parasitics is hard to predict because of the manufacturing approach and irregular outlay.

New attachment techniques are being explored as a result of these deficiencies. In one approach, called *Tape Automated Bonding* (or TAB), the die is attached to a metal lead frame that is printed on a polymer film (typically polyimide) (Figure 8.55a). The connection between chip pads and polymer film wires is made using solder bumps (Figure 8.55b). The tape can then be connected to the package body using a number of techniques. One possible approach is to use pressure connectors.

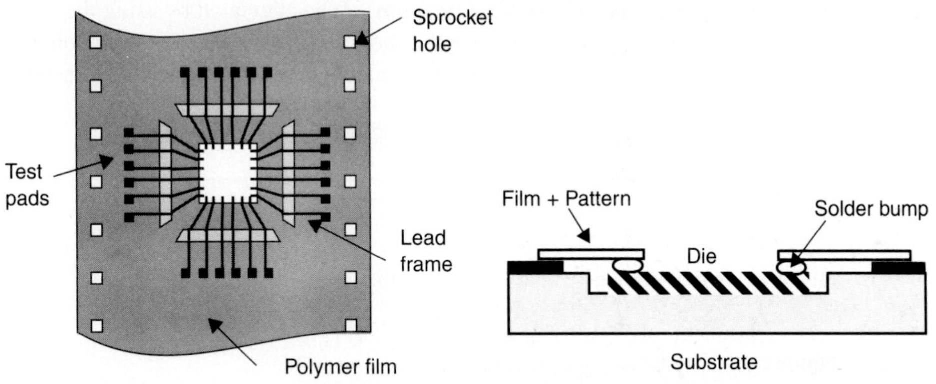

Figure 8.55 Tape-automated bonding (TAB).

The advantage of the TAB process is that it is highly automated. The sprockets in the film are used for automatic transport. All connections are made simultaneously. The printed approach helps to reduce the wiring pitch, which results in higher lead counts. Elimination of the long bonding wires improves the electrical performance. For instance, for a two-conductor layer, 48 mm TAB Circuit, the following electrical parameters hold: $L \approx 0.3$–0.5 nH, $C \approx 0.2$–0.3 pF, and $R \approx 50$–$200\ \Omega$ [Doane93, p. 420].

Another approach is to flip the die upside-down and attach it directly to the substrate using solder bumps. This technique, called *flip-chip* mounting, has the advantage of a superior electrical performance (Figure 8.56). Instead of making all the I/O connections on the die boundary, pads can be placed at any position on the chip. This can help address the power- and clock-distribution problems, since the interconnect materials on the substrate (e.g., Cu or Au) are typically of a better quality than the Al on the chip.

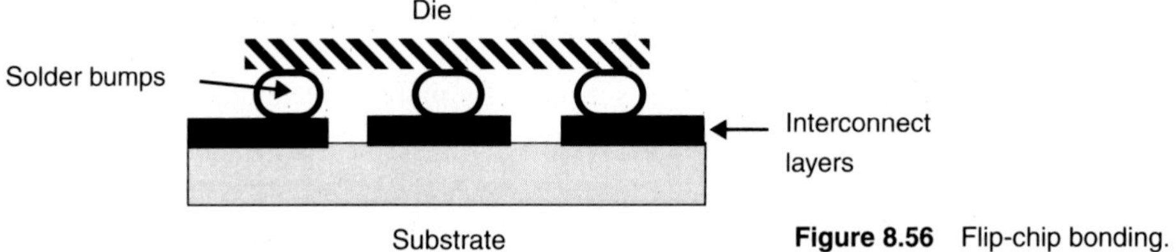

Figure 8.56 Flip-chip bonding.

Interconnect Level 2—Package Substrate to Board

When connecting the package to the PC board, *through-hole mounting* has been the packaging style of choice. A PC board is manufactured by stacking layers of copper and insulating epoxy glass. In the through-hole mounting approach, holes are drilled through the board and plated with copper. The package pins are inserted and electrical connection is made with solder (Figure 8.57a). The favored package in this class was the *dual-in-line* package or DIP (Figure 8.58a). The packaging density of the DIP degrades rapidly when the number of pins exceeds 64. This problem can be alleviated by using the *pin-grid-array* (PGA) package that has leads on the entire bottom surface instead of only on the periphery (Figure 8.58b). PGAs can extend to large pin counts (over 400 pins are possible).

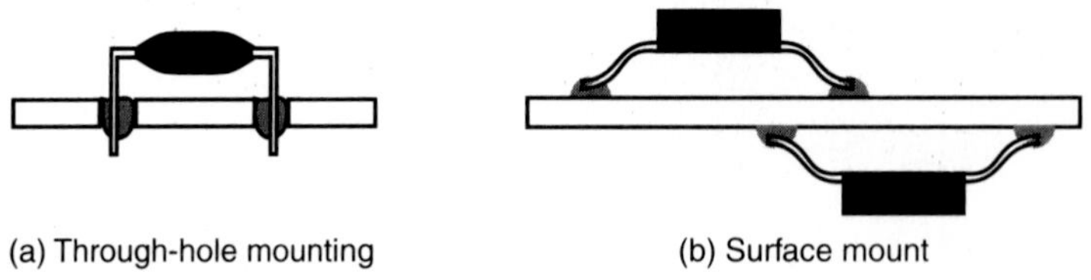

(a) Through-hole mounting

(b) Surface mount

Figure 8.57 Board-mounting approaches.

The through-hole mounting approach offers a mechanically reliable and sturdy connection. However, this comes at the expense of packaging density. For mechanical reasons, a minimum pitch of 2.54 mm between the through-holes is required. Even under those circumstances, PGAs with large numbers of pins tend to substantially weaken the board. Secondly, through-holes limit the board packing density by blocking lines that might otherwise have been routed below them, which results in longer interconnections.

PGAs with large pin counts hence require extra routing layers on the board to connect to the multitudes of pins. Finally, while the PGA has a parasitic capacitance and inductance that are slightly lower than the DIP package, their values are still substantial (Table 8.11).

Many of the shortcomings of the through-hole mounting are solved by using the *surface-mount* technique. A chip is attached to the surface of the board with a solder connection without requiring any through-holes (Figure 8.57b). Packing density is increased for the following reasons: (1) through-holes are eliminated, which provides more wiring space; (2) the lead pitch is reduced; and (3) chips can be mounted on both sides of the board. In addition, the elimination of the through-holes improves the mechanical strength of the board. On the negative side, the on-the-surface connection makes the chip-board connection weaker. Not only is it cumbersome to mount a component on a board, but also more expensive equipment is needed, since a simple soldering iron will not do anymore. Finally, testing of the board is more complex, because the package pins are no longer accessible at the backside of the board. Signal probing becomes hard or even impossible.

A variety of surface-mount packages are currently in use with different pitch and pin-count parameters. Three of these packages are shown in Figure 8.58: the *small-outline package* with gull wings, the *plastic leaded package* (PLCC) with J-shaped leads, and the *leadless chip carrier*. An overview of the most important parameters for a number of packages is given in Table 8.13.

1 Bare die
2 DIP
3 PGA
4 Small-outline IC
5 Quad flat pack
6 PLCC
7 Leadless carrier

Figure 8.58 An overview of commonly used package types.

Multi-Chip Modules—Die-to-Board

Eliminating one layer in the packaging hierarchy by mounting the die directly on the wiring backplanes—board or substrate—offers a substantial benefit when performance or density is a major issue. A number of the previously mentioned die-mounting techniques can be adapted to mount dies directly on the substrate. This includes wire bonding, TAB, and flip-chip, although the latter two are preferable. The substrate itself can vary over a wide range of materials, depending upon the required mechanical, electrical, thermal, and

Table 8.13 Parameters of various types of chip carriers.

Package type	Lead spacing (Typical)	Lead count (Maximum)
Dual-in-line	2.54 mm	64
Pin grid array	2.54 mm	> 300
Small-outline IC	1.27 mm	28
Leaded chip carrier (PLCC)	1.27 mm	124
Leadless chip carrier	0.75 mm	124

economical requirements. Materials of choice are epoxy substrates (similar to PC boards), metal, ceramics, and silicon. Silicon has the advantage of presenting a perfect match in mechanical and thermal properties with respect to the die material.

The main advantages of the MCM approach are the increased packaging density and performance. An example of an MCM module implemented using a silicon substrate (commonly dubbed *silicon-on-silicon*) is shown in Figure 8.59. The module, which implements an avionics processor module and is fabricated by Rockwell International, contains 53 ICs and 40 discrete devices on a 2.2″ × 2.2″ substrate with aluminum polyimide interconnect. The interconnect wires are only an order of magnitude wider than what is typical for on-chip wires, since similar patterning approaches are used. The module itself has 180 I/O pins. Performance is improved by the elimination of the chip-carrier layer with its assorted parasitics, and through a reduction of the global wiring lengths on the die, a result of the increased packaging density. For instance, a solder bump has an assorted capacitance and inductance of only 0.1 pF and 0.01 nH respectively. The MCM technology can

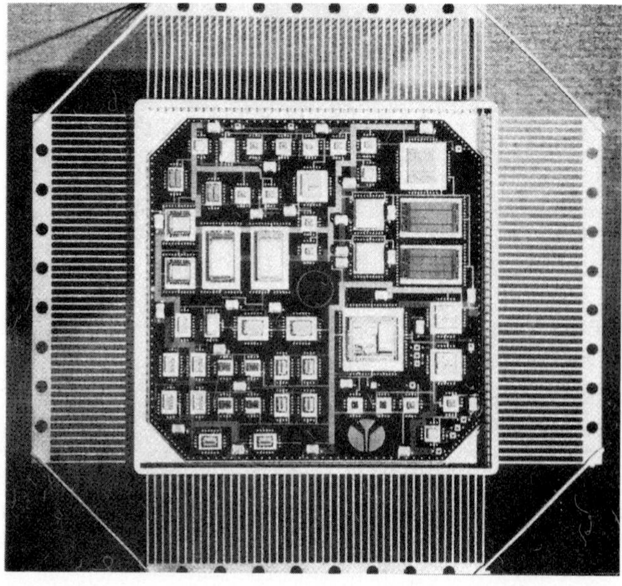

Figure 8.59 Avionics processor module. *Courtesy of Rockwell International.*

also reduce power consumption significantly, since large output drivers—and associated dissipation—become superfluous due to the reduced load capacitance of the output pads. The dynamic power associated with the switching of the large load capacitances is simultaneously reduced.

While MCM technology offers some clear benefits, its main disadvantage is economic. This technology requires some advanced manufacturing steps that make the process expensive. The approach is only justifiable when either dense housing or extreme performance is essential. In the near future, this argument might become obsolete as MCM approaches proliferate; for example, some of the more advanced microprocessors, such as the Intel P6 (Pentium Pro), employ MCM technology.

8.5.3 Thermal Considerations in Packaging

As the power consumption of integrated circuits rises, it becomes increasingly important to efficiently remove the heat generated by the chips. A large number of failure mechanisms in ICs are accentuated by increased temperatures. Examples are leakage in reverse-biased diodes, electromigration, and hot-electron trapping. To prevent failure, the temperature of the die must be kept within certain ranges. The supported temperature range for commercial devices during operation equals 0° to 70°C. Military parts are more demanding and require a temperature range varying from –55° to 125°C.

The cooling effectiveness of a package depends upon the thermal conduction of the package material, which consists of the package substrate and body, the package composition, and the effectiveness of the heat transfer between package and cooling medium. Standard packaging approaches use still or circulating air as the cooling medium. The transfer efficiency can be improved by adding finned metal heat sinks to the package. More expensive packaging approaches, such as those used in mainframes or supercomputers, force air, liquids, or inert gases through tiny ducts in the package to achieve even greater cooling efficiencies.

As an example, a 40-pin DIP has a thermal resistance of 38 °C/W and 25 °C/W for natural and forced convection of air. This means that a DIP can dissipate 2 watts (3 watts) of power with natural (forced) air convection, and still keep the temperature difference between the die and the environment below 75 °C. For comparison, the thermal resistance of a ceramic PGA ranges from 15 ° to 30 °C/W.

Since packaging approaches with decreased thermal resistance are prohibitively expensive, keeping the power dissipation of an integrated circuit within bounds is an economic necessity. The increasing integration levels and circuit performance make this task nontrivial. An interesting relationship in this context has been derived by Nagata [Nagata92]. It provides a bound on the integration complexity and performance as a function of the thermal parameters

$$\frac{N_G}{t_p} \leq \frac{\Delta T}{\theta E} \tag{8.51}$$

where N_G is the number of gates on the chip, t_p the propagation delay, ΔT the maximum temperature difference between chip and environment, θ the thermal resistance between them, and E the switching energy of each gate.

Example 8.18 Thermal Bounds On Integration

For $\Delta T = 100$ °C, $\theta = 2.5$ °C/W and $E = 0.1$ pJ, this results in $N_G/t_p \leq 4 \times 10^5$ (gates/nsec). In other words, the maximum number of gates on a chip, when all gates are operating simultaneously, must be less than 400,000 if the switching speed of each gate is 1 nsec. This is equivalent to a power dissipation of 40 W.

Fortunately, not all gates are operating simultaneously in real systems. The maximum number of gates can be substantially larger, as expressed by the activity factor discussed in Chapter 4. For instance, it was experimentally derived that the ratio between the average switching period and the propagation delay ranges from 20 to 200 in mini- and large-scale computers [Masaki92].

Nevertheless, Eq. (8.51) demonstrates that heat dissipation and thermal concerns present an important limitation on circuit integration. Design approaches for low power that reduce either E or the activity factor are rapidly gaining importance.

8.6 Perspective: When to Consider Interconnect Parasitics

An important question for a designer to answer when analyzing an interconnect network is whether a simple capacitive model is sufficient, or should the effects of *RC* delays be considered, or do transmission line effects come into play? Some simple rules of thumb prove to be very useful here.

1. *RC* delays should only be considered when $t_{pRC} >> t_{pgate}$ (of the driving gate).

This translates into Eq. (8.52), which determines the critical length L of the interconnect wire where *RC* delays become dominant.

$$L >> \sqrt{\frac{t_{pgate}}{0.38rc}} \tag{8.52}$$

Combining this equation with the appropriate technology parameters produces the critical-length values of Table 8.14. From this table, we can deduce that the *RC* delay of a metal wire on a chip should only be considered if it is longer than 2.5 mm in a 1.0 μm technology. Be aware that this table is a generalization and that the critical lengths can vary from technology to technology.

Table 8.14 Critical Length of wires for *RC* delays (from [Weste93]). λ stands for the technology scaling factor lambda and equals 0.5 for a 1 μm CMOS process.

Material	Critical length
Metal1/Metal2/Metal3	> 5000–10,000 λ
Silicides	> 600 λ
Polysilicon	> 200 λ
Diffusion	> 60 λ

2. Transmission line effects should be considered when the rise or fall time of the input signal (t_r, t_f) is smaller than the time-of-flight of the transmission line (t_{flight}).

This leads to the following rule of thumb [Bakoglu90], which determines when transmission line effects should be considered:

$$t_r(t_f) < 2.5t_{flight} = 2.5\frac{L}{v} \tag{8.53}$$

For on-chip wires with a maximum length of 1 cm, one should only worry about transmission line effects when t_r < 150 psec. This is not the case in present-day CMOS design, but is a possibility for fast ECL and GaAs designs. At the board level, where wires can reach a length of up to 50 cm, we should account for the delay of the transmission line when t_r < 8 nsec. This condition is easily achieved with state-of-the-art processes and packaging technologies. Ignoring the inductive component of the propagation delay can easily result in overly optimistic delay predictions.

Although the presented rules are by no means cast in stone, they present a set of useful design guidelines. When in doubt, it is worthwhile to verify the impact of the interconnect parasitics by analyzing a simple test circuit using SPICE. A major challenge from this perspective is that the technology parameters of the interconnect are not always well known or that adequate models are not available. This is the topic of ongoing research [Tewksbury94].

8.7 Chapter Summary

This chapter has discussed the impact of interconnect on the performance and reliability of digital integrated circuits. The parasitics introduced by the interconnect have a dual effect on circuit operation: (1) they introduce noise and (2) they increase the propagation delay.

- Interconnect capacitance consists of capacitance-to-ground and the interwire capacitance. Each of these contain a parallel-plate and a fringing component.
- The interwire capacitance introduces cross talk. Providing the necessary shielding is important for wires such as busses or clock signals.
- Driving large capacitances rapidly in CMOS requires the introduction of a *cascade of buffer stages* that must be carefully sized. Alternative approaches reduce the required signal swing.
- A resistive wire is best modeled as a distributed *rc* line. Approximative models for SPICE were introduced.
- Resistivity affects the reliability of the circuit by introducing *RI* drops. This is especially important for the supply network, where wire sizing is important.
- The extra delay introduced by the *rc* effects can be minimized by partitioning of the wire and by using a better interconnect technology.

- The inductance of the interconnect becomes important at higher switching speeds. The chip package is currently one of the most important contributors of inductance. Novel packaging techniques are gaining importance with faster technologies.
- Ground bounce introduced by the $L\, di/dt$ voltage drop over the supply wires is one of the most important sources of noise in current integrated circuits. Ground bounce can be reduced by providing sufficient supply pins and by controlling the slopes of the off-chip signals.
- Transmission line effects are not yet an issue in CMOS design but are bound to be in the near future. Providing the correct termination is the only means of dealing with the transmission line delay.

8.8 To Probe Further

An excellent overview of the issues involved in analyzing, modeling, and minimizing the impact of interconnect on digital designs can be found in [Bakoglu90]. An interesting collection of papers on interconnect modeling and analysis can be found in [Tewksbury94]. [Johnson93] presents an in-depth analysis of noise issues in high-performance design, while [Doane93] offers a detailed description of the state-of-the-art packaging issues.

REFERENCES

[Abo92] A. Abo and A. Behzad, "Interconnect Driver Design," EE141 Project Report, Univ. of California—Berkeley, 1992.

[Bakoglu90] H. Bakoglu, *Circuits, Interconnections and Packaging for VLSI*, Addison-Wesley, 1990.

[Brews89] J. Brews, "Electrical Modeling of Interconnections," in *SubMicron Integrated Circuits*, ed. R. Watts, John Wiley & Sons, pp. 269–331, 1989.

[Doane93] D. Doane, ed., *Multichip Module Technologies and Alternatives*, Van Nostrand-Reinhold, 1993.

[Dopperpuhl92] D. Dopperpuhl et al., "A 200 MHz 64-b Dual Issue CMOS Microprocessor," *IEEE Journal of Solid State Circuits*, vol. 27, no. 11, pp. 1555–1567, November 1992.

[Elmore48] E. Elmore, "The Transient Response of Damped Linear Networks with Particular Regard to Wideband Amplifiers," *Journal of Applied Physics*, pp. 55–63, January 1948.

[Etter93] D. Etter, "Engineering Problem Solving with Matlab," Prentice Hall, 1993.

[Franzon93] P. Franzon, "Electrical Design of Digital Multichip Modules," in [Doane93], pp 525–568, 1993.

[Heller75] L. Heller et al., "High-Sensitivity Charge-Transfer Sense Amplifier," *Proceedings ISSCC Conf.*, pp. 112-113, 1975.

[Hedenstierna87] N. Hedenstierna et al., "CMOS Circuit Speed and Buffer Optimization," *IEEE Trans. on Computer-Aided Design*, vol CAD-6, no. 2, pp. 270–281, March 1987.

[Horowitz83] M. Horowitz, "Timing Models for MOS Circuits," Ph.D. diss., Stanford University, 1983.

[Johnson93] H. Johnson and M. Graham, *High-Speed Digital Design—A Handbook of Black Magic*, Prentice Hall, 1993.

[Kang87] S. Kang, "Metal-Metal Matrix (M^3) for High-Speed MOS VLSI Layout," *IEEE Trans. Computer-Aided Design*, vol. CAD-6, pp. 886–891, September 1987.

[Landman71] B. Landman and R. Russo, "On a Pin versus Block Relationship for Partitions of Logic Graphs," *IEEE Trans. on Computers*, vol. C-20, pp. 1469–1479, December 1971.

[Masaki92] A. Masaki, "Deep-Submicron CMOS Warms Up to High-Speed Logic," *Circuits and Devices Magazine*, Nov. 1992.

[Mead80] C. Mead and L. Conway, *Introduction to VLSI Systems*, Addison-Wesley, 1980.

[Nagata92] M. Nagata, "Limitations, Innovations, and Challenges of Circuits and Devices into a Half Micrometer and Beyond," *IEEE Journal of Solid State Circuits*, vol. 27, no. 4, pp. 465–472, April 1992.

[Rubinstein83] J. Rubinstein, P. Penfield, and M. Horowitz, "Signal Delay in *RC* Networks," *IEEE Transactions on Computer-Aided Design*, vol. CAD-2, pp. 202–211, July 1983.

[Schaper83] L. Schaper and D. Amey, "Improved Electrical Performance Required for Future MOS Packaging," *IEEE Trans. on Components, Hybrids and Manufacturing Technology*, vol. CHMT-6, pp. 282–289, September 1983.

[Solomon82] P. Solomon, "A Comparison of Semiconductor Devices for High-Speed Logic," *Proc. of the IEEE*, vol. 70, no. 5, May 1982.

[Sorkin87] G. Sorkin, "Asymptotically Perfect Trivial Global Routing: A Stochastic Analysis," *IEEE Trans. on Computer-Aided Design*, vol. CAD-6, p. 820, 1987.

[Steidel83] C. Steidel, "Assembly Techniques and Packaging," in [Sze83], pp. 551–598, 1983.

[Sze81] S. Sze, *Physics of Semi-Conductor Devices*, 2nd ed., Wiley, 1981.

[Sze83] S. Sze, ed., *VLSI Technology*, McGraw-Hill, 1983.

[Tao94] J. Tao, N. Cheung, and C. Hu, "An Electromigration Failure Model for Interconnects under Pulsed and Bidirectional Current Stressing," *IEEE Trans. on Devices*, vol. 41, no. 4, pp. 539–545, April 1994.

[Tewksbury94] S. Tewksbury, ed., *Microelectronics System Interconnections—Performance and Modeling*, IEEE Press, 1994.

[Vaidya80] S. Vaidya, D. Fraser, and A. Sinha, "Electromigration Resistance of Fine-Line Al," *Proc. 18th Reliability Physics Symposium*, IEEE, p. 165, 1980.

[Vdmeijs84] N. Van De Meijs and J. Fokkema, "VLSI Circuit Reconstruction from Mask Topology," *Integration*, vol. 2, no. 2, pp. 85–119, 1984.

[Wada92] O. Wada, T. Kamijoh, and M. Nakamura, "Integrated Optical Connections," *IEEE Circuits and Devices Magazine*, vol. 8, no. 6, pp. 37–42, 1992.

[Weste93] N. Weste and K.Eshragian, *Principles of CMOS VLSI Design: A Systems Perspective*, 2nd ed., Addison-Wesley, 1993.

[Yoshino90] T. Yoshino et al., "A 100 MHz 64-tap FIR Digital Filter in a 0.8 μm BiCMOS Gate Array," *IEEE Journal of Solid State Circuits*, vol. 25, no. 6, pp. 1494–1501, December 1990.

8.9 Exercises and Design Problems

1. [E, None, 8.2] In order to drive a large capacitance (C_L = 20 pF) from a minimum-size gate (with input capacitance, C_i = 10 fF), a designer decides to introduce a two-staged buffer (Figure 8.60). Assume that the propagation delay of a minimum-size inverter (loaded by an identi-

cal gate) is given by t_{p0} = 70 psec. Also, assume that the input capacitance of a gate is proportional to its size. Determine the sizing of the two additional buffer stages that will minimize the propagation delay as well as the value of the minimum delay.

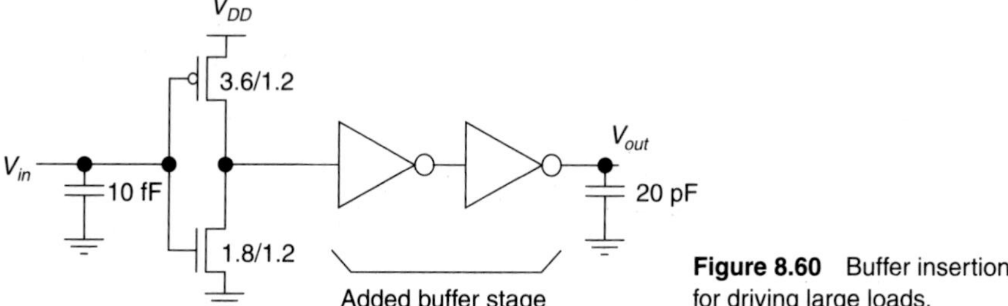

Figure 8.60 Buffer insertion for driving large loads.

2. [E, None, 8.2] A not-so-clever engineer accidentally swaps the *Data* and *Enable* inputs on the tri-state buffer of Figure 8.18, resulting in the new circuit of Figure 8.61.
   a. Does this circuit still function as a tri-state buffer?
   b. For the *In* and *En* waveforms given in the figure, sketch the voltages at nodes *X*, *Y*, and *Out*. Label asymptotic voltages, neglecting body effect.
   c. Why is this circuit dangerous to use in practice?

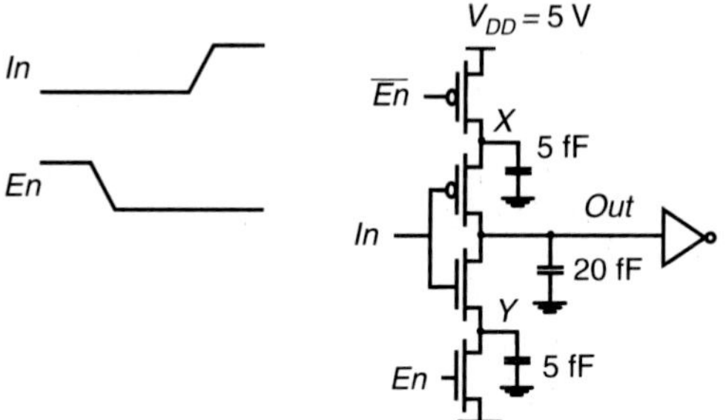

Figure 8.61 Tri-state buffer with swapped inputs.

3. [M, None, 8.2] Stray coupling capacitance can cause signals to be fed directly from gate input to gate output. Consider the parasitic capacitances in the circuit of Figure 8.62. Assume the capacitances shown dominate over other parasitics in the circuit.
   a. For the waveforms in the figure, sketch the voltage waveform at *Out*.
   b. If *Out* drives a transmission gate that is nominally off, is there any danger of improper operation due to the clock feedthrough demonstrated in *a*?
4. [E, None, 8.2–8.3] A two-stage buffer is used to drive a metal wire of 1 cm. The first inverter is a minimum-size inverter with an input capacitance C_i of 10 fF and a propagation delay t_{p0} of 175 psec when loaded with an identical gate. The width of the metal wire is 3.6 μm. The sheet resistance of the metal is 0.08 Ω/□, the capacitance value is 0.03 fF/μm^2, and the fringing-field capacitance is 0.04 fF/μm.
   a. What is the propagation delay t_w over the metal wire?
   b. Compute the optimal size of the second inverter. What is the minimum delay through the buffer?

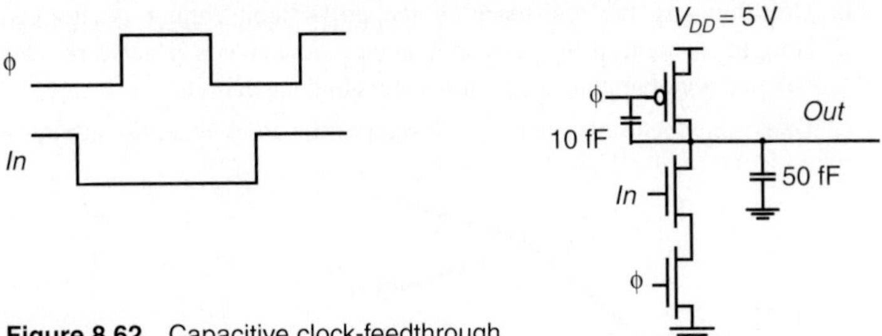

Figure 8.62 Capacitive clock-feedthrough.

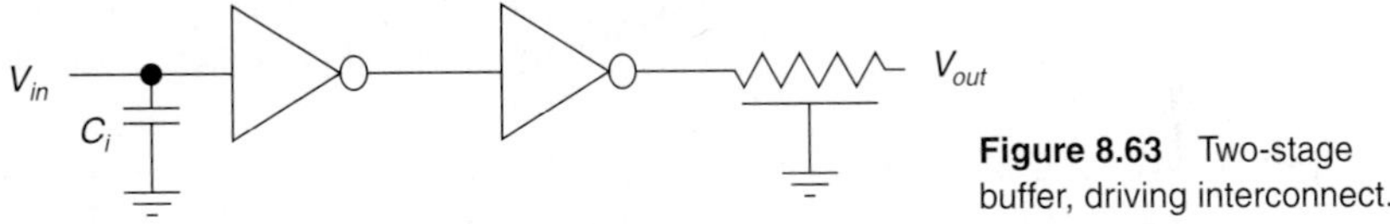

Figure 8.63 Two-stage buffer, driving interconnect.

5. [M, None, 8.2] Consider applying scaling to the two-stage buffer of problem 4. Assume ideal full scaling (i.e., constant electrical-field scaling) with a scaling factor S > 1. More logic is added to the chip, however, so that the length of the metal wire stays constant at 1 cm. All other horizontal and vertical dimensions are scaled. Moreover, the first inverter scales, but the size of the second inverter is adapted so that for each scaling factor, a minimum delay is maintained for the buffer.
   a. How do the interconnect delay and the propagation delay of the buffer scale as a function of S?
   b. At what value of S does the interconnect delay of the metal wire become equal to the propagation delay of the buffer? What are the values of these delays for this value of S?
   c. For the case of fixed-voltage scaling, how does the propagation delay of the buffer scale now? Assume velocity-saturated operation. Repeat a and b under these conditions.
6. [E, None, 8.3] Consider the buffered cascade of transmission gates shown in Figure 8.64. For a total of $N = 30$ transmission gates, find the optimum number of gates between buffers and the total propagation delay of the network. Assume $(W/L)_n = 1.8/1.2$ and $(W/L)_p = 6.6/1.2$ for the transmission gates. Also assume $C = 20$ fF, $t_{inv} = 0.25$ nsec, and $V_{DD} = 5$ V.

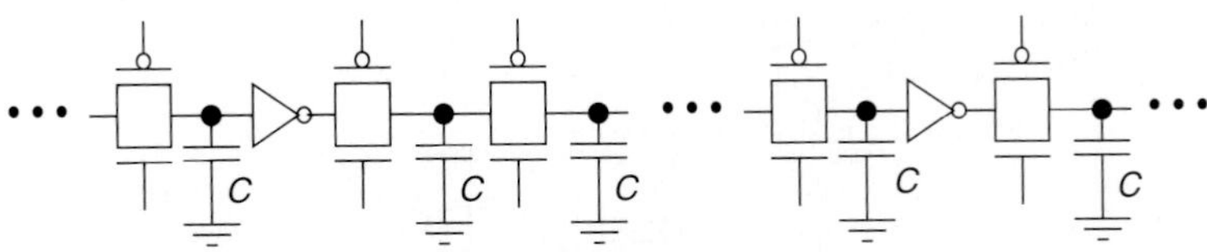

Figure 8.64 Transmission-gate *RC* Network

7. [M, None, 8.3] Figure 8.65 shows a clock-distribution network. Each segment of the clock network (between the nodes) is 5 mm long, 3 μm wide, and is implemented in polysilicon. At each of the terminal nodes (such as R) resides a load capacitance of 100 fF.
   a. Determine the average current of the clock driver, given a voltage swing on the clock lines of 5 V and a maximum delay of 5 nsec between clock source and destination node R. For this part, you may ignore the resistance and inductance of the network

b. Unfortunately the resistance of the polysilicon cannot be ignored. Assume that each straight segment of the network can be modeled as a Π-network. Draw the equivalent circuit and annotate the values of resistors and capacitors.

c. Determine the dominant time-constant of the clock response at node R.

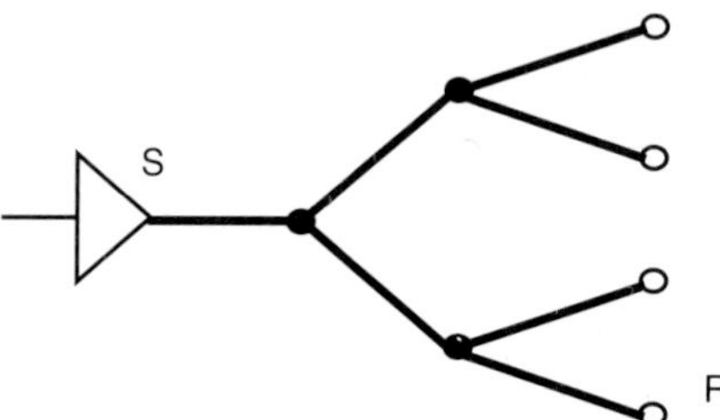

Figure 8.65 Clock-distribution network.

8. [C, SPICE, 8.3] You are designing a clock distribution network in which it is critical to minimize skew between local clocks ($CLK1$, $CLK2$, and $CLK3$). You have extracted the RC network of Figure 8.66, which models the routing parasitics of your clock line. Initially, you notice that the path to $CLK3$ is shorter than to $CLK1$ or $CLK2$. In order to compensate for this imbalance, you insert a transmission gate in the path of $CLK3$ to eliminate the skew.

   a. Write expressions for the time-constants associated with nodes $CLK1$, $CLK2$ and $CLK3$. Assume the transmission gate can be modeled as a resistance R_3.

   b. If $R_1 = R_2 = R_4 = R_5 = R$ and $C_1 = C_2 = C_3 = C_4 = C_5 = C$, what value of R_3 is required to balance the delays to $CLK1$, $CLK2$, and $CLK3$?

   c. For $R = 750\Omega$ and $C = 200\text{fF}$, what (W/L)'s are required in the transmission gate to eliminate skew? Determine the value of the propagation delay.

   d. Simulate the network using SPICE, and compare the obtained results with the manually obtained numbers.

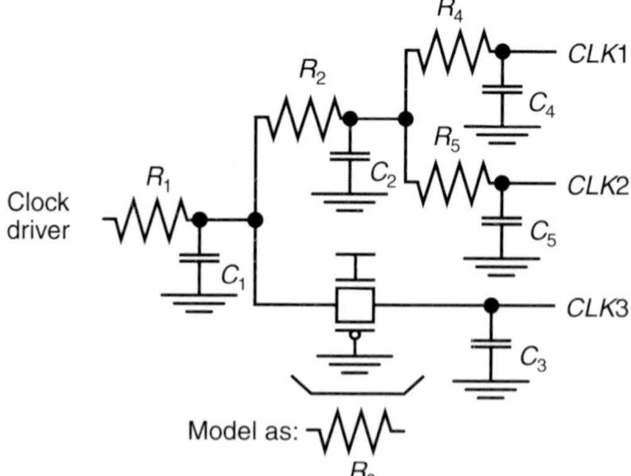

Figure 8.66 *RC* clock-distribution network.

9. [E, None, 8.3] Ohmic voltage drops are particularly troublesome in power distribution networks. Figure 8.67 shows a schematic of such a network that distributes V_{DD} and GND to four processing modules. The numbers annotated on the figure are wire lengths in μm. Assume both supplies are distributed in metal1 wires 5.4 μm in width.

a. Sketch a single-resistor network capable of modeling the actual V_{DD} and *GND* voltages of the four modules. You can model the modules themselves as current sources drawing worst-case currents of I_{peak}.

b. Find the effective worst-case V_{DD} and *GND* voltages seen by each of the four modules using $I_{peak} = 20$ mA.

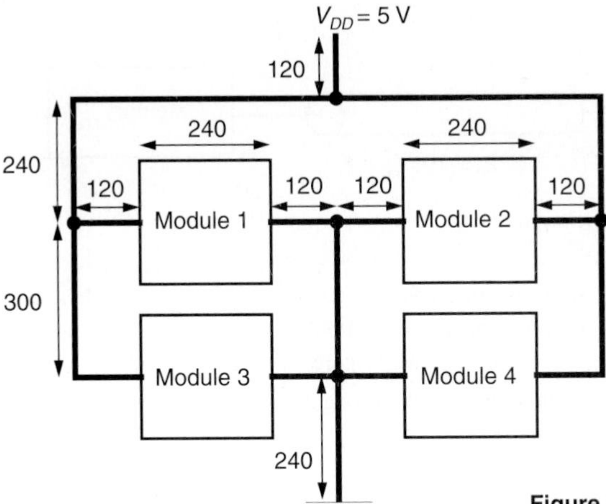

Figure 8.67 Power -distribution network.

10. [E, None, 8.3] Electromigration failure is most common in power and ground wires. Consider again the network of Figure 8.67. The numbers annotated in the figure are the wire lengths in μm. If each module draws a worst-case current of 5 mA, what are the minimum wire widths required for each segment of the network to avoid electromigration failure. Assume the vertical wire thickness is 1 μm, and the maximum current density is 1 mA/μm^2.

11. [M, None, 8.4] Consider again the two-stage buffer design of Problem 1 and Figure 8.60. Assume that each buffer is made five times larger than its predecessor.

a. The *GND* and V_{DD} connections of the buffers are wired to the pins of the package (and the supply) through a bonding wire with an inductance of 3 nH. Approximate the voltage bounce on the supply rail when the input switches from 0 to 1. You may assume that the rise or fall time of a gate equals two times its propagation delay.

b. The voltage bounce on the supply rails is too high. The designer decides to reduce it by a factor of two. Determine the best way of doing this to minimize the effect on the performance.

12. [M, None, 8.4] To connect a processor to an external memory, an off-chip connection is necessary (Figure 8.68) The copper wire on the board, which is 15 cm long, acts as a transmission line with a characteristic impedance of 100 Ω. The memory input pins represent a very high impedance, which can be considered as infinite. The bus driver is a CMOS inverter, consisting of very large devices of (120/1.2) for the NMOS and (360/1.2) for the PMOS. You may also assume that the on-resistance of the minimum-size devices (1.8/1.2 and 5.4/1.2, for NMOS and PMOS, respectively) equals 10 kΩ and scales proportionally with the size of the transistors.

a. Determine the time it takes for a change in signal to propagate from the source to the destination (the time of flight). The wire inductance per unit length equals 75×10^{-8} H/m.

b. Given the driver device sizes defined above, determine how long it will take the output signal to stay within 10% of its final value. You can model the driver as a voltage source with the driving device acting as series source resistance. Determine the reflection coefficients on both source and destination ends, and draw the lattice diagram for the transmission line. Assume a supply and step voltage of 5 V.

c. Resize the device dimensions of the driver to minimize the transmission line delay. Determine that minimum time and derive the sizes for the NMOS and PMOS transistors.

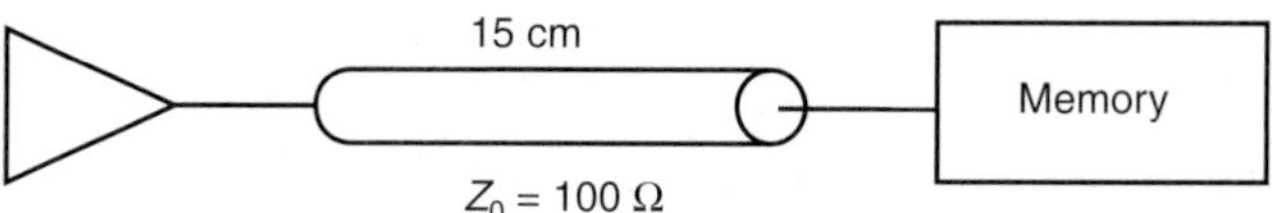

Figure 8.68 Driving memory through a transmission line.

13. [C, None, 8.4] Transmission line effects can become important when considering interchip communications in high-performance CMOS circuits. Figure 8.69 shows an output driver feeding a 2 pF effective fan-out of CMOS gates through a transmission line.

a. Find the line impedance, the wave propagation speed, and the time-of-flight delay for the given transmission line.

b. Sketch V_S and V_L assuming $R_{p,on} = R_{n,on} = Z_0$ for a square wave input. Label critical voltages and times. Assume the input frequency is low enough to allow all voltages to settle between transitions.

c. What benefit is achieved by sizing the driving transistors to meet the criterion $R_{p,on} = R_{n,on} = Z_0$? Hint: Consider the voltage waveform at V_L.

d. Size the driving transistors to satisfy $R_{p,on} = R_{n,on} = Z_0$. Assume $V_{DD} = 5$ V.

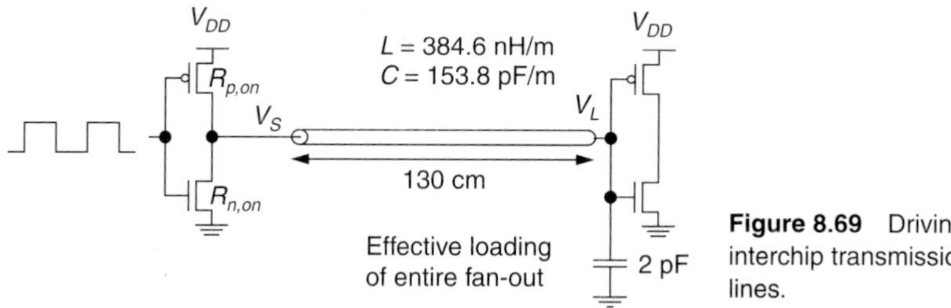

Figure 8.69 Driving interchip transmission lines.

14. [M, SPICE, 8.4] Suppose that you have an inverter made out of two minimum-size transistors driving a capacitance of 40 pF. The power supply inductances can be modeled as two 10 nH inductors. One of them is hooked up between the positive supply rail and the positive node of the inverter (V_+), and one between the negative rail and the negative node (V_-). Draw graphs of V_- and V_+ versus time for a low-to-high-to-low transition of the output. Use $t_R = 2t_{PLH}$ and $t_F = 2t_{PHL}$. Verify with SPICE.

DESIGN PROBLEM

You must send a signal across a poly wire 10 mm long, and then off-chip onto a 20 pF capacitance. Minimize the propagation delay from the input to the output. The signal originates from a minimum-size inverter ($W/L_n = 1.8/1.2$, $W/L_p = 5.4/1.2$). You may place any number of inverters, of any size, at any place along the poly wire. You may also choose the width of the poly wire between each inverter to be 1.2, 1.8, or 2.4 microns. Do not bypass the poly wire with metal. The supply voltage is set at 3 V.

You may use any logic family (static, dynamic, etc.) to implement your buffers as long as the voltage swing at the load capacitor is a full 3 volts. You control the size, number, and position of the buffers along the poly wire. You also choose whether to use 1.2-, 1.8-, or 2.4-micron wide poly wires between inverters.

The design is graded on its energy-delay product, the product of t_p, and the average energy consumed over both low-to-high and high-to-low transitions.

CHAPTER

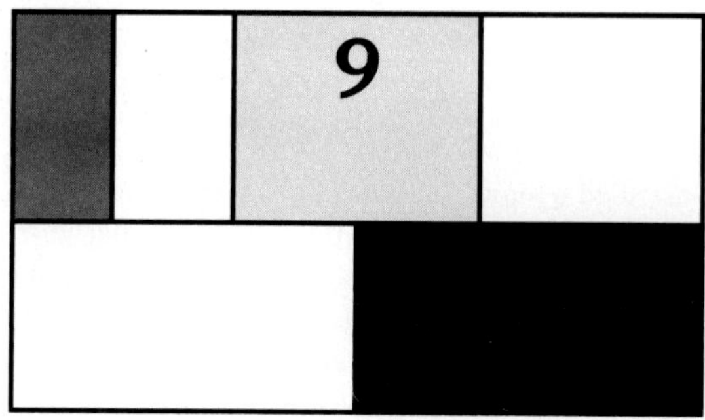

TIMING ISSUES IN DIGITAL CIRCUITS

Impact of clock skew on performance and functionality

Alternative timing methodologies

Synchronization issues in digital IC and board design

Clock generation

9.1 Introduction

All sequential circuits have one property in common—a well-defined ordering of the switching events must be imposed if the circuit is to operate correctly. If this were not the case, wrong data might be written into the memory elements. Most of the current sequential circuits belong to the class of the *synchronous* systems, which means that the latching of data into the memory elements is orchestrated by a number of globally distributed clock signals. The generation of clock signals and their distribution to the memory elements over the chip must adhere to strict regulations; otherwise, the circuit may malfunction.

In this chapter, we first discuss the constraints and requirements imposed by the synchronous design approach. We analyze the impact of delays of the clock signal, called *clock skew*, and introduce approaches to cope with it. One way to avoid the pitfalls of the synchronous design approach is to avoid clocks all together. This *asynchronous design* approach is discussed in the next section. The chapter concludes with a discussion of the interfacing between synchronous and asynchronous partitions of a design and the clock-generation problem.

9.2 Clock Skew and Sequential Circuit Performance

In a synchronous circuit, the clock signal connects to all the registers, flip-flops, and latches as well as the precharge and evaluate transistors of the dynamic gates. This huge fan-out acts as a large capacitive load on the clock line. This load is further increased by the capacitance of the wire itself, which is distributed over the complete chip and may reach a length of many centimeters. The total capacitance of a single clock line easily measures hundreds of pF and can reach into the nF range. For instance, the total clock capacitance of the Alpha microprocessor [Dopperpuhl92] equals 3.25 nF, which is 40% of the total effective switching capacitance on the chip!

The long clock wire also introduces a substantial series resistance, even when routed in metal. A clock line thus behaves as a distributed *RC* line. As the delay of an *RC* line is a function of its length, flip-flops clocked by the "same" clock signal observe different transition times depending on their distance from the clock source. This effect is called *clock skew* and can severely affect the performance of a sequential circuit. One of the goals of a clock signal is to synchronize the updating of the system state. This synchronization is disturbed in the presence of skew, which can result in race conditions and malfunctioning. The scale of the clock skew problem is delineated in Figure 9.1, which plots the delay of a 4 cm clock wire as a function of its width and the size (conductance) of the driver transistor.

We will study the effect of clock skew on circuit functionality and performance for a number of clocking styles. A single reference circuit—the synchronous pipelined datapath of Figure 9.2—will be used throughout this discussion. In this system, each synchronous module is composed of a combinational block *CL* and a latch *R,* and is characterized by six timing parameters.

- The minimum and maximum propagation delays of the register, $t_{r,min}$ and $t_{r,max}$ respectively.

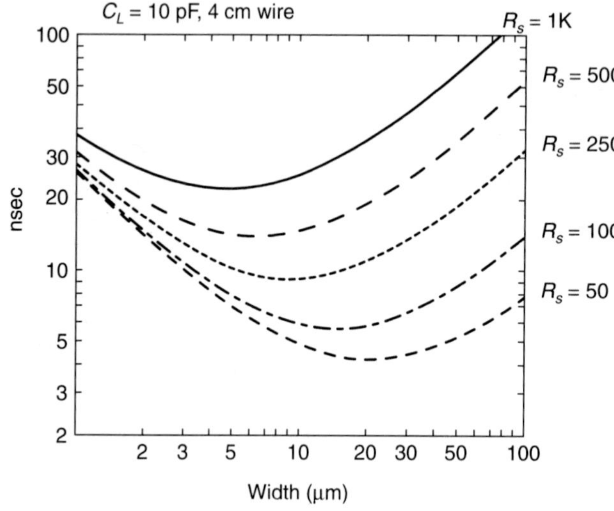

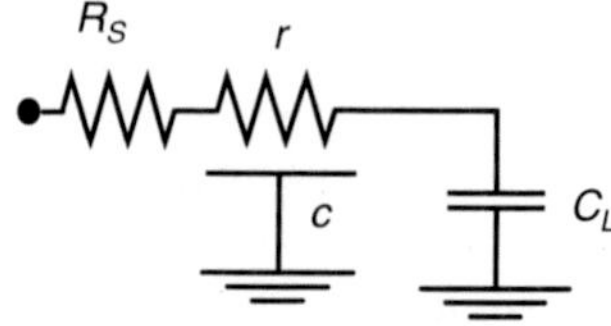

r = 0.07 Ω/□, c = 0.04 fF/μm²
(tungsten wire)

Figure 9.1 Delay of clock wire as a function of width and size of the driver transistor. Notice that this plot only includes wiring delay and ignores fan-out (from [Jacobs90]).

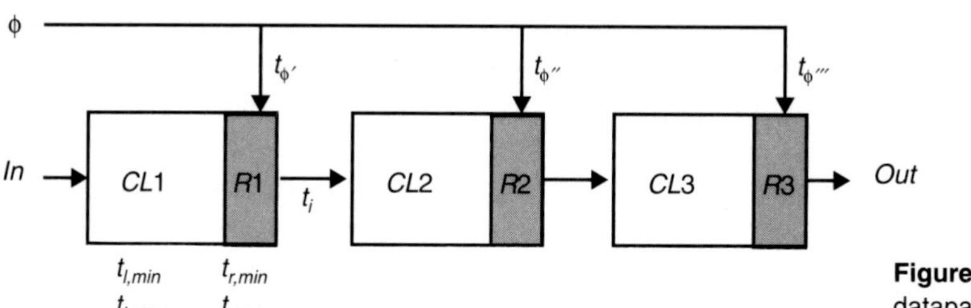

Figure 9.2 Pipelined datapath structure.

- The minimum and maximum delay of the combinational logic, $t_{l,min}$ and $t_{l,max}$.
- The propagation delay of the interconnect t_i.
- The local clock time t_ϕ, which is defined as the skew between the local clock and some global reference that is typically the clock source.

Notice that two delay values are defined for both register and logic. The *maximum propagation delay* corresponds to the traditional definition of the propagation delay and denotes the times it takes for the slowest output signal to respond to a change in the inputs (worst-case situation). This delay determines the maximum allowable clocking speed of the circuit. The *minimum propagation delay* denotes the time it takes for at least one of the outputs to start a transition as a result of a transition on the inputs. This time is generally much shorter than the traditional propagation delay, and is more appropriate for the study of the clock skew, in which the fastest propagation of a signal through the network determines whether or not a race occurs. To simplify the discussion, we have assumed that the setup time of the latches equals zero. The obtained results are easily adapted to address a finite setup time [Hatamian88].

9.2.1 Single-Phase Edge-Triggered Clocking

This clocking style addresses includes both the traditional static *edge-triggered flip-flops* and the dynamic *true single-phase clock registers* introduced in Chapter 6. Consider now the transfer of data between two of those registers $R1$ and $R2$. Due to routing delays, the local clock times are different at $R1$ and $R2$. The difference is called the *clock skew* δ.

$$\delta = t_{\phi''} - t_{\phi'} \tag{9.1}$$

The clock skew can be positive or negative depending upon the routing direction and position of the clock source.

Clock skew can harm the edge-triggered circuit as follows. At the positive (negative) edge of the clock ϕ', data is latched into $R1$ and starts to propagate through the interconnect network and $CL2$ logic block. The earliest time that one of the inputs of $R2$ can make a transition is at $t_{\phi'} + t_{r,min} + t_i + t_{l,min}$. If the local clock of $R2$ is delayed with respect to the clock of $R1$, it might happen that the inputs of $R2$ change before the previous data is latched. A race occurs, and the circuit yields erroneous results, as shown in Figure 9.3a. Condition (9.2) must be satisfied for the circuit to perform properly. This imposes a maximum value on the skew.

$$t_{\phi''} \le t_{\phi'} + t_{r,\,min} + t_i + t_{l,\,min}$$

or (9.2)

$$\delta \le t_{r,\,min} + t_i + t_{l,\,min}$$

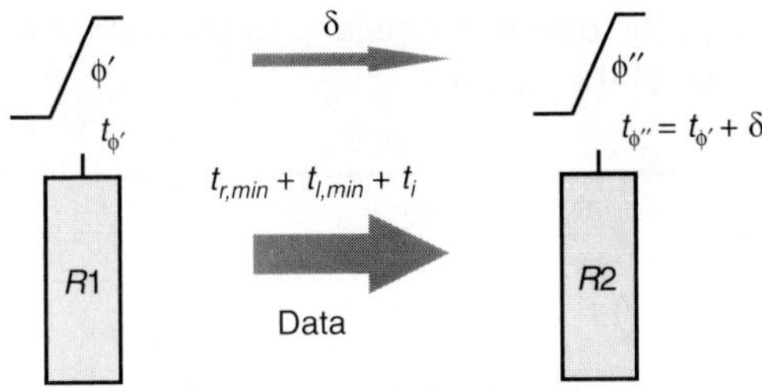

(a) Race between clock and data.

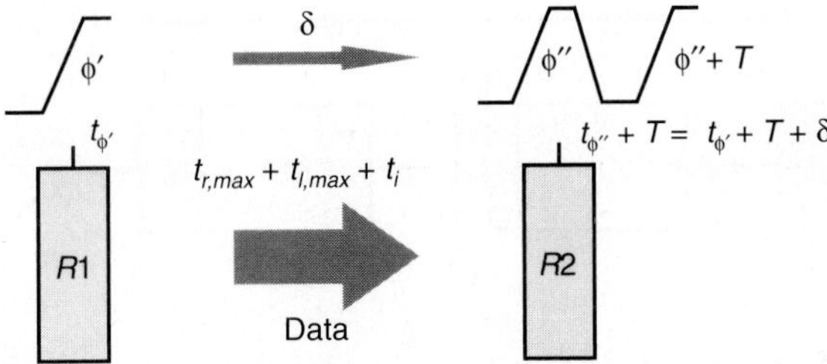

(b) Data should be stable before clock pulse is applied.

Figure 9.3 Constraints on clock skew for edge-triggered clocking.

The correct input data is stable at $R2$ after the worst-case propagation delay or at time $t_{\phi'} + t_{r,max} + t_i + t_{l,max}$. The clock period T must be large enough for the computations to settle. Due to skew, the latching clock edge at $R2$ occurs at time $t_{\phi''} + T$. This translates into a minimum bound on the clock period, as illustrated in Figure 9.3b.

$$t_{\phi 2} + T \geq t_{\phi 1} + t_{r,\,max} + t_i + t_{l,\,max} \quad \text{or} \quad (T \geq t_{r,\,max} + t_i + t_{l,\,max} - \delta) \tag{9.3}$$

Design Consideration

Two scenarios can be envisioned.

- **$\delta > 0$**—The clock is routed in the same direction as the flow of the data through the pipeline (Figure 9.4a). In this case, the skew has to be strictly controlled and satisfy Eq. (9.2). If this constraint is not met, the circuit does malfunction **independent of the clock period**. Therefore, reducing the clock frequency of an edge-triggered circuit does not help get around skew problems! On the other hand, positive skew increases the throughput of the circuit as expressed by Eq. (9.3), because the clock period can be shortened by δ. The extent of this improvement is limited as large values of δ soon provoke violations of Eq. (9.2).
- **$\delta < 0$**—When the clock is routed in the opposite direction of the data (Figure 9.4b), the skew is negative and condition (9.2) is unconditionally met since the propagation delays are always positive. The circuit operates correctly independent of the skew. Unfortunately, negative skew impacts the throughput in a negative way as expressed by Eq. (9.3). The skew reduces the time available for actual computation so that the clock period has to be increased by $|\delta|$. In summary, routing the clock in the opposite direction of the data avoids disasters but hampers the performance of the circuit.

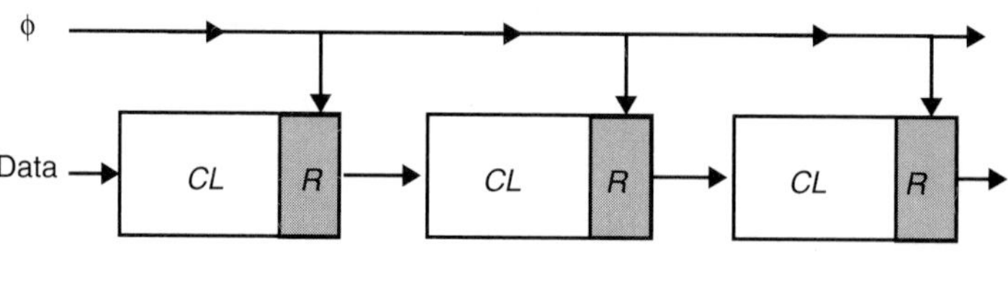

(a) Positive skew

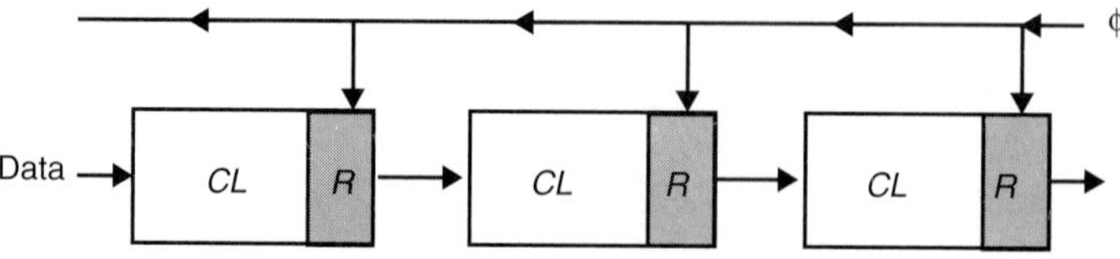

(b) Negative skew

Figure 9.4 Positive and negative clock skew.

The value of the allowable skew is determined by technology parameters such as the delays of interconnect, registers, and logic. Design methodology and architectural style also play an important role. For instance, large skews between connecting registers occur more frequently in a standard-cell approach than in a structured approach such as a bit-sliced datapath—cell placement and interconnect routing are performed automatically in the former, while the cells are carefully placed, and the clock is routed along with the data in the latter. The chance of clock skew harming circuit functionality or performance also increases with decreasing logic depth, caused for instance by the introduction of more pipeline stages.

9.2.2 Two-Phase Master-Slave Clocking

Consider the pipelined circuit of Figure 9.5. Master *M* and slave *S* are transparent latches that could be implemented using the dynamic latch shown in the inset. For the sake of simplicity, we assume that both ϕ_1 and ϕ_2 are routed in the same direction and exhibit the same relative skew. The second register *M2-S2* observes the delayed clocks ϕ_1' and ϕ_2' due to the skew δ. This is illustrated in Figure 9.6 for the case of a positive skew; that is, the clock is routed in the same direction of the data. Some extra definitions are introduced to fully specify the clocks. $T_{\phi 1}$ and $T_{\phi 2}$ stand for the time intervals for which ϕ_1 or ϕ_2 are high. $T_{\phi 12}$ and $T_{\phi 21}$ denote the separation between ϕ_1 and ϕ_2 and vice versa (*nonoverlap* times). The nonoverlap times must be positive for the circuit of Figure 9.5 to operate correctly.

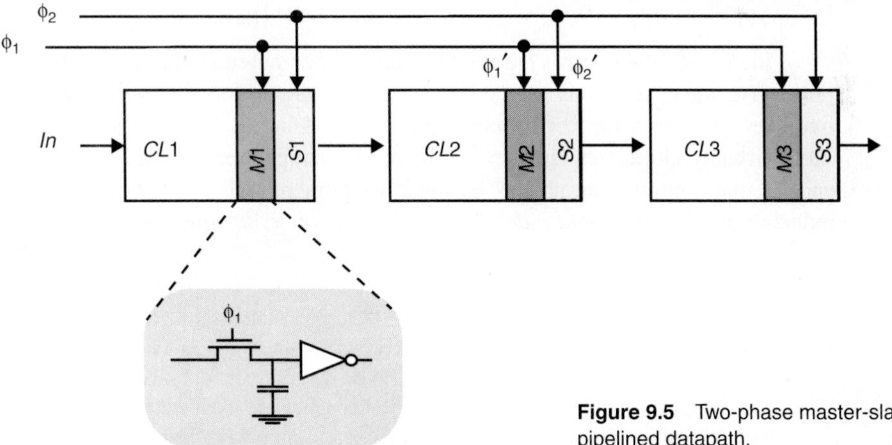

Figure 9.5 Two-phase master-slave pipelined datapath.

New data is applied to the logic block *CL2* at the rising edge of ϕ_2 and is clocked into the register *M2* at the falling edge of ϕ_1'. Due to the skew, ϕ_2 and ϕ_1' might be overlapping, as illustrated in Figure 9.6. If the data signal propagates fast enough through the logic block, it may reach *R2* before ϕ_1' has fallen, thus destroying the data in the latch. To prevent this, the overlap time must be smaller than the fastest propagation time through the logic:

$$\delta - T_{\phi 12} \leq t_{r,min} + t_i + t_{l,min} \tag{9.4}$$

Comparing this to Eq. (9.2), we can see that the nonoverlap time $T_{\phi 12}$ acts as a buffer, partly absorbing the effects of the clock skew. The effects of clock-skew can

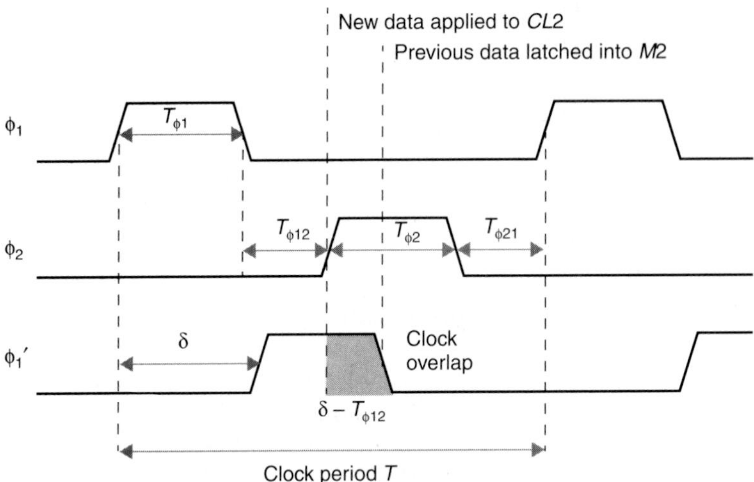

Figure 9.6 Two-phase clocks and clock skew.

always be countered by increasing the nonoverlap time in a two-phase clocking system. This remedy unfortunately has an adverse effect on the performance of the circuit, as expressed by Eq. (9.5), which sets a minimum bound on the clock period:

$$T + \delta - T_{\phi 12} \geq t_{r,\,max} + t_i + r_{l,\,max} \tag{9.5}$$

Increasing the nonoverlap time to battle skew increases the minimum clock period. Otherwise, the conclusions are similar to the ones drawn for the single-phase logic—positive skew increases the performance but poses tighter skew constraints, while negative skew virtually eliminates the skew problem but hampers performance. The crucial difference is that a circuit can always be made to work regardless of the amount of skew by reducing the clock frequency and increasing the nonoverlap time between the clock-phases.

Example 9.1 Clock Skew

Consider the logic shown in Figure 9.7. Determine the minimum and maximum allowable clock skews so that the circuit operates correctly, assuming gate delay t_g, mux delay t_m, latch setup time t_s, latch delay t_l, and clock period T. No differentiation between the maximum and minimum delays of the gates is necessary, since they are identical. Assume that the input signals, other than the one from the latch, arrive early enough so they are stable during the operation. The race analysis puts a max bound on the skew:

$$t_l + t_d + t_m + t_s > \delta$$

The equilibrium requirement at the time of latching imposes a second constraint.:

$$t_l + 5t_d + t_m + t_s < T + \delta$$

These constraints can be combined to define the following bounds on the skew:

$$t_l + t_d + t_m + t_s > \delta > t_l + 5t_d + t_m + t_s - T$$

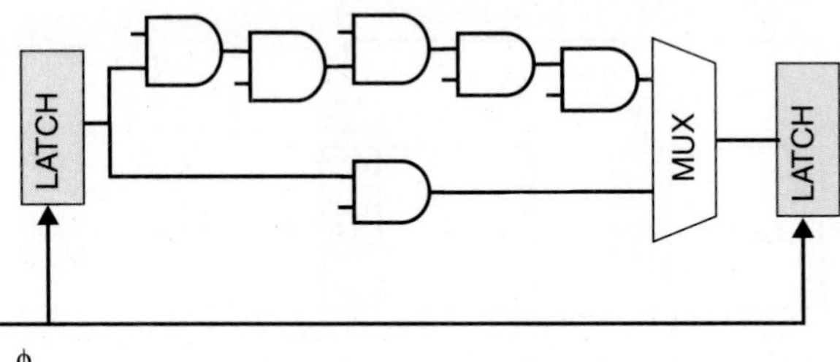

Figure 9.7 Logic network for clock-skew analysis.

CONSIDERATION: Overlap-Insensitive Registers and Clock Skew

In Chapter 6, we established that some register styles such as C^2MOS are insensitive to clock overlap. Does this also mean that these circuits are not affected by clock skew?

Not really. The overlap insensitivity prevents a signal from permeating through multiple register stages all at once. This does not mean that the sampled signal is necessarily the correct one. A C^2MOS register might sample a future value if the overlap between the clock signals is larger than the delay between the registers.

The fact that a circuit style such as C^2MOS allows for clock overlap makes the skew-management problem even more pronounced. The (potentially) negative value of $T_{\phi 12}$ puts some very tight bounds on the skew, as reflected by Eq. (9.4).

9.2.3 Other Clocking Styles

The techniques presented above apply to any other clocking scheme and register arrangement. The important message is that clock skew does not always result in a loss of performance, and can sometimes be used to boost the throughput. This assumes that one is willing to take a considerable risk, since clock skew can produce failure modes that are independent of the clock speed no matter how low it is. Remember also to differentiate between the minimum and maximum propagation delays when analyzing performance and skew. The former determine how much clock skew can be tolerated, while the latter dominate circuit performance.

Problem 9.1 Clock Skew in TSPC (or NORA) Pipelined Logic.

A high-performance datapath often consists of a cascade of modules operating on alternating clock phases (Figure 9.8). Derive expressions for the maximum allowable skew and minimum clock period.

9.2.4 How to Counter Clock Skew Problems

So far, we have learned that clock skew problems can be controlled in two ways.

1. Routing the clock in the opposite direction of the data.
2. Controlling the nonoverlap periods of the clock. This solution is an option only in the case of two-phase clocking.

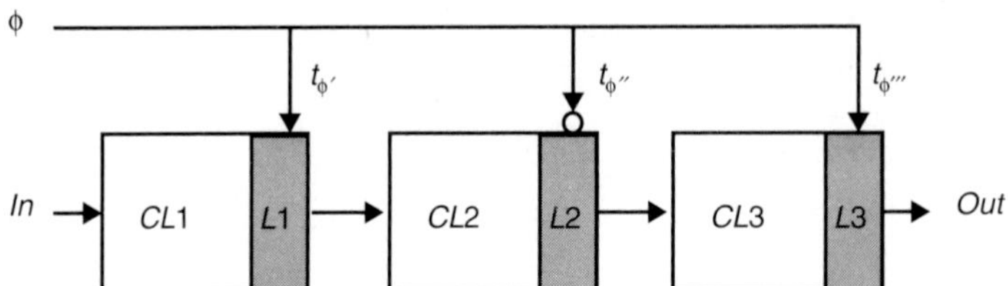

Figure 9.8 Pipelined datapath using alternating phases. *L*1, *L*2, and *L*3 are transparent latches. *L*1 and *L*3 evaluate on ($\phi = 1$), while *L*2 operates on the opposite phase.

Both solutions hamper the performance of the circuit. In addition, the dataflow in a circuit is not always unidirectional. Feedback might occur, as demonstrated in Figure 9.9. The skew can assume both positive and negative values, depending upon the direction of the data transfer. Under these circumstances, the designer has to account for the worst-case skew condition. In general, routing the clock so that only negative skew occurs is feasible for subcircuits such as datapaths. Other approaches must be used at the global chip level.

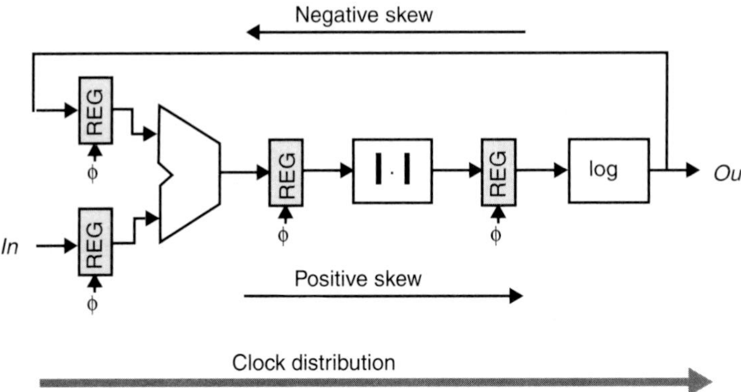

Figure 9.9 Datapath structure with feedback.

The most obvious, and hence most practical, solution to the skew problem is to ensure that the clock skew between communicating registers is bounded. We can do this by careful analysis and design of the clock-distribution network. Note that the absolute value of the skew—between local clock and clock source—is irrelevant. It does not matter if the clock takes 20 nsec to traverse to a given node in the network—what counts is the relative skew between communicating registers.

The designer can govern the clock-distribution network through a number of design parameters.

- Interconnect material used for the clock network
- Shape of the clock-distribution network
- Clock driver and the buffering scheme used
- Load on the clock lines (i.e., the fan-out)

- Rise and fall times of the clock

The last parameter determines the appropriate model for the clock line. Transmission line effects must be considered if the time-of-flight of the clock signal along the line becomes comparable to the rise/fall time of the signal, as discussed in Chapter 8. If not, a distributed *RC* model is sufficient.

A clock-distribution network that achieves the goal of minimizing the skew between communicating elements is the *H-tree network*, illustrated in Figure 9.10 for a 4 × 4 array of cells. This configuration has the property that the clock signal is delayed by an equal amount for each sub-block, since all blocks are equidistant from the clock source. The clock skew is theoretically zero. The *H*-tree configuration is an ideal model and is only useful for regular-array networks in which all elements are identical and the clock can be distributed as a binary tree. However, the underlying concept is more generic. It represents a floorplan that distributes the clock signal so that the interconnections carrying the clock signals to the functional sub-blocks are of equal length. The sub-blocks themselves should be small enough so that the skew within the block is tolerable.

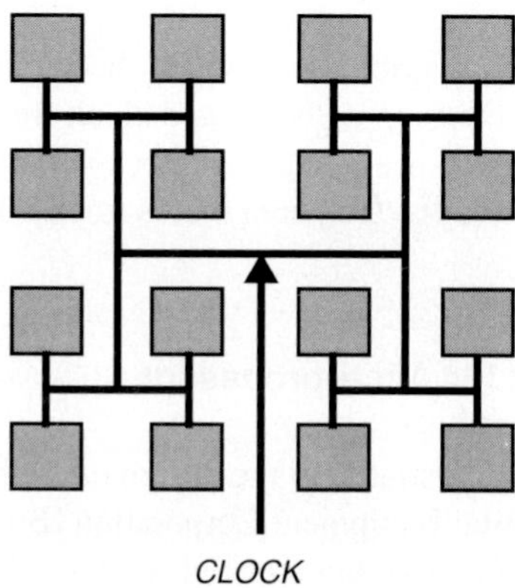

Figure 9.10 Example of an *H*-tree clock-distribution network for a 4 × 4 array of processing elements.

A more general approach is to control the skew by equalizing the local clock delay through a careful routing of the clock signals, combined with a hierarchical clock-buffering scheme. An example of such a scheme is shown in Figure 9.11. This approach does not

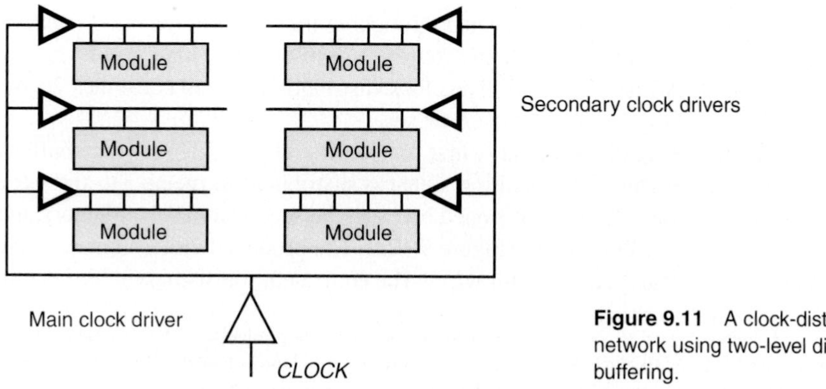

Figure 9.11 A clock-distribution network using two-level distributed buffering.

result in zero skew, but the previous discussion has shown that this is not necessary. The only requirement is that the skew be bounded within a certain limit, which can be quite tolerant in actual designs. The task of the intermediate buffers is to isolate local clock nets from upstream load impedances and to amplify the clock signals degraded by the *rc* network. The buffers also decrease the absolute skew value and produce steep clock slopes. Although a large absolute clock delay does not hamper the performance, it does make it harder to interface different chips and boards. Keeping the clock slopes steep enough is important for design styles using C^2MOS or TSPC latches.[1] The number of intermediate buffering layers depends upon the interconnect material and the size and the fan-out of the clock network, as is apparent from Chapter 8.

It is essential to consider clock distribution in the earlier phases of the design of a complex circuit, since it might influence the shape and form of the chip floorplan. Clock distribution is often only considered in the last phases of the design process, when most of the chip layout is already frozen. This results in unwieldy clock networks and multiple timing constraints that hamper the performance and operation of the final circuit. With careful planning, a designer can avoid many of these problems, and clock distribution becomes a manageable operation.

Despite this observation, be aware that clock distribution is rapidly becoming one of the dominant design problems due to the ever-increasing die sizes and clock frequencies. These problems can be circumvented by avoiding the use of a global clock and by designing circuits in a self-timed manner. This is an active research area that is briefly discussed in the next section.

9.2.5 Case Study—The Digital Alpha 21164 Microprocessor

The importance of controlling clock skew is well illustrated in the design of the high-performance Alpha 21164 microprocessor from Digital Equipment Corporation [Bowhill95]. A careful design of the clock network and a well-thought-out clocking strategy is required to meet the targeted clock frequency of 300 Mhz for a complex microprocessor design containing 9.3 million transistors on a 16.5 mm $\times$ 18.1 mm die in a four-metal-layer 0.55 μm CMOS technology.

First of all, a single-phase clocking methodology was selected. The advantages of this approach are the elimination of the dead times between clock phases and the fact that only one clock signal has to be routed throughout the chip. On the other hand, this clocking methodology combined with an extensive use of dynamic logic results in a substantial load on the clock line of 3.75 nF. The clock-distribution system consumes 20 W, which is 40% of the total dissipation of the processor.

The incoming clock signal is first routed through a single six-stage buffer placed at the center of the chip. The resulting signal is distributed in metal-3 to the left and right banks of final clock drivers, positioned between the secondary cache memory and the outside edge of the execution unit (Figure 9.12a). The produced clock signal is driven onto a single grid of metal-3 and metal-4 wires. The equivalent transistor width of the final driver

[1] Intermediate clock buffering is a controversial issue. A large number of designs use a single, centralized clock driver to drive all clock nodes. This approach results in a larger absolute skew, but is more amenable to design automation and avoids problems with parameter variations over large chips.

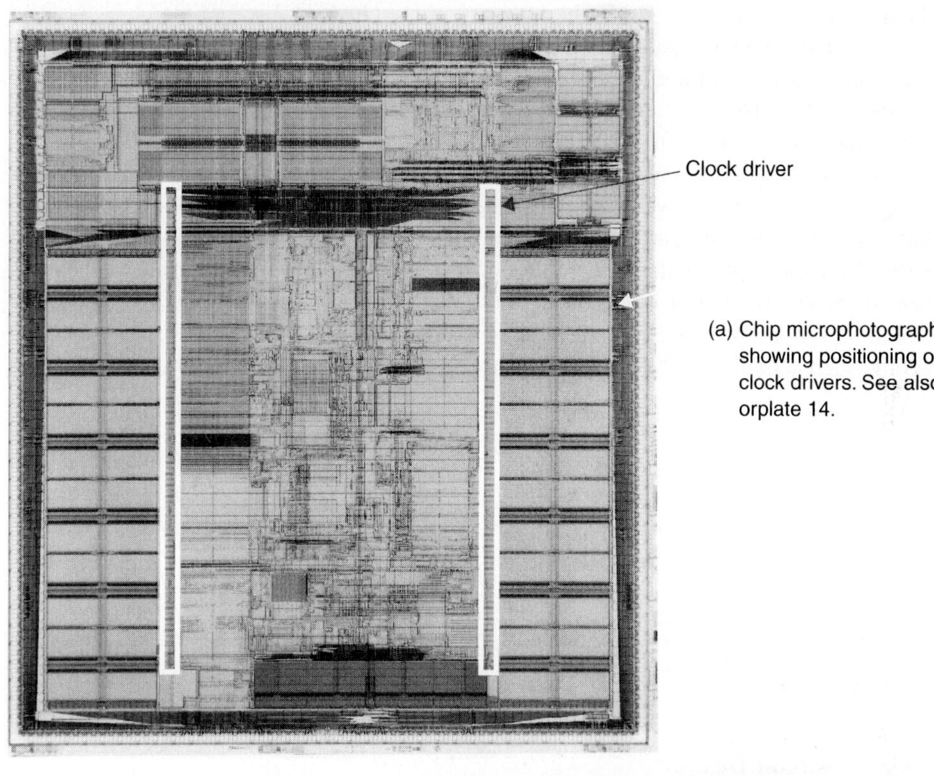

(a) Chip microphotograph, showing positioning of clock drivers. See also Colorplate 14.

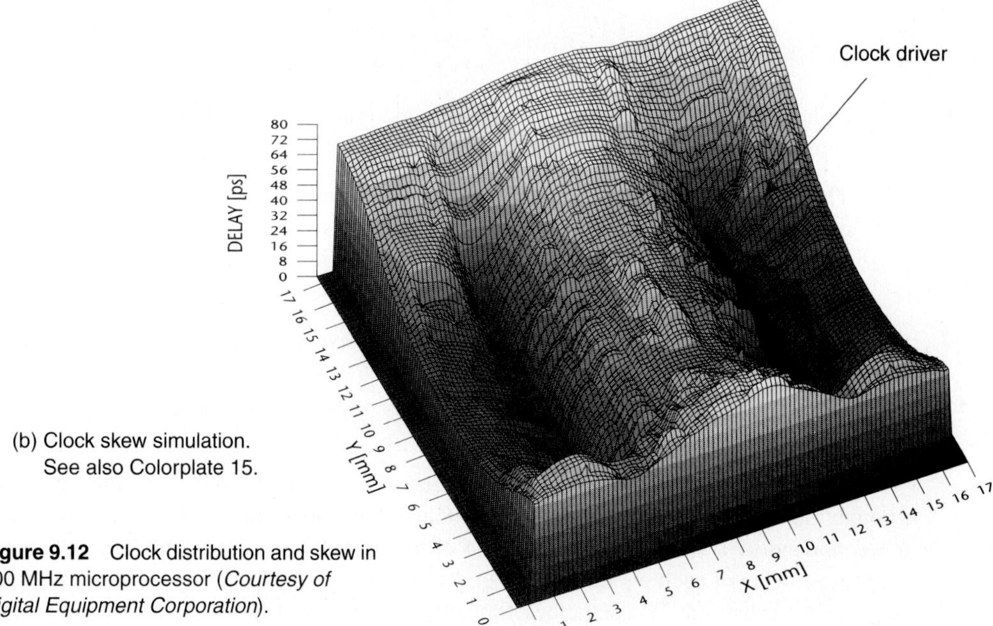

(b) Clock skew simulation. See also Colorplate 15.

Figure 9.12 Clock distribution and skew in 300 MHz microprocessor (*Courtesy of Digital Equipment Corporation*).

inverter equals 58 cm! To ensure the integrity of the clock grid across the chip, the grid was extracted from the layout, and the resulting *RC*-network was simulated using a circuit simulation program. A three-dimensional representation of the simulation results is plotted in Figure 9.12b. The maximum value of the absolute skew is smaller than 90 psec. The critical instruction and execution units all see the clock within 65 psec.

Clock skew and race problems were addressed using a mix-and-match approach. The clock skew problems were eliminated by either routing the clock in the opposite direction of the data at a small expense of performance or by ensuring that the data could not overtake the clock. A standardized library of level-sensitive transmission-gate latches was used for the complete chip. These latches have the advantage of being very fast (10% faster than the TSPC latches used in the preceding 21064 [Dopperpuhl92]) but are susceptible to race-through conditions. The latter problem was addressed by a number of approaches.

- Careful sizing of the local clock buffers so that their skew was minimal.
- At least one gate had to be inserted between connecting latches. This gate, which can be part of the logic function or just a simple inverter, ensures that the signal cannot overtake the clock. Special design verification tools were developed to guarantee that this rule was obeyed over the complete chip.

This example demonstrates that managing clock skew and clock distribution for large, high-performance synchronous designs is a feasible task. However, making such a circuit work in a reliable way requires careful planning and intensive analysis.

9.3 Self-Timed Circuit Design*

9.3.1 Self-Timed Concept

The synchronous design approach advocated in the previous sections assumes that all circuit events are orchestrated by a central clock or clocks. Those clocks have a dual function.

- They insure that the *physical timing constraints* are met. The next clock cycle can only start when all transitions have settled and the system has come to a steady state. This ensures that only legal logical values are applied in the next round of computation. In short, clocks are used to account for the delays of gates and the wiring.
- Clock events serve as a *logical ordering mechanism* for the global system events. A clock provides a time base that determines what will happen and when. On every clock transition, a number of operations are initiated that change the state of the sequential network.

Consider the pipelined datapath of Figure 9.13. In this circuit, the data marches through the operation chain under the command of the clock, like an army column under the command of a shouting sergeant. The clock period is chosen to be larger than the worst-case delay of each pipeline stage, or $T >$ max $(t_{pF1}, t_{pF2}, t_{pF3}) + t_{preg}$. This represents the physical constraint. At each clock transition, a new set of inputs is sampled and computation is started anew. The throughput of the system—which is equivalent to the number

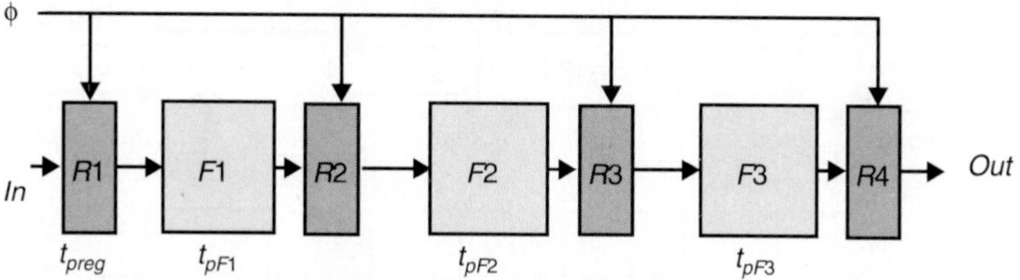

Figure 9.13 Pipelined, synchronous datapath.

of data samples processed per second—is equivalent to the clock rate. When to sample a new input or when an output is available depends upon the *logical ordering* of the system events and is clearly orchestrated by the clock in this example.

The synchronous design methodology has some clear advantages. It presents a structured, deterministic approach to the problem of choreographing the myriad of events that take place in digital designs. The approach taken is to equalize the delays of all operations by making them as bad as the worst of the set. The approach is robust and easy to adhere to, which explains its enormous popularity; however it does have some pitfalls.

- It assumes that all clock events or timing references happen simultaneously over the complete circuit. This is not the case in reality, because of effects such as clock skew.
- The linking of physical and logical constraints has some obvious effects on the performance. For instance, the throughput rate of the pipelined system of Figure 9.13 is directly linked to the worst-case delay of the slowest element in the pipeline. On the average, the delay of each pipeline stage is smaller. The same pipeline could support an average throughput rate that is substantially higher than the synchronous one.

One way to avoid these problems is to opt for an *asynchronous* design approach and to eliminate all the clocks. Designing a purely asynchronous circuit is a nontrivial and potentially hazardous task. Ensuring a correct circuit operation that avoids all potential race conditions under any operation condition and input sequence requires a careful timing analysis of the network. In fact, the logical ordering of the events is dictated by the structure of the transistor network and the relative delays of the signals. This was demonstrated in the *JK* flip-flop circuit of Figure 6.5, which can be considered an asynchronous circuit. Under the condition that $J = K = 1$, this circuit contains a cycle as long as the enable signal ϕ equals 1. If the on-time of ϕ is longer than the propagation delay of the cycle, the output toggles between the on- and off-state creating an unpredictable result. Strict timing constraints must be imposed to make this circuit operate correctly. In this particular case $T_{on,\phi}$ must be smaller than $t_{p,cycle}$. Enforcing these constraints by manipulating the logic structure and the lengths of the signal paths requires an extensive use of CAD tools, and is only recommended when strictly necessary.

A more reliable and robust technique is the self-timed approach, which presents a local solution to the timing problem [Seitz80]. Figure 9.14 uses a pipelined datapath to illustrate how this can be accomplished. Suppose each combinational function has a means of indicating that it has completed a computation. This situation is asserted by raising a *Done* flag. To make the picture complete, an extra input signal called *Start* is needed to act as an initiator for a computation. This mechanism helps establish the precise dura-

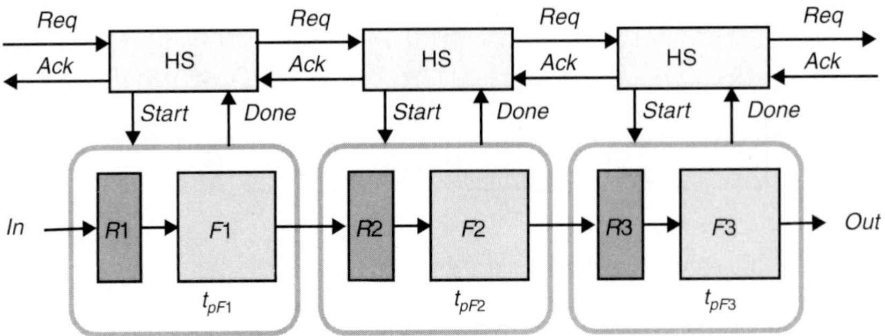

Figure 9.14 Self-timed, pipelined datapath.

tion of a computation in a data-dependent fashion (taking care of the physical constraint). Additionally, the operators must signal each other that they are either ready to receive a next input word or that they have a legal data word at their outputs that is ready for consumption. This signaling ensures the logical ordering of the events and can be achieved with the aid of an extra *Ack(nowledge)* and *Req(uest)* signal. In the case of the pipelined datapath, the scenario could proceed as follows.

1. An input word arrives, and a *Req(uest)* to the block *F*1 is raised. If *F*1 is inactive at that time, it transfers the data and acknowledges this fact to the input buffer, which can go ahead and fetch the next word.

2. *F*1 is enabled by raising the *Start* signal. After a certain amount of time, dependent upon the data values, the *Done* signal goes high indicating the completion of the computation.

3. A *Re(quest)* is issued to the *F*2 module. If this function is available, an *Ack(nowledge)* is raised, the output value is transferred, and *F*1 can go ahead with its next computation.

The self-timed approach effectively separates the physical and logical ordering functions implied in circuit timing. The completion signal *Done* ensures that the physical timing constraints are met and that the circuit is in steady state before accepting a new input. The logical ordering of the operations is ensured by the acknowledge-request scheme, often called *a handshaking protocol.* Both interested parties synchronize with each other by mutual agreement or, if you want, by shaking hands. The ordering protocol described above and implemented in the module HS is only one of many that are possible. The choice of protocol is important, since it has a profound effect on the circuit performance and robustness.

When compared to the synchronous approach, self-timed circuits display some alluring properties.

- In contrast to the global centralized approach of the synchronous methodology, timing signals are generated *locally*. This avoids all problems and overheads of distributing high-speed clocks.

- Since synchronization is achieved locally, there are no global side effects. This results in increased design modularity.

- Separating the physical and logical ordering mechanisms results in a potential increase in performance. In synchronous systems, the period of the clock has to be stretched to accommodate the slowest path over all possible input sequences. In self-timed systems, a completed data-word does not have to wait for the arrival of the next clock edge to proceed to the subsequent processing stages. Since circuit delays are often dependent upon the actual data value, a self-timed circuit proceeds at the *average speed* of the hardware in contrast to the *worst-case* model of synchronous logic. The average period T_{avg} of the pipelined datapath, for instance, is now determined by the max of the average delays of the individual modules: $T_{avg} = \max(t_{avg,F1}, t_{avg,F2}, t_{avg,F3}) + t_{preg}$.
- The automatic shut-down of blocks that are not in use can result in power savings. Additionally, the power consumption overhead of generating and distributing high-speed clocks can be partially avoided. As discussed earlier, this overhead can be substantial.
- Self-timed circuits are by nature robust to variations in manufacturing and operating conditions such as temperature. Synchronous systems are limited by their performance at the extremes of the operating conditions. The performance of a self-timed system is determined by the actual operating conditions.

Unfortunately, these nice properties are not for free; they come at the expense of a substantial circuit-level overhead, which is caused by the need to generate completion signals and the need for handshaking logic that acts as a local traffic agent to order the circuit events (see block HS in Figure 9.14). Both of these topics are treated in more detail in subsequent sections.

9.3.2 Completion-Signal Generation

Adjusting a circuit so that it raises a signal to flag its completion does come at an expense. It actually requires the introduction of redundancy in the data representation to signal that a particular bit is either in a transition or steady-state mode. Consider the redundant data model presented in Table 9.1. Two bits ($B0$ and $B1$) are used to represent a single data bit B. For the data to be valid or the computation to be completed, the circuit must be in a legal 0 ($B0 = 0$, $B1 = 1$) or 1 ($B0 = 1$, $B1 = 0$) state. The ($B0 = 0$, $B1 = 0$) condition signals that the data is nonvalid and the circuit is in either a reset or transition mode. The ($B0 = 1$, $B1 = 1$) state is illegal and should never occur in an actual circuit.

Table 9.1 Redundant signal representation to include transition state.

B	*B0*	*B1*
in transition (or reset)	0	0
0	0	1
1	1	0
illegal	1	1

A circuit that actually implements such a redundant representation is shown in Figure 9.15. Attentive readers will recognize this circuit as a dynamic version of the DCVSL (differential cascode voltage switch logic) logic style, where the clock is replaced by the *Start* signal [Heller84]. DCVSL uses a redundant data representation by nature of its differential dual-rail structure. When the *Start* signal is low, the circuit is precharged by the PMOS transistors, and the output ($B0$, $B1$) goes in the *Reset-Transition* state (0, 0). When the *Start* signal goes high, signaling the initiation of a computation, the NMOS pull-down network evaluates, and one of the precharged nodes is lowered. Either $B0$ or $B1$—but never both—goes high, which raises *Done* and signals the completion of the computation.

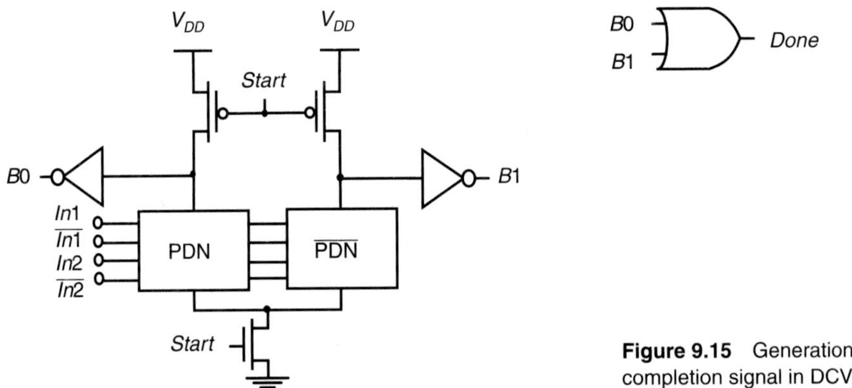

Figure 9.15 Generation of a completion signal in DCVSL.

DCVSL is more expensive in terms of area than a nonredundant circuit due to its dual nature. The completion generation is performed in series with the logic evaluation, and its delay adds directly to the total delay of the logic block. The completion signals of all the individual bits must be combined to generate the completion for an N-bit data word. Completion generation thus comes at the expense of both area and speed. The benefits of the dynamic timing generation often justify this overhead.

The DCVSL approach is not the only circuit technique to offer automatic completion generation. Other dual-rail approaches, such as dynamic pass-transistor logic (CPL), offer similar capabilities. Redundant signal presentations other than the one presented in Table 9.1 can be envisioned. One essential element is the presence of a *transition state* denoting that the circuit is in evaluation mode and the output data is not valid.

When area overhead is a prime issue or when obtaining the average speed is not essential, some of the overhead of the automatic completion generation can be avoided by mimicking the delay of the logic circuit with a separate delay element, as shown in Figure 9.16. In this approach, the delay of the network is set to a fixed number that is larger than the worst-case delay of the logic module. A normal nonredundant circuit style can then be employed. This approach is often used to generate the internal timing of semiconductor memories where self-timing is a commonly used technique. The delay module can be implemented using a variety of approaches, such as an RC network or a chain of inverters.

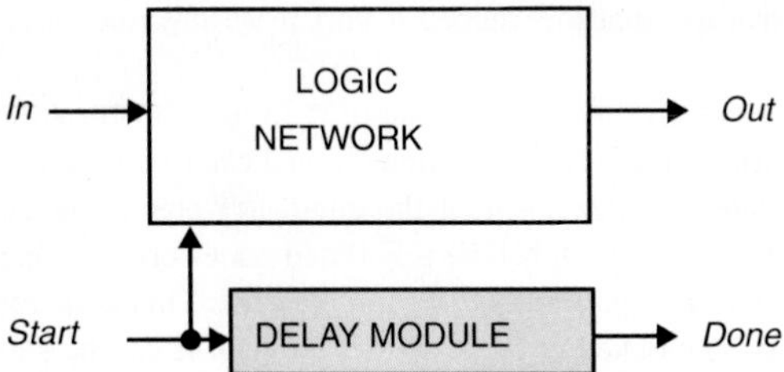

Figure 9.16 Completion-signal generation using delay module.

Example 9.2 Self-Timed Adder Circuit

An efficient implementation of a self-timed adder circuit is shown in Figure 9.17 [Abnous93]. A Manchester-carry scheme is used to boost the circuit performance. The proposed approach is based on the observation that the circuit delay of an adder is dominated by the carry-propagation path. It is consequentially sufficient to use the differential signaling in the carry path only (Figure 9.17a). The completion signal is efficiently derived by combining the carry signals of the different stages (Figure 9.17b). This safely assumes that the sum generation, which

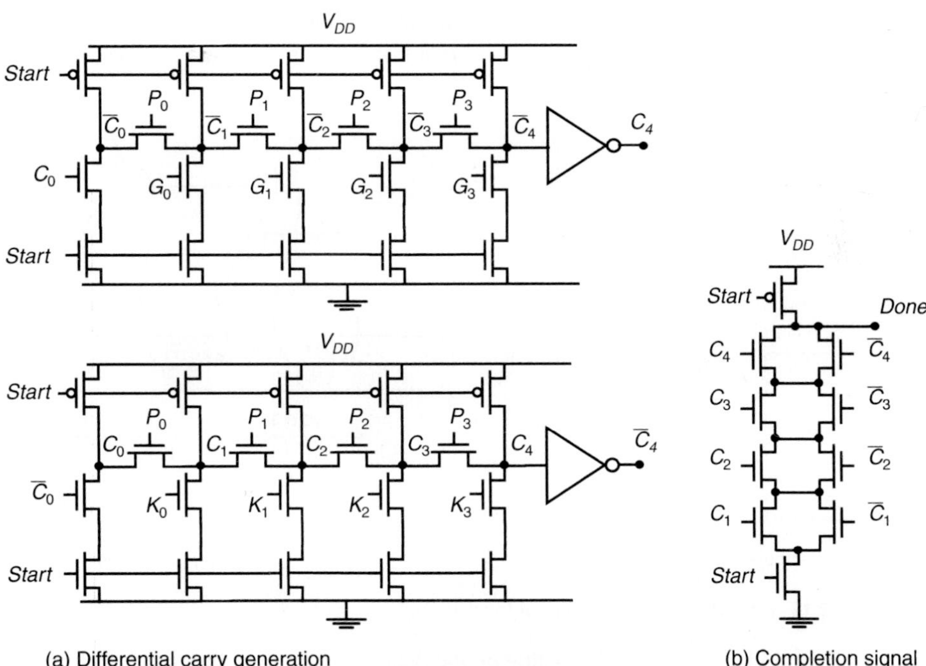

Figure 9.17 Manchester-carry scheme with differential signal representation.

depends upon the arrival of the carry signal, is faster than the completion generation. The benefit of this approach is that the completion generation starts earlier and proceeds in parallel with sum generation, which reduces the critical timing path. All other signals such as *P(ropagate)*, *G(enerate)*, *K(ill)*, and *S(um)* do not require completion generation and can be implemented in single-ended logic. As shown in the circuit schematics, the differential carry paths are virtually identical. The only difference is that the *G(enerate)* signal is replaced by a *K(ill)*.

A simple logic analysis demonstrates that this indeed results in an inverted carry signal and, hence, a differential signaling.

The resulting adder achieves a propagation delay of 0.23 nsec/bit for an area of 4750 μm^2/bit in the 1.2 μm CMOS technology. A synchronous Manchester-carry adder operates at the same speed but is smaller, since it does not need the dual carry chain. The actual performance difference is substantially larger though. The self-timed adder operates at *the average delay* when taken statistically over a large number of input vectors. This is in contrast to the synchronous adder, which must be clocked at the worst-case delay. It can be shown that the average performance of a self-timed ripple-carry adder over all possible input patterns is O(log(*N*)), where *N* equals the number of bits. This is similar to the performance of a synchronous, tree-structured carry-lookahead adder but does not require the complex structure.

9.3.3 Self-Timed Signaling

Besides the generation of the completion signals, a self-timed approach also requires a handshaking protocol to logically order the circuit events avoiding races and hazards. The functionality of the signaling (or handshaking) logic is illustrated by the example of Figure 9.18, which shows a *sender module* transmitting data to a *receiver* ([Sutherland89]).

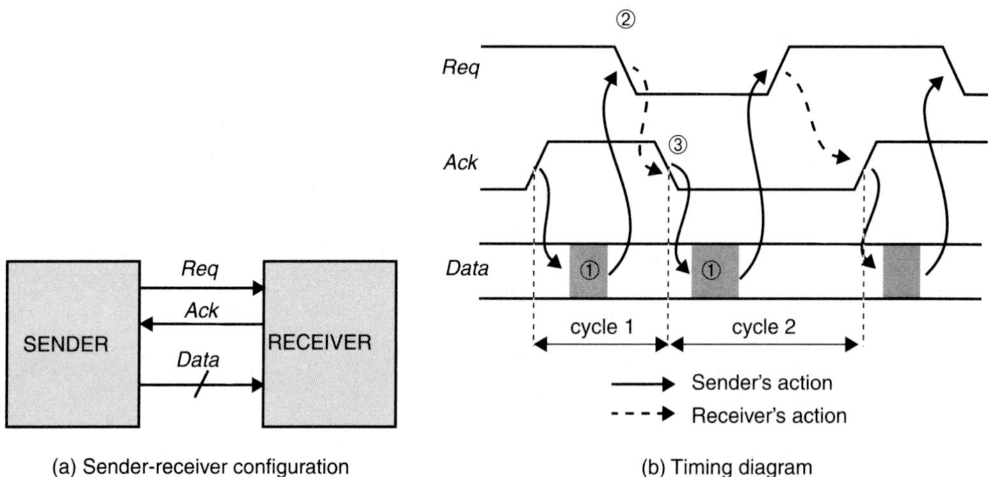

Figure 9.18 Two-phase handshaking protocol.

The sender places the data value on the data bus ① and produces an event on the *Req* control signal by changing the polarity of the signal ②. In some cases the request event is a rising transition; at other times it is a falling one—the protocol described here does not distinguish between them. Upon receiving the request, the receiver accepts the data when possible and produces an event on the *Ack* signal to indicate that the data has been accepted ③. If the receiver is busy or its input buffer is full, no *Ack* event is generated, and the transmitter is stalled until the receiver becomes available by, for instance, freeing space in the input buffer. Once the *Ack* event is produced, the transmitter goes ahead and produces the next data word ①. The four events, *data change, request, data acceptance,* and *acknowledge*, proceed in a cyclic order. Successive cycles may take different amounts of time depending upon the time it takes to produce or consume the data.

This protocol is called *two-phase*, since only two phases of operation can be distinguished for each data transmission—the active cycle of the sender and the active cycle of the receiver. Both phases are terminated by certain events. The *Req* event terminates the active cycle of the sender, while the receiver's cycle is completed by the *Ack* event. The sender is free to change the data during its active cycle. Once the *Req* event is generated, it has to keep the data constant as long as the receiver is active. The receiver can only accept data during its active cycle.

The correct operation of the sender-receiver system requires a strict ordering of the signaling events, as indicated by the arrows in Figure 9.18. Imposing this order is the task of the handshaking logic which, in a sense, performs logic manipulations on events. An essential component of virtually any handshaking module is the *Muller C-element*. This gate, whose schematic symbol and truth table are given in Figure 9.19, performs an AND-operation on events. The output of the C-element is a copy of its inputs when both inputs

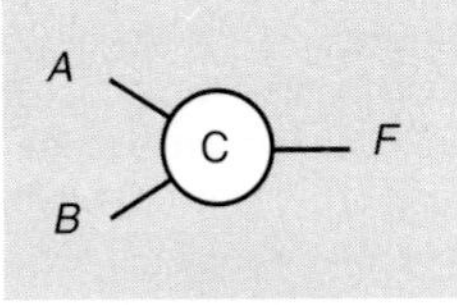

(a) Schematic

A	B	F_{n+1}
0	0	0
0	1	F_n
1	0	F_n
1	1	1

(b) Truth table

Figure 9.19 Muller C-element.

are identical. When the inputs differ, the output retains its previous value. Phrased in a different way—events must occur at both inputs of a Muller C-element for its output to change state and to create an output event. As long as this does not happen, the output remains unchanged and no output event is generated. The implementation of a C-element is centered around a flip-flop, which should be of no surprise, given the similarities in their truth tables. Figure 9.20 displays two potential circuit realizations—a static and a dynamic one respectively. Only a minimal amount of circuitry is required.

Figure 9.21 shows how to use this component to enforce the two-phase handshaking protocol for the example of the sender-receiver. Assume that *Req*, *Ack,* and *Data Ready* are

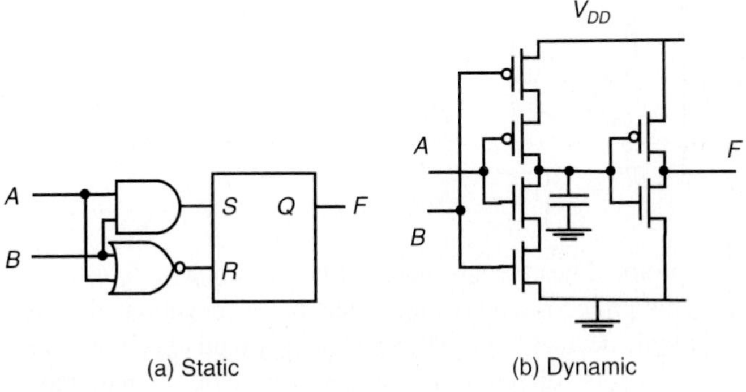

(a) Static (b) Dynamic

Figure 9.20 Implementations of a Muller C-element.

initially 0. When the sender wants to transmit the next word, the *Data Ready* signal is set to 1, which triggers the C-element because both its inputs are at 1. *Req* goes high—this is commonly denoted as *Req*↑. The sender now resides in the wait mode, and control is passed to the receiver. The C-element is blocked, and no new data is sent to the data bus (*Req* stays high) as long as the transmitted data is not processed by the receiver. Once this happens, the *Data accepted* signal is raised. This can be the result of many different actions, possibly involving other C-elements communicating with subsequent blocks. An *Ack*↑ ensues, which unblocks the C-element and passes the control back to the sender. A *Data ready*↓ event, which might already have happened before *Ack*↑, produces a *Req*↓, and the cycle is repeated.

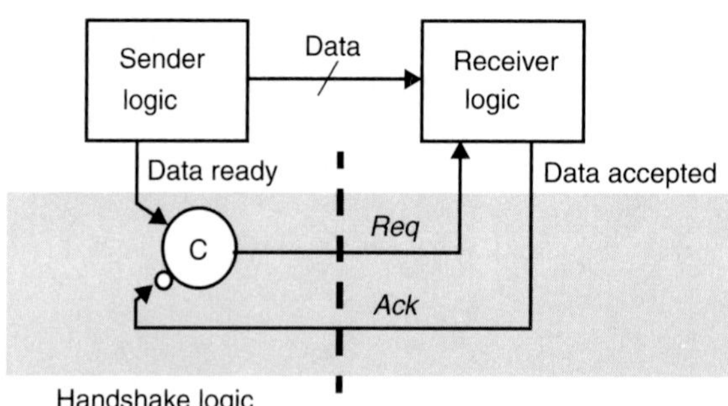

Figure 9.21 A Muller C-element implements a two-phase handshake protocol. The circle at the lower input of the Muller C-element stands for inversion.

Problem 9.2 Two-Phase Self-Timed FIFO

Figure 9.22 shows a two-phase, self-timed implementation of a FIFO (first-in first-out) buffer with three registers. Assuming that the registers accept a data word on both positive- and negative-going transitions of the *En* signals and that the *Done* signal is simply a delayed version of *En*, examine the operation of the FIFO by plotting the timing behavior of all signals of interest. How can you observe that the FIFO is completely empty (full)? (Hint: Determine the necessary conditions on the *Ack* and *Req* signals.)

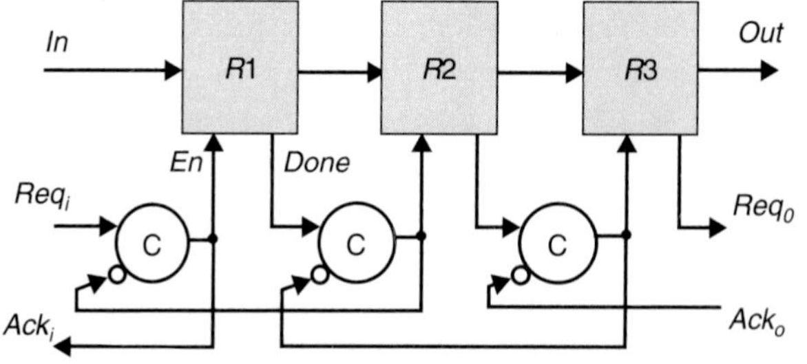

Figure 9.22 Three-stage self-timed FIFO, using a two-phase signaling protocol.

The two-phase protocol has the advantage of being simple and fast. There is some bad news, however. This protocol requires the detection of transitions that may occur in either direction. Most logic devices in the MOS technology tend to be sensitive to levels or to transitions in one particular direction. Event-triggered logic, as required in the two-phase protocol, requires extra logic as well as state information in the registers and the computational elements. Since the transition direction is important, initializing all the

Muller C-elements in the appropriate state is essential. If this is not done, the circuit might dead-lock, which means that all elements are permanently blocked and nothing will ever happen. A detailed study on how to implement event-triggered logic can be found in the text by Sutherland on micropipelines [Sutherland89].

The only alternative is to adopt a different signaling approach, called four-phase signaling, or *return-to-zero* (*RTZ*). This class of signaling requires that all controlling signals be brought back to their initial values before the next cycle can be initiated. Once again, this is illustrated with the example of the sender-receiver. The four-phase protocol for this example is shown in Figure 9.23.

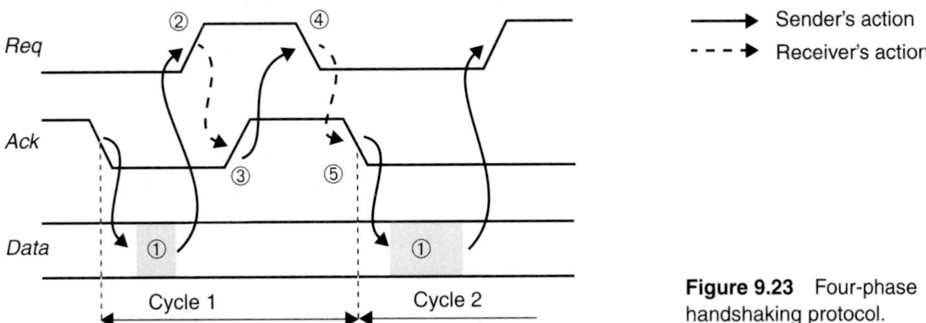

Figure 9.23 Four-phase handshaking protocol.

The protocol presented is initially the same as the two-phase one. Both the *Req* and the *Ack* are initially in the zero-state, however. Once a new data word is put on the bus ①, the *Req* is raised (*Req*↑ or ②) and control is passed to the receiver. When ready, the receiver accepts the data and raises *Ack* (*Ack*↑ or ③). So far, nothing new. The protocol proceeds, now by bringing both *Req* (*Req*↓ or ④) and *Ack* (*Ack*↓ or ⑤) back to their initial state in sequence. Only when that state is reached is the sender allowed to put new data on the bus ①. This protocol is called four-phase because four distinct time-zones can be recognized per cycle: two for the sender; two for the receiver. The first two phases are identical to the two-phase protocol; the last two are devoted to resetting of the state. An implementation of the protocol, based on Muller C-elements, is shown in Figure 9.24. It is interesting to notice that the four-phase protocol requires two C-elements in series (since four states must be represented). The *Data ready* and *Data accepted* signals must be pulses instead of single transitions.

Problem 9.3 Four-Phase Protocol

Derive the timing diagram for the signals shown in Figure 9.24. Assume that the *Data Ready* signal is a pulse and that the *Data Accepted* signal is a delayed version of *Req*.

The four-phase protocol has the disadvantage of being more complex and slower, since two events on *Req* and *Ack* are needed per transmission. On the other hand, it has the advantage of being robust. The logic in the sender and receiver modules does not have to deal with transitions, which can go either way, but only has to consider rising (or falling) transition events or signal levels. This is readily accomplished with traditional logic circuits. For this reason, four-phase handshakes are the preferred implementation approach for most

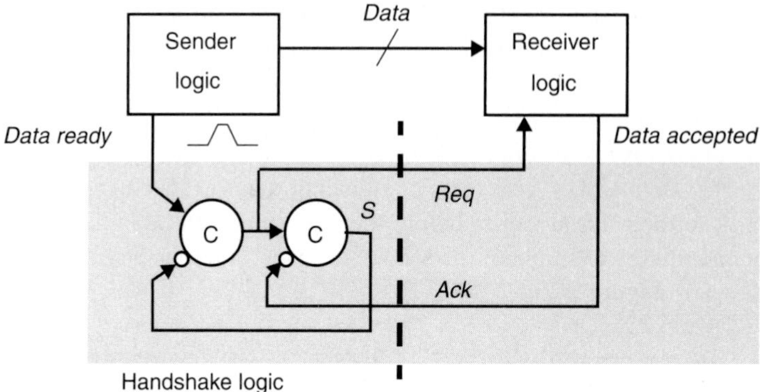

Figure 9.24 Implementation of 4-phase handshake protocol using Muller C-elements.

of the current self-timed circuits. The two-phase protocol is mostly selected when the sender and receiver are far apart and the delays on the control wires (*Ack*, *Req*) are substantial.

Example 9.3 The Pipelined Datapath—Revisited

We have introduced both the signaling conventions and the concepts of the completion-signal generation. Now it is time to bring them all together. We do this with the example of the pipelined data path, which was presented earlier. A view of the self-timed data path, including the timing control, is offered in Figure 9.25. The logic functions *F*1 and *F*2 are implemented using dual-rail, differential logic.

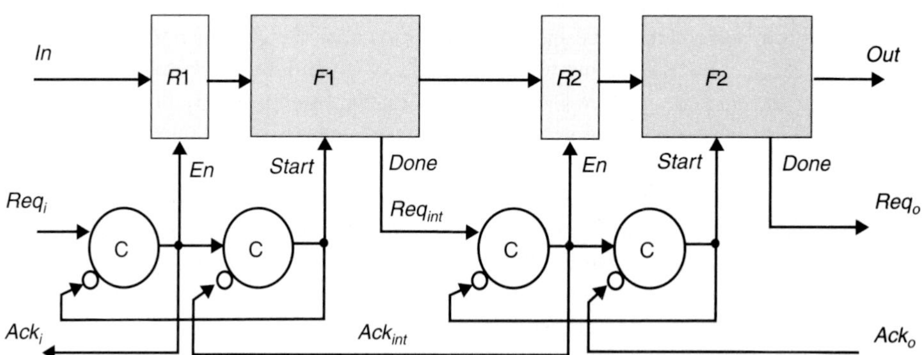

Figure 9.25 Self-timed pipelined datapath—complete composition.

To understand the operation of this circuit, assume that all *Req* and *Ack* signals, including the internal ones, are set to 0, what means there is no activity in the data path. All *Start* signals are low so that all logic circuits are in precharge condition. An input request ($Req_i\uparrow$) triggers the first C-element. The enable signal *En* of *R*1 is raised, effectively latching the input data into the register, assuming a positive edge-triggered or a level-sensitive implementation. $Ack_i\uparrow$ acknowledges the acceptance of the data. The second C-element is triggered as well, since Ack_{int} is low. This raises the *Start* signal and starts the evaluation of *F*1. At its completion, the output data is placed on the bus, and a request is initiated to the second stage ($Req_{int}\uparrow$), which acknowledges its acceptance by raising Ack_{int}.

At this point, stage 1 is still blocked for further computations. However, the input buffer can respond to the $Ack_i\uparrow$ event by resetting Req_i to its zero state ($Req_i\downarrow$). In turn, this lowers *En* and Ack_i. Upon receival of $Ack_{int}\uparrow$, *Start* goes low, the pre-charge phase starts, and *F*1 is ready for new data. Note that this sequence corresponds to the four-phase handshake mechanism described earlier. The dependencies among the events are presented in a more pictorial fashion in the *state transition diagram* (*STG*) shown in Figure 9.26. These STGs can become very complex. Computer tools are often used to derive STGs that ensure proper operation and optimize the performance.

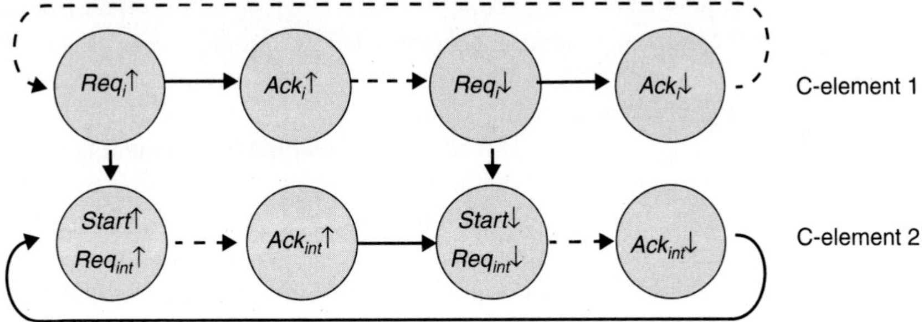

Figure 9.26 State transition diagram for pipeline stage 1. The nodes represent the signaling events, while the arrows express dependencies. Arrows in dashed lines express actions in either the preceding or following stage.

9.4 Synchronizers and Arbiters*

9.4.1 Synchronizers—Concept and Implementation

Even though a complete system may be designed in a synchronous fashion, it must still communicate with the outside world, which is generally asynchronous. An asynchronous input can change value at any time related to the clock edges of the synchronous system, as is illustrated in Figure 9.27.

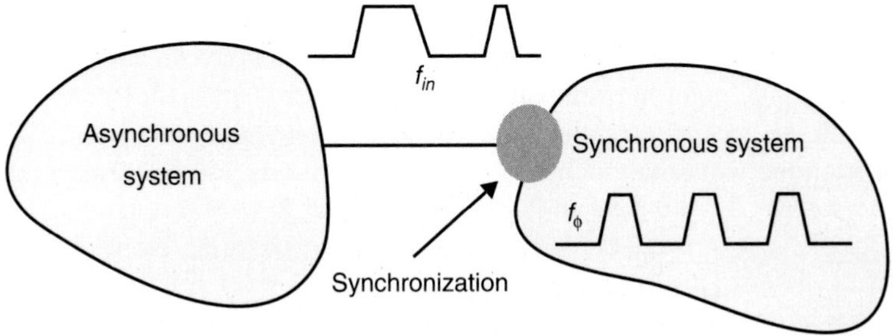

Figure 9.27 Asynchronous-synchronous interface

Consider a typical personal computer. All operations within the system are strictly orchestrated by a central clock that provides a time reference. This reference determines what happens within the computer system at any point in time. This synchronous computer has to communicate with a human through the mouse or the keyboard, who has no knowledge of this time reference and might decide to press a keyboard key at any point in time. The way a synchronous system deals with such an asynchronous signal is to sample or poll it at regular intervals and to check its value. If the sampling rate is high enough, no transitions will be missed—this is known as the Nyquist criterion in the communication community. However, it might happen that the signal is polled in the middle of a transition. The resulting value is neither low or high but undefined. At that point, it is not clear if the key was pressed or not. Feeding the undefined signal into the computer could be the source of all types of trouble, especially when it is routed to different functions or gates that might interpret it differently. For instance, one function might decide that the key is pushed and start a certain action, while another function might lean the other way and issue a competing command. This results in a conflict and a potential crash. Therefore, the undefined state must be resolved in one way or another before it is interpreted further. It does not really matter what decision is made, as long as a unique result is available. For instance, it is either decided that the key is not yet pressed, which will be corrected in the next poll of the keyboard, or it is concluded that the key is already pressed.

Thus, an asynchronous signal must be resolved to be either in the high or low state before it is fed into the synchronous environment. A circuit that implements such a decision-making function is called a *synchronizer*. Now comes the bad news—**building a perfect synchronizer that always delivers a legal answer is impossible!** [Chaney73, Glasser85] A synchronizer needs some time to come to a decision, and in certain cases this time might be arbitrarily long. An asynchronous/synchronous interface is thus always prone to errors called *synchronization failures*. The designer's task is to ensure that the probability of such a failure is small enough that it is not likely to disturb the normal system behavior. Typically, this probability can be reduced in an exponential fashion by waiting longer before making a decision. This is not too troublesome in the keyboard example, but in general, waiting affects system performance and should therefore be avoided to a maximal extent.

To illustrate why waiting helps reduce the failure rate of a synchronizer, consider the synchronizer in Figure 9.28. This circuit is simply a *D* latch, sampling the input signal at the falling edge of the clock ϕ. The sampled signal is likely to be a 0 or 1, but there is a chance that it resides somewhere in the undefined transition zone. The sampled signal eventually evolves into a legal 0 or 1 even in the latter case, since the latch is known to have only two stable states. This is illustrated in Figure 9.29, where the trajectories of the output signal *Out* are plotted as a function of time. If the sampled signal resides above (or below) the metastable point, it eventually converges towards the 1 (or 0) region. The time it takes to reach the acceptable signal zones depends upon the initial distance of the sampled signal from the metastable point. Actually, it might take infinite time if the value of

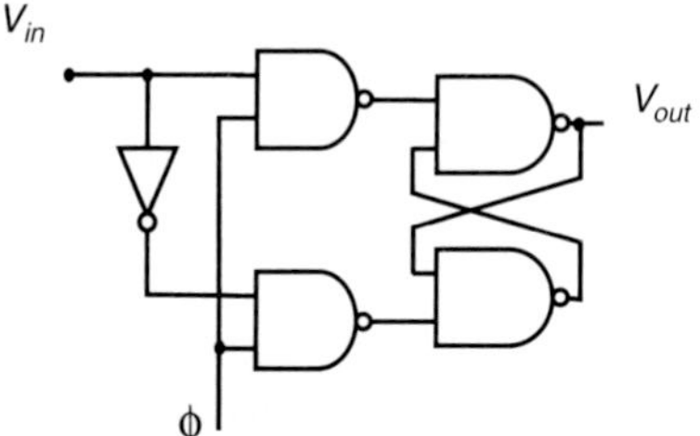

Figure 9.28 A simple synchronizer.

the sampled input signal corresponds exactly to the metastable point of the bistable element! From Figure 9.29, we can see that a longer waiting period increases the probability that the final value has reached a legal 1 or 0 value and has left the undefined zone. It also demonstrates that totally eliminating synchronization failures is not feasible, since it requires an infinite waiting time.

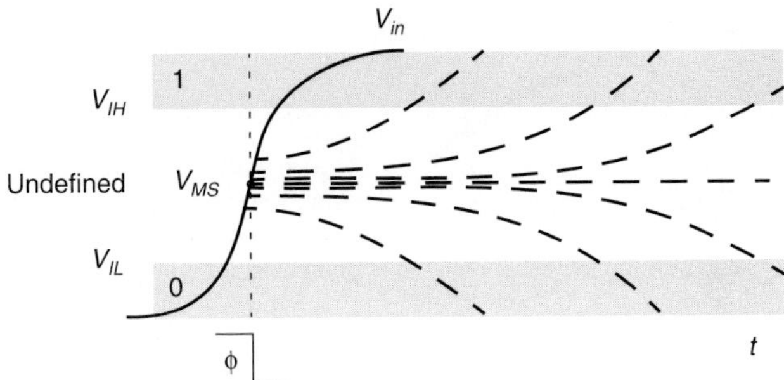

Figure 9.29 Trajectories of the output waveform *Out* after the input signal is sampled as a function of the waiting time. V_{MS} represents the voltage of the metastable point.

In order to determine the required waiting period, let us build a mathematical model of the behavior of the bistable element and use the results to determine the probability of synchronization failure as a function of the waiting period.

The transient behavior around the metastable point of the cross-coupled inverter pair is accurately modeled by a system with a single dominant pole. Under this assumption, the transient response of the bistable element after the turning off of the sampling clock is modeled by Eq. (9.6).

$$v(t) = V_{MS} + (v(0) - V_{MS})e^{t/\tau} \tag{9.6}$$

with V_{MS} the metastable voltage of the latch, $v(0)$ the initial voltage after the sampling clock is turned off, and τ the time-constant of the system.

Example 9.4 Flip-Flop Trajectories

Confirming the validity of the above assumption using manual calculations is a nontrivial process, but it can be established with the aid of simulations. This is demonstrated in Figure 9.30, which plots the simulated trajectories of the output nodes of a cross-coupled static CMOS inverter pair for an initial state close to the metastable point. The inverters are composed of minimum-size devices with an additional load of 20 fF attached to the outputs. The plot also shows the trajectories as predicted by Eq. (9.6). An extremely close match can be observed. The estimated time-constant τ of the system approximately equals 248 psec for the rising output and 310 psec for the falling one. The difference in the time-constants is caused by asymmetries in the flip-flop design.

This model is used to compute the range of values for $v(0)$ that still cause an error, or a voltage in the undefined range, after a waiting period T. A signal is called undefined if its value is situated between V_{IH} and V_{IL}.

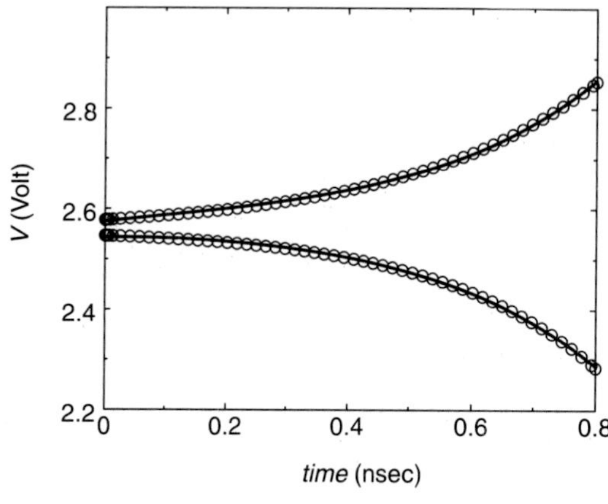

Figure 9.30 Simulated trajectory versus single-pole model for a *D* flip-flop synchronizer. The estimated metastable point is located at 2.57 V.

$$V_{MS} - (V_{MS} - V_{IL})e^{-T/\tau} \le v(0) \le V_{MS} + (V_{IH} - V_{MS})e^{-T/\tau} \qquad (9.7)$$

Eq. (9.7) conveys an important message: The range of input voltages that cause a synchronization error decreases exponentially with the waiting period *T*. Increasing the waiting period from 2τ to 4τ decreases the interval and the chances of an error by a factor of 7.4.

Some information about the asynchronous signal is required in order to compute the probability of an error. Assume that V_{in} is a periodical waveform with an average period T_{signal} between transitions and with identical rise and fall times t_r. Assume also that the slopes of the waveform in the undefined region can be approximated by a linear function (Figure 9.31). Using this model, we can estimate the probability P_{init} that $v(0)$, the value of V_{in} at the sampling time, resides in the undefined region.

$$P_{init} = \frac{\left(\frac{V_{IH} - V_{IL}}{V_{swing}}\right)t_r}{T_{signal}} \qquad (9.8)$$

The chances for a synchronization error to occur depend upon the frequency of the synchronizing clock ϕ. The more sampling events, the higher the chance of running into an error. This means that the average number of synchronization errors per second $N_{sync}(0)$ equals Eq. (9.9) if no synchronizer is used.

$$N_{sync}(0) = \frac{P_{init}}{T_{\phi}} \qquad (9.9)$$

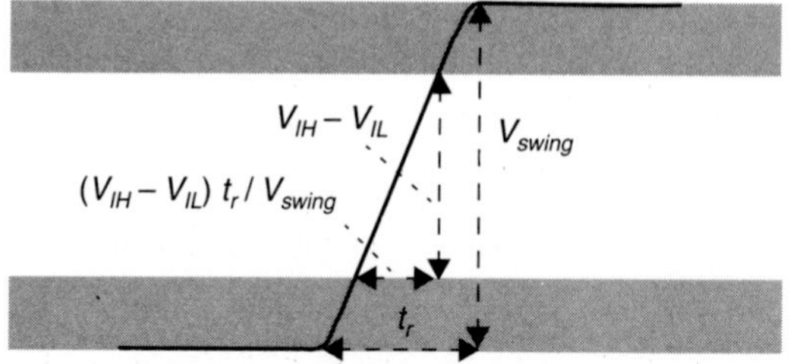

Figure 9.31 Linear approximation of signal slope.

where T_ϕ *is* the sampling period.

From Eq. (9.7) we learned that waiting a time T before observing the output reduces exponentially the probability that the signal is still undefined.

$$N_{sync}(T) = \frac{P_{init}e^{-T/\tau}}{T_\phi} = \frac{(V_{IH} - V_{IL})e^{-T/\tau}}{V_{swing}} \frac{t_r}{T_{signal}T_\phi} \tag{9.10}$$

The robustness of an asynchronous-synchronous interface is hence determined by the following parameters: signal switching rate and rise time, sampling frequency, and waiting time T.

Example 9.5 Synchronizers and Mean Time-to-Failure

Consider the following design example. T_ϕ = 10 nsec, which corresponds to a 100 Mhz clock), $T = T_\phi$ = 10 nsec, T_{signal} = 50 nsec, t_r =1 nsec, and τ = 310 psec (as obtained in Example 9.4). From the VTC of a typical CMOS inverter, it can be derived that $V_{IH} - V_{IL}$ approximately equals 1 V for a voltage swing of 5 V. Evaluation of Eq. (9.10) yields an error probability of 3.9 10^{-9} errors/sec. The inverse of N_{sync} is called the *mean time-to-failure*, or the MTF, and equals 2.6×10^8 sec, or 8.3 years. If no synchronizer was used, the MTF would only have been 2.5 μsec! Doubling the waiting period to 20 nsec, on the other hand, increases the MTF to 8×10^{14} years.

Design Consideration

When designing a synchronous/asynchronous interface, we must keep in mind the following observations:

- The acceptable failure rate of a system depends upon many economic and social factors and is a strong function of its application area.
- The exponential relation in Eq. (9.10) makes the failure rate extremely sensitive to the value of τ. Defining a precise value of τ is not easy in the first place. τ varies from chip to chip and is a function of temperature as well. The probability of an error occurring can thus fluctuate over large ranges even for the same design. A worst-case design scenario is definitely advocated here. If the worst-case failure rate exceeds a certain criterion, it can be reduced by increasing the value of T. A problem occurs when T exceeds the sampling period T_ϕ. This can be avoided by cascading (or pipelining) a number of synchronizers, as shown in Figure 9.32. Each of those synchronizers has a waiting period equal to T_ϕ. Notice that this arrangement requires the ϕ-pulse to be short enough to avoid race conditions. The global waiting period equals the sum of the Ts of all the individual synchronizers. The increase in MTF comes at the expense of an increased latency.

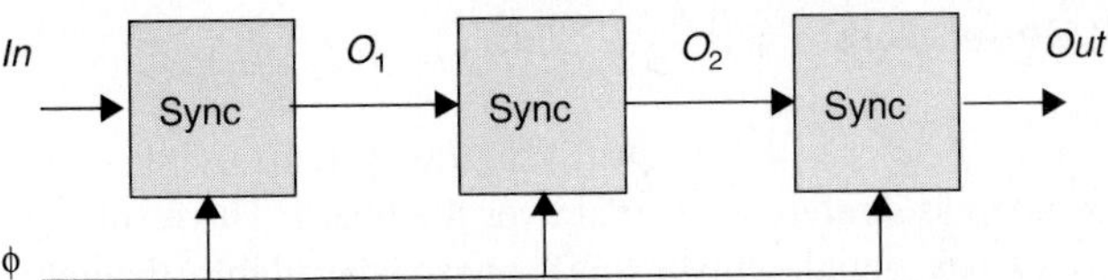

Figure 9.32 Cascading synchronizers reduces the main time-to-failure.

- Synchronization errors are very hard to trace due to their probabilistic nature. Making the mean time-to-failure very large does not preclude errors. The number of synchronizers in a system should therefore be severely restricted. A maximum of one or two per system is advocated.

One might wonder how noise influences the efficiency of a synchronizer. Intuitively, it would seem that noise improves the performance of an arbiter. For instance, noise can move a flip-flop that is trapped in its metastable state out of this precarious position. Along the same lines, one can argue that noise might actually force the flip-flop into its metastable state. From this argument, we realize that low-amplitude noise does not affect the performance of the synchronizer at all—sometimes it helps, and sometimes it hurts.

9.4.2 Arbiters

Finally, a sibling of the synchronizer called the *arbiter,* interlock element, or mutual-exclusion circuit, should be mentioned. An arbiter is an element that decides which of two events has occurred first. Such components for instance allow multiple processors to access a single resource, such as a large shared memory. A synchronizer is actually a special case of an arbiter, since it determines if a signal transition happened before or after a clock event. A synchronizer is thus an arbiter with one of its inputs tied to the clock.

An example of a mutual-exclusion circuit is shown in Figure 9.33. It operates on two input-request signals, that operate on a four-phase signaling protocol; that is, the *Req(uest)* signal has to go back to the reset state before a new *Req(uest)* can be issued. The output consists of two *Ack(nowledge)* signals that should be mutually exclusive. While *Requests* may occur concurrently, only one of the *Acknowledges* is allowed to go high. The operation is most easily visualized starting with both inputs low—neither device issuing a request—nodes *A* and *B* high, and both *Acknowledges* low. An event on one of the inputs (e.g., $Req1\uparrow$) causes the flip-flop to switch, node *A* goes low, and $Ack1\uparrow$. Concurrent events on both inputs force the flip-flop into the metastable state, and the signals *A* and *B* might be undefined for a while. The cross-coupled output structure keeps the output values low until one of the NAND outputs differs from the other by more than a threshold value V_T. This approach eliminates glitches at the output.

9.5 Clock Generation and Synchronization*

9.5.1 Clock Generators

Instead of generating the clock on-chip, most digital systems use an external crystal oscillator that is distributed as a reference signal over the board. The resulting waveform is not directly useful as a clock signal. Only a single phase is available, the duty cycle of the signal does not match the requirements of the logic style, and buffering is needed to drive the

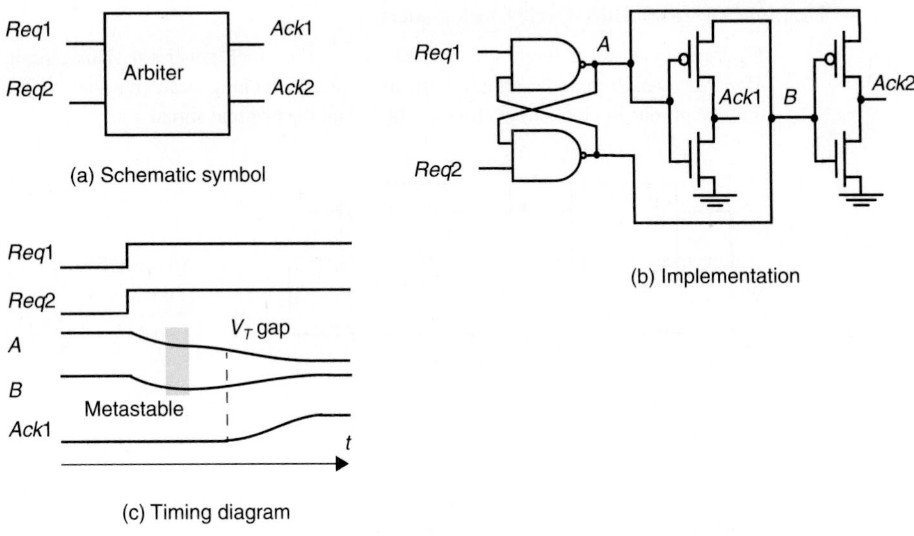

Figure 9.33 Mutual-exclusion element (or arbiter).

large clock-load capacitance. Hence, a clock-generator circuit is required. An example of such a generator is shown in Figure 9.34. It produces a two-phase clock signal with a duty cycle of 25% and a separation of 25% between the clock phases. In order to be functional, it is necessary to initialize all *D* flip-flops to zero at start-up.

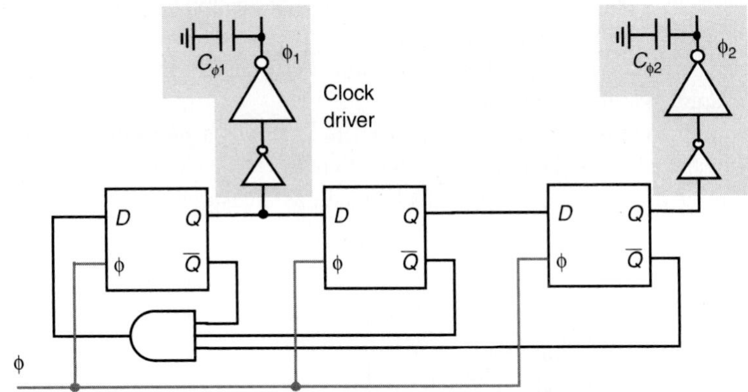

Figure 9.34 Two-phase clock-generator and clock buffers.

A chain of cascaded inverters is added to provide the necessary clock-driving capabilities. The designs of those buffers is quite demanding because the load capacitance can be huge, as was established earlier. Controlling the potential switching noise and keeping the power consumption of the driver within bounds is a challenging design task.

Example 9.6 50% Duty-Cycle Clock Generator.

Figure 9.35 shows a single phase, 50% duty cycle clock generator. This circuit, called the *Moebius counter*, is quite effective in single-phase clocking strategies such as TSPC. Notice that the produced clock is two times slower than the original signal.

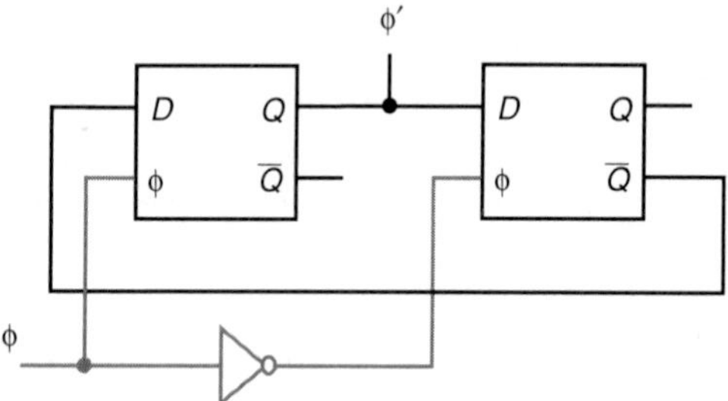

Figure 9.35 50% duty-cycle, single-phase clock generator.

9.5.2 Synchronization at the System Level

The clock-distribution approach presented has one important pitfall. Consider the simple system of Figure 9.36 that combines two communicating integrated circuits on a printed-circuit board. A single reference clock is distributed to both ICs, which derive a number of local clock signals with the aid of on-chip clock generators. Due to the delay of those generators, which include the clock drivers, the derived clock signals are skewed with respect to the reference clock (Figure 9.37a). Other sources for the skew are the delays of the input pads and wiring delay. The value of the skew differs from chip to chip as the clock loading conditions can be substantially different. Since the data signals are sychronized to the local clocks, communicating between the chips becomes hazardous, because input data might be sampled at the wrong time.

One approach to circumvent this problem is to use a *phase-locked loop* (PLL) to synchronize the edges of the internal clock with respect to the edge of the reference clock. The PLL takes as input the internal clock signal, as measured at the output of the clock

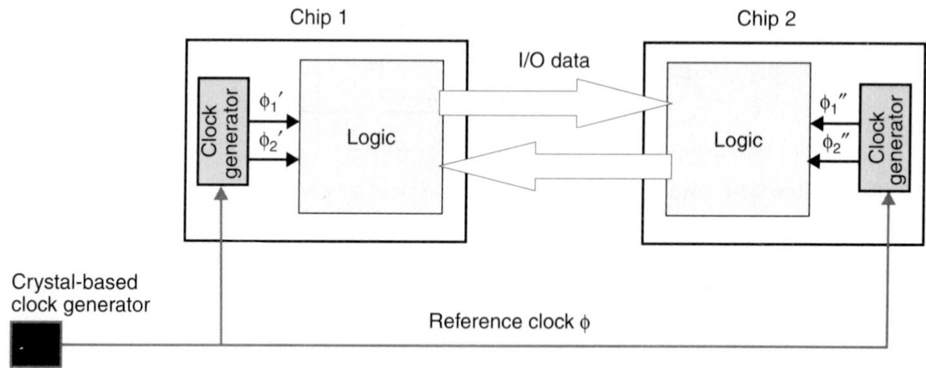

Figure 9.36 Local clock generators create skew.

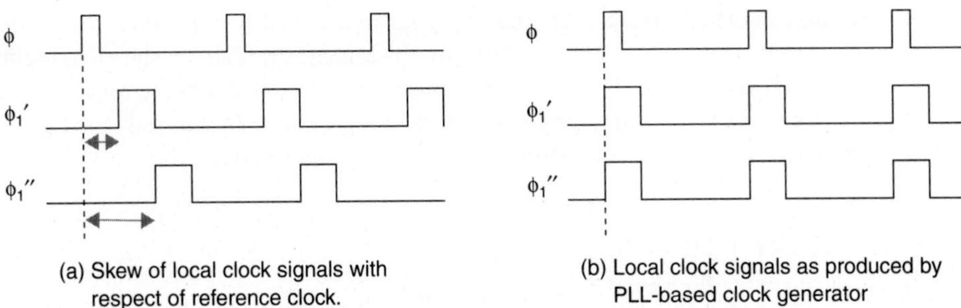

Figure 9.37 Skew of local clock signals with respect to reference clock.

generator or some other important reference point, and the incoming reference clock signal. Its function is to align the phases of both clock signals, so that the internal skew is effectively eliminated. This is illustrated in Figure 9.37b, which shows the clock signals produced by the PLL-based clocked generator. All clock signals are aligned with respect to a particular clock edge.

Although a PLL is a complex, nonlinear circuit, its basic operation is readily understood with the aid of Figure 9.38 [Jeong87].[2] The reference clock and the local clock are passed to a phase detector that compares the phases between the signals and produces an *Up* or *Down* signal when the local clock lags or leads the reference signal. An attentive reader might realize that the phase detector is a close relative of the arbiter circuit discussed earlier. It detects which of the two input signals arrives earlier and produces an appropriate output signal. Next, the *Up* and *Down* signals are fed into a charge pump, which translates the digital encoded control information into an analog voltage [Gardner80]. This translation is necessary because the subsequent *voltage-controlled oscillator* (VCO) uses an analog voltage as its controlling input. An *Up* signal increases the value of the control voltage and speeds up the VCO, which causes the local signal to catch up with the reference clock. A *Down* signal, on the other hand, slows down the oscillator and eliminates the phase lead of the local clock. An example realization of the charge

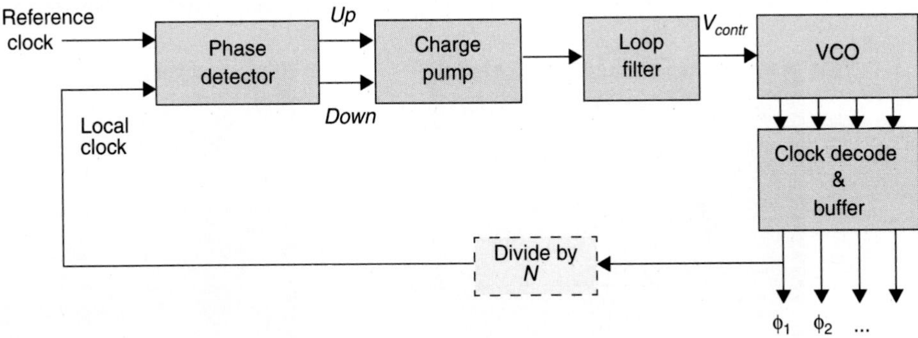

Figure 9.38 Composition of a phase-locked loop (PLL).

[2] The concept of PLLs dates from the 1930s. It has been used in communication systems of many types, particularly in satellite communications. For a compact introduction to PLLs, please refer to [Gray92, pp. 681–709].

pump is shown in Figure 9.39. The VCO, which is typically implemented as a ring oscillator, simultaneously generates the various clock phases needed by the clocking methodology adopted on the IC. Observe that the phase detector compares the reference signals and the generated clock signal probed after the clock drivers. In this way, the PLL also compensates for the delay of the drivers, which can be substantial.

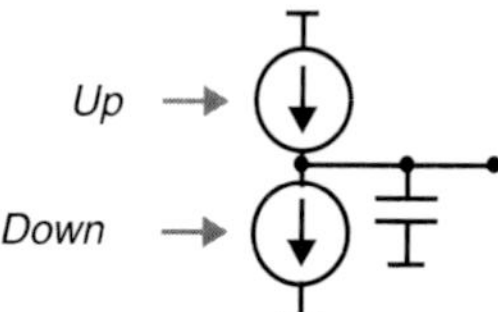

Figure 9.39 Charge pump.

Passing the output of the charge pump directly into the VCO creates a jittery clock signal. The edge of the local clock jumps back and forth instantaneously and oscillates around the targeted position. This effect, called *clock jitter*, is highly undesirable, since it dynamically varies the time available for logic computation. The jitter should hence be kept within a given percentage of the clock period. This is partially accomplished by the introduction of the *loop filter*. This low-pass filter removes the high-frequency components from the VCO control voltage and smooths out its response, which results in a reduction of the jitter. Typically, the order of the loop filter ranges from first to third order. Beware: the PLL structure is a feedback structure and the addition of extra phase shifts, as is done by a high-order filter, may result in instability. Important properties of a PLL are its *lock range*—the range of input frequencies over which the loop can maintain functionality; the *lock time*—the time it takes for the PLL to lock onto a given input signal; and the jitter.

The operation of a PLL-based clock generator is illustrated in Figure 9.40 with an oscilloscope trace of the produced signals. An excellent alignment between the falling edges of reference and local clock can be observed.

Besides providing a mechanism of synchronization between internal and external clocks, the PLL approach offers other potential applications. The introducing of a frequency divider between the local clock and phase detector (indicated by dotted lines in Figure 9.38) makes it possible to operate the internal clock of a chip at a higher rate than the external ref-

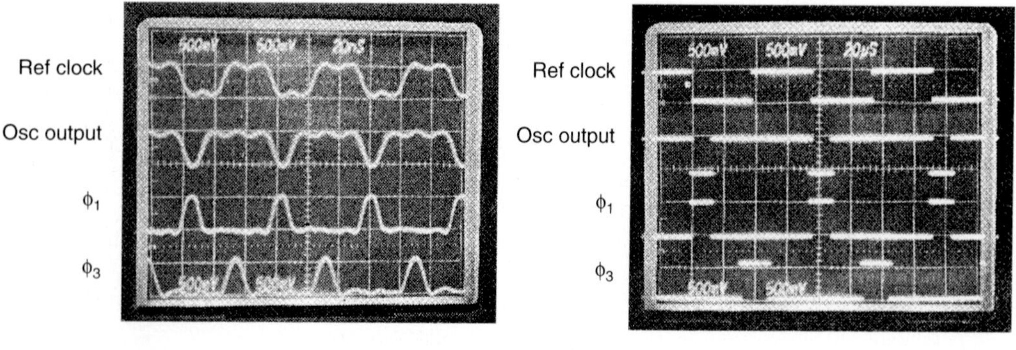

(a) Clock generator output at 18 MHz.

(b) Clock generator output at 15 kHz.

Figure 9.40 Oscilloscope traces of signals in PLL (from [Jeong87]).

erence clock. In other words, the PLL-based clock generator can act as a frequency multiplier as well. This makes sense because it simplifies the design and distribution of the reference clock that is distributed over the complete PC board and is heavily loaded.

Given all these advantages, we may wonder why designers do not flock to put PLLs on their digital integrated circuits. The answer is simple—a PLL is an analog circuit and is inherently sensitive to noise and interference. This is especially true for the loop filter and VCO, where induced noise has a direct effect on the resulting clock jitter. A major source of interference is the noise coupling through the supply rails. We have learned that these are rather ill behaved in a digital environment. Analog circuits with a high supply rejection, such as differential VCOs, are therefore desirable [Kim90]. In summary, integrating a highly sensitive component into a hostile digital environment is nontrivial and requires expert analog design. Completely digital PLLs have been proposed but are prohibitively expensive.

If properly applied, PLLs can become a major part of the digital designer's toolbox. It is interesting to note that the domain of digital communications, which originated the concept of the phase-locked loop, has produced a large number of alternative approaches that might be applicable to the synchronization problem in digital integrated circuits. After all, digital communication systems currently transmit data at GHz rates over very long distances. An examples of a technique that might be applicable to circuit design is the carrier recovery approach that derives the clock from the transmitted data.

9.6 Perspective: Synchronous versus Asynchronous Design

The self-timed approach offers a potential solution to the growing clock-distribution problem. It translates the global clock signal into a number of local synchronization problems. Independence from physical timing constraints is achieved with the aid of completion signals. Handshaking logic is needed to ensure the logical ordering of the circuit events and to avoid race conditions. This requires adherence to a certain protocol, which normally consists of either two or four phases.

Despite all its advantages, self-timing has only been used in a number of isolated cases. Examples of self-timed circuits can be found in signal processing [Jacobs90], fast arithmetic units (such as self-timed dividers) [Williams87], simple microprocessors [Martin89] and memory (static RAM, FIFOs). In general, synchronous logic is both faster and simpler since the overhead of completion-signal generation and handshaking logic is avoided. The design a fool-proof network of handshaking units, that is robust with respect to races, live-lock, and dead-lock, is nontrivial and requires the availability of dedicated design-automation tools.

On the other hand, distributing a clock at high speed becomes exceedingly difficult. This was amply illustrated by the example of the 21164 Alpha microprocessor. Skew management requires extensive modeling and analysis, as well as careful design. It will not be easy to extend this methodology into the next generation of designs. This observation is already reflected in the fact that the routing network for the latest generation of massively parallel supercomputers is completely implemented using self-timing [Seitz92]. For self-timing to become a mainstream design technique however (if it ever will), further innovations in circuit and signaling techniques and design methodologies are needed. Other alternative timing approaches might emerge as well. Possible candidates are fully asynchronous designs or islands of synchronous units connected by an asynchronous network.

9.7 Summary

This chapter has explored the timing of sequential digital circuits.

- An in-depth analysis of the synchronous digital circuits and clocking approaches was presented. Clock skew has a growing impact on the functionality and performance of a system. Important parameters are the clocking scheme used and the nature of the clock-generation and distribution network.
- The increasing complexity of digital ICs makes the fully synchronous approach more and more cumbersome. Therefore, alternative approaches, such as self-timed design, are becoming attractive. Self-timed design uses completion signals and handshaking logic to isolate physical timing constraints from event ordering.
- The connection of synchronous and asynchronous components introduces the risk of synchronization failure. The introduction of synchronizers helps to reduce that risk, but can never eliminate it.
- Phase-locked loops are becoming an important element of the digital-designer's tool box. They help to synchronize signals from different components on a board. The analog nature of the PLL makes its design a real challenge.
- The key message of this chapter is that synchronization and timing are among the most intriguing challenges facing the digital designer of the next decade.

9.8 To Probe Further

While system timing is an important topic, no congruent reference work is available in this area. One of the best discussions so far is the chapter by Chuck Seitz in [Mead80, Chapter 7]. Other in-depth overviews are given in [Bakoglu90, Chapter 8], [Johnson93, Chapter 11], and [Hatamian88]. A collection of papers on clock distribution networks is presented in [Friedman95]. Numerous other publications are available on this topic in the leading journals, some of which are mentioned below.

REFERENCES

[Abnous93] A. Abnous and A. Behzad, "A High-Performance Self-Timed CMOS Adder," in *EE241 Final Class Project Reports*, by J. Rabaey, Univ. of California—Berkeley, May 1993.

[Bakoglu90] H. Bakoglu, *Circuits, Interconnections and Packaging for VLSI*, Addison-Wesley, pp. 338–393, 1980.

[Bowhill95] W. Bowhill et al., "A 300 MHz Quad-Issue CMOS RISC Microprocessor," *Technical Digest of the 1995 ISSCC Conference*, San Fransisco, February 1995.

[Chaney73] T. Chaney and F. Rosenberger, "Anomalous Behavior of Synchronizer and Arbiter Circuits," *IEEE Trans. on Computers*, vol. C-22, April 1973, pp. 421–422.

[Dopperpuhl92] D. Dopperpuhl et al., "A 200 MHz 64-b Dual Issue CMOS Microprocessor," *IEEE Journal on Solid State Circuits*, vol. 27, no. 11, Nov. 1992, pp. 1555–1567.

[Friedman95] E. Friedman, ed., *Clock Distribution Networks in VLSI Circuits and Systems*, IEEE Press, 1995.

[Gardner80] F. Gardner, "Charge-Pump Phase-Locked Loops," *IEEE Trans. on Communications*, vol. COM-28, November 1980, pp. 1849–1858.

[Glasser85] L. Glasser and D. Dopperpuhl, *The Design and Analysis of VLSI Circuits*, Addison-Wesley, 1985, pp. 360–365.

[Gray93] P. Gray and R. Meyer, *Analysis and Design of Analog Integrated Circuits*, 3rd ed., Wiley, 1993.

[Hatamian88] M. Hatamian, "Understanding Clock Skew in Synchronous Systems," in *Concurrent Computations*, ed. S. Tewksbury et al., Plenum Publishing, pp. 86–96, 1988.

[Heller84] L. Heller et al., "Cascade Voltage Switch Logic: A Differential CMOS Logic Family," *IEEE International Solid State Conference Digest*, Feb. 1984, San Francisco, pp. 16–17.

[Jacobs90] G. Jacobs and R. Brodersen, "A Fully Asynchronous Digital Signal Processor," *IEEE Journal on Solid State Circuits*, vol. 25, No 6, December 1990, pp. 1526–1537.

[Jeong87] D. Jeong et al., "Design of PLL-Based Clock Generation Circuits", *IEEE Journal on Solid State Circuits*, vol. SC-22, no 2, April 1987, pp 255–261.

[Johnson93] H. Johnson and M. Graham, *High-Speed Digital Design—A Handbook of Black Magic*, Prentice-Hall, N.J, 1993.

[Kim90] B. Kim, D. Helman, and P. Gray, "A 30 MHz Hybrid Analog/Digital Clock Recovery Circuit in 2 μm CMOS," *IEEE Journal on Solid State Circuits*, vol. SC-25, no. 6, December 1990, pp. 1385–1394.

[Martin89] A. Matrin et. al, "The First Asynchronous Microprocessor: Test Results," *Computer Architecture News*, vol. 17 No 4, June 1989, pp. 95–110.

[Mead80] C. Mead and L. Conway, *Introduction to VLSI Design*, Addison-Wesley, 1980.

[Seitz80] C. Seitz, "System Timing," in [Mead80], pp. 218–262, 1980.

[Seitz92] C. Seitz, "Mosaic C: An Experimental Fine-Grain Multicomputer," in *Future Tendencies in Computer Science, Control and Applied Mathematics*, Proceedings International Conference on the 25th Anniversary of INRIA, Springer-Verlag, Germany, pp. 69–85, 1992.

[Shoji88] M. Shoji, *CMOS Digital Circuit Technology*, Prentice-Hall, 1988.

[Sutherland89] I. Sutherland, "Micropipelines," *Communications of the ACM*, pp. 720–738, June 1989.

[Veendrick80] H. Veendrick, "The Behavior of Flip Flops Used as Synchronizers and Prediction of Their Failure Rates", *IEEE Journal of Solid State Circuits*, vol. SC-15, no 2, April 1980, pp. 169–176.

[Williams87] T. Williams et al., "A Self-Timed Chip for Division," in *Proc. of Advanced Research in VLSI 1987*, Stanford Conf., pp. 75–96, March 1987.

9.9 Exercises and Design Problems

1. [E, None, 9.2] A simple latch is shown in Figure 9.41.
   - **a.** Use a timing diagram to show whether the latch is level-sensitive or edge-triggered.
   - **b.** Determine if races are an issue in designs based on this register. How about skew?
2. [E, None, 9.2] Identify the major problem of the circuit in Figure 9.42 and cure it with the simplest of adjustments.
3. [E, None, 9.2] The inequalities presented in this chapter specifying maximum skew and minimum clock period for single-phase edge-triggered and two-phase master-slave clocking neglect register setup times. Re-derive the four expressions, this time accounting for setup times that can vary between $t_{su,min}$ and $t_{su,max}$.

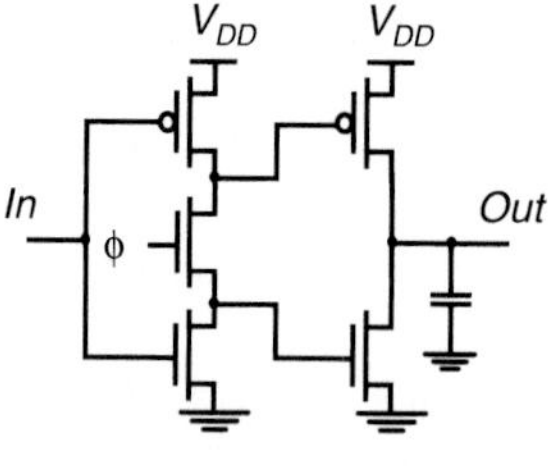

Figure 9.41 Register schematic.

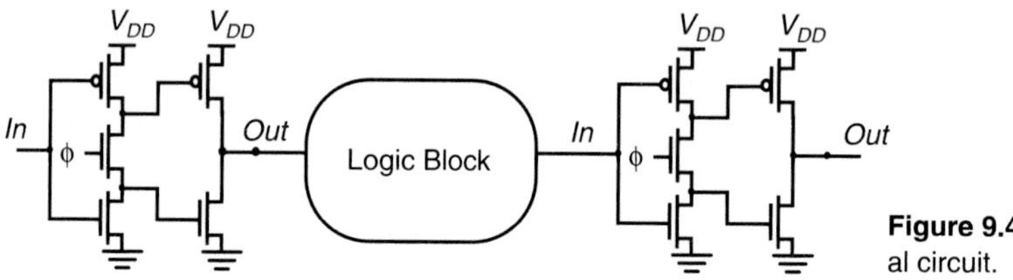

Figure 9.42 Sequenti al circuit.

4. [M, None, 9.2] Consider the circuit in Figure 9.43. Both *D* flip-flops are negative edge-triggered devices, with a zero set-up time and a fixed propagation delay of 0.1 nsec. The inverters separating the FFs have a maximum propagation delay of 0.8 nsec and a minimum delay of 0.2 nsec. The corresponding clock signals are also shown in the figure.
   a. Determine the maximum value of the clock skew, δ_{max}, such that the circuit will still be operational.
   b. Ascertain the minimum clock period, T_{min}, as a function of the skew.
   c. Redraw the circuit to use two-phase dynamic registers instead of edge-triggered flip-flops. Qualitatively discuss how clock skew affects this circuit.

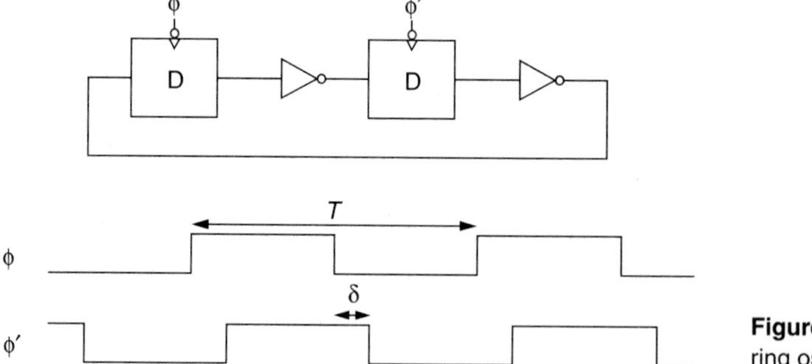

Figure 9.43 Clocked ring oscillator.

5. [C, Spice, 9.2] For the circuit in Figure 9.44, use SPICE to measure t_{max} and t_{min}. Use a minimum-size NAND gate and inverter. Assume no skew and a zero rise/fall time. For the registers, use the following:
   a. A TSPCR, as shown in Figure 6.37a.
   b. A C^2MOS, as shown in Figure 6.24.
6. [C, Spice, 9.2] Introduce clock skew, both positive and negative, to the circuit in problem 5.a. How much skew can the circuit tolerate and still function correctly?
7. [C, Spice, 9.2] Introduce finite rise and fall time to the clocks in the circuit in problem 5.b. Show what can occur and describe why.

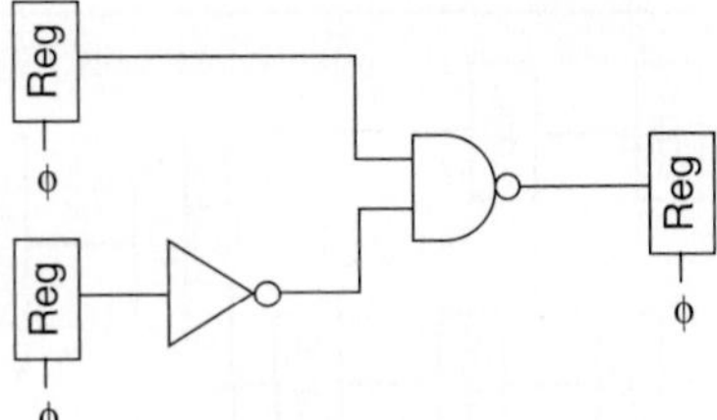

Figure 9.44 Sequential circuit.

8. [M, None, 9.2] Consider the register shown in Figure 9.45. Assume that the delay through an inverter equals t_i, which is also the delay through a pass-transistor (with an appropriate load).

a. Draw the signal waveforms for the signals *X*, *Y*, *Z*, and OUT, given the *IN* and *CLK* signals of Figure 9.46. Logic high shown is equal to the supply voltage and logic low is ground.

b. Does this circuit show a race problem? If yes, say why. If no, determine under what conditions a race would occur.

c. Derive a simple way of making this circuit **absolutely** insensitive to the race problem. Do not add or remove transistors; just reorganize.

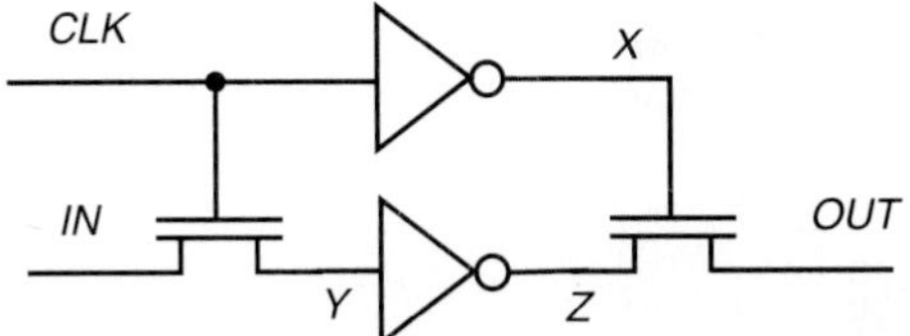

Figure 9.45 Register for problem 9.8.

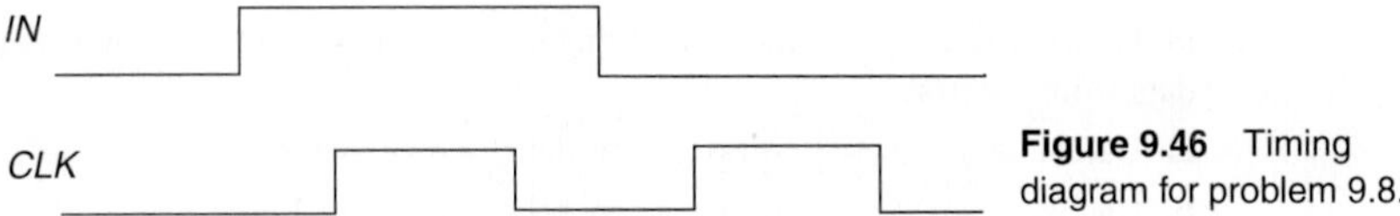

Figure 9.46 Timing diagram for problem 9.8.

9. [C, None, 9.2] For the circuit in Figure 9.47, assume a unit delay through the Register and Logic blocks (i.e., $t_R = t_L = 1$). Assume that the registers, which are positive edge-triggered, have a set-up time t_S of 1. The delay through the multiplexer t_M equals 2 t_R.

a. Determine the minimum clock period. Disregard clock skew.

b. Repeat part a, factoring in a nonzero clock skew: $\delta = t'_\theta - t_\theta = 1$.

c. Repeat part a, factoring in a non-zero clock skew: $\delta = t'_\theta - t_\theta = 4$.

d. Derive the maximum positive clock skew that can be tolerated before the circuit fails.

e. Derive the maximum negative clock skew that can be tolerated before the circuit fails.

10. [M, None, 9.2] For the circuit in Figure 9.48, sketch a clock network to the different processing elements that minimizes the clock skew.

11. [E, None, 9.3] Consider the pipelined datapath of Figure 9.14. Draw the signal-transition graph for the network when employing a four-phase handshaking scheme using Muller-C elements.

12. [M, Spice, 9.3] For the self-timed circuit shown in Figure 9.49, make the following assumptions. The propagation through the NAND gate can be 5 nsec, 10 nsec, or 20 nsec with equal

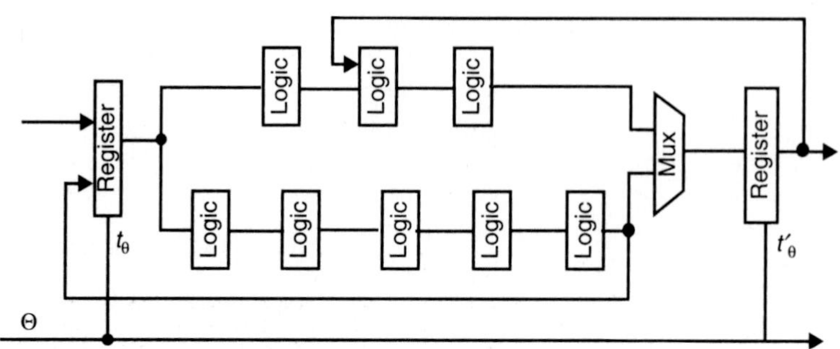

Figure 9.47 Sequential circuit.

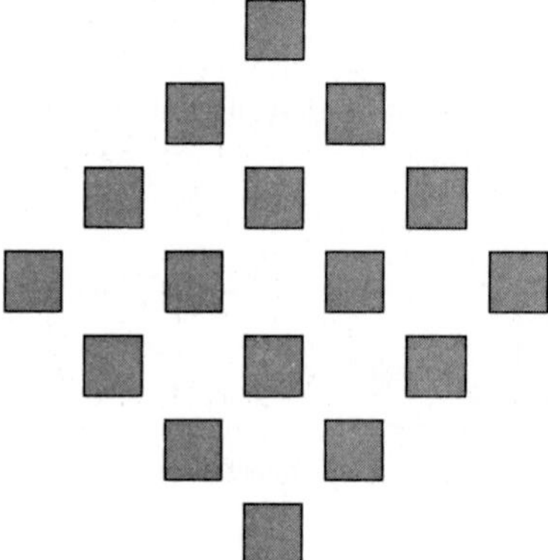

Figure 9.48 Floorplan of sequential circuit.

probability. The logic in the succeeding stages is such that the second stage is always ready for data from the first.

a. Calculate the average propagation delay with t_{hs} = 6 nsec.

b. Calculate the average propagation delay with t_{hs} =12 nsec.

c. If the handshaking circuitry is replaced by a synchronous clock, what is the smallest possible clock frequency?

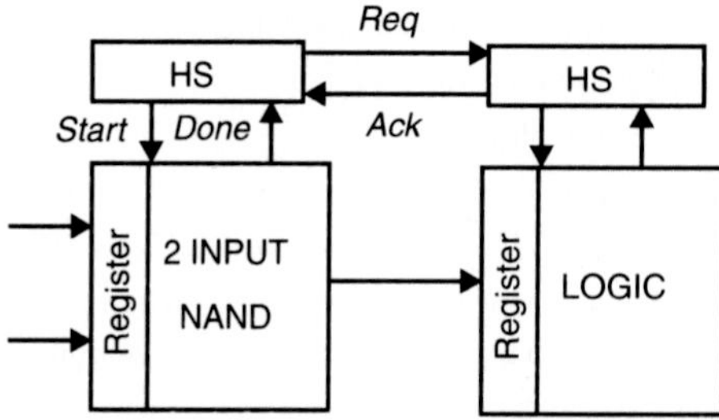

Figure 9.49 Self-timed circuit.

13. [C&D, SPICE & IRSIM, 9.3] Design the circuit in Figure 9.49, using the dynamic Muller element of Figure 9.20b and DCVSL for the NAND gate. Construct an alternative implementation that uses synchronous clocking. Use TSPCR for the registers and minimum-size devices.

a. With a supply voltage of 3.3 V, compare the propagation delay of both circuits using SPICE.

b. Use IRSIM (or any other switch-level simulator) to compare the energy dissipated in both circuits. Scale the voltages so that both circuits have approximately the same delay time. Which of the circuits is more power efficient?

14. [C&D, None, 9.3] Design a self-timed FIFO at the transistor level. It should be six stages deep and have a four-cycle interaction with the outside world. The black-box view of the FIFO is given in Figure 9.50.

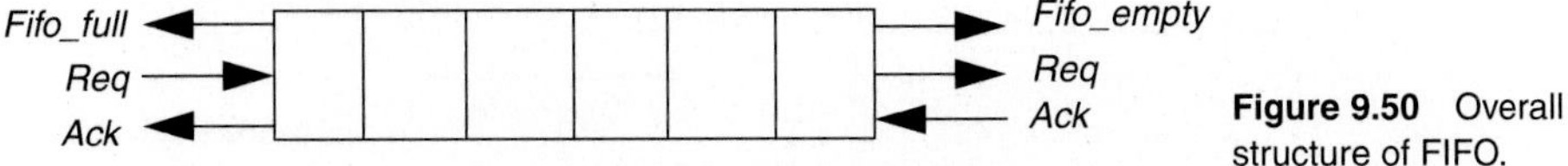

Figure 9.50 Overall structure of FIFO.

15. [E, None, 9.4] Lisa and Marcus Allen have a luxurious symphony hall date. After pulling out of their driveway, they pull up to a four-way stop sign. They pulled up to the sign at the same time as a car on the cross-street. The other car, being on the right, had the right-of-way and proceeded first. On the way they also have to stop at traffic signals. There is so much traffic on the freeway, the metering lights are on. Metering lights regulate the flow of merging traffic by allowing only one lane of traffic to proceed at a time. With all the traffic, they arrive late for the symphony and miss the beginning. The usher does not allow them to enter until after the first movement.

On this trip, Lisa and Marcus proceeded through both synchronizers and arbiters. Please list all and explain your answer.

16. [M, Spice, 9.4] Figure 9.51 shows a simple synchronizer. Assume that the asynchronous input switches at a rate of approximately 10 MHz and that $t_r = 2$ nsec, $f_\phi = 50$ MHz, $V_{IH} - V_{IL} = 1$ V, and $V_{DD} = 5$ V.

a. If all NMOS devices are minimum-size, find $(W/L)p$ required to achieve $V_{MS} = 2.5$ V. Verify with SPICE.

b. Use SPICE to find τ for the resulting circuit.

c. What waiting time T is required to achieve a MTF of 10 years?

d. Is it possible to achieve an MTF of 1000 years (where $T > T_\phi$)? If so, how?

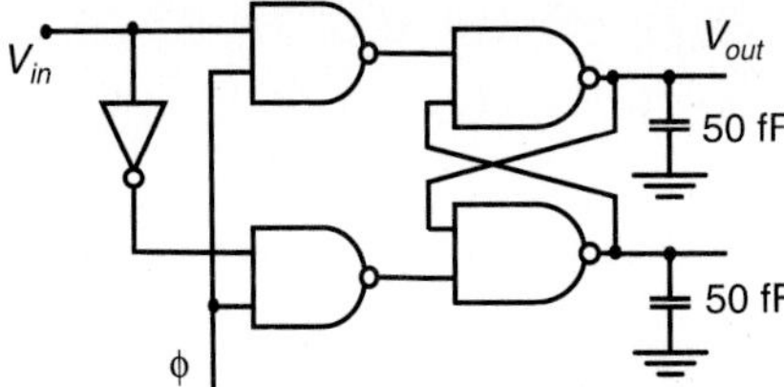

Figure 9.51 Simple synchronizer

17. [M, None, 9.4] An adjustable duty-cycle clock generator is shown in Figure 9.52. Assume the delay through the delay element matches the delay of the multiplexer.

a. Describe the operation of this circuit

b. What is the range of duty-cycles that can be achieved with this circuit.

c. Using an inverter and an additional multiplexer, show how to make this circuit cover the full range of duty cycles.

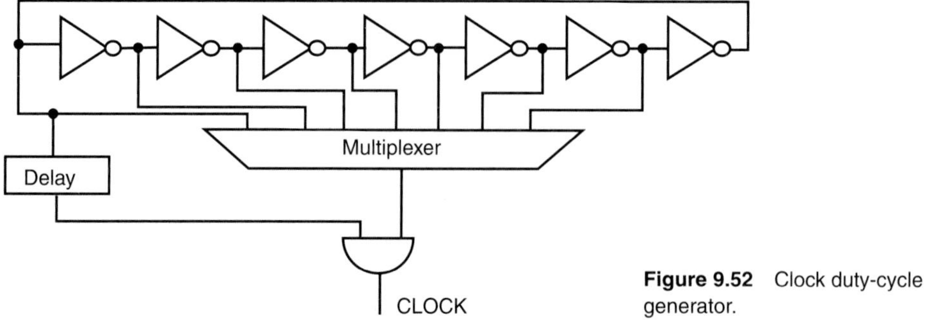

Figure 9.52 Clock duty-cycle generator.

DESIGN PROBLEM

Design an 8-bit, self-timed programmable counter in the standard 1.2 μm CMOS technology (as shown in Figure 9.53). The goal is to maximize the speed while minimizing the power consumption. Grading is based on the *energy-delay product*. Speed is measured by the *average counting rate*. Energy corresponds to the *total energy consumed by a single bit slice of the data path.*

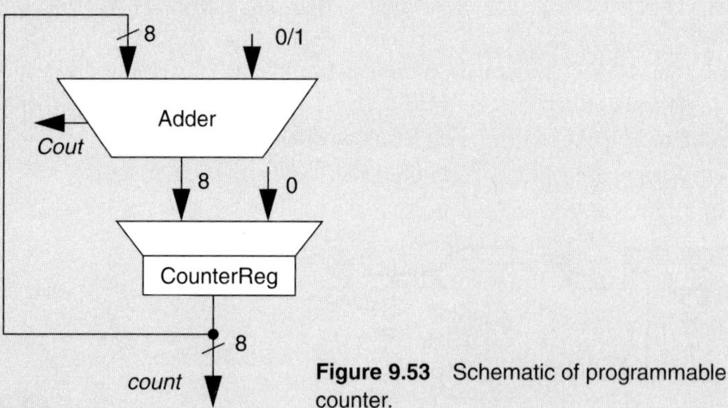

Figure 9.53 Schematic of programmable counter.

The counter should implement the following functions:

if RESET	count ← 0
if INC	count ← count + 1
else	count ← count

When the counter reaches FF it returns back to 00.

You have total freedom in choosing the design style of the registers (static, dynamic). Use the Manchester-carry adder of Figure 9.17. You are free to select the power supply! Lowering the power supply will reduce the energy consumption, but increase the delay.

CHAPTER

10

DESIGNING MEMORY AND ARRAY STRUCTURES

Overview of memory architecture

Data-storage cells for read-only, nonvolatile, and read-write memories

Peripheral circuits such as sense amplifiers, decoders, drivers, and timing generators

Reliability issues in memory design

10.1 Introduction

A large portion of the silicon area of many contemporary digital designs is dedicated to the storage of data values and program instructions. More than half of the transistors in today's high-performance microprocessors are devoted to cache memories, and this ratio is expected to increase in the foreseeable future. The situation is even more dramatic at the system level. High-performance workstations and computers contain more than 64 MBytes of semiconductor memory, a number that is continuously rising. Obviously, dense data-storage circuitry is one of the primary concerns of a digital circuit or system designer.

In Chapter 6, we introduced means of storing Boolean values based on either positive feedback or capacitive storage. The use of the register cells presented as a mass storage medium would lead to excessive area requirements. Memory cells are therefore combined into large arrays, which minimizes the overhead caused by peripheral circuitry and increases the storage density. The sheer size and complexity of these array structures introduces a variety of design problems, some of which are discussed in this chapter.

We first introduce the basic memory architectures and their essential building blocks. Next, we analyze the different memory cells and their properties. The cell structure and topology is mainly driven by the available technology and is somewhat out of the control of the digital designer. On the other hand, the peripheral circuitry has a tremendous impact on the robustness, performance, and power consumption of the memory unit. Therefore, a careful analysis of the options and considerations of the periphery design is appropriate. The chapter concludes with a short discussion on memory reliability.

An interesting aspect of Chapter 10 is that it applies a large number of the circuit techniques introduced in the earlier chapters. In a sense, one can consider memory design as a case study of high-performance, high-density, low-power circuit design.

10.2 Semiconductor Memories—An Introduction

10.2.1 Memory Classification

Electronic memories come in many different formats and styles. The type of memory unit that is preferable for a given application is a function of the required memory size, the time it takes to access the stored data, the access patterns, and the system requirements.

Depending upon the level of abstraction, different means are used to express the size of a memory unit. The *circuit* designer tends to define the size of the memory in terms of *bits* that are equivalent to the number of individual cells (flip-flops or registers) needed to store the data. The *chip* designer expresses the memory size in *bytes* (groups of 8 or 9 bits) or its multiples—kilobytes (Kbyte), megabytes (Mbyte), gigabytes (Gbyte), and ultimately terabytes (Tbyte). The *system* designer likes to quote the storage requirement in terms of *words*, which represent a basic computational entity. For instance, a group of 32 bits represents a word in a computer that operates on 32-bit data.

The timing properties of a memory are illustrated in Figure 10.1. The time it takes to retrieve (*read*) from the memory is called the *read-access time*, which is equal to the delay between the read request and the moment the data is available at the output. This time is

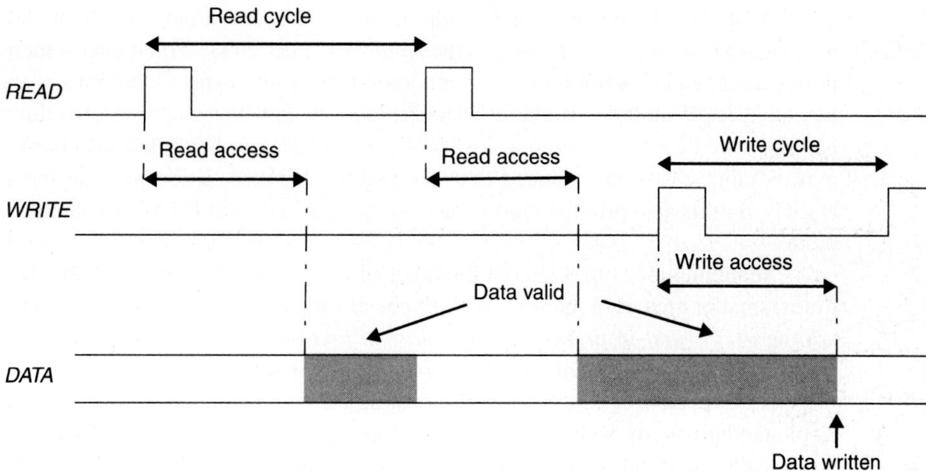

Figure 10.1 Memory-timing definitions.

different from the *write-access time*, which is the time elapsed between a write request and the final writing of the input data into the memory. Finally, another important parameter is the (read or write) *cycle time* of the memory, which is the minimum time required between successive reads or writes. This time is normally greater than the access time for reasons that will become apparent later in the chapter. Read and write cycles do not necessarily have the same length, but their lengths are considered equal for simplicity of system design.

Furthermore, semiconductor memories can be classified on the basis of memory functionality, access patterns, and the nature of the storage mechanism. A distinction is made between *read-only* (ROM) and *read-write* (RWM) memories. The RWM structures have the advantage of offering both read and write functionality with comparable access times and are the most flexible memories. Data is stored either in flip-flops or as a charge on a capacitor. As in the classification introduced in the discussion on sequential circuitry, these memory cells are called *static* and *dynamic* respectively. The former retain their data as long as the supply voltage is retained, while the latter need periodic refreshing to compensate for the charge loss caused by leakage. Since RWM memories use active circuitry to store the information, they belong to the class called *volatile* memories, where the data is lost when the supply voltage is turned off.

Read-only memories, on the other hand, encode the information into the circuit topology—for example, by adding or removing diodes or transistors. Since this topology is hardwired, the data cannot be modified; it can only be read. Furthermore, ROM structures belong to the class of the *nonvolatile* memories. Disconnection of the supply voltage does not result in a loss of the stored data.

The most recent entry in the field are memory modules that can be classified as nonvolatile, yet offer both read and write functionality. Typically, their write operation takes substantially more time than the read. We call them *nonvolatile read-write* (NVRWM) memories. Members of this family are the *EPROM* (erasable programmable read-only memory), *E²PROM* (electrically erasable programmable read-only memory), and *flash* memories.

A second memory classification is based on the order in which data can be accessed. Most memories belong to the *random-access* class, which means memory locations can be read or written in a random order. One would expect memories of this class to be called *RAM* modules (random-access memory). For historical reasons, this name has been reserved for the random-access RWM memories, probably because the RAM acronym is more easily pronounced than the awkward RWM. Be aware that most ROM or NVRWM units also provide random access, but the acronym RAM should not be used for them.

Some memory types restrict the order of access, which results in either faster access times, smaller area, or a memory with a special functionality. Examples of such memories are the FIFO (*first-in first-out*), LIFO (*last-in first-out*, most often used as a stack), *shift register,* and CAM (*contents-addressable memory*) structures.[1]

An overview of the memory classes, as introduced above, is given in Figure 10.2. Implementations for each of the mentioned memory structures will be discussed in subsequent sections. It will be demonstrated how the nature of the memory not only affects the choice of the basic storage cell, but also influences the composition of peripheral units.

RWM		NVRWM	ROM
Random Access	**Non–Random Access**	EPROM E^2PROM FLASH	Mask-programmed programmable (PROM)
SRAM DRAM	FIFO LIFO Shift register CAM		

Figure 10.2 Semiconductor memory classification.

A final classification of semiconductor memories is based on the *number of data input and output ports.* While a majority of the memory units presents only a single port that is shared between input and output, memories with higher bandwidth requirements often have multiple input and output ports. Examples of the latter are the register files used in RISC (reduced instruction set computer) microprocessors. Adding more ports tends to complicate the design of the storage cell.

When massive amounts of storage are needed (hundreds of megabytes and more), semiconductor memories tend to become too expensive. More cost-effective technologies such as tape, magnetic disk, and optical disk should be used. While these provide extensive storage capabilities at a low cost per bit, they tend to be either slow or provide limited access patterns. For instance, a magnetic tape generally allows for serial access only. For these reasons, such memories do not communicate directly with the computing processor, but are interfaced through a number of faster semiconductor memories. They are called *secondary* or *tertiary* memories and are beyond the scope of this textbook.

[1] For lack of space, the latter will only be treated superficially in this text.

10.2.2 Memory Architectures and Building Blocks

When implementing an N-word memory where each word is M bits wide,[2] the most intuitive approach is to stack the subsequent memory words in a linear fashion, as shown in Figure 10.3a. One word at a time is selected for reading or writing with the aid of a *select* bit (S_0 to S_{N-1}), if we assume that this module is a single-port memory. In other words, only one signal S_i can be high at any time. For simplicity, let us temporarily assume that each storage cell is a D flip-flop and that the select signal is used to activate (clock) the cell. While this approach is relatively simple and works well for very small memories, one runs into a number of problems when trying to use it for larger memories.

Assume that we would like to implement a 1-million ($N = 10^6$) by 8 ($M = 8$) memory. The reader should be aware that 1 million is a simplification of the actual memory size, since memory dimensions always come in powers of two. In this particular case, the actual number of words equals $2^{20} = 1024 \times 1024 = 1{,}048{,}576$. For ease of use, it is common practice to denote such a memory as 1 Mword unit.

When implementing this structure using the strategy of Figure 10.3a, we quickly realize that 1 million select signals are needed—one for every word. Since these signals are normally provided from off-chip or from another part of the chip, this translates into insurmountable wiring and/or packaging problems. A *decoder* is inserted to reduce the number of select signals (Figure 10.3b). A memory word is selected by providing a binary encoded *address word* (A_0 to A_{K-1}). The decoder translates this address into $N = 2^K$ select lines, only one of which is active at a time. This approach reduces the number of external

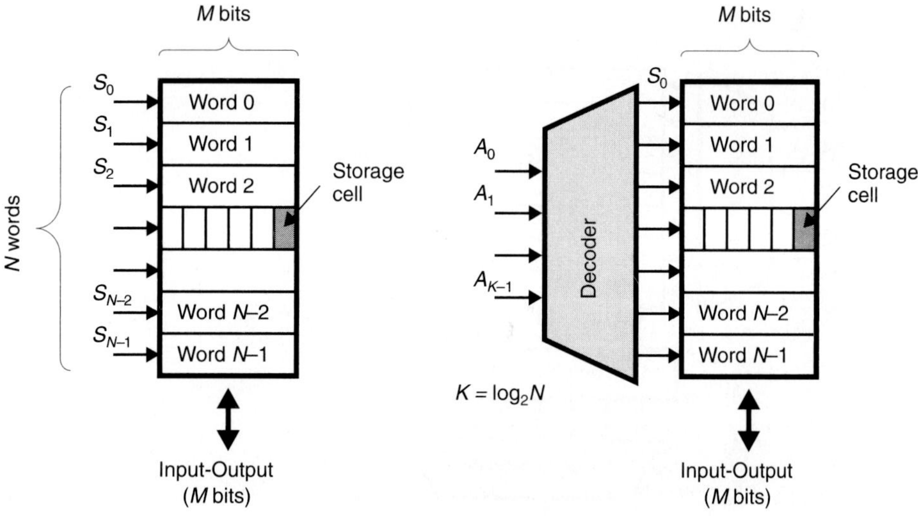

(a) Intuitive architecture for $N \times M$ memory

(b) A decoder reduces the number of address bits

Figure 10.3 Architectures for N-word memory (where each word is M bits).

[2] The length of a word varies between 1 and 128 bits. In commercial memory chips, the word length typically equals 1, 4, or 8 bits.

address lines from 1 million to 20 ($\log_2 2^{20}$) in our example, which virtually eliminates the wiring and packaging problem. The decoder is typically designed so that its dimensions are matched to the size of the storage cell and the connections between the two, in particular the S signals in Figure 10.3b, do not produce any area overhead. The value of this approach can be appreciated by interpreting Figure 10.3b as a physical floorplan of the memory module. By performing the pitch matching between decoder and memory core, the S wires can be very short, and no large routing channel is required.

While this resolves the select problem, it does not address the issue of the memory aspect ratio. Evaluation of the dimensions of the storage array of our token example shows that its height is approximately 128,000 times larger than its width ($2^{20}/2^3$), assuming the shape of the basic storage cell is approximately square which is almost always the case. Obviously, this results in an unacceptable and nonimplementable design. Besides the bizarre shape factor, the resulting design is also extremely slow. The vertical wires connecting the storage cells to the input/outputs become excessively long. Remember that the delay of an interconnect line increases at least linearly with its length.

To address this problem, memory arrays are organized so that the vertical and horizontal dimensions are of the same order of magnitude; thus, the aspect ratio approaches unity. Multiple words are stored in a single row and are selected simultaneously. To route the correct word to the input/output terminals, an extra piece of circuitry called the *column decoder* is needed. The concept is illustrated in Figure 10.4. The address word is partitioned into a *column address* (A_0 to A_{K-1}) and a *row address* (A_K to A_{L-1}). The row address enables one row of the memory for *R/W,* while the column address picks one particular word from the selected row.

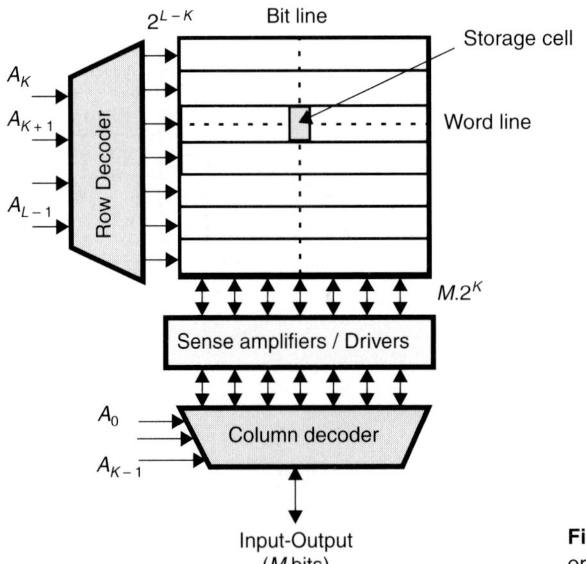

Figure 10.4 Array-structured memory organization.

Example 10.1 Memory Organization

An alternative choice would be to organize the memory core of our example as an array of 4,000 by 2,000 cells (to be more precise, 4096 × 2048), which approaches a square aspect ratio. Each of the 4,000 rows stores 256 8-bit words. This results in a row address of 12 bits, while the column address measures 8 bits. It can be verified that the total address space still equals 20 bits.

Figure 10.4 introduces commonly used terminology. The horizontal select line that enables a single row of cells is called the *word line*, while the wire that connects the cells in a single column to the input/output circuitry is named the *bit line*.

The area of large memory modules is dominated by the size of the memory core. It is thus crucial to keep the size of the basic storage cell as small as possible. We could use one of the register cells introduced in Chapter 6 to implement a R/W memory. Such a cell easily requires more than 10 transistors per bit, and employing it in a large memory would result in excessive area requirements. Semiconductor memory cells therefore reduce the cell area by trading off some desired properties of digital circuits, such as noise margin, logic swing, input-output isolation, fan-out, or speed. While a degradation of some of those properties is allowable within the confined domain of the memory core where noise levels can be tightly controlled, this is not acceptable when interfacing with the external or surrounding circuitry. The desired digital signal properties must be recovered with the aid of peripheral circuitry.

For example, it is common to reduce the voltage swing on the bit lines to a value substantially below the supply voltage. This reduces both the propagation delay and the power consumption. A careful control of the crosstalk and other disturbances is possible within the memory array, ensuring that sufficient noise margin is obtained even for these small signal swings. Interfacing to the external world, on the other hand, requires an amplification of the internal swing to a full rail-to-rail amplitude. This is achieved by the *sense amplifiers* shown in Figure 10.4. The design of those peripheral circuits is discussed in Section 10.4. Relaxation of bounds on a number of the coveted digital properties makes it possible to reduce the transistor count of a single memory cell to between one and six transistors!

The architecture of Figure 10.4 works well for memories up to a range of 64 Kbits to 256 Kbits. Larger memories start to suffer from a serious speed degradation as the length, capacitance, and resistance of the word and bit lines become excessively large. Larger memories have consequently gone one step further and added one extra dimension to the address space, as illustrated in Figure 10.5.

The memory is partitioned into *P* smaller blocks. The composition of each of the individual blocks is identical to one of Figure 10.4. A word is selected on the basis of the row and column addresses that are broadcast to all the blocks. An extra address word called the *block address*, selects one of the *P* blocks to be read or written. This approach has a dual advantage.

1. The length of the local word and bit lines—that is, the length of the lines within the blocks—is kept within bounds, resulting in faster access times.

2. The block address can be used to activate only the addressed block. Nonactive blocks are put in power-saving mode with sense amplifiers and row and column decoders disabled. This results in a substantial power saving that is desirable, since power dissipation is a major concern in very large memories.

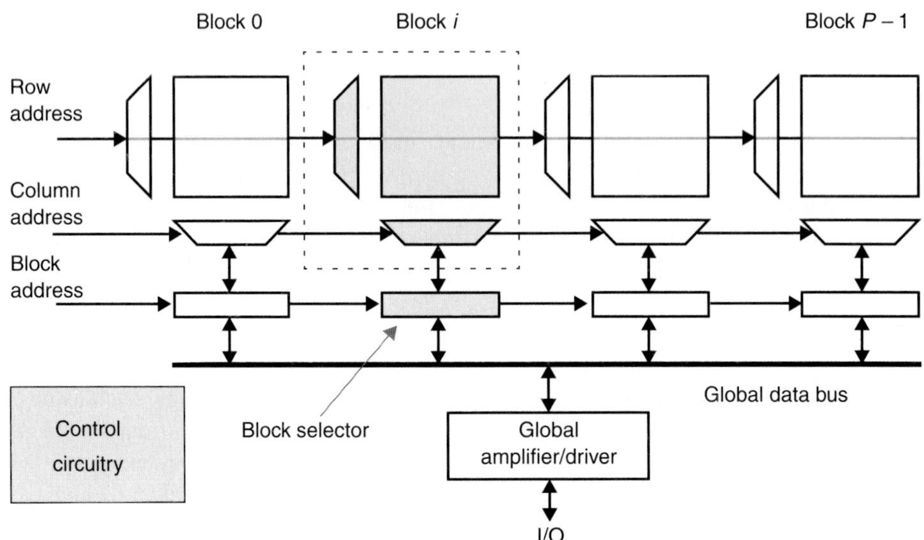

Figure 10.5 Hierarchical memory architecture. The block selector enables a single memory block at a time.

Example 10.2 Hierarchical Memory Architecture

As an example, a 4 Mbit SRAM can be designed [Hirose90] as a composition of 32 blocks, each of which contains 128 Kbits. Each block is structured as an array with 1024 rows and 128 columns. The row address, column address, and block address are 10, 7, and 5 bits wide, respectively.

Multiple variants of the proposed architecture are possible. Variations include the positioning of the sense amplifiers, the partitioning of the word and bit lines, and the styles of the decoders used. The underlying concept is the same—it is advantageous to partition a large memory into smaller subdivisions to combat the delay associated with extra long lines. The gains in performance and power consumption easily outweigh the overhead incurred by the partitioning.

Finally, a component of the memory design that is often overlooked is the input/output interface and control circuitry. The nature of the I/O interface has an enormous impact on the global memory control and timing. This statement is illustrated by comparing the input-output behavior of typical DRAM and SRAM components and the associated timing structure.

Since the early days, DRAM designers have opted for a multiplexed addressing scheme. In this model, the lower and upper halves of the address words are presented sequentially on a single address bus. This approach reduces the number of package pins and has survived through the subsequent memory generations, mostly for reasons of backwards compatibility The presence of a new address word is asserted by raising the a number of strobe signals (Figure 10.6a). Raising the *RAS* (row-access strobe) signal asserts that the msb part of the address is present on the address bus and that the word-decoding process can be initiated. The lsb part of the address is applied next, and the *CAS* (column-access strobe) signal is asserted. To ensure correct the memory operation, a careful timing of the *RAS-CAS* interval is necessary. In fact, the *RAS* and *CAS* signals act as clock inputs to the memory module and are used to synchronize memory events, such as decoding, memory core access, and sensing.

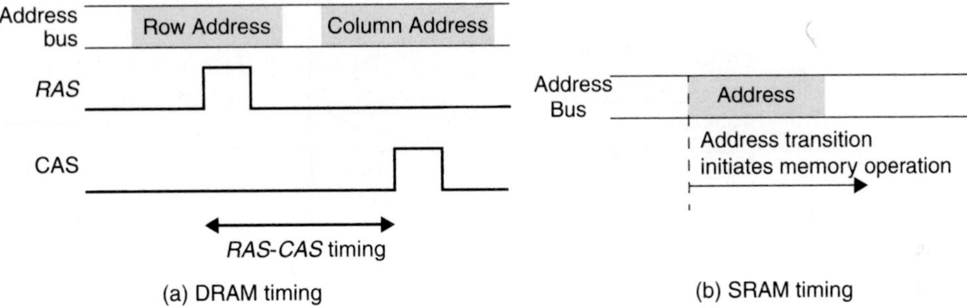

Figure 10.6 Input-output interface of DRAM and SRAM memories and their impact on memory control.

The SRAM designers, on the other hand, have chosen a self-timed approach (Figure 10.6b). The complete address word is presented at once, and circuitry is provided to automatically detect any transitions on that bus. No external timing signals are needed. All internal timing events, such as the enabling of the decoders and sense amplifiers, are derived from the internally generated transition signal.

Designing the control and timing circuitry so that the memory is functional over a wide range of manufacturing tolerances and operating temperatures is a demanding task that requires extensive simulation and design optimization. It is an integral but often overlooked part of the memory-design process and has a major impact on both memory reliability and performance.

10.3 The Memory Core

This section concentrates on the design of the memory core and its composing cells for a variety of semiconductor memory types. While the most compelling issue in designing large memories is to keep the size of the cell as small as possible, this should be done so that other important design-quality measures such as speed and reliability are not fatally affected. In sequence, we discuss ROM, NVRWM, and RAM memory cores.

10.3.1 Read-Only Memories

While the idea of a read-only memory might seem odd at first, a second glance reveals a large number of potential applications. An example is program storage for processors with fixed applications such as washing machines, calculators, and game machines.

ROM Cells—An Overview

The fact that the contents of a ROM cell are permanently fixed considerably simplifies its design. The cell should be designed so that a 0 or 1 is presented to the bit line upon activation of its word line. Figure 10.7 shows several ways to accomplish this.

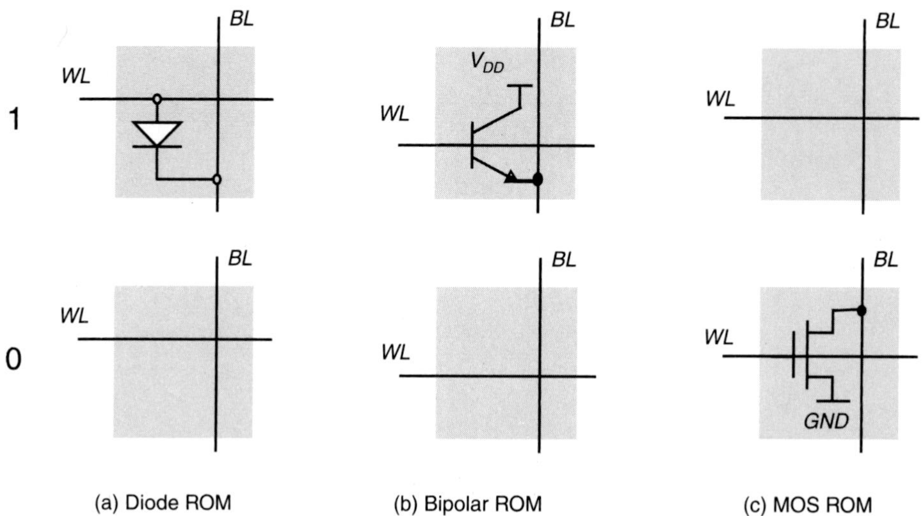

Figure 10.7 1 and 0 ROM cells in different technologies.

Consider first the simplest cell, which is the diode-based ROM cell shown in Figure 10.7a. Assume that the bit line *BL* is resistively clamped to ground, which means that *BL* is pulled low through a resistor connected to ground lacking any other excitations or inputs. This is exactly what happens in the 0 cell. Since no physical connection between the word line *WL* and *BL* exists, the value on *BL* is low, independent of the value of *WL*. On the other hand, when a high voltage V_{WL} is applied to the word line of the 1 cell, the diode is enabled, and the word line is pulled up to $V_{WL} - V_{D(on)}$, resulting in a 1 on the bit line. In other words, the presence or absence of a diode between *WL* and *BL* differentiates between ROM-cells storing a 1 or a 0 respectively.

The disadvantage of the diode cell is that it does not isolate the bit line from the word line. All current required to charge the bit-line capacitance, which can be quite high for large memories, has to be provided through the word line and its drivers; therefore, this approach only works for small memories. A better approach is to use an active device in the cell to provide amplification. This is exemplified in Figure 10.7b, where the diode is replaced by the base-emitter junction of a bipolar device, whose collector is connected to the supply voltage. The operation is identical to that of the diode cell with one major difference—the current to be provided by the word-line driver is substantially reduced due to the current gain of the bipolar transistor. In fact, the combination of ROM cell and resistive load acts as an emitter-follower, which was shown earlier to exhibit excellent capacitance-driving capabilities. The improved isolation comes at the penalty of a more complex cell and a larger area. The latter is caused primarily by the extra supply contact. This contact must be provided in every cell, so that the supply rail must be distributed throughout the array. An example of a 4×4 array is shown in Figure 10.8. Notice how the overhead of the supply lines is reduced by sharing them between neighboring cells. This requires the *mirroring* of the odd cells around the horizontal axis, an approach that is extensively used in memory cores of all styles.

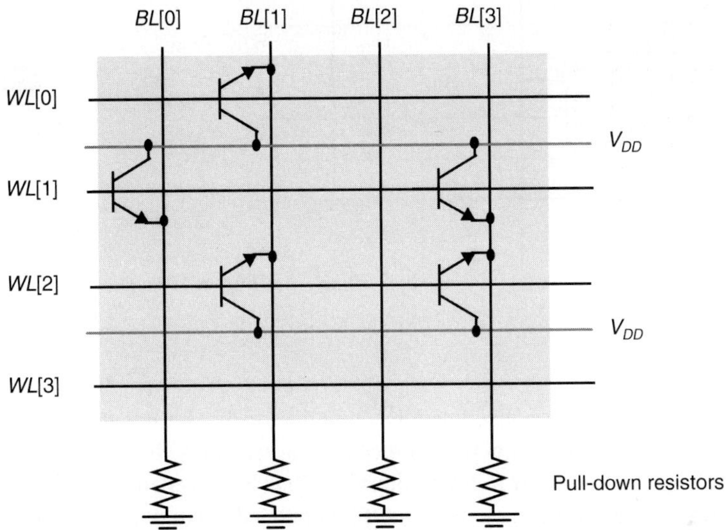

Figure 10.8 A 4×4 bipolar ROM cell array.

Problem 10.1 Bipolar ROM Memory Array

Determine the values of the data stored at addresses 0, 1, 2, and 3 in the ROM of Figure 10.8.

While the bipolar cell offers substantial improvement over the diode ROM, complete isolation between word and bit lines is still not achieved. The MOS cell of Figure 10.7c offers perfect isolation, lower power consumption, and higher density at the cost of reduced speed. To be operational, this cell requires the bit line to be resistively clamped to the supply voltage, or equivalently, the default value at the output must equal 1. The absence of a transistor been *WL* and *BL* hence means that a 1 is stored. The 0-cell is realized by providing an MOS device between bit line and ground. Applying a high voltage on the word line turns on the device, which in turn pulls down the bit line. An example of a 4×4 MOS ROM array is shown in Figure 10.9. Notice how a PMOS load is used to pull up the bit lines in case none of the attached NMOS devices is enabled.

Problem 10.2 MOS NOR ROM Memory Array

Determine the values of the data stored at addresses 0, 1, 2, and 3 in the ROM of Figure 10.9.

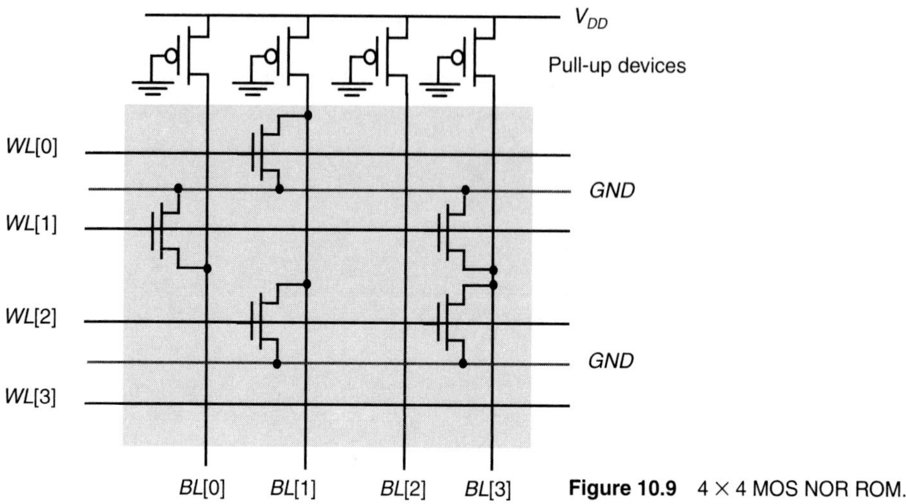

Figure 10.9 4 × 4 MOS NOR ROM.

NOR and NAND ROMs

A careful reader might have noticed that the combination of a bit line, PMOS pull-up, and NMOS pull-downs constitutes nothing other than a pseudo-NMOS NOR gate with the word lines as inputs. An $N \times M$ ROM memory can thus be considered as a combination of M NOR gates with at most N inputs (for a fully populated column). Under normal operating conditions, only one of the word lines goes high, and at most one of the pull-down devices is turned on. This raises some interesting issues regarding the sizing of both the cell and pull-up transistors.

- To keep the cell size and the bit-line capacitance small, the pull-down device should be kept as close as possible to minimum size.
- On the other hand, the resistance of the pull-up device must be larger than the pull-down resistance to ensure an adequate low level. In Chapter 4, we derived a factor of at least four for pseudo-NMOS gates. This large resistance has a detrimental effect on the low-to-high transitions on the bit lines. The bit-line capacitance consists of the contributions of all connected devices and can be in the pF range for larger memories.

This is where memory and logic design differ. For the sake of density and performance, it is possible to relax some of the quality standards imposed on digital gates. In the NOR ROM, we can trade off noise margin for performance by letting the V_{OL} of the bit line stand at a higher voltage (e.g., 1 to 3 V for a 5 V supply). This makes it possible to use a wider pull-up device, which enhances the low-to-high transition. The reduced noise margin is tolerable within the memory core, where the noise conditions and signal interferences can be carefully controlled. Going to the external world requires a restoration of the full voltage swing. This is accomplished by the peripheral devices—in this case, the sense amplifier. For instance, feeding the bit line into a complementary CMOS inverter with an appropriately adjusted switching threshold restores the full signal swing.

Figure 10.10 shows one of the many possible layouts of the 4 × 4 NOR ROM array of Figure 10.9. The array is constructed by repeating the same cell in both the horizontal and vertical directions, mirroring the odd cells around the horizontal axis in order to share the *GND* wire. The memory is programmed by the selective addition of metal-to-diffusion contacts. The presence of a metal contact to the bit line creates a 0-cell, while its absence indicates a 1-cell. Observe that only one mask layer, the contact mask, is used to program the memory array. This makes it possible to mass-fabricate ROM dies, delaying the actual programming of the memory to one of the last process steps.[3]

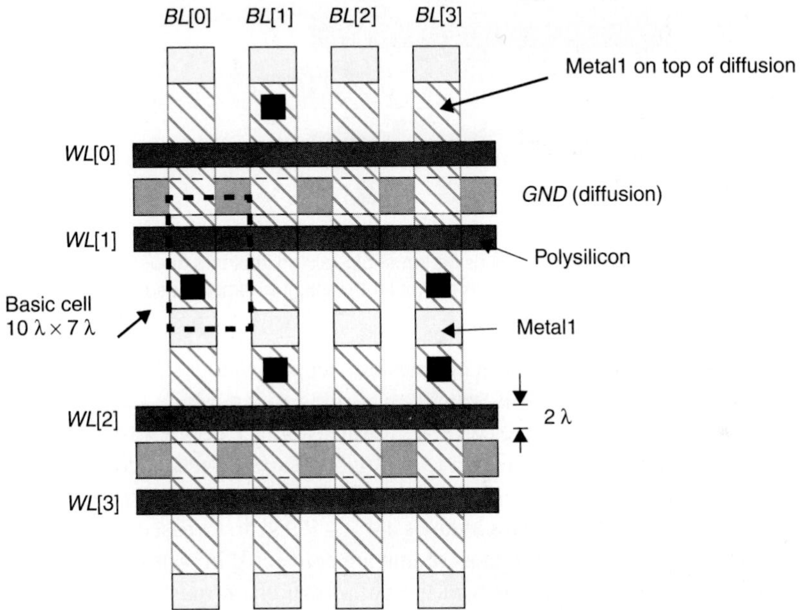

Figure 10.10 Possible layout of a 4 × 4 MOS NOR ROM. The bit lines are implemented in metal-1 and are routed on top of the cell diffusion. *GND* lines are distributed horizontally in diffusion.

Contact-mask programming is not the only possible option. A denser solution is shown in Figure 10.11. In this instance, an extra implant step is introduced in the process. The thresholds of some of the transistors are selectively raised to a value higher than the voltage swing of the word lines—e.g., to 7 V for a 5 V supply voltage—by implanting extra *p*-type impurities. Consequently, it becomes impossible to turn on the device, which is equivalent to eliminating it. The bit-line contact between the cells can now be shared as well, which reduces the cell area to 59.5 λ^2 taking into account the extra spacing required for the implants. The reader should be aware that the proposed layouts by no means yield the minimum possible area. Industrial ROM cells succeed in further reducing the area by using special layout techniques such as 45° lines, or by modifying the manufacturing process.

[3] Single-mask programming is only important for mass-produced commodity ROM devices. This is not an issue when designing custom ROM memory modules that are part of an application-specific processor.

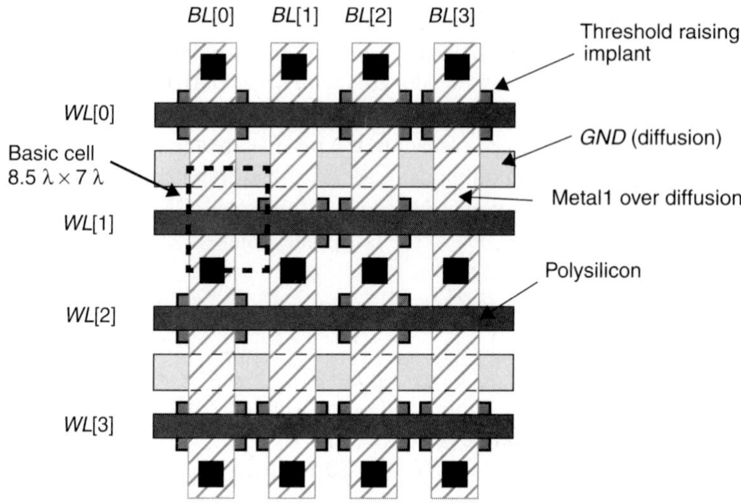

Figure 10.11 4-×-4 MOS NOR ROM using threshold-raising implants to disable transistors. Since the MOSIS design rules do not provide such an implant mask, it is simply assumed that the implant must extend 1 λ on all sides of the transistor.

It is important to note that the transistor occupies only a small ratio of the total cell size, which measures 70 λ^2. It is actually possible to increase its size over the minimum dimensions without affecting the cell size. A transistor size of 4/2 was chosen in the examples of Figure 10.10 and Figure 10.11. A large part of the cell is devoted to the bit-line contact and ground connection. One way to avoid this overhead is to adopt a different memory organization. Figure 10.12 shows a 4 × 4 ROM array based on the NAND configuration. All transistors constituting a column are connected in series. The basic property of a NAND gate is that all transistors in the pull-down chain must be on to produce a low value. To understand the memory operation, we need to be aware that the word lines are

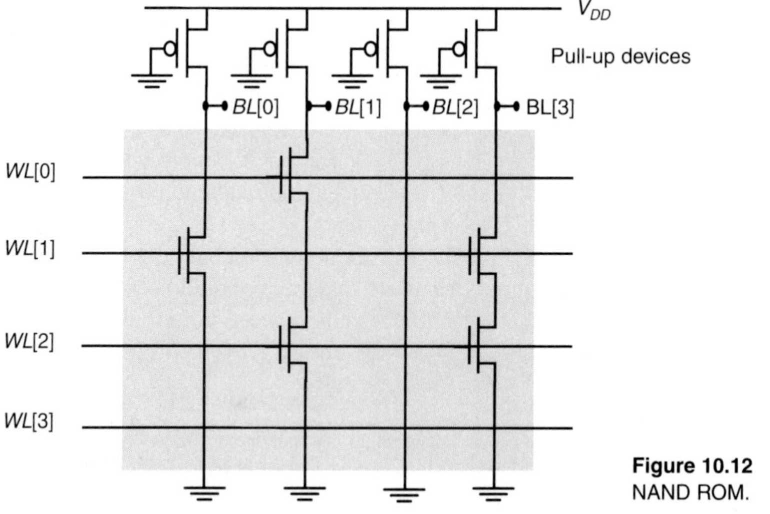

Figure 10.12 4 × 4 MOS NAND ROM.

operated in reverse-logic mode. All word lines are high by default with the exception of the selected row, which is set to 0. All transistors on the nonselected rows are thus turned on. Now suppose that no transistor is present on the intersection between the row and column of interest. Since all other transistors on the series chain are selected, the output is pulled low, and the stored value equals 0. On the other hand, a transistor present at the intersection is turned off when the associated word line is brought low. This results in a high output, which is equivalent to the reading a 1.

Problem 10.3 MOS NAND ROM

Determine the values of the data stored at addresses 0, 1, 2, and 3 in the ROM of Figure 10.12.

The main advantage of the NAND structure is that the basic cell only consists of a transistor (or a lack of a transistor) and that no contact to any of the supply voltages is needed. This reduces the cell size substantially, as illustrated in the layout of Figure 10.13. In this example, we again use an extra implant step to program the memory. In contrast to the NOR ROM, eliminating a transistor means replacing it by a short-circuit. This is accomplished by a threshold-lowering implant using n-type impurities. This turns the device into a depletion transistor, which is always on, regardless of the applied word-line voltage. The resulting cell area of 30 λ^2 is almost two times smaller than the equivalent NOR ROM cell. This comes at a price, however. In the next section, we demonstrate that the NAND configuration results in a considerable loss in performance and is only useful for small memory arrays.

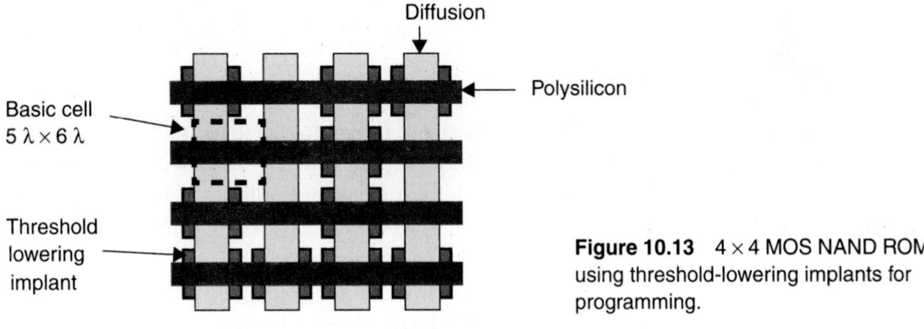

Figure 10.13 4 × 4 MOS NAND ROM using threshold-lowering implants for programming.

Example 10.3 Voltage Swings in NOR and NAND ROMs

Assuming that the layouts of Figure 10.11 and Figure 10.13 are implemented using our standard 1.2 μm CMOS technology, determine the size of the PMOS pull-up device so that the worst-case value of V_{OL} is never higher than 2.5 V (for a 5 V supply voltage). Determine the values for an 8 × 8 and a 512 × 512 array.

1. NOR ROM

Since at most one transistor can be on at a time, the value of V_{OL} is not a function of the array size nor the programming of the array. The low output voltage is computed by analyzing a pseudo-NMOS inverter with a single pull-down device of a (4/2) size. This circuit was analyzed in detail in Chapter 4, which yielded Eq. (4.11). For the sake of simplicity, it is assumed that $V_{Tn} = |V_{Tp}|$.

$$V_{OL} = (V_{DD} - V_T)\left(1 - \sqrt{1 - \frac{k_p}{k_n}}\right)$$

$$= (5 - 0.75)\left(1 - \sqrt{1 - \frac{(W/L)_p}{3(W/L)_n}}\right)$$

Solving for V_{OL} = 2.5 V leads to a required size for the PMOS device of $(W/L)_p = 6.6$. Hint: make sure you take the lateral diffusion into account when deriving this number. It is easily verified that this translates into almost identical high-to-low and low-to-high transitions on the bit lines.

2. NAND ROM

Due to the series chaining, the value of V_{OL} is a function of both the size of the memory (number of rows) and the programming. The worst case occurs when all bits in a column are set to 1, which means *N* transistors are connected in series in the pull-down network. Assuming the *N* transistors can be replaced by an *N*-times longer device, the values of $(W/L)_p$ can be derived for the (8 × 8) and (512 × 512) cases. Observe that the design of Figure 10.13 uses minimum-size devices (3/2).

$$(8 \times 8)\text{: } (W/L)_p = 0.62$$

$$(512 \times 512)\text{: } (W/L)_p = 0.0097$$

While the first case still produces acceptable results for the PMOS device, the second one would require an extremely long pull-up device, which is unacceptable. For this reason, NAND ROM's are rarely used for arrays with more than 8 or 16 rows.

ROM Transient Performance

The transient response of a memory array is defined as the time it takes from the time a word line switches until the point where the bit line has traversed a certain voltage swing ΔV. Since the bit line normally feeds into a sense amplifier, it is not necessary for it to traverse its complete swing. A voltage drop (or raise) of ΔV is sufficient to make the sense amplifier react. Typical values of ΔV range around 0.5 V.

One important difference between the analysis of the propagation delay of a logic gate and that of a memory array is that most of the delay is attributable to the interconnect parasitics. An accurate modeling of these parasitics is therefore of prime importance. This is illustrated in the following example, which extracts the parasitic resistance and capacitance of the word and bit lines of both the NOR and NAND ROM arrays introduced earlier. We cover this example quite extensively because the same approach also holds for other memory styles, such as SRAM and DRAM.

Example 10.4 Word- and Bit-Line Parasitics

In this example, we first derive an equivalent model of the memory arrays of Figure 10.11 and Figure 10.13 respectively. Taking into account all transistors in the array simultaneously quickly leads to intractable equations and models, especially for larger memories. Simplification and abstraction is the obvious approach to be followed. We only consider the case of the (512 × 512) array.

1. NOR ROM

Figure 10.14 shows a model that is appropriate for the analysis of the word- and bit-line delay of the NOR ROM. The word line is best modeled as a distributed *RC* line since it is implemented in polysilicon with a relatively high sheet resistance. The bit line, on the other hand, is implemented in aluminum, and the resistance of the line only comes into play for very long lines. It is reasonable to assume for this example that a purely capacitive model is adequate and that all capacitive loads connected to the wire can be lumped into a single element.

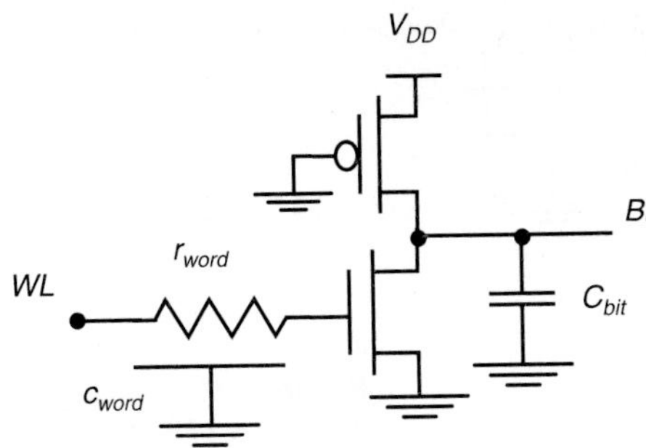

Figure 10.14 Equivalent transient model for the NOR ROM.

Word-line parasitics (for the memory array of Figure 10.11),

Resistance/cell: $(7/2) \times 10\ \Omega/\square = 35\ \Omega$ (using the data of Table 8.9)

Wire capacitance/cell: $(7\lambda \times 2\lambda)\ (0.6)^2\ 0.058 + 2 \times (7\lambda \times 0.6) \times 0.043 = 0.65$ fF (from Table 8.1 and Table 8.2)

Gate capacitance/cell: $(4\lambda \times 2\lambda)\ (0.6)^2\ 1.76 = 5.1$ fF. [4]

Bit-line parasitics:

Resistance/cell: $(8.5/4) \times 0.07\ \Omega/\square = 0.15\ \Omega$ (which is negligible)

Wire capacitance/cell: $(8.5\lambda \times 4\lambda)\ (0.6)^2\ 0.031 + 2 \times (8.5\lambda \times 0.6) \times 0.044 = 0.83$ fF

Drain capacitance/cell:
$((3\lambda \times 4\lambda)\ (0.6)^2 \times 0.3 + 2 \times 3\lambda \times 0.6 \times 0.8) \times 0.375 + 4\lambda \times 0.6 \times 0.43 = 2.6$ fF

The latter term deserves an explanation. The *drain capacitance* contributed by every cell connected to the bit line consists of the bottom-junction, side wall, and overlap capacitances. It may be assumed that all cells, besides the one being switched, are in the off-state, which explains why only the overlap part of the gate capacitance is taken into account. It is furthermore assumed that the bit line swings between 2.5 V and 5 V. The K_{eq} factor evaluates to 0.375 under this condition, as derived in Chapter 3. For all other capacitance values, please refer to Example 3.4.

2. NAND ROM

As in the approach taken for the NOR ROM, we can derive an equivalent model for the analysis of the delay of the NAND structure. While the word-line model is identical, modeling the bit-line behavior is more complex due to the long chain of series-connected transistors. The worst-case behavior occurs when the transistor at the bottom of the chain is switched, and

[4] For simplicity, we assume that the total gate capacitance of the cell appears as a load to the word line. This is not completely accurate, but is appropriate for the intended modeling level.

the column is completely populated with transistors. A model approximating the behavior for that case is shown in Figure 10.15. Each of the series transistors (which are normally in the on-mode) is modeled as a resistance-capacitance combination. The entire chain can be modeled as a distributed *rc*-network for large memories.

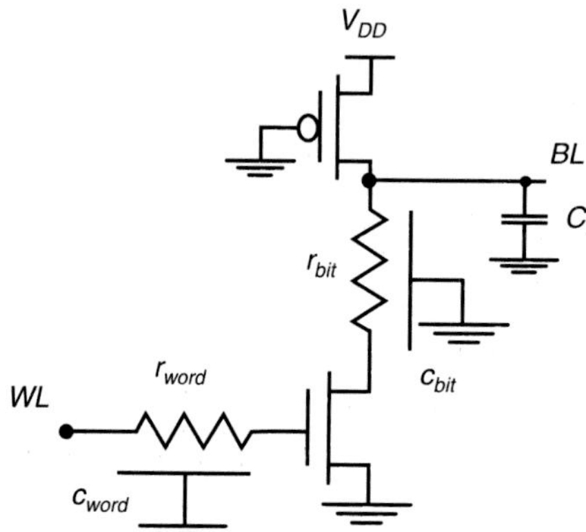

Figure 10.15 Equivalent model for word and bit line of NAND ROM.

Word-line parasitics (for the memory array of Figure 10.13) follow:

Resistance/cell: $(6/2) \times 10\ \Omega/\square = 30\ \Omega$

Wire capacitance/cell: $(6\lambda \times 2\lambda)\ (0.6)^2\ 0.058 + 2 \times (6\lambda \times 0.6) \times 0.043 = 0.56$ fF

Gate Capacitance/cell: $(3\lambda \times 2\lambda)\ (0.6)^2\ 1.76 = 3.8$ fF.

Bit-line parasitics:

Resistance/cell: ~ 10 kΩ, the average transistor resistance over the range of interest.

Wire capacitance/cell: Included in diffusion capacitance

Source/drain capacitance/cell:

$((3\lambda \times 3\lambda)\ (0.6)^2 \times 0.3 + 2 \times 3\lambda \times 0.6 \times 0.8) \times 0.375 + (3\lambda \times 2\lambda)\ (0.6)^2 \times 1.76 = 5.2$ fF

The drain/source capacitance must include the gate-source and gate-drain capacitances, which means the complete gate capacitance. This is in contrast with the NOR case, in which only the drain-source overlap capacitance was included.

Determining the average transistor resistance is more complex. In fact, the resistance varies from device to device depending upon the position in the chain, caused by differing gate-source voltages and body effects. Simulation is the only correct way to derive a meaningful value for this average resistance. An example of such a simulation was shown in Figure 4.24 for an transistor of identical size. It shows a variation of the NMOS resistance between 10 kΩ and 30 kΩ. Experiments and simulations seem to indicate that during transient operation in long resistor chains, the average resistance of the device is close to the minimum value, in this case 10 kΩ. This is the value we will employ in the first-order analysis.

A simpler, but less accurate, approach is to lump all transistors in the pull-down chain into a single, long device and lump all the intermediate capacitors into one load capacitance. This translates into a model with a single pull-down device with a (W/L) of $(3/(2 \times 512))$ and a load capacitance of $(512) \times 5.2$ fF = 2.7 pF.

One might wonder why we bother with models at all, if simulations could readily produce more accurate results. One has to be aware that the memory modules can contain thousands to millions of transistors, making repeated simulations prohibitively slow. Models also help us to understand the behavior of the memory more readily. Computer simulations do not tell where the bottlenecks are located and how to address them.

Using the computed data and the equivalent model, an estimated value of the propagation delay of the memory core and its components can now be derived.

Example 10.5 Propagation Delay of NOR ROM

1. Word-line delay

Once again, we only consider the 512 × 512 case.The delay of the distributed *rc*-line containing M cells can be approximated using the expressions derived in Chapter 8.

$$t_{word} = 0.38\ (r_{word} \times c_{word})\ M^2 = 0.38\ (35\ \Omega \times (0.65 + 5.1)\ \text{fF})\ 512^2 = 20\ \text{nsec}$$

2. Bit-line delay

The response time of the bit line depends upon the transition direction. Assuming a (2.4/1.2) pull-down device and a (6/1.2) pull-up transistor as derived in Example 10.3, we can derive the propagation delay using the familiar techniques. Observe that the bit line switches between 5 V and 2.5 V. The device parameters (k_n, k_p) are identical to those used everywhere else in the book.

$$C_{bit} = 512 \times (2.6 + 0.8)\ \text{fF} = 1.7\ \text{pF}$$

$$I_{avHL} = 1/2\ (2.4/0.9)\ (19.6\ 10^{-6})((4.25)^2/2 + (4.25 \times 3.75 - (3.75)^2/2)) - 1/2\ (6/0.9)\ (5.3\ 10^{-6})\ (4.25 \times 1.25 - (1.25)^2/2) = 0.39\ \text{mA}$$

$$t_{HL} = (1.7\ \text{pF} \times 1.25\ \text{V})\ /\ 0.39\ \text{mA} = 5.5\ \text{nsec}$$

The low-to-high response time can be computed using a similar approach. The computation of the current is left as an exercise.

$$t_{LH} = (1.7\ \text{pF} \times 1.25\ \text{V})\ /\ 0.36\ \text{mA} = 5.9\ \text{nsec}$$

Inspection of the above results shows a large discrepancy between word- and bit-line delay. The former is almost completely due to the large resistance of the polysilicon wire. Some of the techniques to reduce the delay of distributed *RC*-lines come in handy here. One possibility is to partition the line into multiple sections and place buffers between them. A common variant of this approach is to drive the word line from both ends, as shown in Figure 10.16a. This effectively reduces the worst-case delay by a factor of four. Another option is to bypass the line with a metal wire called a *global word line* and connect to the polysilicon line every K cells. Finally, the most obvious approach is to use another interconnect material. Aluminum would be the natural choice. Unfortunately, the extra metal-to-polysilicon contact needed in every cell would translate into an unacceptable increase in area. The almost universally used approach is to employ silicides. For instance, WSi_2 has a resistance that is eight times lower than that of polysilicon. This approach reduces the word-line delay of our example to 3.4 nsec, which is in the same range as the bit-line delay.

If necessary, the bit-line delay can be reduced as well. The approach most often used is to further reduce the voltage swing on the bit line and to let the sense amplifier restore the output signal to the full swing. Voltage swings around 0.5 V are quite common.

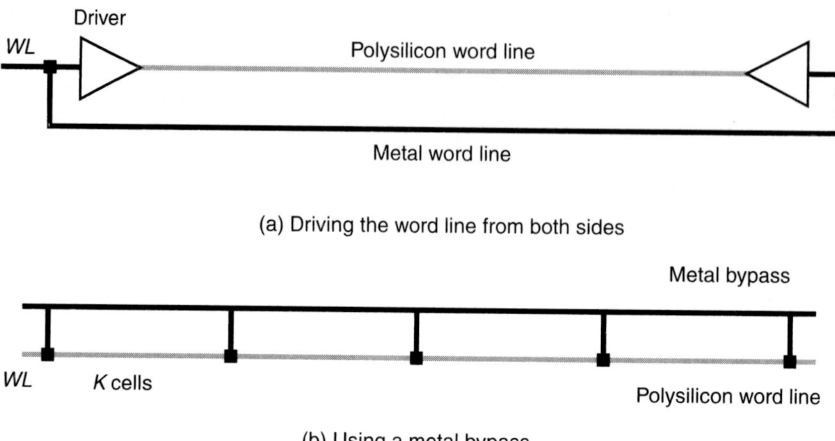

Figure 10.16 Approaches to reduce the word-line delay.

Problem 10.4 Propagation Delay of NAND ROM

Use techniques similar to the ones used in Example 10.5 to determine the word-line and bit-line delay of the 512×512 NAND ROM. Compare the effectiveness and accuracy of the models proposed in Example 10.4 to actual simulation results.

To provide some guidance, Figure 10.17 shows the simulated high-to-low transition of a 16-cell bit line using the full transistor model and lumped model. The simulation using the full model demonstrates how the high-to-low transition is dominated by the propagation delay of the transistor chain that increases quadratically with the length. The low-to-high transition is mainly determined by the high resistance of the pull-up device. These effects are not adequately modeled by the lumped model, as demonstrated by the simulation results. For instance, lumping all the intermediate capacitances in the chain into a single load capacitance

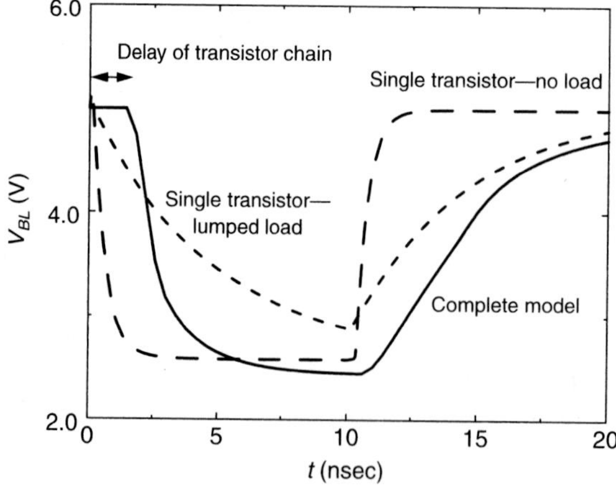

Figure 10.17 Simulated response of 16-word NAND ROM (using a 1.8/4.2 PMOS pull-up): (1) complete model, (2) lumped model replacing the transistor chain by a single device of (1.8/(16 × 0.9 + 0.3) = 1.8/14.7) but no extra load, (3) similar to (2) but with a bit-line load of (16 × 5.2) = 83 fF.

results in a far too pessimistic result, while ignoring the intermediate capacitances yields too optimistic a view.

The distributed *RC*-chain model, on the other hand, might be more accurate if an adequate value of the transistor resistance could be determined. In this particular case, the distributed model predicts a delay of $0.38 \times (10\ k\Omega \times 5.2\ fF) \times 16^2 = 5$ nsec for the high-to-low transition, which is close to the measured value of 4.3 nsec. The simulated low-to-high delay equals 3.7 nsec.

Power Consumption and Precharged Memory Arrays

The proposed NAND and NOR structures inherit all the disadvantages of the pseudo-NMOS gate discussed in Chapter 4.

1. **Ratioed logic**—The V_{OL} is determined by the ratio of the pull-up and pull-down devices. This can result in non-acceptable transistor ratio's as demonstrated earlier in the examples.
2. **Static power consumption**—A static current path exists between the supply rails when the output is low.

The latter can cause severe power dissipation problems. Consider the case of the (512×512) NOR ROM. It is reasonable to assume that 50% of the outputs are low. The standby current for the design of Example 10.4 equals approximately 0.4 mA (for an output voltage of 2.5 V). This translates into a total static dissipation of $(512/2) \times 0.4\ mA \times 5\ V = 0.5\ W$ that is consumed even when nothing happens. This is clearly not acceptable.

To address the two issues raised, one can consult the practices used in designing digital gates. One approach would be to use fully complimentary NAND or NOR gates. The larger number of transistors and the connection to both supply rails makes this approach unattractive from an area perspective. A better approach is to use precharged logic, as shown in Figure 10.18. This approach eliminates the static dissipation as well as the ratioed logic requirements, while keeping the cell complexity the same. Since the logic structure of both the NAND and NOR ROM is simple and is only one level deep, it is possible to ensure that all pull-down paths are off during precharging. This allows us to eliminate the enabling NMOS transistor at the bottom of the pull-down network, keeping the cell simple.

The dynamic architecture makes it possible to control the pull-up and pull-down timing independently. For instance, the PMOS precharge device can be made as large as necessary. Be aware that this transistor loads the clock driver, which might become increasingly hard to design.

The excellent properties of the precharged approach have made it the memory structure of choice. Virtually all large memories currently designed, including NVRWM and RAMs, use dynamic precharging.

ROM Memories—A User Perspective

The reader should be aware that most of the static, dynamic, and power problems raised in the preceding sections are general in nature and apply to other memory architectures as well. Before addressing some of these structures, it is worth discussing the classification

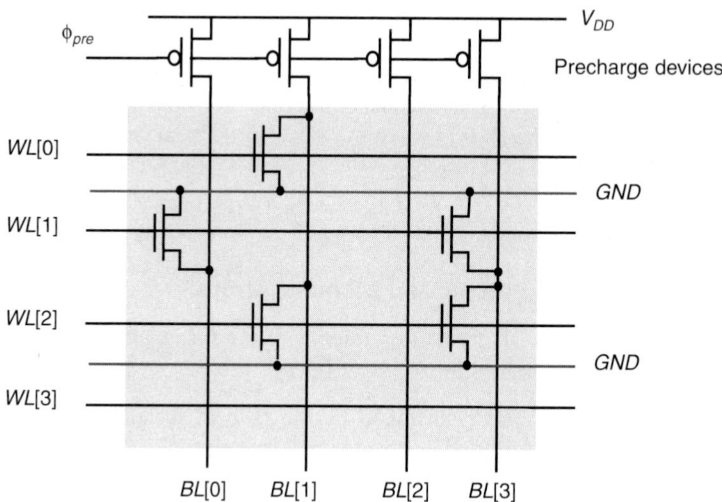

Figure 10.18 Precharged (4 × 4) MOS NOR ROM.

of ROM modules and programming approaches. The first class of ROM modules comprises the so-called *application-specific* ROMs, where the memory module is part of a larger, custom design and programmed for that particular application only. Under these circumstances, the designer has all degrees of freedom and can use any mask layer (or combination thereof) to program the device.

A second, economically more important class are the *commodity* ROM chips, where a vendor mass-produces memory modules that are later customized according to customer specifications. Under these circumstances, it is essential that the number of process steps involved in programming be minimal and that they can be performed as a last phase of the manufacturing process. In this way, large amounts of unprogrammed dies can be preprocessed. This *mask-programmable* approach uses the contact or an extra implant mask to personalize or program the memory, as was shown in some of the examples. The programming of a ROM module involves the manufacturer, which introduces an unwelcome delay in product development. It has consequently become increasingly unpopular.

A more desirable approach is for the client to program the memory at his own facility. One technology that offers such capability is the PROM (Programmable ROM) structure that allows the customer to program the memory one time; hence, it is called a WRITE ONCE device. This is most often accomplished by introducing *fuses* (implemented in nichrome, polysilicon, or other conductors) in the memory cell. During the programming phase, some of these fuses are blown by applying a high current, which disables the connected transistor.

While PROMs have the advantage of being customer-programmable, the single write phase makes them unattractive. For instance, a single error in the programming process or application makes the device unusable. This explains the current preference for devices that can be programmed several times (albeit slowly). The next section explains how this can be achieved.

10.3.2 Nonvolatile Read-Write Memories

The architecture of the NVRW memories is virtually identical to the ROM structure. The memory core consists of an array of transistors placed on a word-line/bit-line grid. The memory is programmed by selectively disabling or enabling some of those devices. In a ROM, this is accomplished by mask-level alterations. In an NVRW memory, a modified transistor that permits its threshold to be altered electrically is used instead. This modified threshold is retained indefinitely (or at least over a long lifetime) even when the supply voltage is turned off. To reprogram the memory, the programmed values must be *erased*, after which a new programming round can be started. The method of erasing is the main differentiating factor between the various classes of reprogrammable nonvolatile memories. The programming of the memory is typically an order of magnitude slower than the reading operation.

We start this section with a description of the floating-gate transistor, which is the device at the heart of all reprogrammable memories. The rest of the section is devoted to a number of alterations of the device, mainly with respect to the erase procedure, which translate into the various NVRWM families.

The Floating-Gate Transistor

Over the years, various attempts have been made to create a device with electrically alterable characteristics and enough reliability to support a multitude of write cycles. For example, the MNOS (Metal Nitride Oxide Semiconductor) transistor held promise, but has been unsuccessful until now. In this device, threshold-modifying electrons are trapped in a Si_3N_4 layer deposited on top of the gate SiO_2 [Chang77]. A more accepted solution is offered by the floating-gate transistor (Figure 10.19), which forms the core of virtually every NVRW memory built today. The structure is similar to a traditional MOS device, except that an extra polysilicon strip is inserted between the gate and channel. This strip is not connected to anything and is called a *floating gate*. The most obvious impact of inserting this extra gate is to double the gate oxide thickness t_{ox}, which results in a reduced device transconductance as well as an increased threshold voltage (see Eq. (2.41) and Eq. (2.48)). Both these properties are not particularly desirable. From other points of view, this device acts as a normal transistor.

More important, this device has the interesting property that its threshold voltage is programmable. Applying a high voltage in the range of 15 to 20 V between the source and

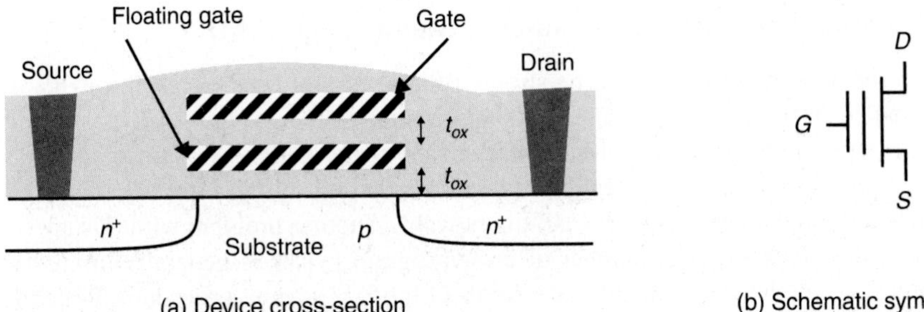

Figure 10.19 Floating-gate transistor (FAMOS).

gate-drain terminals creates a high electric field and causes avalanche injection to occur. Electrons acquire sufficient energy to become "hot" and traverse through the first oxide insulator, so that they get trapped on the floating gate. This phenomenon can occur with oxides as thick as 100 nm, which makes it relatively easy to fabricate the device. In reference to the programming mechanism, the floating-gate transistor is often called a *floating-gate avalanche-injection MOS* (or FAMOS) [Frohman74].

The trapping of electrons on the floating gate effectively drops the voltage on that gate (Figure 10.20a). This process is self-limiting—the negative charge accumulated on the floating gate reduces the electrical field over the oxide so that ultimately it becomes incapable of accelerating any more hot electrons. Removing the voltage leaves the induced negative charge in place, and results in a negative voltage on the intermediate gate (Figure 10.20b). From a device point of view, this translates into an effective increase in threshold voltage. To turn on the device, a higher voltage is needed to overcome the effect of the induced negative charge (Figure 10.20c). Typically, the resulting threshold voltage is somewhere around 7 V; thus, a 5 V gate-to-source voltage is not sufficient to turn on the transistor, and the device is effectively disabled.

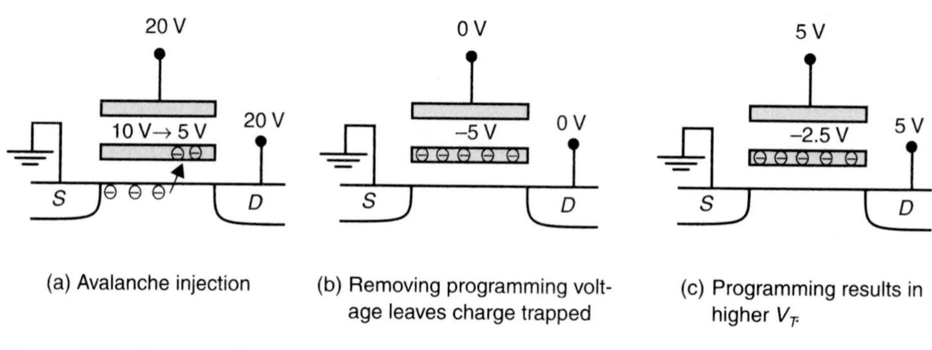

Figure 10.20 Programming the floating-gate transistor.

Since the floating gate is surrounded by SiO_2, which is an excellent insulator, the trapped charge can be stored for many years, even when the supply voltage is removed, which creates a nonvolatile memory. While virtually all nonvolatile memories are currently based on the floating-gate approach, different classes can be discerned, based on the erasure mechanism.

Erasable-Programmable Read-Only Memory (EPROM)

An EPROM is erased by shining strong ultraviolet light on the cells through a transparent window in the package. The UV radiation renders the oxide slightly conductive by the direct generation of electron-hole pairs in the material. The erasure process is slow and can take from seconds to several minutes depending on the intensity of the UV source. Programming takes several (5–10) μsecs/word. Another problem with this approach is the *limited endurance*—the number of erase/program cycles is generally limited to a maximum of one thousand, mainly as a result of the UV erasing procedure. Reliability is also an issue. The device thresholds might vary with repeated programming cycles. Most

EPROM memories therefore contain on-chip circuitry to control the value of the thresholds to within a specified range during programming.

On the other hand, the EPROM cell is extremely simple and dense, making it possible to fabricate large memories at a low cost. EPROM's are therefore attractive in applications that do not require regular reprogramming.

Electrically-Erasable Programmable Read-Only Memory (EEPROM or E²PROM)

The major disadvantage of the EPROM approach is that the erasure procedure has to occur "off-system," which means the memory must be removed from the board and placed in an EPROM programmer for programming. The EEPROM approach avoids this labor-intensive and annoying procedure by providing an electrical-erasure procedure. To achieve this goal, a modified floating-gate device called the FLOTOX (floating-gate tunneling oxide) transistor is used as the programmable device [Johnson80]. A cross-section of the FLOTOX structure is shown in Figure 10.21a. It resembles the FAMOS device, except that a portion of the dielectric separating the floating gate from the channel and drain is reduced in thickness to about 10 nm or less. When a voltage of approximately 10 V (equivalent to an electrical field of around 10^9 V/m) is applied over the thin insulator, electrons travel to and from the floating gate by a mechanism called *Fowler-Nordheim tunneling* [Snow67]. The *I-V* characteristic of the tunneling junction is plotted in Figure 10.21b.

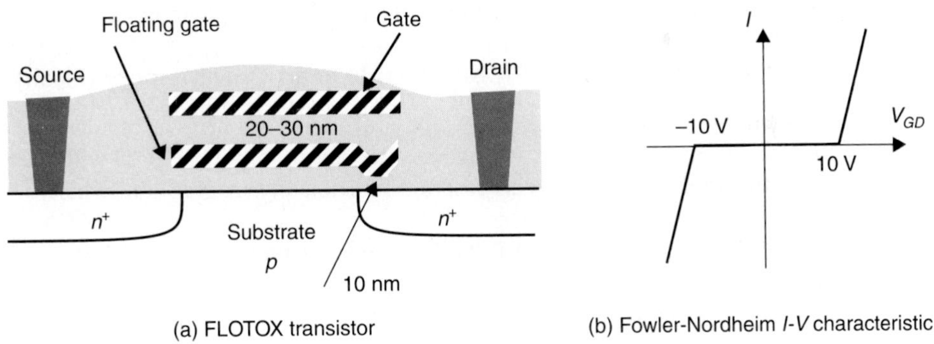

(a) FLOTOX transistor (b) Fowler-Nordheim *I-V* characteristic

Figure 10.21 FLOTOX transistor, programmable using Fowler-Nordheim tunneling.

The main advantage of this programming approach is that it is reversible; that is, erasing is simply achieved by reversing the voltage applied during the writing process. Injecting electrons onto the floating gate raises the threshold, while the reverse operation lowers the V_T. This bidirectionality, however, poses the problem of threshold control. Removing too much charge from the floating gate results in a depletion device that cannot be turned off by the standard word-line signals. Notice that the resulting threshold voltage depends upon the initial charge on the gate. An extra transistor connected in series with the floating-gate transistor is added to the EEPROM cell to remedy this problem. This transistor acts as the access device during the read operation, while the FLOTOX transistor performs the storage function (Figure 10.22). This is in contrast with the EPROM cell, where the FAMOS transistor acts as both the programming and access device.

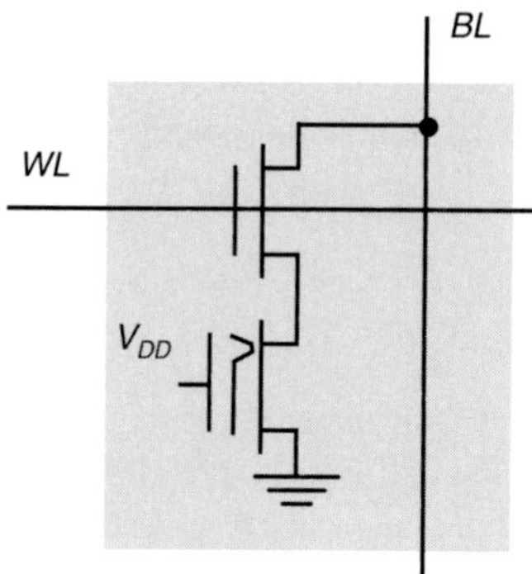

Figure 10.22 EEPROM cell as configured during a read operation. When programmed, the threshold of the FLOTOX device is higher than V_{DD}, effectively disabling it. If not, it acts as a closed switch.

The EEPROM cell with its two transistors is larger than its EPROM counterpart. This area penalty is further aggravated by the fact that the FLOTOX device is intrinsically larger than the FAMOS transistor due to the extra area of the tunneling oxide. Additionally, the fabrication of the very thin oxide is a challenging and costly manufacturing step. EEPROM components thus pack less bits at a higher cost than EPROMs. On the positive side of the balance, EEPROMs offer a higher versatility. They also tend to be more resilient against wear-out and can support up to 10^5 erase/write cycles.[5] Repeated programming causes a drift in the threshold voltages due to permanently trapped charges in the SiO_2. This finally leads to malfunction or the inability to reprogram the device.

Flash Electrically-Erasable Programmable Read-Only Memory (Flash)

The concept of Flash EEPROMs was introduced in 1984 and has rapidly evolved into a popular memory architecture. It combines the density of the EPROM with the versatility of the EEPROM structures, with cost and functionality ranging somewhere between the two.

Technically, the Flash EEPROM is a combination of the EPROM and EEPROM approaches. Most Flash EEPROM devices use the avalanche hot-electron-injection approach to program the devices. Erasure is performed using Fowler-Nordheim tunneling, as for EEPROM cells. The main difference is that erasure is performed in bulk for the complete chip, or for a subsection of the memory. While this represents a reduction in flexibility, it has the advantage that the extra access transistor of the EEPROM cell can be eliminated. Erasing the complete memory core at once allows for a careful monitoring of the device characteristics during erasure, guaranteeing that the unprogrammed transistor acts as an enhancement device. The monitoring control hardware on the memory chip regularly checks the value of the threshold during erasure, hence dynamically adjusting the erasure time. This approach is only practical when erasing large chunks of memory at a time; hence the flash concept. The simpler cell structure results in a substantial reduction in cell size and an increased integration density.

For instance, Figure 10.23 shows the ETOX Flash cell introduced by Intel [Pashley89]. This is only one of the many existing alternatives. It resembles a FAMOS

[5] Thin oxides are not the only way to realize electron tunneling. Other approaches include the use of textured surfaces to locally enhance the surface field [Masuoka91].

gate, except that a very thin tunneling gate oxide is utilized (10 nm). Different areas of the gate oxide are used for programming and erasure. Programming is performed by applying a high voltage (12 V) on gate and drain terminals for a grounded source, while erasure occurs with the gate grounded and the source at 12 V.

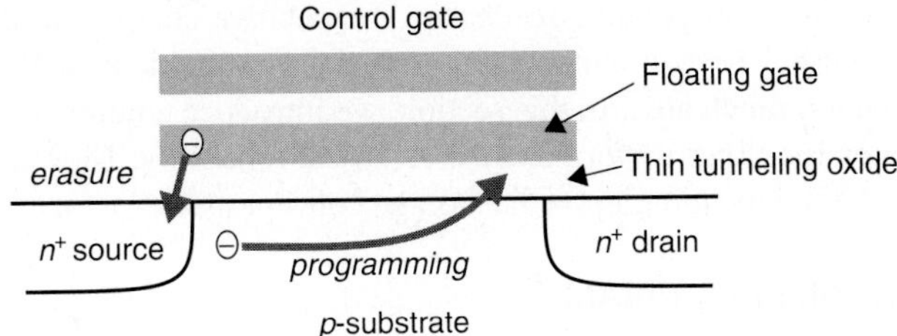

Figure 10.23 ETOX device as used in Flash EEPROM memories.

Nonvolatile Read-Write Memories in Perspective

Some numbers are useful to put the above observations in perspective. Table 10.1 summarizes the essential data for the most recent nonvolatile memories (at the time of writing). The table confirms that the flexibility of the EEPROM structure comes at the expense of density and performance. EPROMs and Flash EEPROM devices are comparable in both density and speed. The versatility of the latter explains its explosive growth in a short time span.

Table 10.1 Characteristics of state-of-the-art nonvolatile memories.

	EPROM [Tomita91]	**EEPROM [Terada89, Pashley89]**	**Flash EEPROM [Jinbo92]**
Memory size	16 Mbit (0.6 μm)	1 Mbit (0.8 μm)	16 Mbit (0.6 μm)
Chip size	7.18×17.39 mm^2	11.8×7.7 mm^2	6.3×18.5 mm^2
Cell size	3.8 μm^2	30 μm^2	3.4 μm^2
Access time	62 nsec	120 nsec	58 nsec
Erasure time	minutes	N.A.	4 sec
Programming time/word	5 μsec	8 msec/word, 4 sec /chip	5 μsec
Erase/write cycles [Pashley89]	100	10^5	10^3–10^5

A large number of design considerations raised in the section on read-only memories are valid for the NVRWMs as well. In addition, the (E)EPROM structures must cope with the extra complexity of the programming and erasure circuitry. Remember that all the proposed structures require the availability of high-voltage signals (12 – 20 V) on word and bit lines during the programming, while standard 3 or 5 V signals are used on the same wires during the read mode. The generation and distribution of those signals requires some interesting circuit design, which is unfortunately beyond the scope of this text.

10.3.3 Read-Write Memories (RAM)

Providing a memory cell with roughly equal read and write performance requires a more complex cell structure. While the contents of the ROM and NVRWM memories is ingrained in the cell topology or programmed into the device characteristics, storage in RAM memories is based on either positive feedback or capacitive charge, similar to the ideas introduced in Chapter 6. These circuits would be perfectly suitable as R/W memory cells but tend to consume too much area. In this section, we introduce a number of simplifications that trade off area for either performance or electrical reliability. They are labeled as either SRAMs or DRAMs depending upon the storage concept used.

Static Random-Access Memory (SRAM)

The generic SRAM cell is (re)introduced in Figure 10.24 and turns out to be virtually identical to a register cell shown in Figure 6.16, deemed of limited use at that time. It requires six transistors/bit. Access to the cell is enabled by the word line, which replaces the clock and controls the two pass-transistors M_5 and M_6. In contrast to the ROM cells, two bit lines transferring both the stored signal and its inverse are required. Although providing both polarities is not a necessity, doing so improves the noise margins during both read and write operations, as will become apparent in the subsequent analysis.

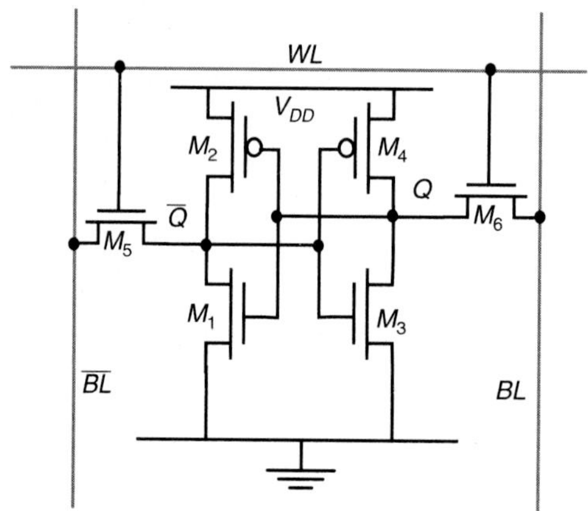

Figure 10.24 Six-transistor CMOS SRAM cell.

Problem 10.5 CMOS SRAM Cell

Does the SRAM cell presented in Figure 10.24 consume stand-by power? Explain. Draw an equivalent pseudo-NMOS implementation. How about the stand-by power in that case?

Operation of SRAM cell. To understand the operation of the memory cell, let us consider the write and read operations in sequence. While doing so, we also derive the transistor-sizing constraints.

Example 10.6 CMOS SRAM Write Operation

In this example, we derive the device constraints necessary to ensure a correct write operation. Assume that a 1 is stored in the cell (or $Q = 1$). A 0 is written in the cell by setting BL to 0 and BL to 1, which is identical to applying a reset pulse. This causes the flip-flop to change state if the devices are sized properly.

During the initiation of a write, the schematic of the SRAM cell can be simplified to the model of Figure 10.25 It is reasonable to assume that the gates of transistors M_1 and M_4 stay at V_{DD} and GND respectively as long as the switching has not commenced. While this condition is violated once the flip-flop starts toggling, the simplified model is more than accurate for hand-analysis purposes. It is sufficient that node Q can be pulled below the switching threshold of the cross-coupled inverter, which is assumed to be located at $V_{DD}/2$, to ensure that the flip-flop will toggle. Node $\overline{Q}$ must be raised above $V_{DD}/2$.

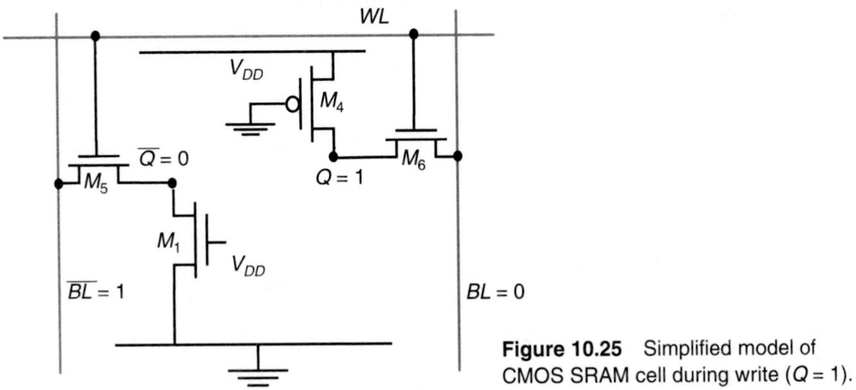

Figure 10.25 Simplified model of CMOS SRAM cell during write ($Q = 1$).

The conditions under which this occurs can be derived by considering the *dc* current equations at the switching threshold point. This is similar to the analysis of Example 6.1.

$$k_{n,M6}\left((V_{DD} - V_{Tn})\frac{V_{DD}}{2} - \frac{V_{DD}^2}{8}\right) = k_{p,M4}\left((V_{DD} - |V_{Tp}|)\frac{V_{DD}}{2} - \frac{V_{DD}^2}{8}\right) \tag{10.1}$$

and

$$\frac{k_{n,M5}}{2}\left(\frac{V_{DD}}{2} - \left(V_{Tn}\left(\text{for } V_{SB} = \frac{V_{DD}}{2}\right)\right)\right)^2 = k_{n,M1}\left((V_{DD} - |V_{Tn}|)\frac{V_{DD}}{2} - \frac{V_{DD}^2}{8}\right) \tag{10.2}$$

The first equation simply expresses that $k_{n,M6}$ should be equal to (or larger than) $k_{p,M4}$. When the cross-coupled inverter is implemented using minimum-size devices, it is acceptable to use a minimum device for the pass-transistor as well. A wider device can help increase the noise margin (as already shown in Figure 6.14).

Evaluating Eq. (10.2) for the 1.2 μm CMOS technology and $V_{DD} = 5$ V yields

$$(W/L)_{n,M5} \geq 10\ (W/L)_{n,M1} \tag{10.3}$$

Pulling up node Q requires a large NMOS pass-transistor, since pulling up a node with a saturated NMOS device is not very effective. This is detrimental in two ways: (1) it increases the size of the cell, and (2) it presents a larger capacitive load to the bit line, which hampers the performance.

Now the good news—the required size of the pass-transistors is set by the combination of both sides of the flip-flop. It is sufficient for one side to start switching in order for the other side to eventually follow. Even when making transistor M_5 equal to the minimum size (or somewhat larger), the flip-flop still switches as long as the combination M_4-M_6 switches. This is the result of the dual bit-line structure. Similar observations are valid when writing a 1. In this case, the switching is initiated from the BL side by the combination M_5-M_3.

This leads to the following observation:

The dual bit-line architecture of the 6T-SRAM cell increases the noise margins, or equivalently, allows for similar noise margins with smaller devices.

Example 10.7 CMOS SRAM—Read Operation

Assume that a 1 is stored at Q. We further assume that both bit lines are precharged to 5 V before the read operation is initiated. The read-cycle is started by asserting the word line, enabling both pass-transistors M_5 and M_6 after the initial word-line delay. During a correct read-event, the values stored in Q and $\overline{Q}$ are transferred to the bit lines by leaving BL at its precharge value and by discharging $\overline{BL}$ through M_1-M_5. Once again, a careful sizing of the transistors is necessary to avoid a malfunctioning of the cell.

This is illustrated in Figure 10.26. Consider the $\overline{BL}$ side of the cell. The bit-line capacitance for larger memories is in the pF range. Consequently, the value of $\overline{BL}$ does not drop instantaneously but stays at the precharged value V_{DD} upon enabling of the read operation ($WL \rightarrow 1$). The combination M_5-M_1 then forms a saturated-load NMOS inverter. The dc value of Q, as imposed by this inverter configuration must stay below the switching point of the inverter M_2-M_4. If not, this could toggle the cross-coupled inverter pair and destroy the value stored in the cell. It is necessary to keep the resistance of transistor M_5 larger than that of M_1 to prevent this from happening.

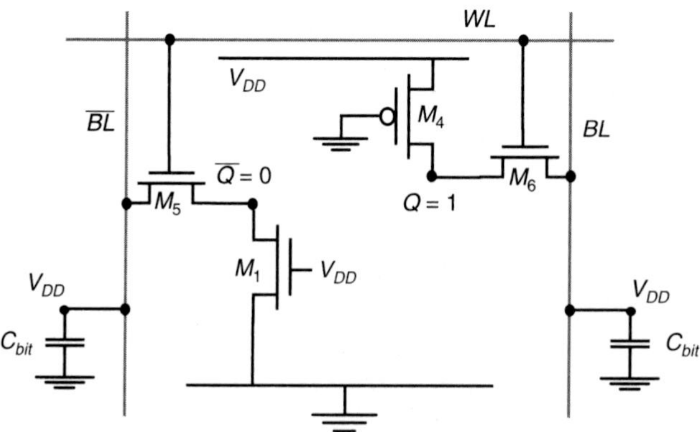

Figure 10.26 Simplified model of CMOS SRAM cell during read ($Q = 1$, $V_{precharge} = V_{DD}$).

The boundary constraints on the device sizes can be derived from solving the current equation at the switching threshold (Eq. (10.4)). Note that this equation is identical to Eq. (10.2).

$$\frac{k_{n,M5}}{2}\left(\frac{V_{DD}}{2} - V_{Tn}\left(\frac{V_{DD}}{2}\right)\right)^2 = k_{n,M1}\left((V_{DD} - |V_{Tn}|)\frac{V_{DD}}{2} - \frac{V_{DD}^2}{8}\right) \tag{10.4}$$

However, the conclusion is just the opposite. To avoid inadvertent switching during the read operation, it is necessary to keep $(W/L)_{n,M5} \leq 10\ (W/L)_{n,M1}$. This condition supersedes Eq. (10.3), which we have already seen to be nonessential.

Notice that the analysis presented is the worst case. The second bit line BL clamps Q to V_{DD}, which makes the inadvertent toggling of the cross-coupled inverter pair difficult. This demonstrates once again the advantage of the dual bit-line architecture.

Beyond adjusting the size of the cell transistors, the erroneous toggling can be prevented by precharging the bit lines to another value, such as 2.5 V. This effectively makes it impossible for Q to reach the switching threshold of the connecting inverter. Precharging to the mid-point of the voltage range has some performance benefits as well, since it limits the voltage swing on the bit lines.

Performance of SRAM cell. When analyzing the transient behavior of the SRAM cell, one realizes that the read operation is the critical one. It requires the (dis)charging of the large bit-line capacitance through the small transistors of the selected cell. For instance, $C_{\overline{BL}}$ has to be discharged through the series combination of M_5 and M_1 in the example of Figure 10.26. The write time is dominated by the propagation delay of the cross-coupled inverter pair, since drivers that bring BL and $\overline{BL}$ to the desired values can be large.

In determining the access times of the SRAM cell array, we can use techniques similar to the ones used in the analysis of the ROM structures. The situation is somewhat complicated by the positive feedback in the cell. It is, however, safe to assume that an equivalent circuit in the style of Figure 10.26 offers a fair approximation.

Problem 10.6 CMOS SRAM—Access Time

Derive an expression for the READ propagation delay of the circuit of Figure 10.26. Assume that BL is precharged to V_{DD}.

Improved MOS SRAM Cells. The six-transistor SRAM cell presented, while simple and reliable, is area-hungry. Besides the devices, it requires the signal routing and connections to two bit lines, a word line, and both supply rails. Figure 10.27 shows a possible layout for such a cell. Its dimensions are dominated by the wiring and interlayer contacts (11.5 of them—the top and bottom ones only count for one half, since they are shared with the neighboring cells).

Designers of large memory arrays have therefore proposed other cell structures whose conception is not only based on revised transistor topologies, but also on the presence of special devices and a more complex technology.

Consider the cell schematic of Figure 10.28, called the *resistive load* SRAM cell (also called the four-transistor SRAM cell). The special feature of this cell is that the cross-coupled CMOS inverter pair is replaced by a pair of resistive-load NMOS inverters. The PMOS transistors are replaced by resistors, and the wiring is simplified. This reduces the SRAM cell size by approximately one third, as illustrated in Table 10.2 for the example of a 1 Mbit SRAM.

We remember from Chapter 4 that this gate topology poses a major dilemma. The resistor value must be as high as possible to retain a reasonable noise margin NM_L and to

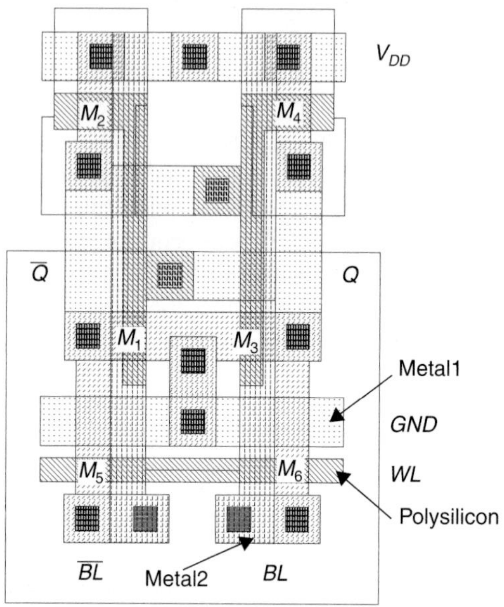

Figure 10.27 Layout of six-transistor CMOS SRAM memory cell (see also Colorplate 16).

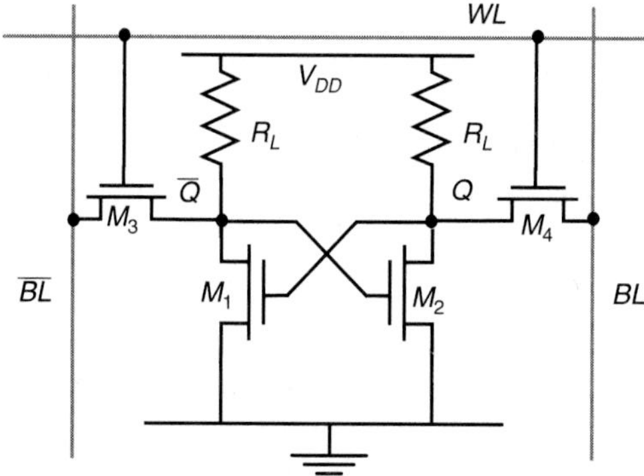

Figure 10.28 Resistive-load SRAM cell.

reduce the static power consumption. On the other hand, too high a resistor value severely deteriorates the low-to-high propagation delay and increases the cell size.

Keeping the static power dissipation per cell as low as possible is the prime design priority in SRAM cells. Consider a 1 Mbit SRAM memory operating at 3 V and using a 10 kΩ resistor as inverter load. With each cell sinking 0.3 mA in static current, a total standby dissipation 300 W can be recorded! Therefore, the only obvious choice is to make the load resistance as large as possible. A very large, yet compact, resistor can be manufactured by employing undoped polysilicon that has a sheet resistance of several TΩ/□ (Tera = 10^{12}!). The propagation-delay problem is addressed by precharging the bit lines to V_{DD}, so that a low-to-high transition on the bit lines only occurs during precharge and never during a read operation. Under those conditions, the resistive loads never provide real current during a transition. Their only goal is to maintain the state of the cell, that is, to compensate for the leakage currents that typically range around 10^{-15} A/cell [Takada91]. The resistor current should be at least two orders of magnitude larger to accomplish that goal, or $I_{load} > 10^{-13}$ A. This puts an upper limit on the resistor value.

The realization that the pull-up devices are only needed for charge-loss compensation has resulted in a revised version of the six-transistor memory cell of Figure 10.24. Instead of using traditional, expensive PMOS devices, the pull-up transistors are realized as parasitic devices deposited on top of the cell structure using a thin-film technology. These PMOS thin-film transistors (TFTs) have inferior properties with respect to normal devices and are characterized by a current of approximately 10^{-8} A and 10^{-13} in the *ON* and *OFF* modes respectively for a 5 V gate-source voltage [Sasaki90, Ootani90]. The complimentary nature of the cell results in an increased cell reliability with less sensitivity to leakage and soft errors,[6] yet at a lower standby current compared to the resistive load cell. This property, combined with a comparable cell size (Table 10.2), makes the TFT-cell the prime candidate for the implementation of ultralarge SRAM memories, especially for portable battery-operated applications.

Table 10.2 Comparison of CMOS SRAM cells used in 1Mbit memory (from [Takada91])

	Complementary CMOS	Resistive Load	TFT cell
Number of transistors	6	4	4 (+2 TFT)
Cell size	58.2 μm^2 (0.7 μm rule)	40.8 μm^2 (0.7 μm rule)	41.1 μm^2 (0.8 μm rule)
Standby current (per cell)	10^{-15} A	10^{-12} A	10^{-13} A

Example 10.8 Bit-Line Precharge

Inspection of the layout of Figure 10.27 reveals the following device sizes:

$$(M_1, M_3) = (2.4/1.2),\ (M_2, M_4) = (1.8/1.8),\ \text{and}\ (M_5, M_6) = (1.8/1.2).$$

From these device sizes, we can conclude that the t_{pLH} of the cell is substantially larger than the t_{pHL}. This implies that the cell is designed for a high precharge value on the bit lines, probably situated at V_{DD} or $V_{DD} - V_T$.

Bipolar SRAM Cells. The availability of very fast SRAMs is essential in the design of high-performance super- and mainframe computers, where cycle times below 5 nsec are required for the cache and control memories. Until recently, bipolar memory was the only option to achieve that performance. The low input impedance of the bipolar transistors effectively eliminates the feasibility of a dynamic bipolar memory cell, as became apparent in Chapter 6. The only option available is to use positive feedback as the storage mechanism.

The design of a bipolar memory cell is further hampered by the lack of a high-quality switch. Such a device is useful for the selection of a single cell for reading or writing, and to isolate all nonselected cells from the bit lines. This is exactly the function of the NMOS pass-transistors M_5 and M_6 in the CMOS SRAM cell of Figure 10.24. Other less-efficient isolation approaches must be employed, including diode isolation and multiple

[6] Soft errors will be discussed later in the chapter in the section on memory reliability.

word lines. One of these techniques is illustrated in Figure 10.29, which shows one of the favored bipolar memory cells called the *Schottky-barrier diode (SBD) load* cell. It consists of a cross-coupled pair of bipolar inverters, whose supply rails are connected to a pair of word lines. The bit lines are resistively clamped to a supply voltage of 1.5 V.

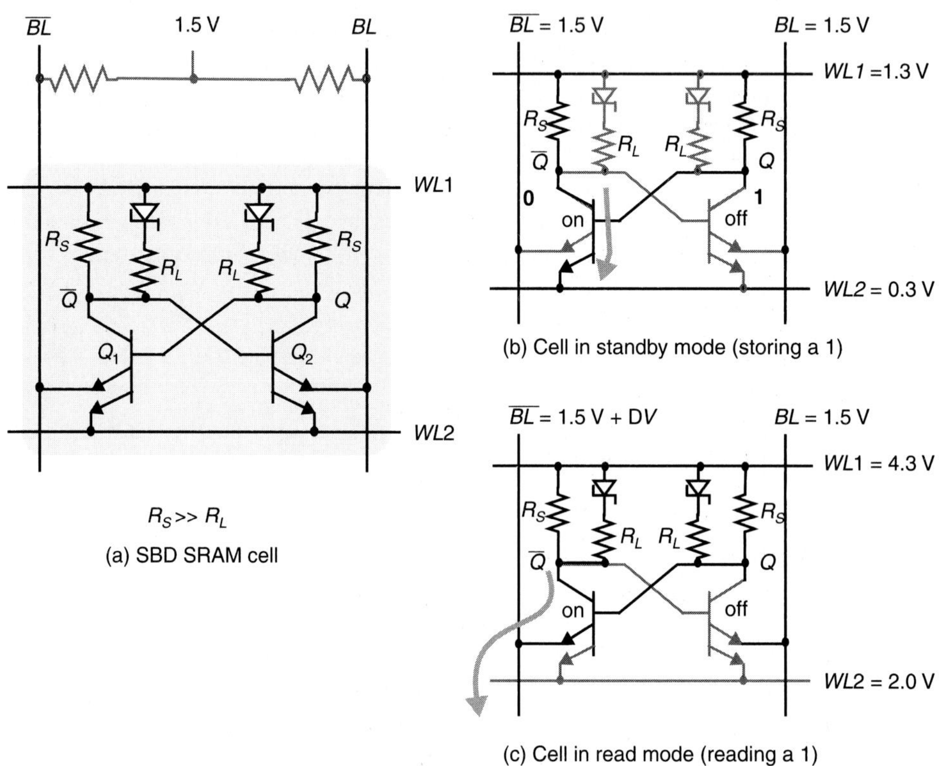

Figure 10.29 Schottky-barrier diode bipolar SRAM cell. The shaded components are inactive in the operation mode under study.

The behavior of the cell is best understood by examining its different operation modes. When not selected (standby mode, Figure 10.29b), the cell is operated on a low supply voltage of approximately 1 V by placing *WL*1 and *WL*2 at 1.3 V and 0.3 V, respectively. This voltage is sufficient to maintain the state of the cell, yet reduces the standby current. Power dissipation is further reduced by disabling the smaller component of the load resistance (R_L) with the aid of a Schottky-barrier diode, leaving only the larger resistance R_S in place. As the bit lines are clamped at 1.5 V, the access diodes—the second emitter of Q_1 and Q_2—are off, which effectively isolates the cell.

In the read mode (Figure 10.29c), both word lines *WL*1 and *WL*2 are pulled high to 4.3 V and 2.0 V respectively. The Schottky-barrier diodes turn on, reducing the load resistance and increasing the available current. The emitters connected to *WL*2 turn off since the bit lines are at a lower value. The current flowing through the on-transistor (in this case Q_1) is diverted to the bit line $\overline{BL}$, which causes a voltage rise on $\overline{BL}$. This change can be detected by the sense amplifier.

To write a 0 into the cell, bit line BL is pulled low (e.g., to 0 V) with both word lines high (4.3 V and 2.0 V, respectively). Transistor Q_2 turns on, and the state of the cell is toggled, after which the word lines can be returned to the standby values.

Problem 10.7 Bipolar Memory Cell

Produce an approximate sketch of the voltage wave-forms of the bit lines during reading and writing of both a 0 and a 1 in the SBD SRAM cell.

Other bipolar memory cells have been devised [Takada91] either to reduce the complexity or increase the speed. The importance of these memory structures has been substantially reduced with the arrival of the BiCMOS technology. A BiCMOS memory combines a dense, area-effective MOS memory array with high-performance, bipolar peripherals such as drivers and sense amplifiers. Access times can be achieved that rival those of the bipolar memories while maintaining the integration density of the MOS memories. For instance, BiCMOS memories have achieved access times of around 5–6 nsec for memory sizes up to 4 Mbit. Purely bipolar memories realize a modest increase in performance (2–3 nsec access) at the expense of a dramatic reduction in density (64 Kbit max) and power consumption (±20 W) [Takada91, Nakamura92].

Dynamic Random-Access Memory (DRAM)

While discussing the resistive-load SRAM cell, we noted that the only function of the load resistors is to replenish the charge lost by leakage. One option is to eliminate these loads completely and compensate for the charge loss by periodically rewriting the cell contents. This *refresh* operation, which consists of a read of the cell contents followed by a write operation, should occur often enough that the contents of the memory cells are never corrupted by the leakage. Typically, refresh should occur every 1 to 4 msec. For larger memories, the reduction in cell complexity more than compensates for the added system complexity imposed by the refresh requirement. These memories are called *dynamic*, since the underlying concept of these cells is based on charge storage on a capacitor.

Three-Transistor Dynamic Memory Cell. The first kind of dynamic cell is obtained by eliminating the load resistors in the schematic of Figure 10.28. The four-transistor cell can be further simplified by observing that the cell stores both the data value and its complement; hence, it contains redundancy. Eliminating one more device (e.g., M_1) removes this redundancy and results in the three-transistor (3T) cell of Figure 10.30 [Regitz70]. This cell formed the core of the first popular MOS semiconductor memories such as the first 1Kbit memory from Intel [Hoff70]. While replaced by more area-efficient cells in the very large memories of today, it is still the cell of choice in many memories embedded in application-specific integrated circuits. This can be attributed to its relative simplicity in both design and operation.

The cell is written to by placing the appropriate data value on $BL1$ and asserting the *write-word line* (WWL). The data is retained as charge on capacitance C_S once WWL is lowered. When reading the cell, the *read-word line* (RWL) is raised. The storage transistor M_2 is either on or off depending upon the stored value. The bit line $BL2$ is either clamped to V_{DD} with the aid of a load device, for example, a grounded PMOS or saturated NMOS transistor, or is precharged to either V_{DD} or $V_{DD} - V_T$. The former approach necessitates

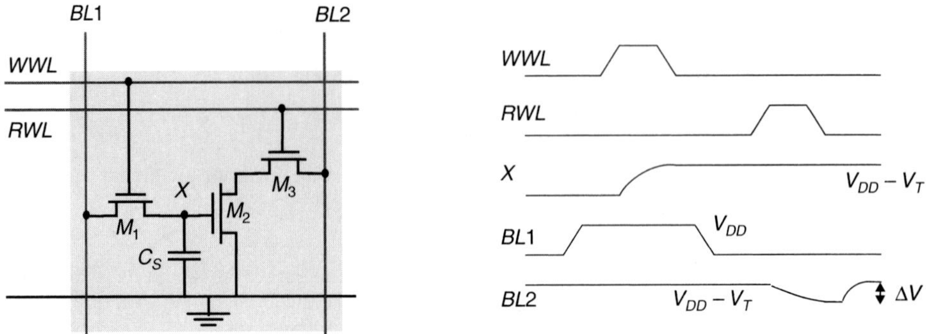

Figure 10.30 Three-transistor dynamic memory cell and the signal waveforms during read and write.

careful transistor sizing and causes static power consumption. Therefore, the precharged approach is generally preferable. The series connection of M_2 and M_3 pulls $BL2$ low when a 1 is stored. $BL2$ remains high in the opposite case. Notice that the cell is inverting; that is, the inverse value of the stored signal is sensed on the bit line. The most common approach to refreshing the cell is to read the stored data, put its inverse on $BL1$, and assert WWL in consecutive order.

The cell complexity is substantially reduced with respect to the static cell. This is illustrated by the example layout of Figure 10.31. The total area of the cell is 576 λ^2, compared to the 1092 λ^2 of the SRAM cell of Figure 10.27. These numbers do not take into account the potential area reduction obtained by sharing with neighboring cells. The area reduction is mainly due to the elimination of contacts and devices.

Further simplifications in the cell structure are possible at the expense of a more complex circuit operation. For instance, bit lines $BL1$ and $BL2$ can be merged into a single

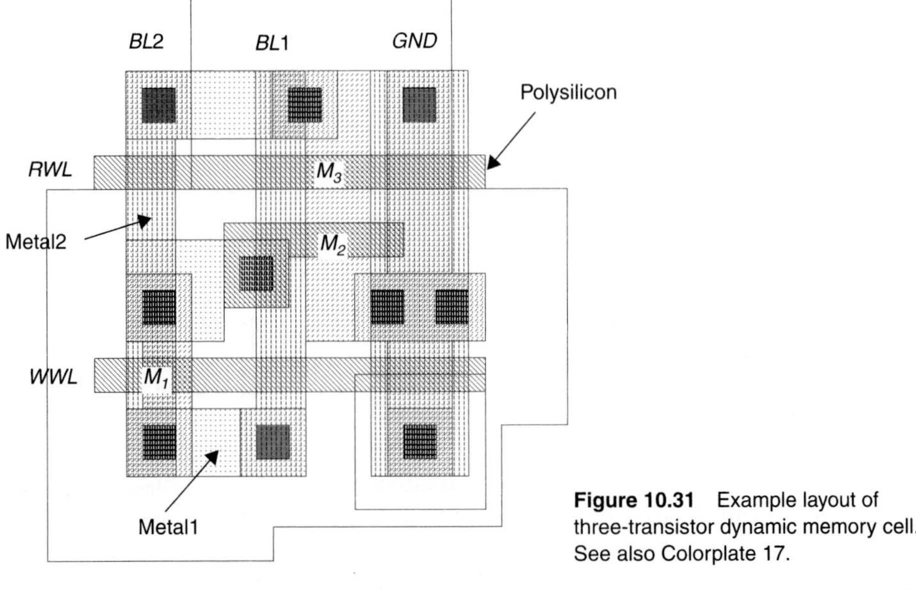

Figure 10.31 Example layout of three-transistor dynamic memory cell. See also Colorplate 17.

wire. The read and write cycles can proceed as before. The read-sense-write refresh cycle must be altered considerably, since the data value read from the cell is the complement of the stored value. This requires the bit line to be driven to both values in a single cycle. Another option is to merge the *RWL* and the *WWL* lines. Once again, this does not significantly change the cell operation. A read operation is automatically accompanied by a refresh of the cell contents. A careful control of the word-line voltage is necessary to prevent a writing of the cell before the actual value is read during refresh.

Finally, the following interesting properties of the 3T-cell are worth mentioning.

1. In contrast to the SRAM cell, no constraints exist on the *device ratios*. This is a common property of dynamic circuits. The choice of device sizes is solely based on performance and reliability considerations. Observe that this statement is not valid when a static bit-line load approach is employed.

2. In contrast to other DRAM cells, reading the 3T-cell contents is *nondestructive;* that is, the data value stored in the cell is not affected by a read.

3. The value stored on the storage node X when writing a 1 equals $V_{WWL} - V_{Tn}$. This threshold loss reduces the current flowing through M_2 during a read operation and increases the read access time. To prevent this, some designs *bootstrap* the word-line voltage, or in other words, raise V_{WWL} to a value higher than V_{DD}.

One-Transistor Dynamic Memory Cell. Another dramatic reduction in cell complexity can be obtained by a further sacrifice in some of the cell properties. The resulting structure, called the one-transistor DRAM cell (1T), is undoubtedly the most pervasive dynamic DRAM cell in commercial memory design.[7] A schematic is shown in Figure 10.32 [Dennard68]. Its basic operational concepts are extremely simple. During a write cycle, the data value is placed on the bit line *BL,* and the word line *WL* is raised. Depending upon the data value, the cell capacitance is either charged or discharged. Before a read operation is performed, the bit line is pre-charged to a voltage V_{PRE}. Upon asserting the

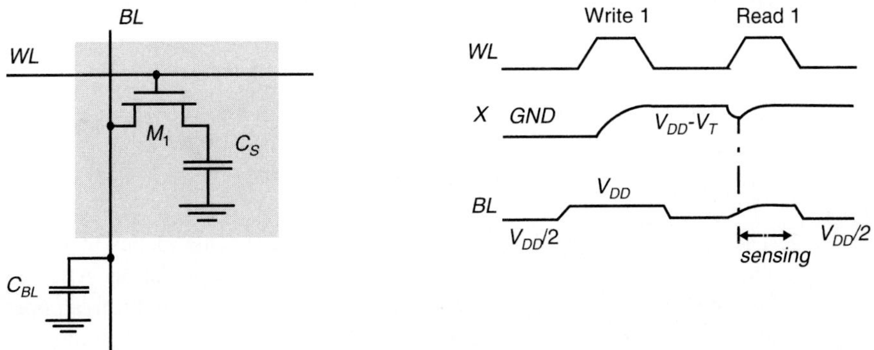

Figure 10.32 One-transistor dynamic RAM cell and the corresponding signal waveforms during read and write.

[7] A DRAM cell containing only two transistors can also be conceived. It offers no substantial advantages over either the 3T and 1T cells and is, therefore, only rarely used.

word line, a charge redistribution takes place between the bit line and storage capacitance. This results in a voltage change on the bit line, the direction of which determines the value of the data stored. The magnitude of the swing is given by the following expression

$$\Delta V = V_{BL} - V_{PRE} = (V_{BIT} - V_{PRE})\frac{C_S}{C_S + C_{BL}} \tag{10.5}$$

where C_{BL} is the bit-line capacitance, V_{BL} the potential of the bit line after the charge redistribution, and V_{BIT} the initial voltage over the cell capacitance C_S. As the cell capacitance is normally one or two orders of magnitude smaller than the bit-line capacitance, this voltage change is very small, typically around 250 mV for state-of-the-art memories [Itoh90]. The ratio $C_S/(C_S + C_{BL})$ is called the *charge-transfer ratio* and ranges between 1% and 10%.

Amplification of ΔV to the full voltage swing is necessary if functionality is to be achieved. This observation marks a first major difference between the 1T and 3T, as well as other, DRAM cells.

1. A 1T DRAM requires the *presence of a sense amplifier* for each bit line to be functional. This is a result of the charge-redistribution based read-out. The read operation of all cells discussed previously relies on current sinking. A sense amplifier is only needed to speed up the read-out, not for functionality considerations. It is also worth noticing that the DRAM memory cells are *single-ended* in contrast to the SRAM cells, which present both the data value and its complement on the bit lines. This complicates the design of the sense amplifier, as will be discussed in the section on periphery.

Example 10.9 1T DRAM Read-out

Assume a bit-line capacitance of 1pF in correspondence to the numbers derived earlier in the chapter, and a bit-line precharge voltage of 2.5 V. The voltage over the cell capacitance C_S (of 50 fF) equals 3.5 V and 0 V for a 1 and 0 respectively. This translates into a charge-transfer efficiency of 4.8% and the following voltage swings on the bit line during a read operation:

$$\Delta V(0) = -2.5V \times \frac{50fF}{50fF + 1pF} = -120mV$$

$$\Delta V(1) = 60mV$$

Other important differences are also worth enumerating.

2. The read-out of the 1T DRAM cell is *destructive*. This means that the amount of charge stored in the cell is modified during the read-operation. After a successful read operation, the original value must be restored. Read and refresh operations are therefore intrinsically intertwined in a 1T-DRAM. Typically, the output of the sense amplifier is imposed onto the bit line during the read-out. Keeping *WL* high ensures that the cell charge is restored during that period. This is illustrated in Figure 10.33, which plots a typical bit-line voltage waveform during read-out.

3. Unlike the 3T cell that relies on charge storage on a gate capacitance, the 1T cell requires the presence of an extra capacitance that must be explicitly included in the

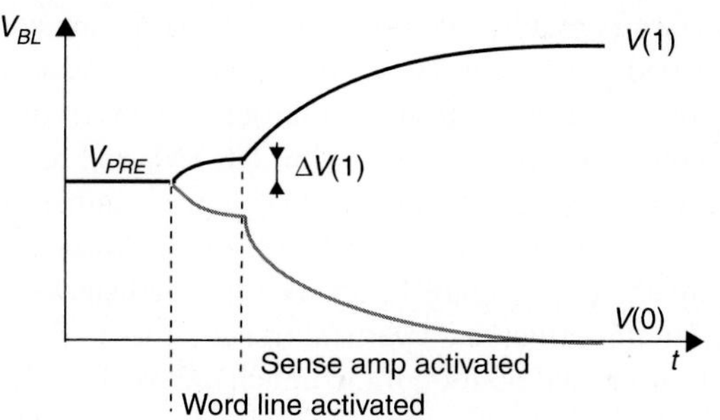

Figure 10.33 Bit-line voltage waveform during read operation (for 1 and 0 data values).

design. For reliability, the charge-transfer ratio is kept large, with the minimum value of the capacitance ranging around 30 fF. Fitting that large of a capacitance in as small an area as possible is one of the key challenges in DRAM designs. Some of the most popular ways to do so are briefly summarized below.

4. Observe that when writing a 1 into the cell, a threshold voltage is lost, which reduces the available charge. This charge loss can be circumvented by bootstrapping the word lines to a value higher than V_{DD}. This is a common practice in state-of-the-art memory design.

Figure 10.34 presents a first approach toward designing a 1T DRAM cell. The main advantage of this design is that it can be realized in a generic CMOS technology. The storage node in this cell is composed of the gate capacitance, sandwiched between a polysilicon plate and an inversion layer, induced into the substrate by applying a positive voltage bias on the polysilicon plate. When writing a 0 in the cell, the potential well of the storage node is filled with electrons, and the capacitor is charged. If a 1 is written in the cell (with a high voltage on the bit line) the electrons are removed from the induced inversion layer, and the surface area is depleted. The voltage over the capacitor is reduced. Observe that this represents the inverse of the scenario described before: the capacitor is charged for a 0 and discharged when storing a 1.

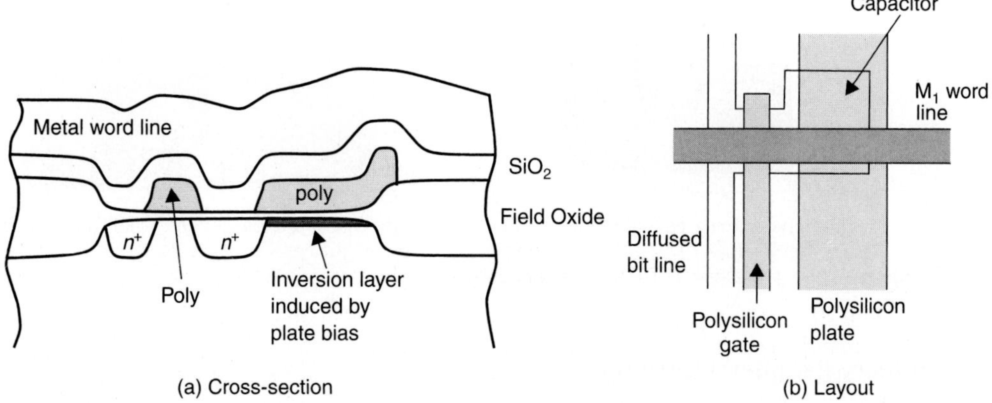

Figure 10.34 1T DRAM cell using a polysilicon-diffusion capacitance as storage node (from [Dillinger88]). The contact between word line and polysilicon gate is accomplished in the neighboring cell.

Implementing denser cells requires modifications in the manufacturing process. A first change is to add a second polysilicon layer, which serves as the second plate of the capacitor, with the first polysilicon layer forming the other plate. In the quest for ever-denser cells, DRAM technology as used in the 16 Mbit DRAMs and beyond, has focused on three-dimensional structures, where the storage capacitance is either implemented vertically in the substrate or on top of the access transistor [Lu89]. Cross-sections of some of the most advanced cells are shown in Figure 10.35. The first cell shows the cross-section of a *trench-capacitor* cell. In this structure, a vertical trench of up to 5 μm deep is etched into the substrate. The side-walls and bottom of the trench are used for the capacitor electrode, which results in a large plate surface that occupies only a small die area. Figure 10.35b shows the cross-section of a *stacked-capacitor cell* (STC), where the capacitance is superimposed on top of the access transistor and bit lines. Up to four polysilicon layers are employed to realize "fin-type" capacitors, reducing the effective cell area. Using these approaches, the cell area in a 64 Mbit DRAM ranges between 1.5 and 2.0 μm^2, yielding a storage capacitance between 20 and 30 fF in a 0.4 μm technology! Obviously, these size reductions are not for free. The production and manufacturing of those esoteric devices to obtain a reasonable yield has become an increasingly difficult and expensive undertaking. More than physics or implementation impediments, economics will determine if there is life beyond 256 Mbit or 1 Gbit semiconductor memories.

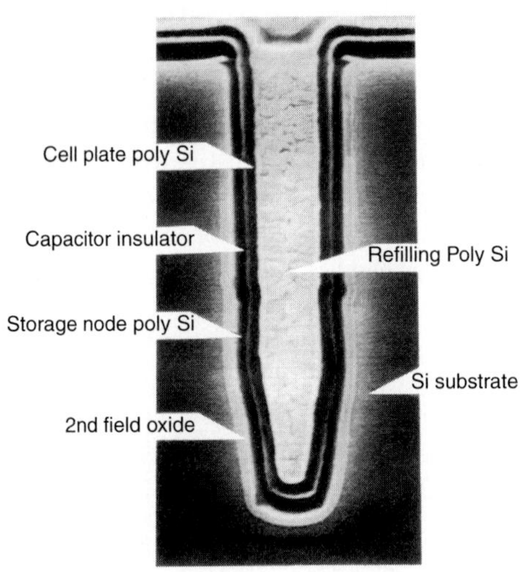

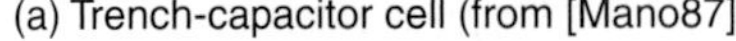

(a) Trench-capacitor cell (from [Mano87]) (b) Stacked-capacitor cell (from [Taguchi91])

Figure 10.35 Advanced 1T DRAM memory cells.

10.4 Memory Peripheral Circuitry

Since the memory core trades performance and reliability for reduced area, memory design relies exceedingly on the peripheral circuitry to recover both speed and electrical integrity. While the design of the core is dominated by technological considerations and is largely beyond the scope of the circuit designer, it is in the design of the periphery where a good designer can make an important difference. In this section, we discuss the address decoders, I/O drivers/buffers, sense amplifiers, and memory timing and control.

10.4.1 The Address Decoders

Whenever a memory allows for random address-based access, address decoders must be present. The design of these decoders has a substantial impact on the speed and power consumption of the memory. In Section 10.2, we introduced two classes of decoders—the row encoders whose task it is to enable one memory row out of 2^M, and the column and block decoders that can be described as 2^K-input multiplexers, where M and K are the widths of the respective fields in the address word. While conceiving these decoders, it is important to keep the global memory in perspective. These units are tightly coupled to the memory core, so that a matching between the cell dimensions of decoders and the core is recommended (*pitch matching*). Failing to do so leads to a dramatic wiring overhead with its associated delay and power dissipation. Examples of pitch-matched decoders and memory arrays are shown in the case studies at the end of this chapter.

Row Decoders

A 1-out-of-2^M decoder is nothing less than a collection of 2^M complex logic gates. Consider a 10-bit address decoder. Each of the outputs WL_i is a logic function of the 10 input address signals (A_0 to A_9). For example, the rows with addresses 0 and 511 are enabled by the following logic functions

$$\begin{aligned} WL_0 &= \bar{A}_0\bar{A}_1\bar{A}_2\bar{A}_3\bar{A}_4\bar{A}_5\bar{A}_6\bar{A}_7\bar{A}_8\bar{A}_9 \\ WL_{511} &= \bar{A}_0 A_1 A_2 A_3 A_4 A_5 A_6 A_7 A_8 A_9 \end{aligned} \tag{10.6}$$

or, using inverting logic, which is necessary when envisioning a single-stage CMOS design

$$\begin{aligned} WL_0 &= \overline{A_0 + A_1 + A_2 + A_3 + A_4 + A_5 + A_6 + A_7 + A_8 + A_9} \\ WL_{511} &= \overline{A_0 + \bar{A}_1 + \bar{A}_2 + \bar{A}_3 + \bar{A}_4 + \bar{A}_5 + \bar{A}_6 + \bar{A}_7 + \bar{A}_8 + \bar{A}_9} \end{aligned} \tag{10.7}$$

In essence, a 10-input NOR gate is needed per row. A total of $11 \times 1024 = 11{,}264$ transistors are needed to realize the decoder in its entirety when implementing this function using pseudo-NMOS or dynamic gates. This number ignores the extra transistors needed for the input drivers and complement generators. Fortunately, the analysis of memory cores—see the section on ROMs—has taught us that these NOR functions can be implemented in a regular and dense fashion. This is illustrated in Figure 10.36, where the transistor diagram and the conceptual layout of a 2-to-4 decoder is depicted. Notice that this structure is geometrically identical to the NOR-ROM array, differing only in the data patterns.

In a similar fashion, one can also implement the decoder as a NAND array, effectively realizing the inverse of the functions of Eq. (10.6). In this case, all the outputs of the array are high by default with the exception of the selected row, which is low. This "active low" signaling is in correspondence with the word-line requirements of the NAND ROM

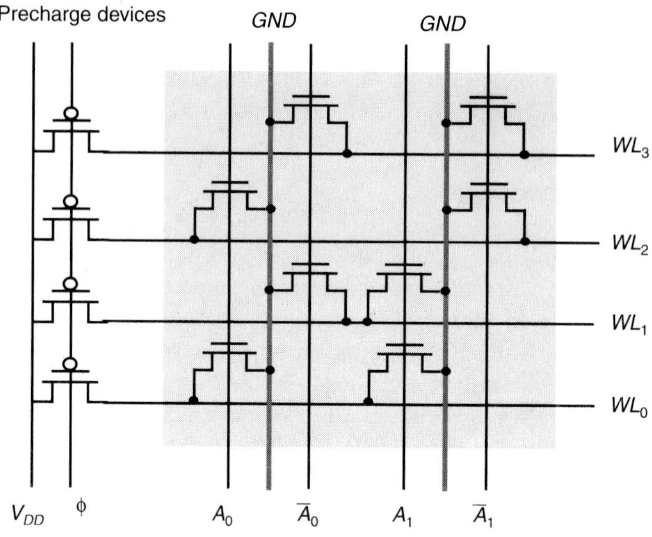

Figure 10.36 Dynamic 2-to-4 NOR decoder.

as discussed in Section 10.3.1 Observe that the interface between decoder and memory often includes a buffer/driver that can be made inverting whenever needed. A 2-to-4 decoder in NAND configuration is shown in Figure 10.37.

All the performance and density considerations raised in the discussion of NAND and NOR ROM arrays are valid. NOR decoders are substantially faster, but consume more area than their NAND counterparts and dramatically more power, as is clear from the fol-

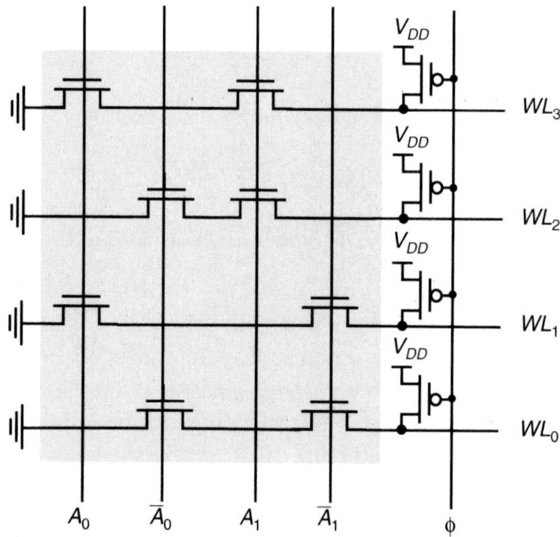

Figure 10.37 A 2-to-4 MOS dynamic NAND decoder. This implementation assumes that all address signals are low during precharge. An alternative approach is to provide evaluate transistors at the bottom of each transistor chain.

lowing observation. Only a single word line is being pulled down after the precharge in a NAND decoder, while only a single wire stays high in the NOR case.

The propagation delay of the decoder is a matter of prime importance, since it adds directly to both read- and write-access times. To realize large, fast decoders, most memory implementations resort to a principle introduced in Chapter 4 to address complex logic gates. Splitting a complex gate into two or more logic layers often produces both a faster and cheaper implementation. This observation has led to the concept of the *predecoder*, which decodes segments of the address in a first logic layer. A second layer of logic gates then produces the final word-line signals.

Consider the case of the 10-input NAND decoder. The expression for WL_0 can be regrouped in the following way

$$\begin{aligned} WL_0 &= \overline{\bar{A}_0\bar{A}_1\bar{A}_2\bar{A}_3\bar{A}_4\bar{A}_5\bar{A}_6\bar{A}_7\bar{A}_8\bar{A}_9} \\ &= \overline{(\overline{A_0+A_1})(\overline{A_2+A_3})(\overline{A_4+A_5})(\overline{A_6+A_7})(\overline{A_8+A_9})} \end{aligned} \tag{10.8}$$

For this particular case, the address is partitioned into sections of 2 bits that are decoded in advance. The resulting signals are then combined using 5-input NAND gates to produce the fully decoded array of word-line signals. The resulting structure is diagrammed in Figure 10.38. The use of a predecoder is advantageous in many ways.

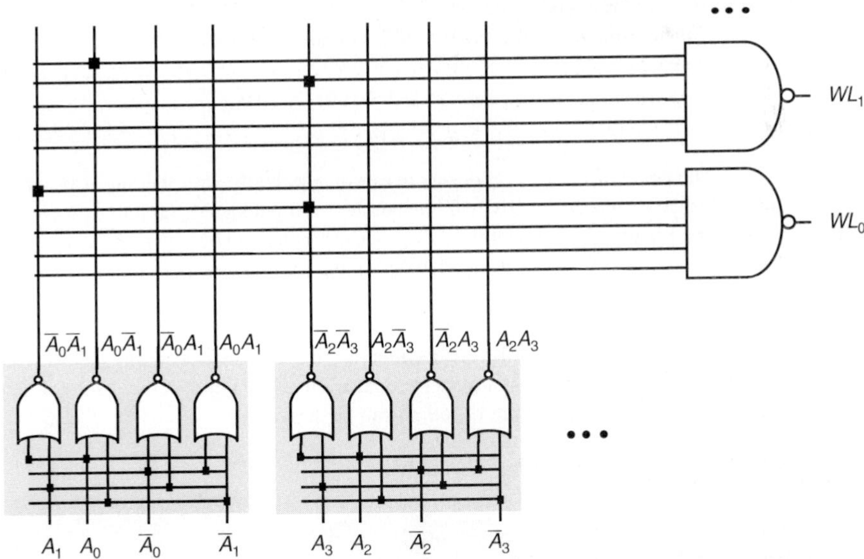

Figure 10.38 A NAND decoder using 2-input predecoders.

- It reduces the number of transistors required. Assuming that the predecoder is implemented in complementary static CMOS, the number of active devices in the 10-input decoder equals (1024 × 6) + (5 × 4 × 4) = 6,224, which is 55% of the original decoder.

- As the number of inputs to the NAND gates is halved, the propagation delay is reduced by approximately a factor of 4. Remember the squared dependency between delay and fan-in.
- The load on the vertical address lines is halved, since only 256 connections are required per line. This reduces the delay and makes the design of the address drivers simpler.

Consequently, all large decoders are realized using the two-layer implementations. This configuration has another advantage. Adding a select signal to each of the pre-decoders makes it possible to disable the decoder when the memory block in question is not selected. This results in important power savings.

Problem 10.8 10-Input NAND Decoder

Determine what partitioning best minimizes the number of devices needed in a 10-input NAND-decoder with a predecoder.

Example 10.10 Speed Optimization of Row Decoders

One of the major advantages of the predecoder approach is the increased performance. An interesting problem is to determine the partitioning between pre-decoder and actual decoder that minimizes the decoding delay.

Consider the case of an *N*-bit NOR decoder and assume that *M* bits are predecoded. The decoder delay equals the sum of the predecoding, decoder-line driving, and the final decoding delays.

$$t_{decode} = t_{pre} + t_{drive} + t_{final} \tag{10.9}$$

The delay of the predecoder increases quadratically with the fan-in *M*, while the delay of final stage increases linearly with the number of input signals (assuming a precharged NOR structure). Finally, the delay of the address-line driving can be assumed to be a linear function of the fan-out. This analysis is summarized in the following expression:

$$t_{decode} \approx M^2 t_{gate} + t_{drive}/M + t_{dc}/M \tag{10.10}$$

where t_{drive} and t_{dc} represent the address-line driving and final-decoding delays without predecoding, and t_{gate} stands for the inverter propagation delay. For specific values of these parameters, the optimal amount of pre-decoding can be derived.

$$M_{opt} = ((t_{drive} + t_{dc})/(2t_{gate}))^{1/3} \tag{10.11}$$

Column and Block Decoders

The functionality of a column and block decoder is best described as a 2^K-input multiplexer, where *K* stands for the size of the address word. Two implementations of that function are in general use. Which one to choose depends upon area, performance, and architectural considerations.

One implementation is based on the CMOS pass-transistor multiplexer introduced in Chapter 4 (Figure 4.22). The control signals of the pass-transistors are generated using a *K*-to-2^K predecoder, realized along the lines described in the previous section. The schematic of a 4-to-1 column decoder is shown in Figure 10.39. The main advantage of this approach is its speed. Only a single pass-transistor is inserted in the signal path, which introduces only a minimal extra resistance. The column-decoding is one of the last actions to be performed in the read-sequence, so that the predecoding can be executed in parallel with other operations, such as the memory access and sensing, and can be performed as soon as the column address is available. Consequently, its propagation delay does not add to the overall memory access time. Slower implementations such as NAND decoders might even be acceptable. The disadvantage of the structure is its large transistor count. $(K+1)2^K + 2^K$ devices are needed for a 2^K-input decoder. For instance, a 1024-to-1 column decoder requires 12,288 transistors. One should also realize that the capacitance and thus the transient response at node D is proportional to the number of inputs of the multiplexer.

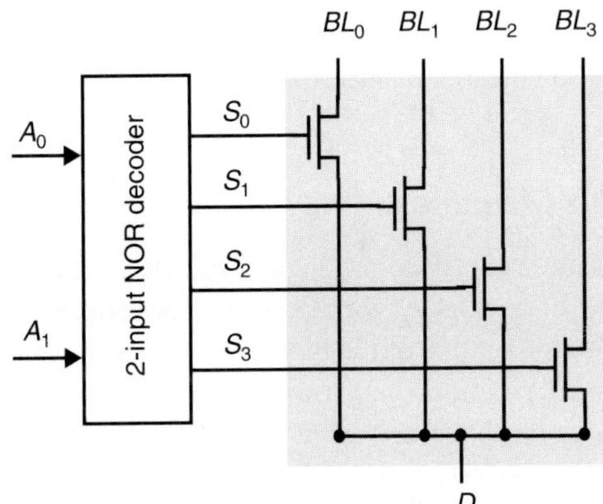

Figure 10.39 Four-input pass-transistor-based column decoder using a NOR predecoder.

A more efficient implementation is offered by a *tree decoder* that uses a binary reduction scheme, as shown in Figure 10.40. Notice that no predecoder is required. The number of devices is drastically reduced, as is derived below for a 2^K-input decoder.

$$N_{tree} = 2^K + 2^{K-1} + \ldots + 4 + 2 = 2 \times (2^K - 1) \tag{10.12}$$

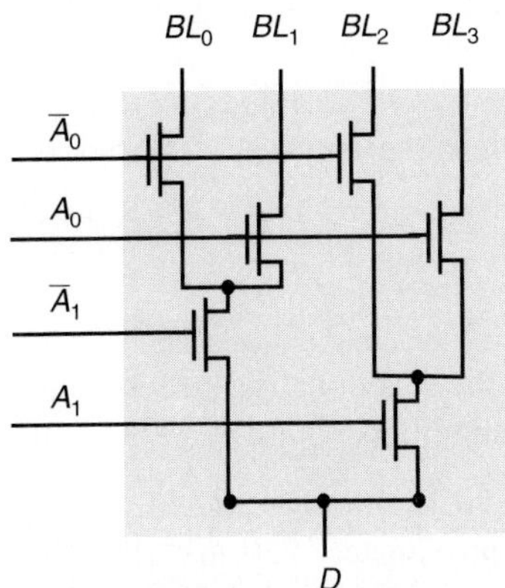

Figure 10.40 A 4-to-1 tree-based column decoder.

This means that a 1024-to-1 decoder requires only 2046 active devices, a reduction by a factor of 6! On the negative side, a chain of K series-connected pass-transistors is inserted in the signal path. Because the delay increases quadratically with the number of sections, the tree approach becomes prohibitively slow for large decoders. This can be remedied by inserting intermediate buffers. A progressive sizing of the transistors is another option, with the transistor size increasing from bottom to top. A final option is to combine the pass-transistor and tree-based approaches. A fraction of the address word is predecoded (for instance, the msb-side), while the remaining bits are tree-decoded. This can reduce both the transistor count and the propagation delay.

Example 10.11 Column Decoders

Consider a 1024-to-1 decoder. Predecoding 5 bits results in the following transistor tally:

$$N_{dec} = N_{pre} + N_{pass} + N_{tree} = 6.\ 2^5 + 2^{10} + 2(2^5 - 1) = 1278!$$

The number of series-connected pass-transistors is reduced to six.

Decoders for Non-Random-Access Memories

Memories that are not of the random-access class do not need a full-fledged decoder. In a serial-access memory, such as a video line memory, the decoder degenerates to an M-bit shift-register with M the number of rows. Only one of the bits is high at a time and is called a *pointer*. The pointer moves to the next position every time an access is performed. An example of such a degenerated decoder implemented using a C^2MOS D-FF is shown in Figure 10.41. Similar approaches can be devised for other memory classes such as FIFOs.

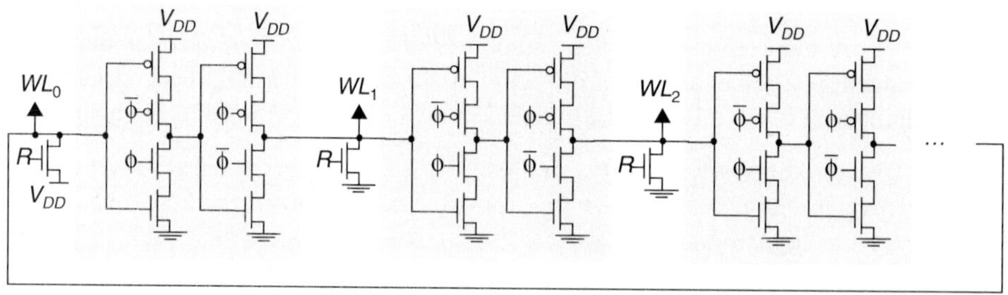

Figure 10.41 Decoder for circular shift-register. The R signal resets the pointer to the first position.

10.4.2 Sense Amplifiers

A number of important functions that influence the functionality, performance, and reliability of the memory are attributed to the sense amplifiers.

- *Amplification*—This task is absolutely essential in the 1T DRAM, where the signal swing would otherwise be restricted to approximately 250 mV [Itoh90]). In other memories, it allows for a reduced voltage swing on the bit lines, which helps to reduce both the delay and the power dissipation.

- *Performance speed-up*—The amplifier compensates for the restricted fan-out-driving capability of the memory cell by accelerating the bit-line transition.
- *Power reduction*—Reducing the signal swing on the bit lines can eliminate a substantial part of the power dissipation related to charging and discharging the bit lines.
- *Signal restoration*—Because the read and refresh functions are intrinsically linked in 1T DRAMs, it is necessary to drive the bit lines to the full signal range after sensing.

What amplifier to use is a strong function of the memory class involved as well as the overall memory architecture. Be aware that amplifiers are essential analog circuits, and their design falls in the realm of analog design. We will also see that the performance of the amplifier is a strong function of the overall timing. A careful optimization is always required.

Differential Sensing

It is generally known that a differential amplifier presents numerous advantages over its single-ended counterpart—one of the most important being the *common-mode rejection*. This is especially true in memories where the exact value of the bit-line signal varies from die to die and even for different locations on a single die. In other words, the absolute value of a 1 or 0 signal is not exactly known and might vary over quite a large range. The picture is further complicated by the presence of multiple noise sources, such as switching spikes on the supply voltages and capacitive cross talk between word and bit lines. The impact of those noise signals can be substantial, especially when we realize that the amplitude of the signal to be sensed is generally small.

The differential approach has the advantage that it only amplifies the difference between its input signals. The signals common to both inputs are suppressed at the output of the amplifier by a ratio called the *common-mode rejection ratio* (CMRR). Similarly, spikes on the power supply are suppressed by a ratio called the *power-supply rejection ratio* (PSRR). Differential sensing is therefore considered the technique of choice.

Unfortunately, the differential approach is only directly applicable to the SRAM memories, since these are only the memory cells that offer a differential output (BL and $\overline{BL}$). A differential sensing scheme is easily conceptualized for these memories, as shown in Figure 10.42a. The bit lines are connected to the inputs x and $\overline{x}$ of the differential amplifier. A read-cycle then proceeds as follows.

1. In the first step, the bit lines are precharged to V_{DD} by pulling $\overline{PC}$ low. Simultaneously, the *EQ*-PMOS transistor is turned on, ensuring that the initial voltages on both bit lines are identical. This operation, called *equalization*, is necessary to prevent the sense amplifier from making erroneous excursions when turned on. In practice, every differential signal in the memory is equalized before performing a read.
2. The read operation is started by disabling the precharge and equalization devices and enabling one of the word lines. One of the bit lines is pulled low by the selected memory cell. Notice that a grounded PMOS load, placed in parallel with the precharge transistor, limits the bit-line swing, preserving power and speeding up the next precharge cycle.

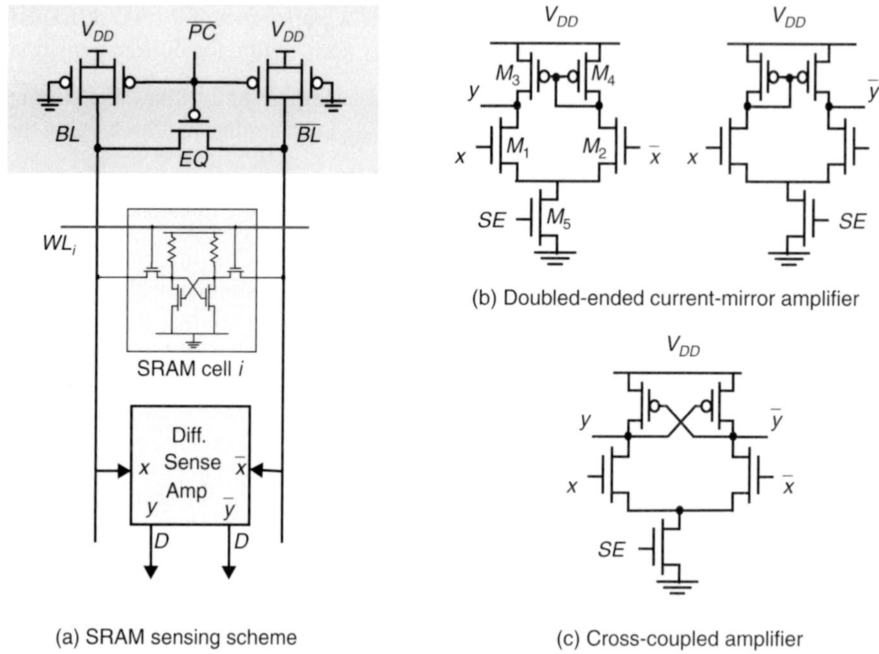

(a) SRAM sensing scheme

(b) Doubled-ended current-mirror amplifier

(c) Cross-coupled amplifier

Figure 10.42 Differential sensing as applied to an SRAM memory column.

3. Once a sufficient signal is built up (typically around 0.5 V), the sense amplifier is turned on by raising *SE*.

Two sense-amplifier circuits are shown in Figure 10.42b and c. While these circuits use PMOS loads and NMOS input devices, the dual configurations with PMOS input devices and NMOS loads are also regularly used, depending upon biasing conditions. These circuits are well known in the analog design world. An in-depth discussion of their operation is beyond the scope of this text. We limit ourselves to a qualitative analysis. The interested reader is referred to standard analog circuit design textbooks, such as [Gray93, pp. 295–302 and 460–466], or [Sedra87, pp. 531–532].

Consider first the current-mirror amplifier of Figure 10.42b. This circuit represents the MOS implementation of the emitter-coupled pair used in ECL and is called a *source-coupled pair.* To boost the gain without incurring an area or power-dissipation penalty, a current mirror acting as an active load replaces the traditional load resistances. In short, one could state that current flowing through transistor M_2 is mirrored by the combination (M_3-M_4) and compared with the drain current of M_1. The current difference is translated into a high voltage gain given by Eq. (10.13).

$$A_{sense} = -g_{m2}(r_{o2} \parallel r_{o3}) \tag{10.13}$$

with g_{m2} the transconductance of the input transistors, and r_o the small-signal device resistance. When operating in the saturation mode, the r_o of the MOS transistor is very high and is determined by the λ value (Chapter 2). The transconductance of the input devices can be increased by either widening the devices or increasing the bias current. The latter also reduces the output resistance of M_2, which limits the usefulness of this approach. A gain around 100 can be achieved. However, the gain of sense amplifiers typically is set to around 10, since gain is secondary to response time. The disadvantage of the current-mirror amplifier is its single-ended output. When a differential output is needed (for instance, when feeding the output into another amplifier stage) a second source-coupled pair with reversed inputs is necessary, as shown in Figure 10.42b.

Example 10.12 Differential Sense Amplifier

Assume an amplifier with the following parameters: I_{bias} = 100 μA, $(W/L)_n = (W/L)_p = 10$. Using the 1.2 μm CMOS parameters, this leads to the following gain.

$$A = \frac{g_m}{(\lambda_n + \lambda_p)(I_{bias}/2)} = \frac{1}{\lambda_n + \lambda_p}\sqrt{\frac{4k'_n(W/L)_n}{I_{bias}}} = 11.2$$

The main goal of the sense amplifier is the rapid production of a differential output signal. From this perspective, it makes sense to replace the current mirror by a cross-coupled pair of PMOS transistors, as shown in Figure 10.42c. Notice how this circuit resembles the DCVSL structure. The positive feedback of the cross-coupled PMOS devices accelerates the sensing speed compared to the current-mirror amplifier. On the negative side, the feedback and the associated high dynamic gain make this structure sensitive to erroneous latching resulting from timing errors or device mismatches [Sasaki90].

One final observation on the scheme of Figure 10.42. A single sense amplifier can be shared between multiple columns by inserting the column decoder pass-transistors between the memory cells and the amplifier. This results in area savings and, potentially, power reduction.

A radically different sensing approach is offered by the circuit of Figure 10.43, where a CMOS cross-coupled inverter pair is used as a sense amplifier. A CMOS inverter exhibits a high gain when positioned in its transient region, as was established in Chapter 3. To act as a sense amplifier, the flip-flop is initialized in its metastable point by equalizing the bit lines. A voltage difference is built over the bit lines in the course of the read process. Once a large enough voltage gap is created, the sense amplifier is enabled by raising *SE*. Depending upon the input, the cross-coupled pair traverses to one of its stable operation points. The transition is swift as a result of the positive feedback.

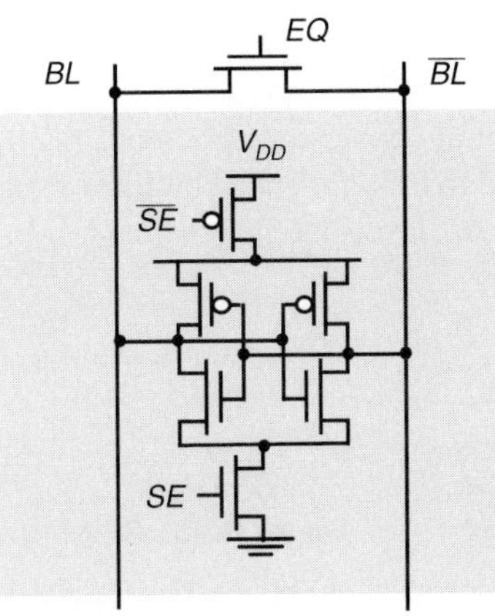

Figure 10.43 Cross-coupled CMOS inverter latch used as sense amplifier.

While the flip-flop sense amplifier is simple and fast, it has the property that inputs and outputs are merged, so that a full rail-to-rail transition is enforced on the bit lines. This is exactly what is needed for a 1T DRAM, where a restoration of the signal levels on the bit lines is necessary for the refresh of the cell contents. The cross-coupled cell is, therefore, almost universally used in DRAM designs. How to turn a single-ended memory structure such as the DRAM cell into a differential one is discussed in a subsequent section. On the other hand, a rail-to-rail excursion of the bit lines is to be avoided in SRAMs and ROMs for power considerations. Shutting off the amplifier after a carefully optimized time interval can help to address this problem.

Single-Ended Sensing

While the differential sensing approach is obviously preferable, memory cells used in ROMs, E(E)PROMs and DRAMs are inherently single-ended. One option is to resort to the single-ended amplification techniques introduced in Section 8.2, "How to drive large capacitances fast." Both the charge-redistribution amplifier (Figure 8.22) and asymmetrical inverter (Figure 8.19) are often used in smaller memory structures. For the sake of completeness, we present the schematics of the charge-redistribution amplifier as it is used in a memory in Figure 10.44. This structure has been very popular in EPROM memories.

Larger memories (>256 kBit) that are exceedingly prone to noise disturbances resort to translating the single-ended sensing problem into a differential one.

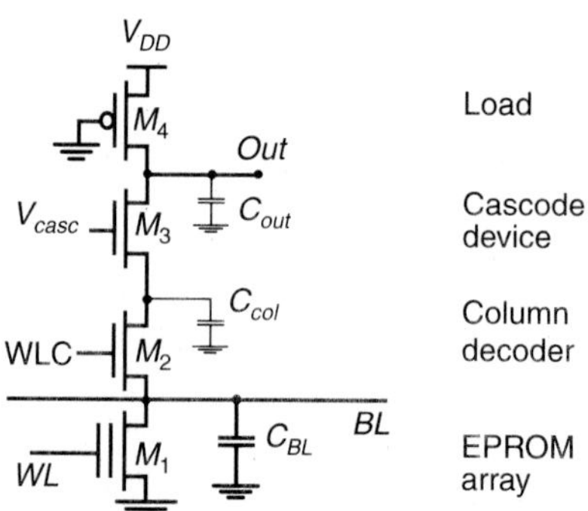

Figure 10.44 Charge-redistribution amplifier as used in EPROM memory.

Single-to-Differential Conversion

The basic concept behind the single-to-differential conversion is demonstrated in Figure 10.45. A differential sense amplifier is connected to a single-ended bit line on one side and a reference voltage, positioned between the 0 and 1 levels, at the other end. Depending on

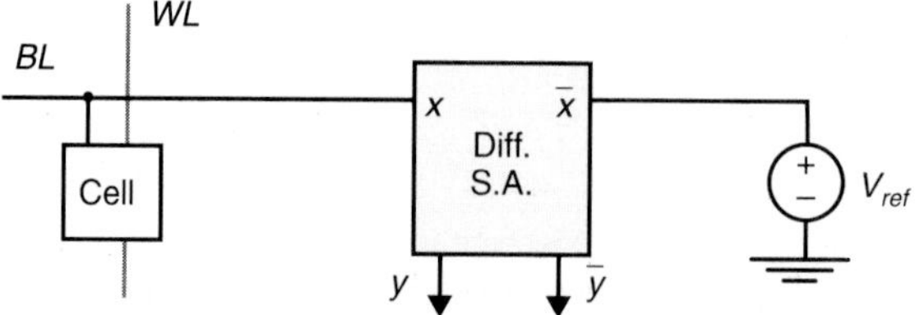

Figure 10.45 Single-to-differential conversion.

the value of *BL*, the amplifier will toggle in one or the other direction. Creating a good reference source is not as easy as it sounds, since the voltage levels tend to vary from die to die or even over a single die. The reference source must therefore track those variations. A popular way of doing so is illustrated in Figure 10.46 for the case of a 1T DRAM. The memory array is divided into two halves, with the differential amplifier placed in the middle. On each side, a column of so-called *dummy cells* is added. These are 1T memory cells that are similar to the others, but whose sole purpose is to serve as reference. This approach is often called the *open bit-line architecture*.

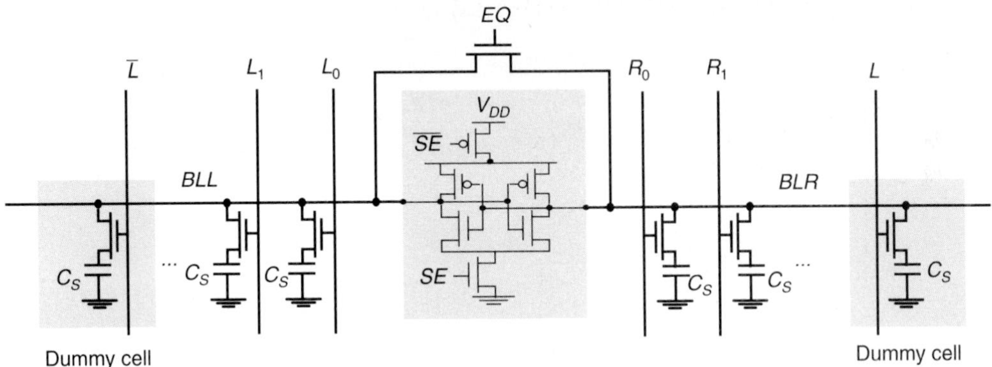

Figure 10.46 Open bit-line architecture with dummy cells.

When the *EQ* signal is raised, both the bit lines *BLL* and *BLR* are precharged to $V_{DD}/2$. Enabling *L* and $\overline{L}$ at the same time ensures that the dummy cells are charged to $V_{DD}/2$. One of the word lines is enabled during the read cycle. Assume that a cell in the left half of the array is selected by raising WL_0, which causes a voltage change on *BLL*. The appropriate voltage reference is created by simultaneously selecting the dummy cell in the other memory half by raising *L*. Under the assumption that the left and right memory sides are perfectly matched, the resulting voltage on *BLR* resides between the 0 and 1 levels and causes the sense latch to toggle. Notice that maintaining perfect symmetry is important. Raising word lines WL_0 and *L* simultaneously turns the capacitive coupling between bit and word lines into a common-mode signal that is effectively eliminated by the sense amplifier. Observe that dividing the bit lines into two halves effectively reduces the bit-line capacitance. This doubles the charge-transfer ratio and improves the signal-to-noise ratio.

Other approaches to create reference voltages using dummy cells can be envisioned, such as using a cell with a halved storage capacitance ($C_S/2$). This reduces the charge injected on the bit line to half, which is sufficient to create the necessary voltage difference. In (E)EPROMs, the dummy cell can be a standard floating-gate cell where the (W/L) of the storage transistor is halved. This causes a current imbalance during reading, which triggers the sense amplifier.

Example 10.13 Sensing in 1T DRAM

The read-operation of a 1T DRAM implemented using the open bit-line architecture, and the latch sense amplifier is simulated using SPICE. The storage capacitance is set to 50 fF, while the bit-line capacitance equals 0.5 pF. This is equivalent to a charge-transfer ratio of 9%.

Assuming that the bit line is precharged to 2.5 V, charge redistribution results in a voltage drop of 220 mV for a 0 signal. Due to the threshold loss over the pass-transistor, the cell voltage associated with a 1 equals 3.6 V, which translates to a voltage increase of only 100 mV on the bit line after the enabling of the word line. These values are confirmed by the simulation, which shows values of 220 mV and 90 mV respectively, before the sense amplifier is turned on (Figure 10.47).

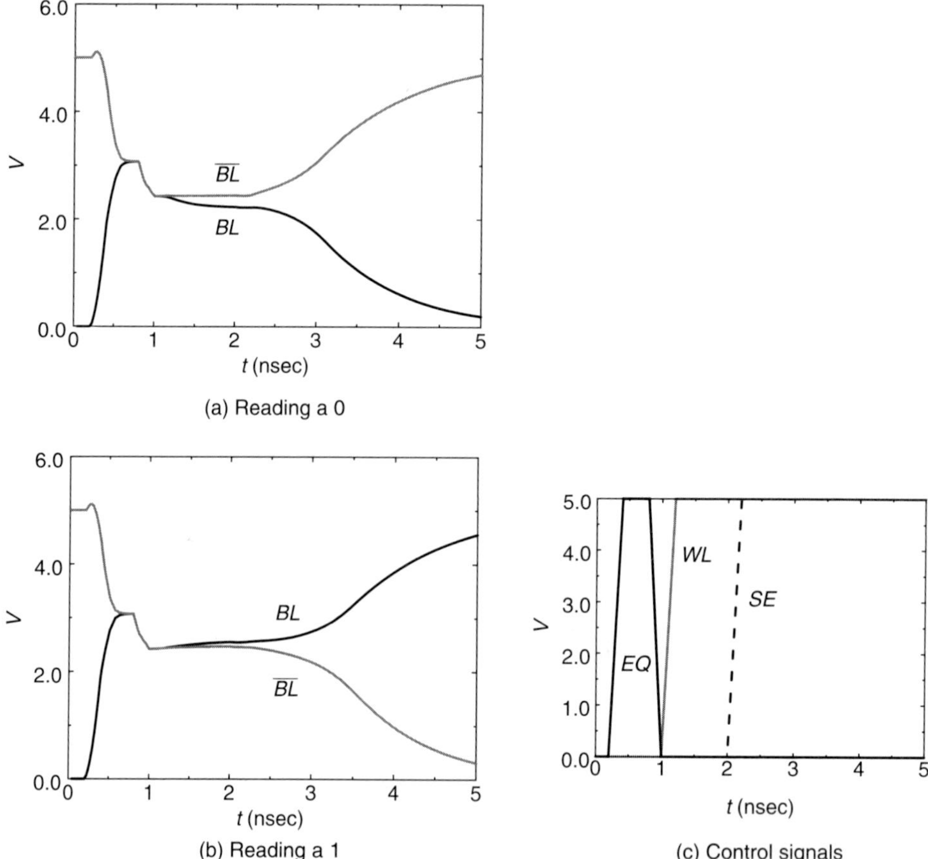

Figure 10.47 Simulation of the read-process in an open bit-line architecture with dummy cell. The dummy cell is connected to $\overline{BL}$ (CS = 50 fF; C_{BL}= 0.5 pF).

While the dummy cell technique presents a reasonable approach to differential sensing, it still suffers from some inadequacies. The positioning of the reference signal in the middle of the voltage range reduces the differential signal presented to the amplifier. This increases the sensing delay. The dummy cell is furthermore placed on the other side of the memory array, so that device mismatches can still cause severe sensing problems. This is particularly so in (E)EPROMs, where the device thresholds vary over time. Some of the larger memories in this class opt for a completely differential implementation, where data is stored in a complementary format. This makes the sensing problem similar to the

SRAM case. Although this approach doubles the number of storage cells, the overhead in area can be limited to approximately 30–50% for the memory core by using clever layout techniques (using so-called *twin* cells).

10.4.3 Drivers/Buffers

The length of word and bit lines increases with increasing memory sizes. Even though some of the associated performance degradation can be alleviated by partitioning the memory array, a large portion of the read and write access time can be attributed to the wire delays. A major part of the memory-periphery area is therefore allocated to the drivers, in particular the address buffers and the I/O drivers. The design of cascaded buffers was discussed at length in Chapter 8. All the issues raised there are valid here as well. If there is one area where the BiCMOS technology has made an impact, it is in the design of fast and compact memory drivers. Itoh [Itoh90] has derived that a BiCMOS driver achieves a 23% lower delay time and a 28% lower power dissipation compared to a CMOS cascade for the same input capacitance and output rise time. BiCMOS is also used in the design of extrasensitive differential sense amplifiers. A 1 Mbit DRAM using BiCMOS drivers and sense amplifiers is on the average 36% faster and consumes 24% less power.

10.4.4 Timing and Control

The above discussions present a picture of the memory module as a complex entity, whose operation is governed by a well-defined sequence of actions such as address-latching, word-line decoding, bit-line precharging and equalization, sense-amplifier enabling, and output driving. To be operational, it is essential that this sequence be adhered to under all operational circumstances and over a range of device and technology parameters. To achieve maximal performance, a careful timing of the different events is necessary. Although the timing and control circuitry only occupies a minimal amount of area, this implies that its design is an integral and defining part of the memory design process and requires careful optimization and multiple long SPICE simulations over a range of operating conditions.

Over time, a number of different memory-timing approaches have emerged that can be largely classified as *clocked* and *self-timed.* A typical example of each class is discussed to illustrate the differences.

DRAM—A Clocked Approach

Since the early days, DRAM memories have opted for a multiplexed addressing scheme where the row address and column address are presented in sequence on the same address bus to save package pins. This approach has survived many memory generations and is still in vogue today, although it is getting increasingly awkward and in disparity with system-level requirements. Some alternative techniques are being promoted.

In the multiplexed addressing scheme, the user must provide two main control signals, *RAS* (row-address strobe) and *CAS* (column-address strobe) that indicate the presence of the row and column addresses respectively (Figure 10.6). Another control signal

(W) indicates if the intended operation is a read or a write. These signals can be interpreted as external clock signals and are used to time the internal memory events. Similar to the synchronous clocking approach, the *RAS* and *CAS* signals must be sufficiently separated so that all the ensued operations have come to completion. Figure 10.48 shows a simplified timing diagram of a 1×4 Mbit DRAM memory and some of the imposed timing constraints. The full specification involves more than 20 timing parameters, all of which have to be held within precise bounds to ensure proper operation. Obviously, generating the correct timing signals for these memories is a nontrivial task!

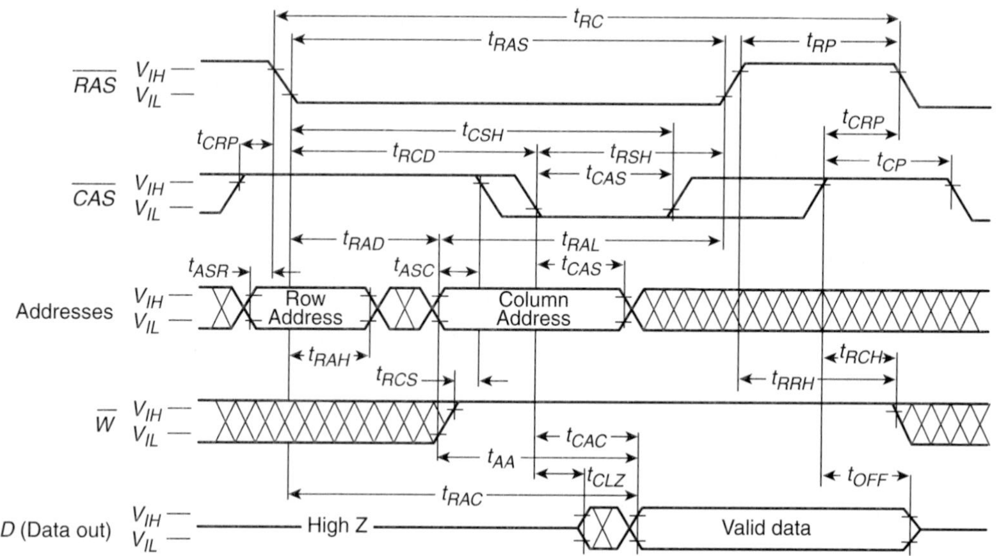

Figure 10.48 Read-cycle timing diagram for 4M × 1 DRAM memory [Mot91]. The minimum length of the read cycle t_{RC} equals 110 nsec. The spacing of the falling edges of *RAS* and *CAS* signals must range between 20 and 40 nsec. A total of 24 timing constraints can be observed.

All the internal timing signals such as the *EQ*, *PC*, and *SE* signals are derivatives of the external control signals. To maximize performance, some of those signals have to fall within very precise intervals. For instance, the *SE* signal cannot be applied too early after word-line decoding is started since this increases the sensitivity to noise. On the other hand, postponing it for too long slows down the memory. The separation of these events is a function of the word-line delay that might vary with temperature and process variations. It is a common practice to adjust the timing interval with operating parameters by including a model of the word line (*dummy word line*) in the timing-generation circuitry. This "delay module" tracks the delay of the actual word line accurately, making the memory more robust while preserving performance.

SRAM—A Self-Timed Approach

In contrast to DRAMs, SRAM memories have historically been designed from a globally static perspective. A memory operation is triggered by an event on the address bus or *R/W* signal (Figure 10.6b). No extra timing or control signals are needed. Such an approach seems to imply a fully static implementation for all circuitry such as decoders and sense amplifiers. A change in one of the input signals (data or address bus, *R/W*) then simply ripples through the subsequent layers of circuitry.

Our previous discussions made it apparent that such a fully static approach is not viable for larger memories for both area and power-dissipation considerations. SRAM memories, therefore, use a clever approach called *address-transition detection* (*ATD*) to automatically generate the internal signals such as *PC* and *SE* upon detection of a change in the external environment.

The *ATD* circuit plays an essential role in the architecture of SRAM and PROM modules. It acts as the source of most timing signals and is an integral part of the critical timing path. Speed is thus of utmost importance. A possible implementation of an *ATD* is shown in Figure 10.49. It consists of a number of transition-triggered one-shots (see Figure 6.47)—one per input bit—connected in a pseudo-NMOS NOR configuration. A transition on any of the input signals causes *ATD* to go low for a period t_d. The resulting pulse acts as the main timing reference for the rest of the memory, which results in a huge fan-out. Adequate buffering is thus recommendable. A more elaborated view on SRAM timing can be found in one of the case studies at the end of this chapter.

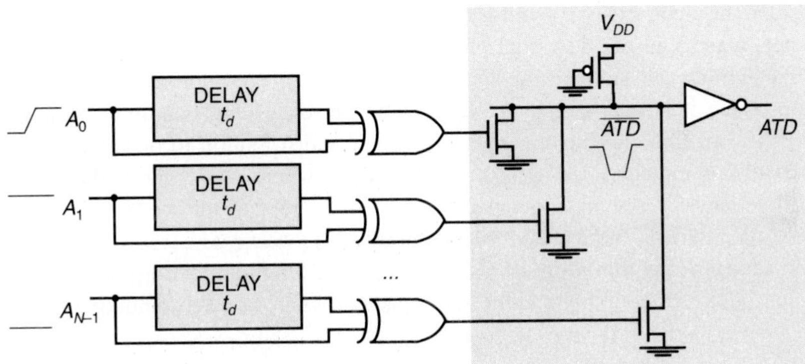

Figure 10.49 Address-transition detection circuitry. The delay lines are typically implemented as an inverter chain.

10.5 Memory Reliability and Yield

Memories are operating under low signal-to-noise conditions. Maximizing the signals while minimizing the noise contributions is essential to achieve stable memory operation. Another problem plaguing memory design is the low yield due to structural and intermittent defects. This section both discusses the nature of some of those problems, as well as potential solutions.

10.5.1 Signal-To-Noise Ratio

A tremendous effort is being made to produce memory cells that generate as large a signal as possible per unit area. Notwithstanding this effort, the produced signal quality decreases gradually with an increase in density. For instance, the DRAM cell capacitance has degraded from approximately 70 fF for the 16K memory to 30 fF for the current generations. Simultaneously, the voltage levels have decreased for both power consumption and reliability purposes. For instance, the memory core of some of the 64 Mbit memories operates at less than 3 V, the main reason being the limited voltage stress that can be endured by the very thin oxides used in cells.

At the same time, the increased integration density raises the noise level, due to the intersignal coupling. While the problem of the word-line-to-bit-line coupling capacitance was already an issue in the early 1980s, closer line spacing has brought the bit-line-to-bit-line coupling into the limelight. Contributing also to the problem are the higher speed requirements that result in an increasing switching noise for every new generation, and α-particle-induced soft errors that can change the state of a high-impedance node at random, and present an additional noise source. The latter problem was traditionally confined to dynamic memories, but now extends to static RAMs as well because the impedance of the SRAM storage nodes is rapidly increasing.

Word-Line-to-Bit-Line Coupling

Consider the open bit-line memory configuration of Figure 10.50. Word line WL_0 gets selected, which causes an amount of charge to be injected into the left bit line. We assume that the dummy cell on the right side, selected by WL_D, does not cause any charge redistribution, since it is precharged to $V_{DD}/2$. The presence of a coupling capacitance C_{WBL} between word line and bit line causes a charge redistribution to occur, whose amplitude approximately equals $\Delta_{WL} \times C_{WBL} / (C_{WBL} + C_{BL})$, where Δ_{WL} is the voltage swing on the word line and C_{BL} the bit-line capacitance. If both sides of the memory array were completely symmetrical, the injected bit-line noise (Δ_{BL}) would be identical on both sides and would appear as a common-mode signal to the sense amplifier. Unfortunately, this is not the case, because both coupling and bit line capacitance can vary substantially over the array.

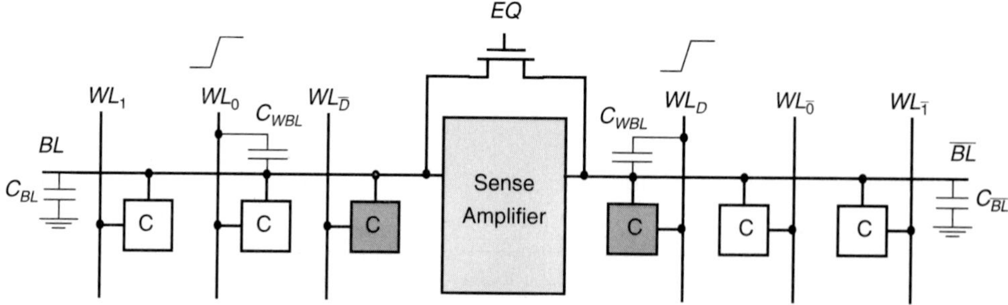

Figure 10.50 Word-line-to-bit-line coupling in open bit-line architecture.

This problem can be addressed by employing the *folded bit-line architecture* shown in Figure 10.51. Placing the sense amplifier at the end of the array and having BL and $\overline{BL}$ routed next to each other ensures a much closer matching between parasitic and bit-line capacitances. The word-line signals (WL_0 and WL_D) cross both bit lines. The cross-coupling noise, hence, appears as a common-mode signal even if the voltage waveforms on

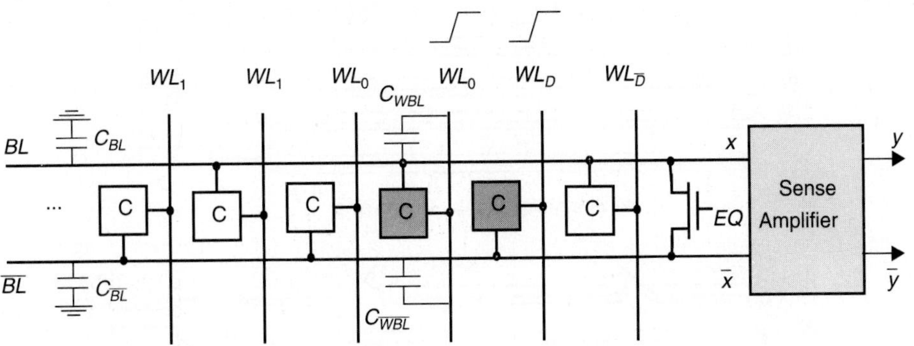

Figure 10.51 Folded bit-line architecture for 1T DRAM.

WL_0 and WL_D differ considerably. While the folded bit-line architecture has an obvious advantage in terms of noise suppression, it tends to increase the length of the bit line and consequently its capacitance. This overhead can be kept to a minimum by a clever interleaving of the cells.

The scenario presented above covers only a part of the total picture. A more convoluted case of word-line-to-bit-line coupling can be described as follows. During a read operation, a bit line *BL* undergoes a voltage transition ΔV. This voltage transition gets coupled to the nonselected word lines that are in precharged mode and thus represent high-impedance nodes. This noise signal, in turn, gets coupled back to other bit lines. The resulting cross talk depends upon the data patterns stored in memory. Suppose that the majority of the bit lines are reading a 0. This causes a negative transition on the nonselected word lines, which in turn pulls down the bit lines. This is especially harmful for the (minority of) bit-lines that are reading a 1 value. The fact that this noise signal is pattern-dependent makes it particularly hard to detect. Testing a memory for functionality requires the application of a wide range of different data patterns. The folded bit-line architecture helps to suppress this type of noise.

Bit-Line-to-Bit-Line Coupling

The impact of interwire cross-coupling increases with reducing dimensions, as was detailed in Chapter 8. This is especially true in the memory array, where the noise-sensitive bit lines run side by side for long distances. The art is to turn the noise signals into a common-mode signal for the sense amplifier. An ingenious way of doing so is represented by the *transposed (or twisted) bit-line architecture* illustrated in Figure 10.52. The first figure represents the straightforward implementation. Both *BL* and *BL* are coupled to the adjoining column (or row) by the capacitance C_{cross}. In the worst case, the signal swing observed at the sense amplifier can be reduced by ΔV_{cross}.

$$\Delta V_{cross} = 2\frac{C_{cross}}{C_{cross} + C_{BL}} V_{swing} \tag{10.14}$$

where V_{swing} is the signal swing on the bit lines. Up to one-fourth of the already weak signal can be lost due to this interference ([Itoh90]).

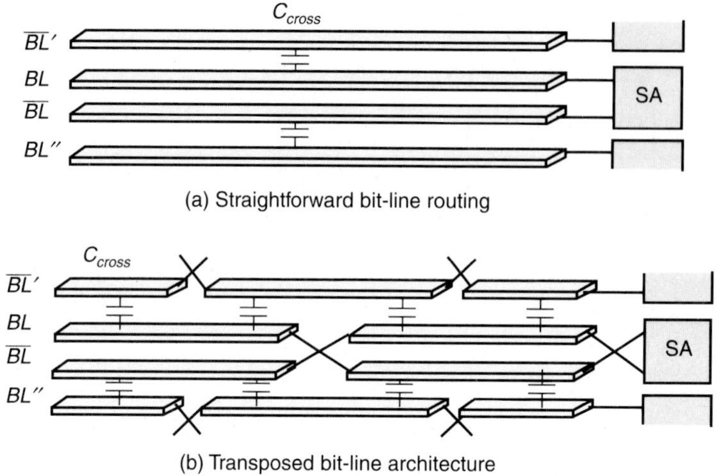

Figure 10.52 Bit-line-to-bit-line coupling.

The *transposed bit-line architecture* (Figure 10.52b) eliminates this source of disturbance by dividing the bit lines into segments that are connected in a cross-coupled fashion. This modification presents the interference signal introduced by a neighboring bit line in approximately the same way to both BL and $\overline{BL}$, turning it into a common-mode signal. An alternative approach is to use the capacitor-plate layer as an extra shielding between the data lines.

Leakage

The last two noise sources to be discussed are specific for dynamic memories or memories that rely on charge storage as the memorization technique. Charge leakage through the junction capacitance causes a gradual degradation in the stored signal charge. The minimum bound on the signal-to-noise ratio sets an upper limit on the refresh time t_{REFmax}. Refresh operations consume power and represent a pure overhead from a systems point of view since they reduce the achievable memory bandwidth. Consequently, it is desirable to keep the refresh period as long as possible. This can be achieved by decreasing the ambient junction temperature and by improving the quality of the fabrication process. Architectural modifications can further reduce the refresh overhead. A typical refresh procedure sequentially toggles each word-line, refreshing all the cells connected to that line simultaneously. It is thus advantageous to keep the number of rows in the memory to a minimum. Dividing the memory into multiple blocks furthermore allows for the simultaneous refreshing of multiple rows.

Critical Charge

Early memory designers were puzzled by the occurrence of *soft errors*, that is, nonrecurrent and nonpermanent errors, in DRAMs that could not be explained by either supply-voltage noise, leakage, or cross-coupling. The impact of soft errors on system design is enormous. If no adequate protection against them is found, they can cause a computer system to crash in a nonreconstructible way making the task of the system debugger virtually hopeless. In a landmark paper, May and Woods identified *alpha particles* as the main culprits [May79].

Alpha particles are He^{2+} nuclei (two protons, two neutrons) emitted from radioactive elements during decay. Traces of such elements are unavoidably present in the device packaging materials. With an emitted energy of 8 to 9 MeV, alpha particles can travel up to 10 μm deep into silicon. While doing so, they interact strongly with the crystalline structure, generating roughly 2×10^6 electron-hole pairs in the substrate. The soft error problem occurs when the trajectory of one of these particles strikes the storage node of a memory cell. Consider the cell of Figure 10.53. When a 1 is stored in the cell, the potential well is empty. Electrons and holes generated by a striking particle diffuse through the substrate. Electrons that reach the edge of the depletion region before recombining are swept into the storage node by the electrical field. If enough electrons are collected, the stored value can change into a 0. The *collection efficiency* measures the percentage of electrons that make it into the cell.

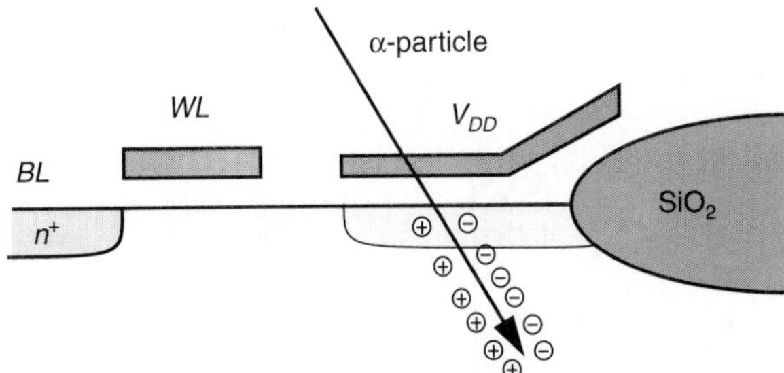

Figure 10.53 Alpha-particle induced soft errors.

The occurrence of soft errors can be reduced by keeping the cell charge larger than a *critical charge* Q_C. This puts a lower bound on the storage capacitance and cell voltage. As an example, a capacitance of 50 fF charge to a potential of 3.5 V holds 1.1×10^6 electrons. A single alpha particle striking such a cell with a collection efficiency of 55% can erase the complete charge. This is one of the main reasons why the cell capacitance of even the densest memories is kept higher than 30 fF. Another approach is to reduce the collection efficiency, which has been observed to be inversely proportional to the diagonal length of the depletion region of the cell.The value of the critical charge thus reduces with technology scaling! For instance, the critical charge equals 30 fC for a diagonal length of 2.5 μm which corresponds to a 64 Mbit memory. A memory die can furthermore be covered with polymide to protect against alpha radiation. It is worth mentioning that alpha particles are also a problem in state-of-the-art SRAMs that are relying more heavily on capacitive storage on high-impedance nodes.

While the occurrence of soft errors can be kept to an absolute minimum by careful design, total elimination is hard to achieve. The occasional occurrence of an error should not necessarily be fatal. System-level techniques such as error correction can help detect and correct most failures. This is briefly discussed in the next section.

10.5.2 Memory yield

With increasing die size and integration density, a reduction in yield is to be expected, notwithstanding the improvements in the manufacturing process. Causes for malfunctioning of a part can be both material defects and process variations. Memory designers use two approaches to combat low yields and to reduce the cost of these complex components: redundancy and error correction.

Redundancy

Memories have the advantage of being extremely regular structures. Providing *redundant hardware* is easily accomplished. Defective bit lines in a memory array can be replaced by redundant ones, and the same holds for word lines (Figure 10.54). When a defective column is detected during testing of the memory part, it is replaced by a redundant one by programming the fuse bank connected to the column decoder. A typical way of doing so is to blow the fuses using a programming laser. A similar approach is followed for the defective word lines. Whenever a failing word line is addressed, the word redundancy system enables a redundant word line.

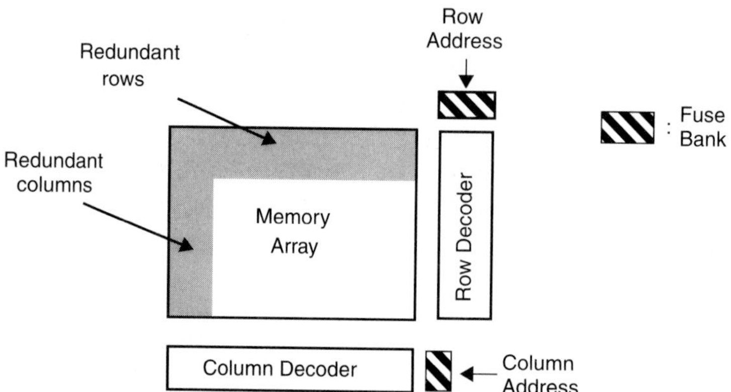

Figure 10.54 Redundancy in memory array increases the yield.

Error Correction

Redundancy helps correct faults that affect a large section of the memory such as defective bit lines or word lines. It is ineffective when dealing with scattered point-errors such as local errors caused by material defects. Achieving a reasonable fault coverage requires too much redundancy under these circumstances and results in a large area overhead. A better approach to address these faults is to use *error correction*. The idea behind this scheme is to use redundancy in the data representation so that an erroneous bit(s) can be detected and even corrected. Adding a parity bit to a data word, for instance, provides a way of detecting but not correcting an error. While a full discussion of error correction techniques would lead us astray, a simple example is used to illustrate the basic concept.

Example 10.14 Error Correction using Hamming Codes

Consider a 4-bit number (B_i), encoded with 3 check bits (P_i),

$$P_1\, P_2\, B_3\, P_4\, B_5\, B_6\, B_7$$

with the P_i chosen such that

$$P_1 \oplus B_3 \oplus B_5 \oplus B_7 = 0$$

$$P_2 \oplus B_3 \oplus B_6 \oplus B_7 = 0$$

$$P_4 \oplus B_5 \oplus B_6 \oplus B_7 = 0$$

Suppose now that an error occurs in B_3. This causes the first two expressions to evaluate to 1 while the last one remains at 0. Binary encoded, this means that bit 011 or 3 is in error. This information is sufficient to correct the fault. In general, single error correction requires that

$$2^k \geq m + k + 1 \tag{10.15}$$

where m and k are the number of data and check bits respectively. To perform single error correction on 64 data bits requires 7 check bits, resulting in a total word length of 71.

An important observation is that error correction not only combats technology-related faults, but is also an effective way of dealing with soft errors and time-variant faults. For instance, error correction is very effective in dealing with threshold variations in EEPROMs.

Error correction and redundancy address different angles of the memory yield problem. To cover all the bases, a combination of both is needed. This is convincingly illustrated in Figure 10.55, which plots the yield percentage for a 16 Mbit DRAM when the yield improvement techniques are used independently or combined [Kalter90].

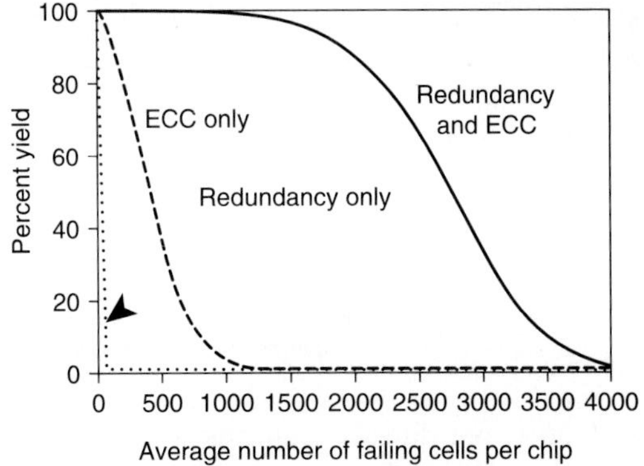

Figure 10.55 Yield curves for 16 Mbit DRAM using ECC and bit-line redundancy [Kalter90].

10.6 Case Studies in Memory Design

While the previous sections have introduced the individual components of a semiconductor memory, it is worthwhile to study a number of examples to understand how it all comes together. We will use two case studies to achieve that goal—the programmable logic array and a 4 Mbit SRAM memory.

10.6.1 The Programmable Logic Array (PLA)

When implementing complex control logic in CMOS, a designer can fall back on two options. The first one is to directly map the multilevel Boolean function into a network of individual CMOS logic gates after performing some optimization on the logic function to minimize either the implementation area or delay. This *multilevel logic* design approach can be automated using the standard cell methodology that is discussed in full in Chapter 11. Another option is to bring the logic function first into a canonical format called the two-level *sum-of-products* representation. This representation can be mapped into a very regular implementation by an automated process. The circuit structure that makes this possible is called the *programmable logic array*, or PLA.

The concept is best explained with the aid of an example. Consider the following logic functions, for which we have transformed the equations into the sum-of-products format using logic manipulations. Such a representation always exists.

$$\begin{aligned} f_0 &= x_0 x_1 + \overline{x_2} \\ f_1 &= x_o x_1 x_2 + \overline{x_2} + \overline{x_0} x_1 \end{aligned} \tag{10.16}$$

An advantage of this representation is that a regular realization is easily conceived, as illustrated in Figure 10.56. A first layer of gates implements the AND operations—also called *product-terms or min-terms*—while a second layer realizes the OR functions, called the *sum-terms*.

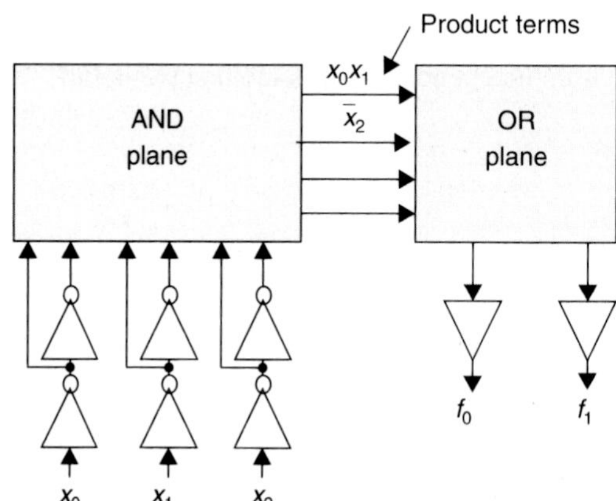

Figure 10.56 Regular two-level implementation of Boolean functions.

The schematic of Figure 10.56 is not directly realizable since single-layer logic functions in CMOS are always inverting. With a few, simple Boolean manipulations, Eq. (10.16) can be rewritten into a NOR-NOR format.

$$\begin{aligned} \overline{f_0} &= \overline{\overline{(\overline{x_0} + \overline{x_1})} + \overline{x_2}} \\ \overline{f_1} &= \overline{\overline{(\overline{x_0} + \overline{x_1} + \overline{x_2})} + \overline{x_2} + \overline{(x_0 + \overline{x_1})}} \end{aligned} \tag{10.17}$$

Problem 10.9 Two-Level Logic Representations

It is equally conceivable to represent Eq. (10.16) in a NAND-NAND format. In general, the NOR-NOR representation is preferred due to the prohibitively slow speed of large fan-in NAND gates. The NAND-NAND configuration is dense however and can help to reduce power consumption. Derive the NAND-NAND representation for the example of Eq. (10.16).

Now follows the reason why the issue of two-level logic implementation is raised in this chapter on memories. When discussing the ROM memory core, we realized that the ROM array is nothing more than a collection of large fan-in NOR (NAND) gates implemented in a very regular format. Similarly, the address decoder is also a collection of NOR (NAND) gates. The same approach can thus be used to implement any two-level logic function. Such a structure is called a PLA. The only difference between a ROM and PLA is that the decoder (AND-plane) of the former enumerates all possible min-terms (m inputs yield 2^m min-terms), while the AND-plane of the latter only realizes a limited set of min-terms, as dictated by the logic equations. Topologically, both structures are identical.

A NOR-NOR implementation of the Boolean functions of Eq. (10.16) using a static pseudo-NMOS circuit style is shown in Figure 10.57. Although compact and fast, its power dissipation makes this realization unattractive for larger PLAs where a dynamic approach is better. However, a direct cascade of the dynamic planes is out of the question. Solutions could be to introduce an inverter between the planes in the DOMINO style or implement the OR-plane with PMOS transistors and use predischarging in the *np*-CMOS fashion. The former solution causes pitch-matching problems, while the latter slows down the structure if minimum-size PMOS devices are used.

Both of those approaches are feasible, although many others have been engineered. We will take this opportunity to examine the memory-timing issue more thoroughly. Figure 10.58 uses a more complex clocking scheme to resolve the plane-connection problem [Weste93]. Assume that the pull-up transistors of the AND plane and OR plane are

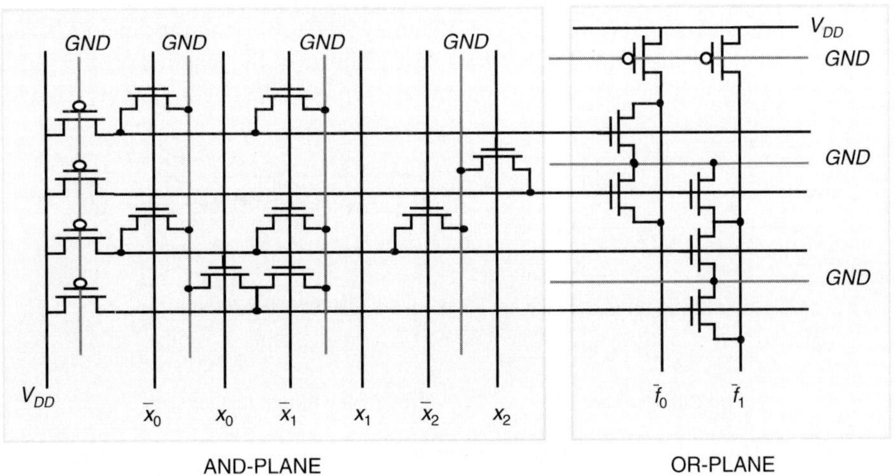

Figure 10.57 Pseudo-NMOS PLA.

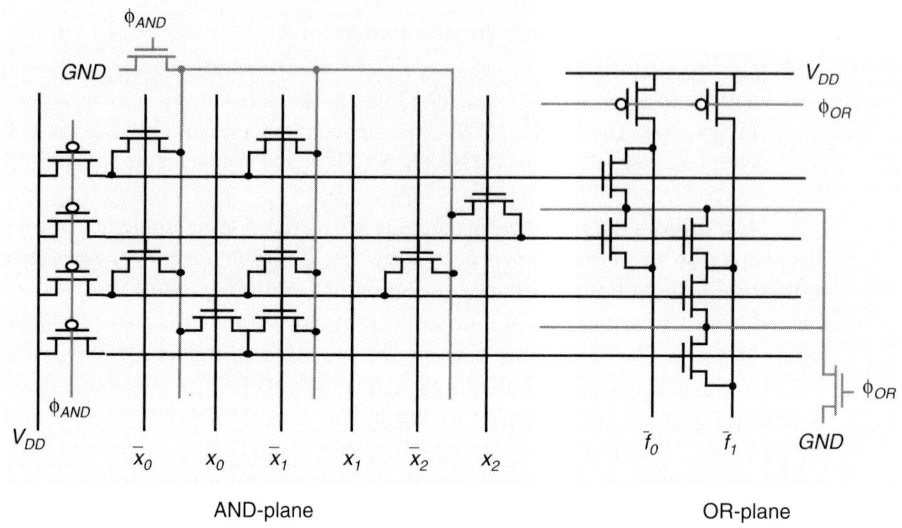

Figure 10.58 Dynamic implementation of PLA.

clocked by signals ϕ_{AND} and ϕ_{OR} respectively. These signals are defined so that a clock cycle starts with the precharging of both planes. After a time interval long enough to ensure that precharging of the AND plane has ended, the AND plane starts to evaluate by raising ϕ_{AND}, while the OR plane remains disabled (Figure 10.59a). Once the AND plane outputs are valid, it is safe to enable the OR plane by raising ϕ_{OR}.

The timing of these events is precarious, since it depends upon the PLA size and programming and technology variations. Although the clock signals can be provided externally, self-timing is the recommended approach to achieve maximum performance at the minimum risk. How to derive the appropriate timing signals from a single clock ϕ is shown in Figure 10.59b. The clock signal ϕ_{AND} is derived from ϕ using a monostable one-shot. The delay element consists of a dummy AND row with maximum loading—a fully populated row that provides the worst-case estimate of the required precharge time. ϕ_{OR} is derived in a similar way using a dummy AND row clocked by ϕ_{AND}. This provides a worst-

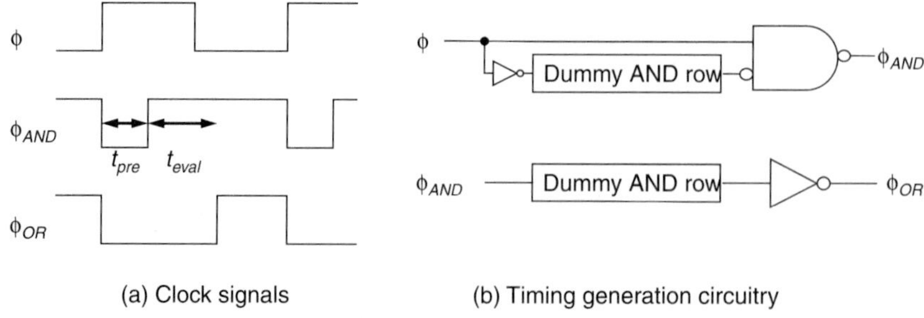

Figure 10.59 Generation of clock-signals for self-timed dynamic PLA. t_{pre} and t_{eval} represent the worst-case precharge and evaluation times of an AND row. These are obtained with the aid of dummy AND rows, which are fully populated; all transistors in the in the NOR network are turned off except one. This represents the worst-case discharge scenario.

case estimate of the discharge time, including some extra safety margin. Although it requires additional logic, this approach ensures reliable operation of the array under a range of conditions.

The PLA approach has the advantage of regularity and ease of automation. In fact, numerous computer-aided design programs have been developed over the years that take as input a set of Boolean equations and produce a minimized layout [Brayton84]. On the negative side, the array-based approach reduces the performance and can be area-ineffective, depending upon the Boolean functions. The use of PLAs has decreased substantially in recent years. Progress in software support for multilevel logic has tilted the balance in that direction, as will be discussed in Chapter 11.

Problem 10.10 PLA Architecture

A PLA implementation of the logic functions described by Eq. (10.16) is shown in Figure 10.60. Derive the circuit diagram and the clocking strategy used. The input and output buffers, which in this example are almost as large as the PLA array, have been omitted to simplify the picture. Notice how metal bypasses are used for the polysilicon wires (word lines and clocks), as well as for the diffusion wires (*GND*).

Explain why the implemented equations are different from the ones in Eq. (10.16). Finally, think about why one of the rows in the array is not used at all.

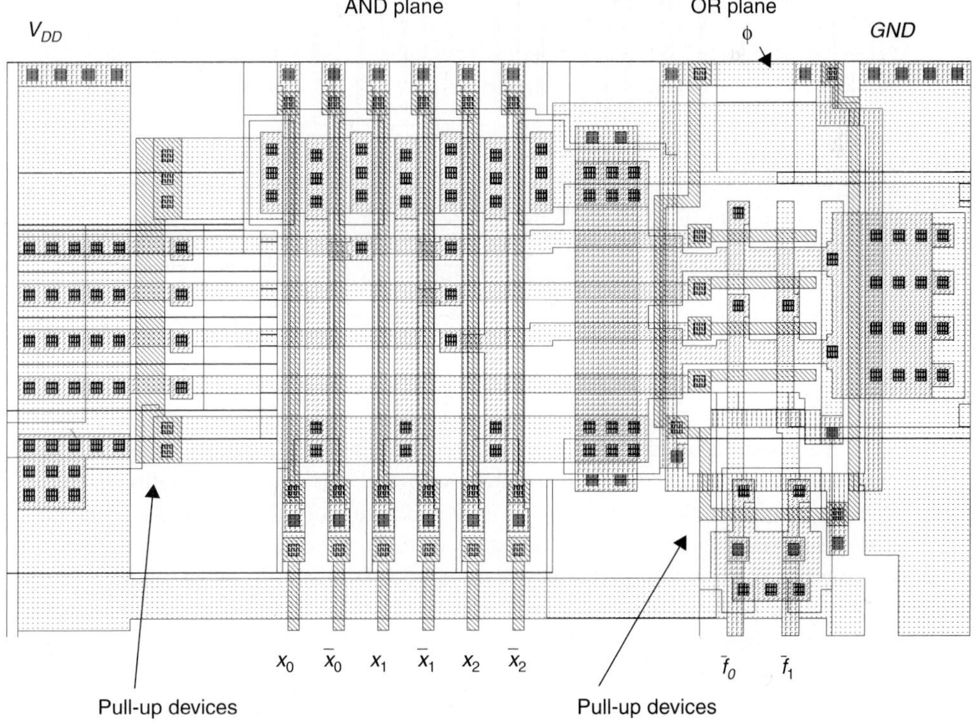

Figure 10.60 PLA layout implementing Eq. (10.16). See also Colorplate 18.

10.6.2 A 4 Mbit SRAM

Figure 10.61 shows the block diagram of a 4 Mbit SRAM memory designed by Mitsubishi [Hirose90]. The memory has an access time of 20 nsec for a single supply voltage of 3.3 V and is fabricated in a quadruple polysilicon, double metal 0.6 μm CMOS technology. The memory is organized as 32 blocks, each block counting 1024 rows and 128 columns. The row-address (*X*), column address (*Y*), and block address (Z) are 10, 7, and 5 bits wide.

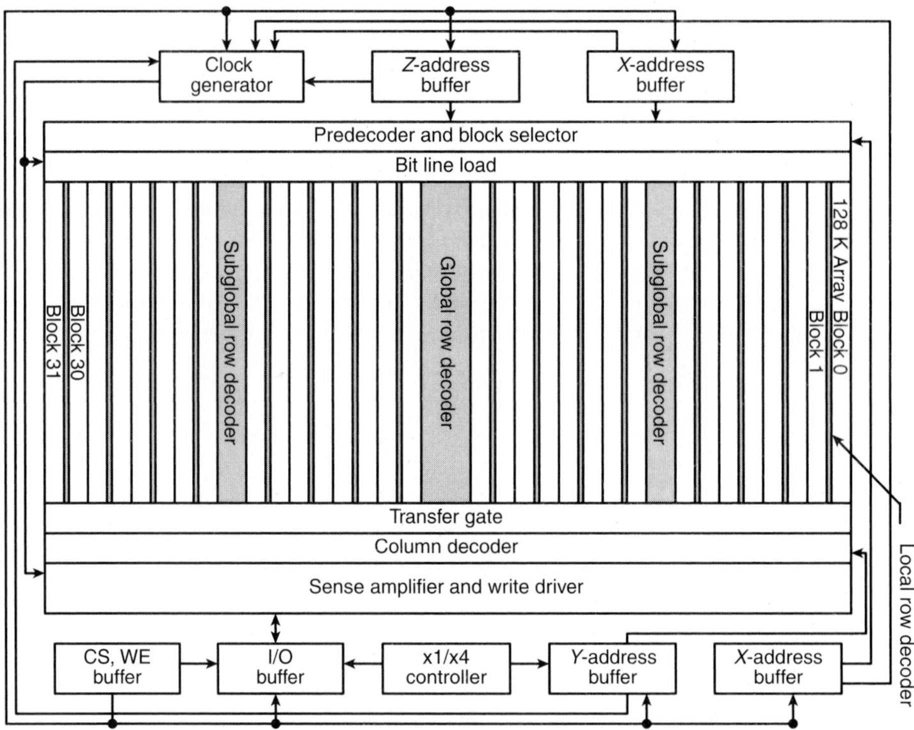

Figure 10.61 Block diagram of 4 Mbit memory (from [Hirose90]).

To provide fast and low-power row decoding, the memory uses an interesting approach called the *hierarchical-word decoding* scheme, shown in Figure 10.62. Instead of broadcasting the decoded *X* address to all blocks in polysilicon, it is distributed in metal and called the *global word line*. The local word line is confined to a single block and is only activated when that particular block is selected using the block address. To further improve the performance and power consumption, an extra word-line hierarchy level called the *subglobal word line* is introduced. This approach results in a row-decoding delay of only 7 nsec.

The bit-line peripheral circuitry is presented in Figure 10.63. The memory cell is a four-transistor resistive-load cell occupying 19 μm^2. The load resistance equals 10 TΩ. A triggering of the *ATD* pulse causes the precharging and the equalizing of the bit lines of the selected block with the aid of the *BEQ* control signal. Notice that the bit-line load is a composition of static and dynamic devices. Lowering *BEQ* starts the read process. One of

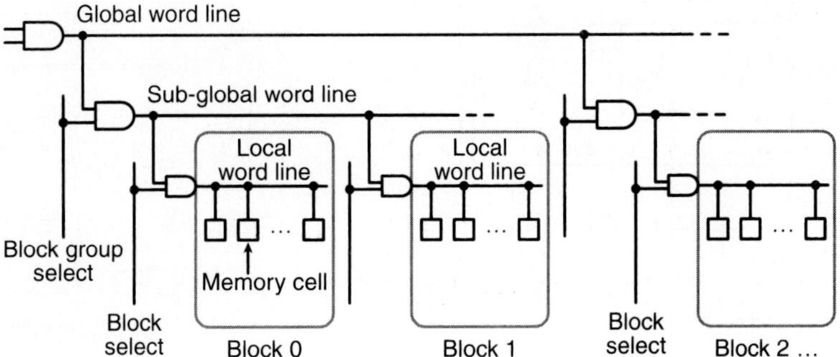

Figure 10.62 Hierarchical word-line selection scheme.

the word lines is enabled, and the appropriate bit lines start to discharge. These are connected to the sense amplifiers after passing through the first layer of the column decoder (*CD*). Only 16 sense amplifiers are needed per block of 128 columns in this scheme. The amplifier consists of two stages (Figure 10.63b). The first is a cross-coupled stage that provides a minimal gain, but also acts as a level-shifter. This permits the second amplifier, which is of the current-mirror type, to operate at the maximum-gain point. A push-pull output stage drives the highly capacitive data lines that lead to the tri-statable input-output drivers. The approximative signal waveforms are displayed in Figure 10.63c. These demonstrate the subtlety of the timing and clocking strategy necessary to operate a large memory at high speed. To write into the memory, the appropriate value and its complement are imposed on the I/O and $\overline{I/O}$ lines, after which the appropriate row, column, and block addresses are enabled. It is worth mentioning that the memory consumes only 70 mA when operated at 40 MHz. The standby current equals 1.5 μA.

10.7 Perspective: Semiconductor Memory Trends and Evolutions

To conclude this chapter, some historical perspectives and evolutionary trends are worth discussing. The first semiconductor memories introduced in the 1960s used bipolar technology and could store no more than 100 bits. A major breakthrough was achieved in the early 1970s when the first MOS DRAM memory was introduced with a capacity of 1 KBit ([Hoff70]). This was the beginning of a dramatic evolution. Currently, DRAMs of 64 Mbit exist, and the 256 MBit memory is on its way. This means that in less than 25 years, the amount of memory that can be integrated on a single die has increased by a factor of 64,000! The evolution of the integration density for a number of memory varieties is plotted in Figure 10.64. It is fair to conclude that the number of memory cells that can be integrated on a single die quadruples approximately every three years.

It is interesting to understand what gave rise to these spectacular improvements. Figure 10.65 plots the evolution of the SRAM cell and die size over the subsequent generations ([Takada91]). Note that the cell size decreases by an approximate factor of 2.6 from generation to generation—in the semiconductor memory business, memory density is upgraded by a factor of four between generations, for example, 1K, 4K, 16K, 64K, etc. At the same time, the die size increases by a factor of 1.5. Combining the two factors suggests

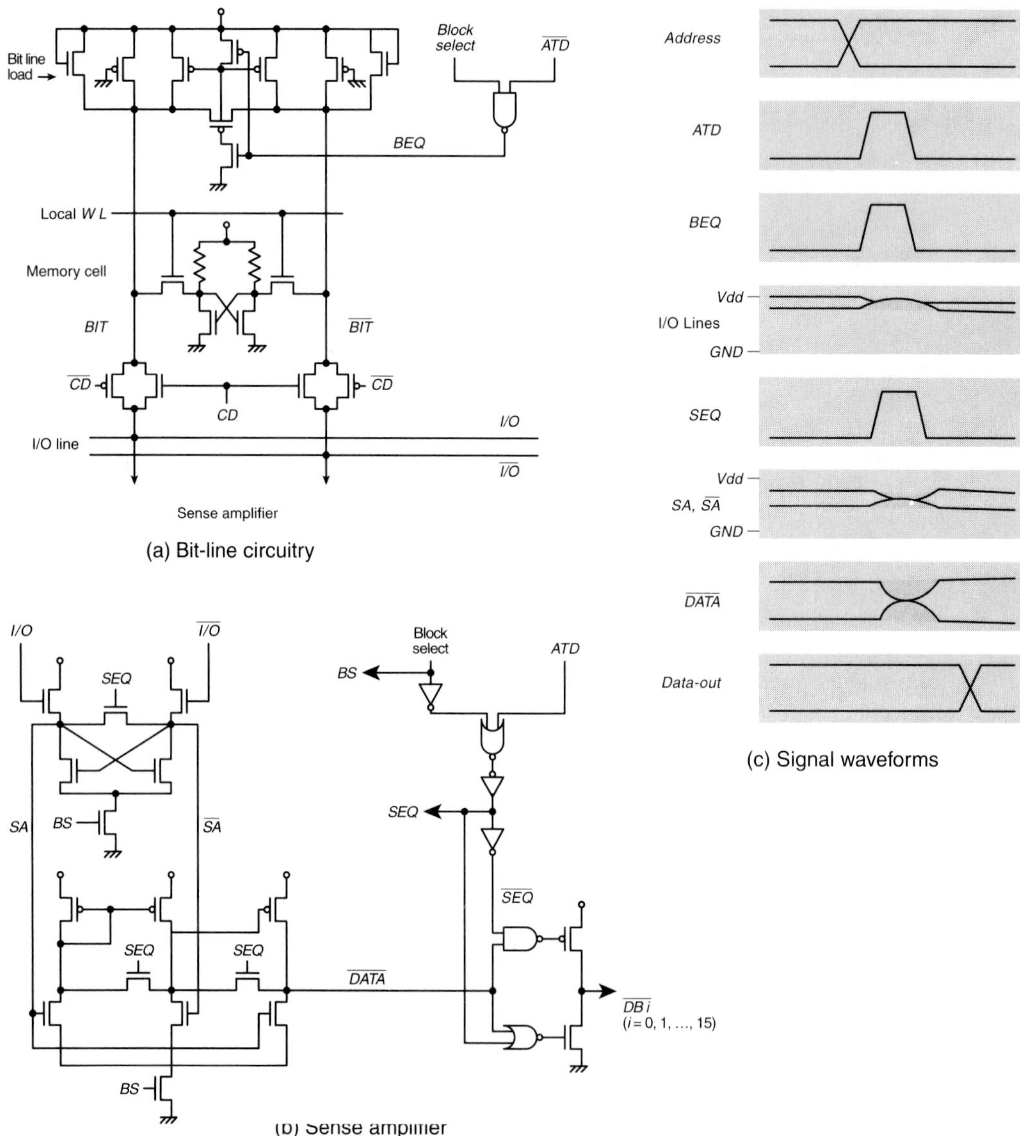

Figure 10.63 The bit-line peripheral circuitry and the associated signal waveforms.

an increase in density by a factor of 3.9 between generations. The remaining discrepancy can be attributed to the increasing ratio between areas devoted to memory cells and peripheral circuitry for larger memories; in other words, the overhead decreases. One would be inclined to attribute the reduction in cell size solely to technology scaling, but this is only partially true. For instance, the L_{eff} of the transistors used in the 4K SRAM memory equaled 3 μm and has decreased from generation to generation so that the 64M SRAM uses transistors with an effective length of 0.25 μm (Figure 10.66). This suggests that scaling reduces the cell size by a factor of $1.5 \times 1.5 = 2.25$ between generations, what

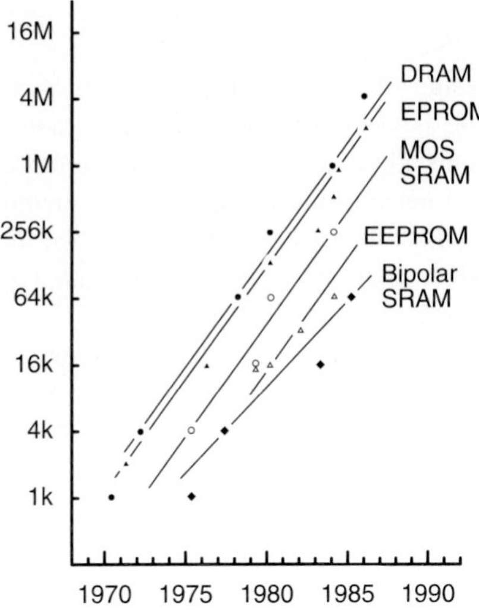

Figure 10.64 Evolution of semiconductor integration density over time (from [Asai86]).

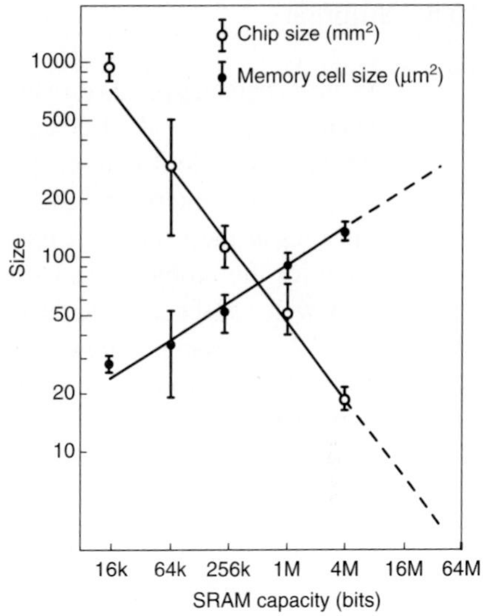

Figure 10.65 Progress in memory cell sizes and change in chip sizes for CMOS SRAMs (from [Takada91])

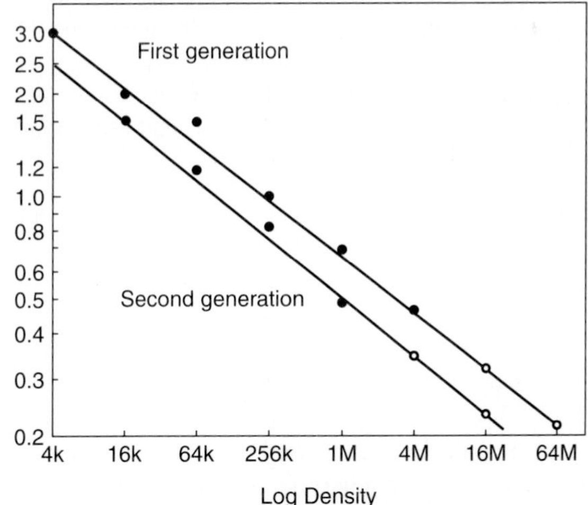

Figure 10.66 Trend of L_{eff} reduction with increasing SRAM density. Notice that a single memory is typically implemented in multiple technologies. After a first generation, a second generation is implemented in a more aggressive technology [Flanagan88].

is not sufficient to account for the observed trends. Additional reductions are obtained by an ever-increasing sophistication in memory cell technology. Important events in this category include the introduction of the polysilicon gate, denser isolation using LOCOS and U-groove, polysilicon load resistors for SRAM cells, and three-dimensional DRAM cells using trench or stacked capacitors. The continuous introduction of such novelties is essential if the current trends in integration are to continue.

10.8 Summary

In this chapter, we have discussed the design of semiconductor memories in extensive detail. While the art of designing memories is vastly different from the traditional logic world, it is obvious that many of the problems addressed by memory designers are similar to those encountered by logic designers. Moreover, due to the fact that memory design operates at the frontier of technological ability, such as the smallest device dimensions, many of the problems and solutions considered in contemporary memory design might very well surface in logic design at a later time.

As an example, power dissipation is one of the major challenges facing high-density memory designers. Low-voltage operation around 1.5 V is one of the solutions being employed in state-of-the-art memories. Other approaches include self-timing and local power-down. Most of these techniques can be applied to the logic design world.

In conclusion, the memory-design problem can be summarized as follows.

- The *memory architecture* has a major impact on the ease of use of the memory, its reliability and yield, its performance, and, power consumption. Memories are organized as arrays of cells. An individual cell is addressed by a block, column, and row address.

- The *memory cell* should be designed so that a maximum signal is obtained in a minimum area. While the cell design is mostly dominated by technological considerations, a clever circuit design can help maximize the signal value and transient response. The cell topologies have not varied much over the last decades. Most of the improvement in density results from technology scaling and advanced manufacturing processes. We discussed cells for read-only, nonvolatile, and read-write memories.

- The *peripheral circuitry* is essential to operate the memory in a reliable way and with a reasonable performance, given the weak signaling characteristics of the cells. Decoders, sense amplifiers, and I/O buffers are an integral part of every memory design. The discussion presented in this chapter only touched the surface. Besides the mentioned functions, other circuitry in the periphery performs functions such as bootstrapping, voltage generation in EEPROMs, and voltage regulation (some of these are discussed in [Nakagome91]). Although interesting and challenging, the discussion of these circuits would have lead us too far. Suffice it to say that the design of memory periphery is still the playground of the hard-core digital circuit designer.

- A memory must operate correctly over a variety of operating and manufacturing conditions. To increase the integration density, memory designers give up on signal-to-noise ratio. This makes the design vulnerable to a whole range of noise signals, which are normally less of an issue in logic design. Identifying the potential sources of malfunction and providing an appropriate model is the first requirement when addressing *memory reliability.* Circuit precautions to deal with potential malfunctions include redundancy and error correction.

10.9 To Probe Further

Although a vast amount of literature on memory design is presented in the leading circuit design journals, not many textbooks are available on this topic. One of the best introductions to memory design can be found in [Hodges88, Chapter9]. A number of interesting tutorial papers have been published over the years, some of which are referenced below. Finally, most ground-breaking innovations in memory design are reported in either the *Journal of Solid State Circuits* (November issue in particular), the ISSCC conference proceedings, and the VLSI circuit symposium.

REFERENCES

[Asai86] S. Asai, "Semiconductor Memory Trends," *Proceedings of the IEEE*, vol. 74, no. 12, pp. 1623–1635, December 1986.

[Brayton84] R. Brayton et al., *ESPRESSO-IIC: Logic Minimization Algorithms for VLSI Synthesis*, Kluwer Academic, The Netherlands, 1984.

[Chang77] J. Chang, "Theory of MNOS Memory Transistor," *IEEE Trans. Electron Devices*, ED-24, pp. 511–518, 1977.

[Dennard68] R. Dennard, "Field-effect transistor memory," U.S. Patent 3 387 286, June 1968.

[Dennard84] R. Dennard, "Evolution of the MOSFET Dynamic RAM—A Personal View," *IEEE Trans. On Electron Devices*, pp. 1549–1555, November 1984.

[Dillinger88] T. Dillinger, *VLSI Engineering*, Chapter 12, Prentice Hall, 1988.

[Flannagan88] S. Flannagan, "Future Technology Trends for Static RAMS," *Proceedings IEDM 1988*, pp. 40–43, 1988.

[Frohman74] D. Frohman, "FAMOS—A New Semiconductor Charge Storage Device," *Solid State Electronics*, vol. 17, pp. 517–529, 1974.

[Gray93] P. Gray and R. Meyer, *Analysis and Design of Analog Integrated Circuits,* 3rd ed., John Wiley and Sons, 1993.

[Hirose90] T. Hirose et al., "A 20-ns 4-Mb CMOS SRAM with Hierarchical Word Decoding Architecture," *IEEE Journal of Solid State Circuits*, vol. 25, no. 5, pp. 1068–1074, October 1990.

[Hodges88] D. Hodges and H. Jackson, *Analysis and Design of Digital Integrated Circuits*, McGraw-Hill, 1988.

[Hoff70] E. Hoff, "Silicon-Gate Dynamic MOS Crams 1,024 Bits on a Chip," *Electronics*, pp.68–73, August 3, 1970.

[Itoh90] K. Itoh, "Trends in Megabit DRAM Circuit Design," *IEEE Journal of Solid State Circuits*, vol. 25, no. 3, pp. 778–798, June 1990.

[Jinbo92] T. Jinbo et al., "A 5V-Only 16 Mbit Flash Memory with Sector Erase Mode," *IEEE Journal of Solid State Circuits,* vol. 27, no. 11, pp. 1547–1554, November 1992.

[Johnson80] W. Johnson et al., "A 16 Kb Electrically Erasable Nonvolatile Memory," *ISSCC Digest of Technical Papers*, pp. 152–153, February 1980.

[Kalter90] H. Kalter et al., "A 50 nsec 16 Mb DRAM with a 10 nsec Data Rate and On-Chip ECC," *IEEE Journal of Solid State Circuits*, vol. 25, no. 5, pp. 1118–1128, October 1990.

[Lu89] N. Lu, "Advanced Cell Structures for Dynamic RAMs," *IEEE Circuits and Devices Magazine*, pp. 27–36, January 1989.

[Mano87] T. Mano et al., "Circuit Technologies for 16 Mbit DRAM's," *ISSCC Digest of Technical Papers*, pp. 22–23, February 1987.

[Masuoka91] F. Masuoka et al., "Reviews and Prospects of Non-Volatile Semiconductor Memories," *IEICE Transactions*, vol. E74, no. 4, pp. 868–874, April 1991.

[May79] T. May and M. Woods, "Alpha-Particle-Induced Soft Errors in Dynamic Memories," *IEEE Transactions on Electron Devices*, ED-26, no. 1, pp 2–9, January 1979.

[Mot91] Motorola, "Memory Device Data," Specification Data Book, 1991.

[Nakagome91] Y. Nakagome and K. Itoh, "Reviews and Prospects of DRAM Technology," *IEICE Transactions*, vol. E74, no. 4, pp. 799–811, April 1991.

[Nakamura92] K. Nakamura et al., "A 6 ns ECL I/O and 8 ns TTL I/O 4-MB BiCMOS SRAM," *IEEE Journal of Solid State Circuits,* vol. 27, no. 11, pp. 1504–1510, November 1992.

[Ootani90] T. Ootani et al., "A 4-Mb CMOS SRAM with PMOS Thin-Film-Transistor Load Cell," *IEEE Journal of Solid State Circuits*, vol. 25, no. 5, pp. 1082–1092, October 1990.

[Pashley89] R. Pashley and S. Lai, "Flash Memories: The Best of Two Worlds," *IEEE Spectrum*, pp. 30–33, December 1989.

[Regitz70] W. Regitz and J. Karp, "A Three-Transistor Cell, 1,024-bit 500 ns MOS RAM," *ISSCC Digest of Technical Papers*, pp 42–43, 1970.

[Sasaki90] K. Sasaki et al., "A 23 ns 4-Mb CMOS SRAM with 0.2 mA Standby Current," *IEEE Journal of Solid State Circuits*, vol. 25, no. 5, pp. 1075–1081, October 1990.

[Sedra87] A. Sedra and K. Smith, *Microelectronic Circuits,* 2nd ed., Holt, Rinehart and Winston, 1987.

[Snow67] E. Snow, "Fowler-Nordheim Tunneling in SiO_2 Films," *Solid State Communications*, vol. 5, pp. 813–815, 1967.

[Takada91] M. Takada and T. Enomoto, "Reviews and Prospects of SRAM Technology," *IEICE Transactions*, vol. E74, no. 4, pp. 827–838, April 1991.

[Terada89] Y. Terada et al., "120 ns 128k × 8bit / 64k × 16bit CMOS EEPROMs," *IEEE Journal of Solid State Circuits*, vol. 24, no. 5, pp. 1244–1249, October 1989.

[Taguchi91] M. Taguchi et al., "A 40-ns 64 Mb DRAM with 64-b Parallel Data Bus Architecture," *IEEE Journal of Solid State Circuits*, vol. 26, no. 11, pp. 1493–1497, November 1991.

[Tomita91] N. Tomita et al., "A 62-ns 16-Mb CMOS EPROM with Voltage Stress Relaxation Technique," *IEEE Journal of Solid State Circuits*, vol. 26, no. 11, pp. 1593–1599, November 1991.

[Weste93] N. Weste and K. Eshraghian, *Principles of CMOS VLSI Design*, 2nd ed., Addison-Wesley, 1993.

10.10 Exercises and Design Problems

1. [E, SPICE, 10.3.1] Use SPICE to compute the access time of the 512 × 512 NOR ROM of Example 10.4. **Use a simplified model** (i.e., do not include all the transistors but model their impact on word and bit lines). Compare the obtained results (row, column, and overall delay) with the results of the hand analysis.
2. [M, SPICE, 10.3.1] Repeat Problem 1 for the 512 × 512 NAND ROM of Example 10.4.
3. [E, None, 10.3.2] Figure 10.67 depicts an EPROM cell with a floating gate. Assume the oxide thickness between the access gate and the floating gate is 30 nm, while below the floating gate there is 20 nm of oxide. Use W/L = 4.8/2.4. Assume, initially, that the potential of the floating gate is 0 V (with all transistor terminals grounded). All other transistor parameters are as normal.

a. Compute the drain current for the transistor with its source grounded and its gate and drain at 5 V. Assume $\phi_{MS} - 2\phi_F = -0.2$ V.

b. Compute the potential of the floating gate *immediately* after applying a programming voltage of 15 V to the transistor gate and drain.

c. Assuming the programming voltage induces a hot electron current of 10^{-10} A, what voltage will the floating gate be after 600 μs of programming?

d. When the transistor gate and drain are returned to 0 V after 1200 μs, what is the final potential on the floating gate?

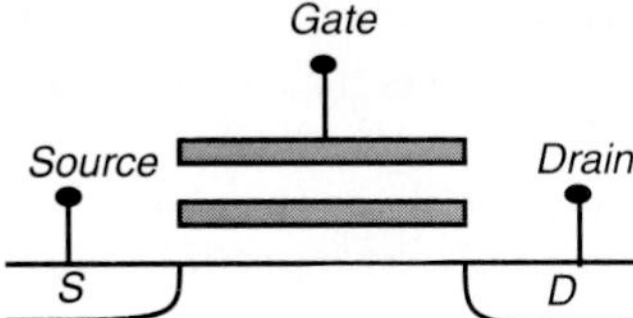

Figure 10.67 EPROM cell with floating gate.

4. [C, None, 10.3.1&10.4] A ROM is designed in a BiCMOS technology. For the MOS transistors, assume the given dimensions are effective, electrical *W/L*'s. For the bipolar components, use the following device data: $V_{BE(max)} = 0.8$ V and $V_{SBD(max)} = 0.4$ V (V_{BE} and V_{SBD} are each 0.1 V less at the edge of conduction); $\tau_F = 0.2$ nsec, $C_{je} = C_{jc} = C_{SBD} = 0.05$ pF, $C_{cs} = 0.1$ pF. Assume these capacitances are fixed, effective values, and neglect base currents.

a. Consider an NMOS ROM array with a bipolar transistor used in the sense amplifier, as shown in Figure 10.68a. Find the dc voltage needed at node 1 to bring node 2 up to 0.7 V.

b. Suppose the word line (node 1) steps from 0 to 5 V (Figure 10.68b). Calculate the propagation delay to node 3. Assume node 2 is the speed-limiting node.

c. Now suppose that node 1 steps from 5 V to 0 V (Figure 10.68b). Calculate the propagation delay to node 3. Again assume that node 2 is the speed limiting node. Also, note that the level-1 k' and λ parameters were extracted with $V_{GS} = V_{DS} = 5$ V, so they do not work well for this problem where V_{DS} is small. Discuss how you can get around this problem.

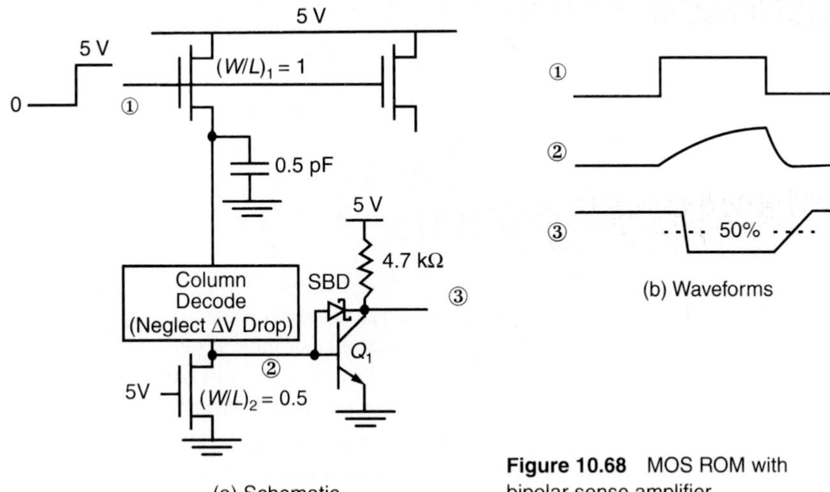

Figure 10.68 MOS ROM with bipolar sense amplifier.

5. [E, None, 10.3.3] For a memory containing a 4096 word × 2048 bit array of SRAM cells with a differential bit-line architecture, assume that the dynamic power consumption is dominated by charging and discharging the bit lines. Assume further that the cells are tiled at a vertical pitch of 24 μm and a horizontal pitch of 15 μm. Also assume that each cell adds a load of 20 fF to BL and $\overline{BL}$. Bit lines are in metal1 and are 3 μm wide.

a. Compute the capacitance loading each bit line. Break it down into contributions from wiring and from memory cells.

b. If the bit lines are precharged to 2.5 V and are allowed to develop a maximum differential voltage of 2 V (symmetric around the precharge voltage) during a read operation, what is the power consumption by the memory while reading at an access rate of 1 MHz.

6. [M, None, 10.3.3] Consider the six-transistor NMOS static memory cell of Figure 10.69. You may ignore the body effect for this problem ($\gamma = 0$). Use $(W/L)_2 = 1.2/24$ and $(W/L)_3 = 2.4/1.2$. The threshold voltage V_{TO} of the depletion transistors equals –2 V.

a. Assume first that node Q is in the 1 state and node Q is 0. In order to write a 0 to node Q, bit line BL is lowered to 0.9 V. Determine the minimal size of transistor M_1 so that the cell just flips when this voltage is reached. Assume that the switching threshold V_M of the NMOS inverter equals 0.92 V.

b. Assume now that $(W/L)_1 = 2.4/1.2$. Determine the maximal precharge voltage on bit line BL so that the cell does not flip during a read operation.

c. Finally, suppose that the bit line has been precharged to 2 V and that the sense amplifier can detect a voltage swing of 0.5 V. Determine how long the read operation would take, given that the bit-line capacitance is 2 pF and the delay of the sense amp itself may be ignored. You may ignore the current through the depletion load M_2 for this part of the problem.

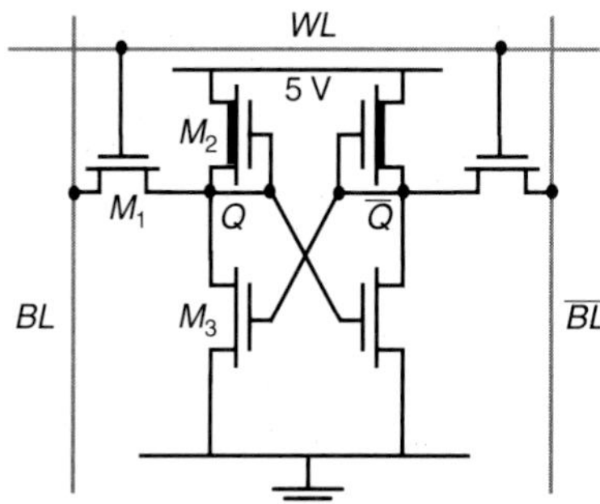

Figure 10.69 6T SRAM cell with NMOS depletion loads.

7. [M, None, 10.3.3] A 5-transistor SRAM cell is shown in Figure 10.70. The bit line is normally precharged to V_{DD}.

a. Describe *the three constraints* that should be imposed on the devices for guaranteeing safe read and write operations. Write down the equations that would help you to size the transistors. Do not solve the equations nor plug in numbers. Using the following variables: V_{DD}, V_T (same for NMOS and PMOS), k_{Mx}, and V_M of an inverter.

b. Based on the above equations, discuss qualitatively the required relative sizing of the transistors in the cell (for instance, transistor M_x must be wider than transistor M_y).

8. [E, None, 10.3.3] Draw the transistor diagram of a two-port SRAM cell. A two-port memory can read or write two independent addresses simultaneously. Discuss qualitatively how this effects the transistor sizing in the cell.

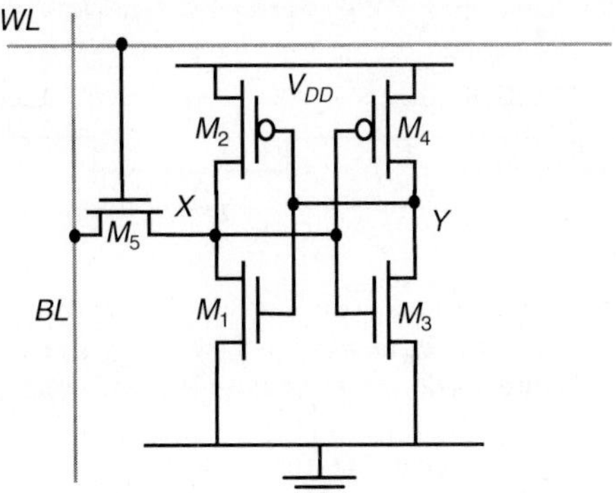

Figure 10.70 Five-transistor SRAM cell.

9. [M, None, 10.3.3] A two-transistor memory cell is shown in Figure 10.71. It uses two identical transistors (M_1 and M_2) with $W/L = 1.8/1.2$. Separate lines are provided for the read select (RS) and write select (WS), which both switch between 0 and 3 V. You may ignore body effect ($\gamma = 0$) and channel-length modulation ($\lambda = 0$) throughout this problem.

a. Explain the operation of the memory. Draw waveforms for WB, RB, WS, and RS for both reads and writes.

b. Determine the maximum possible current flowing into the cell during a read operation. State clearly your assumptions and simplifications.

c. Determine the size (W/L) of transistor M_3 so that the voltage on the bit line RB never drops below 2.5 V during a read operation.

d. Compute the time it takes to achieve a 0.5 V voltage drop on the bit line during a read operation. Assume that $C_c = 50$ fF and $C_b = 2$ pF.

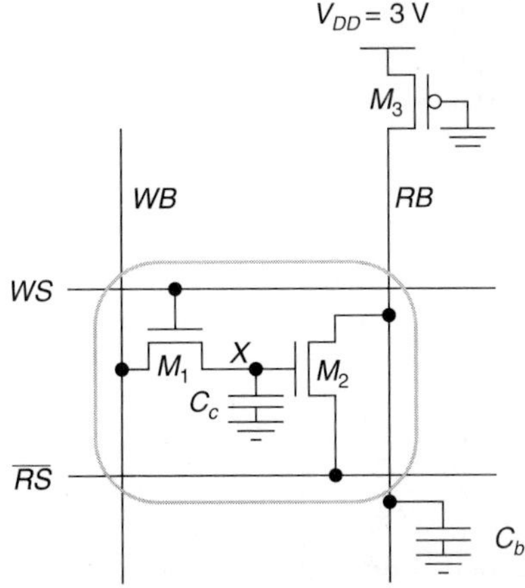

Figure 10.71 2-T memory cell.

10. Figure 10.72 shows a variant of the 1T-DRAM cell.

a. Fill in the timing diagrams for nodes BL and Y when writing a 0 and a 1 into the cell. For WL and PL, use the timing waveforms shown in the figure. Denote the voltage levels in terms of V_{DD} and V_T (ignore body effect). Ignore transient effects.

b. Describe briefly why this is an attractive approach.

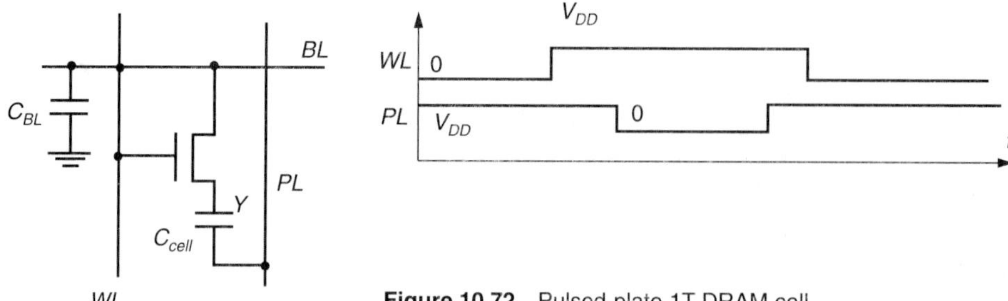

Figure 10.72 Pulsed-plate 1T DRAM cell.

c. Assume that the bit-line capacitance equals 500 fF. The transistor threshold equals 0.75 V (ignore body effect). The supply voltage equals 5 V and the bit line is pre(dis)charged to 0 V. Derive the symbolic equations needed to the derive the bit line voltages after reading a 0 and a 1. Ignore transient effects.

d. Using the results from 10c, derive the minimum cell capacitance so that the voltage difference on the bit line between reading a 0 and a 1 is larger than 250 mV.

11. [C, None, 10.3.3] Figure 10.73 shows an alternative diode-coupled bipolar SRAM cell along with the word-line voltage levels used to access the cell. Use the following bipolar transistor data: $\beta_F = 75$, $V_{BE(max)} = 0.8$ V and $V_{SBD(max)} = 0.4$ V (V_{BE} and V_{SBD} are each 0.1 V less at the edge of conduction). Assume a series diode resistance of 50 Ω.

a. Deduce and explain the operation of this cell, sketching qualitative waveforms for *WL*, *BL*, *BL*, *Q*, and *Q* for the following sequence of operations: write 0, read 0, write 1, and read 1. To write a 1, $\overline{BL}$ is raised to 2.5 V, while *BL* is left floating. Vice versa for writing a 0.

b. Find the range of values for R_5 and R_6 that guarantees correct operation of the cell. Use $R_1 = R_2 = 1.2$ kΩ and $R_3 = R_4 = 60$ kΩ.

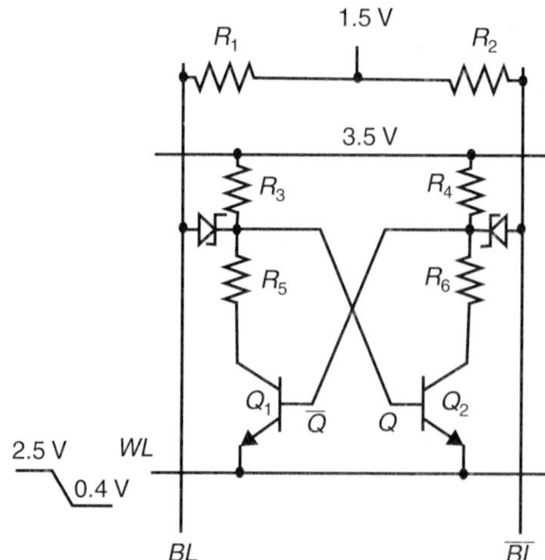

Figure 10.73 Diode-coupled bipolar SRAM cell.

12. [M, None, 10.4.2] Figure 10.74 shows a DRAM with a divided bit line. Each side of the differential sense amplifier connects to 256 DRAM cells and 1 dummy cell. The input capacitance of the sense amp is 50 fF at each input. The *BL* and *BL* lines are in metal for which the resistive and capacitive contributions can be neglected. The lumped junction capacitance of each cell is 8 fF. Ignore body effect.

a. Compute the effective capacitance on each bit line, C_{bit}.

b. Draw the timing diagrams corresponding to reading a 0 and a 1 from cell 1. Draw the waveforms for ϕ_p, ϕ_r, ϕ_d, $\phi_{\bar{r}}$, $\phi_{\bar{d}}$, *BL* and *BL*. Assume the sense amplifier refreshes values after reading them. Discuss the sizing of the capacitor C_d.

c. Compute the minimum values of C_c and C_d so that the sense amp sees a differential voltage of at least 60 mV.

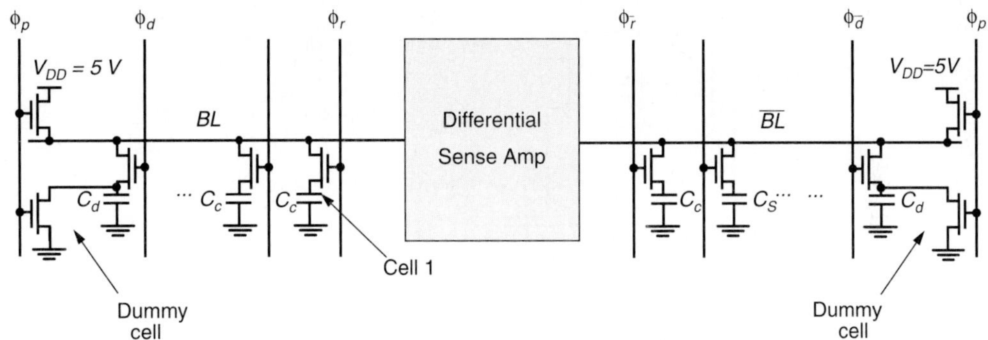

Figure 10.74 DRAM with divided bit-line structure.

13. [M, None, 10.4.1] Figure 10.75 shows a dynamic NOR row decoder (using pseudo two-phase logic) for word line 0 of a memory with 16 word lines.

a. Explain the operation of this row decoder, showing approximate voltage waveforms for all important nodes, including the clocks. What is the purpose of gate 2?

b. Discuss the relative advantages and disadvantages of a dynamic versus a static decoder.

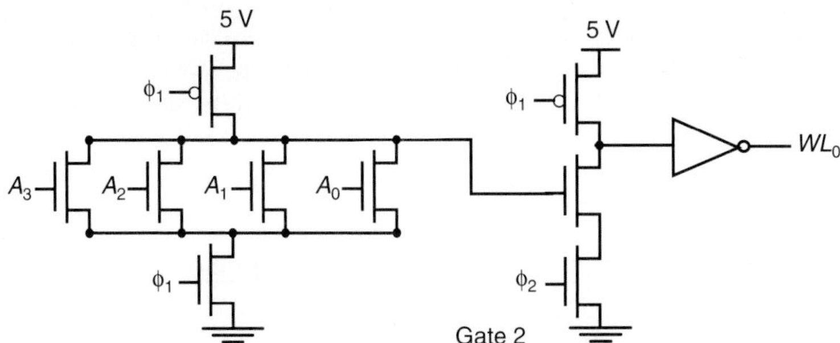

Figure 10.75 Dynamic NOR row-decoder

14. [E, None, 10.4.1] Design a column decoder to select one out of 16 bit lines based on address bits A_3, A_2, A_1, and A_0 and their complements. Minimize the total number of transistors while limiting the maximum number of cascaded pass transistors to three. Draw a transistor-level schematic of your design.

15. [C, None, 10.4.2] A dynamic sense amplifier is shown in Figure 10.76a. The capacitance of each bit line is 1 pF, and average discharging current from the selected RAM cell equals 10 μA. The waveform applied at point *S* is shown in Figure 10.76b. Initially, *S* is high. When $\overline{BL}$ reaches $V_{DD} - 0.5$ V (at time $t = t_1$), *S* is lowered to $V_{DD} - 1.2$ V and reaches this value at time

$t = t_2 = t_1 + 1$ nsec. Determine the sizes of the various transistors M_1 and M_2 so that a delay of less than 3 nsec is obtained. Delay is measured starting at t_1 and is defined as the time when BL reaches 2.1 V. $V_{DD} = 3.3$ V. Complete the waveform for $\overline{BL}$.

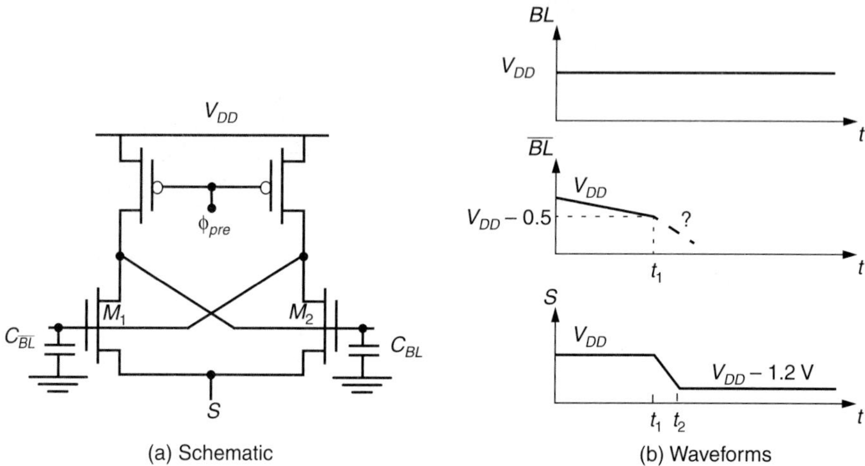

Figure 10.76 Dynamic sense amplifier.

16. [E&D, None, 10.5] Design a hamming-code scheme to correct a single bit in an 11-bit word. How many check-bits are needed? Design the logic needed to perform the error correction (at the gate level). It is sufficient to show the logic for a single bit.

17. [E, None, 10.6] Implement the logic functions described by Eq. (10.16) using a dynamic NAND-NAND PLA. Draw the transistor diagram.

18. [C, SPICE, 10.6] Compute with SPICE the worst-case propagation delay of the PLA of Figure 10.60 (or Colorplate 18). Manually extract the transistor diagram from the layout and estimate the value of the wire parasitics. You may ignore the input and output buffers. The supply voltage is set at 5 V. The input-bits may assumed to be step-functions with a rise/fall time of 0.5 nsec.

DESIGN PROBLEM

Design a 64 × 16 3T DRAM memory down to the layout level. Your goal will be to achieve a worst-case read access time of 20 nsec while minimizing power consumption (both standby and peak) and chip area. You are free to choose supply and voltage levels. As with any large design, your memory can be decomposed into several components: the input drivers, row decoder, the memory cell, the sense amplifiers, the column decoders (if needed), and the output buffers.

CHAPTER

11

DESIGN METHODOLOGIES

Simulation and verification

Implementation and synthesis approaches

Validation and test of manufactured circuits

11.1 Introduction

It was suggested in the introduction of this book that design automation plays an instrumental role in coping with the dramatic increase in complexity of contemporary integrated circuits. Designing a multimillion transistor circuit and ensuring that it operates correctly when the first silicon returns is a daunting task that is virtually impossible without the help of computer aids and well-established design methodologies. In general, the wide range of tools available to a designer can be subdivided into a number of global classes.

- *Analysis* and *verification* tools examine the behavior of a circuit and help determine if the response is within specification bounds.
- *Implementation* and *synthesis* methodologies aid the designer in generating and optimizing the circuit schematics or layout.
- *Testability* techniques provide a combination of design approaches and CAD tools to validate the functionality of a fabricated design.

This chapter presents a brief overview of the leading approaches in each of those categories. Due to the extensive nature of the field, we do not intend to be comprehensive—doing so would require a textbook of its own. Instead we present *a user perspective* that provides a basic perception and insight into what is offered and can be expected from the different design methodologies. This information should be sufficient for a novice designer to get started. Readers interested in an in-depth discussion of the mechanics of the various tools and environments are referred to the textbooks published on this topic, a number of which are listed in the references.

11.2 Design Analysis and Simulation

The primary expectation of a designer with respect to design automation is the availability of accurate and fast analysis tools.The first computer-aided design (CAD) tool to gain wide acceptance was the SPICE circuit simulator, which is undoubtedly the most utilized computer aid for the design of digital circuits at present [Nagel75]. Unfortunately, circuit simulation takes into account all the peculiarities and second-order effects of the semiconductor devices and tends to be time-consuming. It is rapidly becoming unwieldy when designing complex circuits, unless one is willing to spend days of computer time.The designer can address the complexity issue by giving up modeling accuracy and resorting to higher representation levels. A discussion of the different abstraction levels available to the designer and their impact on simulation accuracy is the topic of this section.

At this point, its is useful to differentiate between simulation and verification. In the simulation approach, the value of a design parameter such as noise margin, propagation delay, or dissipated energy is determined by applying a set of excitation vectors to the circuit model of choice and extracting parameters from the obtained signal waveforms. While this approach is very flexible, it has the disadvantage that the results depend strongly upon the choice of the excitations. For instance, the delay of a ripple-carry adder varies widely depending upon the input signal. Identification of the worst-case delay requires a careful choice of the excitation vector, so that the complete ripple-carry path is exercised. Failing to do so produces meaningless results. *Verification*, on the other hand, attempts to extract the system parameters directly from the circuit description. For instance, the critical path of an adder can be recognized from an inspection of the circuit diagram or a model of it. This approach has the advantage that the result is independent of the choice of excitation vectors and is supposedly fool-proof. On the negative side, it requires an inherent understanding of the circuit design style used. For the example of the ripple-carry adder, determination of the propagation delay requires an understanding of the logic operation of the composing circuitry—for example, dynamic or static logic—and a definition of the term *propagation delay*. As a second example, it is necessary to identify the register elements first before the maximum clock speed of a synchronous circuit can be determined. Verification is the topic of the Section 11.3

The best way of differentiating among the myriad of simulation approaches and abstraction levels is to identify how the data and time variables are represented—as analog, continuous variables, as discrete signals, or as abstract data models.

11.2.1 Representing Digital Data as a Continuous Entity

Circuit Simulation

In the course of this textbook, we have used circuit simulation extensively to illustrate the basic concepts of digital circuits and to validate our manual models. The assignments at the end of each chapter also rely heavily on circuit simulation. Hence, you should be very familiar with the capabilities and peculiarities of simulation and analysis at this level of detail. Some important properties of circuit simulation are worth summarizing.

- When analyzing a digital network using a circuit simulator, the resulting voltage and current *signals* are represented as *continuous waveforms*.
- In a transient analysis, *time* seems to be a *continuous* variable, and for all practical purposes can be considered as such. In reality, the simulator executed on a digital computer evaluates only a limited number of time points and obtains the intermediate data points by interpolation.

Executing a transient simulation means solving a set of differential equations at each time point. To make matters worse, the accurate modeling of semiconductor devices requires the introduction of nonlinearities, such as the current equations for bipolar and MOS devices or the diode-capacitance model. For instance, Figure 11.1 shows the network that must be solved when performing a transient analysis on a CMOS inverter. All capacitors and current sources in this model display a strong nonlinear behavior. The analytical solution of the set of nonlinear differential equations describing such a network is generally based on iteration that is computationally expensive. At each time step, an initial guess is made of the node values based on the values of the previous time step. This estimate is iteratively refined until some predefined error criterion is met. The tighter the error-bound, the better the accuracy, but the more iterations needed.

The accuracy of circuit simulation is a direct function of the quality and complexity of the device models employed. A number of those models for bipolar, MOS, and GaAs devices were introduced earlier in the text. While most circuit simulators use similar solu-

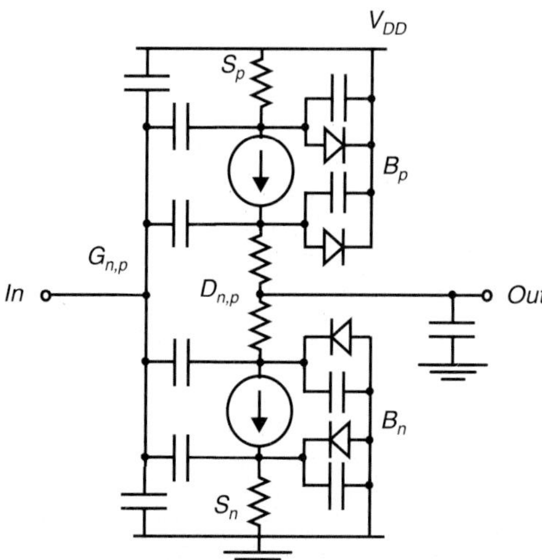

Figure 11.1 Circuit simulation model of CMOS inverter.

tion techniques, the added value is really situated in the robustness, efficiency, and accuracy of these models. For instance, it is possible to come up with very elaborate and accurate models that are completely useless because of poor convergence behavior.

The computational complexity makes circuit simulation impractical for larger circuits. It is, therefore, mostly used for the analysis of the critical parts of a design. Higher-level simulators are employed for the overall analysis.

Timing Simulation

Substantial effort has been invested over time to decrease the computation time at the expense of generality. Consider an MOS digital circuit. Due to the excellent isolation property of the MOS gate, it is often possible to partition the circuit into a number of sections that have limited interaction. A possible approach is to solve each of these partitions individually over a given time period, assuming that the inputs from other sections are known or constant. The resulting waveforms can then be iteratively refined. This *relaxation-based* approach has the advantage of being computationally more effective than the traditional technique, since expensive matrix inversions are avoided, but is restricted to MOS circuits [White87]. When the circuit contains feedback paths, the partitions can become large, and simulation performance degrades. The accuracy of these simulators is generally less than that of the SPICE-class tools, but the execution speed can be one or two orders of magnitude higher.

CAUTION: Novice designers tend to have an almost religious faith in simulation results. It is not rare to see designers optimizing propagation delays of CMOS circuits into the pico-second range. One should be aware that the simulation can diverge from reality due to a number of error sources such as inaccuracies in the device models, deviations in the device parameters, and parasitic resistances and capacitances. The actual and predicted circuit behaviors might further diverge because of process variations over the die or temperature variations. Designers should, therefore, allow a margin of 10% to 20% between the design constraints and simulation results.

11.2.2 Representing Data as a Discrete Entity

When addressing the complexity issue, the only solution is to compromise accuracy. In digital circuits, we are in general not interested in the actual value of the voltage variable, but only the digital value it represents. Therefore, it is possible to envision a simulator where data signals are either in the 0 or 1 range. Signals that do no comply with either of the above are denoted as X, or undefined.

This tertiary representation $\{0,1,X\}$ is used extensively in simulators at both the device and gate level. By augmenting this set of allowable data values, we can obtain more detailed information while retaining the capability of handling complex designs. Possible extensions are the Z value for a tri-state node in the high-impedance state, and R and F for the rising and falling transients. Some commercially offered simulators go as far as offering more than a dozen possible signal states.

While substantial performance improvement is obtained by discretizing the data representation space, similar benefits can be obtained by considering time as a discrete variable. Consider the voltage waveform of Figure 11.2 that represents the signal at the input

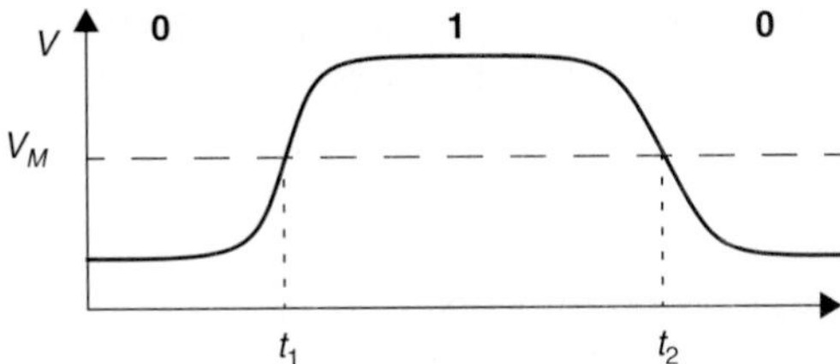

Figure 11.2 Discretizing the time variable.

of an inverter with a switching threshold V_M. It is reasonable to assume that the inverter output changes its value one propagation delay after its input crossed V_M. When one is not strictly interested in the exact shape of the signal waveforms, it is sufficient to evaluate the circuit only at the interesting time points, t_1 and t_2. Similarly, the interesting points of the output waveform are situated at $t_1 + t_{pHL}$ and t_2+t_{pLH}. A simulator that only evaluates a gate at the time an event happens at one of its inputs is called an *event-driven* simulator. The evaluation order is determined by putting projected events on a time queue and processing them in a time-ordered fashion. Suppose that the waveform of Figure 11.2 acts an input waveform to a gate. An event is scheduled to occur at time t_1. Upon processing that event, a new event is scheduled for the fan-out nodes at $t_1 + t_{pHL}$ and is put on the time queue. This event-driven approach is evidently more efficient than the time-step-driven approach of the circuit simulators. To take the impact of fan-out into account, the propagation delay of a circuit can be expressed in terms of an intrinsic delay (t_{in}) and a load-dependent factor (t_l), and can differ over edge transitions.

$$t_{pLH} = t_{inLH} + t_{lLH} \times C_L \tag{11.1}$$

The load C_L can be entered in absolute terms (in pF) or as a function of the number of fan-out gates.

While offering a substantial performance benefit, the above approach still has the disadvantage that events can happen at any point in time. Another simplification could be to discretize the time even further and allow events to happen only at integer multiples of a *unit time* variable. An example of such an approach is the *unit-delay* model, where each circuit is attributed a single delay of one unit. Finally, the simplest model is the *zero-delay model,* in which circuits are assumed to be delay-free. Under this paradigm, time proceeds from one clock event to the next, and all events are assumed to occur instantaneously upon arrival of a clocking event. These concepts can be applied on a number of abstraction levels, resulting in the simulation approaches discussed below.

Switch-Level Simulation

The nonlinear nature of semiconductor devices is one the major impediments to higher simulation speeds. The switch-level model [Bryant81] overcomes this hurdle by approximating the transistor behavior with a linear resistance whose value is a function of the operating conditions. In the off-mode the resistance is set to infinity, while in the on-mode it is set to the average on-resistance of the device (Figure 11.3). The resulting network is a time-variant, linear network of resistors and capacitors that can be more efficiently analyzed. Evaluation of the resistor network determines the steady-state values of the signals and typically employs a {0, 1, X} model. For instance, if the total resistance between a node and *GND* is substantially smaller than the resistance to V_{DD}, the node is set to the 0-state. The timing of the events can be resolved by analyzing the *RC* network. Simpler timing models such as the unit-delay model are also employed.

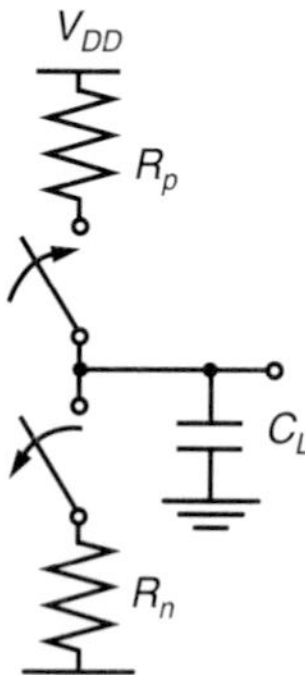

Figure 11.3 Switch-level model of CMOS inverter.

Example 11.1 Switch versus Circuit-Level Simulation

A four-bit adder is simulated using the switch-level simulator IRSIM ([Salz89]). The simulation results are plotted in Figure 11.4. Initially all inputs (*IN*1 and *IN*2) and the carry-input *CIN* are set to 0. After 10 nsec, all inputs *IN*2 as well as *CIN* are set to 1. The display window plots the input signals, the output vector *OUT*[0-3], and the most significant output bits *OUT*[2] and *OUT*[3]. The output converges to the correct value 0000 after a transition period. Notice how the data assumes only 0 and 1 levels. The glitches in the output signals go rail-to-rail, although in reality they might represent only partial excursions. During transients, the signal is marked *X*, which means "undefined." To put this result in perspective, Figure 11.4 plots the SPICE results for the same input vectors. Worth observing are the partial glitches. Also, it shows that the IRSIM timing, which is based on an *RC* model, is relatively accurate and sufficient to get a first-order impression.

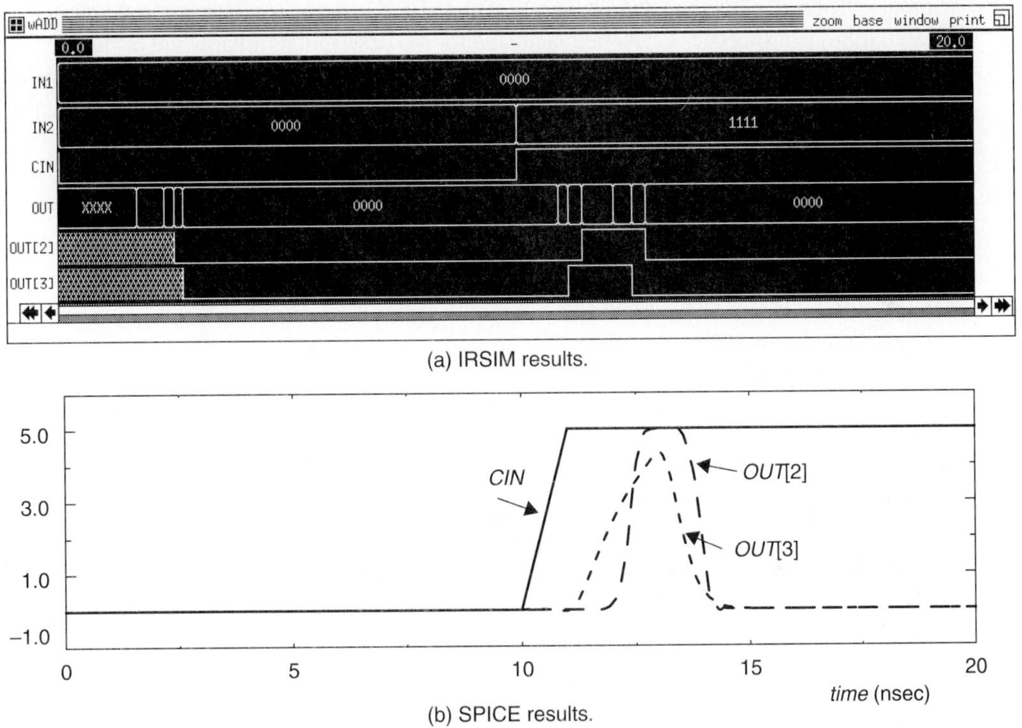

Figure 11.4 Comparison between circuit and switch-level simulations.

Gate-Level (or Logic) Simulation

Gate-level simulators use the same signal values as the switch-level tools, but the simulation primitives are gates rather than transistors. This approach enables the simulation of more complex circuits at the expense of detail and generality. For instance, some common VLSI structures such as tri-state busses and pass-transistors are hard to deal with at this level. Since gate level is the preferred design-entry level for many designers, this simulation approach remained extremely popular until the introduction of logic synthesis tools, which moved the focus to the functional or behavioral abstraction layer. The interest in logic simulation was so great that special and expensive hardware accelerators were developed to expedite the simulation process (e.g., [Agrawal90]).

Functional Simulation

Functional simulation can be considered as a simple extension of logic simulation. The primitive elements of the input description can be of an arbitrary complexity. For instance, a simulation element can be a NAND gate, a multiplier, or an SRAM memory. The functionality of one of these complex units can be described using a modern programming language or a dedicated hardware description language. For instance, the THOR simulator uses the C programming language to determine the output values of a module as a function of its inputs [Thor88]. On the other hand, *VHDL (VHSIC Hardware Description Language)* [VHDL88] is a specially developed language for the description of hardware designs.

In the *structural mode*, VHDL describes a design as a connection of functional modules. Such a description is often called a *netlist*. For instance, Figure 11.5 shows a description of a 16-bit accumulator consisting of a register and adder.

```
entity accumulator is
   port ( -- definition of input and output terminals
      DI: in bit_vector(15 downto 0) -- a vector of 16 bit wide
      DO: inout bit_vector(15 downto 0);
      CLK: in bit
   );
end accumulator;

architecture structure of accumulator is
   component reg -- definition of register ports
      port (
         DI : in bit_vector(15 downto 0);
         DO : out bit_vector(15 downto 0);
         CLK : in bit
      );
   end component;
   component add -- definition of adder ports
      port (
         IN0 : in bit_vector(15 downto 0);
         IN1 : in bit_vector(15 downto 0);
         OUT0 : out bit_vector(15 downto 0)
      );
   end component;
-- definition of accumulator structure
signal X : bit_vector(15 downto 0);
begin
   add1 : add
      port map (DI, DO, X); -- defines port connectivity
   reg1 : reg
      port map (X, DO, CLK);
end structure;
```

Figure 11.5 Functional description of an accumulator in VHDL.

The adder and register can in turn be described as a composition of components such as full-adder or register cells. An alternative approach is to use the *behavioral mode* of the language that describes the functionality of the module as a set of input-output relations regardless of the chosen implementation. As an example, Figure 11.6 describes how the output of the adder is the two's complement sum of its inputs.

```
entity add is
   port (
      IN0 : in bit_vector(15 downto 0);
      IN1 : in bit_vector(15 downto 0);
      OUT0 : out bit_vector(15 downto 0)
   );
end add;

architecture behavior of add is
begin
   process(IN0, IN1)
      variable C : bit_vector(16 downto 0);
      variable S : bit_vector(15 downto 0);
   begin
      loop1:
      for i in 0 to 15 loop
         S(i) := IN0(i) xor IN1(i) xor C(i);
         C(i+1):= IN0(i) and IN1(i) or C(i) and (IN0(i) or IN1(i));
      end loop loop1;
      OUT0 <= S;
   end process;
end behavior;
```

Figure 11.6 Behavioral description of 16-bit adder.

The signal levels of the functional simulator are similar to the switch and logic levels. A variety of timing models can be used. For example, the designer can describe the delay between input and output signals as part of the behavioral description of a module. Most often the zero-delay model is employed, since it yields the highest simulation speed.

11.2.3 Using Higher-Level Data Models

When conceiving a digital system such as a compact-disk player or an embedded microcontroller, the designer rarely thinks in terms of bits. Instead, she envisions data moving over busses as integer or floating-point words, and patterns transmitted over the instruction bus as members of an enumerated set of instruction words such as $\{ACC, RD, WR$ or $CLR\}$. Modeling a discrete design at this level of abstraction has the distinct advantage of being more understandable, and also results in a substantial benefit in simulation speed. Since a 64-bit bus is now handled as a single object, analyzing its value requires only one action instead of the 64 evaluations it formerly took to determine the current state of the bus at the logic level. The disadvantage of this approach is another sacrifice of timing accuracy. Since a bus is now considered to be a single entity, only one global delay can be annotated to it, while the delay of bus-elements can vary from bit to bit at the logic level.

It is common again to distinguish between *functional (or structural)* and *behavioral* descriptions. In a functional-level specification, the description mirrors the intended hardware structure. Behavioral-level specifications only mimic the input-output functionality of a design. Hardware delay loses its meaning, and simulations are normally performed on a per clock-cycle or higher base. For instance, the behavioral models of a microprocessor that are used to verify the completeness and the correctness of the instruction set are performed on a per-instruction base.

The most popular languages at this level of abstraction are the VHDL and VERILOG hardware-description languages. VHDL allows for the introduction of user-defined data types such as 16-bit two's-complement words or enumerated instruction sets. Many designers tend to use traditional programming approaches such as C or C++ for their first-order behavioral models. This approach has the advantage of offering more flexibility but requires the user to define all data types and to essentially write the complete simulation scenario.

Example 11.2 Behavioral-Level VHDL Description

To contrast the functional and behavioral description modes and the use of higher-level data models, consider again the example of the accumulator. This time, we use a fully behavioral description that employs integer data types to describe the module operation (Figure 11.7).

```
entity accumulator is
   port (
      DI : in integer;
      DO : inout integer := 0;
      CLK : in bit
   );
end accumulator;

architecture behavior of accumulator is
begin
   process(CLK)
   variable X : integer := 0; -- intermediate variable
   begin
      if CLK = '1' then
         X <= DO + D1;
         DO <= X;
      end if;
   end process;
end behavior;
```

Figure 11.7 Accumulator for Example 11.2.

Figure 11.8 shows the results of a simulation performed at this level of abstraction. Even for this small example, the simulation performance in terms of CPU time is three times better than what is obtained with the structural description of Figure 11.5.

11.3 Design Verification

Simulation results do not guarantee the correctness and functionality of a hardware design. They only tell how the circuit reacts under a given set of *input excitations*. A charge-redistribution condition in a dynamic logic gate is not detected by a simulation if the exact

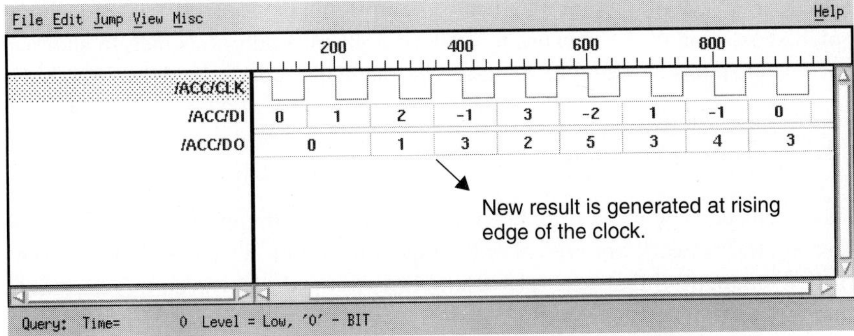

Figure 11.8 Display of simulation results for accumulator example as obtained at the behavioral level. The WAVES display tool (and VHDL simulator) are part of the Synopsis VHDL tool suite (*Courtesy of Synopsys*).

sequence of input patterns that causes the charge sharing is not applied. To avoid this dependence upon excitation vectors, tools have been developed that analyze the design and detect configurations that may lead to circuit failure. This *verification* approach relies on an number of implicit assumptions regarding design techniques and methodologies. The resulting tools are generally restricted in scope and handle only a limited class of circuit styles, such as two-phase synchronous design.

11.3.1 Electrical Verification

Given the transistor schematics of a digital design, it is possible to verify if a number of basic rules are satisfied. Some examples of typical rules help illustrate this concept. Many others can be derived from previous chapters.

- The number of inversions between two C^2MOS gates should be even.
- In a pseudo-NMOS gate, a well-defined ratio between the PMOS pull-up and NMOS pull-down devices is necessary to guarantee a sufficient noise-margin low NM_L.
- To ensure that rise and fall times of the signal waveforms stay within limits, minimum bounds can be set on the sizes of the driver transistors as a function of the fan-out.
- The maximum amount of charge sharing in a dynamic design should be such that the noise-margin high is not violated.

Pure common sense can help define a large set of rules to which a design should always adhere. Applying them requires an in-depth understanding of the circuit structure. An electrical verifier, therefore, starts with the identification of well-known substructures in the overall circuit schematic. Typical templates are simple logic gates, pass-transistors, and registers. The verifier traverses the resulting network on a rule-per-rule base. Since electrical rules tend to be specific for a particular design style, they should be modifiable. For example, rule-based expert systems allow for an easy updating of the rule base

[DeMan85]. The individual rules can be complex and even invoke a circuit simulator for a small subsection of the network to verify if a given condition is met. In summary, electrical verification is a helpful tool and can dramatically reduce the risk of malfunction.

11.3.2 Timing Verification

As circuits become more complex, it is increasingly difficult to define exactly which paths through the network are critical with respect to timing. One solution is to run extensive SPICE simulations that may take a long time to finish. However, this does not guarantee that the identified critical path is the worst case, since the delay path is a function of the applied signal patterns. A timing verifier traverses the electrical network and rank-orders the various paths based on delay. This delay can be determined in a number of ways. One approach is to build an *RC* model of the network and compute bounds on the delay of the resulting passive network using the Penfield-Rubinstein-Horowitz algorithm described in Chapter 8. To obtain more accurate results, many timing verifiers first extract the details of the longest path(s) based on the *RC* model, and perform circuit simulation on the reduced circuit to obtain a better estimation. Examples of well-known timing verifiers are the Crystal [Ousterhout83] and TV [Jouppi84] systems.

One problem that hampered many earlier systems is that they identify *false paths*, that is, critical paths that can never be exercised during normal circuit operation. For example, a false path exists in the carry-bypass adder of Figure 7.15 (also shown in Figure 11.9). From a simple analysis of only the circuit topology, one would surmise that the critical path of the circuit passes through the adder and multiplexer modules as illustrated by the arrow. A closer look at the circuit operation reveals that such a path is not feasible. All individual adder bits must be in the propagate mode for *In* to propagate through the complete adder. However, the *bypass* signal is asserted under these conditions, and the bottom path through the multiplexer is selected instead. The actual critical path is thus shorter than what would be predicted from the first-order analysis. Detecting false timing paths is not easy since it requires an understanding of the logic functionality of the network. Newer timing verifiers are remarkably successful in accomplishing this and have become one of the more important design aids for the high-performance circuit designer [Devadas91].

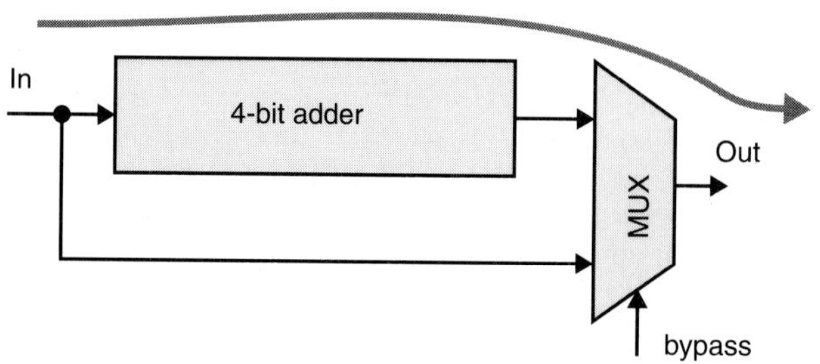

Figure 11.9 Example of false path in timing verification.

Example 11.3 Example of Timing Verification

The output of a static timing verifier, the PathMill tool from EPIC Design, is shown in Figure 11.10. The input to the verification process is a transistor netlist, but gate- and block-level

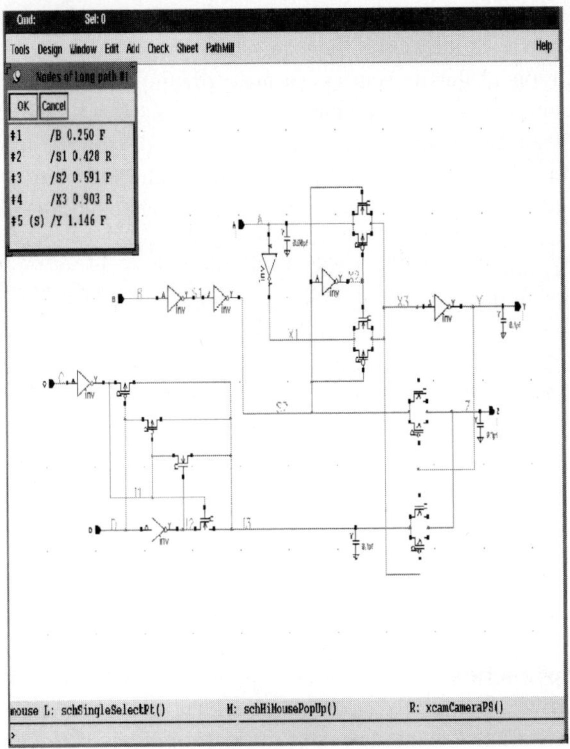

Figure 11.10 Example of static timing verifier response as generated by the PathMill tool (*Courtesy of EPIC Design Technology, Inc.*). The results are displayed using the *Cadence DFII* tool. The critical timing path runs from node *B* to node *Y* via nodes *S*1, *S*2, and *X*3.

models can be included as well. The analysis considers capacitive and resistive parasitics that are obtained from the transistor schematic or layout extraction.

The output of the timing analysis is an ordered list of critical timing paths. For the example of Figure 11.10, the longest path extends from node B (falling edge) to node Y (falling edge) over nodes S1, S2, and X3. The predicted delay is 1.14 nsec. Other paths, close to the critical one are in order of decreasing delay: $B\ (F) \rightarrow Y\ (R)$ (1.11 nsec), $B\ (R) \rightarrow Y\ (F)$ (1.04 nsec), $C\ (R) \rightarrow Z\ (R)$ (1.04 nsec), and $D\ (R) \rightarrow Y\ (R)$ (1.00 nsec). Observe that the circuit contains multiple chained pass-transistors that make the analysis complicated and increase the chances for false paths to occur. An example of the latter is the path $A(F) \rightarrow X1 \rightarrow X3 \rightarrow Z(R)$ for $B = 0$. For a low value on B, node Z can only make a falling transition.

11.3.3 Functional (or Formal) Verification

Each component (transistor, gate, or functional block) of a circuit can be described behaviorally as a function of its inputs and internal state. By combining these component descriptions, an overall circuit model can be generated that symbolically describes the behavior of the complete circuit. Formal verification compares this derived behavior with the designer's initial specification. Although not identical, the two descriptions need to be mathematically equivalent for the circuit to be correct.

Formal verification is the designer's ultimate dream of what design automation should be able to accomplish—*proof that the circuit will work as specified.* Unfortunately, no general and widely accepted verifier has been realized at the time of writing. The complexity of the problem is illustrated by the following argument. One way to prove that two circuit descriptions are identical is to compare the outputs while enumerating all possible input patterns and sequences thereof. This is an intractable approach with a computation time exponential in the number of inputs and states.

This does not mean that formal verification is a fantasy. Techniques have been proposed and successfully implemented for certain classes of circuits. For instance, assuming that a circuit is synchronous helps minimize the search space. Proving the equivalence between state machines has been one of the main research targets that has led to some remarkable advances (e.g., [Coudert90]). Besides fully automatic verification techniques, there exist approaches that involve the user in the proof process, for instance [Claesen91].

Although not yet in the mainstream of the design-automation process, functional verification might become one of the important assets in the designer's tool box. For this to happen, important progress must be made in the fields of design specification and interpretation.

11.4 Implementation Approaches

The viability of a microelectronics design depends upon a number of conflicting factors, such as performance in terms of speed or power consumption, cost, and production volume. For example, a microprocessor has to excel in performance at a low cost to the customer to be competitive in the market. Achieving both goals simultaneously is only possible through large sale volumes. The high development cost associated with high-performance design is then amortized over many parts. Applications such as radar or space systems present another scenario. Since performance is critical, high-performance custom design techniques are desirable. The production volume is small, but the cost of electronic parts is only a small fraction of the overall system and thus not much of an issue. Finally, a majority of the digital designs vie for integration density, not performance. Under these circumstances, the design cost can be reduced substantially by using advanced design-automation techniques that compromise performance but minimize design time. Note that the cost of a semiconductor device is the sum of two components:

- The *nonrecurring expense* (NRE), which is incurred only once for a design and includes the cost of designing the part
- The *production cost per part*, which is a function of the process complexity, design area, and process yield.

These economic considerations have spurred the development of a number of distinct implementation approaches that range from high-performance handcrafted design to fully programmable, medium-to-low performance designs. Figure 11.11 provides an overview of the different methodologies that are discussed in more detail in the remainder of this section.

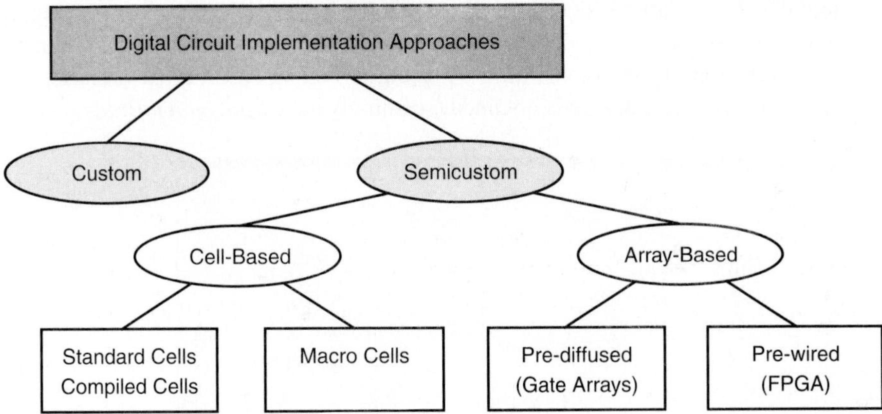

Figure 11.11 Overview of implementation approaches for digital integrated circuits (after [DeMicheli94]).

11.4.1 Custom Circuit Design

When performance or design density is of primary importance, the designer has no other choice than to return to handcrafting the circuit topology and physical design. This approach was the only option in the early days of digital microelectronics. This was adequately demonstrated in the design of the Intel 4004 microprocessor, shown on the back cover. The labor-intensive nature of custom design translates into a high cost and a long *time-to-market*. Therefore, it can only be justified economically under the following conditions:

- The cost can be amortized over a large volume. Microprocessors and semiconductor memories are examples of applications in this class.
- The custom block can be reused many times, for instance as a library cell
- Cost is not the prime design criterion, as it is in space applications

With continuous progress in the design-automation arena, the share of custom design reduces from year to year. Even in high-performance microprocessors, such as the DEC alpha processor shown in Figure 9.12, large portions are designed automatically using semicustom design approaches. Only the most performance-critical modules such as the integer and floating-point execution units are handcrafted.

Even though the amount of design automation in the custom design process is minimal, some design tools have proven to be indispensable. Together with circuit simulators, these programs form the core of every design-automation environment.

Layout Editor

The editor is the premier working tool of the designer and exists primarily for the generation of a physical representation of a design, given a circuit topology. Virtually every design-automation vendor offers an entry in this field. Most well-known is the MAGIC tool developed at the University of California at Berkeley [Ousterhout84], which has been widely distributed. A typical MAGIC display is shown in Figure 11.12 and illustrates the basic function of the layout editor—placing polygons on different mask layers so that a functional physical design is obtained (scathingly called *polygon pushing*).

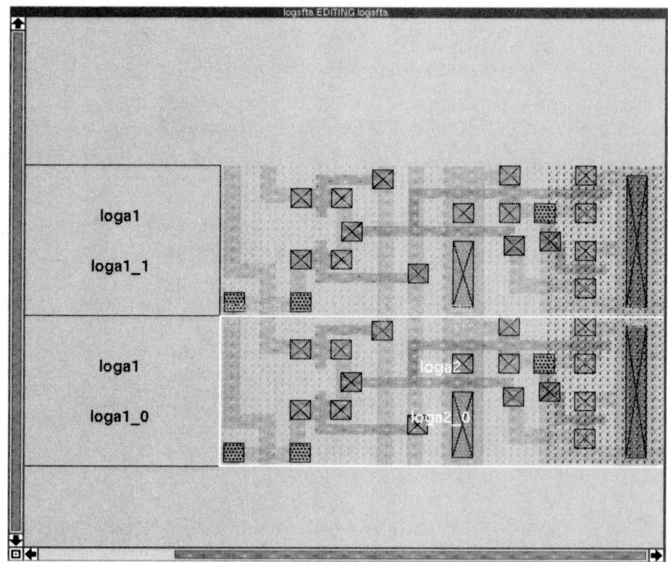

Figure 11.12 View of a MAGIC display window. It plots the layout of a logarithmic shifter with a few of the underlying cells fully expanded. Only outlines are shown of the remaining cells.

Since physical design occupies a major fraction of the design time for a new cell or component, techniques to expedite this process have been in continual demand. The *symbolic-layout* approach has gained popularity over the years. In this design methodology, the designer only draws a shorthand notation for the layout structure. This notation indicates only the *relative* positioning of the various design components (transistors, contacts, wires). The *absolute* coordinates of these elements are determined automatically by the editor using a *compactor* [Hsueh79, Weste93]. The compactor translates the design rules into a set of constraints on the component positions, and solves a constrained optimization problem that attempts to minimize the area or another cost function.

An example of a symbolic notation for a circuit topology, called a *sticks diagram*, is shown in Figure 11.13. The different layout entities are dimensionless, since only positioning is important. You may have noticed that we have made occasional use of such a notation in earlier chapters. The advantage of this approach is that the designer does not have to worry about design rules, because the compactor ensures that the final layout is physically correct. Thus, she can avoid cumbersome polygon manipulations. Another plus of the symbolic approach is that cells can adjust themselves automatically to the environment. For example, automatic pitch-matching of cells is an attractive feature in module generators. Consider the case of Figure 11.14 (from [Croes88]), in which the original cells have different heights, and the terminal positions do not match. Connecting the cells

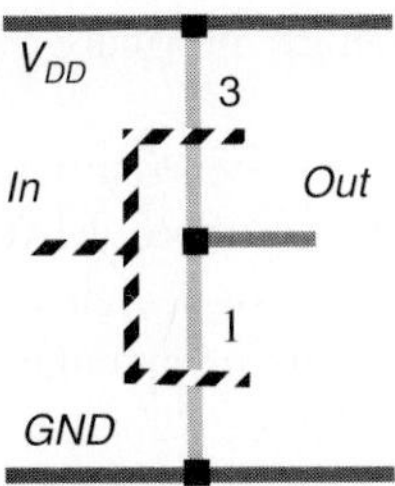

Figure 11.13 Sticks representation of CMOS inverter. The numbers represent the (W/L)-ratios of the transistors.

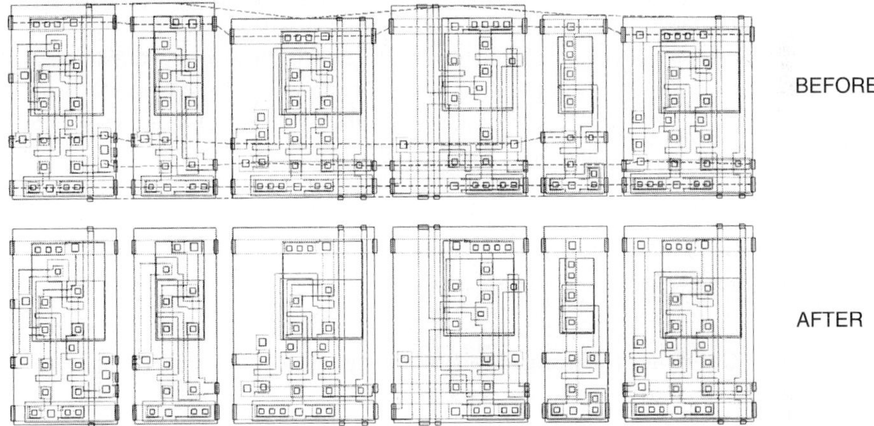

Figure 11.14 Automatic pitch matching of datapath cells based on symbolic layout.

would require extra wiring. The symbolic approach allows the cells to adjust themselves and connect without any overhead.

The disadvantage of the symbolic approach is that the outcome of the compaction phase is often unpredictable. The resulting layout can be less dense than what is obtained with the manual approach. Notwithstanding, symbolic layout tools have improved considerably over the years and are currently a part of the mainstream design process.

Design-Rule Checking

Design rules were introduced in Appendix A as a set of layout restrictions that ensure the manufactured design will operate as desired with no short or open circuits. A prime requirement of the physical layout of a design is that it adhere to these rules. This can be verified with the aid of a *design-rule checker (DRC)*, which uses as inputs the physical layout of a design and a description of the design rules presented in the form of a *technology file*. Since a complex circuit can contain millions of polygons that must be checked against each other, efficiency is the most important property of a good DRC tool. The verification of a large chip can take hours or days of computation time. One way of expediting the process is to preserve the design hierarchy at the physical level. For instance, if a cell is used multiple times in a design, it should be checked only once. Besides speeding up the process, the use of hierarchy can make error messages more informative by retaining knowledge of the circuit structure.

DRC tools come in two formats: (1) The *on-line DRC* runs concurrent with the layout editor and flags design violations during the cell layout. For instance, MAGIC has a built-in design-rule checking facility. (2) *Batch DRC* is used as a post-design verifier, and is run on a complete chip prior to shipping the mask descriptions to the manufacturer.

Circuit Extraction

Another important tool in the custom-design methodology is the circuit extractor, which derives a circuit schematic from a physical layout. By scanning the various layers and their interactions, the extractor reconstructs the transistor network, including the sizes of the devices and the interconnections. The schematic produced can be used to verify that the artwork implements the intended function. Furthermore, the resulting circuit diagram contains precise information on the parasitics, such as the diffusion and wiring capacitances and resistances. This allows for a more accurate simulation and analysis. The complexity of the extraction depends greatly upon the desired information. Most extractors extract the transistor network and the capacitances of the interconnect with respect to *GND* or other network nodes. Extraction of the wiring resistances already comes at a greater cost. For very high speed circuits, extraction of the inductance would be useful as well. Unfortunately, this requires a three-dimensional analysis and is only feasable for small circuits at present.

11.4.2 Cell-Based Design Methodology

Since the custom-design approach proves to be prohibitively expensive, a wide variety of design approaches have been introduced over the years to shorten and automate the design process. This automation comes at the price of reduced integration density and/or performance. The following rule seems to hold—*the shorter the design time, the larger the penalty incurred.* In this section, we discuss a number of design approaches that still require a full run through the manufacturing process for every new design. The *array-based design* approach discussed in the next section cuts the design time and cost even further by requiring only a limited set of extra processing steps or by eliminating processing all together.

The idea behind cell-based design is to reduce the implementation effort by *reusing* a limited library of cells. The advantage is that the cells only need to be designed and verified once for a given technology, and can be reused many times, thus amortizing the design cost. The disadvantage is that the constrained nature of the library reduces the possibility of fine-tuning the design. Cell-based approaches can be partitioned into a number of classes depending upon the granularity of the library elements.

Standard Cell

The standard-cell approach standardizes the design-entry level at the logic gate. A library is provided that contains a wide selection of logic gates over a range of fan-in and fan-out counts. Besides the basic logic functions, such as inverter, AND/NAND, OR/NOR, EXOR/NXOR, and flip-flops, a typical library also contains more complex functions such as AND-OR-INVERT, MUX, full adder, comparator, counter, decoders, and encoders. A design is captured as a schematic containing only cells available in the library. The layout is then automatically generated. This high degree of automation is made possible by placing strong restrictions on the layout options. In the standard-cell philosophy, cells are placed in rows that are separated by routing channels, as illustrated in Figure 11.15. To be effective, this requires that all cells in the library have identical heights. The width of the cell can vary to accommodate for the variation in complexity between the cells. As illustrated in the drawing, the standard-cell technique can be intermixed with other layout approaches to allow for the introduction of modules such as memories and multipliers that do not adapt easily or efficiently to the logic-cell paradigm.

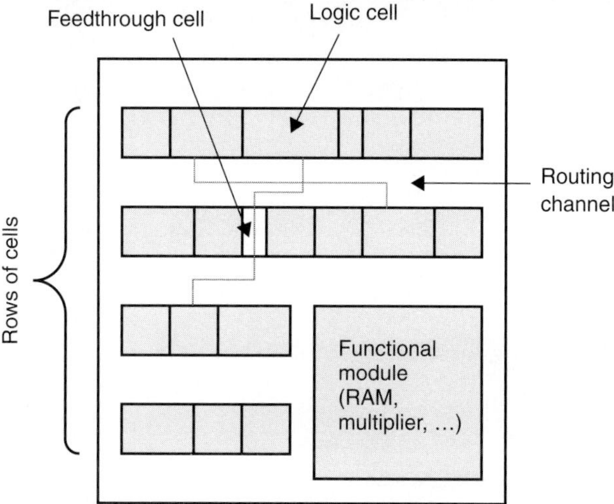

Figure 11.15 Standard-cell layout methodology.

An example of a design implemented in the standard-cell design style is shown in Figure 11.16. A substantial fraction of the area is devoted to signal routing. The minimization of the interconnect overhead is the most important goal of the standard-cell placement and routing tools. One approach to minimize the wire length is to introduce feedthrough cells (Figure 11.15) that make it possible to connect between cells in different rows without having to route around a complete row. The availability of multiple interconnect layers is also a positive development. For instance, in a three-metal layer process, it is possible to avoid feedthrough cells by routing the feedthrough wires on top of the logic cells.

The design of a standard-cell library is a time-intensive undertaking that, fortunately, can be amortized over a large number of designs. Determining the composition of the library is nontrivial. For instance, a pertinent question is if one is better off with a small library in which most cells have a limited fan-in, or if it is more beneficial to have a large library with many versions of every gate (e.g., containing 2, 3, …, 6–input NAND gates). Since the fan-out and load capacitance due to wiring are not known in advance, it is a common practice to ensure that each gate has large current-driving capabilities, that is, employs large output transistors. While this simplifies the design procedure, it has a detrimental impact on area and power consumption. An alternative approach is to provide dif-

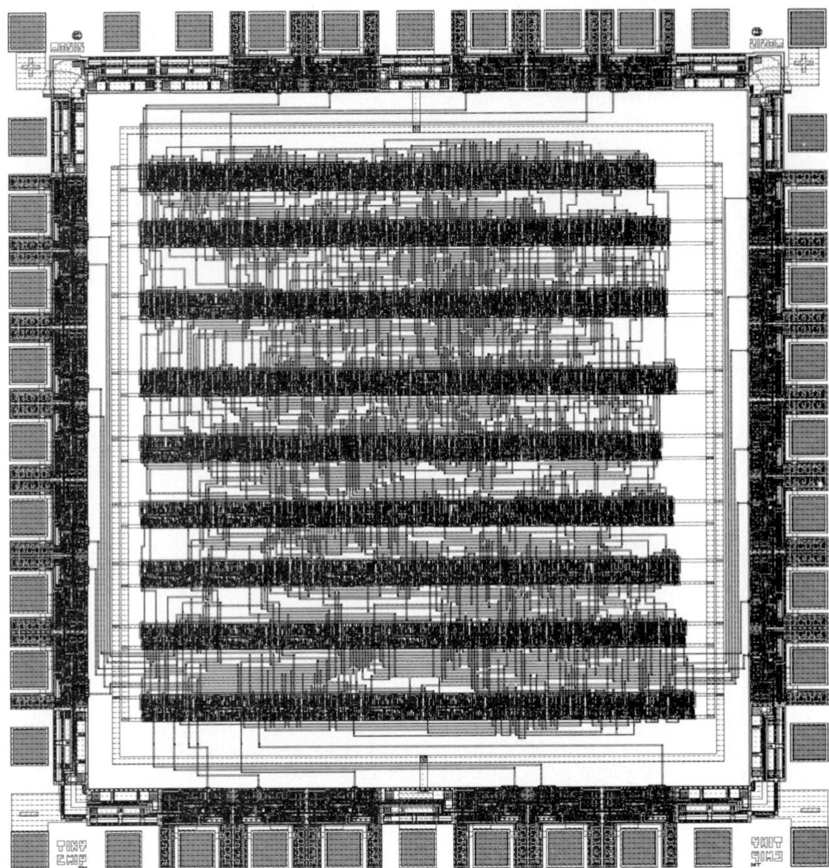

Figure 11.16 Complete standard-cell design (from [Brodersen92]).

ferent versions of each cell with varying sizes of the output devices, ranging from small to medium, and large driving capacity.

To make the library-based approach work, a detailed documentation of the cell library is an absolute necessity. The information should not only contain the layout, a description of functionality and terminal positioning; it must also characterize the delay and power consumption of the cell as a function of load capacitance and the input rise and fall times. Generating this information accounts for a large fraction of the library generation effort.

Example 11.4 A 3-Input NAND Gate

To illustrate some of the above observations, the design and characterization of a 3-input NAND standard cell gate are depicted in Figure 11.17.[1] The NMOS and PMOS transistors in the pull-down (up) networks are sized at 12/2 and 17/2 respectively. Observe how the layout

[1] The Mississippi State University library is accessible from the World-Wide Web at the following address: *http://www.erc.msstate.edu/mpl/libraries/stdcells*.

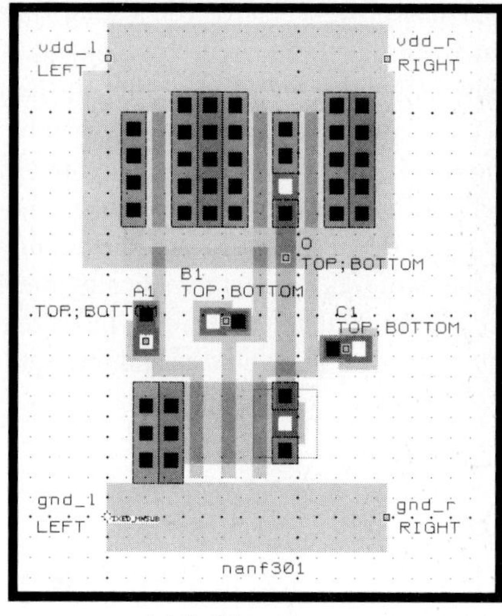

(a) Cell layout

Fanout 4x	0.5 μm	1.0 μm	2.0 μm
A1_tphl	0.595	0.711	0.919
A1_tplh	0.692	0.933	1.360
B1_tphl	0.591	0.739	1.006
B1_tplh	0.620	0.825	1.1.81
C1_tphl	0.574	0.740	1.029
C1_tplh	0.554	0.728	1.026

(b) Cell characterization (delay in nsec) for a fan-out of four and for three different technologies

Figure 11.17 Documentation of 3-input NAND standard cell as taken from the Mississippi State Library (from [Brodersen92]).

strategy follows the approach outlined in Figure C.2 (Appendix C). Supply lines are distributed horizontally and shared between cells in the same row. Input and output signals are brought in vertically using polysilicon or metal2.

The standard-cell approach has become immensely popular for the implementation of random-logic functions and sequential state machines, largely replacing the programmable logic array (PLA) as the premier approach in this area. The main reason for this shift was the advent of more sophisticated *logic-synthesis* tools. The logic-synthesis approach allows for the design to be entered at a high level of abstraction using Boolean equations, state machines or even register transfer languages such as VHDL. The synthesis tools automatically translate this specification into a gate netlist, minimizing a specific cost-function such as area or delay. Early efforts in this domain focused on two-level logic minimization that mapped naturally into PLA implementations. While PLAs have the advantage of being regular structures, they suffer from low silicon utilization and mediocre performance. More recent advances in the area of *multilevel logic synthesis* and place-and-route have tilted the balance towards the standard-cell approach. The majority of the application-specific circuits and even large parts of commodity components such as microprocessors are currently generated using the standard-cell methodology.

Compiled Cells

The cost of implementing and characterizing a library of cells should not be underestimated. In addition, even an extensive library has the disadvantage of being discrete, which means that the number of design options is limited. When targeting performance, customized cells with optimized transistor sizes are attractive. A number of automated approaches have been devised that generate cell layouts on the fly, given the transistor netlists. High-quality automatic cell layout has proven elusive for a long time. Earlier approaches relied on fixed topologies such as the Weinberger topology described in Appendix C (Figure C.1)). Later approaches allow for more flexibility in the transistor placement (e.g [Hill85]). Layout densities close to what can be accomplished by a human designer are now within reach. As an example, Figure 11.18 shows the layout of a random logic function generated by the CLEO cell compiler (from Digital Equipment Corporation). The input of the compiler consists of a gate-level schematic and the required transistor sizes. Notice how the wide PMOS transistors of (80/2) are automatically translated into a pair of parallel devices. This helps reduce the cell height and increase the layout density. Cell compilation is often combined with automatic netlist synthesis and transistor sizing.

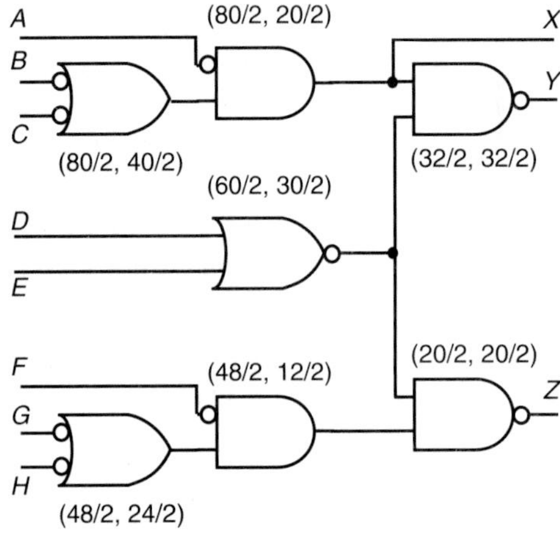

(a) Circuit schematics with annotated transistor sizes (PMOS, NMOS)

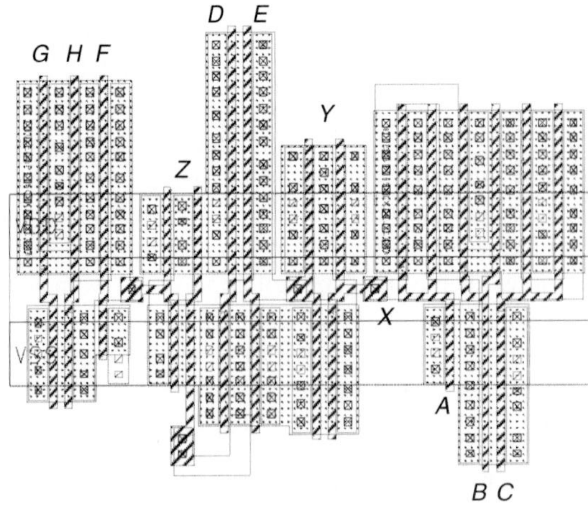

(b) Generated layout

Figure 11.18 Random-logic layout generated by the CLEO cell compiler. *Courtesy of Digital Equipment Corporation.*

Module Generators

Standardizing at the logic gate level is attractive for random logic functions, but turns out to be inefficient for regular structures such as shifters, adders, multipliers, data paths, PLAs or memories. Earlier chapters have demonstrated that a primary design goal is the minimization of parasitic capacitances on the internal nodes of these computational and storage modules. This is hard to achieve in the standard-cell approach, which relies on the use of routing channels to realize the interconnect network. Additionally, the standard-cell technique ignores the inherent regularity that is present in all of the mentioned modules. It is, therefore, advantageous to employ a different layout-generation technique called *structured custom design*. Based on the nature of the regularity in the structure, we discern two types of generators: macrocell generators and datapath compilers.

1. Macrocell Generators

Structures such as PLAs, memories, and multipliers are easily constructed by abutting predesigned leaf-cells in a two-dimensional array topology. All interconnections are made by abutment, and no extra routing is needed if the cells are designed correctly, which minimizes the parasitic capacitance. The carry-save multiplier of Figure 7.32 is an example of such a configuration. The whole array can be constructed with a minimal number of cells, in particular the carry-save and vector-merging adder cells.

The generator itself can be considered as a simple software program that determines the relative positioning of the various leaf-cells in the array. These generators are typically *parameterizable*, which means a single generator can be used to generate various instances of the module such as a 4×4 and a 32×16 multiplier.

2. Datapath Compilers

Datapaths were discussed extensively in Chapter 7. The bit-sliced approach provides regularity in one dimension—an N-bit data path is constructed by repeating the same slice N times. Interconnection between the slices is provided by abutment (Figure 11.19). Using this approach, dense layout structures can be obtained, as was shown in Figure E.1 (Appendix E).

The task of the datapath compiler is to generate the most efficient datapath layout, given the composing modules (adders, multiplexers, registers, etc.) and their interconnections. The generation process includes deriving a linear placement of the modules so that

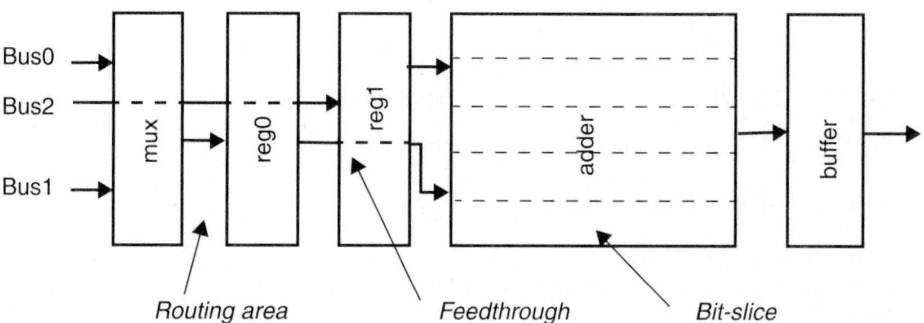

Figure 11.19 Bit-sliced datapath layout methodology.

the number of long interconnections is minimized. Feedthroughs are used as much as possible. Whenever the number of available feedthrough wires is exhausted, it is necessary to stretch the cells to accommodate the extra interconnections. The importance of this technique is illustrated in Figure 11.20, which demonstrates the impact of the availability of feedthroughs on the layout area. The same figure shows that it is advisable to equalize the height of the cells in every bit-slice. Failing to do so results in a *staircase* effect in the routing areas.

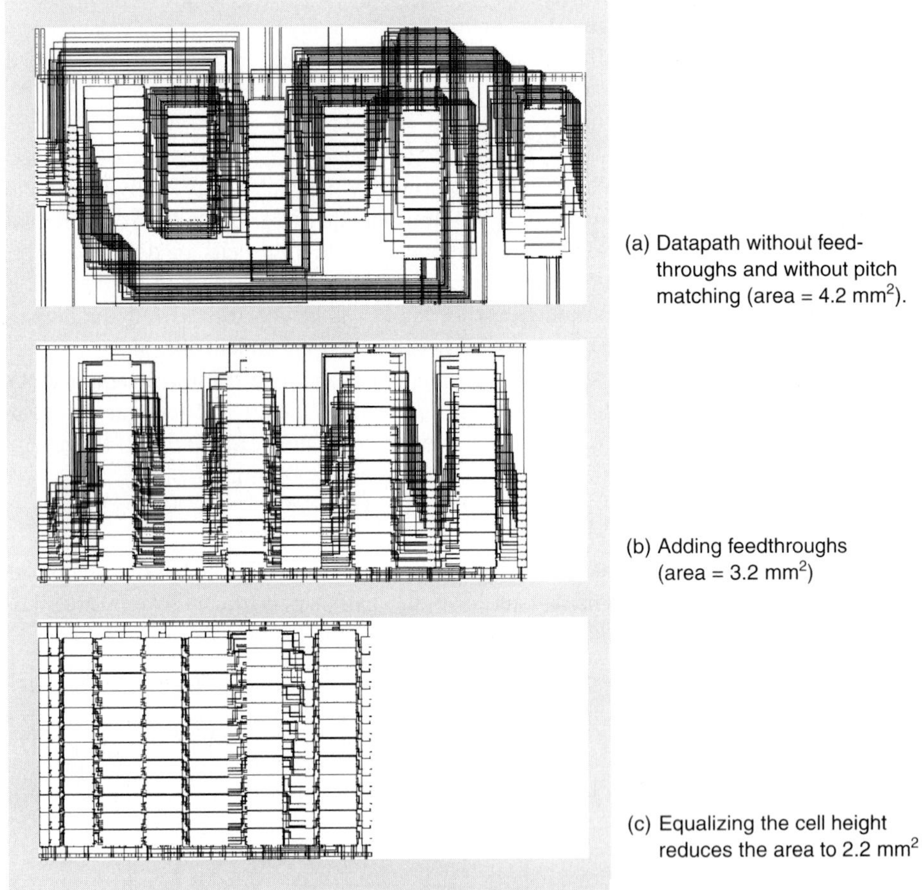

(a) Datapath without feedthroughs and without pitch matching (area = 4.2 mm^2).
(b) Adding feedthroughs (area = 3.2 mm^2)
(c) Equalizing the cell height reduces the area to 2.2 mm^2

Figure 11.20 Impact of wire feedthroughs and pitch matching on datapath layout (from [Brodersen92]).

Macrocell Place and Route

The generators and compilers implement only parts of a design. To assemble the complete artwork and connect the core of the chip to the input-output pads, one last step called *place and route* is required. The challenge is to combine modules of different shapes and aspect ratios and to realize the connections between them so that the silicon area is well utilized and the overall chip area is kept within bounds. The common approach is to accommodate the interconnections in *routing channels* that separate the (generally) rectangular macro-modules. The latter can be generated using standard-cell, module-generator, or even full-custom approaches. An example of such a topology is shown in Figure 11.21, where the gray boxes stand for the macrocells separated by routing channels. The lines connecting the centers of the modules symbolically represent the interconnections to be made and the path that will be followed. The thickness of the line represents the number of wires in that bundle of nets.

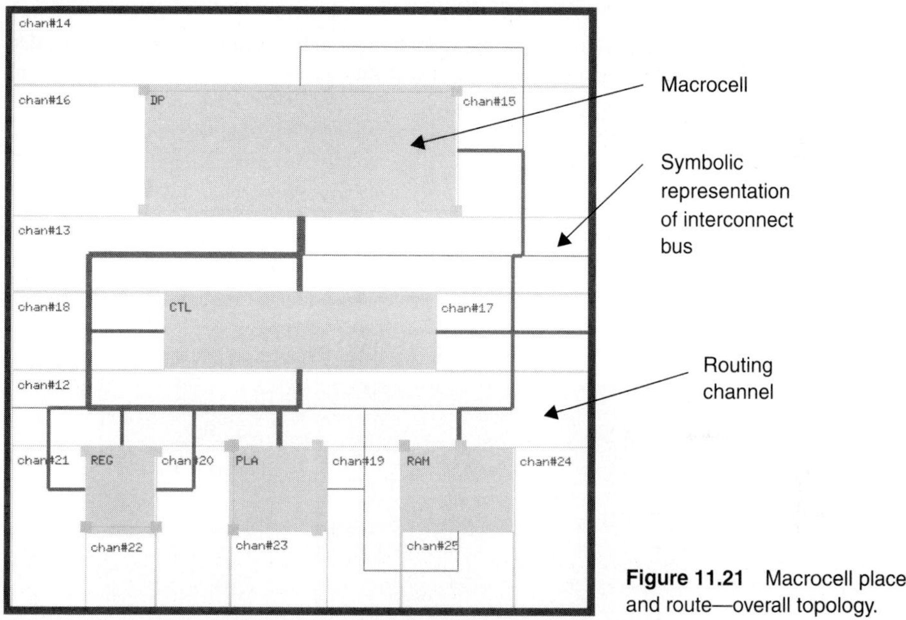

Figure 11.21 Macrocell place and route—overall topology.

To make the routing process tractable, most place-and-route tools currently use an approach called *channel routing*. The concept is illustrated in Figure 11.22 for a manufacturing process with two interconnect layers. Modules are placed on the longitudinal sides of the channel. Wires are implemented as a sequence of orthogonal segments using one interconnect layer for the horizontal sections (e.g., metal2) and the other (e.g., metal1) for the vertical ones. This approach has the advantage that it is easily automated. More

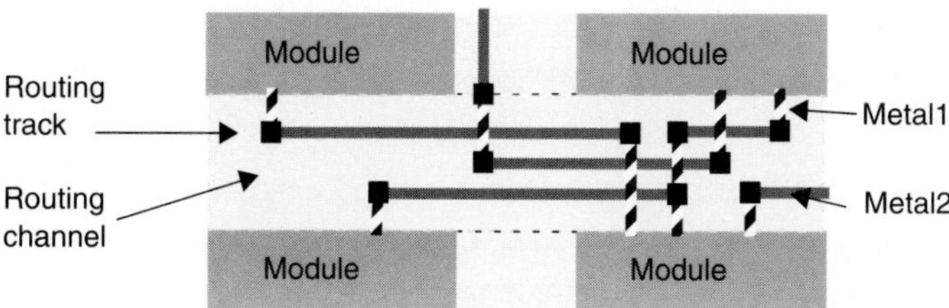

Figure 11.22 Two-layer channel routing.

advanced routers use three or more routing layers, which results in a denser wiring. In general, this means shorter wires, reduced capacitance, and higher performance or lower power. This macrocell place-and-route approach originated when the number of interconnect layers were scarce (two or at most three). Up to six layers are currently available in state-of-the-art manufacturing processes. This makes over-the-block routing an attractive possibility, eliminating wasteful routing channels altogether. Chip place and route will resemble more and more the approaches currently in use for printed-circuit boards (PCB).

Example 11.5 Macrocell Place and Route

Figure 11.23 shows the microphotograph of a video-compression chip generated automatically using a combination of the described techniques [Lidsky94]. The designer only provided a netlist describing the composing modules, their parameters (e.g., the size of the memory unit), and interconnections. The layout was generated using the LAGER *silicon compiler*, so named because of a perceived similarity with software compilers [Brodersen92]. Functions such as the memories are generated using module generators. Standard cells are used for the generation of the controller. The datapath was assembled using a datapath compiler.

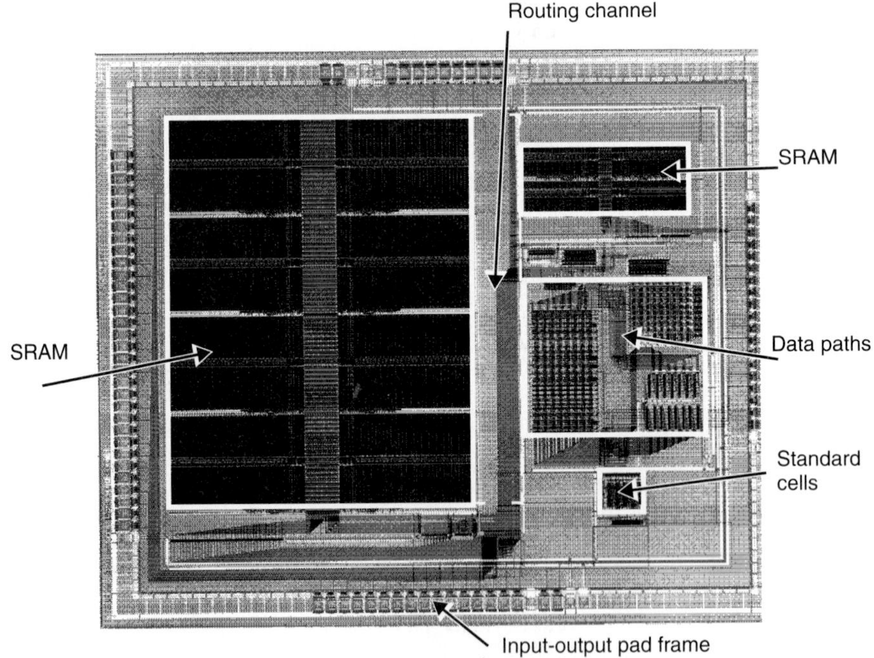

Figure 11.23 Microphotograph of video-compression chip, generated using the LAGER silicon compiler.

The module-based approach has the advantage of automating the layout process to a large degree. The irregular shape of the modules can, however, lead to very inefficient implementations and huge areas devoted to wiring. To address this deficiency, it is important to consider the global topology of the circuit in the early phases of the design process—even before the modules are actually designed. This effort, called *floorplanning* determines the silicon estate attributed to each function, resolves the routing path for the major busses, and provides an initial vision on the supply and clock network.

11.4.3 Array-Based Implementation Approaches

While design automation can help reduce the design time, it does not address the time spent in the manufacturing process. All of the design methodologies discussed so far require a complete run through the fabrication process.This can take from three weeks up to months and can substantially delay the introduction of a product. Additionally, a dedicated process run is expensive, and product economics must determine if this is a viable route.

Consequently, a number of alternative implementation approaches have been devised that do not require a complete run through the manufacturing process or avoid dedicated processing completely. These approaches have the advantage of having a lower NRE (non-recurring expense) and are, therefore, more attractive for small series. This comes at the expense of lower performance, lower integration density, or higher power dissipation.

Prediffused (or Mask-Programmable) Arrays

In this approach, batches of wafers containing arrays of primitive cells or transistors are manufactured by the vendors and stored. All the fabrication steps needed to make transistors are standardized and executed without regard to the final application.

To transform these uncommitted wafers into an actual design, only the desired interconnections have to be added, determining the overall function of the chip with only a few metallization steps. These layers can be designed and applied to the premanufactured wafers much more rapidly, reducing the turn-around time to a week or even days.

This approach is often called the *gate-array* or the *sea-of-gate* approach, depending upon the style of the prediffused wafer. To illustrate the concept, consider the gate-array primitive cell shown in Figure 11.24a. It comprises four NMOS, four PMOS transistors, polysilicon gate connections, and a power and ground rail.There are two possible contact points per diffusion area and two potential connection points for the polysilicon strips. We can turn this cell, which does not implement any logic function so far, into a real circuit by adding some extra wires on the metal layer and contact holes. This is illustrated in Figure 11.24b, where the cell is turned into a four-input NOR gate.

The *gate-array* approach utilizes two metallization layers and places the cells in rows separated by wiring channels, as shown in Figure 11.25. The overall look is similar to the standard-cell technique. With the advent of extra metallization layers, the routing channels can be eliminated, and routing can be performed on top of the primitive cells—occasionally leaving a cell unused. This channelless architecture, called *sea-of-gates,* yields an increased density, and makes it possible to achieve integration levels of more than 100,000 gates on a single die. It is projected that by the end of the century, 500 K to 700 K gates will be available on a single die for a typical gate delay of 0.4 nsec [Ieda91].

The primary challenge when designing a gate-array (or sea-of-gates) template is to determine the composition of the primitive cell and size the individual transistors. A sufficient number of wiring tracks must be provided to minimize the number of cells wasted to interconnect. The cell should be chosen so that the prefabricated transistors can be utilized

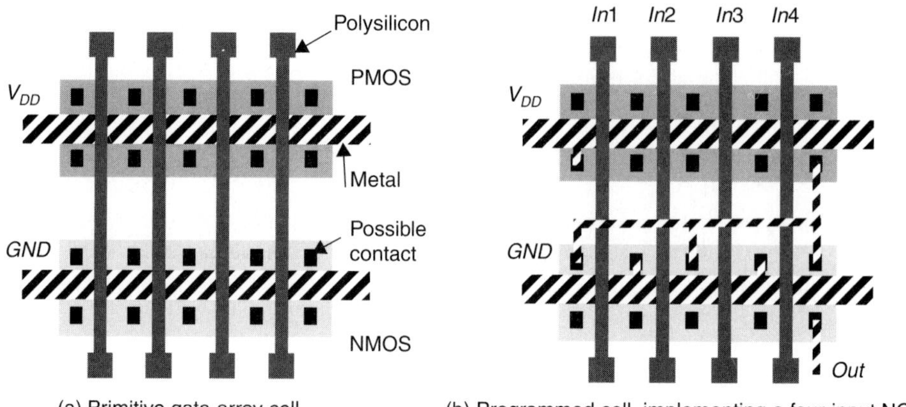

Figure 11.24 An example of the gate-array approach.

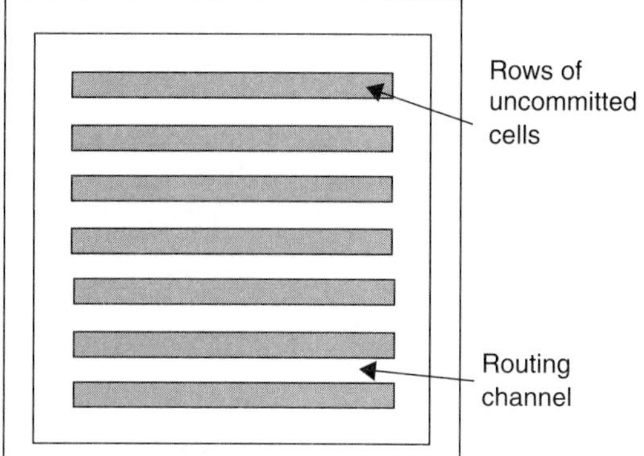

Figure 11.25 Gate-array architecture.

to a maximal extent over a wide range of designs. For instance, the configuration of Figure 11.24 is well suited for the realization of four-input gates, but wastes devices when implementing two-input gates. Multiple cells are needed when implementing a flip-flop. A number of alternative cell structures are pictured in Figure 11.26 in a simplified format. In one approach, each cell contains a limited number of transistors (four to eight). The gates are isolated by means of *oxide isolation*, also called *geometry isolation*. The "dog-bone" terminations on the poly gates allow for denser routing. A second approach provides long rows of transistors, all sharing the same diffusion area. In this architecture, it is necessary to electrically turn off some devices to provide isolation between neighboring gates by tying NMOS and PMOS transistors to *GND* and V_{DD} respectively. This technique is called *gate isolation*.

It is worth observing that the cell of Figure 11.26b provides two rows of smaller NMOS transistors that can be connected in parallel if needed. Smaller transistors come in handy when implementing pass-transistor logic or memory cells. Sizing the transistors in the cells is a clear challenge. Due to the interconnect-oriented nature of the array-based design methodology, the propagation delay is generally dominated by the interconnect capacitance. This seems to favor larger device sizes that cause a larger area loss when unused. On the other hand, it is possible to construct larger transistors by putting several

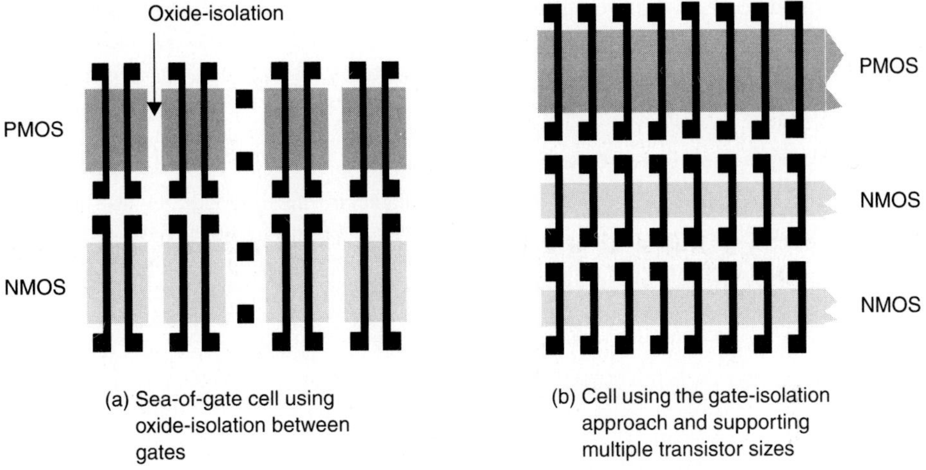

(a) Sea-of-gate cell using oxide-isolation between gates

(b) Cell using the gate-isolation approach and supporting multiple transistor sizes

Figure 11.26 Examples of sea-of-gates primitive cells (from [Veendrick90]).

smaller devices in parallel. One way to address the impact of interconnect capacitance on performance is to make use of the BiCMOS technology. In this technology, a cell can contain a mixture of MOS and bipolar transistors. Depending upon the driving and performance requirements, one can choose between CMOS, BiCMOS, and ECL gate structures. Clock frequencies ranging around 100 MHz have been demonstrated using this approach [Gallia90]. An example of a BiCMOS cell is shown in Figure 11.27.

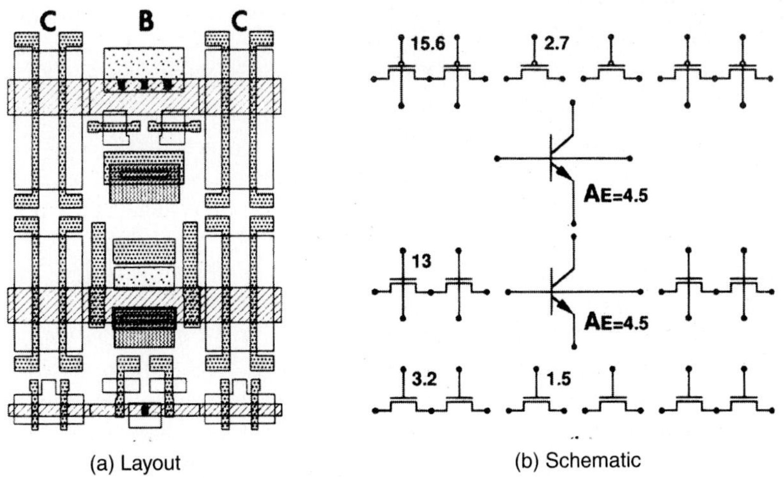

(a) Layout

(b) Schematic

Figure 11.27 BiCMOS sea-of-gate cell ([Gallia90]).

Mapping a logic design onto an array of cells is a largely automated process, involving logic synthesis followed by placement and routing. The quality of these tools has an enormous impact on the final density and performance of a sea-of-gates implementation.

Utilization factors in sea-of-gates structures are a strong function of the type of application being implemented. Utilization factors close to 100% can be obtained for regular structures such as memories. For other applications, utilization factors can be substantially lower (< 75%) largely due to wiring restrictions.

Example 11.6 Sea-of-Gates

An example of a sea-of-gates implementation is shown in Figure 11.28. The array has a maximum capacity of 300 K gates and is implemented in a 0.6 μm HCMOS technology. The upper left part of the array implements a memory subsystem, which results in a regular modular layout. The rest of the array implements random logic.

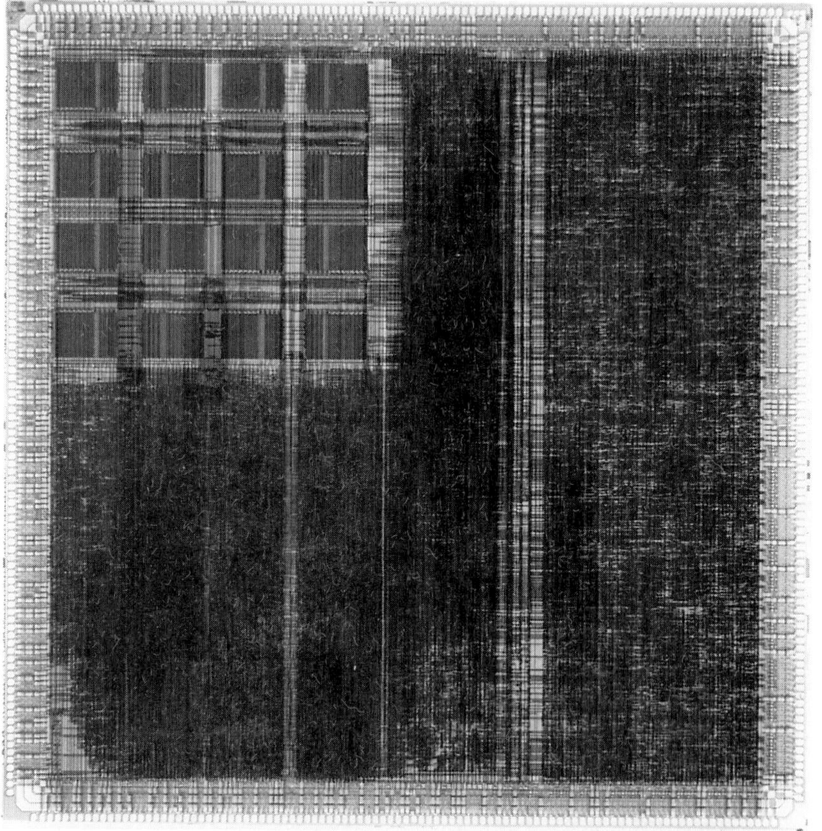

Figure 11.28 Gate-array die microphotograph (LEA300K). *Courtesy of LSI Logic.*

Prewired Arrays

While the prediffused arrays offer a fast road to implementation, it would be even more efficient if dedicated manufacturing steps could be avoided altogether. This leads to the concept of the preprocessed die that can be programmed in the field (i.e., outside the semiconductor foundry) to implement a set of given Boolean functions. Such a programmable prewired array of cells is called a *field-programmable gate array (FPGA)*. The advantage of this approach is that the manufacturing process is completely separated from the implementation phase and can be amortized over a large number of designs. The implementation itself can be performed at the user site with negligible turn-around. The major drawback of this technique is a loss in performance and design density as compared to the more customized approaches.

To understand how such an array can possibly be conceived, we can draw an analogy with the memory world. The gate-array approach is analogous to the mask-programmable ROM (PROM). Both structures require a number of dedicated masks and accompanying processing steps to finalize the functionality of the device. We introduced a number of alternative memory structures in Chapter 10 that circumvent this problem.

1. Programmable ROMs or PROMs use selectively blown *fuses* to discriminate between 0 and 1 cells.
2. In the *nonvolatile* ROM structure, the threshold of the storage transistors is electrically adjusted to reflect the requested storage value.
3. RAM memories store a signal employing either positive feedback or capacitive storage, albeit in a *volatile* format.

The prewired arrays use virtually identical techniques to implement a set of given Boolean functions on top of a regular array of cells without requiring any special processing steps. The various FPGA approaches can thus be classified into similar categories: fuse-based, nonvolatile, and volatile.

Fuse-Based FPGAs. Consider, for instance, the logic structure of Figure 11.29. A circle (o) at an intersection indicates a fusible link that can be blown when no connection is required at that particular point of intersection. An inspection of the circuit reveals that it simply represents a *programmable logic array* (PLA), where both the AND and OR planes can be programmed by blowing fuses if necessary. This approach allows for the implementation of arbitrary logic functions in a two-level *sum-of-products* format. The AND plane creates the required minterms, while the OR plane takes the sum of a selected set of products to form the outputs. The functionality of PLA is restricted by the number of inputs, outputs, and minterms, also called *product terms.*

We can envision variations on this theme, some of which are represented in Figure 11.30. The dot (•) at the intersection of two lines represents a nonfusible, hardwired link. The first structure represents the PROM architecture, in which the AND plane is fixed and enumerates all possible minterms.[2] The second structure, called a *programmable array logic device* (PAL), is located at the other end of the spectrum, where the OR plane is fixed, and the AND plane is programmable. The PLA architecture is the most generic one for the implementation of arbitrary logic functions. The PROM and PAL structures, on the other hand, trade off flexibility for density and performance. Which structure to select depends strongly upon the nature of the Boolean functions to be implemented. All these

[2] Please refer to Chapter 10 for a definition of the ROM and PLA concept and their differences.

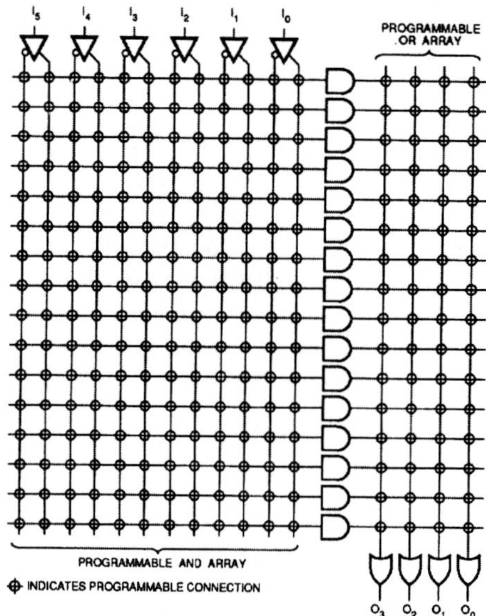

Figure 11.29 Fuse-programmable programmable logic array (PLA). Copyright © Advanced Micro Devices, Inc., 1986. All rights reserved.

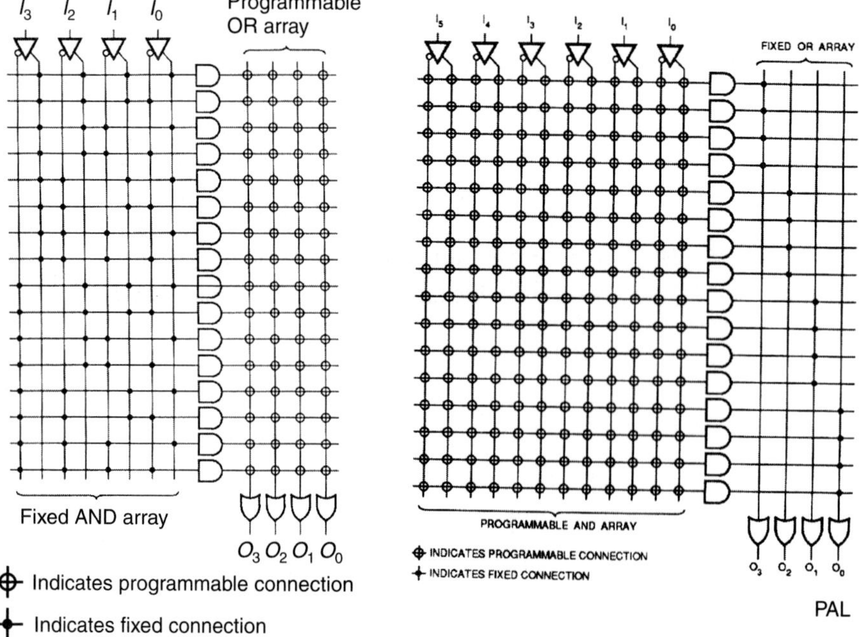

Figure 11.30 Alternative fuse-based programmable logic devices (or PLD's). Copyright © Advanced Micro Devices, Inc., 1986. All rights reserved.

approaches are generally classified under the common term of *programmable logic devices* (PLD).

The sum-of-products approach results in regular structures, but tends to produce designs with a low integration density and sluggish performance. Other approaches can be conceived that are more in line with the multilevel approach favored in the standard-cell and sea-of-gate approaches. An example of such an architecture introduced by Actel [El-Ayat89] is illustrated in Figure 11.31. It consists of rows of logic cells separated by routing channels. Interconnections between channels can be accomplished with the aid of feedthroughs. All logic cells are structurally identical. Their functionality is set at programming time by fixing a number of control inputs. An example of such a logic cell that is used in the Actel architecture is shown in Figure 11.32. It consists of three two-input multiplexers and a two-input NOR gate. The cell can be programmed to realize any two-input and three-input logic functions, some four-input Boolean functions and a latch.

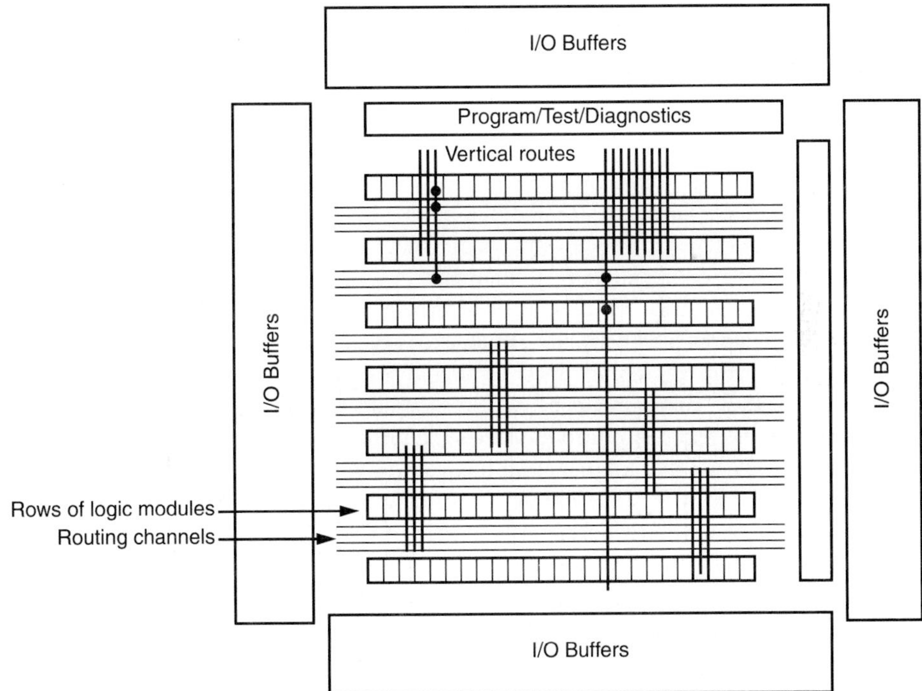

Figure 11.31 Field-programmable gate array modeled after the standard-cell approach

---

Example 11.7 Programmable Logic Cell

It can be verified that the logic cell of Figure 11.32 acts as a two-input XOR under the programming conditions enumerated below. Assume the multiplexers select the bottom input signal when the control signal is high.

$$A = 1; B = 0; C = 0; D = 1; SA = SB = In1; S0 = S1 = In2$$

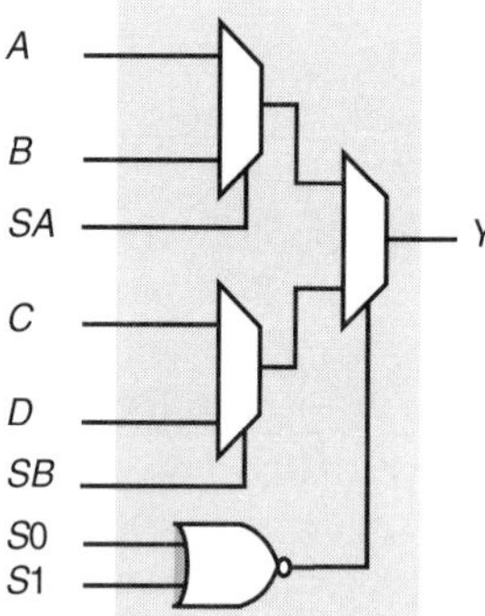

Figure 11.32 Logic cell as used in the Actel fuse-based FPGA.

As an exercise, determine the programming required for the two-input XNOR function. A three-input AND gate can be realized as follows:

$$A = 0;\ B = In1;\ C = 0;\ D = 0;\ SA = In2;\ SB = 0;\ S0 = S1 = In3$$

Finally, the largest function (four-inputs) that can be realized is the four-input multiplexer. *A*, *B*, *C*, and *D* act as inputs, while *SA*, *SB*, and $(S0 + S1)$ are control signals.

The major difference between the PLD approaches and the gate-array techniques is that a considerable part of the programming is related to configuring the interconnections in the latter. One approach would be to configure each routing channel as a fully connected grid of horizontal and vertical interconnect wires and to blow a fuse whenever a connection is not needed. Unfortunately, interconnect networks tend to be sparsely populated, which requires the interruption of an excessive number of switches and results in prohibitively long programming times. To circumvent this problem, the *antifuse* has been devised [El-Ayat89]. This component, which represents an open circuit (or a high resistance > 100 MΩ) in the unprogrammed mode, can be turned into a short-circuit (or small resistance < 500 Ω) during the programming phase. One way to realize an antifuse is to sandwich a dielectric (such as oxide-nitride-oxide) between two conducting layers and cause breakdown in the dielectric during programming by applying a high voltage. This operation is a one-time event and cannot be undone, as was the case for the fuse-based approach.

Antifuses only need to be enabled when a connection is required in the routing channel that represents a small fraction of the overall grid. A substantial reduction in programming time is consequentially obtained. A detailed view on a potential configuration of the routing channel is shown in Figure 11.33. Notice how only two antifuses are needed to set up a connection. Be aware that this figure hides the programming circuitry.

Nonvolatile FPGAs. The fuse-based approach has the important disadvantage of being *one-time programmable*. Circuit corrections or extensions are not possible; thus, new components are required for every design change. In the ROM world, this deficiency has been addressed by the introduction of the erasable (EPROM) and electrically erasable (EEPROM) architecture, both based on the floating-gate transistors. By electrically modifying the threshold of this device, it is possible to program a memory cell to store either a 0 or a 1. Refer to Section 10.3.2 for more information on this topic. An identical approach can be used to commit a programmable logic device to a given logic function. Based on

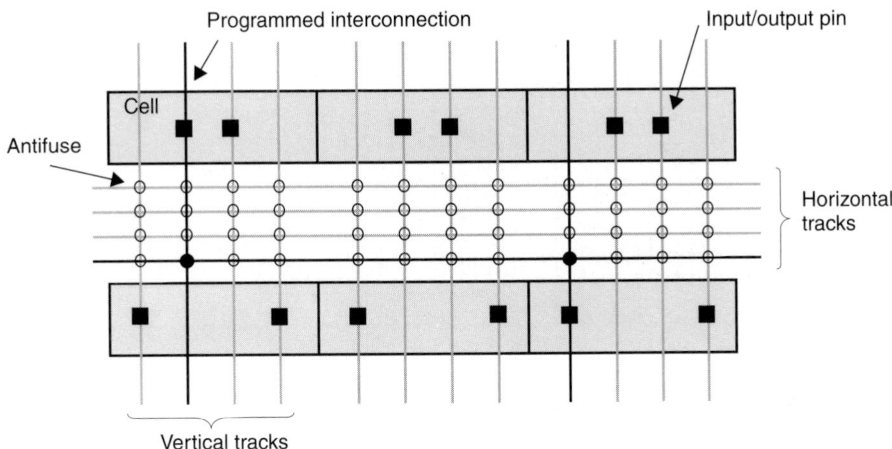

Figure 11.33 Programming interconnect using antifuses.

the programming approach and analogous to the PROM terminology, these devices are either called *erasable programmable logic devices* (EPLD) or *electrically erasable programmable logic devices* (EEPLDs). Ignoring the differences in the overall architecture, it can be stated that the design of these devices proceeds along virtually identical paths as those described for the (E)EPROM structures, and that the same considerations hold regarding area, performance, or power consumption.

The single-array architecture of the PLA, PROM, and PAL structures in Figure 11.29 and Figure 11.30 becomes less attractive in the era of higher integration density. First of all, implementing very complex logic functions on a single, large array results in a loss of programming density and performance. Secondly, the arrays as shown implement only combinational logic. To realize complete, sequential subdesigns, the presence of registers and/or flip-flops is an absolute requirement. These deficiencies can be addressed as follows:

1. partition the array into a number of smaller sections, often called macrocells
2. introduce flip-flops and provide a potential feedback from output signals to the inputs.

One example of how this can be accomplished is shown in Figure 11.34 [Altera92]. The overall architecture consists of eight macrocells, each of which represents an electrically programmable PAL with 18 inputs, eight product terms and a single output. The inputs to the macrocells consist of ten primary inputs and eight macrocell outputs. The combined register/output control block is programmable as well. The register can be configured as a *D*, *T*, *J-K*, or clocked *S-R* flip-flop. The output pin of the macrocell can be connected either to the register or the output of the PAL. Finally, a ninth product term can be used to control the tri-state output buffer.

Based on similar concepts, complex parts have been devised that reach into an equivalent gate range of several thousands [Alford89]. Their ease of use and the avoidance of any manufacturing steps have made these parts competitive with gate arrays in the low-end range. Programming these devices so that the macrocell and interconnect resources are used effectively and a reasonable speed is obtained is nontrivial. All manufacturers

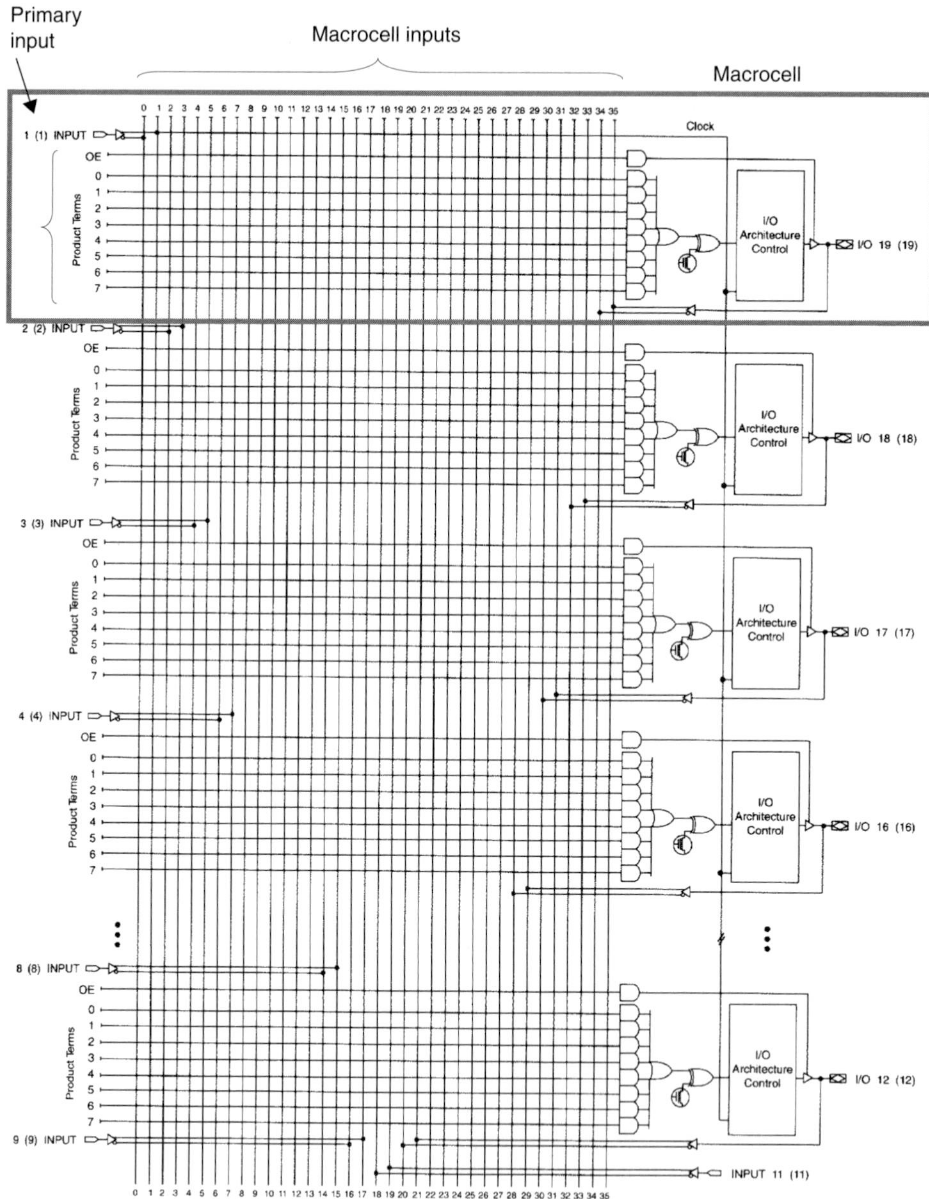

Figure 11.34 EPLD block diagram containing eight macrocells (*Courtesy Altera Corp.*).

provide a suite of software tools and module libraries that help map a set of logic functions onto their architectures. The input specifications for these design tools range from gate schematics, truth tables, and state graphs to complete register-transfer descriptions using languages such as VHDL.

RAM-based (volatile) FPGAs. Fuses and programmable transistors have the advantage that, once programmed, a component retains its functionality even when the supply voltage is removed. This advantage comes at a price either in terms of integration density, cost, or programming turn-around time. Using a dense random-access memory approach instead addresses many of these issues:

1. Programming is fast and can be repeated many times.
2. No high voltages are needed during programming.
3. Integration density is high.

Interrupting the supply voltage results in a loss of the chip's contents. A loading of the "program" from either a nonvolatile memory or over a programming bus connected to a microprocessor board is required at start-up time.

A RAM-based approach that has proven to be particularly successful was introduced by Xilinx, Inc. [Xilinx90]. Similar to the fuse-based approach of Figure 11.31, it is inspired by the gate-array and standard-cell methodologies. The design consists of a large array of programmable logic cells called *configurable logic blocks* (CLBs), an example of which is shown in Figure 11.35. It consists of a five-input, two-output combinatorial section, that can be programmed like a PROM to perform any function of five variables or two functions of four variables. The two *D* flip-flops can be configured for edge-triggered or level-triggered operation. The logic functionality and the configuration of multiplexers and registers is controlled by the RAM state bits. If required, the RAM program memory can double as table look-up for complex logic functions or as a register file.

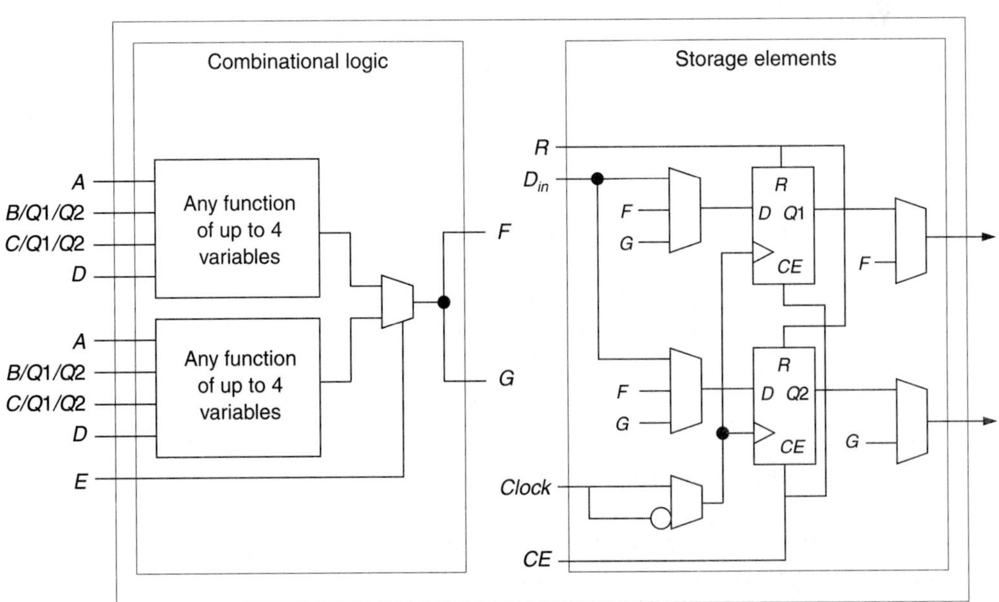

Figure 11.35 RAM-programmable logic block (CLB). *Courtesy of Xilinx, Inc.*

Even more critical to the operation of the device are the programmable interconnect resources. To fully utilize the available logic cells, the interconnect network must be flexible and routing bottlenecks must be avoided. Speed is another prerequisite, since interconnect delay tends to dominate the performance in this style of design. It is worth differentiating between local cell-to-cell interconnections and global signals such as clocks that have to be distributed over the complete chip with low delay. A large number of local interconnections can be accounted for by providing a mesh-like interconnection between neighboring cells. For instance, each cell output can be distributed to its neighbors to the north, east, south and west. To account for interconnections between disjoint cells or to provide global interconnections, routing channels are placed between the cells, containing a fixed number of uncommitted vertical and horizontal routing wires (Figure 11.36). At the junctions of the horizontal and vertical wires, RAM-programmable switching matrices are provided that direct the routing of the data. Cell inputs and outputs can be connected to the global interconnect network by RAM-programmable interconnect points.

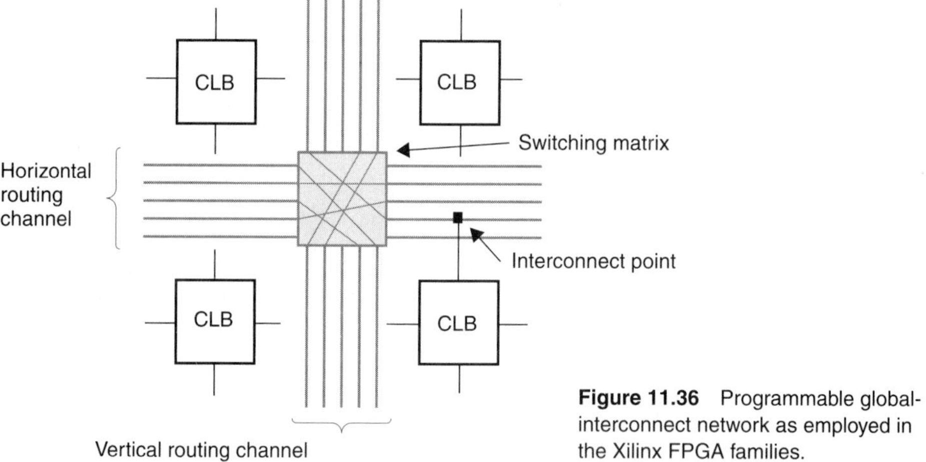

Figure 11.36 Programmable global-interconnect network as employed in the Xilinx FPGA families.

To program the component at start-up time, programming data is shifted serially into the part over a single line (or pin). For all practical purposes, one can consider the FPGA RAM cells to be configured as a giant shift-register during that period. Once all memories are loaded, normal execution is started. The attraction of this approach is that the internals of the chip are modified by changing the software, leaving the hardware untouched. This represents a considerable advantage in the prototyping and debugging phases, but can also come in handy for products that need to undergo field updates. In some sense, this brings a paradigm that was extremely successful in the world of programming (as embodied by the microprocessor) to the domain of logic design.

Example 11.8 FPGA Complexity and Performance

To get an impression of what can be achieved with the volatile field-programmable components, consider the Xilinx 4025. It contains approximately 1000 CLBs organized in a 32× 32 array. This translates into a maximum equivalent gate count of 25,000 gates. The chip contains 422 Kbits of RAM, used mostly for programming. A single CLB is specified to operate at 250

MHz. When taking into account the interconnect network and attempting more complex logic configurations such as adders, clock speeds between 20 and 50 MHz are attainable. To put the integration complexity in perspective, a 32-bit adder requires approximately 62 CLBs. A chip microphotograph of the XC4025 part is shown in Figure 11.37. The horizontal and vertical routing channels are easily recognizable.

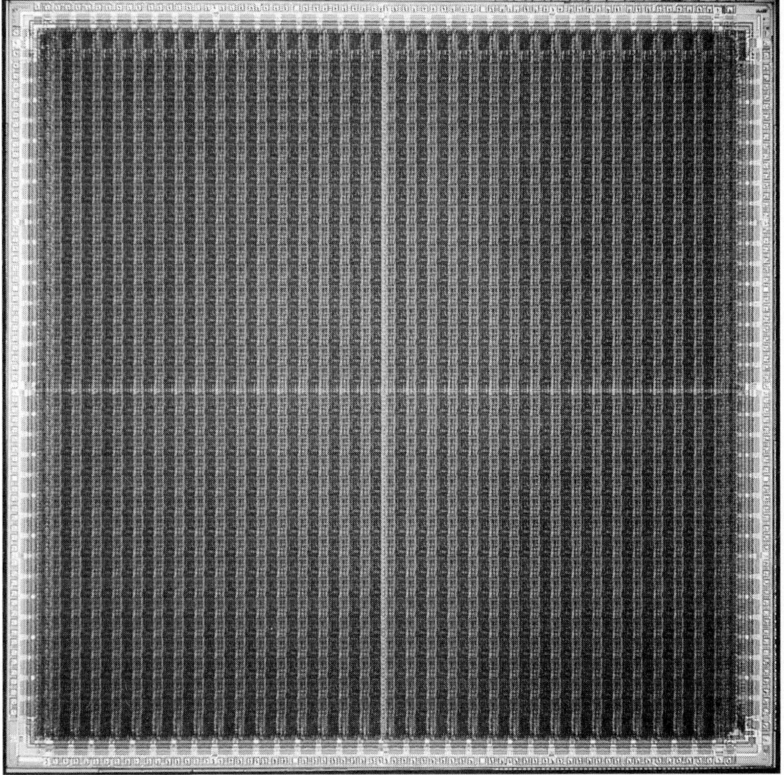

Figure 11.37 Chip microphotograph of XC4025 volatile FPGA. *Courtesy of Xilinx, Inc.*

Prewired logic arrays have rapidly claimed a significant part of the logic component market. Their arrival has effectively ended the era of logic design using discrete components represented by the TTL logic family. It is generally believed that the impact of these components will increase with a further scaling of the technology. To make this approach successful, however, advanced software support in terms of cell placement, signal routing, and synthesis are required. The latter is rather new to the scene and requires further elaboration.

11.5 Design Synthesis

Synthesis can be defined as the transformation between two different design views. Typically it represents a translation from a *behavioral* specification of a design entity into a *structural* description. In laymen's terms, it translates a description of the function a module should perform (the behavior) into a composition, that is, an interconnection of elements (the structure). Synthesis approaches can be defined at each level of abstraction: circuit, logic, and architecture. An overview of the various synthesis levels and their impact is given in Figure 11.38. The synthesis procedures may differ dependent upon the targeted implementation style. For instance, logic synthesis translates a logic description given by a state diagram into an interconnection of gates. The techniques involved in this process strongly depend upon the choice of either a two-level (PLA) or a multilevel (standard-cell or gate-array) implementation style. We briefly describe the synthesis tasks at each of the different modeling levels. Refer to [DeMicheli94] for more information and a deeper insight into design synthesis.

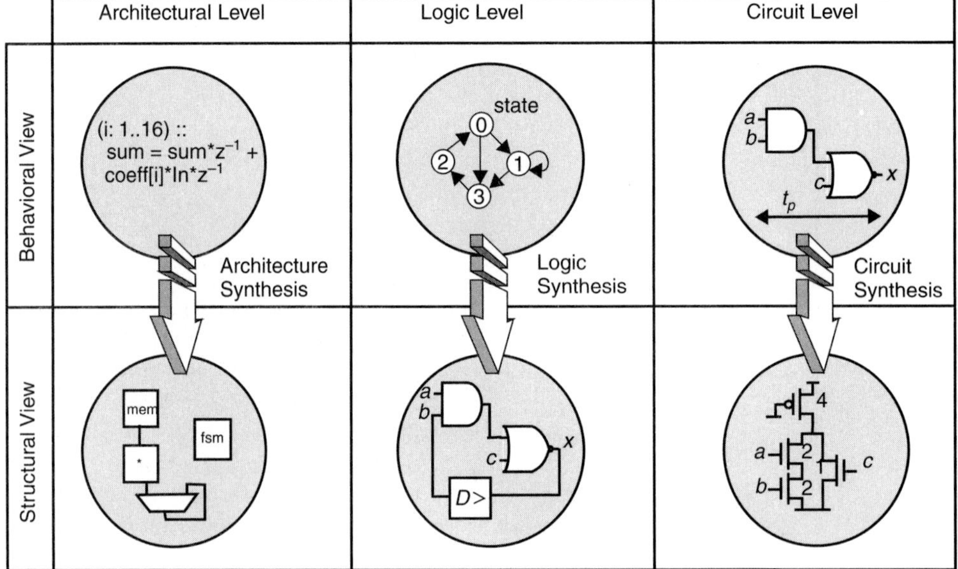

Figure 11.38 A taxonomy of synthesis tasks.

11.5.1 Circuit Synthesis

The task of circuit synthesis is to translate a logic description of a circuit into a network of transistors that meets a set of timing constraints. This process can be divided into two stages.

1. Derivation of *transistor schematics* from the logic equations. This requires the selection of a circuit style (complementary static, pass-transistors, dynamic, DCVSL, etc.) and the construction of the logic network. The former task is usually up to the designer. The latter depends upon the chosen style. For instance, the logic graph technique introduced in Appendix C can be used to derive the complementary pull-down and pull-up networks of a static CMOS gates. Similarly, automated techniques

have been developed to generate the pull-down trees for the DCVSL logic style so that the number of required transistors is minimized [Chu86].

2. *Transistor sizing* to meet performance constraints. This has been a recurring subject throughout this textbook. The choice of the transistor dimensions has a major impact on the area, performance, and power dissipation of a circuit. We have also learned that this is a subtle process. For instance, the performance of a gate is sensitive to a number of layout parasitics, such as the size of the diffusion area, fan-out, and wiring capacitances. Notwithstanding these daunting challenges, some powerful transistor-sizing tools have been developed [e.g., Fishburn85]. The key to the success of these tools is the accurate modeling of the performance of the circuit using *RC* equivalent circuits and a detailed knowledge of the subsequent layout-generation process (see the preceding section on compiled cells). The latter allows for an accurate estimation of the values of the parasitic capacitances.

While circuit synthesis has proven to be a powerful tool, it has not penetrated the design world as much as could be expected. One of the main reasons for this is that the quality of the cell library has a strong influence on the complete design and designers are reluctant to pass this important task to automatic tools that might produce inferior results.

11.5.2 Logic Synthesis

Logic synthesis is the task of generating a structural view of a logic-level model. This model can be specified in many different ways, such as state transition diagrams, finite state machines, schematic diagrams, Boolean equations, truth tables, or HDL descriptions.

The synthesis techniques differ according to the nature of the circuit (combinational or sequential) or the intended implementation architecture (multilevel logic, PLA, or FPGA). The synthesis process consists of a sequence of optimization steps, the order and nature of which depend on the chosen cost function—area, speed, power, or a combination of the above. Typically, logic optimization systems divide the task into two stages:

1. A technology-independent phase, where the logic is optimized using a number of Boolean or algebraic manipulation techniques.

2. A technology-mapping phase that takes into account the peculiarities and properties of the intended implementation architecture. The technology-independent description resulting from the first phase is translated into a gate netlist or a PLA description.

The *two-level minimization* tools were the first logic-synthesis techniques to become widely available. The Espresso program developed at the University of California at Berkeley [Brayton84] is an example of a popular two-level minimization program. For some time, the wide availability of these tools made regular, array-based architectures such as PLAs and PALs the prime choice for the implementation of random logic functions.

At the same time, the groundwork was laid for sequential or state-machine synthesis. Tasks involved include the *state minimization* that aims at reducing the number of machine states, and the *state encoding* that assigns a binary encoding to the states of a finite state machine [DeMicheli94].

The emergence of *multilevel logic synthesis* environments such as the Berkeley MIS tool [Brayton87] swung the pendulum towards the standard-cell and FPGA implementations that offer higher performance or integration density for a majority of random-logic functions.

The combination of these techniques with sequential synthesis has opened the road to complete register-transfer (RTL) synthesis environments that take as an input an HDL description of a sequential circuit and produce a gate netlist. The Synopsy VHDL compiler [Carlson91] is one of the leaders in this domain and spurred a dramatic change in the way digital circuit design is performed.

Example 11.9 Logic Synthesis

To demonstrate the difference between two-level and multilevel logic synthesis, both approaches were applied on the well-known full-adder equations.

$$S = (A \oplus B) \oplus C_i$$
$$C_o = A \cdot B + A \cdot C_i + B \cdot C_i \tag{11.2}$$

The MIS-II logic synthesis environment was employed for both the two-level and multi-level synthesis. The minimized truth table representing the PLA implementation is shown in Table 11.1. It can be verified that the resulting network corresponds to the full-adder equations presented above. The PLA counts three inputs, seven product terms, and two outputs. Observe that no product terms can be shared between the sum and carry outputs. A NOR-NOR implementation in the style of Figure 10.57 requires 26 transistors in the PLA array (17 and 9 in the OR plane and AND planes respectively). This count does not include the input and output buffers.

Table 11.1 Minimized PLA truth table for full adder. The (—) means that the corresponding input does not appear in the product term.

A	B	C_i	S	C_o
1	1	1	1	0
0	0	1	1	0
0	1	0	1	0
1	0	0	1	0
1	1	—	0	1
1	—	1	0	1
—	1	1	0	1

Figure 11.39 shows the multilevel implementation as generated by MIS-II. In the technology-mapping phase, a standard-cell library developed by Mississippi State University [Brodersen92] was targeted. Implementation of the adder requires only six standard cells. This corresponds to 34 (!) transistors in a static CMOS implementation.[3] Observe the usage of complex logic gates such as EXOR and OR-AND-INVERT. For this case study, minimization of the area was selected as the prime optimization target. Other implementations can be

[3] How to implement a static CMOS EXOR gate with only 9 transistors is left as an exercise for the reader.

obtained by targeting performance instead. For instance, the critical timing path from C_i to C_o can be reduced by signal reordering. This requires the designer to identify this path as the most critical, a fact that is not obvious from a simple inspection of the full-adder equations.

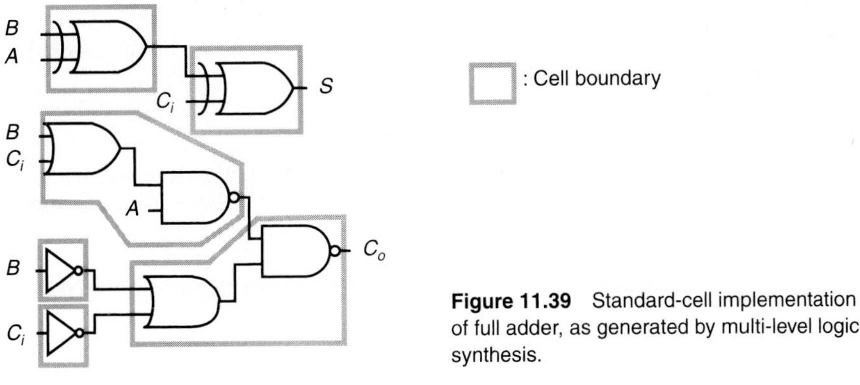

Figure 11.39 Standard-cell implementation of full adder, as generated by multi-level logic synthesis.

11.5.3 Architecture Synthesis

Architecture synthesis is the latest development in the synthesis area. It is also referred to as *behavioral* or *high-level synthesis*. Its task is to generate a structural view of an architecture design, given a behavioral description of the task to be executed, and a set of performance, area, and/or power constraints. This corresponds to determining what architectural resources are needed to perform the task (execution units, memories, busses, and controllers), binding the behavioral operations to hardware resources, and determining the execution order of the operations on the produced architecture. In synthesis jargon, these functions are called *allocation, assignment,* and *scheduling* [Gajski92, DeMicheli94]. While these operations represent the core of architecture synthesis, other steps can have a dramatic impact on the quality of the solution. An example of this are optimizing transformations that manipulate the initial behavioral description so that a superior solution can be obtained in terms of area or speed. *Pipelining* is a typical example of such a transformation. In a sense, this component of the synthesis process is similar to the use of optimizing transformations in software compilers.

While architecture compilers have been extensively researched in the academic community, only recently have they started to make an inroad into the design world, and only in restricted application domains such as digital signal processing. Examples of the latter are the Cathedral-II [DeMan86] and HYPER [Rabaey91] synthesis environments. A number of reasons for this slow penetration can be enumerated:

- A lack of understanding of what a behavioral description means at the architecture level for some domains, such as general-purpose microprocessors.
- Behavioral synthesis assumes the availability of an established synthesis approach at the register transfer level. This has only recently come to maturity.

- For a long time, architecture synthesis has concentrated on a limited aspect of the overall design process. For instance, the impact of interconnect on the overall design cost was long ignored. Also, limitations on the architectural scope resulted in inferior solutions apparent to every experienced designer.

A number of these constraints have been resolved, and architecture synthesis should soon come to fruition.

Example 11.10 Architecture Synthesis

To illustrate the concept and capabilities of architecture synthesis, consider the simple computational flowgraph of Figure 11.40. It describes a program that inputs three numbers a, b, and c from off-chip and produces their sum x at the output.

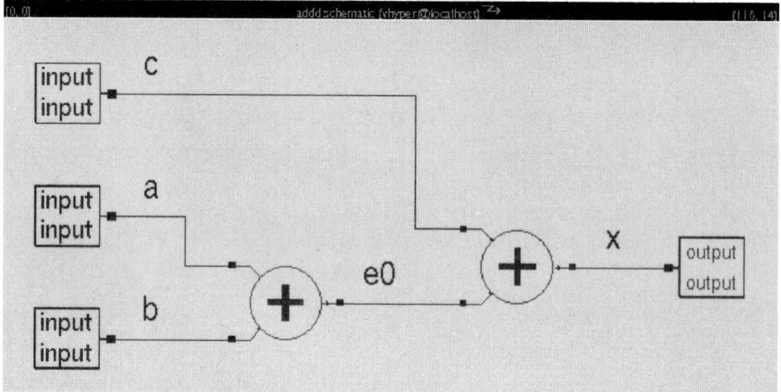

Figure 11.40 Simple program performing the sum of three numbers.

Two possible implementations, as generated by the HYPER synthesis system,[4] are shown in Figure 11.41. The first instance requires four clock cycles and timeshares the input bus as well as the adder. The second architecture performs the program in a single clock cycle. To achieve this performance, it was necessary to pipeline the algorithm; that is, multiple iterations of the computation are overlapping. The increased speed translates as expected to a higher hardware cost—one extra adder, extra registers, and a more dedicated bus architecture, including three input ports. Both architectures were produced automatically, given the behavioral description and the clock-cycle constraint. This includes the pipelining transformation.

11.6 Validation and Testing of Manufactured Circuits

While designers tend to spend numerous hours on the analysis, optimization, and layout of their circuits, one issue is often overlooked—when a component returns from the manufacturing plant, how does one know if it actually works? Does it meet the functionality and performance specifications? The customer expects a delivered component to perform as

[4] A complete tutorial on HYPER containing more complex synthesis examples can be accessed on the world-wide-web at the following address: *http://infopad.eecs.berkeley.edu/~hyper.*

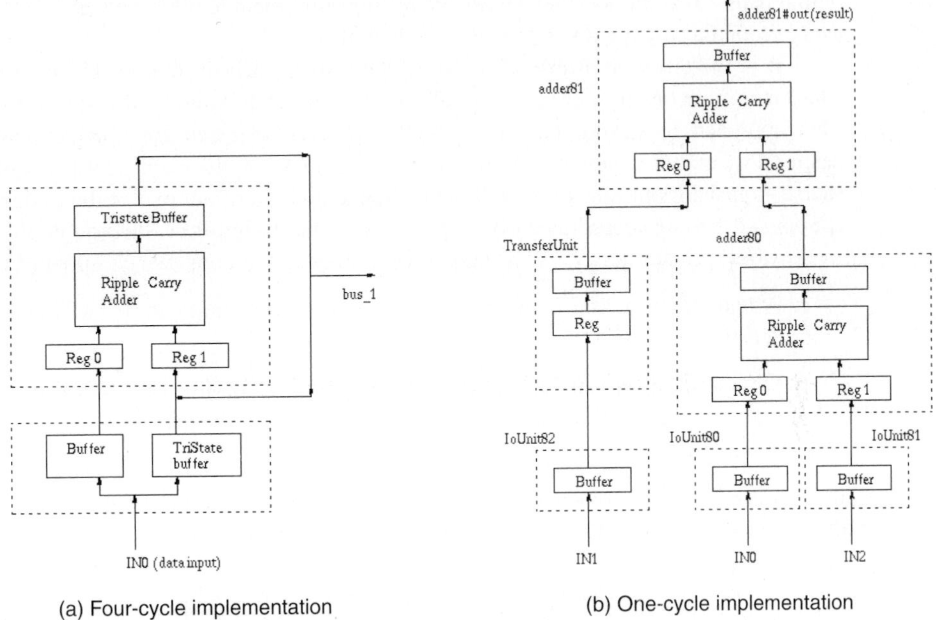

(a) Four-cycle implementation

(b) One-cycle implementation

Figure 11.41 Two alternative architectures implementing the sum program.

described in the specification sheets. Once a part is shipped or deployed in a system, it is expensive to discover that it does not work. The later a fault is detected, the higher the cost of correction. For instance, replacing a component in a sold television set means replacement of a complete board as well as the cost of labor. Shipping a nonworking or partially functional device should be avoided if at all possible.

A correct design does not guarantee that the manufactured component will be operational. A number of manufacturing defects can occur during fabrication, either due to faults in the base material (e.g., impurities in the silicon crystal) or as a result of variations in the process, such as misalignment. Other faults might be introduced during the stress tests that are performed after the manufacturing. These tests expose a part to cycles of temperature and mechanical stress to ensure its operation over a wide range of working conditions. Typical faults include short-circuits between wires or layers and broken interconnections. This translates into network nodes that are either shorted to each other or to the supply rails, or that may be floating.

Making sure a delivered part is operating correctly under all possible input conditions is not as simple as it would seem at a first glance. When analyzing the circuit behavior during the design phase, the designer has unlimited access to all the nodes in the network. He is free to apply input patterns and observe the resulting response at any node he desires. This is not the case once the part is manufactured. The only access one has to the circuit is through the input-output pins. A complex component such as a microprocessor is composed of millions of transistors and contains an uncountable number of possible states. It is a very lengthy process to bring such a component into a particular state and to observe the resulting circuit response through the limited bandwidth offered by the input-output pads—if at all possible. Hardware testing equipment tends to be very expensive and every second a part spends in the tester adds to its price.

It is therefore advisable to consider the testing early in the design process. Some small modifications in a circuit can help make it easier to validate the absence of faults. This approach to design has been dubbed *design-for-testability (DFT).* While often despised by circuit designers who prefer to concentrate on the exciting aspects of design, such as transistor optimization, DFT is an integral and important part of the design process and should be considered as early as possible in the design flow. "If you don't test it, it won't work! (Guaranteed)" [Weste93]. A DFT strategy contains two components:

1. Provide the necessary *circuitry* so that the test procedure can be swift and comprehensive.

2. Provide the necessary *test patterns* (excitation vectors) to be employed during the test procedure. For reasons of cost, it is desirable that the test sequence be as short as possible while covering the majority of possible faults.

In the subsequent sections, we briefly cover some of the most important issues in each of these domains. Before doing so, a short description of a typical test procedure can help to put things in perspective.

11.6.1 Test Procedure

Manufacturing tests fall into a number of categories depending upon the intended goal.

- **The diagnostic test** is used during the debugging of a chip or board and tries to accomplish the following: Given a failing part, identify and locate the offending fault.

- **The functional test** (also called *go/no go* test) determines whether or not a manufactured component is functional. This problem is simpler than the diagnostic test since the only answer expected is yes or no. Since this test must be executed on every manufactured die and has a direct impact on the cost, it should be as simple and swift as possible.

- **The parametric test** checks on a number of nondiscrete parameters, such as noise margins, propagation delays, and maximum clock frequencies, under a variety of working conditions, such as temperature and supply voltage. This requires a different set-up from the functional tests that only deal with 0 and 1 signals. Parametric tests generally are subdivided into static (dc) and dynamic (ac) tests.

A typical manufacturing test proceeds as follows. The predefined test patterns are loaded into the tester that provides excitations to the *device under test* (DUT) and collects the corresponding responses. The test patterns are defined in a *test program* that describes the waveforms to be applied, voltage levels, clock frequency, and expected response. A probe card, or DUT board, is needed to connect the outputs and inputs of the tester to the corresponding pins on the die or package.

A new part is automatically fed into the tester. The tester executes the test program, applies the sequence of input patterns to the DUT, and compares the obtained response with the expected one. If differences are observed, the part is marked as faulty (e.g., with an ink spot). In case of a probe card, the probes are automatically moved to the next die on the wafer. During the scribing process that divides the wafer into the individual dies, spotted parts will be automatically discarded. For a packaged part, the tested component is removed from the test board and placed in a good or faulty bin, depending upon the outcome of the test. The whole procedure takes in the range of a few seconds per part, making it possible for a single tester to handle thousands of parts in an hour.

11.6.2 Design for Testability

Issues in Design for Testability

A high-speed tester that can adequately handle state-of-the-art components comes at an astronomical cost. Reducing the test time for a single component can help increase the throughput of the tester and has an important impact on the testing cost. By considering testing from the early phases of the design process, it is possible to simplify the whole validation process. In this section, we will describe some approaches that achieve that goal. Before detailing these techniques, we should first understand some of the intricacies of the test problem.

Consider the combinational circuit block of Figure 11.42a. The correctness of the circuit can be validated by exhaustively applying all possible input patterns and observing the responses. For an N-input circuit, this requires the application of 2^N patterns. For $N = 20$, this translates into more than 1 million patterns. If the application and observation of a single pattern takes 1 μsec, the total test of the module requires 1 sec. The situation gets

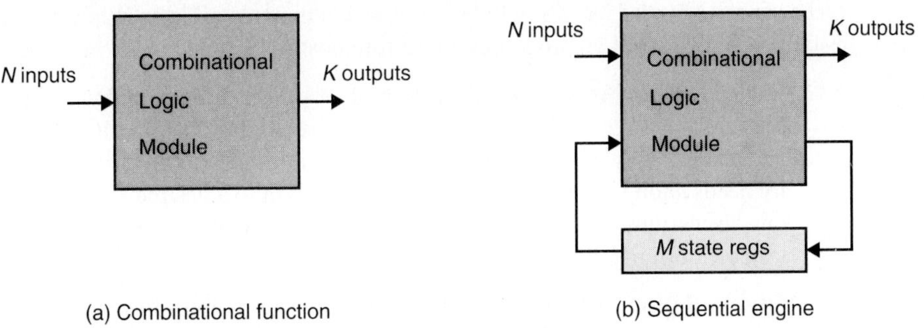

Figure 11.42 Combinational and sequential devices under test.

more dramatic when considering the sequential module of Figure 11.42b. The output of the circuit depends not only upon the inputs applied, but also upon the value of the state. To exhaustively test this finite state machine requires the application of 2^{N+M} input patterns, where M is the number of state registers [Williams83, Weste93]. For a state machine of moderate size (e.g., $M = 10$), this means that 1 billion patterns must be evaluated, which takes 16 minutes on our 1 μsec/pattern testing equipment. Modeling a modern microprocessor as a state machine translates into an equivalent model with over fifty state registers. Exhaustive testing of such an engine would require over a billion years!

Obviously, an alternative approach is required. A more feasible testing approach is based on the following premises.

- An exhaustive enumeration of all possible input patterns contains a substantial amount of *redundancy;* that is, a single fault in the circuit is covered by a number of input patterns. Detection of that fault requires only one of those patterns, while the other patterns are superfluous.
- A substantial reduction in the number of patterns can be obtained by relaxing the condition that all faults must be detected. For instance, detecting the last single percentage of possible faults might require an exorbitant number of extra patterns and the cost of detecting them might be larger than the eventual replacement cost. Typical test procedures only attempt a 95–99% fault coverage.

By eliminating redundancy and providing a reduced fault coverage, it is possible to test most combinational logic blocks with a limited set of input vectors. This does not solve the sequential problem, however. To test a given fault in a state machine, it is not sufficient to apply the correct input excitation; the engine must be brought to the desired state first. This requires a sequence of inputs to be applied. Propagating the circuit response to one of the output pins might require another sequence of patterns. In other words, testing for a single fault in an FSM requires a sequence of vectors. Once again, this might make the process prohibitively expensive.

One way to address the problem is to turn the sequential network into a combinational one by breaking the feedback loop in the course of the test. This is one of the key concepts behind the *scan-test* methodology described below. Another approach is to let the circuit test itself. Such a test does not require external vectors and can proceed at a higher speed. The concept of *self-test* will be discussed in more detail later. When considering the testability of designs, two properties are of foremost importance:

1. **Controllability,** which measures the ease of bringing a circuit node to a given condition using only the input pins. A node is easily controllable if it can be brought to any condition by only a single input vector. A node (or circuit) with low controllability needs a long sequence of vectors to be brought to a desired state. From the previous discussions, it should be clear that a high degree of controllability is desirable in testable designs.

2. **Observability,** which measures the ease of observing the value of a node at the output pins. A node with a high observability can be monitored directly on the output pins. A node with a low observability needs a number of cycles before its state appears on the outputs. Given the complexity of a circuit and the limited number of output pins, a testable circuit should have a high observability. This is exactly the purpose of the test techniques discussed in the coming sections.

Combinational circuits fall under the class of easily observable and controllable circuits, since any node can be controlled and observed in a single cycle. *Design-for-test* approaches for the sequential modules are typically classified in three categories: ad hoc test, scan-based test, and self-test.

Ad Hoc Testing

As suggested by the title, ad hoc testing combines a collection of tricks and techniques that can be used to increase the observability and controllability of a design and that are generally applied in an application-dependent fashion.

An example of such a technique is illustrated in Figure 11.43a, which shows a simple processor with its data memory. Under normal configuration, the memory is only accessible through the processor. Writing and reading a data value into and out of a single memory position requires a number of clock cycles. The controllability and observability of the memory can be dramatically improved by adding multiplexers on the data and address busses (Figure 11.43b). During normal operation mode, these selectors direct the memory ports to the processor. During test, the data and address ports are connected directly to the I/O pins, and testing the memory can proceed more efficiently. The example illustrates some important design-for-testability concepts.

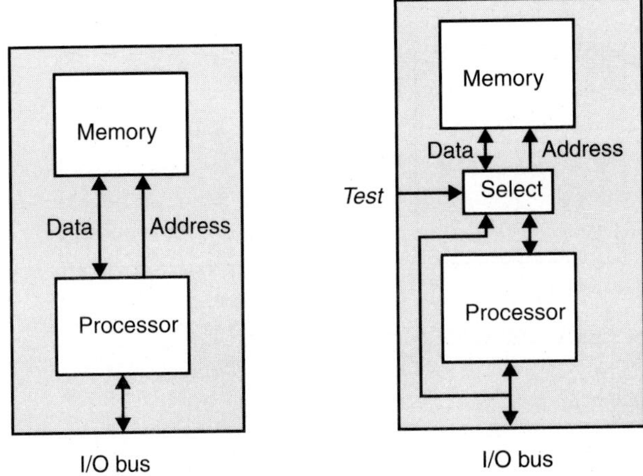

(a) Design with low testability (b) Adding a selector improves testability.

Figure 11.43 Improving testability by inserting multiplexers.

- It is often worthwhile to introduce *extra hardware* that has no functionality except improving the testability. Designers are often willing to incur a small penalty in area and performance if it makes the design substantially more observable or controllable.
- Design-for-testability often means that extra I/O pins must be provided besides the normal functional I/O pins. The *test* port in Figure 11.43b is such an extra pin. To reduce the number of extra pads that would be required, one can multiplex test signals and functional signals on the same pads. For example, the I/O bus in Figure 11.43b serves as a data-bus during normal operation and provides and collects the test patterns during testing.

An extensive collection of ad hoc test approaches can be envisioned. Examples include the partitioning of large state machines, addition of extra test points, provision of reset states, and introduction of test busses. While very effective, the applicability of most of these techniques depends upon the application and architecture at hand. Their insertion into a given design requires expert knowledge and is difficult to automate. This is why structured approaches have been introduced.

Scan-Based Test

One way to avoid the sequential-test problem is to turn all registers into externally loadable and readable elements. This turns the circuit-under-test into a combinational entity. To control a node, an appropriate vector is constructed, loaded into the registers and propagated through the logic. The result of the excitation propagates to the registers and is latched, after which the contents are transferred to the external world. Connecting all the registers in a design to a test bus regrettably introduces an unacceptable amount of overhead. A more elegant approach is offered by the serial-scan approach illustrated in Figure 11.44.

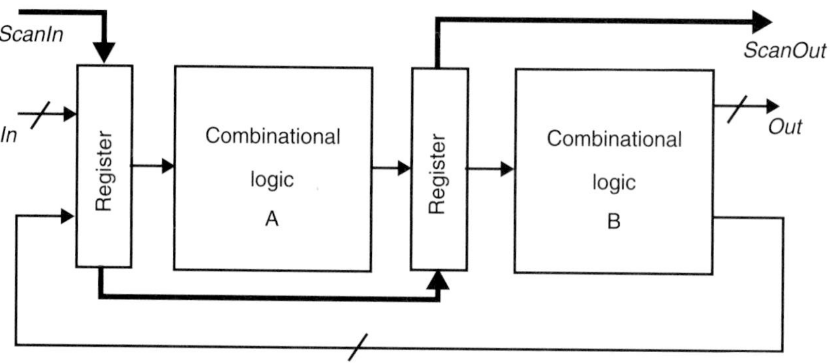

Figure 11.44 Serial-scan test.

In this approach, the registers have been modified to support two operation modes. In the normal mode, they act as N-bit-wide clocked registers. During the test mode, the registers are chained together as a single serial shift-register. A test procedure now proceeds as follows.

1. An excitation vector for logic module A (and/or B) is entered through pin *ScanIn* and shifted into the registers under control of a test clock.
2. The excitation is applied to the logic and propagates to the output of the logic module. The result is latched into the registers by issuing a single system-clock event.
3. The result is shifted out of the circuit through pin *ScanOut* and compared with the expected data. A new excitation vector can be entered simultaneously.

This approach incurs only a minimal overhead. The serial nature of the scan chain reduces the routing overhead. Traditional registers are easily modified to support the scan technique, as is demonstrated in Figure 11.45, which shows a 4-bit register extended with a scan chain. The only addition is an extra multiplexer at the input. When *Test* is low, the circuit is in normal operation mode. Setting *Test* high selects the *ScanIn* input and connects

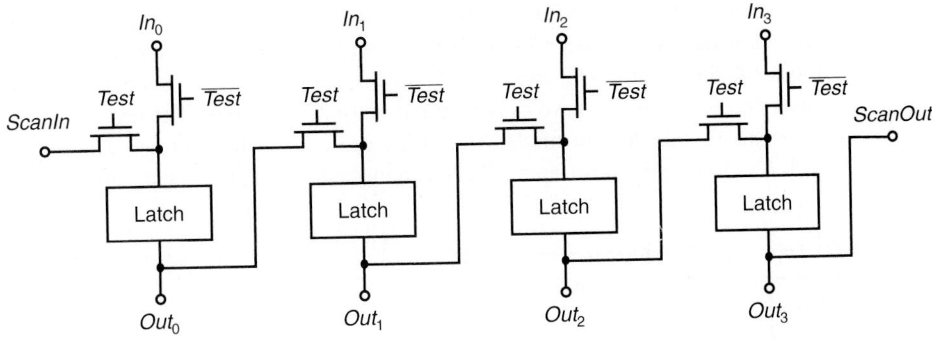

Figure 11.45 Register extended with serial-scan chain.

the registers into the scan chain. The output of the register *Out* connects to the fan-out logic, but also doubles as the *ScanOut* pin that connects to the *ScanIn* of the neighboring register. The overhead in both area and performance is small and can be limited to below 5%.

Problem 11.1 Scan-Register Design

Modify the pseudo-static, two-phase master-slave register of Figure 6.20 to support serial scan.

Figure 11.46 depicts the timing sequence that would be employed for the circuit in Figure 11.44 under the assumption of two-phase clocking approach. For a scan chain *N* registers deep, the *Test* signal is raised, and *N* clock pulses are issued, loading the registers. *Test* is lowered, and a single clock sequence is issued, latching the results from the combinational logic into the registers under the normal circuit-operation conditions. Finally, *N* extra pulses (with *Test* = 1) transfer the obtained result to the output. Note again that the scan-out can overlap with the entering of the next vector.

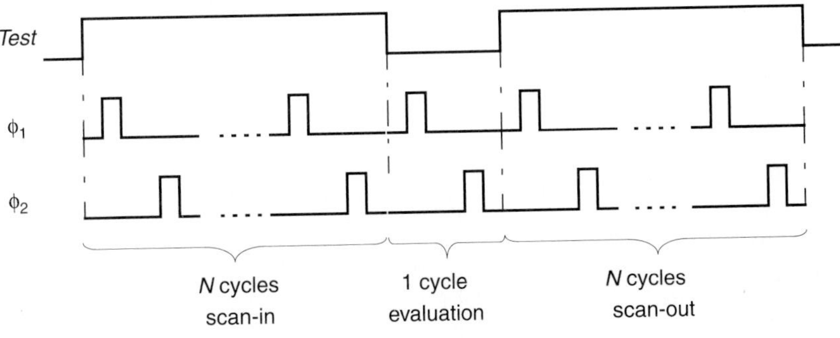

Figure 11.46 Timing diagram of test-sequence. *N* represents the number of registers in the test chain.

Many variants of the serial-scan approach can be envisioned. A very popular one, which was actually the pioneering approach, was introduced by IBM and is called *level-*

sensitive scan design (LSSD) [Eichelberger78]. The basic building block of the LSSD approach is the *shift-register latch* (SRL) shown in Figure 11.47. It consists of two latches *L1* and *L2,* the latter being present only for testing purposes. In normal circuit operation, signals *D*, *Q* ($\overline{Q}$), and *C* serve as latch input, output, and clock. The test clocks *A* and *B* are low in this mode. In scan mode, *SI* and *SO* serve as scan input and output respectively. Clock *C* is low, and clocks *A* and *B* act as nonoverlapping, two-phase test clocks.

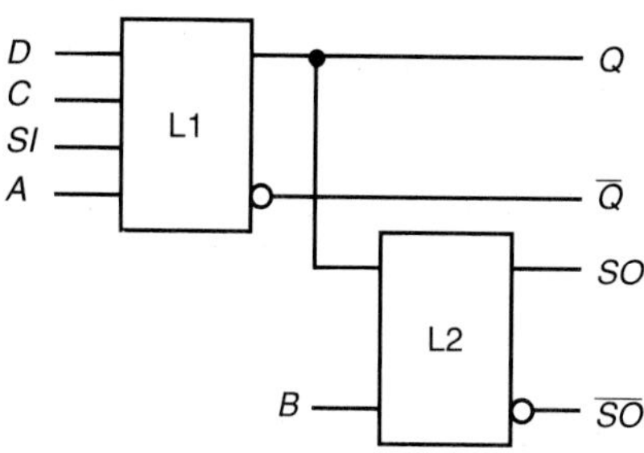

Figure 11.47 Shift-register latch.

The LSSD approach represents not only a test strategy, but also a complete clocking philosophy. By strictly adhering to the rules implied by this methodology, it is possible to automate to a large extent the test generation and the timing verification. This is why the use of LSSD was obligatory within IBM for a long time. The prime disadvantage of the approach is the complexity of the SRL latch.

It is not always necessary to make all the registers in the design scannable. Consider the pipelined datapath of Figure 11.48. The pipeline registers in this design are only present for performance reasons and do not strictly add to the state of the circuit. It is, therefore, meaningful to make only the input and output registers scannable. During test generation, the adder and comparator can be considered together as a single combinational block. The only difference is that during the test execution, two cycles of the clocks are needed to propagate the effects of an excitation vector to the output register. This approach is called *partial scan* and is often employed when performance is of prime interest. The disadvantage is that deciding which registers to make scannable is not always obvious and requires interaction with the designer.

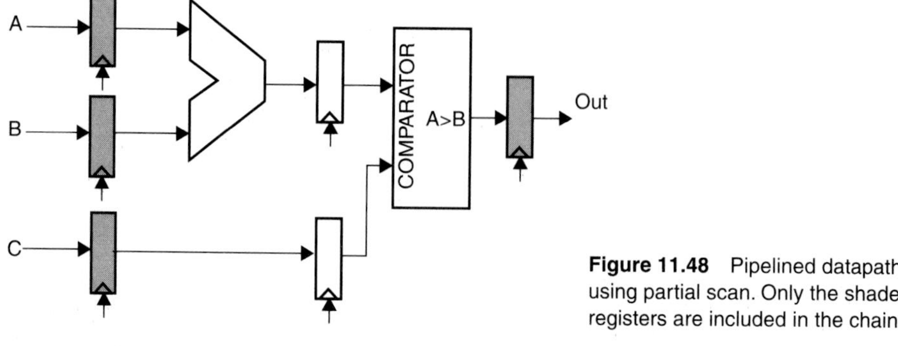

Figure 11.48 Pipelined datapath using partial scan. Only the shaded registers are included in the chain.

Boundary-Scan Design

Until recently, the test problem was most compelling at the integrated circuit level. Testing circuit boards was facilitated by the abundant availability of test points. The through-hole mounting approach made every pin of a package observable at the back-side of the board. For test, it was sufficient to lower the board onto a set of test probes (called "bed-of-nails") and apply and observe the signals of interest. The picture changed with the introduction of advanced packaging techniques such as surface-mount or multichip modules (Chapter 8). Controllability and observability are not as readily available anymore, because the number of probe points is dramatically reduced. This problem can be addressed by extending the scan-based test approach to the component and board levels.

The resulting approach is called *boundary scan* and has been standardized to ensure compatibility between different vendors ([IEEE1149]). In essence, it connects the input-output pins of the components on a board into a serial scan chain, shown in Figure 11.49. During normal operation, the boundary-scan pads act as normal input-output devices. In test mode, vectors can be scanned in and out of the pads, providing controllability and observability at the boundary of the components (hence the name). The test operation proceeds along similar lines as described in the previous paragraph. Various control modes allow for testing the individual components as well as the board interconnect. The overhead incurred includes slightly more complex input-output pads and an extra on-chip test controller (an FSM with 16 states). Boundary scan is now provided in most commodity components.

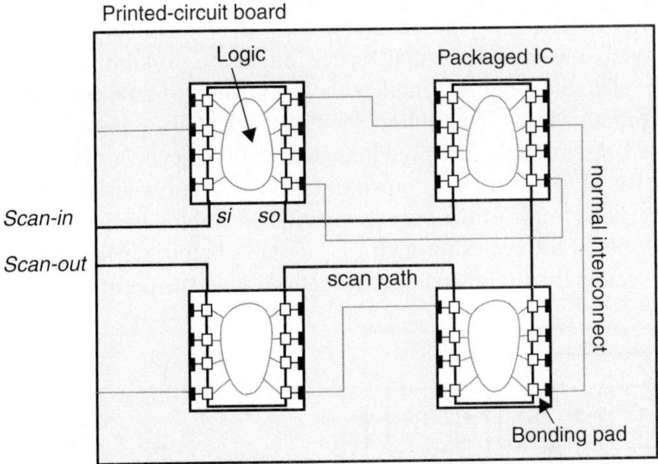

Figure 11.49 The boundary-scan approach to board testing.

Built-in Self-Test (BIST)

An alternative and attractive approach to testability is having the circuit itself generate the test patterns instead of requiring the application of external patterns [Wang86]. Even more appealing is a technique where the circuit itself decides if the obtained results are correct. Depending upon the nature of the circuit, this might require the addition of extra circuitry for the generation and analysis of the patterns. Some of this hardware might already be available as part of the normal operation, and the size overhead of the self-test can be small.

The general format of a built-in self-test design is illustrated in Figure 11.50 ([Kornegay92]). It contains a means for supplying test patterns to the device-under-test and a means of comparing the device's response to a known correct sequence.

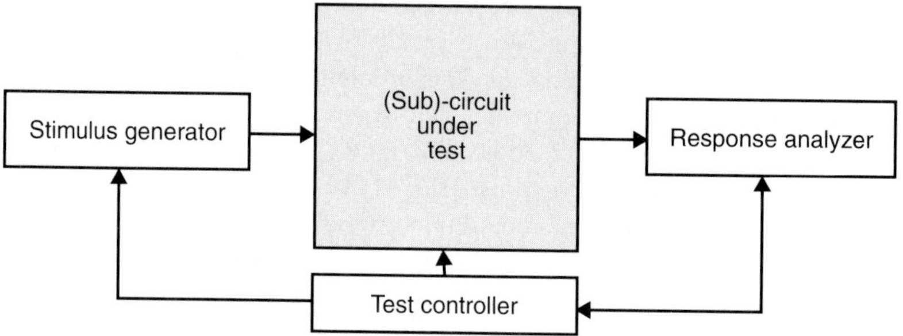

Figure 11.50 General format of built-in self-test structure.

There are many ways to generate stimuli. Most widely used are the *exhaustive* and the *random* approaches. In the exhaustive approach, the test length is 2^N, where N is the number of inputs to the circuit. The exhaustive nature of the test means that all detectable faults will be detected, given the space of the available input signals. An N-bit counter is a good example of an exhaustive pattern generator. For circuits with large values of N, the time to cycle through the complete input space might be prohibitive. An alternative approach is to use random testing that implies the application of a randomly chosen subset of 2^N possible input patterns. This subset should be selected so that a reasonable fault coverage is obtained. An example of a pseudorandom pattern generator is the *linear-feedback shift register* (or LFSR), which is shown in Figure 11.51. It consists of a serial connection of 1-bit registers. Some of the outputs are XOR'd and fed back to the input of the shift register. An N-bit LFSR cycles through $2^N - 1$ states before repeating the sequence, which produces a seemingly random pattern. Initialization of the registers to a given seed

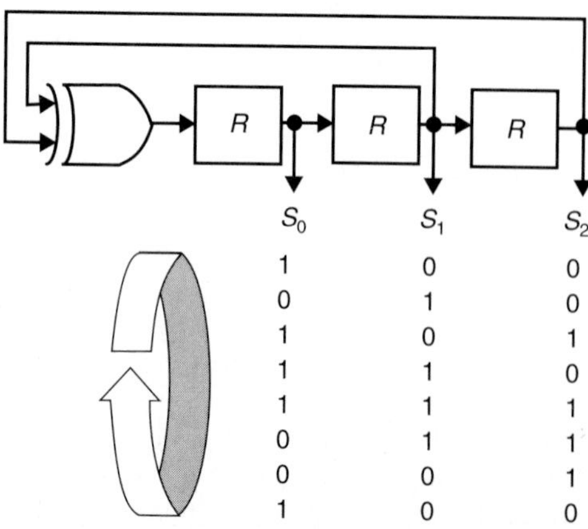

Figure 11.51 Three-bit linear-feedback shift register and its generated sequence.

value (different from 0 for our example circuit) determines what subsequence will be generated.

The response analyzer could be implemented as a comparison between the generated response and the expected response stored in an on-chip memory, but this approach represents too much area overhead to be practical. A cheaper technique is to compress the responses before comparing them. Storing the compressed response of the correct circuit requires only a minimal amount of memory, especially when the compression ratio is high. The response analyzer then consists of circuitry that dynamically compresses the output of the circuit under test and a comparator. The compressed output is often called the *signature* of the circuit, and the overall approach is dubbed *signature analysis*. An example of a signature analyzer that compresses a single bit stream is shown in Figure 11.52.

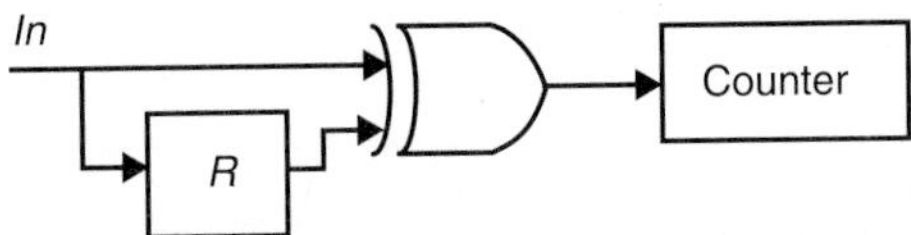

Figure 11.52 Single bit-stream signature analysis.

Inspection reveals that this circuit simply counts the number of $0 \rightarrow 1$ and $1 \rightarrow 0$ transitions in the input stream. This compression does not guarantee that the received sequence is the correct one; that is, there are many different sequences with the same number of transitions. Since the chances of this happening are slim, it may be a risk worth taking if kept within bounds.

Another technique is illustrated in Figure 11.53a. It represents a modification of the linear-feedback shift register and has the advantage that the same hardware can be used for both pattern generation and signature analysis. Each incoming data word is successively XOR'd with the contents of the LFSR. At the end of the test sequence, the LFSR contains the signature, or *syndrome*, of the data sequence, which can be compared with the syndrome of the correct circuit. The circuit not only implements a random-pattern generator and signature analyzer, but can also be used as a normal register and scan register, depending on the values of the control signals B_0 and B_1 (Figure 11.53b). This test approach, which combines all the different techniques, is known as *built-in logic block observation*, or *BILBO* [Koeneman79]. Figure 11.53c illustrates the typical use of BILBO. Using the scan option, the seed is shifted into the BILBO register A while BILBO register B is initialized. Next, registers A and B are operated in the random pattern-generation and signature-analysis modes, respectively. At the end of the test sequence, the signature is read from B using the scan mode.

Finally, it is worth mentioning that self-test is extremely beneficial when testing regular structures such as memories. It is not easy to ensure that a memory, which is a sequential circuit, is fault-free. The task is complicated by the fact that the data value read from or written into a cell can be influenced by the values stored in the neighboring cells because of cross-coupling and other parasitic effects. Memory tests, therefore, include the reading and writing of a number of different patterns into and from the memory using alternating addressing sequences. Typical patterns can be all zeros or ones, or checkerboards of zeros and ones. Addressing schemes can include the writing of the complete memory, followed by a complete read-out or various alternating read-write sequences. With a minimal overhead compared to the size of a memory, this test approach can be built into the integrated

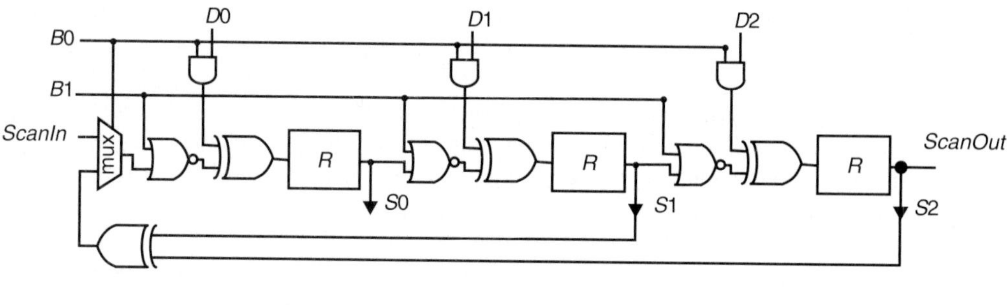

(a) A 3-bit BILBO register

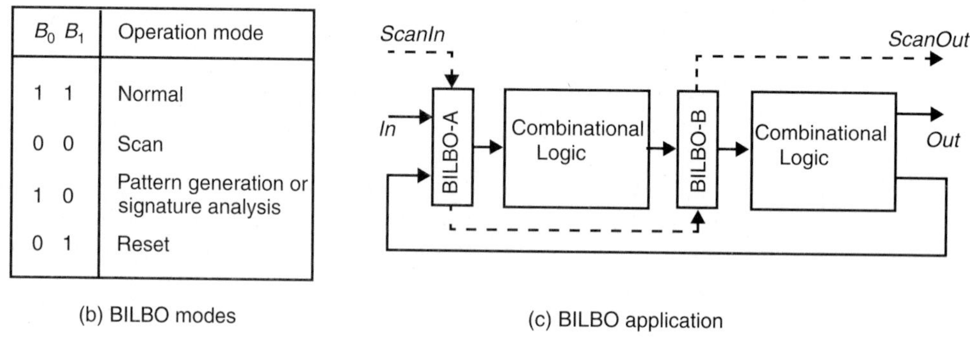

B_0 B_1	Operation mode
1 1	Normal
0 0	Scan
1 0	Pattern generation or signature analysis
0 1	Reset

(b) BILBO modes

(c) BILBO application

Figure 11.53 Built-in logic block observation, or BILBO.

circuit itself, as illustrated in Figure 11.54. This approach significantly improves the testing time and minimizes the external control. Applying self-test is bound to become more important with the increasing complexity of integrated components and the growing popularity of embedded memories.

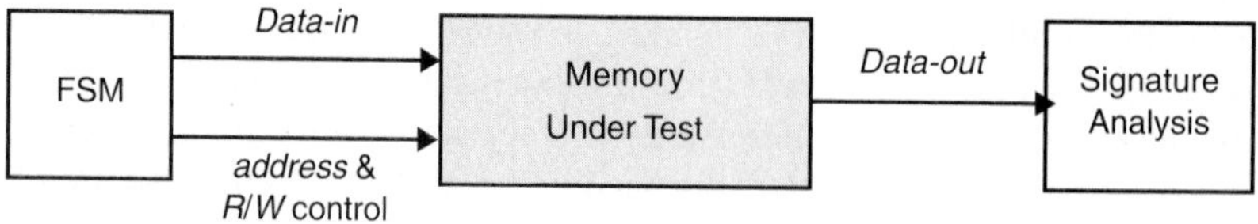

Figure 11.54 Memory self-test.

11.6.3 Test-Pattern Generation

In the preceding sections, we have discussed how to modify a design so that test patterns can be effectively applied. What we have ignored so far is the complex task of determining what patterns should be applied so that a good fault coverage is obtained. This process was extremely problematic in the past, when the test engineer—a different person than the designer—had to construct the test vectors after the design was completed. This invariably required a substantial amount of wasteful reverse engineering that could have been avoided if testing had been considered early in the design flow. An increased sensitivity to design-for-testability and the emergence of automatic test-pattern generation (ATPG) has substantially changed this picture.

In this section, we delve somewhat deeper into the ATPG issue and present techniques to evaluate the quality of a test sequence. Before doing so, we must analyze the fault concept in more detail.

Fault Models

Manufacturing faults can be of a wide variety and manifest themselves as short-circuits between signals, short-circuits to the supply rails, and floating nodes. In order to evaluate the effectiveness of a test approach and the concept of a good or bad circuit, we must relate these faults to the circuit model, or, in other words, derive a *fault model*. The most popular approach is called the *stuck-at* model. Most testing tools consider only the short-circuits to the supplies. These are called the *stuck-at-zero* (sa0) and *stuck-at-one* (sa1) faults for short circuits to *GND* and V_{DD} respectively.

It can be argued that the sa0-sa1 model does not cover the complete range of faults that can occur in a state-of-the-art integrated circuit, and that *stuck-at-open* and *stuck-at-short* faults should also be introduced. However, adding these faults complicates the test-pattern generation process. Moreover, a large number of these faults are covered by the sa0-sa1 model. To illustrate this observation, consider the resistive-load MOS gate of Figure 11.55. All shorts to the supplies are modeled by the introduction of sa0 and sa1 faults at nodes *A*, *B*, *C*, *Z*, and *X*. The figure has been annotated with some stuck-at-open (β) and stuck-at-short faults (α, γ). It can be observed that these faults are already covered by the sa0 and sa1 faults on the various nodes. For example, fault α is covered by A_{sa1}, β is covered by A_{sa0} or B_{sa0}, while γ is equivalent to Z_{sa1}.

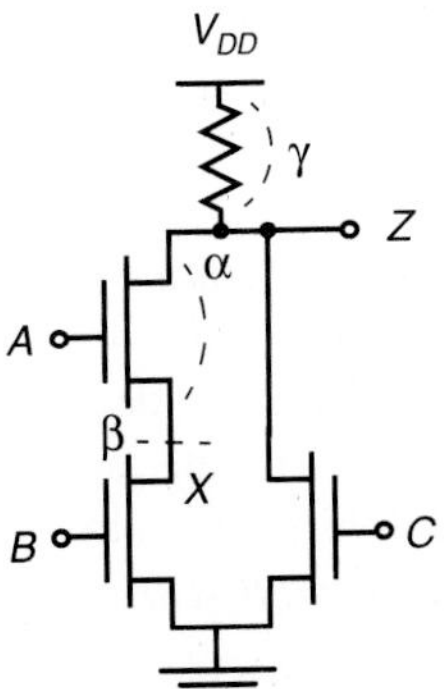

Figure 11.55 Resistive-load gate, annotated with a number of stuck-at-open (β) and stuck-at-short (α,γ) faults.

Even so, shorts and open-circuit faults can cause some interesting artifacts to occur in CMOS circuits that are not covered by the sa0-sa1 model and are worth mentioning. Consider the two-input NAND gate of Figure 11.56, where a stuck-at-open fault α has occurred. The truth table of the faulty circuit is shown in the figure as well. For the combination (A = 1, B = 0), the output node is floating and retains its previous value, while the correct value should be a 1. Depending upon the previous excitation vector, this fault may or may not be detected. In fact, the circuit behaves as a sequential network. To detect this fault, two vectors must be applied in sequence. The first one forces the output to 0 (or $A = 1$ and $B = 1$), while the second applies the $A = 1$, $B = 0$ pattern. Also stuck-at-short faults are troublesome in CMOS circuits since they can cause dc currents to flow between the supply rails for certain input values, which produces undefined output voltages.

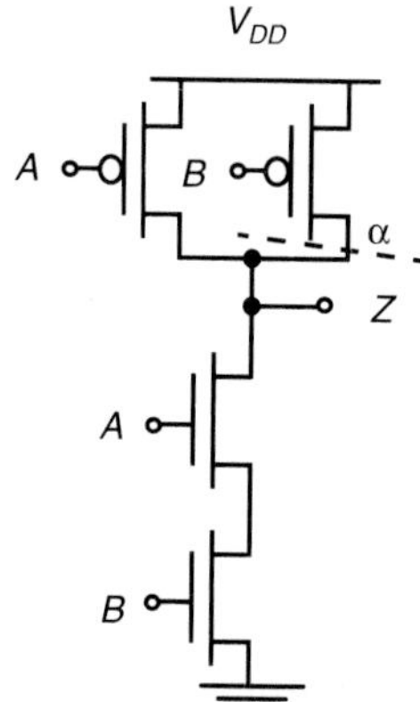

A	B	Z
0	—	1
1	1	0
1	0	Z_{t-1}

Figure 11.56 Two-input complementary CMOS NAND gate and its truth table in the presence of a stuck-at-open fault.

Even though the sa0-sa1 fault model is not perfect, its ease of use and relatively large coverage of the fault space have made it the de facto standard model.

Automatic Test-Pattern Generation (ATPG)

The task of the automatic test-pattern generation (ATPG) process is to determine a minimum set of excitation vectors that cover a sufficient portion of the fault set as defined by the adopted fault model. One possible approach is to start from a random set of test patterns. Fault simulation then determines how many of the potential faults are detected. With the obtained results as guidance, extra vectors can be added or removed iteratively. An alternative and potentially more attractive approach relies on the knowledge of the functionality of a Boolean network to derive a suitable test vector for a given fault. To illustrate the concept, consider the example of Figure 11.57. The goal is to determine the input excitation that exposes an sa0 fault occurring at node U at the output of the network Z. The first requirement of such an excitation is that it should force the fault to occur (*controllability*, again). In this case, we look for a pattern that would set U to 1 under normal circumstances. The only option here is $A = 1$ and $B = 1$. Next, the faulty signal has to propagate to output node Z, so that it can be *observed*. This phase is called *path sensitizing*. For any

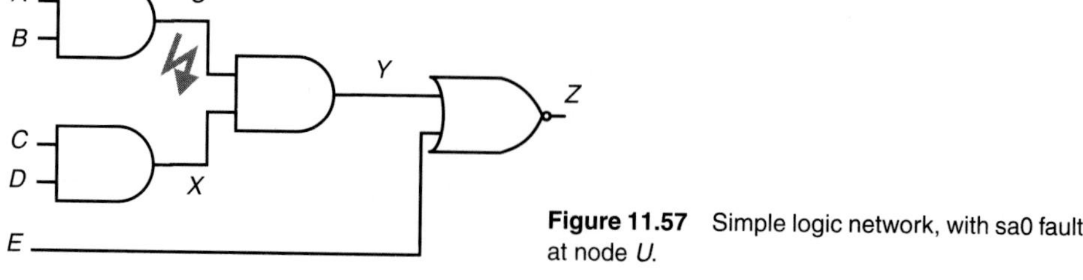

Figure 11.57 Simple logic network, with sa0 fault at node U.

change in node U to propagate, it is necessary for node X to be set to 1 and node E to 0. The (unique) test vector for U_{sa0} can now be assembled: $A = B = C = D = 1$, $E = 0$.

This example is extremely simple, and the derivation of a minimum test-vector set for more complex circuits is substantially more convoluted. A number of excellent approaches to address this problem have been developed. Landmark efforts in this domain are the *D* [Roth66] and PODEM algorithms [Goel81], which underlie many current ATPG tools. It suffices to say that ATPG is currently in the mainstream of design automation, and powerful tools are available from many vendors.

Fault Simulation

A fault simulator measures the quality of a test program. It determines the *fault coverage*, which is defined as the total number of faults detected by the test sequence divided by two times the number of nodes in the network—each node can give rise to an sa0 and sa1 fault. Naturally, the obtained coverage number is only as good as the fault model employed. In an sa0-sa1 model, some of the bridge and short faults are not covered and will not appear in the coverage statistics.

The most common approach to fault simulation is the parallel fault-simulation technique: the correct circuit is simulated concurrently with a number of faulty ones, each of which has a single fault injected. The results are compared, and a fault is labeled as detected for a given test vector set if the outputs diverge. This description is overly simplistic, and most simulators employ a number of techniques such as selecting the faults with a higher chance of detection first, to expedite the simulation process. Hardware fault-simulation accelerators, based on parallel processing and providing a substantial speed-up over pure software-based simulators, are available as well [Agrawal88, pp. 159–240].

11.7 Perspective and Summary

In this chapter, we have briefly scanned the complex world of design methodologies and design automation. In the 1980s, we have witnessed an enormous effort in this domain that has yielded an extensive set of design aids that not only help designers analyze and verify a design, but also optimize and synthesize it and make it testable. These tools and design techniques have had a major impact on the way design is performed today and have made possible the exciting and impressive processors and application-specific circuits to which we have become accustomed. Becoming familiar with these design methodologies is an essential part of the learning experience of the beginning digital designer. We hope this chapter, although compressed, has enticed the reader to further explore the numerous possibilities offered by design automation. We have touched on the following issues in this chapter.

- The simulator is the work-horse of the digital designer. Accuracy can be traded off for execution speed by simplifying the device models and by adapting higher-level data and time abstractions. The following simulation tools are often used by the digital designer: circuit, timing, switch, logic, and functional. Choosing the correct simulation level can reduce the response time by orders of magnitude.

- The verification approach avoids the dependency on excitation vectors that is one of the major pitfalls of the simulation approach. Electrical and timing verification tools are now an integral part of digital design methodology. Formal verification has made important strides in recent years, but needs a solution to the complexity problem before it can become a mainstream approach.
- A wide variety of circuit-implementation methodologies have come into existence, ranging from the high-performance, high-density, custom-design approach on one end to the programmable, fast-turnaround FPGA on the other. In between, we find techniques such as standard-cell, module-generation, gate-arrays and sea-of-gates. What approach to select depends upon the application at hand and is a function of the performance, power, and cost constraints.
- Design synthesis is making a major inroads in the way design is performed and will continue to grow in importance towards the end of the twentieth century. Logic synthesis is already the dominant implementation technique for most application-specific designs, and architectural synthesis is rapidly gaining acceptance. Manual transistor-level design will be mostly confined to the design of subsystems with stringent requirements for either area, power or speed.
- Testability of a design should be considered in the early phases of the design process. A comprehensive design-for-test approach includes techniques to incorporate controllability and observability into the circuit and the accompanying methods to generate a minimum set of test vectors with sufficient fault coverage.

One final observation: Even with the increasing automation of the digital circuit design process, new challenges are continuously emerging—challenges that require the profound insight and intuition only offered by a human designer.

11.8 To Probe Further

The literature on design methodologies and automation for digital integrated circuits has exploded in the last decade. A number of reference works that give an overview of the main activities in the field are worth mentioning.

- General CAD techniques and methodologies—[Ullman84], [Rubin87]
- Implementation approaches—[Preas88], [DeMicheli87], [Brodersen92]
- Synthesis—[Brayton84], [Gajski92], [DeMicheli94]
- Testability—[Agrawal88]

State-of-the-art developments in the design automation domain are generally reported in the *IEEE Transactions on CAD* and the *IEEE Transactions on VLSI Systems*. Premier conferences are, among others, the Design Automation Conference (DAC) and the International Conference on CAD (ICCAD).

REFERENCES

[Alford89] R. Alford, *Programmable Logic Designer's Guide*, Howard Sams & Company, 1989.

[Agrawal88] V. Agrawal and S. Seth, Eds. *Test Generation for VLSI Chips*, IEEE Computer Society Press, 1988.

[Agrawal90] P. Agrawal and W. Dally, "A Hardware Logic Simulation System," *IEEE Trans. Computer-Aided Design*, CAD-9, no. 9, pp. 19–29, January 1990.

[Altera92] *Altera Data Book*, San Jose, 1992.

[Brayton84] R. Brayton et al., *Logic Minimization Algorithms for VLSI Synthesis*, Kluwer Academic Publishers, 1984.

[Brayton87] R. Brayton, R. Rudell, A. Sangiovanni-Vincentelli, and A. Wang, "MIS: A Multi-Level Logic Optimization System," *IEEE Trans. on CAD*, CAD-6, pp. 1062–1081, November 1987.

[Brodersen92]. R. Brodersen, ed., *Anatomy of a Silicon Compiler*, Kluwer Academic Publishers, 1992.

[Bryant81] R. Bryant, *A Switch-Level Simulation Model for Integrated Logic Circuits*, Ph. D. diss., MIT Laboratory for Computer Science, report MIT/LCS/TR-259, March 1981.

[Carlson91] S. Carlson, *Introduction to HDL-Based Design Using VHDL*, Synopsys, 1991.

[Chu86] K. Chu and D. Pulfrey, "Design Procedures for Differential Cascode Logic," *IEEE Journal of Solid State Circuits*, vol. SC-21, no 6, Dec. 1986, pp. 1082-1087.

[Claesen91] L. Claesen, "SFG Tracing: A Method for the Automatic Verification of MOS Transistor-Level Implementations from High-Level Behavioral Specifications," in *Proc. 1991 Int. Workshop Formal Methods in VLSI Design*, 1991.

[Coudert90] O. Coudert and J. Madre, *Proceedings JCCAD 1990*, pp. 126–129, November 1990.

[Croes88] K. Croes, H. De Man, and P. Six, "CAMELEON: A Process-Tolerant Symbolic Layout System," *Journal of Solid State Circuits*, vol. 23 no. 3, pp. 705–713, June 1988.

[DeMan85] H. De Man, I. Bolsens, E. Vandenmeersch, and J. Van Cleynenbreugel, "DIALOG: An Expert Debugging System for MOS VLSI Design," *IEEE Trans. on CAD*, vol. 4, no. 3, pp. 301–311, July 1985.

[DeMan86] H. De Man, J. Rabaey, P. Six, and L. Claesen, "Cathedral-II: A Silicon Compiler for Digital Signal Processing," *IEEE Design and Test*, vol. 3, no. 6, pp. 13–25, December 1986.

[DeMicheli87] G. De Micheli, A. Sangiovanni-Vincentelli, and P. Antognettii, eds., *Design Systems for VLSI Circuits*, Martinus Nijhoff Publishers, 1987.

[DeMicheli94] G. De Micheli, *Synthesis and Optimization of Digital Circuits*, McGraw-Hill, 1994.

[Devadas91] S. Devadas, K. Keutzer, and S. Malik, "Delay Computations in Combinational Logic Circuits: Theory and Algorithms," *Proc. ICCAD-91*, pp. 176–179, Santa Clara, 1991.

[Eichelberger78] E. Eichelberger and T. Williams, "A Logic Design Structure for VLSI Testability," *Journal on Design Automation of Fault-Tolerant Computing*, vol. 2, pp. 165-178, May 1978.

[El-Ayat89]K. El-Ayat, "A CMOS Electrically Configurable Gate Array," *IEEE Journal of Solid State Circuits*, vol. SC-24, no. 3, pp. 752–762, June 1989.

[Fishburn85] J. Fishburn and A. Dunlop, "TILOS: A Polynomial Programming Approach to Transistor Sizing," *Proceedings ICCAD-85*, pp. 326–328, Santa Clara, 1985.

[Gajski92] D. Gajski, N. Dutt, A. Wu, and S. Lin, *High-Level Synthesis—Introduction to Chip and System Design*, Kluwer Academic Publishers, 1992.

[Gallia90] J. Gallia et al., "High-Performance BiCMOS 100 K-Gate Array," *IEEE Journal of Solid State Circuits*, vol. SC-25, no. 1, pp. 142–149, February 1990.

[Goel81] P. Goel, "An Implicit Enumeration Algorithm to Generate Tests for Combinational Logic Circuits," *IEEE Trans. on Computers*, vol. C-30, no. 3, pp. 26–268, June 1981.

[Hill85] D. Hill, "S2C—A Hybrid Automatic Layout System," *Proc. ICCAD-85*, pp. 172–174, November 1985.

[Hsueh79] M. Hsueh and D. Pederson, "Computer-Aided Layout of LSI Building Blocks," *Proceedings ISCAS Conf.*, pp. 474–477, Tokyo, 1979.

[Ieda91] N. Ieda, "Technology Trends in ASIC," *IEICE Transactions,* vol. E74, no. 1, pp. 148–156, January 1991.

[IEEE1149] IEEE Standard 1149.1, "IEEE Standard Test Acess Port and Boundary-Scan Architecture," IEEE Standards Board, New York.

[Jouppi84] N. Jouppi, *Timing Verification and Performance Improvement of MOS VLSI Designs,* Ph. D. diss., Stanford University, 1984.

[Koeneman79] B. Koeneman, J. Mucha, and O. Zwiehoff, "Built-in Logic-Block Observation Techniques," in *Digest 1979 Test Conference*, pp. 37–41, October 1979.

[Kornegay92] K. Kornegay, *Automated Testing in an Integrated System Design Environment,* Ph. D diss. , Mem. No. UCB/ERL M92/104, Sept. 1992.

[Lidsky94] D. Lidsky and J. Rabaey, "Low-Power Design of Memory-Intensive Applications—Case Study: Vector Quantization," *Proc. Symposium on Low-Power Electronics,* San Diego, Oct. 1994.

[Nagel75] L. Nagel, "SPICE2: a Computer Program to Simulate Semiconductor Circuits," Memo ERL-M520, Dept. Elect. and Computer Science, University of California at Berkeley, 1975.

[Ousterhout83] J. Ousterhout, "Crystal: A Timing Analyzer for nMOS VLSI Circuits," *Proc. 3rd Caltech Conf. on VLSI* (Bryant ed.), Computer Science Press, pp. 57–69, March 1983.

[Ousterhout84] J. Ousterhout, G. Hamachi, R. Mayo, W. Scott, and G. Taylor, "Magic: A VLSI Layout System," *Proc. 21st Design Automation Conference*, pp. 152–159, 1984.

[Preas88] M. Preas and M. Lorenzetti, eds, *Physical Design Automation of VLSI Systems,* Benjamin-Cummins, 1988.

[Rabaey91] J. Rabaey, C. Chu, P. Hoang and M. Potkonjak, "Fast Prototyping of Datapath-Intensive Architectures," *IEEE Design and Test*, vol. 8, pp. 40–51, 1991.

[Roth66] J. Roth, "Diagnosis of Automata Failures: A Calculus and a Method," *IBM Journal of Research and Development*, vol. 10, pp. 278–291, 1966.

[Rubin87] S. Rubin, *Computer Aids for VLSI Design*, Addison-Wesley, 1987.

[Salz89] A. Salz and M. Horowitz, "IRSIM: An Incremental MOS Switch-Level Simulator," *Proceedings of the 26th Design Automation Conference*, pp. 173–178, 1989.

[Thor88] R. Alverson et al., "THOR User's Manual," Technical Report CSL-TR-88-348 and 349, Stanford University, January 1988.

[Ullman84] J. Ullman, *Computational Aspects of VLSI,* Computer Science Press, Maryland, 1984.

[VHDL88] VHDL Standards Committee, IEEE Standard VHDL Language Reference Manual, IEEE std 1076-1077, 1978.

[Wang86] L. Wang and E. McCloskey, "Complete Feedback Shift-Register Design for Built-in Self Test," *Proc. ICCAD 1986,* pp. 56–59, November, 1986.

[Weste93] N. Weste and K. Eshraghian, *Principles of CMOS VLSI Design—A Systems Perspective,* Addison-Wesley, 1993.

[Williams83] T. Williams and K. Parker, "Design for Testability—A Survey," *Proceedings IEEE,* vol. 71, pp. 98–112, Jan. 1983.

[White87] J. White and A. Sangiovanni-Vincentelli, *Relaxation Techniques for the Simulation of VLSI Circuits*, Kluwer Academic, 1987.

[Wolf94] W. Wolf, *VLSI Design: A Systems Approach,* Prentice Hall, NJ, 1994.

[Xilinx90] *The Programmable Gate Array Data Book,* Xilinx, Inc., 1990.

11.9 Exercises and Design Problems

1. [E, None, 11.2] Consider a 2-bit full-adder design based on the adder cell of Figure 11.39. Draw the signal waveforms, assuming the following delay models: zero delay, unit delay, and variable delay. For the latter, assume a delay of 1 for the inverters and 2 for the EXOR and AND-OR-INVERT cells. Apply the following input patterns: A = (00, 01, 11), B = (00, 00, 01), (lsb first). Assume a time period of 6 units between consecutive patterns.
2. [E, None, 11.2] Assume that we would like to analyze the power consumption of a CMOS digital circuit. Determine for each simulation level (circuit, switch, gate) if the following power contributions can be included:
   a. Dynamic power consumption
   b. Static power consumption
   c. Direct path current in CMOS
   d. Glitching
3. [M, None, 11.3] Consider the carry-bypass adder of Figure 7.17. Assume a unit-delay model for every cell (full adder, multiplexer, setup, and sum). Find the longest path in the adder. Is this the real worst-case delay of the module? If not, explain and find the worst-case delay.
4. [M, None, 11.4] Use the channel-routing approach to route the channel of Figure 11.58 so that the height H of the channel is minimized (i.e., minimize the number of horizontal tracks). Nodes with the same name should be connected.

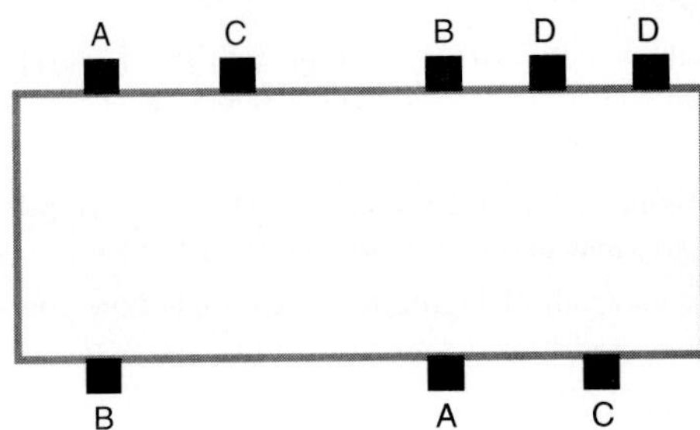

Figure 11.58 Channel-routing problem.

5. [E, None, 11.4] Program the gate-array template of Figure 11.24 to perform the following logical functions:
   a. Four-input NAND gate
   b. $Z = \overline{ABC+DE}$
6. [M, None, 11.4] Implement the four-input NAND gate of problem 5*a* on the gate-isolated cell of Figure 11.26b.
7. [E, None, 11.4] Program the Actel cell of Figure 11.32 to implement a three-input OR gate.
8. [M, None, 11.4] Discuss under what circumstances you would prefer to use the PROM versus the PAL PLD-architectures (Figure 11.30). Give an example for each.
9. [M, None, 11.5] Consider the following logic function (from [Wolf94]):

$$f_1 = (\overline{b} + \overline{c} + d); f_2 = \overline{a}\,\overline{f_1}$$

   a. Find a minimum two-stage implementation.
   b. Find a minimum-cost multilevel implementation given the following cell library: inverter (cost = 1), two-input NOR (cost = 3), two-input NAND (cost = 3).

c. Repeat 9*b* for the following library—inverter (cost = 1), 2-input NOR (cost = 3), 3-input NOR (cost = 4), 2-input NAND (cost = 3), 3-input NAND (cost = 4).

10. [M, None, 11.5] Repeat 9c, but optimize for speed this time. Assume that the delay of a gate is equal to the sum of its fan-in and fan-out. For instance, a two-input NAND with a fan-out of 2 has a delay of 4.

11. [E, None, 11.5] Consider the synthesized architectures of Figure 11.41. For both the 4-cycle and the 1-cycle case, determine when and where each of the operations of the sum-program is executed. In other words, determine the assignment and the schedule. Determine also where the intermediate variables are stored.

12. [M, None, 11.6] Assume that an sa1 fault occurs at the output of $(\overline{A(B+C_i)})$ cell in the full-adder design of Figure 11.39. Determine a test pattern that will propagate this fault to the output. Do the same for an sa0 fault at that node.

13. [C, None, 11.6] Determine a test pattern to detect an sa1 fault at node X of the circuit in Figure 11.59 and to propagate it to the output Z. Is this test pattern unique?

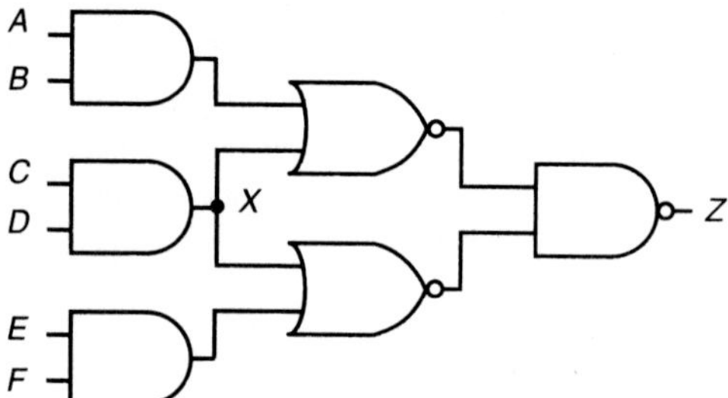

Figure 11.59 Logic circuit with an sa1 fault at node *X*.

14. [C, None, 11.6] Define a minimum set of test vectors to test an 8-bit ripple-carry adder. The vector set should force each carry and sum bit to both the 0 and 1 values.

15. [M, None, 11.6] Consider the datapath of Figure E.2. Determine how you would make this design testable.

a. Using scan-test.

b. Using self-test.

DESIGN PROJECT

To become familiar with the different layout-generation techniques, walk through the tutorial examples of the LAGER silicon compilation system [Brodersen92]. The tutorials can be downloaded from the World-Wide Web page of this book: (http://infopad.eecs.berkeley.edu/~icdesign). If you have a UNIX workstation available, you can actually execute the layout-generation process yourself. The software of LAGER can be downloaded from the same address. Adventurous readers might even want to attempt to design a small system themselves

PROBLEM SOLUTIONS

Problem 2.1—It is easily validated that the short-base model is applicable here:

$$W_n - W_2 = 5\ \mu m - 0.15\ \mu m << L_p = 31\ \mu m$$

$$W_p - W_1 = 0.7\ \mu m - 0.03\ \mu m << L_n = 5\ \mu m$$

With the aid of Eqs. (2.7), (2.8) and (2.10), we find that $I_S = 62 \times 10^{-18}$ A. Finally, $V_D = \phi_T \ln (I_D/I_S)$ = 0.73 V.

Problem 2.2—Similar to the NMOS case, but with all variables negated. A simple approach to avoid sign problems is to use the absolute values of the voltages in the current equations, and adjust the sign of the final result.

Problem 2.3—$\beta_F = 100$, $\beta_R = 1$ (be aware that the latter expresses the relation between I_B and I_E).

Problem 2.4—Similar to Figure 2.40 and Figure 2.41, but all variables are negated.

Problem 2.5—The transient, once again, consists of two components: space charge and base charge. The time to provide the space charge is approximated by the solution of an RC-equivalent circuit:

$$t_{space} = R_B C_j \ln\left(\frac{V_0}{V_0 - V_{BE(on)}}\right) = 5(1.5 \times 20 + 0.68 \times 22)\ln\left(\frac{1}{1 - 0.75}\right) = 311 \text{ psec}$$

The time to provide 90 % of the base charge equals 2.3 times τ_{BF} (similar to the diode model of Eq. (2.33)), or, t_{base} = 2300 psec!

The SPICE simulation shows that it takes approximately 300 psec for the base voltage to reach the 0.7 V point. The collector current reaches its 90% point after 2.35 nsec.

Problem 3.1—The propagation delay is expressed as $t_p = 0.69\ C_L R_{on}$, where 0.69 represents the fraction of the time-constant it takes to reach the 50% point. R_{on} is the average on-resistance of the pull-up or pull-down transistor. For instance, for the low-to-high transition, R_{on} is computed as follows:

$$R_{on} = \frac{1}{2}\left[\frac{V_{DD}}{\frac{k_p}{2}(V_{DD} - |V_{Tp}|)^2} + \frac{V_{DD}/2}{k_p\left((V_{DD} - |V_{Tp}|)\frac{V_{DD}}{2} - \frac{V_{DD}^2}{8}\right)}\right]$$

For V_{DD} = 5 V and $|V_{Tp}|$ this evaluates to $0.44/k_p$. For a (5.4/1.2) device, this means an average on-resistance of 13.6 kΩ. For a load capacitance C_L of 32.6 fF, this translates into a t_{pLH} of 305 psec which is substantially higher than the 200 psec, found using the average current. A better approach is to compute the average resistance, so that the same average current is obtained.

Problem 4.1—$F = \overline{D + A(B + C)}$. To verify that for every input combination a path exists to either V_{DD} or GND (but never to both), enumerate all possible input patterns ands analyze the network.

Problem 4.2—Critical transistors are those on series connections. When N transistors are connectied in series, we multiply their width by N to obtain a composite device with an on-resistance equal to that of a single transistor. Transistors in parallel are not modified, as their worst-case resistance is equal to each of the individual devices. This technique is used recursively for every branch in the network. For instance, the pull-down network consists of two parallel transistors. No size adaptation is needed. The right branch, however, consists of two series devices. Each of these is widened by a factor of 2. The same reasoning can be used for the PUN.

Problem 4.3—NAND logic is preferable, as this results in series chains in the NMOS PDN. This approach is preferable to series chains of PMOS devices.

Problem 4.4—The propagation delay of the current source load equals $C_L R_L/2$ (obtained using the by now traditional propagation-delay formula, with $I_{av} = V_{DD}/R_L$). The delay of the resistive load equals 0.69 $C_L R_L$. The former is 28% smaller than the latter.

Problem 4.5—For simplicity, we assume that V_{swing} is equal to V_{DD}. The average currents are easily derived. The propagation delay is then obtained by using the traditional formula.

$$I_{av}(\mathrm{L} \to \mathrm{H}) = \frac{1}{2}\left[\frac{k_p}{2}(V_{DD} - |V_{Tp}|)^2 + k_p\left((V_{DD} - |V_{Tp}|)\frac{V_{DD}}{2} - \frac{V_{DD}^2}{8}\right)\right]$$

$$I_{av}(\mathrm{H} \to \mathrm{L}) = \frac{1}{2}\left[\frac{k_n}{2}(V_{DD} - V_{Tn})^2 + k_n\left((V_{DD} - V_{Tn})\frac{V_{DD}}{2} - \frac{V_{DD}^2}{8}\right) - k_p\left((V_{DD} - |V_{Tp}|)\frac{V_{DD}}{2} - \frac{V_{DD}^2}{8}\right)\right]$$

$$t_p = \frac{C_L(V_{DD}/2)}{I_{av}}$$

Problem 4.6—In contrast to complementary CMOS, the NOR structure is the preferred topology for pseudo-NMOS. The NOR structure completely avoids series transistors.

Problem 4.7—

- $V_{out} < |V_{tp}|$: NMOS linear, PMOS cutoff.
- $|V_{tp}| < V_{out} < V_{DD} - V_{Tn}$: NMOS linear, PMOS saturated.
- $V_{DD} - V_{Tn} < V_{out}$: NMOS and PMOS saturated.

Problem 4.8—The n-modules have a NM_L equal to V_{Tn}, while the p-modules have an NM_H of $|V_{Tp}|$. Cascading the modules combines both effects. For instance, the output of an n-module feeds into a p-module. This severely increases the noise-sensitivity of that node, as any signal larger than $|V_{Tp}|$ will turn on the PUN of the next stage.

Problem 4.9—These transition probabilities are easily derived from the following observations. The chance that an AND gate is in the 1-state equals the probability that input A is high AND input B is high, or $P_{AND}(1) = P_A \cdot P_B$. From this, the transition probability can be computed. As $P_{0\to1} = (1 - P(1))P(1)$. Similarly, the probability that the output of the XOR gate is high equals the chance that A is high or B is high but not both of them at the same time. This is expressed as follows: $P_{XOR}(1) = P_A + P_B - 2P_AP_B$. Observe that the transition probabilities of the complementary gates (NAND, NOR, NXOR) are identical to their noninverting counterparts.

Problem 4.10—The effective capacitance can be computed using the transition probabilities of Table 4.7.

$$C_{EFF} = C_L(1 - P_AP_B)P_AP_B + C_L(1 - P_A - P_B + 2P_AP_B)(P_A + P_B - 2P_AP_B)$$

The switching energy is then obtained by multiplying C_{EFF} by V_{DD}^2.

Problem 4.11—Trivial, given the solution of Problem 4.9.

Problem 5.1—See Figure P5.1.

Problem 5.2—To make a fair comparison, we equalize the average power consumption of both implementations, or $I_{EF} = (V_{OH} + V_{OL} - 2V_{EE})/2R_B$ (with I_{EF} the value of the current source). This results in the

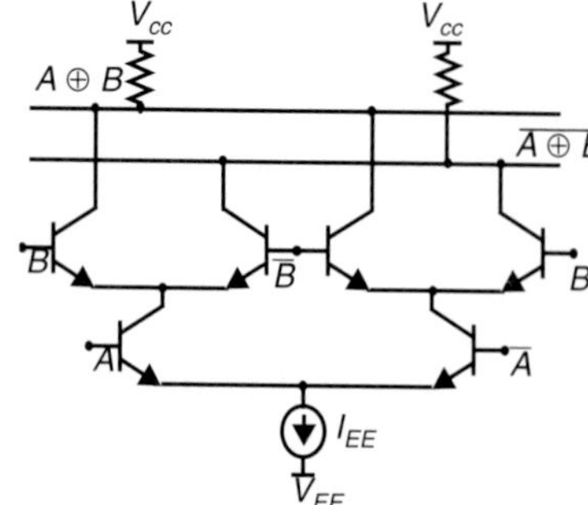

Figure P5.1 The circuit to the left shows the differential solution. The single-ended implementation can be obtained using the wired-OR circuit of Figure 5.3, replacing the A, B, C and D inputs by A, $\overline{B}$, $\overline{A}$, and B, respectively.

following pull-down delay for the emitter-follower (the low-to-high transition is only marginally affected): $t_{pHL} = C_L V_{swing}/2I_{EF} = (V_{swing}/(V_{OH} + V_{OL} - 2V_{EE}))R_B C_L$ (versus 0.69 $R_B C_L$ for the resistive case). For the same power, the current-source implementation generally produces a faster response.

Problem 6.1—See Figure P6.1:

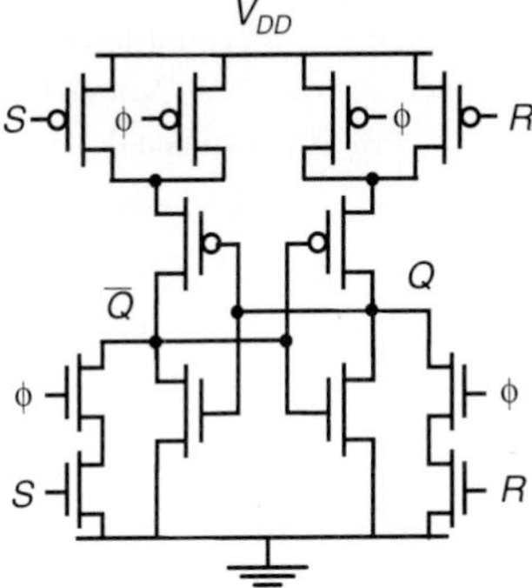

Figure P6.1 The fully complementary implementation eliminates the existence of ratioed inverters during the switching. No transistor sizing is required. This results in a faster switching and smaller transient currents, at the expense of complexity.

Problem 6.2—The access devices only serve as triggers. The ultimate voltage levels are set by the cross-coupled inverter pairs. The voltage loss over the pass-transistors hence does not matter that much.

Problem 6.3—The set-up time equals the time it takes to get the storage capacitor sufficiently charged. Typically we use the 90% point as the measure for this to be true. With R_{on} the linearized on-resistance of the access switch, we can approximate the set-up time by the following expression: $t_{set\text{-}up} = 2.2\ R_{on}\ C_{store}$. The propagation delay of the latch equals the the sum of the delays of the input RC-network and the inverter, or $t_{platch} = 0.69\ R_{on}\ C_{store} + T_{pinv.}$

Problem 6.4—For a low-to-high transition, transistor M_2 only turns on when node X, the output voltage of the saturated-load inverter M_1-M_5 reaches the following value: $V_X = V_{in} - V_{Tn}$. We use this value of V_{in} as an approximation of V_{M-}. Ignoring the body effect, the following expression for V_{M-} can be derived:

$$V_{M-} = \frac{V_{DD} + V_{Tn}\sqrt{k_1/k_5}}{1 + \sqrt{k_1/k_5}}$$

Similarly, we can derive V_{M+}.

$$V_{M+} = \frac{(V_{DD} - |V_{Tp}|)\sqrt{k_4/k_6}}{1 + \sqrt{k_4/k_6}}$$

Problem 6.5—See Figure P6.5.

Problem 7.1—Observe that all nodes in the carry-chain network are precharged high. The NMOS transistors only act as pull-down devices, a task they are well suited for. No threshold voltage is ever lost.

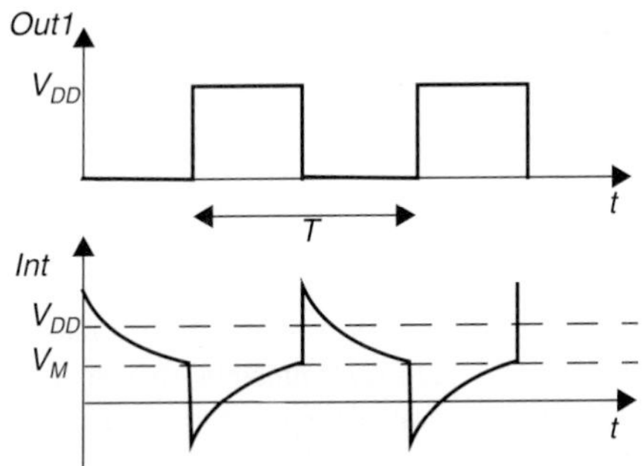

Figure P6.5 The oscillation period is easily derived from the inspection of the waveforms.

Problem 7.2—One of the possible worst-case patterns is given below (msb first). It generates a carry at the lsb-position and propagates it all the way to the msb.

A: 0000 0000 0000 0001

B: 0111 1111 1111 1111

The worst-case delay equals $(1 + 4 + 3 \times 1 + 4 + 1) = 13$. The delay of an equal-length ripple-carry adder would be $(1 + 16 \times 1 + 1) = 18$, or the bypass reduces the delay with 28%. Notice that the shortest delay is observed for the following block organization: 2 3 4 3 2! This yields a delay of only 10!

Problem 7.3—The delay of the 4×4 configuration equals $(1 + 4 + 4 \times 1 + 1) = 10$. A large number of configurations with a delay of 9 can be found. An example is 3 4 5 4.

Problem 7.4—Since the delay of the sum-generation is also proportional to the length of the stage, it is important to make the last stages not too long. Combining this requirement with the equal-arrival-time constraint (at the inputs of the bypass multiplexer) leads to the following optimal structure: stages grow gradually from the lsb, until the middle of the adder, after which they start to shrink again. An example of such a topology is the 2 3 4 3 2 configuration, derived in Problem 7.2. Assuming that the first and last stage have identical lengths, the expression for the propagation delay can be derived

$$\begin{aligned} t_{add} &= t_{setup} + Mt_{carry} + Pt_{bypass} + Mt_{carry} + t_{sum} \\ N &= 2(M + (M+1) + \ldots + (M + P/2 - 1)) \end{aligned}$$

The second expression assumes an even number of stages. For large values of N, a quadratic dependence between N and the number of stages P can once again be observed: $N \approx P^2/4$.

Problem 7.5—Two small changes are sufficient: the upper leftmost adder has to be modified into a FA, taking Y_0X_3 as an extra input (sign extension). The output carry of the lower leftmost adder represents an extra output bit Z_7.

Problem 7.6—In a two's complement notation, the msb represents the sign bit. A shift to the right requires this sign-bit to be repeated. This is exactly what happens in the upper diagonal section of the shift-matrix of the barrel-shifter. The bit A_3 is repeated 1, 2, or 3 times for Sh_1, Sh_2 and Sh_3, respectively.

Problem 7.7—Hint: Use two stages in each cell. The set-up stage compares the bits at position i and has three outputs: >, =, <. The second stage propagates these results upwards in a ripplelike fashion.

Problem 7.8—The pipelining allows us to lower the supply voltage to 2.9 V (just like in the parallel case). Clock frequency remains identical, while the capacitance increases with 15%. This yields the following expression for the power consumption:

$$P_{pipe} = C_{pipe}V^2{}_{pipe}f_{pipe} = (1.15C_{ref})\left(\frac{2.9}{5}V_{ref}\right)^2 f_{ref} \approx 0.39P_{ref}$$

Problem E.1—To conserve area, it is important to minimize the maximum number of feedthroughs per cell. Any other ordering requires more than the one feedthrough per cell that the current solution offers.

Problem 8.1—This result is obtained by replacing the first equation of Eq. (8.15) by

$$t_p = N(u + \alpha)\, t_{p0}$$

and setting the derivative of t_p to 0. The resulting equation is very nonlinear, and numerical equation solving is advisable.

Problem 8.2—The addition of the cascode transistors reduces the swing at the internal nodes in the pull-down network to $V_{ref} - V_T$. The output node still has the full rail-to-rail swing. This means that the DSL logic style is effective when the intrinsic load dominates the extrinsic one. This is the case for complex logic gates with a small-to-medium fan-out. This gate unfortunately consumes static power, since the high logic value of $V_{ref} - V_T$ is too low to turn the connecting pull-up device off, as happens in the DCVSL structure.

Problem 8.3—The correct approach is to scale down the last stage only (b). This is where the largest current change occurs, accounting for most of the *Ldi/dt* effect. The performance impact of scaling on only a single stage is small.

Problem 8.4—See Figure P8.4.

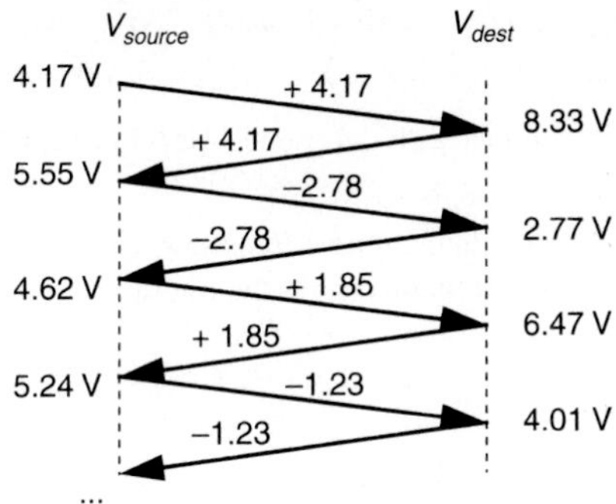

Figure P8.4 The parallel-termination cases proceed along a similar path.

Problem 9.1—Assume, without loss of generality, that the clock ϕ has a 50% duty cycle. Set t_{min} to the shortest timing path between a latch through the logic into the next latch, while t_{max} stands for the worst-case path. The following equations are easily derived using an approach similar to that used for the other clocking methodologies.

$$t_{min} < \delta$$

$$t_{max} < T/2 + \delta$$

We leave it as an exercise for the reader to analyze the case where the duty cycle is different or where a two-phase nonoverlapping clock is used.

Problem 9.2—Assume that the FIFO is initially empty and that *En*1, *En*2, and *En*3 are 0 as well as *Ack*0. We will input new data values without removing any old ones until the FIFO fills up. Notice that each data word automatically moves as deep in the FIFO as it can.

$$Req_i\uparrow \rightarrow En1\uparrow \rightarrow Done1\uparrow \rightarrow En2\uparrow \rightarrow Done2\uparrow \rightarrow En3\uparrow \rightarrow Req0\uparrow$$

$$\rightarrow Ack_i\uparrow \rightarrow Req_i\downarrow \rightarrow En1\downarrow \rightarrow Done1\downarrow \rightarrow En2\downarrow \rightarrow Done2\downarrow$$

$$\rightarrow Ack_i\downarrow \rightarrow Req_i\uparrow \rightarrow En1\uparrow \rightarrow Done1\uparrow$$

$$\rightarrow Ack_i\uparrow \rightarrow Req_i\downarrow$$

We can now see that the FIFO is full if the *Enable* Signals (or the *Ack*'s for the next stage) alternate between 0's and 1's. In the above case, we have *En*1 = 1, *En*2 = 0, *En*3 = 1, and *Ack*0 = 0, while Req_i = 0. On the other hand, the FIFO is empty if all enable signals are equal (either 0 or 1).

Problem 9.3—See Figure P9.3.

Problem 10.1—(0) 0100; (1) 1001; (2) 0101; (3) 0000

Problem 10.2—(0) 1011; (1) 0110; (2) 1010; (3) 1111

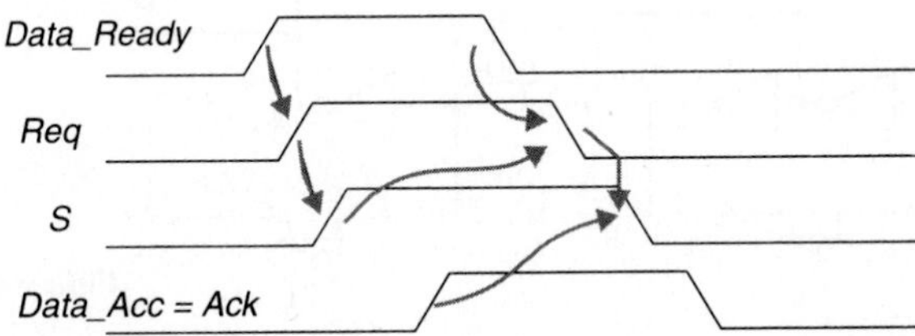

Figure P9.3.

Problem 10.3—(0) 0100; (1) 1001; (2) 0101; (3) 0000

Problem 10.4—Word-line delay:

$$t_{WL} = 0.38\ (30\ \Omega \times (0.56 + 3.8)\ \text{fF}) \times 512 = 12.7\ \text{nsec}$$

Bit-line delay:

$$0.38\ (5.2\ \text{fF} \times 10\ \text{k}\Omega) \times 256^2 \le t_{BL} \le 0.69\ (5.2\ \text{fF} \times 10\ \text{k}\Omega) \times 256^2$$

or t_{BL} is situated between 1.3 and 2.3 μsec!

Problem 10.5—No stand-by power is consumed. This is in contrast with the pseudo-NMOS cell, where there is always conduction in one side of the cell.

Problem 10.6—The pessimistic approach is to lump M1 and M5 into a single transistor, whose length is the sum of the composing devices. The delay can then be computed with the traditional equation: $t_p = C_L V_{swing} / I_{av}$.

Problem 10.7—See Figure P10.7.

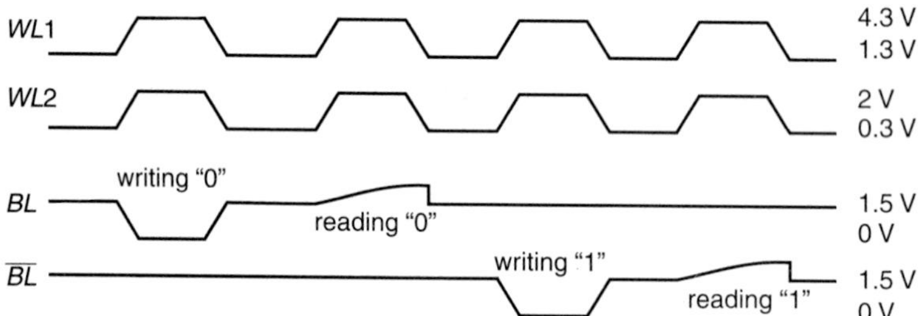

Figure P10.7.

Problem 10.8—The cheapest solution predecodes 5 bits at a time. This requires $(2 \times 32 \times 10) + 1024 \times 3 = 3712$ transistors.

Problem 10.9—

$$\overline{f_0} = \overline{\overline{x_0 x_1} \cdot x_2}$$

$$\overline{f_1} = \overline{\overline{x_0 x_1 x_2} \cdot \bar{x}_2 \cdot \overline{\bar{x}_0 x_1}}$$

Problem 10.10—The OR-plane is implemented in pseudo-NMOS style, while the AND plane is dynamic. The implementation is identical for f_0. The implementation of f_1 is substantially simplified using some logic manipulations. It turns out that $f_1 = x_1 + \bar{x}_2$. The reason one of the rows in the AND plane is not used is related to automatic generation. To make cell mirroring possible, the AND plane is generated two rows at a time. Since the number of min-terms is odd, one of the rows is left idle.

Problem 11.1—See Figure P11.1.

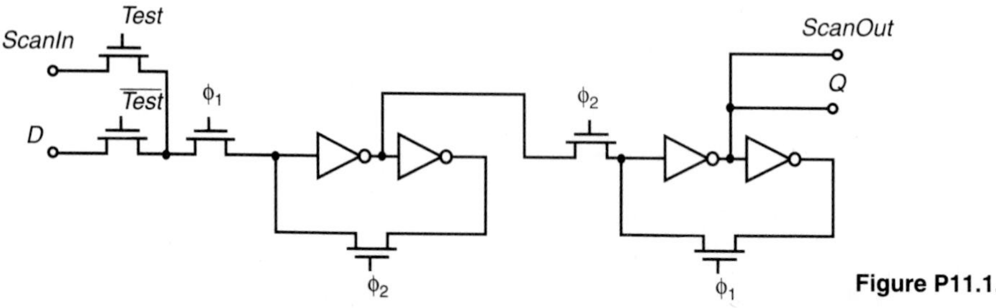

Figure P11.1.

INDEX

SPICE MODELS FOR DEVICES

CMOS (1.2 μm) – Manual Model

NMOS: $V_{T0} = 0.74$ V, $k' = 19.6 \times 10^{-6}$ A/V^2, $\lambda = 0.06$ V^{-1}

PMOS: $V_{T0} = -0.74$ V, $k' = 5.4 \times 10^{-6}$ A/V^2, $\lambda = 0.19$ V^{-1}

CMOS (1.2 μm) – SPICE Level II

```
.MODEL NMOS NMOS LEVEL=2 LD=0.15U TOX=200.0E-10 VTO=0.74 KP=8.0E-05
+ NSUB=5.37E+15  GAMMA=0.54 PHI=0.6 U0=656 UEXP=0.157 UCRIT=31444
+ DELTA=2.34 VMAX=55261 XJ=0.25U LAMBDA=0.037 NFS=1E+12 NEFF=1.001
+ NSS=1E+11 TPG=1.0 RSH=70.00 PB=0.58
+ CGDO=4.3E-10 CGSO=4.3E-10 CJ=0.0003 MJ=0.66  CJSW=8.0E-10 MJSW=0.24

.MODEL PMOS PMOS LEVEL=2 LD=0.15U TOX=200.0E-10 VTO=-0.74 KP=2.70E-05
+ NSUB=4.33E+15  GAMMA=0.58 PHI=0.6 U0=262 UEXP=0.324 UCRIT=65720
+ DELTA=1.79 VMAX=25694 XJ=0.25U LAMBDA=0.061 NFS=1E+12 NEFF=1.001
+ NSS=1E+11 TPG=-1.0 RSH=121 PB=0.64
+ CGDO=4.3E-10 CGSO=4.3E-10 CJ=0.0005 MJ=0.51 CJSW=1.35E-10 MJSW=0.24
```

Bipolar – Manual Model

NPN: $\beta_F = 100$, $I_S = 1.E\text{-}17$ A, $V_{BE(on)} = 0.7$ V, $V_{BE(sat)} = 0.8$ V, $V_{CE(sat)} = 0.1$ V

Bipolar (2 mm × 3.75 A_E npn) – SPICE

```
.MODEL NPN NPN BF=100 BR=1 IS=1.E-17 VAF=50
+ TF=10E-12 TR=5E-9 IKF=2E-2 IKR=0.5
+ RE=20 RC=75 RB=120
+ CJE=20E-15 VJE=0.8 MJE=0.5 CJC=22E-15 VJC=0.7 MJC=0.33
+ CJS=47E-15 VJS=0.7 MJS=0.33
```

GaAs (1 μm) – HSPICE

```
.MODEL ENH NJF
+ VT0=0.23 BETA=250u LAMBDA=0.2 ALPHA=6.5 UCRIT=0 GAMDS=0
+ LDEL=-0.4u WDEL=-0.15u RSH=210 N=1.16 IS=0.5m
+ LEVEL=3 SAT=0 ACM=1 CAPOP=1

.MODEL DP NJF
+ VT0=-0.825 BETA=190u LAMBDA=0.065 ALPHA=3.5 UCRIT=0 GAMDS=0
+ LDEL=-0.4u WDEL=-0.15u RSH=210 n=1.18 IS=10m
+ LEVEL=3 SAT=0 ACM=1 CAPOP=1
```

INTERCONNECT MODELS

Capacitance to substrate (for 1 μm CMOS process)

Interconnect layer	Area capacitance (fF/μm²)	Fringing capacitance (fF/μm)
Polysilicon	0.058 ± 0.004	0.043 ± 0.004
Metal1	0.031 ± 0.001	0.044 ± 0.001
Metal2	0.015 ± 0.001	0.035 ± 0.001
Metal3	0.010 ± 0.001	0.033 ± 0.001
n^+ Diffusion (@ 0 Volt)	0.36 ± 0.02	
p^+ Diffusion (@ 0 Volt)	0.46 ± 0.06	

Interlayer capacitance

Interconnect layer	Area capacitance (fF/μm²)	Fringing capacitance (fF/μm)
Metal1-to-polysilicon	0.055	0.049
Metal2-to-polysilicon	0.022	0.040
Metal2-to-metal1	0.035	0.046

Sheet resistance

Interconnect layer	Sheet Resistance (Ω/□)
n^+ Diffusion	10 ± 2
p^+ Diffusion	10 ± 2
n-well (under field)	1150 ± 250
Polysilicon ($H = 0.33$ μm)	10 ± 2
Metal1, Metal2	0.07 ± 0.006